MECHATRONICS
A Foundation Course

MECHATRONICS
A Foundation Course

Clarence W. de Silva

CRC Press
Taylor & Francis Group
Boca Raton London New York

CRC Press is an imprint of the
Taylor & Francis Group, an **informa** business

CRC Press
Taylor & Francis Group
6000 Broken Sound Parkway NW, Suite 300
Boca Raton, FL 33487-2742

© 2010 by Taylor and Francis Group, LLC
CRC Press is an imprint of Taylor & Francis Group, an Informa business

No claim to original U.S. Government works

Printed in the United States of America on acid-free paper
10 9 8 7 6 5 4 3 2 1

International Standard Book Number: 978-1-4200-8211-1 (Hardback)

Library of Congress Cataloging-in-Publication Data

De Silva, Clarence W.
   Mechatronics : a foundation course / Clarence W. de Silva.
       p. cm.
   Includes bibliographical references and index.
   ISBN 978-1-4200-8211-1 (alk. paper)
   1. Mechatronics. I. Title.

TJ163.12.D449 2010
621--dc22   2009050495

Visit the Taylor & Francis Web site at
http://www.taylorandfrancis.com

and the CRC Press Web site at
http://www.crcpress.com

To Professor Jim A.N. Poo

for his friendship and continuous support.

"We live in a society exquisitely dependent on science and technology, in

which hardly anyone knows anything about science and technology."

Carl Sagan

Contents

Preface

This is an introductory book on the subject of mechatronics. It can serve as both a textbook and a reference book for engineering students and practicing professionals, respectively. As a textbook, it is suitable for undergraduate or entry-level graduate courses in mechatronics, mechatronic devices and components, sensors and actuators, electromechanical systems, system modeling and simulation, and control system instrumentation.

Mechatronics concerns the synergistic and concurrent use of mechanics, electronics, computer engineering, and intelligent control systems in modeling, analyzing, designing, developing, implementing, and controlling smart electromechanical products. As modern machinery and electromechanical devices are typically being controlled using analog and digital electronics and computers, the technologies of mechanical engineering in such a system can no longer be isolated from those of electronic and computer engineering. For example, in a robotic system or a micromachine, mechanical components are integrated and embedded with analog and digital electronic components and microcontrollers to provide single functional units or products. Similarly, devices with embedded and integrated sensing, actuation, signal processing, and control have many practical advantages. In the framework of mechatronics, a unified approach is taken to integrate different types of components and functions, both mechanical and electrical, in modeling, analysis, design, implementation, and control, with the objective of harmonious operation that meets a desired set of performance specifications in a rather "optimal" manner, resulting in benefits with regard to performance, efficiency, reliability, cost, and environmental impact.

Mechatronics has emerged as a bona fide field of practice, research, and development, and simultaneously as an academic discipline in engineering. Historically, the approach taken in learning a new field of engineering has been to first concentrate on a single branch of engineering, such as electrical, mechanical, civil, chemical, or aerospace engineering, in an undergraduate program and then learn the new concepts and tools during practice, graduate study, or research. Since the discipline of mechatronics involves electronic and electrical engineering, mechanical and materials engineering, and control and computer engineering, a more appropriate approach would be to acquire a strong foundation in the necessary fundamentals from these various branches of engineering in an integrated manner in a single and unified undergraduate curriculum. In fact, many universities in the United States, Canada, Europe, Asia, and Australia have established both undergraduate and graduate programs in mechatronics. This book is focused toward an integrated education and practice as related to electromechanical systems.

Scope of the Book

The book is an outgrowth of my experience in integrating key components of mechatronics into senior-level courses for engineering students, and in teaching graduate and professional courses in mechatronics and related topics. The material in the book has acquired an application orientation through industrial experience I have gained at organizations such as IBM Corporation, Westinghouse Electric Corporation, Bruel and Kjaer, and NASA's Lewis and Langley Research Centers. To maintain clarity and the focus and to maximize the usefulness of the book, I have presented the material in a manner that will be convenient and useful to anyone with a basic engineering background, be it electrical, mechanical, aerospace, control, or computer engineering. Case studies, detailed worked examples, and exercises are provided throughout the book. Complete solutions to the end-of-chapter problems are presented in a "Solutions Manual," which will be available to instructors who take up a detailed study of the book.

The book consists of 10 chapters and 4 appendices. The chapters are devoted to presenting the fundamentals in electrical and electronic engineering, mechanical engineering, control engineering, and computer engineering, which are necessary for forming the foundation of mechatronics. In particular, they cover mechanical components, electrical and electronic components, modeling, analysis, instrumentation, sensors, transducers, signal processing, actuators, control, and system design and integration. The book uniformly incorporates the underlying fundamentals into analytical methods, modeling approaches, and design techniques in a systematic manner throughout the main chapters. The practical application of the concepts, approaches, and tools presented in the introductory chapters are demonstrated through a wide range of practical examples and a comprehensive set of case studies. The background theory and techniques that are not directly useful to present the fundamentals of mechatronics are provided in a concise manner in the appendices. Also, in the Solutions Manual, a curriculum is suggested for an undergraduate degree in mechatronics.

Main Features of the Book

The following are the main features of the book, which will distinguish it from other books on the same topic:

- Readability and convenient reference are given priority in the presentation and formatting of the book.
- Key concepts and formulas developed and presented in the book are summarized in windows, tables, and lists in a user-friendly format for easy reference and recollection.
- A large number of worked examples are included and are related to real-life situations and the practice of mechatronics.
- Numerous problems and exercises, most of which are based on practical situations and applications, and convey additional useful information on mechatronics, are provided at the end of each chapter.
- The use of MATLAB®, Simulink®, LabVIEW®, and associated toolboxes are described, and a variety of illustrative examples are provided for their use. Many

problems in the book are cast for solution using these computer tools. However, the main goal of the book is not simply to train the students in the use of software tools. Instead, a thorough understanding of the core and foundation of the subject, as facilitated by the book, will enable the student to learn the fundamentals and engineering methodologies behind the software tools: the choice of proper tools to solve a given problem, interpret the results generated by them, assess the validity and correctness of the results, and understand the limitations of the available tools.

- The material is presented in a manner so that users from diverse engineering backgrounds (mechanical, electrical, computer, control, aerospace, and material) will be able to follow and benefit from it equally.

- Useful material that could not be conveniently integrated into the main chapters is provided separately in four appendices at the end of the book.

- An Instructor's Manual (Solutions Manual) is available, which provides suggestions for curriculum planning and development, and gives detailed solutions to all the end-of-chapter problems in the book.

A Note to Instructors

A curriculum for a 4-year bachelor's degree in mechatronics is given in the Instructor's Manual (Solutions Manual). This book will be suitable as a text for several courses in such a curriculum, as listed below:

Mechatronics

Mechatronic devices and components

Sensors and actuators

Electromechanical systems

System modeling and simulation

Control system instrumentation

Mechatronic system instrumentation

Further material on these topics can be found in the following textbooks (with solutions manuals):

de Silva, C.W., *Sensors and Actuators—Control System Instrumentation*, CRC Press/Taylor & Francis, Boca Raton, FL, 2007.
de Silva, C.W., *Modeling and Control of Engineering Systems*, CRC Press/Taylor & Francis, Boca Raton, FL, 2009.

Windows® and Word® are software products of Microsoft® Corporation, Redmond, WA.

For MATLAB® and Simulink® product information, please contact

The MathWorks, Inc.
3 Apple Hill Drive
Natick, MA, 01760-2098 USA
Tel: 508-647-7000
Fax: 508-647-7001
E-mail: info@mathworks.com
Web: www.mathworks.com

LabVIEW™ is a product of National Instruments, Inc, Austin, TX.

I have personally used these software tools for teaching and for the development of this book.

Clarence W. de Silva
Vancouver, British Columbia, Canada

Acknowledgments

Many individuals have assisted in the preparation of this book, but it is not practical to acknowledge all such assistance here. First, I wish to recognize the contributions, both direct and indirect, of my graduate students, research associates, and technical staff. Particular mention should be made of my PhD student, Roland H. Lang, whose research assistance has been very important. I am particularly grateful to Jonathan W. Plant, senior editor, CRC Press/Taylor & Francis, for his interest, enthusiasm, and strong support throughout the project. Others at CRC Press and its affiliates, in particular, Christine Selvan, Joette Lynch, Anithajohny Mariasusai, Jessica Vakili, and others for their fine effort in producing this book. I also wish to acknowledge the advice and support of various authorities in the field—particularly, Professor Devendra Garg of Duke University; Professor Madan Gupta of the University of Saskatchewan; Professor Mo Jamshidi of the University of Texas (San Antonio, Texas); Professors Ben Chen, Tong-Heng Lee, and Kok-Kiong Tan of the National University of Singapore; Professor Maoqing Li of Xiamen University; Professor Max Meng of the Chinese University of Hong Kong; Dr. Daniel Repperger of the U.S. Air Force Research Laboratory; and Professor David N. Wormley of the Pennsylvania State University. Finally, I owe an apology to my wife and children for the unintentional "neglect" that they may have faced during the latter stages of the preparation of this book and am grateful for all their support.

Author

Clarence W. de Silva, PE, fellow ASME, fellow IEEE, fellow Royal Society of Canada, is a professor of mechanical engineering at the University of British Columbia, Vancouver, Canada, and occupies the Tier 1 Canada Research Chair professorship in mechatronics and industrial automation. Prior to this he occupied the Natural Sciences and Engineering Research Council–British Columbia Packers Research Chair in Industrial Automation since 1988. He served as a faculty member at Carnegie Mellon University (1978–1987) and as a Fulbright visiting professor at the University of Cambridge, Cambridge, England (1987–1988).

Professor de Silva received two PhD degrees from the Massachusetts Institute of Technology (1978) and the University of Cambridge, England (1998), and an honorary DEng from the University of Waterloo, Waterloo, Ontario, Canada (2008). Dr. de Silva also occupied the Mobil Endowed Chair professorship in the Department of Electrical and Computer Engineering at the National University of Singapore and the honorary chair professorship of the National Taiwan University of Science and Technology.

Other fellowships include the Fellow Canadian Academy of Engineering, Lilly Fellow, NASA-ASEE Fellow, Senior Fulbright Fellow to Cambridge University, Fellow of the Advanced Systems Institute of BC, Killam Fellow, and Erskine Fellow.

Professor de Silva has received the Paynter Outstanding Investigator Award and Takahashi Education Award, ASME Dynamic Systems & Control Division; Killam Research Prize; Outstanding Engineering Educator Award, IEEE Canada; and the Lifetime Achievement Award, World Automation Congress. He has also received the IEEE Third Millennium Medal; Meritorious Achievement Award, Association of Professional Engineers of BC; Outstanding Contribution Award, IEEE Systems, Man, and Cybernetics Society. He has given 20 keynote addresses.

He has served on 14 journals including *IEEE Transactions on Control System Technology*; the *Journal of Dynamic Systems, Measurement & Control*; and *Transactions of ASME*; editor in chief, the *International Journal of Control and Intelligent Systems* and the *International Journal of Knowledge-Based Intelligent Engineering Systems*. Other editorial duties include senior technical editor, *Measurements and Control*; and regional editor, North America, *Engineering Applications of Artificial Intelligence—IFAC International Journal of Intelligent Real-Time Automation*.

Professor de Silva has published 19 technical books, 18 edited books, 44 book chapters, about 200 journal articles, and over 215 conference papers.

In the areas of research and development, he has been involved in industrial process monitoring and automation, intelligent multi-robot cooperation, mechatronics, intelligent control, sensors, actuators, and control system instrumentation, with funding of about $7 million, as principal investigator, during the past 15 years.

1

Mechatronic Engineering

Study Objectives

- Introduction to mechatronics
- Modeling and design in mechatronics
- Mechatronic technologies
- Application areas
- Study of mechatronics

1.1 Introduction

The subject of mechatronics concerns the synergistic application of mechanics, electronics, controls, and computer engineering in the development of electromechanical products and systems through an integrated design approach. A mechatronic system will require a multidisciplinary approach for its modeling, design, development, and implementation. In the traditional development of an electromechanical system, the mechanical components and electrical components are designed or selected separately and then integrated, possibly with other components and hardware and software. In contrast, in the mechatronic approach, the entire electromechanical system is treated concurrently in an integrated manner by a multidisciplinary team of engineers and other professionals. Naturally, a system formed by interconnecting a set of independently designed and manufactured components will have a lower level of performance than that of a mechatronic system, which employs an integrated approach for design, development, and implementation. The main reason is straightforward. The best match and compatibility between component functions can be achieved through an integrated and unified approach to design and development, and the best performance is possible through an integrated implementation. Generally, a mechatronic product will be more efficient and cost effective, more precise and accurate, more reliable, more flexible and functional, less mechanically complex, safer, and more environment friendly than a non-mechatronic product requiring a similar level of effort in its development. The performance of a non-mechatronic system can be improved through sophisticated control, but this is achieved at an additional cost of sensors, instrumentation, and control hardware and software, and with added complexity.

Mechatronic products and systems include modern automobiles and aircraft, smart household appliances, medical robots, space vehicles, and office automation devices. In this chapter, the subject of mechatronics is introduced, important issues in modeling, design, and the development of a mechatronic product or system are highlighted, and the associated technology areas and applications are indicated.

1.2 Mechatronic Systems

A typical mechatronic system consists of a mechanical skeleton, actuators, sensors, controllers, signal conditioning/modification devices, computer/digital hardware and software, interface devices, and power sources. Different types of sensing, information acquisition, and transfer are involved among all these various types of components. For example, a servomotor, which is a motor with the capability of sensory feedback for accurate generation of complex motions, consists of mechanical, electrical, and electronic components (see Figure 1.1). The main mechanical components are the rotor, stator, and the bearings. The electrical components include the circuitry for the field windings and rotor windings (not in the case of permanent-magnet rotors) and the circuitry for power transmission and commutation (if needed). Electronic components include those needed for sensing (e.g., an optical encoder for displacement and speed sensing and a tachometer for speed sensing). The overall design of a servomotor can be improved by taking a mechatronic approach.

The humanoid robot shown in Figure 1.2a is a more complex and "intelligent" mechatronic system. It may involve many servomotors and a variety of mechatronic components, as is clear from the sketch in Figure 1.2b. A mechatronic approach can greatly benefit the analysis/modeling, design, and development of a complex electromechanical system of this nature.

In the computer industry, hard-disk drives (HDD; see Figure 1.3), devices for disk retrieval, access and ejection, and other electromechanical components can considerably benefit from high-precision mechatronics. The impact goes

FIGURE 1.1
A servomotor is a mechatronic device. (Courtesy of Danaher Motion, Rockford, IL.)

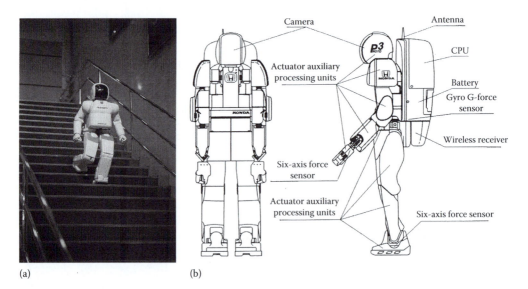

FIGURE 1.2
(a) A humanoid robot is a complex and "intelligent" mechatronic system; (b) components of a humanoid robot. (Courtesy of American Honda Motor Co. Inc., Torrance, CA.)

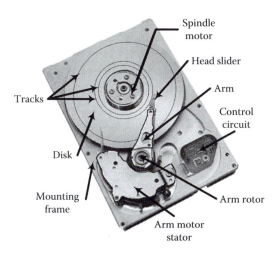

FIGURE 1.3
An HDD unit of a computer.

further because digital computers are integrated into a vast variety of other devices and mechatronic applications.

Technology issues of a general mechatronic system are indicated in Figure 1.4. It is seen that they span the traditional fields of mechanical engineering, electrical and electronic engineering, control engineering, and computer engineering. Each aspect or issue within

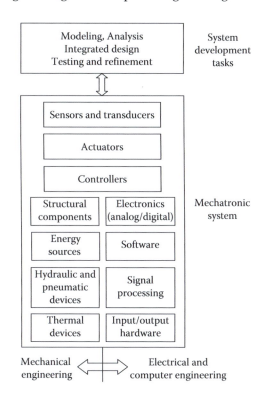

FIGURE 1.4
Concepts and technologies of a mechatronic system.

the system may take a multi-domain character. For example, as noted before, an actuator (e.g., dc servo motor) itself may represent a mechatronic device within a larger mechatronic system such as an automobile or a robot.

The study of mechatronic engineering should include all stages of modeling, design, development, integration, instrumentation, control, testing, operation, and maintenance of a mechatronic system.

1.3 Modeling and Design

A model is a representation of a real system, and the subject of model development (modeling) is important in mechatronics (see Chapter 3). Modeling and design can go hand-in-hand in an iterative manner. Of course, in the beginning of the design process, the desired system does not exist. In this context, a model of the anticipated system can be very useful. In view of the complexity of a design process, particularly when striving for an optimal design, it is useful to incorporate system modeling as a tool for design iteration particularly because prototyping can become very costly and time consuming.

In the beginning, by knowing some information about the system (e.g., intended functions, performance specifications, past experience, and knowledge of related systems) and by using the design objectives, it is possible to develop a model of sufficient (low to moderate) detail and complexity. By analyzing and carrying out computer simulations of the model, it will be possible to generate useful information that will guide the design process (e.g., the generation of a preliminary design). In this manner, design decisions can be made and the model can be refined using the available (improved) design. This iterative link between modeling and design is schematically shown in Figure 1.5.

It is expected that the mechatronic approach will result in a higher quality of the products and services, improved performance, and increased reliability while approaching some form of optimality. This will enable the development and production of

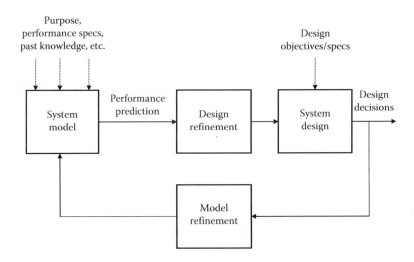

FIGURE 1.5
Link between modeling and design.

electromechanical systems efficiently, rapidly, and economically. When performing the integrated design of a mechatronic system, the concepts of energy and power present a unifying thread. The reasons are clear. First, in an electromechanical system, ports of power and energy exist that link electrical dynamics and mechanical dynamics. Hence, the modeling, analysis, and optimization of a mechatronic system can be carried out using a hybrid system (or multi-domain system) formulation (a model) that integrates mechanical aspects and electrical aspects of the system. Second, an optimal design will aim for minimal energy dissipation and maximum energy efficiency. There are related implications; for example, greater dissipation of energy will mean reduced overall efficiency and increased thermal problems, noise, vibration, malfunctions, wear and tear, and increased environmental impact. Again, a hybrid model that presents an accurate picture of the energy/power flow within the system will present an appropriate framework for the mechatronic design. (*Note*: Refer to linear graph models in particular, as discussed in Chapter 3.)

A design may use excessive safety factors and worst-case specifications (e.g., for mechanical loads and electrical loads). This will not provide an optimal design or may not lead to the most efficient performance. Design for optimal performance may not necessarily lead to the most economical (least costly) design, however. When arriving at a truly optimal design, an objective function that takes into account all important factors (performance, quality, cost, speed, ease of operation, safety, environmental impact, etc.) has to be optimized. A complete design process should generate the necessary details for the construction or assembly of the system.

1.4 Mechatronic Design Concept

In a true mechatronic sense, the design of a multi-domain multicomponent system will require the simultaneous consideration and integrated design of all its components, as indicated in Figure 1.4. Such an integrated and "concurrent" design will call for a fresh look at the design process itself and also a formal consideration of information and energy transfer between the components within the system.

In an electromechanical system, there exists an interaction (or coupling) between electrical dynamics and mechanical dynamics. Specifically, electrical dynamics affect the mechanical dynamics and vice versa. Traditionally, a "sequential" approach has been adopted to the design of multi-domain (or mixed) systems such as electromechanical systems. For example, first the mechanical and structural components are designed, next the electrical and electronic components are selected or developed and interconnected, then a computer is selected and interfaced with the system, subsequently a controller is added, and so on. The dynamic coupling between various components of a system dictates, however, that an accurate design of the system should consider the entire system as a whole rather than designing the electrical/electronic aspects and the mechanical aspects separately and sequentially. When independently designed components are interconnected, several problems can arise as follows:

1. When two independently designed components are interconnected, the original characteristics and operating conditions of the two will change due to the loading or dynamic interactions (see Chapter 4).

2. A perfect matching of two independently designed and developed components will be practically impossible. As a result, a component can be considerably under-utilized or overloaded, in the interconnected system, both conditions being inefficient and undesirable.

3. Some of the external variables in the components will become internal and "hidden" due to interconnection, which can result in potential problems that cannot be explicitly monitored through sensing and cannot be directly controlled.

The need for an integrated and concurrent design for electromechanical systems can be identified as a primary motivation for the developments in the field of mechatronics.

1.4.1 Coupled Design

An uncoupled design is where each subsystem is designed separately (and sequentially), while keeping the interactions with the other subsystems constant (i.e., ignoring the dynamic interactions). Mechatronic design involves an integrated or "coupled" design. The concept of mechatronic design may be illustrated using an example of an electro-mechanical system, which can be treated as a coupling of an electrical subsystem and a mechanical subsystem. An appropriate model for the system is shown in Figure 1.6a. Note that the two subsystems are coupled using a loss-free (pure) energy transformer while the losses (energy dissipation) are integral with the subsystems. In this system, assume that under normal operating conditions the energy flow is from the electrical subsystem to the mechanical subsystem (i.e., the electrical subsystem behaves like a motor rather than a generator). At the electrical port that connects to the energy transformer, there exists a current i (a "through" variable) flowing in and a voltage v (an "across" variable) with the shown polarity (the concepts of through and across variables and the related terminology are explained in Chapter 3). The product vi is the electrical power, which is positive out of the electrical subsystem and into the transformer. Similarly, at the mechanical port that comes out of the energy transformer, there exists a torque τ (a through variable) and an angular speed ω (an across variable) with the sign convention indicated in Figure 1.6a. Accordingly, a positive mechanical power $\omega\tau$ flows out of the transformer and into the mechanical subsystem. The ideal transformer implies that

$$vi = \omega\tau \tag{1.1}$$

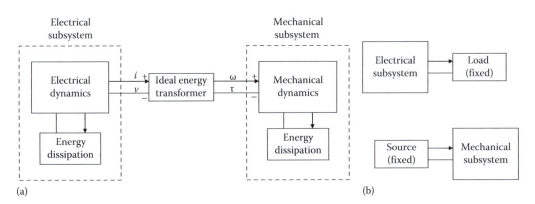

FIGURE 1.6
(a) An electromechanical system; (b) conventional design.

In a conventional uncoupled design of the system, the electrical subsystem is designed by treating the effects of the mechanical subsystem as a fixed load, and the mechanical subsystem is designed by treating the electrical subsystem as a fixed energy source, as indicated in Figure 1.6b. Suppose that, in this manner, the electrical subsystem achieves an optimal "design index" of I_{ue} and the mechanical subsystem achieves an optimal design index of I_{um}.

Note: The design index is a measure of the degree to which the particular design satisfies the design specifications (design objectives).

When the two uncoupled designs (subsystems) are interconnected, there will be dynamic interactions. As a result, neither the electrical design objectives nor the mechanical design objectives will be satisfied at the levels dictated by I_{ue} and I_{um}, respectively. Instead, they will be satisfied at lower levels as given by the design indices I_e and I_m. A truly mechatronic design will attempt to bring I_e and I_m as close as possible to I_{ue} and I_{um}, respectively. This may be achieved, for example, by minimizing the quadratic cost function:

$$J = \alpha_e (I_{ue} - I_e)^2 + \alpha_m (I_{um} - I_m)^2 \tag{1.2}$$

subject to

$$\begin{bmatrix} I_e \\ I_m \end{bmatrix} = D(p) \tag{1.3}$$

where
   D denotes the transformation that represents the design process
   p denotes information including system parameters that is available for the design

Even though this formulation of the mechatronic design problem appears rather simple and straightforward, the reality is otherwise. In particular, the design process, as denoted by the transformation D, can be quite complex and typically nonanalytic. Furthermore, minimization of the cost function J is by and large an iterative practical scheme and undoubtedly a knowledge-based and nonanalytic procedure. This complicates the process of mechatronic design. In any event, the design process will need the information represented by p.

1.4.2 Mechatronic Design Quotient

Mechatronic systems are complex and require multiple technologies in multiple domains. Their optimal design may call for multiple performance indices. The problem of mechatronic design may be treated as a maximization of a "mechatronic design quotient" or MDQ. In particular, an alternative formulation of the optimization problem given by (1.2) and (1.3) would be the maximization of the MDQ:

$$MDQ = \frac{\alpha_e I_e^2 + \alpha_m I_m^2}{\alpha_e I_{ue}^2 + \alpha_m I_{um}^2} \tag{1.4}$$

subject to (1.3).

Even though Equation 1.4 is formulated for two categories of technologies or devices m and e (and the corresponding indices I_e and I_m), the MDQ may be generalized for three or more categories such as: reliability, maintainability, efficiency, cost effectiveness, power and efficiency, size and geometry, control friendliness, and level of intelligence. The corresponding indices may be qualitative or nonanalytic and may have correlations or interactions. Then, more sophisticated representations (e.g., the use of fuzzy measures) and optimization techniques (e.g., evolutionary computing or genetic programming or GP) may be employed in the design process.

For example, in the use of genetic algorithms (GA) for mechatronic design, we start with a group (population) of initial chromosomes (embryos) where an individual chromosome is one possible design. An individual gene in a chromosome corresponds to an element of information in a design (e.g., system component, connection structure, set of parameters, design attribute). Alleles are possible values of a gene (e.g., available choices for a particular component). The "fitness function" of the GA represents the "value," "goodness," or "fitness" of a design. In the present context, the fitness function is the MDQ, which is computable for a given design once the element information of the design is known. Then the problem of design optimization becomes:

$$\text{Maximize MDQ}(p_1, p_2, \ldots, p_m) \tag{1.5}$$

where, p_i is the ith design aspect.

The strength and applicability of the MDQ approach stem from the possibility that the design process may be hierarchically separated. Then, an MDQ may be optimized for one design layer involving two more technology groups in that layer before proceeding to the next lower design layer where each technology group is separately optimized by considering several technology/component groups within that group together with an appropriate MDQ for that lower-level design problem. For example, an upper layer may optimize the actuator type for the particular application (e.g., hydraulic, dc, induction, stepper; see Chapter 7) with an appropriate MDQ. The next lower level may optimize the motor selection (e.g., select a motor from an available set of dc motors) with another MDQ. In this manner, a complex design optimization may be achieved through several design optimizations at different design levels. The final design may not be precisely optimal, yet intuitively adequate for practical purposes; say in a conceptual design.

1.4.3 Design Evolution

Traditionally, the online monitoring of responses/outputs of a system may be used to detect and diagnose the faults and malfunctions (existing or impending) of a system. We believe that such monitoring may also be used to improve the design of an existing mechatronic system. In particular, just like how a health monitoring system can pinpoint a defective component in a system, it should be possible for the same system to at least identify the possible regions (sites) of design weakness in the system. This is the premise of the approach for "design evolution", as outlined below.

A model of the existing system (whose design needs to be improved) and evolutionary computing (GP) may facilitate the approach of "evolutionary" design improvement through online monitoring. A possible framework for implementing this approach is indicated in Figure 1.7.

The relevant steps are as follows:

1. Develop a model of the existing system.
2. Establish (using a machine health monitoring system and an expert system) which aspects or segments of the original system (and its model) may be modified/improved using information monitored from the system. These will provide "modifiable sites" for the existing system/model.

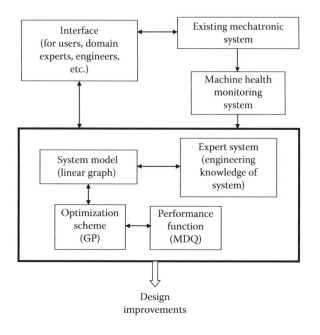

FIGURE 1.7
Structure of a system for evolutionary design.

3. Formulate a performance function to represent the "goodness" of the design. This is the MDQ.

4. Use an optimization method (GP) to evolve the model so as to maximize the performance function.

5. Implement, in the existing system, the design changes represented by the evolved model.

The optimization scheme will gradually improve the original model of the system so as to produce better performance (as judged by the MDQ). This will require the comparison of the monitored response of the original system and the simulated response of the model as it evolves (improves), with respect to the MDQ. In Figure 1.7, in addition to the initial model of the system, the evolutionary computing approach, and online monitoring, we have shown an expert system as well for "intelligent" decision making associated with design/model improvement. This expert system may be generated from the knowledge/expertise of the existing system, its design, and engineering know how.

1.5 Evolution of Mechatronics

Mechanical engineering products and systems that employ some form of electrical engineering principles and devices have been developed and used since the early part of the twentieth century. These systems included the automobile, electric typewriter, aircraft, and elevator. Some of the power sources used in these systems were not necessarily electrical, but there were batteries and/or a conversion of thermal power into electricity through generators. These "electromechanical" systems were not "mechatronic" systems because they did not use an integrated approach characterizing mechatronics for their analysis, design, development, and implementation.

Rapid advances in electromechanical devices and systems were possible particularly due to developments in control engineering, which began for the most part in the early 1950s, and still more rapid advances in digital computer and communication as a result of integrated circuit (IC) and microprocessor technologies, starting from the late 1960s. With these advances, engineers and scientists felt the need for a multidisciplinary approach to design and hence a "mechatronic" approach. In 1969, Yasakawa Electric in Japan was the first to coin the term mechatronics, for which the company obtained a trademark in 1972. Subsequently, in 1982, the company released the trademark rights. Even though a need for mechatronics was felt even in those early times, no formal discipline or educational programs existed for the engineers to be educated and trained in this area. The research and development activities mainly in automated transit systems and robotics in the 1970s and the 1980s undoubtedly paved the way for the evolution of the field of mechatronics. With today's sophisticated technologies of mechanics and materials, analog and digital electronics, sensors, actuators, controllers, electromechanical design, and microelectromechanical systems (MEMS) with embedded sensors, actuators, and microcontrollers, the field of mechatronics has attained a high degree of maturity. Now many universities around the world offer undergraduate and graduate programs in mechatronic engineering, which have become highly effective and popular among students, instructors, employees, and employers alike.

1.6 Application Areas

The application areas of mechatronics are numerous and involve those that concern multidomain (mixed) systems and particularly electromechanical systems. These applications may involve the following:

1. Modifications and improvements to conventional designs by using a mechatronic approach
2. The development and implementation of original and innovative mechatronic systems

In either category, the applications may employ sensing, actuation, control, signal conditioning, component interconnection and interfacing, and communication, generally using tools of mechanical, electrical, electronic, computer, and control engineering. Some important areas of application are indicated below.

Transportation is a broad area where mechatronic engineering has numerous applications. In ground transportation, particularly automobiles, trains, and automated transit systems use mechatronic devices. They include airbag deployment systems, antilock braking systems (ABS), cruise control systems, active suspension systems, and various devices for monitoring, toll collection, navigation, warning, and control in intelligent vehicular highway systems (IVHS). In air transportation, modern aircraft designs with advanced materials, structures, electronics, and control benefit from the concurrent and integrated approach of mechatronics to develop improved designs of flight simulators, flight control systems, navigation systems, landing gear mechanisms, traveler comfort aids, and the like.

Manufacturing and production engineering is another broad field that uses mechatronic technologies and systems. Factory robots (for welding, spray painting, assembly, inspection, and so on), automated guided vehicles (AGVs), modern computer-numerical control (CNC) machine tools, machining centers, rapid (and virtual) prototyping systems, and micromachining systems are examples of mechatronic applications. High-precision motion control is particularly important in these applications.

In medical and health care applications, robotic technologies for patient examination, surgery, rehabilitation, drug dispensing, and general patient care are being developed and used. Mechatronic technologies are being applied for patient transit devices, various diagnostic probes and scanners, beds, and exercise machines.

In a modern office environment, automated filing systems, multifunctional copying machines (copying, scanning, printing, FAX, and so on), food dispensers, multimedia presentation and meeting rooms, and climate control systems incorporate mechatronic technologies.

In household applications, home security systems, robotic caregivers and helpers, robotic vacuum cleaners, washers, dryers, dishwashers, garage door openers, and entertainment centers use mechatronic devices and technologies.

The computer industry can considerably benefit from mechatronics. The impact goes further because digital computers are integrated into a vast variety of other devices and applications.

In civil engineering applications, cranes, excavators, and other machinery for building, earth removal, mixing, and so on will improve their performance by adopting a mechatronic design approach.

In space applications, mobile robots such as NASA's Mars exploration Rover, space-station robots, and space vehicles are fundamentally mechatronic systems.

It is noted that there is no end to the type of devices and applications that can incorporate mechatronics. In view of this, the traditional boundaries between engineering disciplines will become increasingly fuzzy, and the field of mechatronics will grow and evolve further through such merging of disciplines.

1.7 Study of Mechatronics

Mechatronics is a multidisciplinary field that is concerned with the integrated modeling, analysis, design, manufacture, control, testing, and operation of smart electromechanical products and systems. Hence, one should not use a "compartmentalized" approach in studying this field. Specifically, rather than using a conventional approach to learning standard subjects separately in a disjointed manner, they need to be integrated into a common "mechatronics" framework, along with other specialized subjects.

The study of mechatronics requires a good foundation of such core subjects as mechanics, electronics, modeling, control, signal processing and conditioning, communication, and computer engineering, and specialized subjects like electrical components, mechanical components, sensors and transducers, instrumentation, drives and actuators, intelligent control, interfacing hardware, software, testing, performance evaluation, and cost-benefit analysis. In mechatronics, all these subjects are unified through an integrated approach of modeling, analysis, design, and implementation for multi-domain systems. A traditional undergraduate curriculum in engineering does not provide such a broad and multidisciplinary foundation. A more realistic approach would be to follow a traditional engineering curriculum in the first 2 years of a 4 year undergraduate program, and then get into an integrated mechatronic curriculum in the next 2 years. What is presented in this book is the necessary material in mechatronics that is not traditionally covered in the first 2 years of an undergraduate engineering program.

1.8 Organization of the Book

The book consists of 10 chapters and 4 appendices. The chapters are devoted to presenting the fundamentals in electrical and electronic engineering, mechanical engineering, control engineering, and computer engineering, which are necessary for forming the core of mechatronics. In particular, they cover modeling, analysis, mechanics, electronics, instrumentation, sensors, transducers, signal processing, actuators, drive systems, computer engineering, control, and system design and integration. The book uniformly incorporates the underlying fundamentals into analytical methods, modeling approaches, design techniques, and control schemes in a systematic manner throughout the main chapters. The practical application of the concepts, approaches, and tools presented in the introductory chapters are demonstrated through numerous illustrative examples and a comprehensive set of case studies. The background theory and techniques that are not directly useful to present the fundamentals of mechatronics are given in a concise manner in the appendices.

Chapter 1 introduces the field of mechatronics. The evolution of the field is given. The underlying design philosophy of mechatronics and how it relates to modeling is described. This introductory chapter sets the tone for the study, which spans the remaining nine chapters.

Chapter 2 deals with mechanical components, which are important constituents of the mechatronic system. It also studies electronic components, which form another class of important constituents of a mechatronic system. Electronic material and both passive and active electronic components are discussed. Common practical uses of these components are indicated.

Chapter 3 deals with the modeling and analysis of dynamic systems. Mechanical, electrical, fluid, and thermal systems and mixed systems such as electromechanical systems are studied. The usual techniques of modeling are presented, while emphasizing those methods that are particularly appropriate for mechatronic systems. Analysis in both time domain and frequency domain is introduced, while discussing response analysis and computer simulation.

Chapter 4 presents the component interconnection and signal conditioning, which is in fact a significant unifying subject within mechatronics. Impedance considerations of component interconnection and matching are studied. Amplification, filtering, analog-to-digital conversion, digital-to-analog conversion, bridge circuits, and other signal conversion and conditioning techniques and devices are discussed.

Chapter 5 covers the performance analysis of a mechatronic device or component. Methods of performance specification are addressed, both in the time domain and the frequency domain. Common instrument ratings that are used in industry and generally in the engineering practice are discussed. Related analytical methods are given. Instrument bandwidth considerations are highlighted and a design approach based on component bandwidth is presented. Errors in digital devices, particularly resulting from signal sampling, are discussed from the analytical and practical points of view.

Chapter 6 presents important types, characteristics, and operating principles of sensors and transducers. Particular attention is given to sensors that are commonly used in mechatronic systems. Motion sensors, force, torque and tactile sensors, optical sensors, ultrasonic sensors, temperature sensors, pressure sensors, and flow sensors are discussed. Analytical basis, selection criteria, and application areas are indicated. Unlike analog sensors, digital transducers generate pulses or digital outputs. These devices have clear advantages, particularly when used in computer-based digital systems. They do possess quantization errors, which are unavoidable in a digital representation of an analog quantity. The related issues of accuracy and resolution are also addressed in Chapter 6.

Chapter 7 studies the actuators for mechatronic systems. In particular, stepper motors that produce incremental motions are studied. Under satisfactory operating conditions, they have the advantage of being able to generate a specified motion profile in an open-loop manner without requiring motion sensing and feedback control. Continuous-drive actuators such as dc motors, ac motors, hydraulic actuators, and pneumatic actuators are also covered in Chapter 7. The operating principles, analytical methods, sizing and selection considerations, drive systems, and control techniques are described. The advantages and drawbacks of various types of actuators on the basis of the nature and the needs of an application are discussed and practical examples are given.

Chapter 8 covers digital logic and hardware, microprocessors, and microcontrollers, which fall within the area of electronic and computer engineering. Logic devices and ICs are widely used in mechatronic systems for such purposes as sensing, signal conditioning, and control. The basic principles of digital components and circuits are presented in this chapter. The types and applications of logic devices are discussed. The technology of ICs is introduced. The microcontroller and the embedded microprocessor have become standard components in a large variety of mechatronic devices. A microprocessor together with memory, software, and interface hardware (i.e., a microcontroller) provides an effective and economical miniature digital computer in mechatronic applications. Smart sensors, actuators, controllers, and other essential components of a mechatronic system can immensely benefit from the programmability, flexibility, and processing power of a microcontroller.

Chapter 9 deals with the control of mechatronic systems. Both time-domain techniques and frequency-domain techniques of control are covered. In particular, performance specification, stability analysis, and control schemes are presented. Underlying analytical methods are described. The popular approach of intelligent control, known as fuzzy logic control, is presented. Popular advanced techniques of control are outlined.

Chapter 10 concludes the main body of the book by presenting the design approach of mechatronics and by giving extensive case studies of practical mechatronic systems. The techniques covered in the previous chapters come together and are consolidated in these case studies. Several design exercises and practical projects in mechatronics are given.

The four appendices provide some useful fundamentals, techniques, and tools for the study of mechatronics in a concise and condensed form. Appendix A presents the basic theory of solid mechanics and elasticity. Appendix B gives useful techniques of the Laplace transform and Fourier transform. Appendix C outlines the basics of probability and statistics. Appendix D presents several useful software tools. In particular, Simulink® and MATLAB® toolboxes of control systems and fuzzy logic are outlined. The LabVIEW®

program development environment, which is an efficient tool for laboratory experimentation (particularly, data acquisition and control) is described.

Problems

1.1 The following have been claimed as benefits of the mechatronic design of a system:

- Optimality and better component matching
- Ease of system integration and enhancement
- Compatibility and ease of cooperation with other systems
- Increased efficiency and cost effectiveness
- Improved controllability
- Improved maintainability
- Improved reliability and product life
- Reduced environmental impact

Briefly justify each of these claims.

1.2 You are a mechatronic engineer who has been assigned the task of designing and instrumenting a mechatronic system. In the final project report, you will have to describe the steps of establishing the design/performance specifications for the system, selecting and sizing sensors, transducers, actuators, drive systems, controllers, signal conditioning and interface hardware, and software for the instrumentation and component integration of this system. Keeping this in mind, write a project proposal given the following information:

1. Select a process (plant) as the system to be developed. Describe the plant indicating the purpose of the plant, how the plant operates, what is the system boundary (physical or imaginary), what are important inputs (e.g., voltages, torques, heat transfer rates, flow rates), response variables (e.g., displacements, velocities, temperatures, pressures, currents, voltages), and what are important plant parameters (e.g., mass, stiffness, resistance, inductance, conductivity, fluid capacity). You may use sketches.

2. Indicate the performance requirements (or operating specifications) for the plant (i.e., how the plant should behave under control). You may use any available information on such requirements as accuracy, resolution, speed, linearity, stability, and operating bandwidth.

3. Give any constraints related to cost, size, weight, environment (e.g., operating temperature, humidity, dust-free or clean room conditions, lighting, wash-down needs), etc.

4. Indicate the type and the nature of the sensors and transducers present in the plant and what additional sensors and transducers might be needed for properly operating and controlling the system.

5. Indicate the type and nature of the actuators and drive systems present in the plant and which of these actuators have to be controlled. If you need to add new actuators (including control actuators) and drive systems, indicate such requirements in sufficient detail.

6. Mention what types of signal modification and interfacing hardware would be needed (i.e., filters, amplifiers, modulators, demodulators, ADC, DAC, and other data acquisition and control needs). Describe the purpose of these devices. Indicate any software (e.g., driver software) that may be needed along with this hardware.

7. Indicate the nature and operation of the controllers in the system. State whether these controllers are adequate for your system. If you intend to add new controllers, briefly give their nature, characteristics, objectives, etc. (e.g., analog, digital, linear, nonlinear, hardware, software, control bandwidth).

8. Describe how the users and/or operators interact with the system, and the nature of the user interface requirements (e.g., graphic user interface or GUI).

The following plants/systems may be considered:

1. A hybrid electric vehicle
2. A household robot
3. A smart camera
4. A smart airbag system for an automobile
5. Rover mobile robot for Mars exploration developed by NASA
6. An AGV for a manufacturing plant
7. A flight simulator
8. A hard disk drive for a personal computer
9. A packaging and labeling system for a grocery item
10. A vibration testing system (electrodynamic or hydraulic)
11. An active orthotic device to be worn by a person to assist a disabled or weak hand (which has some sensation, but is not fully functional)

Further Reading

This book has relied on many publications, directly and indirectly, in its development and evolution. Many of these publications are based on the work of the author and his coworkers. Also, there are some excellent books the reader may refer to for further information and knowledge. Some selected books are listed below.

Auslander, D.M. and Kempf, C.J., *Mechatronics Mechanical System Interfacing*, Prentice Hall, Upper Saddle River, NJ, 1996.

Bolton, W., *Mechatronics*, 2nd edn., Longman, Essex, England, 1999.

Cetinkunt, S., *Mechatronics*, John Wiley & Sons, Hoboken, NJ, 2007.

Chen, B.M., Lee, T.H., and Venkataramenan, V., *Hard Disk Drive Servo Systems*, Springer-Verlag, London, England, 2002.

Histand, M.B. and Alciatore, D.G., *Introduction to Mechatronics and Measurement Systems*, WCB McGraw-Hill, New York, 1999.

Jain, L. and de Silva, C.W. (Eds.), *Intelligent Adaptive Control: Industrial Applications*, CRC Press, Boca Raton, FL, 1999.

Karray, F. and de Silva, C.W., *Soft Computing and Intelligent Systems Design*, Addison Wesley, Pearson, New York, 2004.

Necsulescu, D., *Mechatronics*, Prentice Hall, Upper Saddle River, NJ, 2002.

Shetty, D. and Kolk, R.A., *Mechatronics System Design*, PWS Publishing Co., Boston, MA, 1997.

de Silva, C.W. and Wormley, D.N., *Automated Transit Guideways: Analysis and Design*, D.C. Heath and Co., Simon & Schuster, Lexington, MA, 1983.

de Silva, C.W. *Dynamic Testing and Seismic Qualification Practice*, D.C. Heath and Co., Simon & Schuster, Lexington, MA, 1983.

de Silva, C.W., *Control Sensors and Actuators*, Prentice-Hall, Englewood Cliffs, NJ, 1989.

de Silva, C.W. and MacFarlane, A.G.J., *Knowledge-Based Control with Application to Robots*, Springer-Verlag, Berlin, Germany, 1989.

de Silva, C.W., *Intelligent Control—Fuzzy Logic Applications*, CRC Press, Boca Raton, FL, 1995.

de Silva, C.W. (Ed.), *Intelligent Machines: Myths and Realities*, Taylor & Francis/CRC Press, Boca Raton, FL, 2000.

de Silva, C.W., *Mechatronics—An Integrated Approach*, Taylor & Francis/CRC Press, Boca Raton, FL, 2005.

de Silva, C.W. (Ed.), *Vibration and Shock Handbook*, Taylor & Francis/CRC Press, Boca Raton, FL, 2005.

de Silva, C.W., *Vibration Fundamentals and Practice*, 2nd edn., CRC Press, Boca Raton, FL, 2007.

de Silva, C.W., *Sensors and Actuators—Control System Instrumentation*, Taylor & Francis/CRC Press, Boca Raton, FL, 2007.

de Silva, C.W. (Ed.), *Mechatronic Systems—Devices, Design, Control, Operation, and Monitoring*, Taylor & Francis/CRC Press, Boca Raton, FL, 2007.

Smaili, A. and Mrad, F., *Applied Mechatronics*, Oxford University Press, Oxford, England, 2008.

Tan, K.K., Lee, T.H., Dou, H, and Huang, S., *Precision Motion Control*, Springer-Verlag, London, England, 2001.

2

Basic Elements and Components

Study Objectives

- Physical (constitutive) equations for basic elements in a mechatronic system from which other components and systems can be modeled, analyzed, and built
- Basic elements: mechanical, fluid, thermal, electrical and electronic (passive and active)
- Material properties that give rise to useful element/component characteristics
- Some useful components/devices and applications
- Important results in solid mechanics (also see Appendix A)

2.1 Introduction

The field of mechatronics is primarily concerned with the integration of mechanics and electronics. In a mechatronic product, mechanical, electrical, and electronic components play a crucial role. They can serve functions of structural support or load bearing, mobility, transmission of motion and power or energy, actuation, manipulation, sensing, switching, signal conditioning, drive circuitry, and control. The mechanical components have to be designed to be integral with electronics, controls, and so on to satisfy such desirable characteristics as high accuracy, light weight, high strength, high speed, low noise and vibration, long design life, fewer moving parts, high reliability, low-cost production and distribution, and infrequent and low-cost maintenance.

Even in an integrated electromechanical system, there are good reasons why a distinction has to be made between the mechanical components and the electronic and computer (hardware and software) components. One reason relates to the energy (or power) conversion. The types of energy that are involved will differ in these different types of components (or functions). The level of energy (or power) can differ greatly as well. For example, digital electronic circuits and computer hardware typically use low levels of power and voltage. Analog devices such as amplifiers and power supplies, however, can accommodate high voltages and power. Motors and other actuators (e.g., ac motors and hydraulic actuators in particular) can receive high levels of electrical power and can generate similar high levels of mechanical power. Analog-to-digital conversion (ADC) and digital-to-analog conversion (DAC) involve relatively low levels of power. But, drive (power) amplifiers of electrical motors, pumps, and compressors of hydraulic and pneumatic systems typically deal with much higher levels of power. It follows that the level of power needed for a task and the nature of energy conversion that is involved can separate mechanical components from others in a mechatronic system.

Another important consideration that separates a mechanical component from electronic and computing components (hardware/software) is the bandwidth (speed, time constant, etc.). Typically, mechanical components have lower time constants than electronic components. Accordingly, their speeds of operation will differ and furthermore, the bandwidth (useful frequency content) of the associated signals will differ as well. For example, process plants can have time constants as large as minutes and robotic devices and machine tools have time constants in the ms range. The time constants of analog electrical circuitry can be quite low (μs range). Software-based computer devices can conveniently generate digital actions in the kHz rate (i.e., ms time scale). If faster speeds are needed, one will have to go for faster processors, efficient computing algorithms, and computers with faster operation cycles. In order to carry out digital control and other digital actions at much faster speeds (MHz speed, μs cycle time), one will have to rely on hardware (not software) solutions using dedicated analog and digital electronics.

This chapter first identifies the basic elements of mechanical, electrical, and electronic systems. Then it studies some fundamental characteristics and principles of mechanical, electrical, and electronic components.

2.2 Mechanical Elements

The three basic elements in a mechanical system are mass (or inertia element), spring (or flexibility element), and damper (or dissipative element). The input elements (or source elements) of a mechanical system are the force source, where its force is the independent variable, which is not affected by the changes in the system (while the associated velocity variable—the dependent variable—will be affected) and the velocity source, where its velocity is the independent variable, which is not affected by the changes in the system (while the associated force variable—the dependent variable—will be affected). These are "ideal" sources since in practice the source variable will be affected to some extent by the dynamics of the system and is not completely "independent." Now we express the physical equations (i.e., constitutive equations) of the three basic mechanical elements, relating velocity (an across variable) and force or torque (a through variable).

2.2.1 Mass (Inertia) Element

Consider the mass element shown in Figure 2.1a. The constitutive equation (the physical law) of the element is given by Newton's second law:

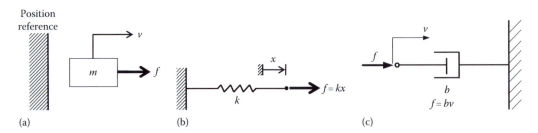

FIGURE 2.1
Basic mechanical elements: (a) Mass (inertia); (b) spring (stiffness element); (c) damper (dissipating element).

$$m\frac{dv}{dt} = f \tag{2.1}$$

Here

v denotes the velocity of mass m, measured relative to an inertial (fixed on earth) reference

f is the force applied "through" the mass

Since power $= fv =$ rate of change of energy, by substituting Equation 2.1, the energy of the element may be expressed as

$$E = \int fv\,dt = \int m\frac{dv}{dt}v\,dt = \int mv\,dv$$

or

$$\text{Energy, } E = \frac{1}{2}mv^2 \tag{2.2}$$

This is the well-known *kinetic energy*. Now by integrating Equation 2.1, we have

$$v(t) = v(0^-) + \frac{1}{m}\int_{0^-}^{t} f\,dt \tag{2.3}$$

By setting $t = 0^+$ in Equation 2.3, provided force f is finite, we see that

$$v(0^+) = v(0^-) \tag{2.4}$$

Note that 0^- denotes the instant just before $t = 0$ and 0^+ denotes the instant just after $t = 0$. In view of these observations, we may state the following:

1. Inertia is an energy storage element (kinetic energy).
2. Velocity can represent the state of an inertia element. This is justified by two reasons: first, from Equation 2.3, the velocity at any time t can be completely determined with the knowledge of the initial velocity and the applied force during the time interval 0 to t. Second, from Equation 2.2, the energy of an inertia element can be represented by the variable v alone.
3. Velocity across an inertia element cannot change instantaneously unless an infinite force is applied to it.
4. A finite force cannot cause an infinite acceleration (or step change in velocity) in an inertia element. Conversely, a finite instantaneous (step) change in velocity will need an infinite force. Hence, v is a natural output (or response) variable for an inertia element, which can represent its dynamic state (i.e., state variable), and f is a natural input variable for an inertia element.
5. Since its state variable, velocity, is an across variable, inertia is an *A*-type element.

Since its dynamic state is represented by velocity, an across variable, mass (or inertia element) is termed an *A*-type element.

2.2.2 Spring (Stiffness) Element

Consider the spring element (linear) shown in Figure 2.1b. The constitutive equation (physical law) for a spring is given by Hook's law:

$$\frac{df}{dt} = kv \tag{2.5}$$

Here, k is the stiffness of the spring.

Note: We have differentiated the familiar force-deflection Hooke's law in order to be consistent with the response/state variable (velocity) that is used for its counterpart, the inertia element.

Now, following the same steps as for the inertia element, the energy of a spring element may be expressed as

$$E = \int fv\, dt = \int f \frac{1}{k} \frac{df}{dt}\, dt = \int \frac{1}{k} f\, df$$

or

$$\text{Energy, } E = \frac{1}{2} \frac{f^2}{k} \tag{2.6}$$

This is the well-known (elastic) *potential energy.*
Also,

$$f(t) = f(0^-) + k \int_{0^-}^{t} v\, dt \tag{2.7}$$

We see that as long as the applied velocity is finite,

$$f(0^+) = f(0^-) \tag{2.8}$$

In summary, we have

1. A spring (stiffness element) is an energy storage element (elastic potential energy).
2. Force can represent the state of a stiffness (spring) element. This is justified by two reasons: first, from Equation 2.7, the force of a spring at any general time t may be completely determined with the knowledge of the initial force and the applied velocity from time 0 to t. Second, from Equation 2.6, the energy of a spring element can be represented in terms of the variable f alone.
3. Force through a stiffness element cannot change instantaneously unless an infinite velocity is applied to it.
4. Force f is a natural output (response) variable, which can represent its dynamic state (i.e., state variable), and v is a natural input variable for a stiffness element.
5. Since its state variable, force, is a through variable, a spring is a *T*-type element.

Since its dynamic state is represented by force (or torque), a through variable, spring (or flexibility element) is termed a *T*-type element.

2.2.3 Damping (Dissipation) Element

Consider the mechanical damper (linear viscous damper or dashpot) shown in Figure 2.1c. It is termed a *D*-type element (energy dissipating element). The constitutive equation (physical law) is

$$f = bv \tag{2.9}$$

where b is the damping constant. Equation 2.9 is an algebraic equation. Hence, either f or v can serve as the natural output variable for a damper, and either one can determine its state. However, since the state variables v and f are established by an independent inertial element and an independent spring element, respectively, a damper will not introduce a new state variable.

In summary:

1. A mechanical damper is an energy dissipating element (*D*-type element).
2. Either force f or velocity v may represent its state.
3. No new state variable is defined by this element.

2.3 Fluid Elements

In a fluid component, pressure (P) is the across variable and the volume flow rate (Q) is the through variable. The three basic fluid elements are shown in Figure 2.2 and are discussed below. Note the following:

1. The elements are usually distributed, but lumped-parameter approximations are used here.
2. The elements are usually nonlinear (particularly, the fluid resistor), but linear models are used here.

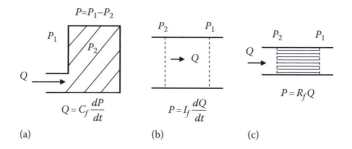

FIGURE 2.2
Basic fluid elements: (a) Capacitor; (b) inertor; (c) resistor.

The input elements (or source elements) of a fluid system are the pressure source, where its pressure is the independent variable, which is not affected by the changes in the system (while the associated flow rate variable—the dependent variable—will be affected); and the flow source, where its flow rate is the independent variable, which is not affected by the changes in the system (while the associated pressure variable—the dependent variable—will be affected).

2.3.1 Fluid Capacitor or Accumulator (*A*-Type Element)

Consider a rigid container with a single inlet through which fluid is pumped in at the volume flow rate Q, as shown in Figure 2.2a. The pressure inside the container with respect to the outside is P. We can write the linear constitutive equation as

$$Q = C_f \frac{dP}{dt} \tag{2.10}$$

where C_f is the fluid capacitance (capacity). Several special cases of fluid capacitance will be discussed later.

A fluid capacitor stores potential energy, given by $(1/2)C_f P^2$. Hence, this element is like a fluid spring. The appropriate state variable is the pressure difference (across variable) P. Contrast here that the mechanical spring is a *T*-type element.

2.3.2 Fluid Inertor (*T*-Type Element)

Consider a conduit carrying an accelerating flow of fluid, as shown in Figure 2.2b. The associated linear constitutive equation may be written as

$$P = I_f \frac{dQ}{dt} \tag{2.11}$$

where I_f is the fluid inertance (inertia).

A fluid inertor stores kinetic energy, given by $(1/2)I_f Q^2$. Hence, this element is a fluid inertia. The appropriate state variable is the volume flow rate (through variable) Q. Contrast here that the mechanical inertia is an *A*-type element. An energy exchange between a fluid capacitor and a fluid inertor leads to oscillations (e.g., water hammer) in fluid systems, analogous to oscillations in mechanical and electrical systems.

2.3.3 Fluid Resistor (*D*-Type Element)

Consider the flow of fluid through a narrow element such as a thin pipe, orifice, or valve. The associated flow will result in energy dissipation due to fluid friction. The linear constitutive equation is (see Figure 2.2c)

$$P = R_f Q \tag{2.12}$$

2.3.4 Derivation of Constitutive Equations

We now indicate the derivation of the constitutive equations for fluid elements.

2.3.4.1 *Fluid Capacitor*

The capacitance in a fluid element may originate from the following:

1. Bulk modulus effects of liquids
2. Compressibility effects of gases
3. Flexibility of the fluid container itself
4. Gravity head of a fluid column

The derivation of the associated constitutive equations is outlined below.

2.3.4.1.1 *Bulk Modulus Effect of Liquids*

Consider a rigid container. A liquid is pumped in at the volume flow rate Q. An increase of the pressure in the container will compress the liquid volume, thereby letting in more liquid (see Figure 2.3a). From calculus we have

$$\frac{dP}{dt} = \frac{\partial P}{\partial V}\frac{dV}{dt}$$

where V is the control volume of liquid. Now, the volume flow rate (into the container) is given by

$$Q = -\frac{dV}{dt}$$

The bulk modulus of liquid is defined by

$$\beta = -V\frac{\partial P}{\partial V} \tag{2.13}$$

Hence,

$$\frac{dP}{dt} = \frac{\beta}{V}Q \quad \text{or} \quad Q = \frac{V}{\beta}\frac{dP}{dt} \tag{2.14}$$

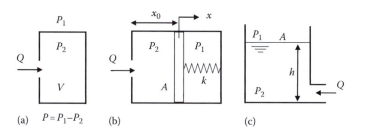

(a) $P = P_1 - P_2$ (b)   (c)

FIGURE 2.3
Three types of fluid capacitance: (a) Bulk modulus or compressibility; (b) flexibility of container; (c) gravity head of fluid column.

and the associated capacitance is

$$C_{bulk} = \frac{V}{\beta}$$

(2.15)

2.3.4.1.2 Compression of Gases

Consider a perfect (ideal) gas, which is governed by the gas law

$$PV = mRT$$

(2.16)

where
P is the pressure (units are Pascal's: $1\,Pa = 1\,N/m^2$)
V is the volume (units are m^3)
T is the absolute temperature (units are °K or degrees Kelvin)
m is the mass (units are kg)
R is the specific gas constant (units: kJ/kg/°K where $1\,J = 1$ joule $= 1\,N \cdot m$; $1\,kJ = 1000\,J$).

Isothermal case: consider a slow flow of gas into a rigid container (see Figure 2.3a) so that the heat transfer is allowed to maintain the temperature constant (isothermal). Differentiate (2.16) keeping T constant (i.e., RHS is constant):

$$P\frac{dV}{dt} + V\frac{dP}{dt} = 0$$

Noting that $Q = -(dV/dt)$ and substituting into this the above equation and (2.16), we get

$$Q = \frac{V}{P}\frac{dP}{dt} = \frac{mRT}{P^2}\frac{dP}{dt}$$

(2.17)

Hence, the corresponding capacitance is given by

$$C_{comp} = \frac{V}{P} = \frac{mRT}{P^2}$$

(2.18)

Adiabatic case: consider a fast flow of gas (see Figure 2.3a) into a rigid container so that there is no time for heat transfer (adiabatic $\Rightarrow$ zero heat transfer). The associated gas law is known to be

$$PV^k = C \quad \text{with} \quad k = \frac{C_p}{C_v}$$

(2.19)

where
C_p is the specific heat when the pressure is maintained constant
C_v is the specific heat when the volume is maintained constant
C is the constant
k is the ratio of specific heats

By differentiating (2.19), we get

$$PkV^{k-1}\frac{dV}{dt} + V^k\frac{dP}{dt} = 0$$

Divide by V^k

$$\frac{Pk}{V}\frac{dV}{dt} + \frac{dP}{dt} = 0$$

Now use $Q = -(dV/dt)$ as usual, and also substitute (2.16):

$$Q = \frac{V}{kP}\frac{dP}{dt} = \frac{mRT}{kP^2}\frac{dP}{dt} \qquad (2.20)$$

The corresponding capacitance is

$$C_{comp} = \frac{V}{kP} = \frac{mRT}{kP^2} \qquad (2.21)$$

2.3.4.1.3 Effect of Flexible Container

Without loss of generality, consider a cylinder of cross-sectional area A with a spring-loaded wall (stiffness k) as shown in Figure 2.3b. As a fluid (assumed incompressible) is pumped into the cylinder, the flexible wall will move through x.

Conservation of flow:

$$Q = \frac{d(A(x_0 + x))}{dt} = A\frac{dx}{dt} \qquad (i)$$

Equilibrium of spring:

$$A(P_2 - P_1) = kx \quad \text{or} \quad x = \frac{A}{k}P \qquad (ii)$$

By substituting (ii) in (i), we get

$$Q = \frac{A^2}{k}\frac{dP}{dt} \qquad (2.22)$$

The corresponding capacitance is

$$C_{elastic} = \frac{A^2}{k} \qquad (2.23)$$

Note: For an elastic container and a fluid having bulk modulus, the combined capacitance will be additive:

$$C_{eq} = C_{bulk} + C_{elastic}$$

A similar result holds for a compressible gas and an elastic container.

2.3.4.1.4 *Gravity Head of a Fluid Column*

Consider a liquid column (tank) having area of a cross section A, height h, and mass density ρ, as shown in Figure 2.3c. A liquid is pumped into the tank at the volume rate Q. As a result, the liquid level rises.

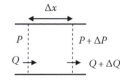

FIGURE 2.4
A fluid flow element.

Relative pressure at the foot of the column $P = P_2 - P_1 = \rho g h$

Flow rate $Q = \dfrac{d(Ah)}{dt} = A\dfrac{dh}{dt}$

Direct substitution gives

$$Q = \frac{A}{\rho g}\frac{dP}{dt} \tag{2.24}$$

The corresponding capacitance is

$$C_{grav} = \frac{A}{\rho g} \tag{2.25}$$

2.3.4.2 Fluid Inertor

First, assume a fluid flow in a conduit with a uniform velocity distribution across it. Along a small element of length Δx of fluid, as shown in Figure 2.4, the pressure will change from P to $P + \Delta P$, and the volume flow rate will change from Q to $Q + \Delta Q$.

Mass of the fluid element $= \rho A \Delta x$
Net force in the direction of flow $= -\Delta P A$
Velocity of flow $= Q/A$

where
  ρ is the mass density of the fluid
  A is the area of the cross section
  Assuming A to be constant, we have

$$\text{Fluid acceleration} = \frac{1}{A}\frac{dQ}{dt}$$

Hence, Newton's second law gives

$$-\Delta P A = (\rho A \Delta x)\frac{1}{A}\frac{dQ}{dt}$$

or

$$-\Delta P = \frac{\rho \Delta x}{A}\frac{dQ}{dt} \tag{2.26}$$

Hence,

$$\text{Fluid inertance, } I_f = \frac{\rho \Delta x}{A} \tag{2.27a}$$

where a nonuniform cross section, $A = A(x)$, is assumed. Then, for a length L, we have

$$I_f = \int_0^L \frac{\rho}{A(x)} dx \qquad (2.27b)$$

For a circular cross section and a parabolic velocity profile, we have

$$I_f = \frac{2\rho \Delta x}{A} \qquad (2.27c)$$

or, in general

$$I_f = \alpha \frac{\rho \Delta x}{A} \qquad (2.27)$$

where α is a suitable correction factor.

2.3.4.3 Fluid Resistor

For the ideal case of viscous, laminar flow we have (Figure 2.2c)

$$P = R_f Q \qquad (2.28)$$

with
$R_f = 128(\mu L / \pi d^4)$ for a circular pipe of diameter d.
$R_f = 12(\mu L / wb^3)$ for a pipe of a rectangular cross section (width w and height b) with
  $b \ll w$

where
  L is the length of pipe segment
  μ is the absolute viscosity of fluid (dynamic viscosity)
  Note: Fluid stress $= \mu(du/dy)$, where du/dy is the velocity gradient across the pipe.
    $\upsilon = (\mu / \rho) =$ kinematic viscosity
    Reynold's number, $R_e = uL/\upsilon = \rho uL/\mu$
    u is the fluid velocity along the pipe

For turbulent flow, the resistance equation will be nonlinear, as given by

$$P = K_R Q^n \qquad (2.29)$$

2.4 Thermal Elements

Thermal systems have temperature (T) as the across variable, as it is always measured with respect to some reference (or as a temperature difference across an element), and heat transfer (flow) rate (Q) as the through variable. The heat source and temperature source are the two types of source elements (inputs). The former type of source is more common. The latter type of source may correspond to a large reservoir whose temperature is hardly affected by heat transfer into or out of it. There is only one type of energy (thermal energy) in a thermal system. Hence, there is only one type (*A*-type) of energy storage element (a thermal capacitor) with the associated state variable, temperature. There is no *T*-type element in a thermal system (i.e., there are no thermal inductors). As a direct result of the absence of two different types of energy storage elements (unlike the case of mechanical, electrical, and fluid systems), a pure thermal system cannot exhibit natural oscillations. It can exhibit "forced" oscillations, however, when excited by an oscillatory input source.

The constitutive equations in a thermal system are the physical equations for thermal capacitors (*A*-type elements) and thermal resistors (*D*-type elements). There are no *T*-type elements. There are three types of thermal resistance: conduction, convection, and radiation.

2.4.1 Thermal Capacitor

Consider a control volume of an object with various heat transfer processes Q_i taking place at the boundary of the object (see Figure 2.5). The level of thermal energy in the object $= \rho\, VcT$,

where
 T is the temperature of the object (assumed uniform)
 V is the volume of the object
 ρ is the mass density of the object
 c is the specific heat of the object

Since the net heat inflow is equal to the rate of change (increase) of thermal energy, the associated constitutive relation is

$$\sum Q_i = \rho Vc \frac{dT}{dt} \tag{2.30}$$

where ρVc is assumed to be constant. We write this as

$$Q = C_h \frac{dT}{dt} \tag{2.31}$$

where
 $C_h = \rho Vc = mc =$ thermal capacitance
 Here, $m = \rho V$ is the mass of the element
 Note: Thermal capacitance means the "capacity" to store thermal energy in a body

2.4.2 Thermal Resistor

A thermal resistor provides resistance to heat transfer in a body or a medium. There are three general types of thermal resistance:

- Conduction
- Convection
- Radiation

We will now give constitutive relations for each of these three types of thermal resistors.

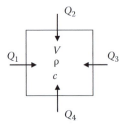

FIGURE 2.5
A control volume of a thermal system.

2.4.2.1 Conduction

The heat transfer in a medium takes place by conduction when the molecules of the medium itself do not move to transfer the heat. Heat transfer takes place from a point of higher temperature to one of lower temperature. Specifically, the heat conduction rate is proportional to the negative temperature gradient and is given by the *Fourier equation*:

$$Q = -kA \frac{\partial T}{\partial x} \tag{2.32}$$

where
  x is the direction of heat transfer
  A is the area of the cross section of the element along which heat transfer takes place
  k is the thermal conductivity

The (Fourier) equation 2.32 is a "local" equation. If we consider a finite object of length Δx and cross section A with temperatures T_2 and T_1 at the two ends, as shown in Figure 2.6a, the one-dimensional heat transfer rate Q can be written according Equation 2.32 as

$$Q = kA \frac{(T_2 - T_1)}{\Delta x} \tag{2.33a}$$

or

$$Q = \frac{1}{R_k} (T_2 - T_1) \tag{2.33b}$$

where

$$R_k = \frac{\Delta x}{kA} = \text{conductive thermal resistance} \tag{2.34}$$

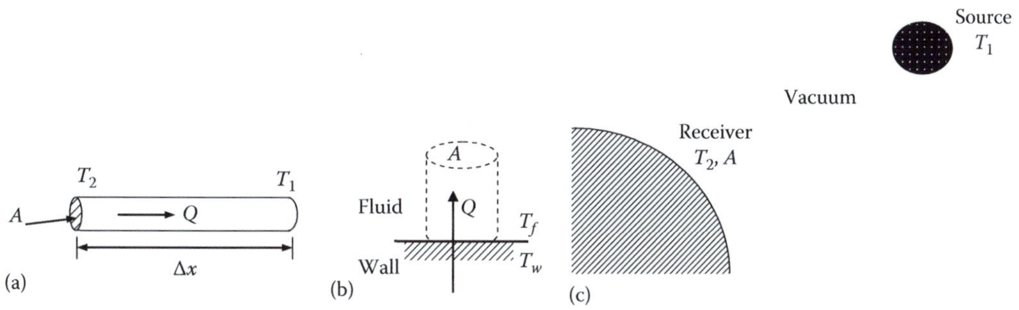

FIGURE 2.6
Three types of thermal resistance: (a) An element of 1-D heat conduction; (b) a control volume for heat transfer by convection; (c) heat transfer by radiation.

2.4.2.2 Convection

In convection, the heat transfer takes place by the physical movement of the heat-carrying molecules in the medium. An example is the case of fluid flowing against a wall, as shown in Figure 2.6b. The constitutive equation is

$$Q = h_c A(T_w - T_f) \tag{2.35a}$$

where

T_w is the wall temperature
T_f is the fluid temperature at the wall interface
A is the area of the cross section of the fluid control volume across which heat transfer
   Q takes place
h_c is the convection heat transfer coefficient

In practice, h_c may depend on the temperature itself, and hence Equation 2.35a is nonlinear in general. But, by approximating to a linear constitutive equation, we can write

$$Q = \frac{1}{R_c}(T_w - T_f) \tag{2.35b}$$

where

$$R_c = \frac{1}{h_c A} = \text{convective thermal resistance} \tag{2.36}$$

In *natural convection*, the particles in the heat transfer medium move naturally. In *forced convection*, they are moved by an actuator such as a fan or pump.

2.4.2.3 Radiation

In radiation, the heat transfer takes place from a higher temperature object (source) to a lower temperature object (receiver) through energy radiation, without needing a physical medium between the two objects (unlike in conduction and convection), as shown in Figure 2.6c. The associated constitutive equation is the Stefan–Boltzman law:

$$Q = \sigma c_e c_r A(T_1^4 - T_2^4) \tag{2.37a}$$

where

A is the effective (normal) area of the receiver
c_e is the effective emmissivity of the source
c_r is the shape factor of the receiver
σ is the Stefan–Boltzman constant ($=5.7 \times 10^{-8}$ W/m²/°K⁴)

This corresponds to a nonlinear thermal resistor.
   The heat transfer rate is measured in watts (W), the area is measured in square meters (m²), and the temperature is measured in degrees Kelvin (°K). The relation (2.37a) is nonlinear, which may be linearized as

$$Q = \frac{1}{R_r}(T_1 - T_2) \tag{2.37b}$$

where R_r = radiation thermal resistance.

Since the slope $\partial Q / \partial T$ at an operating point may be given by $4\sigma c_e c_r A \overline{T}^3$, where $\overline{T}$ is the representative temperature (which is variable) at the operating point, we have

$$R_r = \frac{1}{4\sigma c_e c_r A \overline{T}^3} \tag{2.38a}$$

Alternatively, since $T_1^4 - T_2^4 = (T_1^2 + T_2^2)(T_1 + T_2)(T_1 - T_2)$, we may use the approximate expression

$$R_r = \frac{1}{\sigma c_e c_r A (\overline{T}_1^2 + \overline{T}_2^2)(\overline{T}_1 + \overline{T}_2)} \tag{2.38b}$$

where the over-bar denotes a representative (operating point) temperature.

2.4.3 Three-Dimensional Conduction

Conduction heat transfer in a continuous three-dimensional medium is represented by a distributed-parameter model. In this case, the Fourier equation (2.32) is applicable in each of the three orthogonal directions (x, y, z). To obtain a model, the thermal capacitance equation (2.30) has to be applied as well.

Consider the small three-dimensional model element of sides dx, dy, and dz in a conduction medium, as shown in Figure 2.7. First consider heat transfer into the bottom ($dx \times dy$) surface in the z direction, which according to (2.32) is $-k\,dx\,dy(\partial T/\partial z)$.

Since the temperature gradient at the top ($dx \times dy$) surface is $(\partial T/\partial z) + (\partial^2 T/\partial z^2 dx$ (from calculus), the heat transfer out of this surface is $k\,dx\,dy((\partial T/\partial z) + (\partial^2 T/\partial z^2)dz)$. Hence, the net heat transfer into the element in the z direction is $k\,dx\,dy(\partial^2 T/\partial z^2)dz$ or $k\,dx\,dy\,dz(\partial^2 T/\partial z^2)$. Similarly, the net heat transfer in the x and y directions are $k\,dx\,dy\,dz(\partial^2 T/\partial x^2)$ and $k\,dx\,dy\,dz(\partial^2 T/\partial y^2)$, respectively.

The thermal energy of the element is $\rho\,dx\,dy\,dz\,cT$ where $\rho\,dx\,dy\,dz$ is the mass of the element and c is the specific heat (at constant pressure). Hence, the capacitance equation (2.30) gives

$$k\,dx\,dy\,dz\left(\frac{\partial^2 T}{\partial x^2} + \frac{\partial^2 T}{\partial y^2} + \frac{\partial^2 T}{\partial z^2}\right) = \rho\,dx\,dy\,dz\,c\,\frac{\partial T}{\partial t}$$

or

$$\frac{\partial^2 T}{\partial x^2} + \frac{\partial^2 T}{\partial y^2} + \frac{\partial^2 T}{\partial z^2} = \frac{1}{\alpha}\frac{\partial T}{\partial t} \tag{2.39}$$

where $\alpha = (k/\rho c) =$ thermal diffusivity.

Equation 2.39 is called the *Laplace equation*. Note that partial derivatives are used because T is a function of many variables; and derivatives with respect to x, y, z, and t would be needed. Hence, in general, distributed-parameter models have spatial

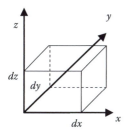

FIGURE 2.7
A 3-D heat conduction element.

variables (x, y, z) as well as the temporal variable (t) as independent variables and are represented by partial differential equations.

2.4.4 Biot Number

This is a nondimesional parameter giving the ratio: (conductive resistance)/(convective resistance). Hence, from Equations 2.34 and 2.36, we have

$$\text{Biot number} = \frac{R_k}{R_c} = \frac{\Delta x h_c A}{kA} = \frac{h_c \Delta x}{k} \tag{2.40}$$

This parameter may be used as the basis for approximating the distributed-parameter model (2.39) by a lumped parameter one. Specifically, divide the conduction medium into slabs of thickness Δx. If the corresponding Biot number ≤ 0.1, a lumped-parameter model may be used for each slab.

2.5 Mechanical Components

Common mechanical components in a mechatronic system may be classified into some useful groups, as follows:

1. Load bearing/structural components (strength and surface properties)
2. Fasteners (strength)
3. Dynamic isolation components (transmissibility)
4. Transmission components (motion conversion)
5. Mechanical actuators (generated force/torque)
6. Mechanical controllers (controlled energy dissipation)

In each category, we have indicated within parentheses the main property or attribute that is characteristic of the function of that category. The analytical results in solid mechanics, as given in Appendix A, are particularly useful in the design and development of these components.

In load bearing or structural components, the main function is to provide structural support. In this context, mechanical strength and surface properties (e.g., hardness, wear resistance, friction) of the component are crucial. The component may be rigid or flexible and stationary or moving. Examples of load bearing and structural components include bearings, springs, shafts, beams, columns, flanges, and similar load-bearing structures.

Fasteners are closely related to load bearing/structural components. The purpose of a fastener is to join two mechanical components. Here as well, the primary property of importance is the mechanical strength. Examples are bolts and nuts, locks and keys, screws, rivets, and spring retainers. Welding, bracing, and soldering are processes of fastening and will fall into the same category.

Dynamic isolation components perform the main task of isolating a system from another system (or environment) with respect to motion and forces. These involve the "filtering" of motions and forces/torques. Hence, motion transmissibility and force transmissibility are the key considerations in these components. Springs, dampers, and inertia elements may form the isolation element. Shock and vibration mounts for machinery, inertia blocks, and the suspension systems of vehicles are examples of isolation dynamic components.

Transmission components may be related to isolation components in principle, but their functions are rather different. The main purpose of a transmission component is the conversion of motion (in magnitude and from). In the process, the force/torque of the input member is also converted in magnitude and form. In fact, in some applications, the modification of the force/torque may be the primary requirement of the transmission component. Examples of transmission components are gears, lead screws and nuts (or power screws), racks and pinions, cams and followers, chains and sprockets, belts and pulleys (or drums), differentials, kinematic linkages, flexible couplings, and fluid transmissions.

Mechanical actuators are used to generate forces (and torques) for various applications. The common actuators are electromagnetic in form (i.e., electric motors) and are not purely mechanical. Since the magnetic forces are "mechanical" forces that generate mechanical torques, electric motors may be considered as electromechanical devices. Other types of actuators that use fluids for generating the required effort may be considered in the category of mechanical actuators. Examples are hydraulic pistons and cylinders (rams), hydraulic motors, their pneumatic counterparts, and thermal power units (prime movers) such as steam/gas turbines. Of particular interest in mechatronic systems are the electromechanical actuators and hydraulic and pneumatic actuators.

Mechanical controllers perform the task of modifying dynamic response (motion and force/torque) in a desired manner. Purely mechanical controllers carry out this task by the controlled dissipation of energy. These are not as common as electrical/electronic controllers and hydraulic/pneumatic controllers. In fact, hydraulic/pneumatic servovalves may be treated in the category of purely mechanical controllers. Furthermore, mechanical controllers are closely related to transmission components and mechanical actuators. Examples of mechanical controllers are clutches and brakes.

In selecting a mechanical component for a mechatronic application, many engineering aspects have to be considered. The foremost are the capability and performance of the component with respect to the design requirements (or specifications) of the system. For example, motion and torque specifications, flexibility and deflection limits, strength characteristics including stress–strain behavior, failure modes and limits, fatigue life, surface and material properties (e.g., friction, nonmagnetic, noncorrosive), operating range, and design life will be important. Other factors such as size, shape, cost, and commercial availability can be quite crucial.

The foregoing classification of mechanical components is summarized in Figure 2.8. It is not within the scope of this chapter to study all the types of mechanical components that are summarized here. Rather, we select for further analysis a few important mechanical components that are particularly useful in mechatronic systems.

2.5.1 Transmission Components

Transmission devices are indispensable in mechatronic applications. We will discuss a few representative transmission devices here. It should be noted that in the present treatment, a transmission is isolated and treated as a separate unit. In an actual application, however, a transmission device works as an integral unit with other components, particularly the actuator, electronic drive unit, and the load of the system. Hence, a transmission design

Mechanical components

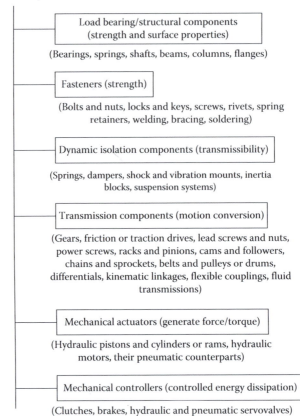

FIGURE 2.8
Classification of mechanical components.

or selection should involve an integrated treatment of all interacting components. This should be clear in the subsequent chapters (see Chapters 3, 4, 7, and 10 in particular).

Perhaps the most common transmission device is a gearbox. In its simplest form, a gearbox consists of two gear wheels, which contain teeth of identical pitch (tooth separation) and of unequal wheel diameter. The two wheels are meshed (i.e., the teeth are engaged) at one location. This device changes the rotational speed by a specific ratio (gear ratio) as dictated by the ratio of the diameters (or radii) of the two gear wheels. In particular, by stepping down the speed (in which case the diameter of the output gear is larger than that of the input gear), the output torque can be increased. Larger gear ratios can be realized by employing more than one pair of meshed gear wheels. Gear transmissions are used in a variety of applications including automotive, industrial-drive, and robotics. Specific gear designs range from conventional spur gears to harmonic drives, as discussed later in this section.

Gear drives have several disadvantages. In particular, they exhibit "backlash" because the tooth width is smaller than the tooth space of the mating gear. Some degree of backlash is necessary for proper meshing. Otherwise, jamming will occur. Unfortunately, backlash is a nonlinearity, which can cause irregular and noisy operation with brief intervals of zero torque transmission. It can lead to rapid wear and tear and even instability. The degree of backlash can be reduced by using proper profiles (shapes) for the gear teeth. Backlash can be eliminated through the use of spring-loaded gears. Sophisticated feedback control may be used as well to reduce the effects of gear backlash.

Conventional gear transmissions, such as those used in automobiles with standard gearboxes, contain several gear stages. The gear ratio can be changed by disengaging the drive-gear wheel (pinion) from a driven wheel of one gear stage and engaging it with another wheel of a different number of teeth (different diameter) of another gear stage, while the power source (input) is disconnected by means of a clutch. Such a gearbox provides only a few fixed gear ratios. The advantages of a standard gearbox include relative simplicity of design and the ease with which it can be adapted to operate over a reasonably wide range of speed ratios, albeit in a few discrete increments of large steps. There are many disadvantages: Since each gear ratio is provided by a separate gear stage, the size, weight, and complexity (and associated cost, wear, and unreliability) of the transmission increases directly with the number of gear ratios provided. Also, the drive source has to be disconnected by a clutch during the shifting of gears, the speed transitions are generally not smooth, and operation is noisy. There is also dissipation of power during the transmission steps, and wear and damage can be caused by inexperienced operators. These shortcomings can be reduced or eliminated if the transmission is able to vary the speed ratio continuously rather than in a stepped manner. Furthermore, the output speed and corresponding torque can be matched to the load requirements closely and continuously for a fixed input power. This results in a more efficient and smoother operation and many other related advantages. A continuously variable transmission, which has these desirable characteristics, will be discussed later in this section. First, we will discuss a power screw, which is a converter of angular motion into rectilinear motion.

2.5.2 Lead Screw and Nut

A lead-screw drive is a transmission component that converts rotatory motion into rectilinear motion. Lead screws, power screws, and ball screws are rather synonymous. Lead screw and nut units are used in numerous applications including positioning tables, machine tools, gantry and bridge systems, automated manipulators, and valve actuators. Figure 2.9 shows the main components of a lead-screw unit. The screw is rotated by a motor, and as a result, the nut assembly moves along the axis of the screw. The support block, which is attached to the nut, provides means for supporting the device that has to be moved using the lead-screw drive. The screw holes that are drilled on the support block may be used for this purpose. Since there can be backlash between the screw and the nut as a result of the assembly clearance and/or wear and tear, a keyhole is provided in the nut to apply a

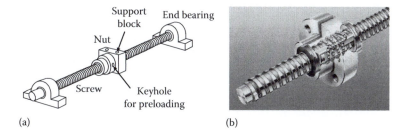

(a)   (b)

FIGURE 2.9
A lead screw and nut unit: (a) Assembly; (b) component details. (Courtesy of Deutsche Star GmbH, Schweinfurt, Germany.)

preload through some form of a clamping arrangement that
is designed into the nut. The end bearings support the moving load. Typically, these are ball bearings that can carry axial
loads as well by means of an angular-contact thrust bearing
design.

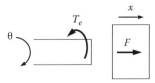
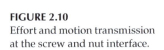

FIGURE 2.10
Effort and motion transmission
at the screw and nut interface.

The basic equation for the operation of a lead-screw drive
is obtained now. As shown in Figure 2.10, suppose that a
torque T_R is provided by the screw at (and reacted by) the
nut. Note that this is the net torque after deducting the inertia torque (due to the inertia of the motor rotor and the lead
screw) and the frictional torque of the bearings from the motor (magnetic) torque. Torque
T_R is not completely available to move the load that is supported on the nut. The reason is
the energy dissipation (friction) at the screw and nut interface. Suppose that the net force
available from the nut to drive the load in the axial direction is F. Denote the screw rotation
by θ and the rectilinear motion of the nut by x.

When the screw is rotated (by a motor) through $\delta\theta$, the nut (which is restrained from
rotating due to the guides along which the support block moves) will move through Δx
along the axial direction. The work done by the screw is $T_R \cdot \delta\theta$ and the work done in moving the nut (with its load) is $F \cdot \delta x$. The lead screw efficiency e is given by

$$e = \frac{F \cdot \delta x}{T_R \cdot \delta\theta} \tag{2.41}$$

Now $r\Delta\theta = \Delta x$, where the transmission parameter of the lead screw is r (axial distance
moved per one radian of screw rotation). The "lead" l of the lead screw is the axial distance
moved by the nut in one revolution of the screw, and it satisfies

$$l = 2\pi r \tag{2.42}$$

In general, the lead is not the same as the "pitch" p of the screw, which is the axial distance
between two adjacent threads. For a screw with n threads,

$$l = np \tag{2.43}$$

Substituting r in Equation 2.41, we have

$$F = \frac{e}{r} T_R = \frac{2\pi e}{l} T_R \tag{2.44}$$

This result is the representative equation of a lead screw and may be used in the design
and selection of components in a lead-screw drive system.

For a screw of mean diameter d, the helix angle α is given by

$$\tan\alpha = \frac{l}{\pi d} = \frac{2r}{d} \tag{2.45}$$

Assuming square threads, we obtain a simplified equation for the screw efficiency in terms
of the coefficient of friction μ. First, for a screw of 100% efficiency ($e = 1$), from Equation 2.44,
a torque T_R at the nut can support an axial force (load) of T_R/r. The corresponding frictional

force F_f is $\mu T_R/r$. The torque required to overcome this frictional force is $T_f = F_f d/2$. Hence, the frictional torque is given by

$$T_f = \frac{\mu d}{2r} T_R \tag{2.46}$$

The screw efficiency is

$$e = \frac{T_R - T_f}{T_R} = 1 - \frac{\mu d}{2r} = 1 - \frac{\mu}{\tan \alpha} \tag{2.47}$$

For threads that are not square (e.g., for slanted threads such as Acme threads, Buttress threads, modified square threads), Equation 2.47 has to be appropriately modified.

It is clear from Equation 2.46 that the efficiency of a lead screw unit can be increased by decreasing the friction and increasing the helix angle. Of course, there are limits. For example, typically the efficiency will not increase by increasing the helix angle beyond 30°. In fact, a helix angle of 50° or more will cause the efficiency to drop significantly. The friction can be decreased by a proper choice of material for the screw and nut and through surface treatments, particularly lubrication. Typical values for the coefficient of friction (for identical mating material) are given in Table 2.1. Not that the static (starting) friction will be higher (as much as 30%) than the dynamic (operating) friction. An ingenious way to reduce friction is by using a nut with a helical track of balls instead of threads. In this case, the mating between the screw and the nut is not through threads but through ball bearings. Such a lead screw unit is termed a *ball screw*. A screw efficiency of 90% or greater is possible with a ball screw unit.

In the driving mode of a lead screw, the frictional torque acts in the opposite direction to (and has to be overcome by) the driving torque. In the "free" mode where the load is not driven by an external torque from the screw, it is likely that the load will try to "back-drive" the screw (say, due to gravitational load). Then, however, the frictional torque will change direction and the back motion has to overcome it. If the back-driving torque is less than the frictional torque, motion will not be possible and the screw is said to be self-locking.

Example 2.1

A lead screw unit is used to drive a load of mass up an incline of angle θ, as shown in Figure 2.11. Under quasi-static conditions (i.e., neglecting inertial loads), determine the drive torque needed by the motor to operate the device. The total mass of the moving unit (load, nut, and fixtures) is *m*. The efficiency of the lead screw is *e* and the lead is *l*. Assume that the axial load (thrust) due to

TABLE 2.1

Some Useful Values for Coefficient of Friction

Material	Coefficient of Friction
Steel (dry)	0.2
Steel (lubricated)	0.15
Bronze	0.10
Plastic	0.10

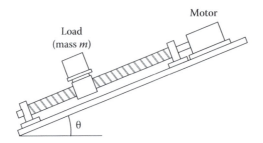

FIGURE 2.11
A lead-screw unit driving an inclined load.

gravity is taken up entirely by the nut (in practice, a significant part of the axial load is supported by the end bearings, which have the thrust-bearing capability).

Solution
The effective load that has to be acted upon by the net torque (after allowing for friction) in this example is

$$F = mg\sin\theta$$

Substitute into Equation 2.44. The required torque at the nut is

$$T_R = \frac{mgr}{e}\sin\theta = \frac{mgl}{2\pi e}\sin\theta \tag{2.48}$$

2.5.3 Harmonic Drives

Usually, motors run efficiently at high speeds. Yet, in many practical applications, low speeds and high toques are needed. A straightforward way to reduce the effective speed and increase the output torque of a motor is to employ a gear system with high gear reduction. Gear transmission has several disadvantages, however. For example, backlash in gears would be unacceptable in high-precision applications. Frictional loss of torque, wear problems, and the need for lubrication must also be considered. Furthermore, the mass of the gear system consumes energy from the actuator (motor) and reduces the overall torque/mass ratio and the useful bandwidth of the actuator.

A harmonic drive is a special type of transmission device that provides very large speed reductions (e.g., 200:1) without backlash problems. Also, a harmonic drive is comparatively much lighter than a standard gearbox. The harmonic drive is often integrated with conventional motors to provide very high torques, particularly in direct-drive and servo applications. The principle of operation of a harmonic drive is shown in Figure 2.12. The rigid circular spline of the drive is the outer gear and it has internal teeth. An annular flexispline has external teeth that can mesh with the internal teeth of the rigid spline in a limited region when pressed in the radial direction. The external radius of the flexispline is slightly smaller than the internal radius of the rigid spline. As the name implies, the flexispline undergoes some elastic deformation during the meshing process. This results in a tight mesh without any clearance between the meshed teeth, and hence the motion is backlash free.

In the design shown in Figure 2.12, the rigid spline is fixed and may also serve as the housing of the harmonic drive. The rotation of the flexispline is the output of the drive;

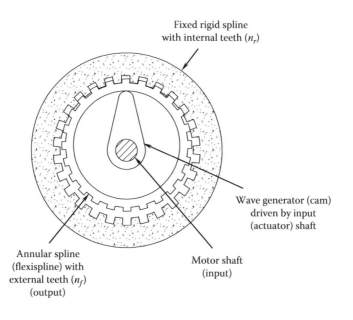

Fixed rigid spline with internal teeth (n_r)

Wave generator (cam) driven by input (actuator) shaft

Annular spline (flexispline) with external teeth (n_f) (output)

Motor shaft (input)

FIGURE 2.12
The principle of operation of a harmonic drive.

hence, it is connected to the driven load. The input shaft (motor shaft) drives the wave generator (represented by a cam in Figure 2.12). The wave generator motion brings about controlled backlash-free meshing between the rigid spline and the flexispline.

Suppose that

n_r is the number of teeth (internal) in the rigid spline

n_f is the number of teeth (external) in the flexispline

It follows that

the tooth pitch of the rigid spline $= 2\pi/n_r$ rad

the tooth pitch of the flexispline $= 2\pi/n_f$ rad

Furthermore, suppose that n_r is slightly smaller than n_f. Then, during a single tooth engagement, the flexispline rotates through $(2\pi/n_r - 2\pi/n_f)$ radians in the direction of rotation of the wave generator. During one full rotation of the wave generator, there will be a total of n_r tooth engagements in the rigid spline (which is stationary in this design). Hence, the rotation of the flexispline during one rotation of the wave generator (around the rigid spline) is

$$n_r \left(\frac{2\pi}{n_r} - \frac{2\pi}{n_f} \right) = \frac{2\pi}{n_f}(n_f - n_r)$$

It follows that the gear reduction ratio (r:1) representing the ratio: input speed/output speed is given by

$$r = \frac{n_f}{n_f - n_r} \qquad (2.49)$$

We can see that by making n_r very close to n_f, very high gear reductions can be obtained. Furthermore, since the efficiency of a harmonic drive is given by

$$\text{Efficiency, } e = \frac{\text{Output power}}{\text{Input power}} \qquad (2.50)$$

we have

$$\text{Output torque} = \frac{e n_f}{(n_f - n_r)} \times \text{Input torque} \qquad (2.51)$$

This result illustrates the torque amplification capability of a harmonic drive.

An inherent shortcoming of the harmonic drive sketched in Figure 2.12 is that the motion of the output device (flexispline) is eccentric (or epicyclic). This problem is not serious when the eccentricity is small (which is the case for typical harmonic drives) and is further reduced because of the flexibility of the flexispline. For improved performance, however, this epicyclic rotation has to be reconverted into a concentric rotation. This may be accomplished by various means, including flexible coupling and pin-slot transmissions. The output device of a pin-slot transmission is a flange that has pins arranged on the circumference of a circle centered at the axis of the output shaft. The input to the pin-slot transmission is the flexispline motion, which is transmitted through a set of holes on the flexispline. The pin diameter is smaller than the hole diameter, the associated clearance being adequate to take up the eccentricity in the flexispline motion. This principle is shown schematically in Figure 2.13. Alternatively, pins could be attached to the flexispline and the slots on the output flange. The eccentricity problem can be eliminated altogether by using a double-ended cam in place of the single-ended cam wave generator shown in Figure 2.12. With this new arrangement, meshing takes place at two diametrical ends simultaneously, and the flexispline is deformed elliptically in doing this. The center of rotation of the flexispline now coincides with the center of the input shaft. This double-mesh design is more robust and is quite common in industrial harmonic drives.

Other designs of harmonic drive are possible. For example, if $n_f < n_r$ then r in Equation 2.49 will be negative and the flexispline will rotate in the opposite direction to the wave generator (input shaft). Also, as indicated in the example below, the flexispline may be fixed and the rigid spline may serve as the output (rotating) member.

Traction drives (or friction drives) employ frictional coupling to eliminate backlash and overloading problems. These are not harmonic drives. In a traction drive, the drive member (input roller) is frictionally engaged with the driven member (output roller). The disadvantages of traction drives include indeterminacy of the speed ratio under slipping (overload) conditions and large size and weight for a specified speed ratio.

Example 2.2

An alternative design of a harmonic drive is sketched in Figure 2.14a. In this design, the flexispline is fixed. It loosely fits inside the rigid spline and is pressed against the internal teeth of the rigid spline at diametrically opposite locations. Tooth meshing occurs at these two locations only. The rigid spline is the output member of the harmonic drive (see Figure 2.14b).

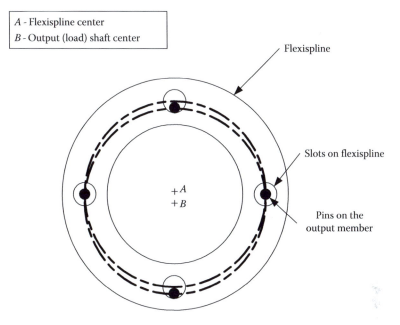

FIGURE 2.13
The principle of a pin-slot transmission.

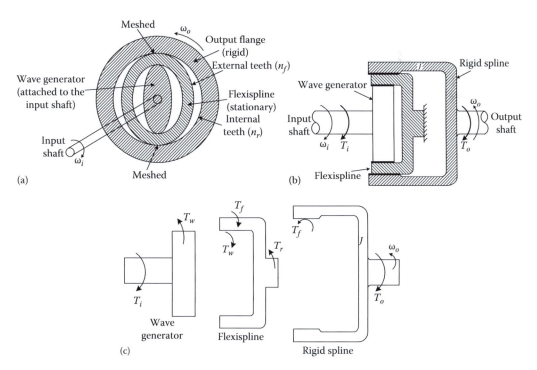

FIGURE 2.14
(a) An alternative design of harmonic drive; (b) torque and speed transmission of the harmonic drive; (c) free-body diagrams.

1. Show that the speed reduction ratio is given by

$$r = \frac{\omega_i}{\omega_f} = \frac{n_r}{(n_r - n_f)} \qquad (2.49b)$$

Note that if $n_f > n_r$, the output shaft will rotate in the opposite direction to the input shaft.

2. Now consider the free-body diagram shown in Figure 2.14c. The axial moment of inertia of the rigid spline is J. Neglecting the inertia of the wave generator, write approximate equations for the system. The variables shown in Figure 2.14c are defined as the following:

   T_i is the torque applied on the harmonic drive by the input shaft
   T_o is the torque transmitted to the driven load by the output shaft (rigid spline)
   T_f is the torque transmitted by the flexispline to the rigid spline
   T_r is the reaction torque on the flexispline at the fixture
   T_w is the torque transmitted by the wave generator

Solution

Part 1:
Suppose that n_r is slightly larger than n_f. Then, during a single tooth engagement, the rigid spline rotates through $(2\pi/n_f - 2\pi/n_r)$ radians in the direction of rotation of the wave generator. During one full rotation of the wave generator, there will be a total of n_f tooth engagements in the flexispline (which is stationary in the present design). Hence, the rotation of the rigid spline during one rotation of the wave generator (around the flexispline) is

$$n_f \left(\frac{2\pi}{n_f} - \frac{2\pi}{n_r} \right) = \frac{2\pi}{n_r} (n_r - n_f)$$

It follows that the gear reduction ratio (r:1) representing the ratio: input speed/output speed, is given by

$$r = \frac{n_r}{n_r - n_f} \qquad (2.49c)$$

It should be clear that if $n_f > n_r$, the output shaft will rotate in the opposite direction to the input shaft.

Part 2:
Equations of motion for the three components are as follows:

1. Wave generator
   Here, since inertia is neglected, we have

$$T_i - T_w = 0 \qquad (2.52a)$$

2. Flexispline
   Here, since the component is fixed, the equilibrium condition is

$$T_w + T_f - T_r = 0 \qquad (2.52b)$$

3. Rigid spline
   Newton's second law gives

$$T_f - T_o = J \frac{d\omega_o}{dt} \qquad (2.52c)$$

2.6 Passive Electrical Elements and Materials

A passive element does not rely on an external power source to exhibit its governing characteristics. In electrical systems, the capacitor is the *A*-type element with the voltage (across the variable) as its state variable and the inductor is the *T*-type element with current (through variable) as its state variable. These are energy storage elements and their constitutive equations are differential equations. The resistor is the energy dissipater (*D*-type element) and as usual, with an algebraic constitutive equation, it does not define a new state variable. These three elements are shown in Figure 2.15.

The input elements (or source elements) of an electrical system are the voltage source where its voltage is the independent variable, which is not affected by the changes in the system (while the associated current variable—the dependent variable—will be affected) and the current source, where its current is the independent variable, which is not affected by the changes in the system (while the associated voltage variable—the dependent variable—will be affected). These are "ideal" sources since in practice the source variable will be affected to some extent by the dynamics of the system and is not completely "independent."

2.6.1 Resistor (Dissipation) Element

Consider the resistor element shown in Figure 2.15a. It is a *D*-type element (energy dissipating element). The constitutive equation (physical law) is the well-known Ohm's law:

$$v = Ri \tag{2.53a}$$

where voltage applied across the conductor is v (volts), the current that will flow through the conductor is i (amperes), and R is the resistance of the resistor (measured in units of "ohm," denoted by Ω). Equation 2.53a is an algebraic equation. Hence, either v or i can serve as the natural output variable for a resistor and either one can determine its state. However, since the state variables v and i are established by an independent capacitor element and an independent inductor element, respectively, a damper will not introduce a new state variable.

In summary,

1. An electrical resistor is an energy dissipating element (*D*-type element)
2. Either current i or voltage v may represent its state
3. No new state variable is defined by this element

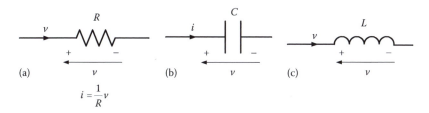

FIGURE 2.15
Basic passive electrical elements: (a) Resistor (dissipating element); (b) capacitor; (c) inductor.

2.6.1.1 Conductance and Resistance

The flow electrons or electric current in a linear conductive is governed by Ohm's law (2.53a), which can also be expressed as

$$i = Gv \qquad \text{(2.53b)}$$

where the conductance of the conductor is G. It is measured in the units of "mho" or "siemen" (denoted by S). Resistance is the inverse of conductance; thus

$$R = \frac{1}{G} \qquad \text{(2.54)}$$

Silver, copper, gold, and aluminum are good conductors of electricity.

2.6.1.2 Resistivity

The resistance increases with the length (L) of the conductor and decreases with the area of the cross section (A). The corresponding relationship is

$$R = \frac{\rho L}{A} \qquad \text{(2.55)}$$

The constant of proportionality ρ is the resistivity of the conducting material. Hence, resistivity may be defined as the resistance of a conductor of unity length and unity cross-sectional area. It may be expressed in the units $\Omega \cdot cm^2/cm$ or $\Omega \cdot cm$. A larger unit would be $\Omega \cdot m^2/m$ or $\Omega \cdot m$.

Alternatively, resistivity may be defined as the resistance of a conductor of unity length and unity diameter. According to this definition,

$$R = \frac{\hat{\rho} L}{d^2} \qquad \text{(2.56)}$$

where
  d represents the wire diameter
  $\hat{\rho}$ is the resistivity in the new units

Resistivities of several common materials are given in Table 2.2.

Example 2.3

Determine the conversion factor between ρ expressed in $\Omega \cdot m$ and $\hat{\rho}$ expressed in diameters of 1 cm and lengths of 1 m. How would you determine the resistance of a conductor that is 5.0 mm in diameter and 2.0 m in length using these two resistivity values?

Solution

Wire diameter = 1.0 cm = 1 × 10⁻² m

Wire length = 1.0 m

Wire resistance, $R = \rho \times 1.0 \left/ \frac{\pi}{4} (1 \times 10^{-2})^2 \right. \ \Omega$

or $R = \rho \times 1.273 \times 10^4 \ \Omega$

TABLE 2.2

Resistivities of Some Useful Materials

Material	Resistivity ρ ($\Omega \cdot$ m) at 20°C (68°F)
Aluminum	2.8×10^{-8}
Carbon	4000.0×10^{-8}
Constantan	44.0×10^{-8}
Copper	1.7×10^{-8}
Ferrite (manganese-zinc)	20.0
Gold	2.4×10^{-8}
Graphite carbon	775.0×10^{-8}
Iron	10.0×10^{-8}
Lead	9.6×10^{-8}
Magnesium	45.8×10^{-8}
Mercury	20.4×10^{-8}
Nichrome	112.0×10^{-8}
Polyester	1×10^{10}
Polystyrene	1×10^{16}
Porcelain	1×10^{16}
Silver	1.6×10^{-8}
Steel	15.9×10^{-8}
Tin	11.5×10^{-8}
Tungsten	5.5×10^{-8}

Hence, multiply ρ by 1.273×10^4 to obtain the resistivity $\hat{\rho}$ in $\Omega \cdot$ cm-diam/m.
Now, according to Equation 2.56, the resistance of the given conductor is

$$R = \frac{\hat{\rho} 2.0}{(0.5)^2} \, \Omega = 8.0 \times \hat{\rho} \, \Omega = 8.0 \times 1.273 \times 10^4 \times \rho \, \Omega$$

Hence, multiply $\hat{\rho}$ by 8.0 or ρ by $8.0 \times 1.273 \times 10^4$ to determine the resistance of the given conductor.

2.6.1.3 Effect of Temperature on Resistance

The resistance of a typical metal increases with temperature. The resistance decreases with temperature for many nonmetals and semiconductors. Typically, temperature effects on hardware have to be minimized in precision equipment and temperature compensation or calibration would be necessary. On the other hand, high temperature sensitivity of resistance in some materials is exploited in temperature sensors such as resistance temperature detectors (RTDs) and *thermistors*. The sensing element of a RTD is made of a metal such as nickel, copper, platinum, or silver. If the temperature variation is not too large, the following linear relationship is valid:

$$R = R_0(1 + \alpha \cdot \Delta t) \tag{2.57}$$

where
   R is the final resistance
   R_0 is the initial resistance
   ΔT is the change in temperature
   α is the temperature coefficient of resistance

Values of α for several common materials are given in Table 2.3. Each of these values can be expressed in ppm/°C (parts per million per degree centigrade) by multiplying by 10^6. Note that graphite has a negative temperature coefficient and nichrome has a very low temperature coefficient of resistance. A platinum RTD can operate accurately over a wide temperature range and possesses a high sensitivity (typically $0.4\,\Omega/°C$).

Thermistors are made of semiconductor material such as oxides of cobalt, copper, manganese, and nickel. Their resistance decreases with temperature. The relationship is nonlinear and is given approximately by

$$R = R_0 e^{-\beta\left(\frac{1}{T_0} - \frac{1}{T}\right)} \tag{2.58}$$

where
   the temperatures T and T_0 are in absolute degrees (°K or °R)
   R and R_0 are the corresponding resistances
   the parameter β is a material constant

2.6.1.4 Effect of Strain on Resistance

The property of resistance change with strain in materials is termed *piezoresistivity*. This property is used in strain gages, in particular. A foil strain gage uses a metallic foil (e.g., a copper-nickel alloy called constantan) as its sensing element. A semiconductor strain

TABLE 2.3

Temperature Coefficients of Resistance for Several Materials

Material	Temperature Coefficient of Resistance α (per °C) at 20°C (68°F)
Aluminum	0.0040
Brass	0.0015
Copper	0.0039
Gold	0.0034
Graphite carbon	−0.0005
Iron	0.0055
Lead	0.0039
Nichrome	0.0002
Silver	0.0038
Steel	0.0016
Tin	0.0042
Tungsten	0.0050

gage uses a semiconductor element (e.g., silicon with the trace impurity boron) in place of a metal foil. An approximate relationship for a strain gage is

$$\frac{\Delta R}{R} = S_s \varepsilon \tag{2.59}$$

where
  ΔR is the change in resistance due to strain ε
  R is the initial resistance
  S_s is the sensitivity (gage factor) of the strain gage

The gage factor is on the order of 4.0 for a metal-foil strain gage and can range from 40.0 to 200.0 for a semiconductor strain gage.

Temperature effects have to be compensated for in the high-precision measurement of strain. Compensation circuitry may be employed for this purpose. In a semiconductor strain gage, self-compensation for temperature effects can be achieved due to the fact that its temperature coefficient of resistance varies nonlinearly with the concentration of the dope material. The temperature coefficient curve of a p-type semiconductor strain gage is shown in Figure 2.16.

2.6.1.5 Superconductivity

The resistivity of some materials drops virtually to zero when the temperature is decreased close to absolute zero, provided that the magnetic field strength of the environment is less than some critical value. Such materials are called superconducting materials. The superconducting temperature T (absolute) and the corresponding critical magnetic field strength H are related through

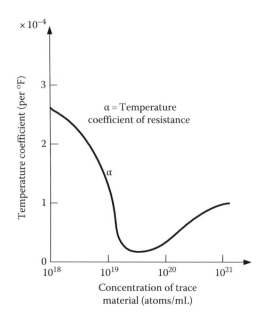

FIGURE 2.16
The temperature coefficient of resistance of a p-type semiconductor strain gage.

TABLE 2.4

Superconductivity Constants for
Several Materials

Material	T_c (°K)	H_0 (A/m)
Aluminum	1.2	0.8×10^4
Gallium	1.1	0.4×10^4
Indium	3.4	2.3×10^4
Lead	7.2	6.5×10^4
Mercury	4.0	3.0×10^4
Tin	3.7	2.5×10^4
Vanadium	5.3	10.5×10^4
Zinc	0.9	0.4×10^4

$$H = H_0 \left(1 - \frac{T}{T_c} \right)^2 \qquad (2.60)$$

where

H_0 is the critical magnetic field strength for a superconducting temperature of absolute zero

T_c is the superconducting temperature at zero magnetic field

The constants H_0 and T_c for several materials are listed in Table 2.4.

Superconducting elements can be used to produce high-frequency (e.g., 1×10^{11} Hz) switching elements (e.g., Josephson junctions), which can generate two stable states (e.g., zero voltage and a finite voltage or zero magnetic field and a finite magnetic field). Hence, they are useful as computer memory elements. Other applications of superconductivity include powerful magnets with low dissipation (for medical imaging, magnetohydrodynamics, fusion reactors, particle accelerators, etc.), actuators (for motors, magnetically levitated vehicles, magnetic bearings, etc.), and sensors.

2.6.1.6 Color Code for Fixed Resistors

Carbon, wound metallic wire, and conductive plastics are commonly used as commercial resistors. A wire-wound resistor element is usually encapsulated in a cylindrical casing made of an insulating material such as porcelain or bakelite. Axial or radial leads are provided for external connection. The outer surface of a cylindrical resistor element is color coded for the purpose of its specification. Four color stripes (bands) are marked on a cylindrical resistor element for coding. The first stripe gives the first digit of a two-digit number and the second stripe gives the second digit. The third stripe specifies a multiplier that should be included with the two-digit number to give the resistance value of the element in ohms. These three bands are equally spaced. The fourth stripe, which is spaced somewhat farther from the first three, gives the percentage tolerance of the resistance value. This color code is given in Table 2.5. According to the notation given in the table, the resistance value is given by

$$R = mn \times 10^p \ \Omega \ \pm e\% \qquad (2.61)$$

TABLE 2.5

Color Code for Fixed Resistor Elements

	First Stripe	Second Stripe	Third Stripe	Fourth Stripe
Color	First Digit (m)	Second Digit (n)	Power of 10 (p)	Tolerance (e %)
Silver	—	—	−2	±10
Gold	—	—	−1	±5
Black	0	0	0	—
Brown	1	1	1	±1
Red	2	2	2	±2
Orange	3	3	3	—
Yellow	4	4	4	—
Green	5	5	5	—
Blue	6	6	6	—
Violet	7	7	7	—
Gray	8	8	8	—
White	9	9	9	—

If there is no fourth band, the tolerance of the element is considered to be ±20%.

Example 2.4

A cylindrical wire-lead resistor element has the following bands printed on it: red, green, blue, and silver. What is the resistance of the element?

Solution
According to Equation 2.61,

$$R = 25 \times 10^6 \ \Omega \ \pm 10\%$$

$$= 22.5 \times 10^6 - 27.5 \times 10^6 \ \Omega$$

In addition to the common cylindrical wire-lead elements, resistors are available in other packages that are particularly useful in assembling a printed circuit board. They include a single in-line package, which has a single array of pins projecting downward, and a dual in-line package (DIP), which has two parallel arrays of pins projecting downward. Each such package has multiple resistors, as determined by the number of pins. Also, there are flat surface-mount packages without pins or leads but with solder tabs as the connection points. Metal-film resistors provide high precision. Since the tolerance of such a component is typically less than 1%, the code printed on the package takes a somewhat different meaning. Specifically, the first three stripes (not just the first two) are used to denote a three-digit number (coefficient) and the fourth stripe gives the power of 10 of the multiplier.

2.6.2 Dielectric Material and Capacitor Element

Components made of dielectric material exhibit properties of electrical capacitance. The "capacity" to storage electrical charge is the underlying basis. Dielectric materials are insulators, having resistivities larger than $1 \times 10^{12} \ \Omega \cdot m$ and containing less than 1×10^6 mobile electrons per m^3. When a voltage is applied across a medium of dielectric material that is sandwiched between two electrode plates, a charge polarization takes place at the two electrodes. The resulting charge depends on the capacitance of the capacitor formed in this manner.

Consider the linear capacitor element shown in Figure 2.15b. Its constitutive equation is

$$q = Cv \tag{2.62a}$$

where
   v is the applied voltage (volts or V)
   q is the stored charge (coulombs or C)
   C is the capacitance (farads or F)

Since current (i) is the rate of change of charge (dq/dt), we can write the following proper form of the constitutive equation:

$$C\frac{dv}{dt} = i \tag{2.62b}$$

Since power is given by the product iv, by substituting Equation 2.62b, the energy in a capacitor may be expressed as

$$E = \int iv\,dt = \int C\frac{dv}{dt}v\,dt = \int Cv\,dv$$

or

$$\text{Energy, } E = \frac{1}{2}Cv^2 \tag{2.63}$$

This is the familiar *electrostatic energy* of a capacitor.
   Also,

$$v(t) = v(0^-) + \frac{1}{C}\int_{0^-}^{t} i\,dt \tag{2.64}$$

Hence, for a capacitor with a finite current, we have

$$v(0^+) = v(0^-) \tag{2.65}$$

We summarize:

1. A capacitor is an energy storage element (electrostatic energy).

2. Voltage is an appropriate (natural) response variable (or state variable) for a capacitor element. This is justified by two reasons: First, from Equation 2.16, the voltage at any time t can be completely determined with the knowledge of the initial voltage and the applied current during the time interval 0 to t. Second, from Equation 2.63, the energy of a capacitor element can be represented by the variable v alone.

3. Voltage across a capacitor cannot change instantaneously unless an infinite current is applied.

4. Voltage is a natural output variable and current is a natural input variable for a capacitor.

5. Since its state variable, voltage, is an across variable, a capacitor is an *A*-type element.

2.6.2.1 Permittivity

Consider a capacitor made of a dielectric plate of thickness *d* sandwiched between two conducting plates (electrodes) of common (facing) area *A*. Its capacitance is given by

$$C = \frac{\varepsilon A}{d} \tag{2.66}$$

where ε is the permittivity (or, dielectric constant) of the dielectric material.

The relative permittivity (or dielectric constant) ε_r is defined as

$$\varepsilon_r = \frac{\varepsilon}{\varepsilon_0} \tag{2.67}$$

where ε_0 is the permittivity of vacuum (approx. 8.85×10^{-12} F/m or 8.85 pF/m).

The relative permittivities of some materials are given in Table 2.6.

TABLE 2.6

Dielectric Constants of Some Common Materials

Material	Relative Permittivity, ε_r
Air	1.0006
Carbon dioxide gas	1.001
Ceramic (high permittivity)	8000.0
Cloth	5.0
Common salt	5.9
Diamond	5.7
Glass	6.0
Hydrogen (liquid)	1.2
Mica	6.0
Oil (mineral)	3.0
Paper (dry)	3.0
Paraffin wax	2.2
Polythene	2.3
PVC	6.0
Porcelain	6.0
Quartz (SiO_2)	4.0
Vacuum	1.0
Water	80.0
Wood	4.0

2.6.2.2 Capacitor Types

The capacitance of a capacitor increases with the common surface area of the electrode plates. This increase can be achieved without compromising the compact size of the capacitor by employing a rolled-tube construction. Here, a dielectric sheet (e.g., paper or a polyester film) is placed between two metal foils and the composite is rolled into a tube. Axial or radial leads are provided for an external connection. If the dielectric material is not flexible (e.g., mica), a stacked-plate construction may be employed in place of the rolled construction to obtain compact capacitors having a high capacitance. High-permittivity ceramic disks are used as the dielectric plates in miniature, single-plate, high-capacitance capacitors. Electrolytic capacitors can be constructed using the rolled-tube method using a paper soaked in an electrolyte in place of the dielectric sheet. When a voltage is applied across the capacitor, the paper becomes coated with a deposit of dielectric oxide, which is formed through electrolysis. This becomes the dielectric medium of the capacitor. Capacitors having low capacitances of the order of 1×10^{-12} F (1 pF) and high capacitances of the order of 4×10^{-3} F are commercially available.

TABLE 2.7

Approximate Dielectric Strengths of Several Materials

Material	Dielectric Strength (V/mil)
Air	25
Ceramics	1000
Glass	2000
Mica	3000
Oil	400
Paper	1500

An important specification for a capacitor is the breakdown voltage—the voltage at which discharge will occur through the dielectric medium (i.e., the dielectric medium ceases to function as an insulator). This is measured in terms of the dielectric strength, which is defined as the breakdown voltage for a dielectric element of thickness 1 mil (1×10^{-3} in.). Approximate dielectric strengths of several useful materials are given in Table 2.7.

2.6.2.3 Color Code for Fixed Capacitors

Color codes are used to indicate the specifications of paper capacitors or ceramic capacitors. The code consists of a colored end followed by a series of four dots printed on the outer surface of the capacitor. The end color gives the temperature coefficient of the capacitance in ppm/°C. The first two dots specify a two-digit number. The third dot specifies a multiplier, which together with the two-digit number, gives the capacitance value of the capacitor in pF. The fourth dot gives the tolerance of the capacitance. This code is shown in Table 2.8.

2.6.2.4 Piezoelectricity

When subjected to stress (strain), some materials produce an electric charge. These are termed piezoelectric materials and the effect is called piezoelectricity. Most materials that possess a nonsymmetric crystal structure are known to exhibit the piezoelectric property. Some examples are barium titanate, cadmium sulfide, lead titanate, quartz, and rochelle salt. The reverse piezoelectric effect (i.e., the material deforms in an electric field) is also useful in practice.

The piezoelectric characteristic of a material may be represented by its piezoelectric coefficient k_p, which is defined as

$$k_p = \frac{\text{Change in strain} \, (\text{m/m})}{\text{Change in electric field strength} \, (\text{V/m})}$$

TABLE 2.8

Color Code for Ceramic and Paper Capacitors

Color	End Color Temperature Coefficient (ppm/°C)	First Dot First Digit	Second Dot Second Digit	Third Dot Multiplier	Fourth Dot Tolerance For ≤10 pF	For >10 pF
Black	0	0	0	1	±2 pF	±20%
Brown	−30	1	1	10	±0.1 pF	±1%
Red	−80	2	2	1×10^2	—	±2%
Orange	−150	3	3	1×10^3	—	±2.5%
Yellow	−220	4	4	1×10^4	—	—
Green	−330	5	5	—	±0.5 pF	±5%
Blue	−470	6	6	—	—	—
Violet	−750	7	7	—	—	—
Grey	30	8	8	0.01	±0.25 pF	—
White	100	9	9	0.1	±1 pF	±10%

TABLE 2.9

Piezoelectric Coefficients of Some Materials

Material	Piezoelectric Coefficient k_p (m/V)
Barium titanate	2.5×10^{-10}
Lead zirconate titanate (PZT)	6.0×10^{-10}
Quartz	0.02×10^{-10}
Rochelle salt	3.5×10^{-10}

with no applied stress. The piezoelectric coefficients of some common materials are given in Table 2.9. Quartz and Rochelle salt are naturally occurring piezoelectric materials. PZT and barium titanate are synthetic materials whose piezoelectric properties are achieved by heating and gradually cooling them in a strong magnetic field.

A piezoelectric element may be treated as a capacitor, which satisfies Equation 2.62. A piezoelectric element has a very high impedance, particularly at low frequencies. For example, a quartz crystal may present an impedance of several megohms at 100 Hz. For this reason (and due to charge leakage), in particular, piezoelectric sensors do not function properly at low frequencies.

The sensitivity of a piezoelectric crystal may be represented either by its *charge sensitivity* or by its *voltage sensitivity*. Charge sensitivity is defined as

$$S_q = \frac{\partial q}{\partial F} \tag{2.68}$$

where
q denotes the generated charge
F denotes the applied force

For a crystal with surface area A, Equation 2.68 may be expressed as

$$S_q = \frac{1}{A} \frac{\partial q}{\partial p} \tag{2.69}$$

where p is the stress (normal or shear) or pressure applied to the crystal surface. Voltage sensitivity S_v is given by the change in voltage due to a unit increment in pressure (or stress) per unit thickness of the crystal. Thus, in the limit, we have

$$S_v = \frac{1}{d}\frac{\partial v}{\partial p} \tag{2.70}$$

where d denotes the crystal thickness.

For a capacitor, from Equation 2.62a, we have $\Delta q = C\Delta v$. Then, by using Equation 2.66 for a capacitor element, the following relationship between charge sensitivity and voltage sensitivity is obtained:

$$S_q = \varepsilon S_v \tag{2.71}$$

Note that ε is the dielectric constant (permittivity) of the crystal (capacitor).

Example 2.5

A barium titanate crystal has a charge sensitivity of 150.0 picocoulombs per newton (pC/N). (*Note*: $1\,pC = 1 \times 10^{-12}$ C; coulombs = farads × volts). The dielectric constant (permittivity) of the crystal is 1.25×10^{-8} farads per meter (F/m). What is the voltage sensitivity of the crystal?

Solution
The voltage sensitivity of the crystal is given by

$$S_v = \frac{150.0\ \text{pC/N}}{1.25 \times 10^{-8}\ \text{F/m}} = \frac{150.0 \times 10^{-12}\ \text{C/N}}{1.25 \times 10^{-8}\ \text{F/m}} = 12.0 \times 10^{-3}\ \text{V} \cdot \text{m/N} = 12.0\ \text{mV} \cdot \text{m/N}$$

The sensitivity of a piezoelectric element is dependent on the direction of loading. This is because the sensitivity depends on the crystal axis. Sensitivities of several piezoelectric materials along their most sensitive crystal axis are listed in Table 2.10.

Applications of piezoelectric materials include actuators for ink-jet printers, miniature step motors, force sensors, precision shakers, high-frequency oscillators, and acoustic amplifiers. Note that large k_p values are desirable in piezoelectric actuators. For instance, PZT is used in microminiature step motors. On the other hand, small k_p values are desirable in piezoelectric sensors (e.g., quartz accelerometers).

TABLE 2.10

Sensitivities of Several Piezoelectric Material

Material	Charge Sensitivity S_q (pC/N)	Voltage Sensitivity S_v (mV · m/N)
PZT	110	10
Barium titanate	140	6
Quartz	2.5	50
Rochelle salt	275	90

2.6.3 Magnetic Material and Inductor Element

Components made of magnetic material are useful in a range of applications. They include actuators (e.g., motors, magnetically levitated vehicles, tools, magnetic bearings), sensors and transducers, relays, resonators, and cores of inductors and transformers.

2.6.3.1 Magnetism and Permeability

When electrons move (or spin), a magnetic field is generated. The combined effect of such electron movements is the cause of magnetic properties of a material. In the linear range of operation of a magnetic element, we can write

$$B = \mu H \tag{2.72}$$

where
B is the magnetic flux density (Wb/m² or T)
H is the magnetic field strength (A/m)
μ is the permeability of the magnetic material

The relative permeability μ_r of a magnetic material is defined as

$$\mu = \frac{\mu}{\mu_0} \tag{2.73}$$

where μ_0 is the permeability of a vacuum (approximately $4\pi \times 10^{-7}$ H/m). (*Note*: 1 tesla = 1 weber per square meter; 1 henry = 1 weber per ampere.)

2.6.3.2 Hysteresis Loop

The B versus H curve of a magnetic material is not linear and exhibits a hysteresis loop as shown in Figure 2.17. It follows that μ is not a constant. Initial μ values (when magnetization is started at the demagnetized state of $H=0$ and $B=0$) are usually specified. Some representative values are given in Table 2.11.

Properties of magnetic materials can be specified in terms of parameters of the hysteresis curve. Some important parameters are shown in Figure 2.17; specifically,
H_c is the coercive field or coercive force (A/m)
B_r is the remnant flux density (Wb/m² or T)
B_{sat} is the saturation flux density (T)

The magnetic parameters of a few permanent-magnetic materials are given in Table 2.12. Note that high values of H_c and B_r are desirable for high-strength permanent magnets. Furthermore, high values of μ are desirable for core materials whose purpose is to concentrate magnetic flux.

2.6.3.3 Magnetic Materials

Magnetic characteristics of a material can be imagined as if contributed by a matrix of microminiature magnetic dipoles. Paramagnetic materials (e.g., platinum and tungsten) have their magnetic dipoles arranged in a somewhat random manner. These materials

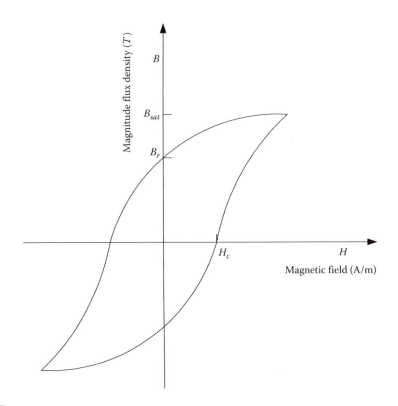

FIGURE 2.17
Hysteresis curve (magnetization curve) of a magnetic material.

TABLE 2.11

Initial Relative Permeability (Approximate) of Some Materials

Material	Relative Permeability, μ_r
Alnico (Fe$_2$ Ni Al)	6.5
Carbon steel	20
Cobalt steel (35% Co)	12
Ferrite (manganese-zinc)	800–10,000
Iron	200
Permalloy (78% Ni, 22% Fe)	3,000
Silicon iron (grain oriented)	500–1,500

TABLE 2.12

Parameters of Some Magnetic Materials

Material	H_c (A/m)	B_r (Wb/m^2)
Alnico	4.6×10^4	1.25
Ferrites	14.0×10^4	0.65
Steel (carbon)	0.4×10^4	0.9
Steel (35% Co)	2.0×10^4	1.4

have a μ_r value approximately equal to 1 (i.e., no magnetization). Ferromagnetic materials (e.g., iron, cobalt, nickel, and some manganese alloys) have their magnetic dipoles aligned in one direction (parallel) with virtually no cancellation of polarity. These materials have a high μ_r (on the order of 1000). Antiferromagnetic materials (e.g., chromium and manganese) have their magnetic dipoles arranged in parallel, but in an alternately opposing manner, thereby virtually canceling the magnetization ($\mu_r = 1$). Ferrites have parallel magnetic dipoles arranged alternately opposing, as in antiferromagnetic materials, but the adjacent dipoles have unequal strengths. Hence, there is a resultant magnetization (μ_r is on the order of 1000).

2.6.3.4 Piezomagnetism

When a stress (or strain) is applied to a piezomagnetic material, the degree of magnetization of the material changes. Conversely, a piezomagnetic material undergoes deformation when the magnetic field in which the material is situated is changed.

2.6.3.5 Hall-Effect Sensors

Suppose that a dc voltage v_{ref} is applied to a semiconductor element that is placed in a magnetic field in an orthogonal direction, as shown in Figure 2.18. A voltage v_o is generated in the third orthogonal direction, as indicated in the figure. This is known as the

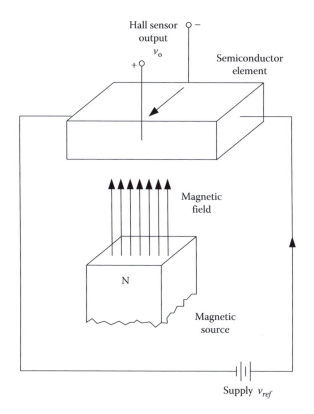

FIGURE 2.18
A Hall-effect sensor.

Hall effect. Hall-effect sensors use this phenomenon. For example, the motion of a ferromagnetic element can be detected in this manner, since the magnetic field in which the sensor is mounted would vary as a result of the motion of the ferromagnetic element. Hall-effect sensors are useful as position sensors, speed sensors, and commutation devices for motors.

2.6.3.6 Magnetic Bubble Memories

Consider a film of magnetic material such as gadolinium gallium garnet $(Gd_3Ga_5O_{12})$ deposited on a nonmagnetic garnet layer (substrate). The direction of magnetization will be perpendicular to the surface of the film. Initially, some regions of the film will be N poles and the remaining regions will be S poles. An external magnetic field can shrink either the N regions or the S regions, depending on the direction of the field. The size of the individual magnetic regions can be reduced to the order of $1\,\mu m$ in this manner. These tiny magnetic bubbles are the means by which information is stored in a magnetic bubble memory.

2.6.3.7 Reluctance

Suppose that a conducting coil having n turns is wound around a doughnut-shaped core of ferromagnetic material (e.g., soft iron). When a current i is applied to the coil, magnetic flux ϕ (Wb) will be generated, which will flow through closed magnetic path of the core. This forms a simple magnetic circuit, as shown in Figure 2.19.

The magnetomotive force (mmf) F, which enables the flow of magnetic flux, is given by

$$F = ni \tag{2.74}$$

The equation for the magnetic circuit (analogous to Ohm's law) is given by

$$F = \Re\phi \tag{2.75}$$

in which $\Re$ is the *reluctance* of the magnetic circuit. Now suppose that the mean length of the closed core (magnetic circuit) is L and the area of the cross section is A. Then, in Equation 2.72, $B = \phi/A$ and $H = ni/L$, which gives

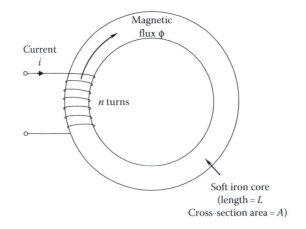

FIGURE 2.19
A magnetic circuit.

$$\mathfrak{R} = \frac{L}{\mu A} \tag{2.76}$$

This shows that the reluctance of a magnetic circuit is inversely proportional to the permeability of the circuit material. Compare Equation 2.76 with Equation 2.55 for electrical resistance.

2.6.3.8 Inductance

Suppose that a conducting coil having n turns is placed in a magnetic field of flux ϕ(Wb). The resulting flux linkage is $n\phi$. If the flux linkage is changed, a voltage is induced in the coil. This induced voltage (v) is given by

$$v = \frac{d(n\phi)}{dt} = \frac{nd\phi}{dt} \tag{2.77}$$

If the change in magnetic flux is brought about by a change in current (i), we can write

$$v = \frac{Ldi}{dt} \tag{2.78}$$

where L = inductance of the coil (H). The linear inductor element is shown in Figure 2.15c. As before, using Equation 2.78, it can be easily shown that energy in an inductor is given by

$$E = \frac{1}{2} L i^2 \tag{2.79}$$

This is the well-known *electromagnetic energy* of an inductor.

Also, by integrating (2.78,) we get

$$i(t) = i(0^-) + \frac{1}{L} \int_{0^-}^{t} v \, dt \tag{2.80}$$

Hence, for an inductor with a finite voltage, we have

$$i(0^+) = i(0^-) \tag{2.81}$$

We summarize:

1. An inductor is an energy storage element (electromagnetic energy).
2. Current is an appropriate response variable (or state variable) for an inductor. This is justified by two reasons: First, from Equation 2.20, the current at any time t can be completely determined with the knowledge of the initial current and the applied current during the time interval 0 to t. Second, from Equation 2.19, the energy of an inductor element can be represented by the variable i alone.

3. Current through an inductor cannot change instantaneously unless an infinite voltage is applied.

4. Current is a natural output variable and voltage is a natural input variable for an inductor.

5. Since its state variable, current, is a through variable, an inductor is a *T*-type element.

2.7 Active Electronic Components

An active device depends on an external power source to activate its behavior. Active components made of semiconductor junctions and field effect components are considered in this section. Junction diodes, bipolar junction transistors (BJTs), and field-effect transistors (FETs) are of particular interest here. Active components are widely used in the monolithic (integrated-circuit) form as well as in the form of discrete elements. They are extensively used in both analog and digital electronic devices including sensors, actuator circuits, controllers, interface hardware, and signal conditioning circuitry. In particular, the fields of consumer electronics and digital computers have been revolutionized due to the advances in active components. An understanding of the characteristics of the discrete components is important as well in the application of the corresponding monolithic devices.

2.7.1 Diodes

A semiconductor diode is formed by joining a *p*-type semiconductor with an *n*-type semiconductor. A diode offers much less resistance to current flow in one direction (forward) than in the opposite direction (reverse). There are many varieties of diodes. Zener diodes, voltage variable capacitor (VVC) diodes, tunnel diodes, microwave power diodes, pin diodes, photodiodes, and light-emitting diodes (LED) are examples. First, we need to understand *pn* junctions.

2.7.1.1 PN Junctions

Semiconductors exhibit properties that border conductors and insulators. A pure semiconductor can be "doped" by mixing a small quantity of special material to form either a *p*-type semiconductor or an *n*-type semiconductor. A *pn* junction is formed by joining a *p*-type semiconductor element and an *n*-type semiconductor element. This subject is explored now.

2.7.1.2 Semiconductors

Semiconductor materials have resistivities that are several million times larger than those of conductors and several billion times smaller than those of insulators. Crystalline materials such as *silicon*, *germanium*, and *cadmium sulfide* are semiconductors. For example, the resistivity of pure silicon is about 5×10^{10} times that of silver. Similarly, the resistivity of pure germanium is about 5×10^{7} times that of silver. Typically, semiconductors have resistivities ranging from 10^{-4} to $10^{7}\ \Omega \cdot m$. Other examples of semiconductor materials are gallium arsenide, cadmium sulfide, and selenium.

A pure (intrinsic) semiconductor material has some *free electrons* (negative charge carriers) and *holes* (positive charge carriers). Note that a hole is formed in an atom when an electron is removed. Strictly, the holes cannot move. But, suppose that an electron shared by two atoms (a *covalent electron*) enters an existing hole in an atom, leaving behind a hole at the point of origin. The resulting movement of the electron is interpreted as a movement of a hole in the direction opposite to the actual movement of the covalent electron.

The number of free electrons in a pure semiconductor is roughly equal to the number of holes. The number of free electrons or holes in a pure semiconductor can be drastically increased by adding traces of impurities in a controlled manner (*doping*) into the semiconductor during crystal growth (e.g., by alloying in a molten form and by solid or gaseous diffusion of the trace). An atom of a pure semiconductor that has four electrons in its outer shell will need four more atoms to share in order to form a stable covalent bond. These covalent bonds are necessary to form a crystalline lattice structure of atoms, which is typical of semiconductor materials. If the trace impurity is a material such as *arsenic, phosphorus,* or *antimony* whose atoms have five electrons in the outer shell (a donor impurity), a free electron will be left over after the formation of a bond with an impurity atom. The result will be an *n*-type semiconductor having a very large number of free electrons. If, on the other hand, the trace impurity is a material such as *boron, gallium, aluminum,* or *indium* whose atoms have only three electrons in the outer shell (an acceptor impurity), a hole will result upon the formation of a bond. In this case, a *p*-type semiconductor, consisting of a very large number of holes, will result. Doped semiconductors are termed extrinsic.

It is clear that doping can change the current-carrying characteristics of a semiconductor. These characteristics can be altered as well by temperature (the principle of a temperature), stress (the principle of a strain gage), magnetic field (in Hall effect devices, for example), and light (the principle of a photosensor).

2.7.1.3 Depletion Region

When a *p*-type semiconductor is joined with an *n*-type semiconductor, a *pn* junction is formed. A *pn* junction exhibits the diode effect; much larger resistance to current flow in one direction than in the opposite direction across the junction. As a *pn* junction is formed, electrons in the *n*-type material in the neighborhood of the common layer will diffuse across into the *p*-type material. Similarly, the holes in the *p*-type material near the junction will diffuse into the opposite side (strictly, the covalent electrons will diffuse in the opposite direction). The diffusion will proceed until an equilibrium state is reached. But, as a result of the loss of electrons and the gain of holes on the *n* side and the opposite process on the *p* side, a potential difference is generated across the *pn* junction with a negative potential on the *p* side and a positive potential on the *n* side. Due to the diffusion of carriers across the junction, the small region surrounding the common area will be virtually free of carriers (free electrons and holes). Hence, this region is called the *depletion region*. The potential difference that exists in the depletion region is mainly responsible for the diode effect of a *pn* junction.

2.7.1.4 Biasing

The forward biasing and the reverse biasing of a *pn* junction are shown in Figure 2.20. In the case of forward biasing, a positive potential is connected to the *p* side of the junction

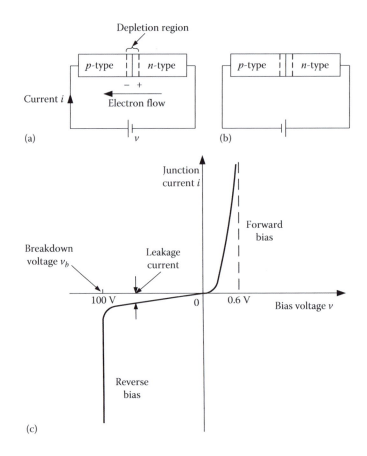

FIGURE 2.20
A *pn*-junction diode: (a) Forward biasing; (b) reverse biasing; (c) characteristic curve.

and a negative potential is connected to the *n* side. The polarities are reversed for reverse biasing. Note that in forward biasing, the external voltage (bias voltage *v*) complements the potential difference of the depletion region (Figure 2.20a). The free electrons that crossed over to the *p* side from the *n* side will continue to flow toward the positive terminal of the external supply, thereby generating a current (junction current *i*). The junction current increases with the bias voltage, as shown in Figure 2.20c.

In forward biasing, the potential in the depletion region is opposed by the bias voltage (Figure 2.20b). Hence, the diffusion of free electrons from the *n* side into the *p* side is resisted. Since there are some (very few) free electrons in the *p* side and some holes in the *n* side, the reverse bias will reinforce the flow of these minority electrons and holes. This will create a very small current (about 10^{-9} A for silicon and 10^{-6} A for germanium at room temperature), known as the *leakage current*, in the opposite direction to the forward-bias current. If the reverse bias is increased, at some voltage (*breakdown voltage v_b* in Figure 2.20c) the junction will break down generating a sudden increase in the reverse current. There are two main causes of this breakdown. First, the intense electric field of the external voltage can cause electrons to break away from neutral atoms in large numbers. This is known as *zener breakdown*. Second, the external voltage will accelerate the minority free electrons on the *p* side (and minority holes on the *n* side) creating collisions that will cause electrons on the outer shells of neutral atoms to break away in large numbers. This is known as

TABLE 2.13

Typical Breakdown Voltage of *pn* Junction at Room Temperature

	Breakdown Voltage (Volts)	
Semiconductor	Dope Concentration $=10^{15}$ Atoms/cm³	Dope Concentration $=10^{17}$ Atoms/cm³
Germanium	400	5.0
Silicon	300	11.0
Gallium arsenide	150	16.0

the *avalanche breakdown*. In some applications (e.g., rectifier circuits), junction breakdown is detrimental. In some other types of applications (e.g., as constant voltage sources and in some digital circuits), the breakdown state of specially designed diodes is practically utilized. Typical breakdown voltages of *pn* junctions made of three common semiconductor materials are given in Table 2.13. Note that the breakdown voltage decreases with the concentration of the trace material.

The current through a reverse-biased *pn* junction will increase exponentially with temperature. For a forward-biased *pn* junction, current will increase with temperature at low to moderate voltages and will decrease with temperature at high levels of voltage.

2.7.1.5 Zener Diodes

Zener diodes are a particular type of diodes that are designed to operate in the neighborhood of the reverse breakdown (both zener and avalanche breakdowns). In this manner, somewhat constant voltage output (the breakdown voltage) can be generated. This voltage depends on the concentration of the trace impurity. By varying the impurity concentration, output voltages in the range of 2–200 V may be realized from a zener diode. Special circuits would be needed to divert large currents that are generated at the breakdown point of the diode. The rated power dissipation of a zener diode should take into consideration the current levels that would be possible in the breakdown region. Applications of zener diodes include constant voltage sources, voltage clipper circuits, filter circuits for voltage transients, digital circuits, and two-state devices.

Example 2.6

A simple circuit for regulated voltage supply is shown in Figure 2.21. An unregulated dc source of voltage v_s is connected through a series resistor R to a zener diode in reverse bias. Assuming that v_s is larger than the breakdown voltage of the diode, the output voltage v_o of the circuit is maintained rather constant, as is clear from the characteristic curve shown in Figure 2.20c. Obtain an expression for the fluctuations in the output voltage in terms of the fluctuations in the voltage source.

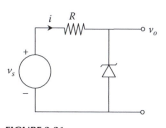

FIGURE 2.21
A voltage regulator using a zener diode.

Solution

The voltage summation in the circuit loop gives

$$v_s = Ri + v_o \qquad \text{(i)}$$

where *i* is the current through *R*, which is the same as the current through the diode, assuming that output is in open circuit (or the load resistance is quite high). The signal fluctuations are expressed as differentials in this equation; thus

$$\delta v_s = R\,\delta i + \delta v_o \qquad\qquad (ii)$$

Also, for the diode we have

$$\frac{\delta v_o}{\delta i} = R_z \qquad\qquad (iii)$$

where R_z is the resistance provided by the zener diode. Substitute (iii) in (ii) to obtain

$$\delta v_o = \frac{R_z}{R_z + R}\,\delta v_s$$

Since R_z is quite small compared with *R*, following the breakdown of the diode, as is clear from the slope of the left-hand curve of Figure 2.20c beyond breakdown, it is clear that Δv_o will be quite small compared with Δv_s.

2.7.1.6 VVC Diodes

VVC diodes use the property of a diode that, in reverse bias, the diode capacitance decreases (nonlinearly) with the bias voltage. The depletion region of a *pn* junction is practically free of carriers (free electrons and holes) and hence behaves like the dielectric medium of a capacitor. The adjoining *p* region and *n* region serve as the two plates of the capacitor. The width of the depletion region increases with the bias voltage. Consequently, the capacitance of a reverse biased *pn* junction decreases as the bias voltage is increased. The obtainable range of capacitance can be varied by changing the dope concentration and also by distributing the dope concentration nonuniformly along the diode. For example, a capacitance variation of 5–500 pF may be obtained in this manner (*Note*: $1\,\text{pF} = 1 \times 10^{-12}$ F). VVC diodes are also known as *varactor diodes* and *varicaps* and are useful in voltage-controlled tuners and oscillators.

2.7.1.7 Tunnel Diodes

The depletion of a *pn* junction can be made very thin by using very high dope concentrations (in both *p* and *n* sides). The result is a tunnel diode. Since the depletion region is very narrow, charge carriers (free electrons and holes) in the *n* and *p* sides of the diode can tunnel through the region into the opposite side, on application of a relatively small voltage. The voltage–current characteristic of a tunnel diode is quite linear at low (forward and reverse) voltages. When the forward bias is further increased, however, the behavior becomes quite nonlinear; the junction current peaks, then drops (a negative conductance) to a minimum (valley) and finally rises again as the voltage is increased. Due to the linear behavior of the tunnel diode at low voltages, almost instantaneous current reversal (i.e., very low reverse recovery time) can be achieved by switching the bias voltage. Tunnel diodes are useful in high-frequency switching devices, sensors, and signal conditioning circuits.

2.7.1.8 PIN Diodes

The width of the depletion region of a conventional *pn* junction varies with many factors, primarily the applied (bias) voltage. The capacitance of a junction depends on this width and varies due to such factors. A diode with practically a constant capacitance is obtained by adding a layer of silicon between the *p* and *n* elements. The sandwiched silicon layer is called the *intrinsic layer* and the diode is called a *pin* diode. The resistance of a *pin* diode varies inversely with the junction current. Pin diodes are useful as current-controlled resistors at constant capacitance.

2.7.1.9 Schottky Barrier Diodes

Most diodes consist of semiconductor–semiconductor junctions. An exception is a Schottky barrier diode that consists of a metal–semiconductor (*n*-type) junction. A metal such as *gold, silver, platinum,* or *palladium* and a semiconductor such as *silicon* or *gallium arsenide* may be used in the construction. Since no holes exist in the metal, a depletion region cannot be formed at the metal–semiconductor junction. Instead, an *electron barrier* is formed by the free electrons from the *n*-type semiconductor. Consequently, the junction capacitance will be negligible and the reverse recovery time will be very small. For this reason, Schottky diodes can handle very high switching frequencies (10^9 Hz range). Since by using a reverse bias, the electron barrier is easier to penetrate than a depletion region, Schottky diodes exhibit much lower breakdown voltages. Operating noise is also lower than for semiconductor–semiconductor diodes.

2.7.1.10 Thyristors

A thyristor, also known as a *silicon-controlled rectifier,* a *solid-state controlled rectifier,* a *semiconductor-controlled rectifier,* or simply an SCR, possesses some of the characteristics of a semiconductor diode. It consists of four layers (*pnpn*) of semiconductor and has three terminals—the *anode,* the *cathode,* and the *gate*—as shown in Figure 2.22a. The circuit symbol for a thyristor is shown in Figure 2.22b. The thyristor current is denoted by *i*, the external voltage is *v*, and the gate potential is v_g. The characteristic curve of a thyristor is shown in Figure 2.22c. Note that a thyristor cannot conduct in either direction (*i* almost zero) until either the reverse voltage reaches the reverse breakdown voltage (v_b) or the forward voltage reaches the forward breakover voltage (v_{fb}). The forward breakover is a bistable state and once this voltage is reached, the voltage drops significantly and the thyristor begins to conduct like a forward-biased diode. When v_g is less than or equal to zero with respect to the cathode, v_{fb} becomes quite high. When v_g is made positive, v_{fb} becomes small and v_{fb} will decrease as the gate current (i_g) is increased. A small positive v_g can make v_{fb} very small and then the thyristor will conduct from anode to cathode but not in the opposite direction (i.e., it behaves like a diode). It follows that a thyristor behaves like a voltage-triggered switch; a positive firing signal (a positive v_g) will close the switch. The switch will be opened when both *i* and v_g are made zero. When the supply voltage *v* is dc and nonzero, the thyristor will not be able to turn itself off. In this case, a *commutating circuit* that can make the trigger voltage v_g slightly negative has to be employed. Thyristors are commonly used in control circuits for dc and ac motors.

Parameter values for diodes are given in data sheets provided by the manufacturer. Commonly used variables and characteristic parameters in association with diodes are described in Table 2.14. For thyristors, as mentioned before, several other quantities such

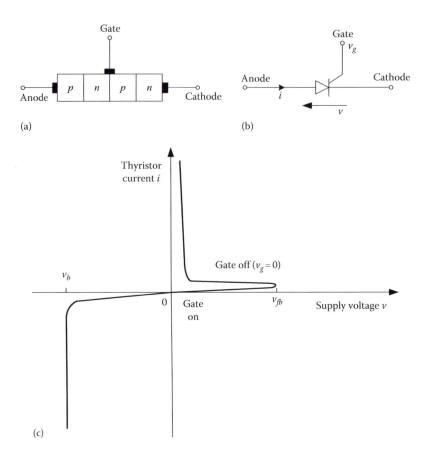

FIGURE 2.22
The thyristor: (a) Schematic representation; (b) circuit symbol; (c) characteristic curve.

TABLE 2.14

Characteristic Variables and Parameters for Diodes

Diode Variable/Parameter	Description
Forward bias (v_f)	A positive external voltage at p with respect to n
Reverse bias (v_r)	A positive external voltage at n with respect to p
Breakdown voltage (v_b)	The minimum reverse bias that will break down the junction resistance
Junction current (i_j)	Forward current through a forward-biased diode
Leakage current (i_r)	Reverse current through a reverse-biased diode
Transition capacitance (C_t)	Capacitance (in the depletion region) of a reverse-biased diode
Diffusion capacitance (C_d)	Capacitance exhibited while a forward biased diode is switched off
Forward resistance (R_f)	Resistance of a forward-biased diode
Reverse recovery time (t_{rr})	Time needed for the reverse current to reach a specified level when the diode is switched from forward to reverse
Operating temperature range (T_A)	Allowable temperature range for a diode during operation
Storage temperature range (T_{srg})	Temperature that should be maintained during storage of a diode
Power dissipation (P)	The maximum power dissipation allowed for a diode at a specified temperature

as v_{fb}, v_g, and i_g should be included. The time required for a thyristor to be turned on by the trigger signal (turn-on time) and the time for it to be turned off through commutation (*turn-off time*) determine the maximum switching frequency (*bandwidth*) for a thyristor. Another variable that is important is the *holding current* or *latching current*, which denotes the small forward current that exists at the breakover voltage.

2.7.2 Transistors

2.7.2.1 Bipolar Junction Transistors

A BJT has two junctions that are formed by joining *p* regions and *n* regions. Two types of transistors, *npn* and *pnp*, are possible with this structure. A BJT has three terminals, as indicated in Figure 2.23a. The middle (sandwiched) region of a BJT is thinner than the end regions, and this region is known as the *base*. The end regions are termed the *emitter* and the *collector*. Under normal conditions, the emitter-base junction is forward biased and the collector–base junction is reverse biased, as shown in Figure 2.23b.

To explain the behavior of a BJT, consider an *npn* transistor under normal biasing. The forward bias at the emitter–base junction will cause free electrons in the emitter to flow into the base region, thereby creating the emitter current (i_e). The reverse bias at the collector–base junction will increase the depletion region there. The associated potential difference at the collector–base junction will accelerate the free electrons in the base into the collector and will form the collector current (i_c). Holes that are created in the base, for recombination with some free electrons that entered the base, will form the base current (i_b). Usually, i_c is slightly smaller than i_e. Furthermore, i_b is much smaller than i_c.

2.7.2.1.1 Transistor Characteristics

The *common-emitter* connection is widely used for transistors in amplifier applications. In this configuration, the emitter terminal will be common to the input side and the output side of the circuit. Transistor characteristics are usually specified for this configuration. Figure 2.24 shows typical characteristic curves for a junction transistor in the common-emitter connection. In this configuration, both *voltage gain* (output voltage/input voltage) and

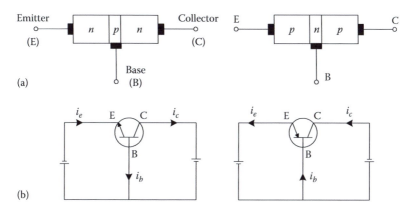

FIGURE 2.23
BJTs: (a) *npn* and *pnp* transistors; (b) circuit symbols and biasing.

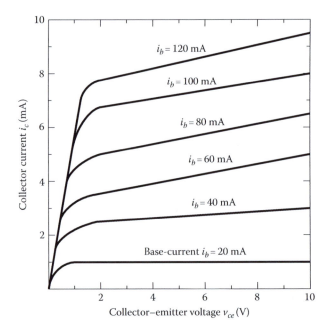

FIGURE 2.24
Characteristic curves of a common emitter BJT.

current gain (collector current/base current) will be greater than unity, thereby providing a voltage amplification as well as a current amplification. Note that from Figure 2.24 the control signal is the base current (i_b) and the characteristic of the transistor depends on i_b. This is generally true for any BJT; a BJT is a *current-controlled* transistor. In the common-base configuration, the base terminal is common to both the input and the output.

The maximum frequency of operation and allowable switching rate for a transistor are determined by parameters such as *rise time, storage time,* and *fall time.* These and some other useful ratings and characteristic parameters for BJTs are defined in Table 2.15. Values for these parameters are normally given in the manufacturer's data sheet for a particular transistor.

2.7.2.1.2 Fabrication Process of Transistors

The actual manufacturing process for a transistor is complex and delicate. For example, an *npn* transistor can be fabricated by starting with a crystal of *n*-type silicon. This starting element is called the *wafer* or *substrate.* An *npn* transistor is formed in the top half of the substrate by using the *planar diffusion method,* as follows: The substrate is heated to about 1000°C. A gas stream containing a *donor-type* impurity (which forms *n*-type regions) is impinged on the crystal surface. This produces an *n*-type layer on the crystal. Next, the crystal is oxidized by heating to a high temperature. The resulting layer of *silicon dioxide* acts as an insulating surface. A small area of this layer is then dissolved off to form a window using *hydrofluoric acid.* The crystal is again heated to 1000°C and a gas stream containing an *acceptor-type* impurity (which forms *p*-type regions) is impinged on the window. This produces a *p* region under the window, on top of the *n* region, which was formed earlier.

Oxidation is repeated to cover the newly formed *p* region. Using hydrofluoric acid, a smaller window is cut on the latest silicon dioxide layer and a new *n* region is formed

TABLE 2.15

Rating Parameters for Transistors

Transistor Parameter	Description
Collector to base voltage (v_{cb})	Voltage limit across collector and base with emitter open
Collector to emitter voltage (v_{ce})	Voltage limit across collector and emitter with base connected to emitter
Emitter to base voltage (v_{eb})	Voltage limit across emitter and base with collector open
Collector cutoff current (i_{co})	Reverse saturation current at collector with either emitter open (i_{ceo}) or base open (i_{cbo})
Transistor dissipation (P_T)	Power dissipated by the transistor at rated conditions
Input impedance (h_i)	Input voltage/input current with output voltage=0 (defined for both common emitter and common base configurations, h_{ie}, h_{ib})
Output admittance (h_o)	Output current/output voltage with input current=0 (h_{oe}, h_{ob} are defined)
Forward current transfer ratio (h_f)	Output current/input current with output voltage=0 (h_{fe}, h_{fb} are defined)
Reverse voltage transfer ratio (h_r)	Input voltage/output voltage with input current=0 (h_{re}, h_{rb} are defined)
Rise time (t_r)	Time taken to reach the full current level for the first time when turned on
Storage time (t_s)	Time taken to reach the steady current level when turned on
Fall time (t_f)	Time taken for the current to reach zero when turned off

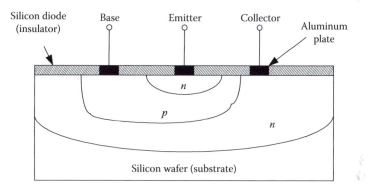

FIGURE 2.25
An *npn* transistor manufactured by the planar diffusion method.

as before, on top of the *p* region. The entire manufacturing process has to be properly controlled so as to control the properties of the resulting transistor. Aluminum contacts have to be deposited on the uppermost *n* region, the second *p* region (in a suitable annular window cut on the silicon dioxide layer), and on the *n* region below it or on the crystal substrate. A pictorial representation of an *npn* transistor fabricated in this manner is shown in Figure 2.25.

2.7.2.2 Field-Effect Transistors

A FET, unlike a BJT, is a *voltage-controlled* transistor. The electrostatic field generated by a voltage applied to the gate terminal of a FET controls the behavior of the FET. Since the device is voltage controlled at very low input current levels, the *input impedance* is very high and the input power is very low. Other advantages of an FET over a BJT are that the former is cheaper and requires significantly less space on a chip in the monolithic form. FETs are somewhat slower (in terms of switching rates) and more nonlinear than BJTs, however.

There are two primary types of FETs; *metal-oxide-semiconductor field-effect transistors* (MOSFETs) and *junction field-effect transistors* (JFETs). Even though the physical structure of the two types is somewhat different, their characteristics are quite similar. Insulated gate FET (or IGFET) is a general name given to a MOSFET.

2.7.2.3 The MOSFET

An *n*-channel MOSFET is produced using a *p*-type silicon substrate, and a *p*-channel MOSFET is produced by an *n*-type substrate. An *n*-channel MOSFET is shown in Figure 2.26a. During manufacture, two heavily doped *n*-type regions are formed on the substrate. One region is termed *source* (S) and the other region is termed *drain* (D). The two regions are connected by a moderately doped and narrow *n* region called *channel*. A metal coating deposited over an insulating layer of silicon dioxide, which is formed on the channel, is the gate (G). The source lead is usually joined with the substrate lead. This is a *depletion-type* MOSFET (or D-MOSFET). Another type is the *enhancement-type* MOSFET (or E-MOSFET). In this type, a channel linking the drain and the source is not physically present in the substrate, but is induced during the operation of the transistor.

Consider the operation of the *n*-channel D-MOSFET shown in Figure 2.26a. Under normal operation, the drain is positively biased with respect to the source. Drain current i_d is considered the output of a MOSFET (analogous to the collector current of a BJT). The control signal of a MOSFET is the gate voltage v_{gs} with respect to the source (analogous to the base current of a BJT). It follows that a MOSFET is a *voltage-controlled* device. Since the source terminal is used as the reference for both input (gate voltage) and output (drain), this connection is called the *common-source* configuration. Suppose that the gate voltage is negative with respect to the source. This will induce holes in the channel, thereby decreasing the free electrons there through recombination. This in turn will reduce the concentration of free electrons in the drain region and hence will reduce the drain current i_d. Clearly, if the magnitude of the negative voltage at the gate is decreased, the drain current will increase, as indicated by the characteristic curves in Figure 2.26b. A positive bias at the gate will further increase the drain current of an *n*-channel MOSFET as shown. The opposite will be true for a *p*-channel MOSFET.

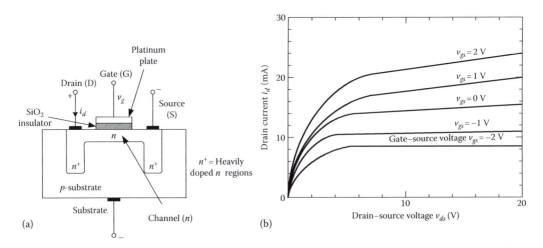

FIGURE 2.26
A metal-oxide semiconductor FET: (a) An *n*-channel depletion-type MOSFET; (b) D-MOSFET characteristics.

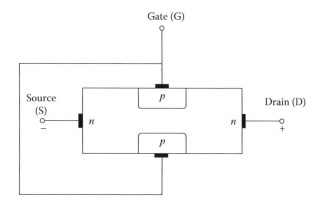

FIGURE 2.27
An *n*-channel JFET.

2.7.2.4 The Junction Field Effect Transistor

A JFET is different in physical structure to a MOSFET, but similar in characteristic. The structure of an *n*-channel JFET is shown in Figure 2.27. It consists of two *p*-type regions formed inside an *n*-type region. The two *p* regions are separated by a narrow *n* region called *channel*. The channel links two *n*-type regions called *source* (S) and *drain* (D). The two *p* regions are linked by a common terminal and form the *gate* (G). As in the case of a MOSFET, drain current i_d is considered the output of the JFET and gate voltage v_{gs} with respect to the source is considered the control signal. For normal operation, the drain is positively biased with respect to the source, as for an *n*-channel MOSFET, and the *common-source* configuration is used.

To explain the operation of a JFET, consider the *n*-channel JFET shown in Figure 2.27. Depletion regions are present at the two *pn* junctions of the JFET (as for a semiconductor diode). If the gate voltage is made negative, the resulting field will weaken the *p* regions. As a result, the depletion regions will shrink. Some of the free electrons from the drain will diffuse toward the channel to occupy the growing *n* regions due to the shrinking depletion regions. This will reduce the drain current. It follows that drain current decreases as the magnitude of the negative voltage at the gate is increased. This behavior is similar to that of a MOSFET. A *p*-channel JFET has two *n* regions representing the gate and two *p* regions forming the source and the drain, which are linked by a *p*-channel. Its characteristic is the reverse of an *n*-channel JFET.

Common types of transistors are summarized in Table 2.16. Semiconductor devices have numerous uses. A common use is as a switching device or as a two-state element. Typical two-state elements are schematically illustrated in Figure 2.28.

Example 2.7

The temperature dependence of a BJT allows it to be used as a temperature sensor. The relevant relation (Ebers–Moll model) is

$$v_{eb} = \frac{kT}{q}\ln\frac{i_c}{I_s} \tag{2.82}$$

TABLE 2.16

Common Transistor Types

Transistor Type		Description
Abbreviation	**Name**	**Description**
BJT	Bipolar junction Transistor	A three-layer device (*npn* or *pnp*) Current controlled Control = base current Output = collector current
FET	Field-effect transistor	A physical or induced channel (*n*-channel or *p*-channel) voltage controlled Control = gate voltage Output = drain current
MOSFET	Metal-oxide semiconductor FET	*n*-channel or *p*-channel
D-MOSFET	Depletion-type MOSFET	A channel is physically present
E-MOSFET	Enhancement-type MOSFET	A channel is induced
VMOS	V-shaped gate MOSFET or VFET	An E-MOSFET with increased power handling capacity
DG-MOS	Dual-gate MOSFET	A secondary gate is present between main gate and drain (lower capacitance)
D-MOS	Double-diffused MOSFET	A channel layer is formed on a high-resistivity substrate and then source and drain are formed (by diffusion). High breakdown voltage
CMOS	Complementary symmetry MOSFET	Uses two E-MOSFETs (*n*-channel and *p*-channel). Symmetry is used to save space on chip. Cheaper and lower power consumption.
GaAs	Gallium arsenide MOSFET	Uses gallium arsenide, aluminum gallium arsenide, (AlGaAs), indium gallium arsenide phosphide (InGaAsP), etc. in place of silicon substrate. Faster operation
JFET	Junction FET	*p*-channel or *n*-channel. Has two (*n* or *p*) regions in a (*p* or *n*) region linked by a channel (*p* or *n*) Control = gate voltage Output = drain current

where

T is the absolute temperature (°K)

k is the Boltzmann constant (1.38×10^{-23} J/K)

q is the electron charge magnitude (1.6×10^{-19} C)

I_s is the saturation current

The difficulty here is that I_s also varies with temperature. This problem can be overcome by using two identical transistors with a common base and maintaining their collector currents (i_c) at a fixed ratio. An appropriate circuit is shown in Figure 2.29. Obtain an expression for the output v_o for the circuit and show that this varies linearly with T.

Solution

For the two transistors, which are identical, we have

$$v_{eb2} = \frac{kT}{q} \ln \frac{i_{c2}}{I_s}$$

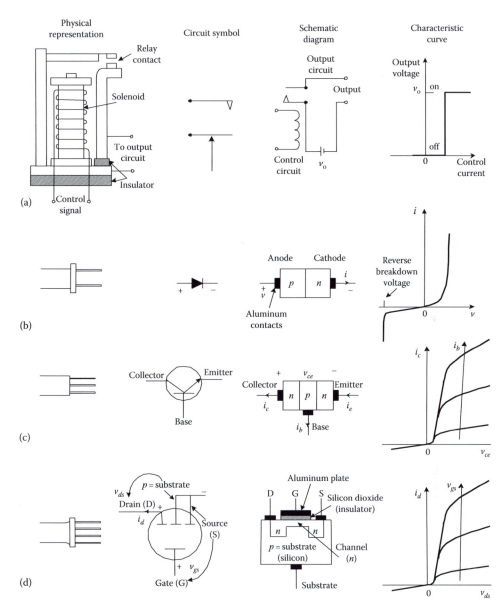

FIGURE 2.28
Discrete switching (two-state) elements: (a) Electromagnetic relay; (b) zener diode; (c) BJT (*npn*); (d) *n*-channel MOSFET.

$$v_{eb1} = \frac{kT}{q} \ln \frac{i_{c1}}{I_s}$$

Subtracting the second equation from the first, we get

$$v_{eb2} - v_{eb1} = \frac{kT}{q} \ln \frac{i_{c2}}{i_{c1}}$$

Note that the temperature-dependent term I_s has been eliminated. Thus,

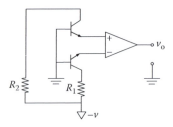

$$v_o = \frac{kT}{q} \ln \frac{i_{c2}}{i_{c1}} \qquad (2.83)$$

It is seen that the output voltage varies linearly with temperature. The ratio of the collector currents (i_{c2}/i_{c1}) has to be maintained at a constant when taking sensor readings.

FIGURE 2.29
A circuit for a semiconductor temperature sensor.

2.7.2.4.1 Switching Elements

Semiconductor devices are particularly useful as high-frequency switching elements in circuits. For example, in pulse width modulation (PWM), the average voltage and power supplied by a drive circuit (e.g., amplifier connected to a motor) are varied by varying the duty ratio. Here the switching frequency and the output voltage level are kept constant.

The duty ratio is defined as

$$d = \frac{t_{on}}{t_{on} + t_{off}} \qquad (2.84)$$

where
t_{on} is the on time of the switch
t_{off} is the off time of the switch within a switching cycle

The switching frequency is given by

$$f_s = \frac{1}{t_{on} + t_{off}} \qquad (2.85)$$

2.8 Light Emitters and Displays

Visible light is part of the *electromagnetic spectrum*. Electromagnetic waves in the *wavelength* range of 390–770 nm (*Note*: 1 nm = 1 × 10⁻⁹ m) form the visible light. Ultraviolet rays and x-rays are also electromagnetic waves, but have lower wavelengths (higher frequencies). Infrared rays, microwaves, and radio waves are electromagnetic waves with higher wavelengths. Table 2.17 lists the wavelengths of several types of electromagnetic waves. Visible light occupies a broad range of wavelengths. In optical coupling applications, for example, the narrower the wave spectrum, the clearer (noise free) the coupling process. Consequently, it is advantageous to use light sources with particular spectral characteristics in applications of that nature. In particular, since visible light can be contaminated by environmental light, thereby introducing an error signal into the optical device, it is also useful to consider electromagnetic waves that are different from what is commonly present in operating environments, in applications such as sensing, optical coupling, switching, and signal processing. Lighting is also crucial in a variety of automated industrial

TABLE 2.17

Wavelengths of Several Selected Components of the
Electromagnetic Spectrum

Wave Type	Approximate Wavelength Range (μm)
Radio waves	$1 \times 10^6 – 5 \times 10^6$
Microwaves	$1 \times 10^3 – 1 \times 10^6$
Infrared rays	$0.8 – 1 \times 10^3$
Visible light	$0.4 – 0.8$
Ultraviolet rays	$1 \times 10^{-2} – 0.4$
X-rays	$1 \times 10^{-6} – 5 \times 10^{-2}$

processes for object recognition and measurement, product inspection and grading, defect recognition and quality control, image analysis, computer vision, and visual servoing of robotic manipulators.

2.8.1 Light-Emitting Diodes

Semiconductor-based light sources such as LEDs are integral components of optoelectronic devices. The basic components of an LED are shown in Figure 2.30a. The element symbol that is commonly used in electrical circuits is shown in Figure 2.30b. The main component of an LED is a *semiconductor diode* element, typically made of gallium compounds (e.g., *gallium arsenide* or GaAs and *gallium arsenide phosphide* or GaAsP). When a voltage is applied in the *forward-bias* direction to this semiconductor element, it emits visible light (and also other electromagnetic wave components, primarily *infrared*). An LED needs a somewhat higher voltage (about 2 V) for its activation (in forward bias) than an ordinary silicon diode. In the forward-bias configuration, electrons are injected into the *p* region of the diode and are recombined with holes. Radiation energy (including visible light) is released spontaneously in this process. This is the principle of operation of an LED. Suitable doping with trace elements such as *nitrogen* will produce the desired effect. The radiation energy generated at the junction of a diode has to be directly transmitted to a window of the diode in order to reduce absorption losses. Two types of construction are commonly used: *edge emitters*, which emit radiation along the edges of the *pn* junction and *surface emitters*, which emit radiation normal to the junction surface.

Infrared light-emitting diodes (IRED) are LEDs that emit infrared radiation at a reasonable level of power. GaAs, *gallium aluminum arsenide* (GaA/As), and *indium gallium arsenide phosphide* (InGaAsP) are the commonly used IRED material. Gallium compounds and not

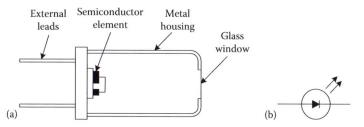

(a)   (b)

FIGURE 2.30
An LED: (a) Physical construction; (b) circuit symbol.

TABLE 2.18

Wavelength Characteristics of Common LEDs

LED Type	Wavelength at Peak Intensity	Color
Gallium arsenide	5,500	Green
	9,300	Infrared
Gallium arsenide phosphide	5,500	Green
	7,000	Red
Gallium phosphide	5,500	Green
Gallium aluminum arsenide	8,000	Red
	8,500	Infrared
Indium gallium arsenide phosphide	13,000	Infrared

Note: $1\,\text{Å} = 1 \times 10^{-10}$ m.

silicon or germanium are used in LEDs for reasons of efficiency of energy conversion and intensity characteristics (gallium compounds exhibit sharp peaks of spectral output in the desired frequency bands). Table 2.18 gives the wavelength characteristics of common LED and IRED types ($1\,\text{Å} = 1 \times 10^{-10}$ m $= 0.1$ nm). *Note*: denotes the unit "angstrom." These diodes come in small sizes (e.g., a few mm in diameter) and use small currents (20–100 mA) at low voltages (2 V).

LEDs are widely used in optical electronics because of their energy conversion efficiency, spectral composition, speed, size, durability, and low cost. They can be constructed in miniature sizes, they have small time constants and low impedances, they can provide high switching rates (typically over 1000 Hz), and they have a much longer component life than incandescent lamps. They are useful as both light sources and displays.

2.8.2 Lasers

A laser (light amplification by stimulated emission of radiation) is a light source that emits a concentrated beam of light, which will propagate typically at one or two frequencies (wavelengths) and in phase. Usually the frequency band is extremely narrow (i.e., *monochromatic*) and the waves in each frequency are in phase (i.e., *coherent*). Furthermore, the energy of a laser is highly concentrated (power densities of the order of 1 billion watts/cm^2). Consequently, a laser beam can travel in a straight line over a long distance with very little dispersion. Hence, it is a structured light source that is particularly useful in imaging, gauging, and aligning applications. Lasers can be used in a wide variety of sensors (e.g., motion sensors, tactile sensors, laser-doppler velocity sensors, 3D imaging) that employ photosensing and fiber optics. Also, lasers are used in medical applications, microsurgery in particular. Lasers have been used in manufacturing and material removal applications such as precision welding, cutting, and drilling of different types of materials including metals, glass, plastics, ceramics, rubber, leather, and cloth. Lasers are used in inspection (detection of faults and irregularities) and gauging (measurement of dimensions) of parts. Other applications of lasers include heat treatment of alloys, holographic methods of nondestructive testing, communication, information processing, and high-quality printing.

Lasers may be classified as a *solid, liquid, gas,* or *semiconductor*. In a solid laser (e.g., *ruby* laser, *glass* laser), a solid rod with reflecting ends is used as the laser medium. The laser medium of a liquid laser (e.g., *dye* laser, *salt-solution* laser) is a liquid such as an organic solvent with a dye or an inorganic solvent with dissolved salt compound. Very high peak power levels are possible with liquid lasers. Gas lasers (e.g. *helium-neon* or He-Ne laser, *helium-cadmium* or He-Cd laser, *carbon dioxide* or CO_2 laser) use a gas as the laser medium. Semiconductor lasers (e.g., *gallium arsenide* laser) use a semiconductor diode similar to an edge-emitting LED. Some lasers have their main radiation components outside the visible spectrum of light. For example, CO_2 lasers (with a wavelength of about 110,000 Å) primarily emit infrared radiation.

In a conventional laser unit, the laser beam is generated by first originating an excitation to create a light flash. This will initiate a process of emitting photons from molecules within the laser medium. This light is then reflected back and forth between two reflecting surfaces before the light beam is finally emitted as a laser. These waves will be limited to a very narrow frequency band (*monochromatic*) and will be in phase (*coherent*). For example, consider the He-Ne laser unit schematically shown in Figure 2.31. The helium and neon gas mixture in the *cavity resonator* is heated by a filament lamp and ionized using a high dc voltage (2000 V). Electrons released in the process will be accelerated by the high voltage and will collide with the atoms, thereby releasing *photons* (light). These photons will collide with other molecules, releasing more photons. This process is known as lasing. The light generated in this manner is reflected back and forth by the silvered surface and the partially reflective lens (beam splitter) in the cavity resonator, thereby stimulating it. This is somewhat similar to a resonant action. The stimulated light is concentrated into a narrow beam by a glass tube and emitted as a laser beam, through the partially silvered lens.

A *semiconductor laser* is somewhat similar to an LED. The laser element is typically made of a *pn* junction (diode) of semiconductor material such as gallium arsenide (GaAs) or indium gallium arsenide phosphide (InGaAsP). The edges of the junction are reflective (naturally or by depositing a film of silver). As a voltage is applied to the semiconductor laser, the ionic injection and spontaneous recombination that take place near the *pn* junction will emit light as in an LED. This light will be reflected back and forth between

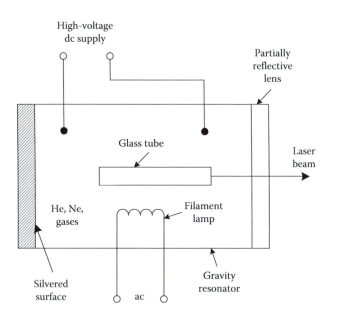

FIGURE 2.31
Helium-neon (He-Ne) laser.

TABLE 2.19

Properties of Several Types of Lasers

Laser Type	Wavelength (Å)	Output Power (W)
Solid		
Ruby	7,000	0.1–100
Glass	1,000	0.1–500
Liquid		
Dye	4,000–10,000	0.001–1
Gas		
Helium-neon	6,330	0.001–2
Helium-cadmium	4,000	0.001–1
Carbon dioxide	110,000	$1–1 \times 10^4$
Semiconductor		
GaAs	9,000	0.002–0.01
InGaAsP	13,000	0.001–0.005

Note: $1 \text{Å} = 1 \times 10^{-10}$ m.

the reflective surfaces, passing along the depletion region many times and creating more photons. The stimulated light (laser) beam is emitted through an edge of the *pn* junction. Semiconductor lasers are often maintained at very low temperatures in order to obtain a reasonable component life. Semiconductor lasers can be manufactured in very small sizes. They are lower in cost and require less power in comparison with the conventional lasers. The wavelength and power output characteristics of several types of lasers are given in Table 2.19.

2.8.3 Liquid Crystal Displays

A liquid crystal display (LCD) consists of a medium of liquid crystal material (e.g., organic compounds such as *cholesteryl nonanote* and *p-azoxyanisole*) trapped between a glass sheet and a mirrored surface, as shown in Figure 2.32. Pairs of transparent electrodes (e.g.,

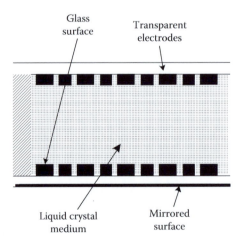

FIGURE 2.32
An LCD element.

indium tin oxide) arranged in a planar matrix are deposited on the inner surfaces of the sandwiching plates. In the absence of an electric field across an electrode pair, the atoms of liquid crystal medium in that region will have a parallel orientation. As a result, any light that falls on the glass sheet will first travel through the liquid crystal, then will be reflected back by the mirrored surface and finally will return unscattered. Once an electrode pair is energized, the molecular alignment of the entrapped medium will change, causing some scattering. As a result, a dark region in the shape of the electrode will be visible. Alphanumeric characters and other graphic images can be displayed in this manner by energizing a particular pattern of electrodes.

Other types of LCD construction are available as well. In one type, polarized glass sheets are used to entrap the liquid crystal. In addition, a special coating is applied on the inner surfaces of the two sheets that will *polarize* the liquid crystal medium in different directions. This polarization structure is altered by an electric field (supplied by an electrode pair), thereby displaying an image element. LCDs require external light to function. But they need significantly low currents and power levels to operate. For example, an LED display might need a watt of power whereas a comparable LCD might require just a small fraction of a mW. Similarly, the current requirement for an LCD is in the μA range. LCDs usually need an ac biasing, however. An image resolution on the order of 5 lines/mm is possible with an LCD.

2.8.4 Plasma Displays

A plasma display is somewhat similar to an LCD in construction. The medium used in a plasma display is an ionizing gas (e.g., *neon* with traces of *argon* or *xenon*). A planar matrix of electrode pairs is used on the inner surfaces of entrapping glass. When a voltage above the ionizing voltage of the medium is applied to the electrode pair, the gas will breakdown and a discharge will result. The electron impacts generated at the cathode as a result will cause further release of electrons to sustain the discharge. A characteristic *orange glow* will result. The pattern of energized electrodes will determine the graphic image.

The electrodes could be either *dc coupled* or *ac coupled*. In the case of the latter, the electrodes are coated with a layer of dielectric material to introduce a capacitor at the gas interface. The *power efficiency* of a plasma display is higher than that of an LED display. A typical image resolution of 2 lines/mm is obtainable.

2.9 Light Sensors

Semiconductor-based light sensors as well as light sources are needed in optoelectronics. A light sensor (also known as a *photodetector* or *photosensor*) is a device that is sensitive to light. Usually, it is a part of an electrical circuit with associated signal conditioning (amplification, filtering, etc.) so that an electrical signal representative of the intensity of light falling on the photosensor is obtained. Some photosensors can serve as energy sources (*cells*) as well. A photosensor may be an integral component of an optoisolator or other optically coupled system. In particular, a commercial optical coupler typically has an LED light source and a photosensor in the same package, with leads for connecting it to other circuits, together with power leads.

By definition, the purpose of a photodetector or photosensor is to sense visible light. But there are many applications where sensing of adjoining bands of the electromagnetic spectrum, namely *infrared* radiation and *ultraviolet* radiation, would be useful. For instance, since objects emit reasonable levels of infrared radiation even at low temperatures, infrared sensing can be used in applications where imaging of an object in the dark is needed. Applications include infrared photography, security systems, and missile guidance. Also, since infrared radiation is essentially *thermal energy*, infrared sensing can be effectively used in thermal control systems. Ultraviolet sensing is not as widely applied as infrared sensing.

Typically, a photosensor is a resistor, diode, or transistor element that brings about a change (e.g., generation of a potential or a change in resistance) in an electrical circuit in response to light that is falling on the sensor element. The power of the output signal may be derived primarily from the power source that energizes the electrical circuit. Alternatively, a photocell can be used as a photosensor. In this latter case, the energy of the light falling on the cell is converted into the electrical energy of the output signal. Typically, a photosensor is available as a tiny cylindrical element with a sensor head consisting of a circular window (lens). Several types of photosensors are described below.

2.9.1 Photoresistors

A photoresistor (or *photoconductor*) has the property of decreasing its electrical resistance (increasing the conductivity) as the intensity of light falling on it increases. Typically, the resistance of a photoresistor could change from very high values (megohms) in the dark to reasonably low values (less than $100\,\Omega$) in bright light. As a result, very high sensitivity to light is possible. Some photocells can function as photoresistors because their impedance decreases (output increases) as the light intensity increases. Photocells used in this manner are termed *photoconductive cells*. The circuit symbol of a photoresistor is given in Figure 2.33a. A photoresistor may be formed by sandwiching a photoconductive crystalline material such as *cadmium sulfide* (CdS) or *cadmium selenide* (CdSe) between two electrodes. Lead sulfide (PbS) or lead selenide (PbSe) may be used in infrared photoresistors.

2.9.2 Photodiodes

A photodiode is a *pn* junction of semiconductor material that produces electron–hole pairs in response to light. The symbol for a photodiode is shown in Figure 2.33b. Two types of photodiodes are available. A *photovoltaic* diode generates a sufficient potential

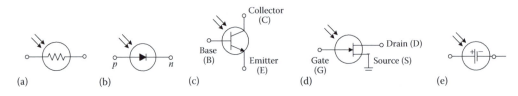

FIGURE 2.33
Circuit symbols of some photosensors: (a) Photoresistor; (b) photodiode; (c) phototransistor (*npn*); (d) phot-FET (*n*-channel); (e) photocell.

at its junction in response to light (photons) falling on it. Hence, an external bias source is not necessary for a photovoltaic diode. A *photoconductive* diode undergoes a resistance change at its junction in response to photons. This type of photodiode is operated in reverse-biased form; the *p*-lead of the diode is connected to the negative lead of the circuit and *n*-lead is connected to the positive lead of the circuit. The breakdown condition may occur at about 10 V and the corresponding current will be nearly proportional to the intensity of light falling on the photodiode. Hence, this current can be used as a measure of the light intensity. The sensitivity of a photodiode is rather low particularly due to the reverse-bias operation. Since the output current level is usually low (a fraction of a milliampere), amplification might be necessary before using it in the subsequent application (e.g., signal transmission, actuation, control, display). Semiconductor materials such as silicon, germanium, cadmium sulfide, and cadmium selenide are commonly used in photodiodes. The response speed of a photodiode is high. A diode with an intrinsic layer (a *pin diode*) can provide a faster response than with a regular *pn* diode.

2.9.3 Phototransistor

Any semiconductor photosensor with amplification circuitry built into the same package (chip) is popularly called a phototransistor. Hence, a photodiode with an amplifier circuit in a single unit might be called a phototransistor. Strictly, a phototransistor is manufactured in the form of a conventional *BJT* with *base* (B), *collector* (C), and *emitter* (E) leads.

A symbolic representation of a phototransistor is shown in Figure 2.33c. This is an *npn* transistor. The base is the central (*p*) region of the transistor element. The collector and the emitter are the two end regions (*n*) of the element. Under the operating conditions of the phototransistor, the collector–base junction is *reverse biased* (i.e., a positive lead of the circuit is connected to the collector and a negative lead of the circuit is connected to the base of an *npn* transistor). Alternatively, a phototransistor may be connected as a two-terminal device with its base terminal floated and the collector terminal properly biased (positive for an *npn* transistor). For a given level of source voltage (usually applied between the emitter lead of the transistor and load, the negative potential being at the emitter lead), the collector current (current through the collector lead) i_c is nearly proportional to the intensity of the light falling on the collector–base junction of the transistor. Hence, i_c can be used as a measure of the light intensity. Germanium or silicon is the semiconductor material that is commonly used in phototransistors.

2.9.4 Photo-Field Effect Transistor

A photo-field effect transistor (FET) is similar to a conventional FET. The symbol shown in Figure 2.33d is for an *n*-channel photo-FET. This consists of an *n*-type semiconductor element (e.g., *silicon* doped with *boron*) called *channel*. A much smaller element of *p*-type material is attached to the *n*-type element. The lead on the *p*-type element forms the *gate* (G). The *drain* (D) and the *source* (S) are the two leads on the channel. The operation of a FET depends on the electrostatic fields created by the potentials applied to the leads of the FET.

Under the operating conditions of a photo-FET, the gate is reverse-biased (i.e., a negative potential is applied to the gate of an *n*-channel photo-FET). When light is projected at the gate, the drain current i_d will increase. Hence, the drain current (current at the D lead) can be used as a measure of light intensity.

2.9.5 Photocells

Photocells are similar to photosensors except that a photocell is used as an electricity source rather than a sensor of radiation. Solar cells, which are more effective in sunlight, are commonly available. A typical photocell is a semiconductor junction element made of a material such as single-crystal silicon, polycrystalline silicon, and cadmium sulfide. Cell arrays are used in moderate-power applications. The typical power output is 10 mW/cm^2 of surface area with a potential of about 1.0 V. The circuit symbol of a photocell is given in Figure 2.33e.

2.9.6 Charge-Coupled Device

A charge-coupled device (CCD) is an integrated circuit (a *monolithic device*) element of semiconductor material. A CCD made from silicon is schematically represented in Figure 2.34. A silicon wafer (*p*-type or *n*-type) is oxidized to generate a layer of SiO$_2$ on its surface. A matrix of metal electrodes is deposited on the oxide layer and is linked to the CCD output leads. When light falls onto the CCD element (from an object), *charge packets* are generated within the substrate *silicon wafer*. Now if an external potential is applied to a particular electrode of the CCD, a *potential well* is formed under the electrode and a charge packet is deposited here. This charge packet can be moved across the CCD to an output circuit by sequentially energizing the electrodes using pulses of external voltage. Such a charge packet corresponds to a *pixel* (a picture element) of the image of the object. The circuit output is the video signal of the image. The pulsing rate could be higher than 10 MHz. CCDs are commonly used in imaging applications, particularly in video cameras. A typical CCD element with a facial area of a few square centimeters may detect 576 × 485 pixels, but larger elements (e.g., 4096 × 4096 pixels) are available for specialized applications. A *charge injection device* (CID) is similar to a CCD. In a CID, however, there is a matrix of semiconductor capacitor pairs. Each capacitor pair can be directly addressed through voltage pulses. When a particular element is addressed, the potential well there will shrink, thereby injecting minority carriers into the substrate. The corresponding signal, tapped from the substrate, forms the video signal. The signal level of a CID is substantially smaller than that of a CCD, as a result of higher capacitance.

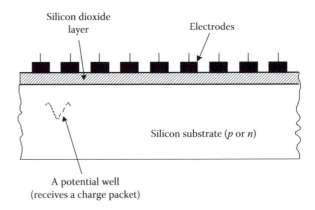

FIGURE 2.34
A CCD element.

2.9.7 Applications of Optically Coupled Devices

One direct application is in the isolation of electric circuitry. When two circuits are directly connected through hard electrical connections (cables, wires, etc.) a two-way path for the electrical signals is created at the interface. In other words, signals in circuit A will affect circuit B and signals in circuit B will affect circuit A. This interaction means that "noise" in one circuit will directly affect the other. Furthermore, there will be loading problems; the source will be affected by the load. Both these situations are undesirable. If the two circuits are optically coupled, however, there is only a one-way interaction between the two circuits (see Figure 2.35). Variations in the *output circuit* (*load circuit*) will not affect the *input circuit*. Hence, the input circuit is *isolated* from the output circuit. The connecting cables in an electrical circuit can introduce noise components such as *electromagnetic interference, line noise*, and *ground-loop noise*. The likelihood of these noise components affecting the overall system is also reduced by using optical coupling. In summary, isolation between two circuits and isolation of a circuit from noise can be achieved by optical coupling. For these reasons, optical coupling is widely used in communication networks (telephones, computers, etc.) and in circuitry for high-precision signal conditioning (e.g., for sophisticated sensors and control systems).

The medium through which light passes from the light source to the photosensor can create noise problems, however. If the medium is open (see Figure 2.35), then ambient lighting conditions will affect the output circuit, resulting in an error. Also, environmental impurities (dust, dirt, smoke, moisture, etc.) will affect the light received by the photosensor. Hence, a more controlled medium of transmission would be desirable. Linking the light source and the photosensor using *optical fibers* is a good way to reduce problems due to ambient conditions in optically coupled systems.

Optical coupling may be used in *relay circuits* where a low-power circuit is used to operate a high-power circuit. If the relay that operates the high-power circuit is activated using an optical coupler, reaction effects (noise and loading) on the low-power circuit can be eliminated. Optical coupling is used in *power electronics* and *control systems* in this manner.

Many types of sensors and transducers that are based on optical methods do indeed employ optical coupling. (e.g., optical encoders, fiberoptic tactile sensors). Optical sensors are widely used in industry for parts counting, parts detection, and level detection. In these sensors, a light beam is projected from a source to a photodetector, both units being stationary. An interruption of the beam through the passage of a part will generate a pulse at the detector, and this pulse is read by a counter or a parts detector. Furthermore, if the

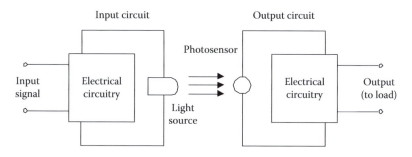

FIGURE 2.35
An optically coupled device.

light beam is located horizontally at a required height, its interruption when the material filled into a container reaches that level may be used for filling control in the packaging industry. Note that the light source and the sensor can be located within a single package if a mirror is used to reflect light from the source back onto the detector. Further applications include computer disk drive systems, e.g., to detect the write protect notch as well as the position of the recording head.

Problems

2.1 (a) What are through variables in mechanical, electrical, fluid, and thermal systems?

(b) What are across variables in mechanical, electrical, fluid, and thermal systems?

(c) Can the velocity of a mass change instantaneously?

(d) Can the voltage across a capacitor change instantaneously?

(e) Can the force in a spring change instantaneously?

(f) Can the current in an inductor change instantaneously?

(g) Can purely thermal systems oscillate?

2.2 Consider a hollow cylinder of length l, inside diameter d_i, and outside diameter d_o. If the conductivity of the material is k, what is the conductive thermal resistance of the cylinder in the radial direction?

2.3 In a lead-screw unit, the coefficient of friction μ was found to be greater than $\tan \alpha$, where α is the helix angle. Discuss the implications of this condition.

2.4 The nut of a lead-screw unit may have means of preloading, which can eliminate backlash. What are the disadvantages of preloading?

2.5 A load is moved in a vertical direction using a lead-screw drive, as shown in Figure P2.5. The following variables and parameters are given:

T is the motor torque

J is the overall moment of inertia of the motor rotor and the lead screw

m is the overall mass of the load and the nut

a is the upward acceleration of the load

r is the transmission ratio (rectilinear motion/angular motion) of the lead screw

e is the fractional efficiency of the lead screw

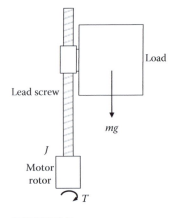

Show that

$$T = \left(J + \frac{mr^2}{e}\right)\frac{a}{r} + \frac{r}{e}mg$$

In a particular application, the system parameters are $m = 500\,\text{kg}$, $J = 0.25\,\text{kg}\cdot\text{m}^2$, and the screw lead is 5.0 mm. In view of the static friction, the starting efficiency is 50% and the operating efficiency is 65%. Determine the torque required to start the load and then move it upwards at an acceleration of 3.0 m/s². What is the torque required

FIGURE P2.5
Moving a vertical load using a lead-screw drive.

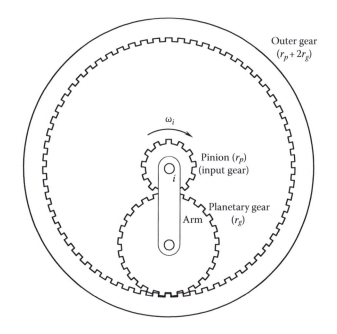

FIGURE P2.6
A planetary gear unit.

to move the load downwards at the same acceleration? Show that in either case much of the torque is used in accelerating the rotor (J). Note that in view of this observation it is advisable to pick a motor rotor and a lead screw with the least moment of inertia.

2.6 Consider the planetary gear unit shown in Figure P2.6. The pinion (pitch-circle radius r_p) is the input gear and it rotates at an angular velocity Ω_i. If the outer gear is fixed, determine the angular velocities of the planetary gear (pitch-circle radius r_g) and the connecting arm. Note that the pitch-circle radius of the outer gear is $r_p + 2r_g$.

2.7 List some advantages and shortcomings of conventional gear drives in speed transmission applications. Indicate ways to overcome or reduce some of the shortcomings.

2.8 A motor of torque T and moment of inertia J_m is used to drive an inertial load of moment of inertia J_L through an ideal (loss free) gear of motor-to-load speed ratio r:1, as shown in Figure P2.8. Obtain an expression for the angular acceleration $\ddot{\theta}_g$ of the load. Neglect the flexibility of the connecting shaft. Note that the gear inertia may be incorporated into the terms J_m and J_L.

2.9 In drive units of mechatronic systems, it is necessary to minimize backlash. Discuss the reasons for this. Conventional techniques for reducing backlash in gear drives include preloading (or spring loading), the use of bronze bearings that automatically compensate for tooth wear, and the use of high-strength steel and other alloys that can be machined accurately to obtain tooth profiles of low backlash and that have minimal wear problems. Discuss the shortcomings of some of the conventional methods of backlash reduction. Discuss the operation of a drive unit that has virtually no backlash problems.

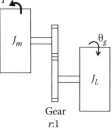

FIGURE P2.8
An inertial load driven by a motor through a gear transmission.

2.10 In some types of (indirect-drive) robotic manipulators, joint motors are located away from the joints and torques are transmitted to the joints through transmission devices such as gears, chains, cables, and timing belts. In some other types of (direct-drive) manipulators, joint motors are located at the joints themselves, the rotor being on one link and the stator being on the joining link. Discuss the advantages and disadvantages of these two designs.

2.11 In the harmonic drive configuration shown in Figure 2.12, the outer rigid spline is *fixed* (stationary), the wave generator is the input member, and the flexispline is the output member. Five other possible combinations of harmonic drive configurations are tabulated below. In each case, obtain an expression for the gear ratio in terms of the standard ratio (for Figure 2.12) and comment on the drive operation.

Case	Rigid Spline	Wave Generator	Flexispline
1	Fixed	Output	Input
2	Output	Input	Fixed
3	Input	Output	Fixed
4	Output	Fixed	Input
5	Input	Fixed	Output

2.12 Figure P2.12 shows a picture of an induction motor connected to a flexible shaft through a flexible coupling. Using this arrangement, the motor may be used to drive a load that is not closely located and also not oriented in a coaxial manner with respect to the motor. The purpose of the flexible shaft is quite obvious in such an arrangement. Indicate the purpose of the flexible coupling. Could a flexible coupling be used with a rigid shaft instead of a flexible shaft?

FIGURE P2.12
An induction motor linked to flexible shaft through a flexible coupling.

2.13 Backlash is a nonlinearity, which is often displayed by robots having gear transmissions. Indicate why it is difficult to compensate for backlash by using sensing and feedback control. What are the preferred ways to eliminate backlash in robots?

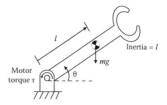

FIGURE P2.14
A single-link robot (inverted pendulum).

2.14 A single-degree-of-freedom robot arm (inverted pendulum) moving in a vertical plane is shown in Figure P2.14. The centroid of the arm is at a distance l from the driven joint. The mass of the arm is m and the moment of inertia about the drive axis is I. A direct-drive motor (without gears) whose torque is τ is used to drive the arm. The angle of rotation of the arm is θ, as measured from a horizontal axis. The dissipation at the joint is represented by a linear viscous damping coefficient b and a Coulomb friction constant c. Obtain an expression for the drive torque τ (which may be used in control).

2.15 Consider a single joint of a robot driven by a motor through gear transmission, as shown in Figure P2.15. The joint inertia is represented by an axial load of inertia J_l whose angular rotation is θ_l. The motor rotation is θ_m and the inertia of the motor rotor is J_m. The equivalent viscous damping constant at the load is b_l, and that at the motor rotor is b_m. The gear reduction ratio is r (i.e., θ_m: $\theta_l = r$:1). The fractional efficiency of the gear transmission is e (Note: $0 < e < 1$). Derive an expression for the drive motor torque τ, assuming that the motor speed $\dot{\theta}_m$ and the acceleration $\ddot{\theta}_m$ are known. What are the overall moment of inertia and the overall viscous damping constant of the system, as seen from the motor end?

2.16 Consider the resistivity ρ of a material, given in the units $\Omega \cdot m$. If the wire diameter is 1 mil (or 1/1000 in.), the wire area would be 1 circular mil (or cmil). Furthermore, if the wire length is 1 ft, the units of resistivity would be $\Omega \cdot cmil/ft$. Determine the factor by which the resistivity given in $\Omega \cdot m$ should be multiplied by in order to obtain the resistivity in $\Omega \cdot cmil/ft$.

2.17 Two wire-lead resistor elements having the following color codes are connected in parallel:

Red, brown, red, gold

Red, black, brown, gold

What is the color code of the equivalent resistance?

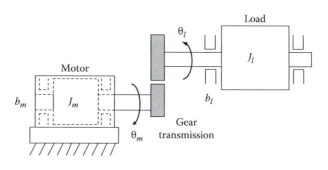

FIGURE P2.15
A geared robot joint.

2.18 The output voltage Δv_o of a strain-gage bridge circuit is related to the measured strain ε through the relation

$$\frac{\delta v_o}{v_{ref}} = \frac{R}{(R+R_c)} S_s K \varepsilon$$

where
  R is the resistance of the active strain gage
  R_c is the compensating resistor connected to the supply lead
  S_s is the strain gage sensitivity (gage factor)
  v_{ref} is the supply voltage
  K is the bridge constant

Assume that K does not change with temperature. The temperature coefficient of resistance α and the temperature coefficient of sensitivity β are defined by

$$R = R_o(1 + \alpha \cdot \Delta T)$$

$$S_s = S_{so}(1 + \beta \cdot \Delta T)$$

where ΔT denotes the temperature increase. Determine an expression for the R_c so that the circuit would possess self-compensation for temperature changes (i.e., the bridge output would not change due to ΔT). Under what conditions will this be feasible?

2.19 Charge sensitivity (or voltage sensitivity) of a piezoelectric material is expressed with respect to two axes—the axis along/about which the force or stress is applied and the axis along which the generated charge (or voltage) is measured. The three orthogonal (Cartesian) axes are denoted by 1, 2, and 3 and the corresponding rotational (torsion) directions are denoted by 4, 5, and 6, respectively. Consider a piezoelectric element of charge sensitivity $S_{q26} = 460.0\,\text{pC/N}$. What this means is, if a torque is applied about axis 3 (i.e., in the rotational direction 6) creating a torsional shear stress of $1.0\,\text{N/m}^2$, then a charge of $460.0\,\text{pC/m}^2$ is generated along axis 2. Suppose that the relative permittivity of the material is 500.0. If the thickness of the piezoelectric element along axis 2 is $1.0 \times 10^{-3}\,\text{m}$, determine the voltage drop along this axis for a stress of $1.0\,\text{N/m}^2$ applied in direction 6.

2.20 Compare the parameter *reluctance* in a magnetic circuit element with the parameter resistance in an electrical circuit element. In particular, highlight the analogy by giving the associated variables and parameters.

   Consider the magnetic circuit shown in Figure P2.20. The source coil has n turns and carries a current i, producing a magnetic flux ϕ. This flux is branched into $\phi1$ and $\phi2$ in the circuit as shown. If the reluctances of these two branched paths are $\mathfrak{R}_1$ and $\mathfrak{R}_2$, respectively, show that the overall reluctance $\mathfrak{R}$ of the circuit, as felt at the source, is given by $1/\mathfrak{R} = 1/\mathfrak{R}_1 + 1/\mathfrak{R}_2$. Compare this relation with what governs the parallel connection of the two resistors.

2.21 Consider the voltage regulator circuit shown in Figure 2.21. From the information sheet for the diode, it is known that at the operating value of the voltage regulator, which is 20V and is beyond the breakdown voltage, the dynamic resistance of the diode is $10\,\Omega$. Also, the series resistance is $200\,\Omega$. If the fluctuations in the voltage source are in the range $\pm 5.0\,\text{V}$, estimate the voltage fluctuations in the output of the regulator. Express this as a percentage of the output voltage.

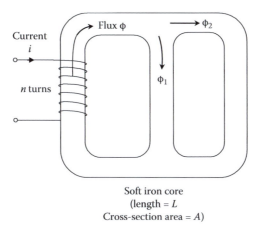

Current
i

n turns

Flux ϕ

ϕ_2

ϕ_1

Soft iron core
(length = L
Cross-section area = A)

FIGURE P2.20
A magnetic circuit with two branches.

2.22 Explain why it is not practical to use a single diode or a single BJT as a temperature sensor. The saturation current I_s of a BJT is dependent on the absolute temperature (T) according to the relation

$$I_s = aT^3 \exp\left(-\frac{qV_g}{kT}\right)$$

where
    a is a constant of the semiconductor material
    k is the Boltzmann constant
    q is the electron charge magnitude
    V_g is the band gap voltage

Obtain an expression for the sensitivity to temperature changes of the emitter-to-base current v_{eb}.

2.23 A BJT may be used as a switching element for an electronic circuit. A simple arrangement that may be used is shown in Figure P2.23.

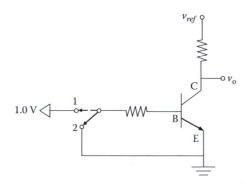

FIGURE P2.23
A switching circuit using a BJT.

The switching positions correspond to 1 and 2 for the base lead, as shown. Explain the operation of this circuit.

2.24 The duty ratio of a transistor switch used in a power width modulation (PWM) amplifier is d. If the output voltage level (high) of the amplifier is v_{ref} (the low level of the output $= 0$), what is the average voltage supplied by the amplifier to its load?

2.25 Digital transistor-to-transistor logic (TTL) signals use voltage levels between 0 and 0.4 V to represent binary 0 and voltage levels between 2.4 and 5 V to represent binary 1. Develop a simple circuit that uses a BJT to switch on and off an LED using a TTL trigger at these digital voltage levels. Assume that the LED resistance is $10\,\Omega$ and the current required by the LED to generate the required brightness is close to (but not more than) 40 mA. Also assume that an *npn* BJT and a 10 V dc power supply are available. Also, a current limiting resistor of $20\,k\Omega$ is available for the base of the BJT and a resistor of $200\,\Omega$ is available in series with the power supply and the LED. Indicate a practical application of this device.

3

Modeling of Mechatronic Systems

Study Objectives

- Understanding the steps of modeling
- Mechanical, electrical/electronic, fluid, and thermal domains
- Across variables, through variables, and analogies among various domains
- Development of state-space models
- Use of linear graphs as a modeling tool
- Transfer functions, block diagrams, and frequency-domain models
- Generalized/unified use of Thevenin and Norton equivalent circuits
- Response analysis by time-domain and Laplace transform approaches
- Computer simulation and Simulink®
- Laplace and Fourier transforms (also see Appendix B)

3.1 Introduction

The design, development, modification, and control of a mechatronic system require an understanding and a suitable "representation"—a "model" of the system. Properties established and results derived are associated with the model rather than the actual system, whereas, in experimental modeling, the excitations are applied to and the output responses are measured from the actual system. When using model-based approaches, it is important to keep in mind the possible sources of error:

1. Errors in the model itself (model errors)
2. Inaccurate or incomplete information on the inputs to the model (i.e., incompletely or inaccurately known inputs; presence of input noise or disturbances)
3. Incompletely known (e.g., immeasurable) or inaccurately processed or interpreted outputs

A mechatronic system may consist of several different types of components, and it is termed a *multi-domain* (or *mixed*) *system*. Furthermore, it may contain *multi-functional components*, for example, a piezoelectric component that can function as both a sensor and an actuator. It is useful to use analogous procedures for modeling such components. Then the component models (see Chapter 2 for elements or basic components in the mechanical, fluid, thermal, and electrical/electronic domains) can be conveniently and systematically integrated to obtain the overall model. Analytical models may be developed for mechanical, electrical, fluid, and thermal systems in a rather analogous manner, because some clear analogies are present among these four types of systems. In view of the *analogy*, then, a unified approach may be adopted in the analysis, design, and control of engineering systems. Emphasized here are model types; the tasks of "understanding" and analytical representation (i.e., analytical modeling) of mechanical, electrical, fluid, and thermal systems; identification of lumped elements (inputs/sources, and equivalent capacitor, inductor, and resistor elements; considerations of the associated variables (e.g., through and across variables; state variables); the development of models and input–output models; use of linear graphs as a modeling tool; frequency-domain models; the generalized/unified use of Thevenin and Norton equivalent circuits; and response analysis.

3.2 Dynamic Systems and Models

Each interacted component or element of an engineering system will possess an *input–output* (or *cause–effect*, or *causal*) relationship. An analytical model (or mathematical model) comprises equations (e.g., differential equations) or an equivalent set of information, which represents the system to some degree of accuracy. Alternatively, a set of curves, digital data (e.g., arrays or tables) stored in a computer, and other numerical data—rather than a set of equations—may be termed a model, strictly a numerical model (or experimental model) from which a representative analytical model can be established or "identified" (the related topic is called "model identification" or "system identification" in the subject of automatic control).

3.2.1 Terminology

A general representation of a dynamic system is given in Figure 3.1. Some useful terms are defined below.

System: Collection of interacting components of interest that are demarcated by a system boundary.

Dynamic system: A system whose rates of changes of response/state variables cannot be neglected.

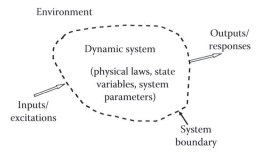

FIGURE 3.1
Nomenclature of a dynamic system.

TABLE 3.1

Examples of Dynamic Systems

System	Typical Input	Typical Outputs
Human body	Neuroelectric pulses	Muscle contraction, body movements
Company	Information	Decisions, finished products
Power plant	Fuel rate	Electric power, pollution rate
Automobile	Steering wheel movement	Front wheel turn, direction of heading
Robot	Voltage to joint motor	Joint motions, effector motion

Plant or process: The system to be controlled.

Inputs: Excitations (known or unknown) applied to the system.

Outputs: Responses of the system.

State variables: A minimal set of variables that completely identify the "dynamic" state of the system. *Note*: If the state variables at one state in time and the inputs from that state up to a future state in time are known, the future state can be completely determined.

Control system: The system that includes at least the plant and its controller. It may include other subsystems and components (e.g., sensors, signal conditioning, and modification).

Dynamic systems are not necessarily engineering, physical, or man-made systems. Some examples of dynamic systems with their representative inputs and outputs are given in Table 3.1. Try to identify several known and deliberately applied inputs; unknown and/or undesirable inputs (e.g., disturbances); desirable outputs and undesirable outputs for each of these systems.

3.2.2 Dynamic Models

3.2.2.1 Model Complexity

It is unrealistic to attempt to develop a "universal model" that will incorporate all conceivable aspects of the system. For example, an automobile model that will simultaneously represent ride quality, power, speed, energy consumption, traction characteristics, handling, structural strength, capacity, load characteristics, cost, safety, and so on is not very practical and can be intractably complex. The model should be as simple as possible, and may address only a few specific aspects of interest in the particular study or application. Approximate modeling and model reduction are relevant topics in this context.

3.2.2.2 Model Types

In general, models may be grouped into the following categories:

1. Physical models (prototypes)
2. Analytical models
3. Computer (numerical) models (data tables, arrays, curves, programs, files, etc.)
4. Experimental models (use input–output experimental data for model "identification")

The main advantages of analytical models (and computer models) over physical models are the following:

1. Modern, high-capacity, high-speed computers can handle complex analytical models at high speed and low cost
2. Analytical/computer models can be modified quickly, conveniently, and with high speed at low cost
3. There is high flexibility of making structural and parametric changes
4. Directly applicable in computer simulations
5. Analytical models can be easily integrated with computer/numerical/experimental models, to generate "hybrid" models
6. Analytical modeling can be conveniently done well before a prototype is built (in fact this step can be instrumental in deciding whether to prototype)

3.2.2.3 Types of Analytical Models

There are many types of analytical models. They include the following:

1. Time-domain model: Differential equations with time t as the independent variable
2. Transfer-function model: Laplace transform of the output variable divided by the Laplace transform of the input variable (algebraic equation with the Laplace variable s as the independent variable)
3. Frequency-domain model: Frequency transfer function (or frequency response function) which is a special case of the Laplace transfer function, with $s = j\omega$. The independent variable is frequency ω
4. Nonlinear model: Nonlinear differential equations (principle of superposition does not hold)
5. Linear model: Linear differential equations (principle of superposition holds)
6. Distributed (or continuous)-parameter model: Partial differential equations (dependent variables are functions of time and space)
7. Lumped-parameter model: Ordinary differential equations (dependent variables are functions of time, not space)
8. Time-varying (or nonstationary or nonautonomous) model: Differential equations with time-varying coefficients (model parameters vary with time)
9. Time-invariant (or stationary or autonomous) model: Differential equations with constant coefficients (model parameters are constant)
10. Random (stochastic) model: Stochastic differential equations (variables and/or parameters are governed by probability distributions)
11. Deterministic model: Non-stochastic differential equations
12. Continuous-time model: Differential equations (time variable is continuously defined)
13. Discrete-time model: Difference equations (time variable is defined as discrete values at a sequence of time points)
14. Discrete transfer function model: z-transform of the discrete-time output divided by the z-transform of the discrete-time input

3.2.2.4 Principle of Superposition

All practical systems can be nonlinear to some degree. If the nonlinearity is negligible, for the situation being considered, the system may be assumed linear. Since linear systems/ models are far easier to handle (analyze, simulate, design, control, etc.) than nonlinear systems/models, linearization of a nonlinear model, which may be valid for a limited range or set of conditions of operation, might be required. This subject will be studied in detail in another chapter.

All linear systems (models) satisfy the principle of superposition. A system is linear if and only if the principle of superposition is satisfied. This principle states that, if y_1 is the system output when the input to the system is u_1, and y_2 is the output when the input is u_2, then $\alpha_1 \cdot y_1 + \alpha_2 \cdot y_2$ is the output when the input is $\alpha_1 \cdot u_1 + \alpha_2 \cdot u_2$ where α_1 and α_2 are any real constants. This property is graphically represented in Figure 3.2a.

Another important property that is satisfied by linear systems is the interchangeability in series connection. This is illustrated in Figure 3.2b. Specifically, sequentially connected linear systems (or subsystems or components or elements) may be interchange without affecting the output of the overall system for a given input. Note that interchangeability in parallel connection is a trivial fact, which is also satisfied.

In this book, we will study/employ the following modeling techniques for response analysis, simulation, and control of a mechatronic system:

1. **State models:** They use state variables (e.g., position and velocity of lumped masses, force and displacement in springs, current through an inductor, voltage across a capacitor) to represent the state of the system, in terms of which the system response can be expressed. These are time-domain models, with time t as the independent variable

2. **Linear graphs:** They use line graphs where each line represents a basic component of the system, with one end as the point of action and the other end as the point of reference. They are particularly useful in the development of a state model.

3. **Transfer-function models** including **frequency-domain models**

3.2.3 Lumped Model of a Distributed System

In a lumped-parameter model, various characteristics of the system are lumped into representative elements located at a discrete set of points in a geometric space. The corresponding analytical models are ordinary differential equations. In most physical systems,

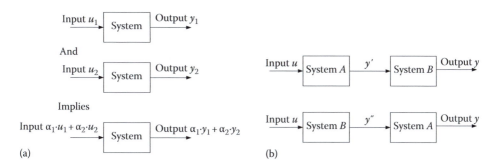

(a)

(b)

FIGURE 3.2
Properties of a linear system: (a) Principle of superposition; (b) interchangeability in series connection.

the properties are continuously distributed in various components or regions; they have distributed-parameter (or continuous) components. To represent the system parameters that are continuously distributed in space, we need spatial coordinates. These dynamic systems have time (t) and space coordinates (e.g., x, y, and z) as the independent variables. The corresponding analytical models are partial differential equations. For analytical convenience, we may attempt to approximate such distributed-parameter models into lumped-parameter ones. The accuracy of the model can be improved by increasing the number of discrete elements in such a model, for example, by using finite element techniques. In view of their convenience, the lumped-parameter models are more commonly employed than continuous-parameter models. Continuous-parameter elements may be included into otherwise lumped-parameter models in order to improve the model accuracy.

3.2.3.1 Heavy Spring

A coil spring has a mass, an elastic (spring) effect, and an energy-dissipation characteristic, each of which is distributed over the entire coil. The distributed mass of the spring has the capacity to store *kinetic energy* by acquiring velocity. The stored kinetic energy can be recovered as work done through a process of deceleration. Furthermore, in view of the distributed flexibility of the coil, each small element in the coil has the capacity to store *elastic potential energy* through reversible (elastic) deflection. If the coil was moving in the vertical direction, there would be changes in the *gravitational potential energy*, but we can disregard this in dynamic response studies if the deflections are measured from the static equilibrium position of the system. The coil will undoubtedly get warmer, make creaking noises, and over time will wear out at the joints, clear evidence of its capacity to dissipate energy. A further indication of damping is provided by the fact that when the coil is pressed and released, it will eventually come to rest; the work done by pressing the coil is completely dissipated. Often, a discrete or lumped-parameter model is sufficient to predict the system response to a forcing function. For instance, if the maximum kinetic energy is small in comparison with the maximum elastic potential energy in general (particularly true for light stiff coils, and at low frequencies of oscillation), and if in addition the rate of energy dissipation is relatively small (determined with respect to the time span of interest), the coil can be modeled by a discrete (lumped) stiffness (spring) element only. These are modeling decisions.

In an analytical model, the individual distributed characteristics of inertia, flexibility, and dissipation of a heavy spring can be approximated by a separate mass element, a spring element, and a damper element, which are interconnected in some parallel-series configuration—a lumped-parameter model. Since a heavy spring has its mass continuously distributed throughout its body, it has an infinite number of degrees of freedom. A single coordinate cannot represent its motion. But, in many practical purposes, a lumped-parameter approximation with just one lumped mass to represent the inertial characteristics of the spring would be sufficient. Such an approximation may be obtained by using one of several approaches, for example, energy equivalence and natural frequency equivalence.

3.2.3.1.1 Kinetic Energy Equivalence

Consider the uniform, heavy spring shown in Figure 3.3, with one end fixed and the other end moving at velocity v. Note that: m_s = mass of spring; k = stiffness of spring; and l = length of spring.

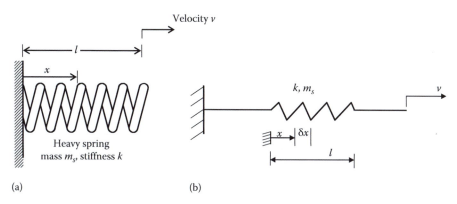

FIGURE 3.3
(a) A uniform heavy spring; (b) analytical representation.

In view of the linear distribution of the speed along the spring, with zero speed at the fixed end and v at the free end (Figure 3.3b), the local speed of an infinitesimal element δx of the spring is given by $(x/l)v$. Element mass $= (m_s/l)\delta x$. Hence, the element kinetic energy

$$KE = \frac{1}{2} \frac{m_s}{l} \delta x \left(\frac{x}{l} v \right)^2.$$

In the limit time, we have $\delta x \to dx$. Accordingly, by performing the necessary integration, we get

$$\text{Total } KE = \int_0^l \frac{1}{2} \frac{m_s}{l} dx \left(\frac{x}{l} v \right)^2 = \frac{1}{2} \frac{m_s v^2}{l^3} \int_0^l x^2 dx = \frac{1}{2} \frac{m_s v^2}{3}$$

Hence,

$$\text{Equivalent lumped mass concentrated at the free end} = \frac{1}{3} \times \text{spring mass} \qquad (3.1)$$

Note: This derivation assumes that one end of the spring is fixed and, furthermore, the conditions are uniform along the spring.

3.2.3.1.2 *Natural Frequency Equivalence*

Here we derive an equivalent lumped-parameter model by equating the fundamental (lowest) natural frequency of the distributed-parameter system to the natural frequency of the lumped-parameter model (in the one-degree-of-freedom case). We will illustrate our approach by using an example.

A heavy spring of mass m_s and stiffness k_s with one end fixed and the other end attached to a sliding mass m, is shown in Figure 3.4a. If the mass m is sufficiently large than m_s, then at relatively high frequencies the mass will virtually stand still. Under these conditions, we have the configuration shown in Figure 3.4b where the two ends of the spring are fixed. Also, approximate the distributed mass by an equivalent mass m_e at the mid point of the

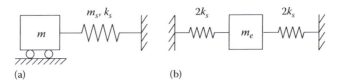

FIGURE 3.4
(a) A lumped mass connected to distributed-parameter system; (b) a lumped-parameter model of the system.

spring: each spring segment has double the stiffness of the original spring. Hence, the overall stiffness is $4k_s$. The natural frequency of the lumped-model is

$$\omega_e = \sqrt{\frac{4k_s}{m_e}} \tag{3.2}$$

It is known from a complete analysis of a heavy spring (which is beyond the present scope) that the natural frequency for the fixed–fixed configuration is

$$\omega_s = \pi n \sqrt{\frac{k_s}{m_e}} \tag{3.3}$$

where n is the mode number. Then, for the fundamental (first) mode (i.e., $n=1$), the natural frequency equivalence gives $\sqrt{4k_s/m_e} = \pi\sqrt{k_s/k_s}$

or,

$$m_e = \frac{4}{\pi^2} m_s \approx 0.4 m_s \tag{3.4}$$

Note: Since the effect of inertia decreases with increasing frequency, it is not necessary to consider the case of high frequencies.

The natural frequency equivalence may be generalized as an *eigenvalue* equivalence (*pole* equivalence) for any dynamic system. In this general approach, the eigenvalues of the lumped-parameter model are equated to the corresponding eigenvalues of the distributed-parameter system, and the model parameters are determined accordingly.

3.3 Lumped Elements and Analogies

A system may possess various physical characteristics incorporating multiple domains, for example, mechanical, electrical, thermal, and fluid components and processes. The procedure of model development will be facilitated if we understand the similarities of these various domains and in the characteristics of the different types of components. This issue is addressed in the present section. The basic system elements in an engineering system can be divided into two groups: energy-storage elements and energy-dissipation elements. The dynamic "state" of a system is determined by its independent energy storage elements and the associated state variables. Depending on the element we can use either an across variable or a through variable as its state variable.

3.3.1 Across Variables and through Variables

An across variable is measured across an element, as the difference in the values at the two ends. Velocity, voltage, pressure, and temperature are across variables. A through variable represents a property that appears to flow through an element, unaltered. Force, current, fluid flow rate, and heat transfer rate are through variables. As noted in Chapter 2, if the across variable of an element is the appropriate state variable for that element, it is termed an *A*-type element. Alternatively, if the through variable of an element is the appropriate state variable for that element, it is termed a *T*-type element. Analogies exist among mechanical, electrical, hydraulic, and thermal systems/processes. An integrated and unified development of a model is desirable, where all domains are modeled together using similar approaches. Then analogous procedures are used to model all components, in developing an analytical model.

Table 3.2 summarizes the linear constitutive relationships, which describe the behavior of translatory-mechanical, electrical, thermal, and fluid elements. The analogy used in Table 3.2 between mechanical and electrical elements is known as the force–current analogy. This follows from the fact that both force and current are *through variables*, which are analogous to fluid flow through a pipe, and furthermore, both velocity and voltage are *across variables*, which vary across the flow direction, as in the case of fluid pressure along a pipe. This analogy appears more logical than a force–voltage analogy, as is clear from Table 3.3. The correspondence between the parameter pairs given in Table 3.3 follows from

TABLE 3.2

Some Linear Constitutive Relations

	Constitutive Relation for		
	Energy Storage Elements		**Energy Dissipating Elements**
System **Type**	**A-Type (Across) Element**	**T-Type (Through) Element**	**D-Type (Dissipative) Element**
Translatory-mechanical	Mass	Spring	Viscous damper
	$m\dfrac{dv}{dt} = f$	$\dfrac{df}{dt} = kv$	$f = bv$
v = velocity	(Newton's second law)	(Hooke's law)	
f = force	m = mass	k = stiffness	b = damping constant
Electrical	Capacitor	Inductor	Resistor
v = voltage	$C\dfrac{dv}{dt} = i$	$L\dfrac{di}{dt} = v$	$Ri = v$
i = current	C = capacitance	L = inductance	R = resistance
Thermal	Thermal capacitor	None	Thermal resistor
T = temperature difference	$C_t\dfrac{dT}{dt} = Q$		$R_t Q = T$
Q = heat transfer rate	C_t = thermal capacitance		R_t = thermal resistance
Fluid	Fluid capacitor	Fluid inertor	Fluid resistor
P = pressure difference	$C_f\dfrac{dP}{dt} = Q$	$I_f\dfrac{dQ}{dt} = P$	$R_f Q = P$
Q = volume flow rate	C_f = fluid capacitance	I_f = inertance	R_f = fluid resistance

TABLE 3.3

Force–Current Analogy

System Type	Mechanical	Electrical
System-response variables:		
Through variables	Force f	Current i
Across variables	Velocity v	Voltage v
System parameters	m	C
	k	$1/L$
	b	$1/R$

the relations in Table 3.2. A rotational (rotary) mechanical element possesses constitutive relations between torque and angular velocity, which can be presented as a generalized force and a generalized velocity, in Table 3.2.

3.3.2 Natural Oscillations

Mechanical systems can produce natural (free) oscillatory responses (or, free vibrations) because they can possess two types of energy (kinetic and potential energies). When one type of stored energy is converted into the other type repeatedly, back and forth, the resulting response is oscillatory. Of course, some of the energy will dissipate (through the dissipative mechanism of a D-type element or damper) and the free natural oscillations will decay as a result. Similarly, electrical circuits and fluid systems can exhibit free, natural oscillatory responses due to the presence of two types of energy storage mechanisms, where energy can "flow" back and forth repeatedly between the two types of elements. But, thermal systems have only one type of energy storage element (A-type) with only one type of energy (thermal energy). Hence, purely thermal systems cannot naturally produce oscillatory responses, unless forced by external means, or integrated with other types of systems that can produce natural oscillations (e.g., fluid systems).

3.4 Analytical Model Development

We have been able to make the following observations concerning analytical dynamic models:

- A dynamic model is a representation of a dynamic system.
- It is useful in analysis, simulation, design, modification, and control of a system.
- In view of the multi-domain nature of practical engineering systems, integrated and unified development of models is desirable. Then all domains are modeled together using similar approaches.
- It is desirable to use analogous procedures to model all components in a system.
- Capability to incorporate multifunctional devices (e.g., piezoelectric elements that work as both sensors and actuators) into the modeling framework is desirable.
- Analogies exist in mechanical, electrical, fluid, and thermal systems.

A systematic procedure for the development of a lumped-parameter analytical model of a dynamic system involves formulating three types of equations:

1. Constitutive equations (physical laws for the lumped elements)
2. Continuity equations (or node equations or equilibrium equations) for the through variables
3. Compatibility equations (or loop equations or path equations) for the across variables

Among these, the constitutive equations have been studied in Chapter 2.

A continuity equation is the equation written for the through variables at a junction (i.e., node) connecting several lumped elements in the system. It dictates the fact that there cannot be any accumulation (storage) or disappearance (dissipation) or generation (source) of the through variables at a junction (i.e., what comes in must go out), because node is not an element but a junction that connects elements. Summation of forces (force balance or equilibrium), currents (Kirchhoff's current law), fluid flow rates (flow continuity equation), or heat transfer rates at a junction to zero provides a continuity equation. Note that source elements, which generate inputs to the system should be included as well in writing these equations.

A compatibility equation is the equation written for the across variables around a closed path (i.e., loop) connecting several lumped elements in the system. It dictates the fact that at a given instant, the value of the across variable at a point in the system should be unique (i.e., cannot have two or more different values). This guarantees the requirement that a closed path is indeed a closed path; there is no breakage of the loop (i.e., compatible). Summation of velocities, voltages (Kirchhoff's voltage law), pressures, or temperatures to zero around a loop of elements provides a compatibility equation. Again, source elements, which generate inputs to the system, should be included as well in writing these equations.

3.4.1 Steps of Model Development

Development of a suitable analytical model for a large and complex system requires a systematic approach. Tools are available to aid this process. The process of modeling can be made simple by following a systematic sequence of steps. The main steps are listed below:

1. Identify the system of interest by defining its *purpose* and the system *boundary*.
2. Identify or specify the *variables* of interest. These include inputs (forcing functions or excitations) and outputs (response).
3. Approximate (or model) various segments (components or processes or phenomena) in the system by *ideal elements* that are suitably interconnected.
4. Draw a *free-body diagram* for the system where the individual elements are isolated or separated, as appropriate.
5. a. Write *constitutive equations* (physical laws) for the elements.
   b. Write *continuity* (or conservation) equations for through variables (equilibrium of forces at joints; current balance at nodes, fluid flow balance, etc.) at junctions (nodes) of the system.

    c. Write *compatibility* equations for across (potential or path) variables around closed paths linking elements. These are loop equations for velocities (geometric connectivity), voltage (potential balance), pressure drop, etc.

    d. Eliminate *auxiliary* variables that are redundant and not needed to define the model.

6. Express the system *boundary conditions* and response *initial conditions* using system variables.

These steps should be self-explanatory, and should be integral with the particular modeling technique that is used. The associated procedures will be elaborated in the subsequent sections and chapters where many illustrative examples will be provided as well.

3.4.2 Input–Output Models

More than one variable may be needed to represent the response of a dynamic system. Furthermore, there may be more than one input variable in a system. Then we have a *multivariable* system or a multi-input–multi-output (MIMO) system. A time-domain analytical model may be developed as a set of differential equations relating the response variables to the input variables. This is specifically a multivariable *input–output model*. Generally, this set of system equations is coupled, so that more than one response variable appears in each differential equation, and each equation cannot be analyzed, solved, or computer simulated separately.

3.4.3 State-Space Models

A particularly useful time-domain representation for a dynamic system is a state-space model. The state variables are a minimal set of variables that can define the dynamic state of a system. In the state-space representation, the dynamics of an *n*th-order system is represented by *n* first-order differential equations that generally are coupled. This is called a *state-space model* or simply a *state model*. An entire set of state equations can be written as a single vector-matrix state equation.

    The choice of state variables is not unique: many choices are possible for a given system. Proper selection of state variables is crucial in developing an analytical model (state model) for a dynamic system. A general approach that may be adopted is to use across variables of the independent *A*-type (or, across-type) energy storage elements and the through variables of the independent *T*-type (or, through-type) energy storage element as the state variables. Note that if any two elements are not independent (e.g., if two spring elements are directly connected in series or parallel) then only a single state variable should be used to represent both elements. Separate state variables are not needed to represent *D*-type (dissipative) elements because their response can be represented in terms of the state variables of the energy storage elements (*A*-type and *T*-type). State-space models and their characteristics are discussed in more detail now.

3.4.3.1 Linear State Equations

Nonlinear state models are difficult to analyze and simulate. Often linearization is necessary, through various forms of approximations and assumptions. An *n*th-order linear, state model is given by the state equations (differential):

$$\dot{x}_1 = a_{11}x_1 + a_{12}x_2 + a_{13}x_3 + \cdots + a_{1n}x_n + b_{11}u_1 + b_{12}u_2 + \cdots + b_{1r}u_r$$

$$\dot{x}_2 = a_{21}x_1 + a_{22}x_2 + a_{23}x_3 + \cdots + a_{2n}x_n + b_{21}u_1 + b_{22}u_2 + \cdots + b_{2r}u_r \quad (3.5a)$$

$$\vdots$$

$$\dot{x}_n = a_{n1}x_1 + a_{n2}x_2 + a_{n3}x_3 + \cdots + a_{nn}x_n + b_{n1}u_1 + b_{n2}u_2 + \cdots + b_{nr}u_r$$

and the output equations (algebraic):

$$y_1 = c_{11}x_1 + c_{12}x_2 + c_{13}x_3 + \cdots + c_{1n}x_n + d_{11}u_1 + d_{12}u_2 + \cdots + d_{1r}u_r$$

$$y_2 = c_{21}x_1 + c_{22}x_2 + c_{23}x_3 + \cdots + c_{2n}x_n + d_{21}u_1 + d_{22}u_2 + \cdots + d_{2r}u_r \quad (3.6a)$$

$$\vdots$$

$$y_m = c_{m1}x_1 + c_{m2}x_2 + c_{m3}x_3 + \cdots + c_{mn}x_n + d_{m1}u_1 + d_{m2}u_2 + \cdots + d_{mr}u_r$$

Here, $\dot{x} = dx/dt$; $x_1, x_2, \ldots, x_n$ are the n *state variables*; $u_1, u_2, \ldots, u_r$ are the r *input variables*; and $y_1, y_2, \ldots, y_m$ are the m *input variables*. Equation 3.5a simply says that a change in any of the n variables and the r inputs of the system will affect the rate of change of any given state variable. In general, in addition to the state variables, the output variables are needed as well to represent the output variables, as indicated in Equation 3.6a. More often, however, the input variables are not present in this set of output equations (i.e., the coefficients d_{ij} are all zero).

This state model may be rewritten in the vector-matrix form as

$$\dot{x} = Ax + Bu \quad (3.5b)$$

$$y = Cx + Du \quad (3.6b)$$

The bold-type upper-case letter indicates that the variable is a *matrix*; a bold-type lower-case letter indicates a *vector*, typically a column vector. Specifically,

$$x = \begin{bmatrix} x_1 \\ x_2 \\ \vdots \\ x_n \end{bmatrix}; \quad \dot{x} = \begin{bmatrix} \dot{x}_1 \\ \dot{x}_2 \\ \vdots \\ \dot{x}_n \end{bmatrix}; \quad A = \begin{bmatrix} a_{11} & a_{12} & \cdots & a_{1n} \\ a_{21} & a_{22} & \cdots & a_{2n} \\ \vdots & \vdots & & \vdots \\ a_{n1} & a_{n2} & \cdots & a_{nn} \end{bmatrix}; \quad B = \begin{bmatrix} b_{11} & b_{12} & \cdots & b_{1r} \\ b_{21} & b_{22} & \cdots & b_{2r} \\ \vdots & \vdots & & \vdots \\ b_{n1} & b_{n2} & \cdots & b_{nr} \end{bmatrix};$$

$$C = \begin{bmatrix} c_{11} & c_{12} & \cdots & c_{1n} \\ c_{21} & c_{22} & \cdots & c_{2n} \\ \vdots & \vdots & & \vdots \\ c_{m1} & c_{m2} & \cdots & c_{mn} \end{bmatrix}; \quad D = \begin{bmatrix} d_{11} & d_{12} & \cdots & d_{1r} \\ d_{21} & d_{22} & \cdots & d_{2r} \\ \vdots & \vdots & & \vdots \\ d_{m1} & d_{m2} & \cdots & d_{mr} \end{bmatrix}$$

where
$x = [x_1\ x_2 \ldots x_n]^T$ is the state vector (nth order)
$u = [u_1\ u_2 \ldots u_r]^T$ is the input vector (rth order)
$y = [y_1\ y_2 \ldots y_m]^T$ is the output vector (mth order)

A is the system matrix $(n \times n)$
B is the input distribution matrix $(n \times r)$
C is the output (or measurement) gain matrix $(m \times n)$
D is the feed-forward input gain matrix $(m \times r)$

Note that $[]^T$ denotes the transpose of a matrix or vector. The system matrix A tells us how the system responds naturally without any external input, and B tells us the input u affects (i.e., how it is amplified and distributed when reaching) the system.

Summarizing

- The state vector is a least (minimal) set of variables that completely determines the dynamic state of system $\rightarrow$ A state variable cannot be expressed as a linear combination of the remaining state variables.
- State vector is not unique; many choices are possible for a given system.
- Output (response) variables can be completely determined from any such choice of state variables.
- State variables may or may not have a physical interpretation.

A linear system is time-invariant if the matrices A, B, C, and D (in Equations 3.5 and 3.6) are constant.

Steps for state model development

Note: Inputs (u) and outputs (y) are given

Step 1: State variable (x) selection

Across variables for independent A-type energy storage elements

Through variables for independent T-type energy storage elements

Step 2: Write constitutive equations for all the elements (both energy storage and dissipative elements)

Step 3: Write compatibility equations and continuity equations

Step 4: Eliminate redundant variables

Step 5: Express the outputs in terms of state variables

Example 3.1

The rigid output shaft of a diesel engine prime mover is running at known angular velocity $\Omega(t)$. It is connected through a friction clutch to a flexible shaft, which in turn drives a hydraulic pump (see Figure 3.5a). A linear model for this system is shown schematically in Figure 3.5b. A viscous rotatory damper of damping constant B_1 represents the clutch (units: torque/angular velocity). The stiffness of the flexible shaft is K (units: torque/rotation). The pump is represented by a wheel of moment of inertia J (units: torque/angular acceleration) and viscous damping constant B_2.

(a) Write down the two state equations relating the state variables T and ω to the input Ω.
where
T is the torque in the flexible shaft
ω is the pump speed

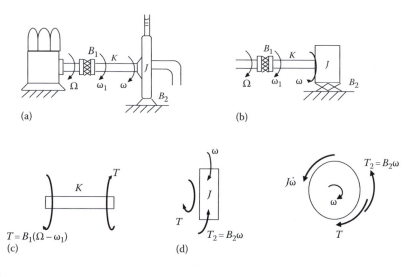

FIGURE 3.5
(a) Diesel engine; (b) linear model; (c) free body diagram of the shaft; (d) free body diagram of the wheel.

Hints:

1. Free body diagram for the shaft is given in Figure 3.5c, where ω_1 is the angular speed at the left end of the shaft.
2. Write down the "torque balance" and "constitutive" relations for the shaft, and eliminate ω_1.
3. Draw the free body diagram for the wheel J and use D'Alembert's principle.
4. Comment on why the compatibility equations and continuity equations are not explicitly used in the development of the state equations.
   (a) Express the state equations in the vector-matrix form.
   (b) To complete the state-space model, determine the output equation for: (i) Output = ω; (ii) Output = T; (iii) Output = ω_1.
   (c) Which one of the translatory systems given in Figure 3.6 is the system in Figure 3.5b analogous to?

Solution
(a)
Constitutive relation for K:

$$\frac{dT}{dt} = K(\omega_1 - \omega) \tag{i}$$

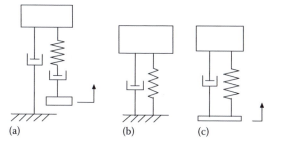

FIGURE 3.6
Three translatory mechanical systems.

Constitutive relation for B_1:

$$T = B_1(\Omega - \omega_1) \tag{ii}$$

Substitute (ii) into (i):

$$\frac{dT}{dt} = -\frac{K}{B_1}T - K\omega + K\Omega \tag{iii}$$

This is one state equation.
   Constitutive equation for J (D'Alembert's principle, see Figure 3.5d):

$$J\dot{\omega} = T - T_2 \tag{iv}$$

Constitutive relation for B_2:

$$T_2 = B_2\omega \tag{v}$$

Substitute (v) in (iv):

$$\frac{d\omega}{dt} = -\frac{B_2}{J}\omega + \frac{1}{J}T \tag{vi}$$

This is the second state equation.

Note: For this example, it is not necessary to write the continuity and compatibility equations because they are implicitly satisfied by the particular choice of variables as given in Figure 3.5.

Important comments

1. Generally, some of the continuity equations (node equations) and compatibility equations (loop equations) are automatically satisfied by the choice of the variables. Then, we do not have to write the corresponding equations.
2. In mechanical systems, typically, compatibility equations are automatically satisfied. This is the case in the present example. In particular, from Figure 3.7a we have:

$$\text{Loop 1 equation} : \omega + (-\omega) = 0$$

$$\text{Loop 2 equation:} \ \omega + (\omega_1 - \omega) + (\Omega - \omega_1) + (-\Omega) = 0$$

3. Node equations may be written in further detail by introducing other auxiliary variables into the free-body diagram; and furthermore, the constitutive equations for the damping elements may be written separately. Specifically, from Figure 3.7b, we can write
   Node 1: $T_{d1} - T = 0$
   Node 2: $T - T_{J1} = 0$
   Node 3: $T_{J2} - T_{d2} = 0$
   Damper B_1: $T_{d1} = B_1(\Omega - \omega_1)$
   Damper B_2: $T_{d2} = B_2\omega$

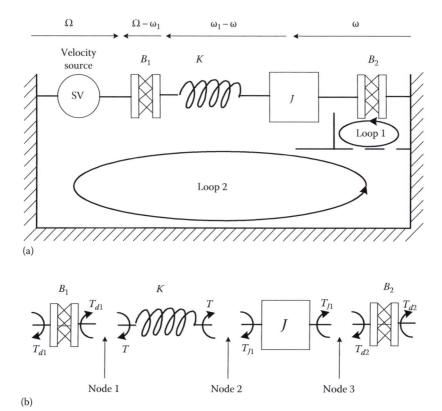

(a)

(b)

FIGURE 3.7
Model details for writing (a) loop equations; (b) node equations.

(b) Vector-matrix form of the state equations (iii) and (iv) is

$$
\begin{bmatrix} \dfrac{dT}{dt} \\[2ex] \dfrac{d\omega}{dt} \end{bmatrix}
=
\begin{bmatrix} \dfrac{-K}{B_1} & -K \\[2ex] \dfrac{1}{J} & -\dfrac{B_2}{J} \end{bmatrix}
\begin{bmatrix} T \\[1ex] \omega \end{bmatrix}
+
\begin{bmatrix} K \\[1ex] 0 \end{bmatrix}\Omega
$$

with the state vector $x = [T\ \omega]^T$ and the input $u = [\Omega]$

(c)
    (i) $C = [0\ 1]$; $D = [0]$

    (ii) $C = [1\ 0]$; $D = [0]$

    (iii) Here, we use the continuity equation to express the output as

$$
\omega_1 = \Omega - \frac{T}{B_1}
$$

Then, the corresponding matrices are

$$
C = [-1/B_1 \quad 0]; \quad D = [1]
$$

In this case, we notice a direct "feed-forward" of the input Ω into the output ω_1 through the clutch B_1. Furthermore, now the system transfer function will have its numerator order equal to the denominator order (=2). This is a characteristic of systems with direct feed-forward of inputs into the outputs.

(d) The translatory system in Figure 3.6a is analogous to the given rotatory system.

Example 3.2

Consider two water tanks joined by a horizontal pipe with an on–off valve. With the valve closed, the water levels in the two tanks were initially maintained unequal. When the valve was suddenly opened, some oscillations were observed in the water levels of the tanks. Suppose that the system is modeled as two gravity-type capacitors linked by a fluid resistor. Would this model exhibit oscillations in the water levels when subjected to an initial-condition excitation? Clearly explain your answer.

A centrifugal pump is used to pump water from a well into an overhead tank. This fluid system is schematically shown in Figure 3.8a. The pump is considered as a pressure source $P_s(t)$ and the water level h in the overhead tank is the system output. P_a denotes the ambient pressure. The following system parameters are given:

L_v, d_v = length and internal diameter of the vertical segment of pipe
L_h, d_h are the length and internal diameter of the horizontal segment of pipe
A_t is area of cross section of overhead tank (uniform)
ρ is mass density of water
μ is dynamic viscosity of water
g is acceleration due to gravity

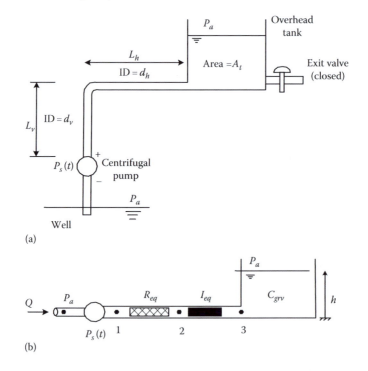

FIGURE 3.8
(a) A system for pumping water from a well into an overhead tank; (b) a lumped-parameter model of the fluid system.

Suppose that this fluid system is approximated by the lumped-parameter model shown in Figure 3.8b.

(a) Give expressions for the equivalent linear fluid resistance of the overall pipe (i.e., combined vertical and horizontal segments) R_{eq}, the equivalent fluid inertance within the overall pipe I_{eq}, and the gravitational fluid capacitance of the overhead tank C_{grv}, in terms of the system parameters defined as above.

(b) Treating $\mathbf{x} = [P_{3a}\ Q]^T$ as the state vector,

where
P_{3a} is the pressure head of the overhead tank
Q is the volume flow rate through the pipe

develop a complete state-space model for the system. Specifically, obtain the matrices $\mathbf{A}$, $\mathbf{B}$, $\mathbf{C}$, and $\mathbf{D}$.

(c) Obtain the input–output differential equation of the system.

Solution

Since the inertia effects are neglected in the model, and only two capacitors are used as the energy storage elements, there exists only one type of energy in this system. Hence this model cannot provide an oscillatory response to an initial condition excitation (i.e., natural oscillations are not possible). However, the actual physical system has fluid inertia, and hence the system can exhibit an oscillatory response.

(a) Assuming a parabolic velocity profile, the fluid inertance in a pipe of uniform cross-section A and length L, is given by

$$I = \frac{2\rho L}{A}$$

Since the same volume flow rate Q is present in both segments of piping (continuity) we have, for series connection,

$$I_{eq} = \frac{2\rho L_v}{\frac{\pi}{4} d_v^2} + \frac{2\rho L_h}{\frac{\pi}{4} d_h^2} = \frac{8\rho}{\pi}\left[\frac{L_v}{d_v^2} + \frac{L_h}{d_h^2}\right]$$

The linear fluid resistance in a circular pipe is

$$R = \frac{128\mu L}{\pi d^4}$$

where d is the internal diameter.

Again, since the same Q exists in both segments of the series-connected pipe,

$$R_{eq} = \frac{128\mu}{\pi}\left[\frac{L_v}{d_v^4} + \frac{L_h}{d_h^4}\right]$$

Also,

$$C_{grv} = \frac{A_t}{\rho g}$$

(b) State-space shell:

$$C_{grv} \frac{dP_{3a}}{dt} = Q$$

$$I_{eq} \frac{dQ}{dt} = P_{23}$$

Remaining constitutive equation:

$$P_{12} = R_{eq}Q$$

Note: Constitutive (node) equations are automatically satisfied.

Compatibility (loop) equations:

$$P_{1a} = P_{12} + P_{23} + P_{3a} \quad \text{with } P_{1a} = P_s(t) \text{ and } P_{3a} = \rho g h$$

Now eliminate the auxiliary variable P_{23} in the state-space shell, using the remaining equations. We get

$$P_{23} = P_{1a} - P_{12} - P_{3a}$$
$$= P_s(t) - R_{eq}Q - P_{3a}$$

Hence, the state-space model is given by

State equations:

$$\frac{dP_{3a}}{dt} = \frac{1}{C_{grv}} Q \tag{i}$$

$$\frac{dQ}{dt} = \frac{1}{I_{eq}} \left[P_s(t) - P_{3a} - R_{eq}Q \right] \tag{ii}$$

Output equation:

$$h = \frac{1}{\rho g} P_{3a} \tag{iii}$$

Corresponding matrices are

$$A = \begin{bmatrix} 0 & 1/C_{grv} \\ -1/I_{eq} & -R_{eq}/I_{eq} \end{bmatrix}; \quad B = \begin{bmatrix} 0 \\ 1/I_{eq} \end{bmatrix}$$

$$C = \begin{bmatrix} \frac{1}{\rho g} & 0 \end{bmatrix}; \quad D = 0$$

(c) Substitute equation (i) into (ii):

$$I_{eq}C_{grv} \frac{d^2 P_{3a}}{dt^2} = P_s(t) - P_{3a} - R_{eq}C_{grv} \frac{dP_{3a}}{dt}$$

Now substitute equation (iii) for P_{3a}:

$$I_{eq}C_{grv}\frac{d^2h}{dt^2} + R_{eq}C_{grv}\frac{dh}{dt} + h = \frac{1}{\rho g}P_s(t)$$

This is the input–output model.

3.5 Model Linearization

Real systems are nonlinear and they are represented by nonlinear analytical models consisting of nonlinear differential equations. Linear systems (models) are in fact idealized representations, and are represented by linear differential equations. Clearly, it is far more convenient to analyze, simulate, design, and control linear systems. In particular, the *principle of superposition* holds for linear systems, thereby making the analytical procedures far simpler. For these reasons, nonlinear systems are often approximated by linear models.

If the input/output relations are nonlinear algebraic equations, it represents a *static nonlinearity*. Such a situation can be handled simply by using nonlinear calibration curves, which will linearize the device without introducing nonlinearity errors. If, on the other hand, the input/output relations are nonlinear differential equations, analysis usually becomes quite complex. This situation represents a *dynamic nonlinearity*. Common manifestations of nonlinearities in devices and systems are saturation, dead zone, hysteresis, the jump phenomenon, limit cycle response, and frequency creation.

It is not possible to represent a highly nonlinear system by a single linear model in its entire range of operation. For small "changes" in the system response, however, a linear model may be developed, which is valid in the neighborhood of an operating point of the system about which small response changes take place. In this section, we will study linearization of a nonlinear system/model in a restricted range of operation, about an operating point. First linearization of both analytical models, particularly state-space models and input–output models will be treated. Then linearization of experimental models (experimental data) is addressed.

3.5.1 Linearization about an Operating Point

Linearization is carried out about some operating point—typically the normal operating condition of the system, by necessity (which is a steady state or the equilibrium state). In a steady state, by definition, the rates of changes of the system variables are zero. Hence, the steady state (equilibrium state) is determined by setting the time-derivative terms in the system equations to zero and then solving the resulting algebraic equations. This may lead to more than one solution, since the steady state (algebraic) equations themselves are nonlinear. The steady state (equilibrium) solutions can be

1. Stable (given a slight shift, the system response will eventually return to the original steady state)
2. Unstable (given a slight shift, the system response will continue to move away from the original steady state)
3. Neutral (given a slight shift, the system response will remain in the shifted condition)

Consider a nonlinear function $f(x)$ of the independent variables x. Its Taylor series approximation about an operating point $()_o$, up to the first derivative, is given by

$$f(x) \approx f(x_o) + \frac{df(x_o)}{dx}\delta x \quad \text{with } x = x_o + \delta x \tag{3.7a}$$

Note that δx represents a small change from the operating point.

Now denote operating condition by $(^-)$ and a small increment about that condition by $(\hat{\ })$. We have

$$f(\bar{x} + \hat{x}) \approx f(\bar{x}) + \frac{df(\bar{x})}{dx}\hat{x} \tag{3.7b}$$

A graphical illustration of this approach to linearization is given in Figure 3.9.

From Equation 3.7 it is seen that the increment of the function, due to the increment in its independent variable, is

$$\delta f = f(x) - f(x_o) \approx \frac{df(x_o)}{dx}\delta x \tag{3.8a}$$

or

$$\hat{f} = f(\bar{x} + \hat{x}) - f(\bar{x}) \approx \frac{df(\bar{x})}{dx}\hat{x} \tag{3.8b}$$

The error resulting from this approximation is

$$\text{Error } e = f(\bar{x} + \hat{x}) - \left[f(\bar{x}) + \frac{df(\bar{x})}{dx}\hat{x} \right] \tag{3.9}$$

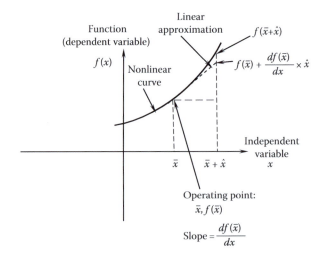

FIGURE 3.9
Linearization about an operating point.

This error can be decreased by

1. Making the nonlinear function more linear
2. Making the change $\hat{x}$ from the operating point smaller

Note: If the term is already linear, as for example, is the case of

$$f = ax$$

where a is the coefficient (constant) of the term, the corresponding linearized incremental term is

$$\delta f = a \, \delta x \qquad (3.10\text{a})$$

or

$$\hat{f} = a\hat{x} \qquad (3.10\text{b})$$

This is obvious because, for a linear term, the first derivative (slope) is simply its coefficient.

Furthermore, the following incremental results hold for the time derivatives of the variables:

$$\delta \dot{x} = \frac{d\hat{x}}{dt} = \dot{\hat{x}} \qquad (3.11)$$

$$\delta \ddot{x} = \frac{d^2\hat{x}}{dt^2} = \ddot{\hat{x}} \qquad (3.12)$$

3.5.2 Function of Two Variables

The process of linearization, as presented above, can be easily extended to functions of more than one independent variable. For illustration consider a nonlinear function $f(x, y)$ of two independent variables x and y. The first order Taylor series approximation is

$$f(x,y) \approx f(x_o,y_o) + \frac{\partial f(x_o,y_o)}{\partial x}\delta x + \frac{\partial f(x_o,y_o)}{\partial y}\delta y \quad \text{with} \ x = x_o + \delta x, \ y = y_o + \delta y \qquad (3.7\text{a})$$

or

$$f\left(\overline{x} + \hat{x}, \overline{y} + \hat{y}\right) \approx f(\overline{x},\overline{y}) + \frac{\partial f(\overline{x},\overline{y})}{\partial x}\hat{x} + \frac{\partial f(\overline{x},\overline{y})}{\partial y}\hat{y} \qquad (3.7\text{b})$$

where ($^-$) denotes the operating condition and ($^\wedge$) denotes a small increment about that condition, as usual. In this case, for the process of linearization, we need two local slopes $\partial f(\overline{x},\overline{y})/\partial x$ and $\partial f(\overline{x},\overline{y})/\partial y$ along the two orthogonal directions of the independent variables x and y.

It should be clear now that the linearization of a nonlinear system is carried out by replacing each term in the system equation by its increment, about an operating point. We summarize below the steps of local linearization about an operating point:

1. Select the operating point (or reference condition). This is typically a steady state that can be determined by setting the time derivative terms in the system equations to zero and solving the resulting nonlinear algebraic equations.
2. Determine the slopes (first-order derivatives) of each nonlinear term (function) in the systems equation at the operating point, with respect to (along) each independent variable.
3. Consider each term in the system equation. If a term is nonlinear, replace it by its slope (at the operating point) times the corresponding incremental variable. If a term is linear, replace it by its coefficient (which is indeed the constant slope of the linear term) times the corresponding incremental variable.

3.5.3 Reduction of System Nonlinearities

Under steady conditions, system nonlinearities can be removed through calibration. Under dynamic conditions, however, the task becomes far more difficult. The following are some of the precautions and procedures that can be taken to remove or reduce nonlinearities in dynamic systems:

1. Avoid operating the device over a wide range of signal levels
2. Avoid operation over a wide frequency band
3. Use devices that do not generate large mechanical motions
4. Minimize Coulomb friction and stiction (e.g., through lubrication)
5. Avoid loose joints, gear coupling, etc. that can cause backlash
6. Use linearizing elements such as resistors and amplifiers
7. Use linearizing feedback

Next, we will illustrate model linearization and operating point analysis using several examples, which involve state-space models and input–output models.

Example 3.3

The robotic spray painting system of an automobile assembly plant employs an induction motor and pump combination to supply paint, at an overall peak rate of 15 gal/min, to a cluster of spray-paint heads in several painting booths. The painting booths are an integral part of the production line in the plant. The pumping and filtering stations are in the ground level of the building and the painting booths are in an upper level. Not all booths or painting heads operate at a given time. The pressure in the paint supply line is maintained at a desired level (approximately 275 psi or 1.8 MPa) by controlling the speed of the pump, which is achieved through a combination of voltage control and frequency control of the induction motor. An approximate model for the paint pumping system is shown in Figure 3.10.

The induction motor is linked to the pump through a gear transmission of efficiency η and speed ratio 1:r, and a flexible shaft of torsional stiffness k_p. The moments of inertia of the motor rotor and the pump impeller are denoted by J_m and J_p, respectively. The gear inertia is neglected (or lumped with J_m). The mechanical dissipation in the motor and its bearings is modeled as a linear viscous

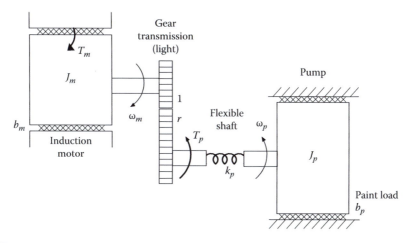

FIGURE 3.10
A model for a paint pumping system in an automobile assembly plant.

damper of damping constant b_m. The load on the pump (the paint load plus any mechanical dissipation) is also modeled by a viscous damper, and the equivalent damping-constant is b_p. The magnetic torque T_m generated by the induction motor is given by

$$T_m = \frac{T_0 q \omega_0 (\omega_0 - \omega_m)}{(q\omega_0^2 - \omega_m^2)}$$ (3.8)

in which ω_m is the motor speed. The parameter T_0 depends directly (quadratically) on the phase voltage (ac) supplied to the motor. The second parameter ω_0 is directly proportional to the line frequency of the ac supply. The third parameter q is positive and greater than unity, and this parameter is assumed constant in the control system.

(a) Comment about the accuracy of the model shown in Figure 3.10.
(b) Taking the motor speed ω_m, the pump-shaft torque T_p, and the pump speed ω_p as the state variables, systematically derive the three state equations for this (nonlinear) model. Clearly explain all steps involved in the derivation. What are the inputs to the system?
(c) What do the motor parameters ω_0 and T_0 represent, with regard to motor behavior? Obtain the partial derivatives $\partial T_m/\partial \omega_m$, $\partial T_m/\partial T_0$ and $\partial T_m/\partial \omega_0$ and verify that the first of these three expressions is negative and the other two are positive. *Note:* under normal operating conditions $0 < \omega_m < \omega_0$.
(d) Consider the steady-state operating point where the motor speed is steady at $\bar{\omega}_m$. Obtain expressions for ω_p, T_p, and T_0 at this operating point, in terms of $\bar{\omega}_m$ and $\bar{\omega}_0$.
(e) Suppose that $\partial T_m/\partial \omega_m = -b$, $\partial T_m/\partial T_0 = \beta_1$, and $\partial T_m/\partial \omega_0 = \beta_2$ at the operating point given in part (d). *Note:* Voltage control is achieved by varying T_0 and frequency control by varying ω_0. Linearize the state model obtained in part (b) about the operating pint and express it in terms of the incremental variables $\hat{\omega}_m$, $\hat{T}_p$, $\hat{\omega}_p$, $\hat{T}_0$, and $\hat{\omega}_0$. Suppose that the (incremental) output variables are the incremental pump speed $\hat{\omega}_p$ and the incremental angle of twist of the pump shaft. Express the linear state-space model in the usual form and obtain the associated matrices **A**, **B**, **C**, and **D**.
(f) For the case of frequency control only (i.e., $\hat{T}_0 = 0$) obtain the input–output model (differential equation) relating the incremental output $\hat{\omega}_p$ and the incremental input $\hat{\omega}_0$. Using this equation, show that if $\hat{\omega}_0$ is suddenly changed by a step of $\Delta\hat{\omega}_0$ then $d^3\hat{\omega}_p/dt^3$ will immediately change by a step of $(\beta_2 r k_p / J_m J_p)\Delta\hat{\omega}_0$, but the lower derivatives of $\hat{\omega}_p$ will not change instantaneously.

Solution

(a) Backlash and inertia of the gear transmission have been neglected in the model shown. This is not accurate in general. Also, the gear efficiency η, which is assumed constant here, usually varies with the gear speed.

- Usually, there is some flexibility in the shaft (coupling), which connects the gear to the drive motor.
- Energy dissipation (in the pump load and in various bearings) has been lumped into a single linear viscous-damping element. In practice, this energy dissipation is nonlinear and distributed.

(b) Motor speed $\omega_m = \dfrac{d\theta_m}{dt}$

$$\text{Load (pump) speed } \omega_p = \frac{d\theta_p}{dt}$$

where
θ_m = motor rotation
θ_p = pump rotation

Let T_g = reaction torque on the motor from the gear. By definition, gear efficiency is given by

$$\eta = \frac{T_p r\, \omega_m}{T_g\, \omega_m} = \frac{\text{Output power}}{\text{Input power}}$$

Since r is the gear ratio, $r\omega_m$ is the output speed of the gear. Also, power = torque × speed. We have

$$T_g = \frac{r}{\eta} T_p \tag{i}$$

The following three constitutive equations can be written as
Newton's second law (torque = inertia × angular acceleration) for the motor:

$$T_m - T_g - b_m\omega_m = J_m\dot{\omega}_m \tag{ii}$$

Newton's second law for the pump:

$$T_p - b_p\omega_p = J_p\dot{\omega}_p \tag{iii}$$

Hooke's law (torque = torsional stiffness × angle of twist) for the flexible shaft:

$$T_p = k_p(r\theta_m - \theta_p) \tag{iv}$$

Equations (ii) through (iv) provide the three state equations. Specifically,
Substitute (i) into (ii):

$$J_m\dot{\omega}_m = T_m - b_m\omega_m - \frac{r}{\eta} T_p \tag{v}$$

Differentiate (iv):

$$\dot{T}_p = k_p(r\omega_m - \omega_p) \tag{vi}$$

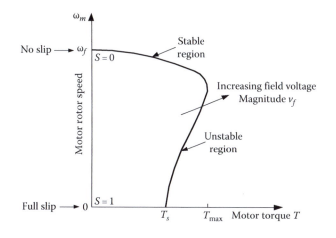

FIGURE 3.11
Torque–speed characteristic curve of an induction motor.

Equation (iii):

$$J_p \dot{\omega}_p = T_p - b_p \omega_p \qquad \text{(vii)}$$

Equations (v) through (vii) are the three state equations. This is a nonlinear model with the state vector $[\omega_m \ T_p \ \omega_p]^T$. The input is T_m. Strictly there are two inputs, in view of the torque–speed characteristic curve of the motor as given by Equation 3.8 and sketched in Figure 3.11, where the fractional slip S of the induction motor is given by

$$S = \frac{(\omega_0 - \omega_m)}{\omega_0} \qquad (3.9)$$

The two inputs are
ω_0, the speed of the rotating magnetic field, which is proportional to the line frequency
T_0, which depends quadratically on the phase voltage

(c) From Equation 3.8:
When $\omega_m = 0$ we have $T_m = T_0$. Hence
T_0 = starting torque of the motor.
When $T_m = 0$, we have $\omega_m = \omega_0$. Hence,
ω_0 = no-load speed.

This is the synchronous speed—under no-load conditions, there is no slip in the induction motor (i.e., actual speed of the motor is equal to the speed ω_0 of the rotating magnetic field).
Differentiate Equation 3.8 with respect to T_0, ω_0, and ω_m. We have

$$\frac{\partial T_m}{\partial T_0} = \frac{q\omega_0(\omega_0 - \omega_m)}{(q\omega_0^2 - \omega_m^2)} = \beta_1 \text{ (say)} \qquad (3.10)$$

$$\frac{\partial T_m}{\partial \omega_0} = \frac{T_0 q\left[(q\omega_0^2 - \omega_m^2)(2\omega_0 - \omega_m) - \omega_0(\omega_0 - \omega_m)2q\omega_0\right]}{(q\omega_0^2 - \omega_m^2)^2}$$

$$= \frac{T_0 q\omega_m\left[(\omega_0 - \omega_m)^2 + (q - 1)\omega_0^2\right]}{(q\omega_0^2 - \omega_m^2)^2} = \beta_2 \text{ (say)} \qquad (3.11)$$

Furthermore, the following somewhat general observations can be made from this example:

1. Mechanical damping constant b_m comes from bearing friction and other mechanical sources of the motor
2. Electrical damping constant b_e comes from the electromagnetic interactions in the motor
3. The two must occur together (e.g., in model analysis, simulation, design, and control). For example, whether the response is underdamped or overdamped depends on the sum $b_m + b_e$ and not the individual components ← electromechanical coupling

3.5.4 Linearization Using Experimental Operating Curves

In some situations, an accurate analytical model may not be readily available for an existing physical system. Yet experiments may be conducted on the system to gather operating curves for the system. These operating curves are useful in deriving a linear model, which can be valuable, for example, in controlling the system. This approach is discussed now, taking an electric motor as the example system.

3.5.4.1 Torque–Speed Curves of Motors

The speed versus torque curves of motors under steady conditions (i.e., steady-state operating curves), are available from the motor manufacturers. These curves have the characteristic shape that they decrease slowly up to a point and then drop rapidly to zero. An example of an ac induction motor is given in Figure 3.11. The operating curves of direct current (dc) motors take a similar, not identical characteristic form. The shape of the operating curve depends on how the motor windings (rotor and stator) are connected and excited. The torque at zero speed is the "braking torque" or "starting torque" or "stalling torque." The speed at zero torque is the "no-load speed" which, for an ac induction motor, is also the "synchronous speed." Typically, these experimental curves are obtained as follows. The supply voltage to the motor windings is maintained constant, a known load (torque) is applied to the motor shaft, and once the conditions become steady (i.e., constant speed) the motor speed is measured. The experiment is repeated for increments of torques within an appropriate range. This gives one operating curve, for a specified supply voltage. The experiment is repeated for other supply voltages and a series of curves are obtained.

It should be noted that the motor speed is maintained steady in these experiments as they represent "steady" operating conditions. That means the motor inertia (inertial torque) is not accounted for in these curves, while mechanical damping is. Hence, motor inertia has to be introduced separately when using these curves to determine a "dynamic" model for a motor. Since mechanical damping is included in the measurements, it should not be introduced again. Of course, if the motor is connected to an external load, the damping, inertia, and flexibility of the load all have to be accounted for separately when using experimental operating curves of motors in developing models for motor-integrated dynamic systems.

3.5.4.2 Linear Models for Motor Control

Consider an experimental set of steady-state operating curves for a motor, each obtained at a constant supply/control voltage. In particular consider one curve measured at voltage v_c and the other measured at voltage $v_c + \Delta v_c$, where ΔT_m is the voltage increment from one operating to the other, as shown in Figure 3.12.

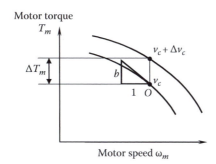

FIGURE 3.12
Two steady-state operating curves of a motor at constant input voltage.

Draw a tangent to the first curve at a selected point (operating point O). The slope of the tangent is negative, as shown. Its magnitude b is given by

$$\text{Damping constant } b = -\left.\frac{\partial T_m}{\partial \omega_m}\right|_{v_c = \text{constant}} = \text{slope at } O \qquad (3.13)$$

It should be clear that b represents an equivalent rotary damping constant (torque/angular speed) that includes both electromagnetic and mechanical damping effects in the motor. The included mechanical damping comes primarily from the friction of the motor bearings and aerodynamic effects. Since a specific load is not considered in the operating curve, load damping is not included.

Draw a vertical line through the operating point O to intersect the other operating curve. We get:

$$\Delta T_m = \text{torque intercept between the two curves}$$

Since a vertical line is a constant speed line, we have

$$\text{Voltage gain } k_v = \left.\frac{\partial T_m}{\partial v_c}\right|_{\omega_m = \text{constant}} = \frac{\Delta T_m}{\Delta v_c} \qquad (3.14)$$

Since the motor torque T_m is a function of both motor speed ω_m and the input voltage v_c (i.e., $T_m = T_m(\omega_m, v_c)$) we write from basic calculus:

$$\delta T_m = \left.\frac{\partial T_m}{\partial \omega_m}\right|_{v_c} \partial \omega_m + \left.\frac{\partial T_m}{\partial v_c}\right|_{\omega_m} \partial v_c \qquad (3.15a)$$

or

$$\delta T_m = -b \delta \omega_m + k_v \delta v_c \qquad (3.15b)$$

where the motor damping constant b and the voltage gain k_v are given by Equations 3.13 and 3.14, respectively.

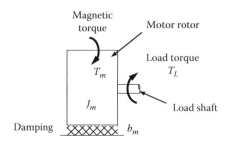

FIGURE 3.13
Mechanical system of the motor.

Equation 3.15 represents a linearized model of the motor. The torque needed to drive rotor inertia of the motor is not included in this equation (because steady-state curves are used in determining parameters). The inertia term should be explicitly present in the mechanical equation of the motor rotor, as given by Newton's second law (see Figure 3.13), in the linearized (incremental) form:

$$J_m \frac{d\delta\omega_m}{dt} = \delta T_m - \delta T_L \tag{3.16}$$

where
 J_m is the moment of inertia of the motor rotor
 T_L is the load torque (equivalent torque applied on the motor by the load that is driven by the motor)

Note that mechanical damping of the motor, as shown in Figure 3.13, is not included in Equation 3.16 because it (and electromagnetic damping) is already included in Equations 3.15.

3.6 Linear Graphs

Among the graphical tools for developing and representing a model of a mechatronic system, linear graphs take an important place. In particular, state-space models of lumped-parameter dynamic systems (mechanical, electrical, fluid, thermal, or multi-domain or mixed) can be conveniently developed by *linear graphs*. Interconnected line segments (called branches) are used in a linear graph to represent a dynamic model. The term "linear graph" stems from this use of line segments, and does not mean that the system itself has to be linear. Particular advantages of using linear graphs for model development and representation are: they allow visualization of the system structure (even before formulating an analytical model); they help identify similarities (structure, performance, etc.) in different types of systems; they are applicable for multi-domain systems (the same approach is used in any domain); and they provide a unified approach to model multifunctional devices (e.g., a piezoelectric device that can function as both a sensor and an actuator.

3.6.1 Variables and Sign Convention

Each branch in the linear graph model has one *through variable* and one *across variable* associated with it. Their product is the power variable. It is important to adhere to standard and uniform conventions so that that there will not be ambiguities in a given linear graph representation. In particular, a standard sign convention must be established.

3.6.1.1 Sign Convention

Consider Figure 3.14 where a general basic element (strictly, a single-port element, as will be discussed later) of a dynamic system is shown. In the linear graph representation, as shown in Figure 3.14b, the element is shown as a branch (i.e., a line segment). One end of any branch is selected as the *point of reference* and the other end automatically becomes the *point of action* (see Figure 3.14a and c). The choice is somewhat arbitrary, and may reflect the physics of the actual system. An *oriented* branch is one to which a direction is assigned, using an arrowhead, as in Figure 3.14b. The arrowhead denotes the positive direction of power flow at each end of the element. By convention, the positive direction of power is taken as "into" the element at the point of action, and "out of" the element at the point of reference. According to this convention, the arrowhead of a branch is always pointed toward the point of reference. In this manner, the reference point and the action point are easily identified.

The across variable is always given relative to the point of reference. It is also convenient to give the through variable f and the across variable v as an ordered pair $(f \cdot v)$ on one side of the branch, as in Figure 3.14b. Clearly, the relationship between f and v (the constitutive relation or physical relation) can be linear or nonlinear. The parameter of the element (e.g., mass, capacitance) is shown on the other side of the branch. It should be noted that the direction of a branch does not represent the positive direction of f or v. For example, when the positive directions of both f and v are changed, as in Figure 3.14a and c, the linear graph remains unchanged, as in Figure 3.14b, because the positive direction of power flow remains the same. In a given problem, the positive direction of any one of the two variables f and v should be preestablished for each branch. Then the corresponding positive direction of the other variable is automatically determined by the convention used to

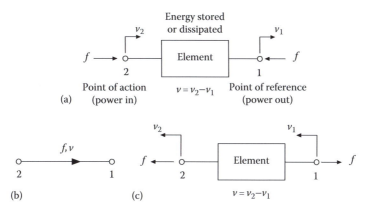

FIGURE 3.14
Sign convention for a linear graph: (a) A basic element and positive directions of its variables; (b) linear graph branch of the element; (c) an alternative sign convention.

orient linear graphs. It is customary to assign the same positive direction for f (and v) and the power flow at the point of action (i.e., the convention shown in Figure 3.14a is customary, not Figure 3.14c). Then the positive directions of the variables at the point of reference are automatically established.

Note that in a branch (line segment), the through variable (f) is transmitted through the element with no change in value; it is the "through" variable. The absolute value of the across variable, however, changes across the element (from v_2 to v_1, in Figure 3.14a). In fact, it is this change ($v = v_2 - v_1$) across the element that is called the across variable. For example, v_2 and v_1 may represent electric potentials at the two ends of an electric element (e.g., a resistor) and then v represents the voltage across the element. Accordingly, the across variable, is measured relative to the point of reference of the particular element.

According to the sign convention shown in Figure 3.14, the work done on the element at the point of action (by an external device) is positive (i.e., power flows in), and work done by the element at the point of reference (on an external load or environment) is positive (i.e., power flows out). The difference in the work done on the element and the work done by the element (i.e., the difference in the work flow at the point of action and the point of reference) is either stored as energy (e.g., kinetic energy of a mass; potential energy of a spring; electrostatic energy of a capacitor; electromagnetic energy of an inductor), which has the capacity to do additional work; or dissipated (e.g., mechanical damper; electrical resistor) through various mechanisms manifested as heat transfer, noise, and other phenomena.

In summary,

1. An element (a single-port element) is represented by a line segment (branch). One end is the point of action and the other end is the point of reference.

2. The through variable f is the same at the point of action and the point of reference of an element; the across variable differs, and it is this difference (value relative to the point of reference) that is called the across variable v.

3. The variable pair (f, v) of the element is shown on one side of the branch. Their relationship (constitutive relation) can be linear or nonlinear. The parameter of the element is shown on the other side of the branch.

4. Power flow p is the product of the through variable and the across variable. By convention, at the point of action, f and p are taken to be positive in the same direction; at the point of reference, f is positive in the opposite direction.

5. The positive direction of power flow p (or energy or work) is into the element at the point of action; and out of the element at the point of reference. This direction is shown by an arrow on the linear graph branch (an oriented branch).

6. The difference in the energy flows at the two ends of the element is either stored (with capacity to do further work) or dissipated, depending on the element type.

Linear graph representation is particularly useful in understanding rates of energy transfer (power) associated with various phenomena, and dynamic interactions in a physical system (mechanical, electrical, fluid, etc.) can be interpreted in terms of power transfer. Power is the product of a through variable (a generalized force or current) and the corresponding across variable (a generalized velocity or voltage). For example, consider a mechanical system. The total work done on the system is, in part, used as stored energy (kinetic and potential) and the remainder is dissipated. The stored energy can be completely recovered when the system is brought back to its original state (i.e., when the cycle is completed). Such a process is *reversible*. On the other hand, dissipation corresponds to irreversible energy transfer that cannot be recovered by returning the system to its initial state. (A fraction of the mechanical energy lost in this manner could be recovered, in principle, by operating a heat engine, but we shall not go into these thermodynamic details, which are beyond the present scope). Energy dissipation may appear in many forms including temperature rise (a molecular phenomenon), noise (an acoustic phenomenon), or work used up in wear mechanisms.

3.6.2 Linear Graph Elements

We will discuss two types of basic elements in the categories of single-port elements and two-port elements. Analogous elements in these categories exist across the domains (mechanical, electrical, fluid, and thermal) for the most part.

3.6.2.1 Single-Port Elements

Single-port (or, *single energy port*) elements are those that can be represented by a single branch (single line segment) of the linear graph. These elements possess only one power (or energy) variable; hence the name "single-port." They have two terminals. The general form of these elements is shown in Figure 3.14b.

In modeling mechanical systems, we require three passive single-port elements, as shown in Figure 3.15. Although translatory mechanical elements are presented in Figure 3.15,

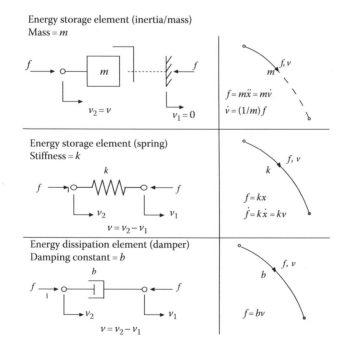

FIGURE 3.15
Single-port mechanical system elements and their linear graph representations.

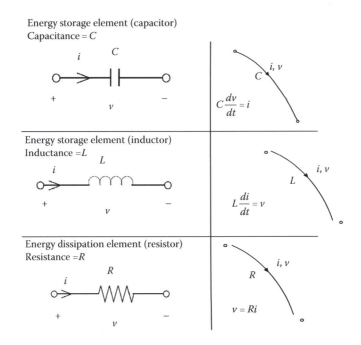

FIGURE 3.16
Single-port electrical system elements and their linear graph representations.

corresponding rotary elements are easy to visualize: f denotes an applied torque and v the relative angular velocity in the same direction. Note that the linear graph of an inertia element has a broken line segment. This is because, through inertia, the force (inertia force) does not physically travel from one end of its linear graph branch to the other end, but rather is "felt" at the two ends.

Analogous single-port electrical elements may be represented in a similar manner. These are shown in Figure 3.16.

3.6.2.1.1 Source Elements

In linear-graph models, system *inputs* are represented by *source elements*. There are two types of sources, as shown in Figure 3.17.

(a) *T*-type source (e.g., force source, current source):
The independent variable (i.e., the source output, which is the system input) is the through variable f. The arrowhead indicates the positive direction of f.

Note: For a *T*-type source, the sign convention that the arrow gives the positive direction of f still holds. However, the sign convention that the arrow is from the point of action to the point of reference (or the direction of the drop in the across variable) does not hold.

(b) *A*-type source (e.g., velocity source, voltage source):

The independent variable is the across variable v. The arrowhead indicates the positive direction of the "drop" in v. *Note*: + and − signs are indicated as well, where the drop in v occurs from + to − terminals.

Note: For an *A*-type source, the sign convention that the arrow is from the point of action to the point of reference (or the direction of the drop in the across variable) holds.

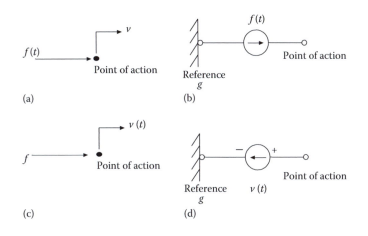

FIGURE 3.17
(a) *T*-type source (through variable input); (b) linear graph representation of a *T*-type source; (c) *A*-type source; (d) linear graph representation of an *A*-type source.

However, the sign convention that the arrow gives the positive direction of *f* does not hold.

An ideal force source (a through-variable source) is able to supply a force input that is not affected by interactions with the rest of the system. The corresponding relative velocity across the force source, however, will vary as determined by the dynamics of the overall system. It should be clear that the direction of *f*(*t*) as shown in Figure 3.17a is the applied force. The reaction on the source would be in the opposite direction. An ideal velocity source (across-variable source) supplies a velocity input independent of the system to which it is applied. The corresponding force is, of course, determined by the system dynamics.

3.6.2.2 Two-Port Elements

A two-port element has two energy ports and two separate, yet coupled, branches corresponding to them. These elements can be interpreted as a pair of single-port elements whose net power is zero. A transformer (mechanical, electrical, fluid, etc.) is a two-port element. Also, a mechanical gyrator is a two-port element. Examples of mechanical transformers are a lever and pulley for translatory motions and a meshed pair of gear wheels for rotation. A gyrator is typically an element that displays gyroscopic properties. We shall consider only the linear case, that is, ideal transformers and gyrators only. The extension to the nonlinear case should be clear.

3.6.2.2.1 Transformer

In an ideal transformer, the across variables in the two ports (branches) are changed without dissipating or storing energy in the process. Hence, the through variables in the two ports will also change. Examples of mechanical, electrical, and fluid transformers are shown in Figure 3.18a through d. The linear graph representation of a transformer is given in Figure 3.18e.

In Figure 3.18e, as for a single-port passive element, the arrows on the two branches (line segments) give the positive direction of power flow (i.e., when the product of the through variable and the across variable for that segment is positive). One of the two ports may be considered the input port and the other the output port. Let

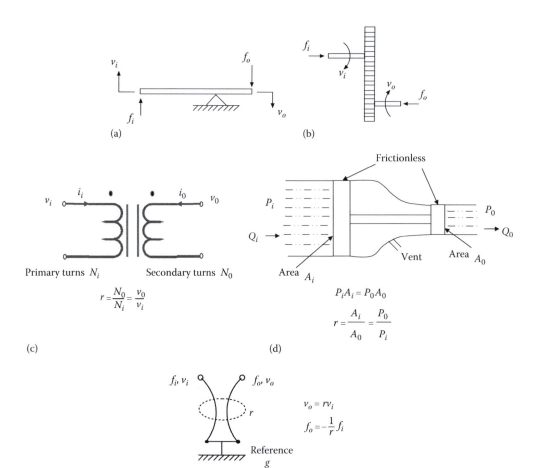

FIGURE 3.18
Transformer: (a) Lever; (b) meshed gear wheels; (c) electrical transformer; (d) fluid transformer; (e) linear graph representation.

v_i and f_i are across and through variables at the input port
v_o and f_o are across and through variables at the output port

The (linear) transformation ratio r of the transformer is given by

$$v_o = r\,v_i \tag{3.17}$$

Due to the conservation of power, we have

$$f_i v_i + f_o v_o = 0 \tag{3.18}$$

By substituting Equation 3.17 into Equation 3.18 gives

$$f_o = -\frac{1}{r} f_i \tag{3.19}$$

Here r is a dimensional parameter. The two constitutive relations for a transformer are given by Equations 3.17 and 3.19.

3.6.2.2.2 Electrical Transformer

As shown in Figure 3.18c, an electrical transformer has a primary coil, which is energized by an ac voltage (v_i), a secondary coil in which an ac voltage (v_o) is induced, and a common core, which helps the linkage of magnetic flux between the two coils. Note that a transformer converts v_i to v_o without making use of an external power source. Hence, it is a passive device, just like a capacitor, inductor, or resistor. The turn ratio of the transformer:

$$r = \frac{\text{Number of turns in the secondary coil } (N_o)}{\text{Number of turns in the primary coil } (N_i)}$$

In Figure 3.18c, the two dots on the top side of the two coils indicate that the two coils are wound in the same direction.

In a pure and ideal transformer, there will be full flux linkage without any dissipation of energy. Then, the flux linkage will be proportional to the number of turns. Hence,

$$\lambda_o = r\lambda_i \tag{3.20}$$

where λ denotes the flux linkage in each coil. Differentiation of Equation 3.20, noting that the induced voltage in coil is given by the rate of charge of flux, gives

$$v_o = rv_i \tag{3.17}$$

For an *ideal transformer*, there is no energy dissipation and also the signals will be in phase. Hence, the output power will be equal to the input power; thus,

$$v_o i_o = v_i i_i \tag{3.18b}$$

Hence, the current relation becomes

$$i_o = \frac{1}{r} i_i \tag{3.19b}$$

3.6.2.2.3 Gyrator

An ideal gyroscope is an example of a mechanical *gyrator* (Figure 3.19a). It is simply a spinning top that rotates about its own axis at a high angular speed ω (positive in the x direction) and is assumed to remain unaffected by other small motions that may be present. If the moment of inertia about this axis of rotation (x in the shown configuration) is J, the corresponding angular momentum is $h = J\omega$, and this vector is also directed in the positive x direction, as shown in Figure 3.19b.

Suppose that the angular momentum vector h is given an incremental rotation $\delta\theta$ about the positive z axis, as shown. The free end of the gyroscope will move in the positive y direction as a result. The resulting change in the angular momentum vector is $\delta h = J\omega\delta\theta$ in the positive y direction, as shown in Figure 3.19b. Hence, the rate of change of angular momentum is

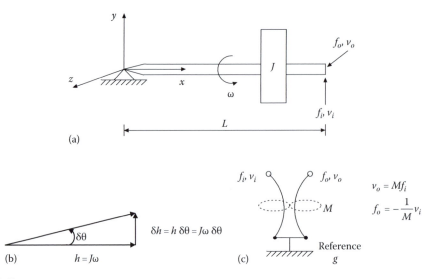

FIGURE 3.19
(a) Gyrator (gyroscope or spinning top)—a two-port element; (b) derivation of the constitutive equations; (c) linear graph representation.

$$\frac{\delta h}{\delta t} = \frac{J\omega\,\delta\theta}{\delta t}$$

where δt is the time increment of the motion. Hence, in the limit, the rate of change of angular momentum is

$$\frac{dh}{dt} = J\omega\frac{d\theta}{dt}$$

If the velocity given to the free end of the gyroscope, in the positive y direction, to generate this motion is v_i (which will result in a force f_i at that point, in the positive y direction) the corresponding angular velocity about the positive z axis is

$$\frac{d\theta}{dt} = \frac{v_i}{L}$$

in which L is the length of the gyroscope. Substitute (iii) in (ii). The rate of change of angular momentum is

$$\frac{dh}{dt} = \frac{J\omega v_i}{L} \tag{3.21}$$

about the positive y direction. By Newton's second law, to sustain this rate of change of angular momentum, it will require a torque equal to $J\omega v_i/L$ in the same direction. If the corresponding force at the free end of the gyroscope is denoted by f_o in the positive z-direction, the corresponding torque is $f_o L$ acting about the negative y-direction. It follows that

$$-f_o L = \frac{J \omega v_i}{L} \tag{3.22}$$

This may be expressed as

$$f_o = -\frac{1}{M} v_i \tag{3.23}$$

By the conservation of power (Equation 3.18) for an ideal gyroscope, it follows from Equation 3.23 that

$$v_o = M f_i \tag{3.24}$$

in which, the gyroscope parameter

$$M = \frac{L^2}{J \omega} \tag{3.25}$$

Note: M is a "mobility" parameter (velocity/force).

Equations 3.23 and 3.24 are the constitutive equations of a gyrator. The linear graph representation of a gyrator is shown in Figure 3.19c.

3.6.3 Linear Graph Equations

Three types of equations have to be written to obtain an analytical model from a linear graph:

1. Constitutive equations for all the elements that are not sources (inputs)
2. Compatibility equations (loop equations) for all the independent closed paths
3. Continuity equations (node equations) for all the independent junctions of two or more branches

Constitutive equations of elements have been discussed earlier. In some examples, compatibility equations and continuity equations are not used explicitly because the system variables are chosen to satisfy these two types of equations. In modeling of complex dynamic systems, systematic approaches, which can be computer-automated, will be useful. In that context, approaches are necessary to explicitly write the compatibility equations and continuity equations. The related approaches and issues are discussed next.

3.6.3.1 Compatibility (Loop) Equations

A loop in a linear graph is a closed path formed by two or more branches. A loop equation (compatibility equation) is obtained by summing all the across variables along the branches of the loop is zero. This is a necessary condition because, at a given point in the linear graph there must be a unique value for the across variable, at a given time. For example, a mass and a spring connected to the same point must have the same velocity at a particular time, and this point must be intact (i.e., does not break or snap); hence, the system is "compatible."

3.6.3.1.1 Sign Convention

1. Go in the counterclockwise direction of the loop.
2. In the direction of a branch arrow the across variable drops. This direction is taken to be positive (except in a *T*-source, where the arrow direction indicates an increase in its across variable, which is the negative direction).

The arrow in each branch is important, but we need not (and indeed cannot) always go in the direction of the arrows in the branches that form a loop. If we do go in the direction of the arrow in a branch, the associated across variable is considered positive. When we go opposite to the arrow, the associated across variable is considered negative.

3.6.3.1.2 Number of "Primary" Loops

Primary loops are a "minimal" set of loops from which any other loop in the linear graph can be determined. A primary loop set is an "independent" set. It will generate all the independent loop equations.

Note: Loops closed by broken-line (inertia) branches should be included as well in counting the primary loops.

3.6.3.2 Continuity (Node) Equations

A node is the point where two or more branches meet. A node equation (or, continuity equation) is created by equating to zero the sum of all the through variables at a node. This holds in view of the fact that a node can neither store nor dissipate energy; in effect saying, "What goes in must come out." Hence, a node equation dictates the continuity of the through variables at a node. For this reason, one must use proper signs for the variables when writing either node equations or loop equations. The sign convention that is used is the through variable "into" the node is positive.

The meaning of a node equation in the different domains is indicated below.

Mechanical systems: Force balance, equilibrium equation, Newton's third law, etc.

Electrical systems: Current balance, Kirchoff's current law, conservation of charge, etc.

Hydraulic systems: Conservation of matter

Thermal systems: Conservation of energy

3.6.4 State Models from Linear Graphs

We can obtain a state model of a dynamic system from its linear graph. Each branch in the linear graph is a "model" of an actual system element of the system, with an associated "constitutive relation." For a mechanical system, it has been justified to use the velocities of independent inertia elements and the forces through independent stiffness (spring) elements as state variables. Similarly, for an electrical system, voltages across independent capacitors and currents through independent inductors are appropriate state variables. In general then, in the linear graph approach we use as *state variables*: across variables of independent *A*-type elements and through variables of independent *T*-type elements.

3.6.4.1 System Order

It is known that *A*-type elements and *T*-type elements are energy storage elements. The *system order* is given by the number of independent energy-storage elements in the system. This is also equal to the number state variables, the order of the state-space model, the number of initial conditions required to solve the response of the analytical model, and the order of the input–output differential equation model.

The total number of energy storage elements in a system can be greater than the system order because some of these elements might not be independent.

3.6.4.2 Sign Convention

The important first step of developing a state-space model using linear graphs is indeed to draw a linear graph for the considered system. A sign convention should be established, as discussed before. The sign convention that we use is as follows:

1. Power flows into the action point and out of the reference point of an element (branch). The branch arrow (which is an oriented branch) shows this direction. *Exception*: In a source element, power flows out of the action point.
2. Through variable (f), across variable (v), and power flow (fv) are positive in the same direction at an action point. At reference point, v is positive in the same direction given by the linear-graph arrow, but f is taken positive in the opposite direction.
3. In writing node equations: Flow into a node is positive.
4. In writing loop equations: Loop direction is counterclockwise. A potential (*A*-variable) "drop" is positive (same direction as the branch arrow. *Exception*: In a *T*-source the arrow is in the direction in which the *A*-variable increases).

Note: Once the sign convention is established, the actual values of the variables can be positive or negative depending on their actual direction.

Steps of obtaining a state model

1. Choose as state variables: across variables for independent *A*-type elements and through variables for independent *T*-type elements.
2. Write constitutive equations for independent energy storage elements. This will give the *state-space shell*.
3. Do similarly for the remaining elements (dependent energy storage elements and dissipation (*D*-type) elements, transformers, etc.).
4. Write compatibility equations for the primary loops.
5. Write continuity equations for the primary nodes (total number of nodes − 1).
6. In the state-space shell, retain state and input variables only. Eliminate all other variables using the loop and node equations and extra constitutive equations.

3.6.4.3 General Observation

Now some general observations are made with regard to a linear graph in terms of its geometric (topological) characteristics (nodes, loops, branches), elements, unknown and known variables, and relevant equations (constitutive, compatibility, and continuity). First let

$$\text{Number of sources} = s$$

$$\text{Number of branches} = b$$

Since each source branch has 1 unknown variable (because one variable is the known input to the system—the source output) and all other passive branches have 2 unknown variables each, we have

$$\text{Total number of unknown variables} = 2b - s \tag{3.26}$$

Since each branch other than a source branch provides 1 constitutive equation, we have

$$\text{Number of constitutive equations} = b - s \tag{3.27}$$

Let

$$\text{Number of primary loops} = \ell$$

Since each primary loop gives a compatibility equation, we have

$$\text{Number of loop (compatibility) equations} = \ell$$

Let

$$\text{Number of nodes} = n$$

Since one of these nodes does not provide an extra node equation, we have

$$\text{Number of node (continuity) equations} = n - 1 \tag{3.28}$$

Hence,

$$\text{Total number of equations} = (b - s) + \ell + (n - 1) = b + \ell + n - s - 1$$

To uniquely solve the analytical model, we must have

$$\text{Number of unknowns} = \text{Number of equations or } 2b - s = b + \ell + n - s - 1$$

Hence, we have the result

$$\ell = b - n + 1 \tag{3.29}$$

This topological result must be satisfied by any linear graph.

Example 3.4

A robotic sewing system consists of a conventional sewing head. During operation, a panel of garment is fed by a robotic hand into the sewing head. The sensing and control system of the robotic hand ensures that the seam is accurate and the cloth tension is correct in order to guarantee the quality of the stitch. The sewing head has a frictional feeding mechanism, which pulls the fabric in a cyclic manner away from the robotic hand, using a toothed feeding element. When there is slip between the feeding element and the garment, the feeder functions as a *force source* and the applied force is assumed cyclic with a constant amplitude. When there is no slip, however, the feeder functions as a *velocity source*, which is the case during normal operation. The robot hand has inertia. There is some flexibility at the mounting location of the hand on the robot. The links of the robot are assumed rigid and some of its joints can be locked to reduce the number of degrees of freedom, when desired.

Consider the simplified case of a single-degree-of-freedom robot. The corresponding robotic sewing system is modeled as in Figure 3.20. Here the robot is modeled as a single moment of inertia J_r, which is linked to the hand with a light rack-and-pinion device with its speed transmission parameter given by

$$\frac{\text{Rack tanslatory movement}}{\text{Pinion rotatory movement}} = r$$

The drive torque of the robot is T_r and the associated rotatory speed is ω_r. Under conditions of slip the feeder input to the cloth panel is force f_r and with no slip the input is the velocity v_r. Various energy dissipation mechanisms are modeled as linear viscous damping of damping constant b (with corresponding subscripts). The flexibility of various system elements is modeled by linear springs with stiffness k. The inertia effects of the cloth panel and the robotic hand are denoted by the lumped masses m_c and m_h, respectively, having velocities v_c and v_h, as shown in Figure 3.20.

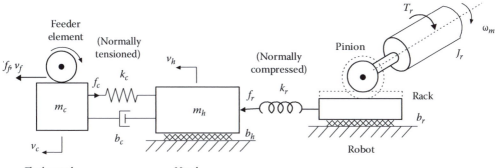

FIGURE 3.20
A robotic sewing system.

Note: The cloth panel is normally in tension with tensile force f_c. In order to push the panel, the robotic wrist is normally in compression with the compressive force f_r.

First, consider the case of the feeding element with slip:

(a) Draw a linear graph for the model shown in Figure 3.20, orient the graph, and mark all the element parameters, through variables and across variables on the graph.
(b) Write all the constitutive equations (element physical equations), independent node equations (continuity), and independent loop equations (compatibility). What is the order of the model?
(c) Develop a complete state-space model for the system. The outputs are taken as the cloth tension f_c, and the robot speed ω_r, which represent the two variables that have to be measured to control the system. Obtain the system matrices A, B, C, and D.
    Now consider the case where there is no slip at the feeder element:
(d) What is the order of the system now? How is the linear graph of the model modified for this situation? Accordingly, modify the state-space model obtained earlier to represent the present situation and from that obtain the new model matrices A, B, C, and D.
(e) Generally comment on the validity of the assumptions made in obtaining the model shown in Figure 3.20 for a robotic sewing system.

Solution

(a) Linear graph of the system is drawn as in Figure 3.21. Since in this case the feeder input to the cloth panel is force f_f, a T-source, the arrow of the source element should be retained but the $+$ and $-$ signs (used for an A-source) should be removed.
(b) In the present operation, ff is an input. This case corresponds to a 5th order model, as will be clear from the development given below.

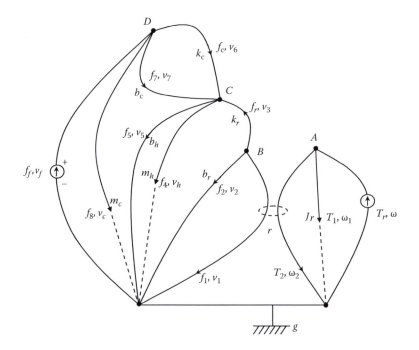

FIGURE 3.21
Linear graph of the robotic sewing system.

Constitutive equations:

$$\left.
\begin{aligned}
J_r \frac{d\omega_r}{dt} &= T_1 \\[8pt]
\frac{df_r}{dt} &= k_r v_3 \\[8pt]
m_h \frac{dv_h}{dt} &= f_4 \\[8pt]
\frac{df_c}{dt} &= k_c v_6 \\[8pt]
m_c \frac{dv_c}{dt} &= f_8
\end{aligned}
\right\} \text{State-space shell}$$

$$v_1 = r\omega_2$$

$$f_1 = -\frac{1}{r} T_2$$

$$f_2 = -b_r v_2$$

$$f_5 = -b_h v_5$$

$$f_7 = -b_c v_7$$

Continuity equations (Node equations):

Node A: $T_r - T_1 - T_2 = 0$
Node B: $-f_1 - f_2 - f_r = 0$
Node C: $f_r + f_c + f_7 - f_5 - f_4 = 0$
Node D: $-f_c + f_f - f_8 - f_7 = 0$

Compatibility equations (Loop equations):

$$-\omega + \omega_1 = 0$$

$$-\omega_1 + \omega_2 = 0$$

$$-v_1 + v_2 = 0$$

$$-v_1 + v_3 + v_h = 0$$

$$-v_h + v_5 = 0$$

$$-v_6 + v_7 = 0$$

$$-v_h - v_7 + v_c = 0$$

$$-v_c + v_f = 0$$

(c) Eliminate unwanted variables as follows:

$$T_1 = T_r - T_2 = T_r + rf_1 = T_r + r(-f_2 - f_r)$$
$$= T_r - rb_r v_2 - rf_r = T_r - rb_r v_1 - rf_r$$
$$= T_r - rb_r r\omega_2 - rf_r$$
$$= T_r - r^2 b_r \omega_2 - rf_r$$

$$v_3 = v_1 - v_h = r\omega_2 - v_h = r\omega_r - v_h$$

$$f_4 = f_r + f_c + f_7 - f_5 = f_r + f_c + b_c v_7 - b_h v_5$$
$$= f_r + f_c + b_c(v_c - v_h) - b_h v_h$$

$$v_6 = v_7 = v_c - v_h$$

$$f_8 = f_f - f_c - f_7 = f_f - f_c - b_c v_7 = f_f - f_c - b_c(v_c - v_h)$$

State–space model:

$$J_r \frac{d\omega_r}{dt} = -r^2 b_r \omega_r - rf_r + T_r$$

$$\frac{df_r}{dt} = k_r(r\omega_r - v_h)$$

$$m_h \frac{dv_h}{dt} = f_r - (b_c + b_h)v_h + f_c + b_c v_c$$

$$\frac{df_c}{dt} = k_c(-v_h + v_c)$$

$$m_c \frac{dv_c}{dt} = b_c v_h - f_c - b_c v_c + f_f$$

with $\boldsymbol{x} = \begin{bmatrix} \omega_r & f_r & v_h & f_c & v_c \end{bmatrix}^T$; $\boldsymbol{u} = \begin{bmatrix} T_r & f_f \end{bmatrix}^T$; $\boldsymbol{y} = \begin{bmatrix} f_c & \omega_r \end{bmatrix}^T$

$$\boldsymbol{A} = \begin{bmatrix} -r^2 b_r/J_r & -r/J_r & 0 & 0 & 0 \\ rk_r & 0 & -k_r & 0 & 0 \\ 0 & 1/m_h & -(b_c + b_h)/m_h & 1/m_h & b_c/m_h \\ 0 & 0 & -k_c & 0 & k_c \\ 0 & 0 & b_c/m_c & -1/m_c & -b_c/m_c \end{bmatrix} ; \quad \boldsymbol{B} = \begin{bmatrix} 1/J_r & 0 \\ 0 & 0 \\ 0 & 0 \\ 0 & 0 \\ 0 & 1/m_c \end{bmatrix} ;$$

$$\boldsymbol{C} = \begin{bmatrix} 0 & 0 & 0 & 1 & 0 \\ 1 & 0 & 0 & 0 & 0 \end{bmatrix} ; \quad \boldsymbol{D} = 0$$

(a) In this case, v_f is an input, which is an A-source. The corresponding element in the linear graph given in Figure 3.21 should be modified to account for this. Specifically, the direction of the arrow of this source element should be reversed (because it is an A-source) and the + and − signs (used for an A-source) should be retained. Furthermore, the inertia element m_c ceases to influence the dynamics of the overall system because, $v_c = v_f$ in this case and is completely specified. This results from the fact that any elements connected in parallel with an A-source have no effect on the rest of the system. Accordingly, the branch representing the m_c element should be removed from the linear graph.

Hence, we now have a fourth order model, with

$$\text{State vector } \boldsymbol{x} = \begin{bmatrix} \omega_r & f_r & v_h & f_c \end{bmatrix}^T; \quad \text{Input vector } \boldsymbol{u} = \begin{bmatrix} T_r & v_f \end{bmatrix}^T$$

State model:

$$J_r \frac{d\omega_r}{dt} = -r^2 b_r \omega_r - rf_r + T_r$$

$$\frac{df_r}{dt} = k_r(r\omega_r - v_h)$$

$$m_h \frac{dv_h}{dt} = f_r - (b_c + b_h)v_h + f_c + b_c v_f$$

$$\frac{df_c}{dt} = k_c(-v_h + v_f)$$

The corresponding model matrices are

$$\boldsymbol{A} = \begin{bmatrix} -r^2 b_r/J_r & -r/J_r & 0 & 0 \\ rk_r & 0 & -k_r & 0 \\ 0 & 1/m_h & -(b_c + b_h)/m_h & 1/m_h \\ 0 & 0 & -k_c & 0 \end{bmatrix}; \quad \boldsymbol{B} = \begin{bmatrix} 1/J_r & 0 \\ 0 & 0 \\ 0 & b_c/m_h \\ 0 & k_c \end{bmatrix};$$

$$\boldsymbol{C} = \begin{bmatrix} 0 & 0 & 0 & 1 \\ 1 & 0 & 0 & 0 \end{bmatrix}; \quad \boldsymbol{D} = 0$$

3.6.4.3.1 Amplifiers

An amplifier is a common component, primarily in an electrical system or electrical subsystem. Purely mechanical, fluid, and thermal amplifiers have been developed and envisaged as well. The common characteristics of an amplifier are

1. They accomplish tasks of signal amplification
2. They are active devices (i.e., they need an external power to operate)
3. They are not affected (ideally) by the load which they drive (i.e., loading effects are small)
4. They have a decoupling effect on systems (i.e., the desirable effect of reducing dynamic interactions between components)

Electrical signals voltage, current, and power are amplified using voltage amplifiers, current amplifiers, and power amplifiers, respectively. Operational amplifiers (op-amps) are the basic building block in constructing these amplifiers. Particularly, an op-amp, with feedback provides the desirable characteristics of very high input impedance, low output impedance, and stable operation. For example, due to its impedance characteristics, the device (load) that is connected to its output does not affect the output characteristics of a good amplifier. In other words, electrical loading errors are negligible.

Analogous to electrical amplifiers, a mechanical amplifier can be designed to provide force amplification (a *T*-type amplifier) or a fluid amplifier can be designed to provide pressure amplification (an *A*-type amplifier). Amplifiers are typically active devices—an external power source is needed to operate the amplifier (e.g., to drive a motor–mechanical load combination).

3.6.4.4 Linear Graph Representation

In its linear graph representation, an amplifier is considered as a "dependent source" element or a "modulated source" element. Specifically, the amplifier output depends on (modulated by) the amplifier input, and is not affected by the dynamics of any devices that are connected to the output of the amplifier (i.e., the load of the amplifier). This is the ideal case. In practice some loading error will be present (i.e., the amplifier output will be affected by the load which it drives).

The linear graph representations of an across-variable amplifier (e.g., voltage amplifier, pressure amplifier) and a through-variable amplifier (e.g., current amplifier, force amplifier) are shown in Figure 3.22a and b, respectively. The pertinent constitutive equations in the general and linear cases are given as well in the figures.

3.6.4.4.1 DC Motor

The dc motor is a commonly used electrical actuator. It converts dc electrical energy into mechanical energy. The principle of operation is based on the fact that when a conductor carrying current is placed in a magnetic field, a force is generated (Lorentz's law). It is this force that results from the interaction of two magnetic fields, which is presented as the magnetic torque in the rotor of the motor.

A dc motor has a stator and a rotor (armature) with windings that are excited by a field voltage v_f and an armature voltage v_a, respectively. The equivalent circuit of a dc motor is shown in Figure 3.23a, where the field circuit and the armature circuit are shown separately, with the corresponding supply voltages. This is the separately excited case. If the stator filed is provided by a permanent magnet, then the stator circuit that is shown in Figure 3.23a is simply an equivalent circuit, where the stator current i_f can be assumed constant. Similarly, if the rotor is a permanent magnet, what is shown in Figure 3.23a is an

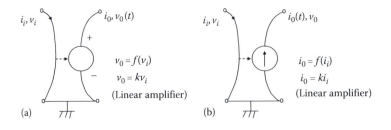

FIGURE 3.22
Linear graph representation of: (a) An across-variable amplifier (*A*-type amplifier); (b) a through-variable amplifier (*T*-type amplifier).

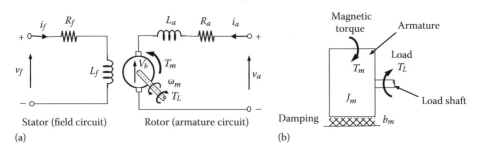

FIGURE 3.23
(a) Equivalent circuit of a dc motor (separately excited); (b) armature mechanical loading.

equivalent circuit where the armature current i_a can be assumed constant. The magnetic torque of the motor is generated by the interaction of the stator field (proportional to i_f) and the rotor field (proportional to i_a) and is given by

$$T_m = k i_f i_a \tag{3.30}$$

A back-electromotive force (back e.m.f) is generated in the rotor (armature) windings to oppose its rotation when these windings rotate in the magnetic field of the stator (Lenz's law). This voltage is given by

$$v_b = k' i_f \omega_m \tag{3.31}$$

where
  i_f is the field current
  i_a is the armature current
  ω_m is the angular speed of the motor

Note: For perfect transfer of electrical energy to mechanical energy in the rotor we have

$$T_m \omega_m = i_a v_b \tag{3.32}$$

This is an electromechanical transformer.
  The field circuit equation is

$$v_f = R_f i_f + L_f \frac{di_f}{dt} \tag{3.33}$$

where
  v_f is the supply voltage to stator
  R_f is the resistance of the field windings
  L_f is the inductance of the field windings

The armature (rotor) circuit equation is

$$v_a = R_a i_a + L_a \frac{di_a}{dt} + v_b \tag{3.34}$$

where

   v_a is the armature supply voltage
   R_a is the armature winding resistance
   L_a is the armature leakage inductance

Suppose that the motor drives a load whose equivalent torque is T_L. Then from Figure 3.23b, The mechanical (load) equation is

$$J_m \frac{d\omega_m}{dt} = T_m - T_L - b_m \omega_m \tag{3.35}$$

where

   J_m is the moment of inertia of the rotor
   b_m is the equivalent (mechanical) damping constant for the rotor
   T_L is the load torque

In field control of the motor, the armature supply voltage v_a is kept constant and the field voltage v_f is controlled. In armature control of the motor, the field supply voltage v_f is kept constant and the armature voltage v_a is controlled.

Example 3.5

A classic problem in robotics is the case of robotic hand gripping and turning a doorknob to open a door. The mechanism is schematically shown in Figure 3.24a. Suppose that the actuator of the robotic hand is an armature-controlled dc motor. The associated circuit is shown in Figure 3.24b. The field circuit provides a constant magnetic field to the motor, and is not important in the present problem. The armature (with motor rotor windings) circuit has a back e.m.f. v_b, a leakage inductance L_a, and a resistance R_a. The input signal to the robotic hand is the armature voltage $v_a(t)$ as shown. The rotation of the motor (at an angular speed ω_m) in the two systems of magnetic field generates a torque T_m (which is negative as marked in Figure 3.24b during normal operation). This torque (magnetic torque) is available to turn the doorknob, and is resisted by the inertia force (moment of inertia J_d), the friction (modeled as linear viscous damping of damping constant b_d) and the spring (of stiffness k_d) of the hand-knob-lock combination. A mechanical model is shown in Figure 3.24c. The dc motor may be considered as an ideal electromechanical transducer that is represented by a linear-graph transformer. The associated equations are

$$\omega_m = \frac{1}{k_m} v_b \tag{3.36}$$

$$T_m = -k_m i_b \tag{3.37}$$

Note: The negative sign in Equation 3.37 arises due to the specific sign convention. The linear graph may be easily drawn, as shown in Figure 3.24d, for the electrical side of the system.

Answer the following questions:

(a) Complete the linear graph by including the mechanical side of the system.
(b) Give the number of branches (*b*), nodes (*n*), and the independent loops (*l*) in the completed linear graph. Verify your answer.

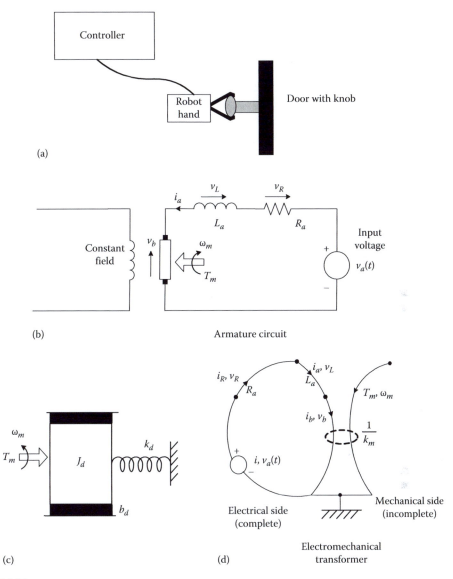

(a)

(b) Armature circuit

(c)

(d) Electrical side (complete)   Mechanical side (incomplete)

Electromechanical transformer

FIGURE 3.24
(a) Robotic hand turning a doorknob; (b) armature-controlled dc motor of the robotic hand; (c) mechanical model of the hand-doorknob system; (d) incomplete linear graph.

(c) Take current through the inductor (i_a), speed of rotation of the doorknob (ω_d), and the resisting torque of the spring within the door lock (T_k) as the state variables, the armature voltage $v_a(t)$ as the input variable, and ω_d and T_k as the output variables. Write the independent node equations, independent loop equations, and the constitutive equations for the completed linear graph. Clearly show the state-space shell. Also, verify that the number of unknown variables is equal to the number of equations obtained in this manner.

(d) Eliminate the auxiliary variables and obtain a complete state-space model for the system, using the equations written in part (c) above.

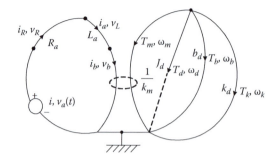

FIGURE 3.25
The complete linear graph of the system.

Solution

(a) The complete linear graph is shown in Figure 3.25.
(b) $b=8$, $n=5$, $l=4$ for this linear graph. It satisfies the topological relationship $l=b-n+1$
(c) **Independent node equations:**

$$i - i_R = 0$$

$$i_R - i_a = 0$$

$$i_a - i_b = 0$$

$$-T_m - T_d - T_b - T_k = 0$$

Independent loop equations:

$$v_a(t) - v_R - v_L - v_b = 0$$

$$\omega_m - \omega_d = 0$$

$$\omega_d - \omega_b = 0$$

$$\omega_b - \omega_k = 0$$

Constitutive equations:

$$\left.\begin{aligned} L_a \frac{di_a}{dt} &= v_L \\ J_d \frac{d\omega_d}{dt} &= T_d \\ \frac{dT_k}{dt} &= k_d\omega_k \end{aligned}\right\} \text{State-space shell}$$

$$\left.\begin{aligned} v_R &= R_a i_R \\ T_b &= b_d\omega_b \end{aligned}\right\} \text{Auxiliary constitutive equations}$$

$$\left.\begin{aligned} \omega_m &= \frac{1}{k_m}v_b \\ T_m &= -k_m i_b \end{aligned}\right\} \text{Electromechanical transformer}$$

Note: There are 15 unknown variables (i, i_R, i_a, i_b, T_m, T_d, T_b, T_k, v_R, v_L, v_b, ω_m, ω_d, ω_b, and ω_k) and 15 equations.

Number of unknown variables $= 2b - s = 2 \times 8 - 1 = 15$
Number of independent node equations $= n - 1 = 5 - 1 = 4$
Number of independent loop equations $= l = 4$
Number of constitutive equations $= b - s = 8 - 1 = 7$
Check: $15 = 4 + 4 + 7$

(d) Eliminate the auxiliary variables from the state-space shell, by substitution:

$$v_L = v_a(t) - v_R - v_b = v_a(t) - R_a i_a - k_m \omega_m$$

$$= v_a(t) - R_a i_a - k_m \omega_d$$

$$T_d = -T_k - T_m - T_b = -T_k + k_m i_b - b_d \omega_b$$

$$= k_m i_a - b_d \omega_d - T_k$$

$$\omega_k = \omega_b = \omega_d$$

Hence, we have the state-space equations:

$$L_a \frac{di_a}{dt} = -R_a i_a - k_m \omega_d + v_a(t)$$

$$J_d \frac{d\omega_d}{dt} = k_m i_a - b_d \omega_d - T_k$$

$$\frac{dT_k}{dt} = k_d \omega_d$$

With $x = [i_a \ \omega_d \ T_k]^T$, $u = [v_a(t)]$, and $y = [\omega_d \ T_k]^T$ we have the state-space model

$$\dot{x} = Ax + Bu$$

$$y = Cx + Du$$

The model matrices are

$$A = \begin{bmatrix} -R_a/L_a & -k_m/L_a & 0 \\ k_m/J_d & -b_d/J_d & -1/J_d \\ 0 & k_d & 0 \end{bmatrix}; \quad B = \begin{bmatrix} 1/L_a \\ 0 \\ 0 \end{bmatrix};$$

$$C = \begin{bmatrix} 0 & 1 & 0 \\ 0 & 0 & 1 \end{bmatrix}; \quad D = 0$$

Note 1: This is a multi-domain (electromechanical model).
Note 2: Multifunctional devices (e.g., a piezoelectric device that serves as both actuator and sensor) may be modeled similarly, using an electromechanical transformer (or, through the use of the "reciprocity principle").

3.6.5 Linear Graphs of Thermal Systems

Thermal systems have temperature (T) as the across variable, as it is always measured with respect to some reference (or as a temperature difference across an element), and heat transfer (flow) rate (Q) as the through variable. Heat source and temperature source are the two types of source elements. The former is more common. The latter may correspond to a large reservoir whose temperature is virtually not affected by heat transfer into or out of it. There is only one type of energy (thermal energy) in a thermal system. Hence, there is only one type (A-type) energy storage element with the associated state variable, temperature. There is no T-type element in a thermal system.

Example 3.6

A traditional Asian pudding is made by blending roughly equal portions by volume of treacle (a palm honey similar to maple syrup), coconut milk, and eggs, spiced with cloves and cardamoms, and baking in a special oven for about 1 h. The traditional oven uses charcoal fire in an earthen pit that is well insulated, as the heat source. An aluminum container half-filled with water is placed on fire. A smaller aluminum pot containing the dessert mixture is placed inside the water bath and covered fully with an aluminum lid. Both the water and the dessert mixture are well stirred and assumed to have uniform temperatures. A simplified model of the oven is shown in Figure 3.26a.

Assume that the thermal capacitances of the aluminum water container, dessert pot, and the lid are negligible. Also, the following equivalent (linear) parameters and variables are defined:

C_r = thermal capacitance of the water bath
C_d = thermal capacitance of the dessert mixture
R_r = thermal resistance between the water bath and the ambient air
R_d = thermal resistance between the water bath and the dessert mixture
R_c = thermal resistance between the dessert mixture and the ambient air, through the covering lid

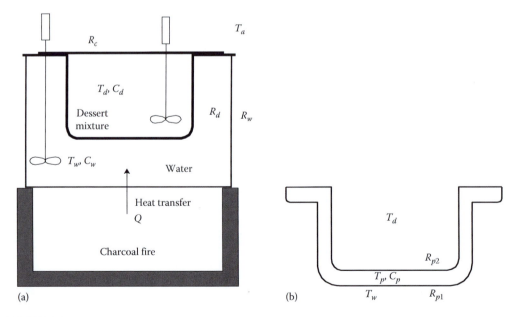

(a)   (b)

FIGURE 3.26
(a) A simplified model of an Asian dessert oven; (b) an improved model of the dessert pot.

T_r=temperature of the water bath
T_d=temperature of the dessert mixture
T_s=ambient temperature
Q=input heat flow rate from the charcoal fire into the water bath

(a) Assuming that T_d is the output of the system, develop a complete state-space model for the system. What are the system inputs?
(b) In part (a), suppose that the thermal capacitance of the dessert pot is not negligible, and is given by C_p. Also, as shown in Figure 3.26b, thermal resistances R_{p1} and R_{p2} are defined for the two interfaces of the pot. Assuming that the pot temperature is maintained uniform at T_p show how the state-space model of part (a) should be modified to include this improvement. What parameters do R_{p1} and R_{p2} depend on?
(c) Draw the linear graphs for the systems in (a) and (b). Indicate in the graph only the system parameters, input variables, and the state variables.

Solution

(a) For the water bath:

$$C_w \frac{dT_w}{dt} = Q - \frac{1}{R_w}(T_w - T_a) - \frac{1}{R_d}(T_w - T_d) \tag{i}$$

For the dessert mixture:

$$C_d \frac{dT_d}{dt} = \frac{1}{R_d}(T_w - T_d) - \frac{1}{R_c}(T_d - T_a) \tag{ii}$$

Equations (i) and (ii) are the state equations with

State vector $x = [T_w, T_d]^T$
Input vector $u = [Q, T_a]^T$
Output vector $y = [T_d]^T$

The corresponding matrices of the state-space model are

$$A = \begin{bmatrix} -\frac{1}{C_w}\left(\frac{1}{R_w}+\frac{1}{R_d}\right) & \frac{1}{C_w R_d} \\ \frac{1}{C_d R_d} & -\frac{1}{C_d}\left(\frac{1}{R_d}+\frac{1}{R_c}\right) \end{bmatrix}; \quad B = \begin{bmatrix} \frac{1}{C_w} & \frac{1}{C_w R_w} \\ 0 & \frac{1}{C_d R_c} \end{bmatrix};$$

$$C = \begin{bmatrix} 0 & 1 \end{bmatrix}; \quad D = \begin{bmatrix} 0 & 0 \end{bmatrix}$$

(b) For the dessert pot:

$$C_p \frac{dT_p}{dt} = \frac{1}{R_{p1}}(T_w - T_p) - \frac{1}{R_{p2}}(T_p - T_d) \tag{iii}$$

Equations (i) and (ii) have to be modified as

$$C_w \frac{dT_w}{dt} = Q - \frac{1}{R_w}(T_w - T_a) - \frac{1}{R_{p1}}(T_w - T_p) \tag{i*}$$

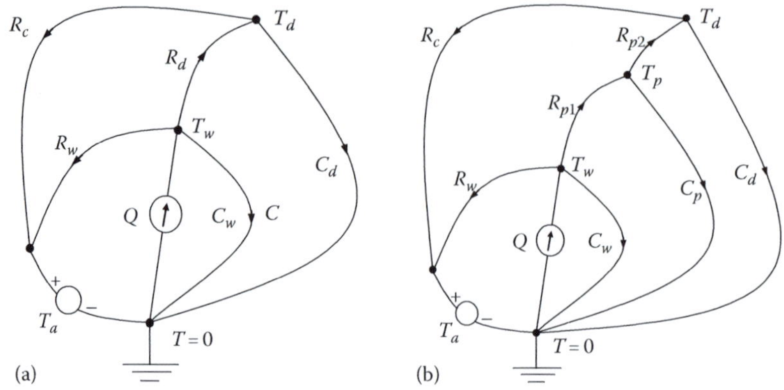

FIGURE 3.27
Linear graph of the: (a) simplified model; (b) improved model.

$$C_d \frac{dT_d}{dt} = \frac{1}{R_{p2}}(T_w - T_d) - \frac{1}{R_c}(T_d - T_a) \qquad \text{(ii*)}$$

The system has become third order now, with the state equations (i*), (ii*), and (iii) and the corresponding state vector:

$$x = \begin{bmatrix} T_w & T_d & T_p \end{bmatrix}^T$$

But u and y remain the same as before. Matrices A, B, and C have to be modified accordingly.

The resistance R_{pi} depends on the heat-transfer area A_i and the heat transfer coefficient h_i. Specifically,

$$R_{pi} = \frac{1}{h_i A_i}$$

(c) The linear graph for case (a) is shown in Figure 3.27a. The linear graph for case (b) is shown in Figure 3.27b.

3.7 Transfer Functions and Frequency-Domain Models

Transfer-function models (strictly, Laplace transfer functions) are based on the Laplace transform, and are versatile means of representing linear systems with constant (time-invariant) parameters. Frequency-domain models (or frequency transfer functions) are a special category of Laplace-domain models, and they are based on the Fourier transform. They are interchangeable—a Laplace-domain model can be converted into the corresponding frequency-domain model in a trivial manner, and *vice versa*. Similarly, linear, constant-coefficient (time-invariant) time-domain model (e.g., input–output differential equation or a state-space model) can be converted into a transfer function, and *vice versa*, in a simple and straightforward manner. A system with just one input (excitation) and one output (response) can be represented uniquely by one transfer function. When a system has two or more inputs (i.e., an input vector) and/or two or more outputs (i.e., and output vector), its representation needs several transfer functions (i.e., a transfer function matrix is needed). The response characteristics at a given location (more correctly, in a given degree of freedom) can be determined using a single frequency-domain transfer function.

Basics of Laplace and Fourier transforms are given in Appendix B. Some useful results are summarized below:

Laplace transform of time derivatives:

$$\mathcal{L}\frac{d^n y(t)}{dt^n} = s^n Y(s) - s^{n-1} y(0) - s^{n-2}\dot{y}(0) - \cdots - \frac{d^{n-1}y}{dt^{n-1}}(0) \tag{3.38}$$

Note: With zero ICs, we have

$$\mathcal{L}\frac{d^n y(t)}{dt^n} = s^n Y(s) \tag{3.39}$$

Laplace transform of integrator:

$$\mathcal{L}\int_0^t y(\tau)d\tau = \frac{1}{s} Y(s) \tag{3.40}$$

Laplace transform:

> Time domain $\rightarrow$ Laplace (complex frequency) domain

> Time derivative $\rightarrow$ Laplace variable s

> Differential equations $\rightarrow$ Algebraic equations(easier math)

> Time integration $\rightarrow$ $1/s$

Fourier transform:

> Time domain $\rightarrow$ Frequency domain

Conversion from Laplace to Fourier (one-sided): Set $s = j\omega$.

In using techniques of Laplace transform, the general approach is to first convert the time-domain problem into an s-domain problem (conveniently, by using Laplace transform tables); perform the necessary analysis (algebra rather than calculus) in the s-domain; and convert the results back into the time domain (again, conveniently using Laplace transform tables). Further discussion, techniques and Laplace tables are found in Appendix B.

3.7.1 Transfer Function

Consider the nth-order linear, constant-parameter system given by

$$a_n \frac{d^n y}{dt^n} + a_{n-1}\frac{d^{n-1}y}{dt^{n-1}} + \cdots + a_0 y = b_0 u + b_1 \frac{du}{dt} + \cdots + b_m \frac{d^m u}{dt^m} \tag{3.41}$$

Note: We will assume $m < n$, or at worst $m \leq n$ when the corresponding systems are said to be *physically realizable*. For systems that possess dynamic delay (i.e., systems whose response does not tend to feel the excitation either instantly or ahead of time, or systems whose excitation or its derivatives are not directly fed forward to the output, we will have $m < n$. These are the systems that concern us most in real applications.

Use the result (3.39) in (3.41), assuming zero ICs. We obtain the transfer function:

$$\frac{Y(s)}{U(s)} = G(s) = \frac{b_0 + b_1 s + \cdots + b_m s^m}{a_0 + a_1 s + \cdots + a_n s^n} \tag{3.42}$$

It should be clear from (3.41) and (3.42) that the transfer function corresponding to a system differential equation can be written simply by inspection, without requiring any knowledge of Laplace-transform theory. Conversely, once the transfer function is given, the corresponding time-domain (differential) equation should be immediately obvious.

Note: The dominator polynomial of a transfer function is called the *characteristic polynomial*, and the corresponding equation is called the *characteristic equation*: $a_0 + a_1 s + L + a_n s^n = 0$.

3.7.2 Frequency-Domain Models

If the output and input of a system are expressed in the frequency domain, the frequency transfer function of the system is given by the ratio of the Fourier transforms of the output to the input. Frequency-domain representations are particularly useful in the analysis, design, control, and testing of mechatronic systems. The signal waveforms encountered in such a system can be interpreted and represented as a series of sinusoidal components. Indeed, any waveform can be so represented, and sinusoidal excitation is often used in testing of equipment and components. It is usually easier to obtain frequency-domain models than the associated time-domain models by testing.

3.7.2.1 Frequency Transfer Function (Frequency Response Function)

Consider the time-domain system (3.41) whose transfer function (in the Laplace-domain) is given by (3.42). Suppose that a harmonic (sinusoidal) input, given in the complex form:

$$u = u_o e^{j\omega t} = u_o(\cos \omega t + j \sin \omega t) \tag{3.43}$$

is applied to the system. After the conditions settle down (i.e., at steady state) the output (response) of the system will also be harmonic, and is given by

$$y = y_o e^{j\omega t} = y_o(\cos \omega t + j \sin \omega t) \tag{3.44}$$

By substituting Equations 3.43 and 3.44 in Equation 3.41 and canceling the common term $e^{j\omega t}$, we get

$$y_o = \left[\frac{b_m(j\omega)^m + b_{m-1}(j\omega)^{m-1} + \cdots + b_0}{a_n(j\omega)^n + a_{n-1}(j\omega)^{n-1} + \cdots + a_0} \right] u_o \tag{3.45a}$$

or, in view of (3.42),

$$y_o = G(j\omega)u_o \qquad (3.45b)$$

(*Note*: $de^{j\omega t}/dt = j\omega e^{j\omega t}$)

Here, the *frequency transfer function* (or, *frequency response function*) is given by

$$G(j\omega) = G(s)\big|_{s=j\omega} = \frac{b_0 + b_1(j\omega) + \cdots + b_m(j\omega)^m}{a_0 + a_1(j\omega) + \cdots + a_n(j\omega)^n} \qquad (3.46)$$

Note: Angular frequency variable (rad/s) $\omega = 2\pi f$ where f is the cyclic frequency variable (Hz).

Also, directly from the Laplace-domain result (3.42) we have the frequency-domain result:

$$G(j\omega) = \frac{Y(j\omega)}{U(j\omega)} \qquad (3.46b)$$

where $Y(j\omega) = \mathcal{F}y(t)$ and $U(j\omega) = \mathcal{F}u(t)$ with $\mathcal{F}$ denoting the Fourier transform operator.

3.7.2.1.1 Magnitude (Gain) and Phase

Let us denote the magnitude of $G(j\omega)$ as

$$|G(j\omega)| = M \qquad (3.47a)$$

and the phase angle of $G(j\omega)$ as

$$\angle G(j\omega) = \phi \qquad (3.47b)$$

Then, we can write

$$G(j\omega) = M\cos\phi + jM\sin\phi = Me^{j\phi} \qquad (3.47c)$$

and from (3.44) and (3.45b):

$$y = u_o M e^{j(\omega t + \phi)} \qquad (3.48)$$

Observations

When a harmonic input of frequency ω is applied to the system:

1. The output is magnified by $M = |G(j\omega)|$
2. The output has a *phase lead* w.r.t. input by $\phi = \angle G(j\omega)$.

Note: For practical systems $\angle G(j\omega)$ is typically a negative phase lead (i.e., output usually lags input).

It follows that $G(j\omega)$ constitutes a complete model for a linear, constant-parameter system, as does $G(s)$.

3.7.2.2 Bode Diagram (Bode Plot) and Nyquist Diagram

The frequency transfer function $G(j\omega)$ is in general a complex function of frequency ω (which is a real variable). From the result (3.47) it should be clear that applying a harmonic (i.e., sinusoidal) excitation and measuring the amplitude gain and the phase change at the output (response) for a series of frequencies, is a convenient method of experimental determination of a system model. This approach of "experimental modeling" is termed *model identification*. Either a *sine-sweep* or a *sine-dwell* excitation may be used with these tests. Specifically, a sinusoidal excitation is applied (i.e., input) to the system and the amplification factor and the phase-lead angle of the resulting response are determined at steady state. The frequency of excitation is varied continuously for a sine-sweep, and in steps for a sine-dwell. The sweep rate should be sufficiently slow, or dwell times should be sufficiently long, to guarantee achieving a steady-state response in these methods. The results are usually presented as either a pair of curves:

$|G(j\omega)|$ versus ω

$\angle G(f)$ versus ω

with log axes for magnitude (e.g., decibels) and frequency (e.g., decades). This pair of curves is called the *Bode plot* or *Bode diagram*.

If the same information is plotted on the complex $G(j\omega)$ plane with the real part plotted on the horizontal axis and the imaginary part on the vertical axis, the resulting curve is termed *Nyquist diagram* or *argand plot* or *polar plot*.

In a Bode diagram, the frequency is shown explicitly on one axis, whereas in a Nyquist plot the frequency is a parameter on the curve, and is not explicitly shown unless the curve itself is calibrated. In Bode diagrams, it is customary and convenient to give the magnitude in decibels ($20 \log_{10} |G(j\omega)|$) and scale the frequency axis in logarithmic units (typically factors of 10 or decades). Since the argument of a logarithm should necessarily be a dimensionless quantity, $Y(j\omega)$ and $U(j\omega)$ should have the same units, or the ratio of $G(j\omega)$ with respect to some base value, such as $G(0)$ should be used.

The arrow on the Nyquist curve indicates the direction of increasing frequency. Only the part corresponding to positive frequencies is actually shown. The frequency response function corresponding to negative frequencies is obtained by replacing ω by $-\omega$ or, equivalently, $j\omega$ by $-j\omega$. The result is clearly the complex conjugate of $G(j\omega)$, and is denoted by $G^*(j\omega)$:

$$G^*(j\omega) = \left. |G(s)| \right|_{s=-j\omega} \tag{3.49}$$

Since, in complex conjugation, the magnitude does not change and the phase angle changes sign, it follows that the Nyquist plot for $G^*(j\omega)$ is the mirror image of that for $G(j\omega)$ about the real axis. In other words, the Nyquist plot for the entire frequency range ω $[-\infty, +\infty]$ is symmetric about the real axis.

Consider the damped simple oscillator shown in Figure 3.28. The shape of these plots for a simple oscillator is shown in Figure 3.29.

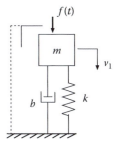

FIGURE 3.28
Damped simple oscillator.

3.7.3 Transfer Functions of Electromechanical Systems

The significance of frequency transfer function as a dynamic model can be explained by considering the simple oscillator (i.e., a single degree-of-freedom mass-spring-damper system) shown in Figure 3.28. Its force–displacement transfer function, in the frequency domain, can be written as

$$G(j\omega) = \frac{1}{ms^2 + bs + k} \quad \text{with } s = j\omega \qquad (3.50)$$

in which m, b, and k denote mass, damping constant, and stiffness, respectively of the oscillator. When the excitation frequency ω is small in comparison to the system natural frequency $\sqrt{k/m}$, the terms ms^2 and bs can be neglected with respect to k and the system behaves as a simple spring. When the excitation frequency ω is much larger than the system natural frequency, the terms bs and k can be neglected in comparison to ms^2. In this case, the system behaves like a simple mass element. When the excitation frequency ω is very close to the natural frequency (i.e., $s = j\omega \approx j\sqrt{k/m}$), it is seen from (3.50) that the term $ms^2 + k$ in the denominator of the transfer function (i.e., the characteristic polynomial) becomes almost zero, and can be neglected. Then the transfer function can be approximated by $G(j\omega) = 1/(bs)$ with $s = j\omega$.

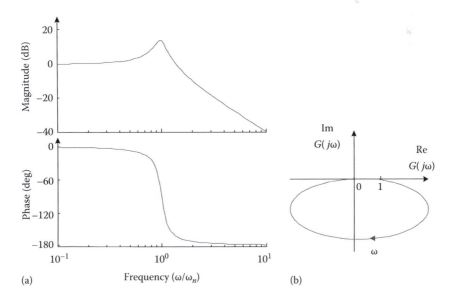

(a)   (b)

FIGURE 3.29
Frequency-domain model of a simple oscillator: (a) Bode plot; (b) Nyquist plot.

In summary:

1. In the neighborhood of a resonance or natural frequency (i.e., for intermediate values of excitation frequencies), the system damping becomes the most important parameter.
2. At low excitation frequencies, the system stiffness is the most significant parameter.
3. At high excitation frequencies, the mass is the most significant parameter.

Note: In these observations, instead of the physical parameters m, k, and b, we could use natural frequency $\omega_n = \sqrt{k/m}$ and the damping ratio $\zeta = b/(2\sqrt{mk})$ as the system parameters. Then the number of system parameters reduces to two, which is an advantage in parametric and sensitivity studies.

3.7.3.1 Mechanical Transfer Functions

Any type of force or motion variable may be used as input and output variables in defining a transfer function of a mechanical system. We can define several versions of frequency transfer functions that may be useful in the modeling and analysis of mechanical systems, as given in Table 3.4.

In the frequency domain:

$$\text{Acceleration} = (j\omega) \times (\text{Velocity})$$

$$\text{Displacement} = \text{Velocity}/(j\omega)$$

In view of these relations, many of the alternative types of transfer functions as defined in Table 3.4 are related to mechanical impedance and mobility through a factor of $j\omega$; specifically:

$$\text{Dynamic stiffness} = \text{Force}/\text{Displacement} = \text{Impedance} \times j\omega$$

$$\text{Receptance} = \text{Displacement}/\text{Force} = \text{Mobility}/(j\omega)$$

$$\text{Dynamic inertia} = \text{Force}/\text{Acceleration} = \text{Impedance}/(j\omega)$$

$$\text{Accelerance} = \text{Acceleration}/\text{Force} = \text{Mobility} \times j\omega$$

TABLE 3.4

Definitions of Useful Mechanical Transfer Functions

Transfer Function	Definition (in Laplace or Frequency Domain)
Dynamic stiffness	Force/displacement
Receptance (dynamic flexibility or compliance)	Displacement/force
Mechanical impedance (Z)	Force/velocity
Mobility (M)	Velocity/force
Dynamic inertia	Force/acceleration
Accelerance	Acceleration/force
Force transmissibility (T_f)	Transmitted force/applied force
Motion transmissibility (T_m)	Transmitted velocity/applied velocity

In these definitions the variables force, acceleration and displacement should be interpreted as the corresponding Fourier spectra.

3.7.3.1.1 Interconnection Laws

Once the transfer functions of the system components are known, the interconnection laws may be used to determine the overall transfer function of the system. Two types of interconnection are useful:

1. Series connection
2. Parallel connection

The determination of the interconnection laws is straightforward in view of the fact that:

1. For series-connected elements, the through variable is common and the across variables add.
2. For parallel-connected elements, the across variable is common and the through variables add.

Since mobility is given by an across variable (velocity) divided by a through variable (force), it is clear (by dividing throughout by the common through variable) that for series-connected elements the mobilities add (or, the inverse of impedance will be additive). Since mechanical impedance is given by a through variable (force) divided by an across variable (velocity), it is clear (by dividing throughout by the common across variable) that for parallel-connected elements the mechanical impedances add (or, the inverse of mobility will be additive). These interconnection laws are presented in Table 3.5.

Since electrical impedance is given by an across variable (voltage) divided by a through variable (current), it is clear (by dividing throughout by the common through variable) that for series-connected elements the electrical impedances add (or, the inverse of admittance

TABLE 3.5

Interconnection Laws for Mechanical Impedance (Z) and Mobility (M)

Series Connection	Parallel Connection

$$v = v_1 + v_2$$

$$\frac{v}{f} = \frac{v_1}{f} + \frac{v_2}{f}$$

$$M = M_1 + M_2$$

$$\frac{1}{Z} = \frac{1}{Z_1} + \frac{1}{Z_2}$$

$$f = f_1 + f_2$$

$$\frac{f}{v} = \frac{f_1}{v} + \frac{f_2}{v}$$

$$Z = Z_1 + Z_2$$

$$\frac{1}{M} = \frac{1}{M_1} + \frac{1}{M_2}$$

will be additive). Since admittance is given by a through variable (current) divided by an across variable (voltage), it is clear (by dividing throughout by the common across variable) that for parallel-connected elements the admittances add (or, the inverse of electrical impedance will be additive). These interconnection laws for electrical are presented in Table 3.6.

3.7.3.1.2 A-Type Transfer Functions and T-Type Transfer Functions

Electrical impedance and mechanical mobility are "A-type transfer functions" because they are given by [across variable/through variable]. They follow the same interconnection laws (compare Tables 3.5 and 3.6). Electrical admittance and mechanical impedance are "T-type transfer functions" because they are given by [through variable/across variable]. They follow the same interconnection laws (compare Tables 3.5 and 3.6).

TABLE 3.6

Interconnection Laws for Electrical Impedance (Z) and Admittance (W)

Series Connections	Parallel Connections
$v = v_1 + v_2$	$i = i_1 + i_2$
$\dfrac{v}{i} = \dfrac{v_1}{i} + \dfrac{v_2}{i}$	$\dfrac{i}{v} = \dfrac{i_1}{v} = \dfrac{i_2}{v}$
$Z = Z_1 + Z_2$	$W = W_1 + W_2$
$\dfrac{1}{W} = \dfrac{1}{W_1} + \dfrac{1}{W_2}$	$\dfrac{1}{Z} = \dfrac{1}{Z_1} + \dfrac{1}{Z_2}$

3.7.3.2 Transfer Functions of Basic Elements

Since a complex system can be formed through series and parallel interconnections of basic elements, it is possible to systematically generate the transfer function of a complex system by using the transfer functions of the basic elements.

The frequency transfer functions are obtained by substituting $j\omega$ or $j2\pi f$ for s. In this manner, the transfer functions of the basic (linear) mechanical elements: mass, spring, and damper may be obtained, as given in Table 3.7.

Similarly, the transfer functions of the basic (linear) electrical elements may be obtained, as given in Table 3.8.

Example 3.7: Ground-Based Mechanical Oscillator

Consider the simple oscillator shown in Figure 3.30a. Its mechanical circuit representation is given in Figure 3.30b. The input is the force $f(t)$; accordingly, the source element is a force source (a through-variable source or *T*-source). The output (response) of the system is the velocity v. In this situation, the transfer function $V(j\omega)/F(j\omega)$ is a mobility function. On the other hand, if the

TABLE 3.7

Mechanical Impedance and Mobility of Basic Mechanical Elements

Element	Time-Domain Model	Impedance	Mobility (Generalized Impedance)
Mass m	$m\dfrac{dv}{dt} = f$	$Z_m = ms$	$M_m = \dfrac{1}{ms}$
Spring k	$\dfrac{df}{dt} = kv$	$Z_k = \dfrac{k}{s}$	$M_k = \dfrac{s}{k}$
Damper b	$f = bv$	$Z_b = b$	$M_b = \dfrac{1}{b}$

TABLE 3.8

Impedance and Admittance of Basic Electrical Elements

Element	Time-Domain Model	Impedance (Z)	Admittance (W)
Capacitor C	$C\dfrac{dv}{dt} = i$	$Z_C = \dfrac{1}{Cs}$	$W_C = Cs$
Inductor L	$L\dfrac{di}{dt} = v$	$Z_L = Ls$	$W_L = \dfrac{1}{Ls}$
Resistor R	$Ri = v$	$Z_R = R$	$W_R = \dfrac{1}{R}$

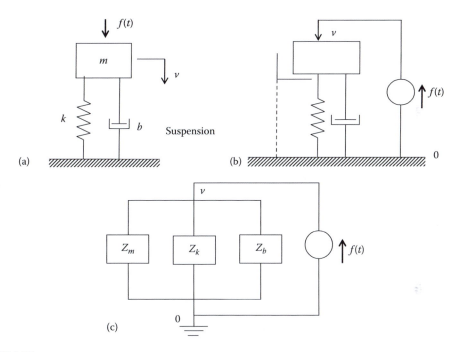

FIGURE 3.30
(a) Ground-based mechanical oscillator; (b) schematic mechanical circuit; (c) impedance circuit.

input is the velocity $v(t)$, the source element is a velocity source; and if force f is exerted on the environment, it is the output, and the corresponding transfer function $F(j\omega)/V(j\omega)$ is an impedance function.

Suppose that using a force source, a known forcing function is applied to this system (with zero initial conditions) and the velocity response is measured. If we were to move the mass exactly at this predetermined velocity (using a velocity source), the force generated at the source would be identical to the originally applied force. In other words, mobility is the reciprocal (inverse) of impedance, as noted earlier. This reciprocity should be intuitively clear because we are dealing with the same system and same initial conditions. Due to this property, we may use either the impedance representation or the mobility representation, depending on whether the elements are connected in parallel or in series, irrespective of whether the input is a force or a velocity. Once the transfer function is determined in one form, its reciprocal gives the other form.

In summary:

From the viewpoint of analysis/modeling of a linear system, it is immaterial as to what type of transfer function is used. In particular, mechanical impedance or mobility may be used without affecting the analytical outcomes. From the physical point of view, however, one transfer function may not be realizable while another is.

In the present example, the three elements are connected in parallel. Hence, as is clear from the impedance circuit shown in Figure 3.30c, the impedance representation (rather than the mobility representation) is more convenient. The overall impedance function of the system is

$$Z(j\omega) = \frac{F(j\omega)}{V(j\omega)} = Z_m + Z_k + Z_b = ms + \frac{k}{s} + b\bigg|_{s=j\omega} = \frac{ms^2 + bs + k}{s}\bigg|_{s=j\omega} \tag{3.51a}$$

The mobility function is the inverse of $Z(j\omega)$:

$$M(j\omega) = \frac{V(j\omega)}{F(j\omega)} = \frac{s}{ms^2 + bs + k}\bigg|_{s=j\omega} \tag{3.51b}$$

Note that if, physically, the input to the system is the force, the mobility function governs the system behavior. In this case, the characteristic polynomial of the system is $s^2 + bs + k$, which corresponds to a simple oscillator and, accordingly, the (dependent) velocity response of the system would be governed by this characteristic polynomial. If, on the other hand, physically, the input is the velocity, the impedance function governs the system behavior. The characteristic polynomial of the system, in this case, is s, which corresponds to a simple integrator ($1/s$). The (dependent) force response of the system would be governed by an integrator type behavior. To explore this behavior further, suppose the velocity source has a constant value. The inertia force will be zero. The damping force will be constant. The spring force will increase linearly. Hence, the net force will have an integration (linearly increasing) effect. If the velocity source provides a linearly increasing velocity (constant acceleration), the inertia force will be constant, the damping force will increase linearly, and the spring force will increase quadratically. In fact, the mobility function (as given above), not the impedance function, is the physically realizable transfer function for the oscillator example in Figure 3.30.

3.8 Equivalent Circuits and Linear Graph Reduction

We have observed that transfer function approaches are more convenient than the differential equation approaches, in dealing with linear systems. This stems primarily from the fact that transfer function approaches use algebra rather than calculus. Also we have noted that when dealing with circuits (particularly, impedance and mobility circuits), transfer function approaches are quite natural. Since the circuit approaches are extensively used in electrical systems, and as a result, quite mature procedures are available in that context, it is useful to consider extending such approaches to mechanical systems (and hence, to electromechanical and mechatronic systems). In particular, circuit reduction is convenient using Thevenin's equivalence and Norton's equivalence for electrical circuits. In particular, linear graphs can be simplified by using transfer function (frequency domain) approaches and circuit reduction.

3.8.1 Thevenin's Theorem for Electrical Circuits

Thevenin's theorem provides a powerful approach to reduce a complex circuit segment into a simpler equivalent representation. Two types of equivalent circuits are generated by this theorem:

1. Thevenin equivalent circuit (consists of a voltage source and an impedance, Z_e in series)
2. Norton equivalent circuit (consists of a current source and an impedance, Z_e in parallel)

The theorem provides means to determine the equivalent source and the equivalent impedance for either of these two equivalent circuits.

Consider a (rather complex) segment of a circuit, consisting of impedances and source elements, as represented in Figure 3.31a. According to the Thevenin's theorem, this circuit segment can be represented by the Thevenin equivalent circuit, as shown in Figure 3.31b or the Norton equivalent circuit, as shown in Figure 3.31c so that for either equivalent circuit, the voltage v and the current i are identical to those at the output port of the considered circuit segment.

Note: The circuit segment of interest (Figure 3.31a) is isolated by "virtually" cutting (separating) a complex circuit into the complex segment of interest and a quite simple (and fully known) segment which is connected to the complex segment. The "virtual" cut is made at the two appropriate terminals linking the two parts of the circuit. The two terminal ends formed by the virtual cutting is the "virtual" output port of the isolated circuit segment. Actually, these terminals are not in open-circuit condition because the cut is "virtual," and a current flows through them.

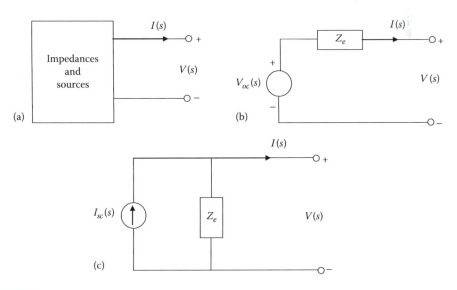

FIGURE 3.31
(a) Circuit segment with impedances and sources; (b) Thevenin equivalent circuit; (c) Norton equivalent circuit.

$V(s)$ = Voltage across the cut terminals when the entire circuit is complete

$I(s)$ = Current through the cut terminals when the entire circuit is complete

$V_{oc}(s)$ = Open-circuit voltage at the cut terminals (i.e., voltage with the terminals open)

$I_{sc}(s)$ = Short-circuit current at the cut terminals (i.e., current when the terminals are shorted)

Z_e = Equivalent impedance of the circuit segment with the source killed (i.e., voltage sources shorted and current source opened) = Thevenin resistance

Note 1: Variables are expressed in the Laplace (or frequency domain) using the Laplace variable *s*

Note 2: For a circuit segment with multiple sources, use superposition (linear system), by taking one source at a time.

Example 3.8: Illustrative Example for Thevenin's Theorem

As usual, we will use electrical impedances, in the Laplace domain. Consider the circuit in Figure 3.32a. We cut it as indicated by the dotted line and determine the Thevenin and Norton equivalent circuits for the left-side portion.

Determination of the equivalent impedance Z_e

First we kill the two sources (i.e., open the current source and short the voltage source so that the source signals become zero). The resulting circuit is shown in Figure 3.32b.

Note the series element and two parallel elements. Since the impedances add in series and inversely in parallel, we have

$$Z_e = Z_L + \frac{Z_R Z_C}{Z_R + Z_C} = Ls + \frac{R/(Cs)}{R + 1/(Cs)} = Ls + \frac{R}{RCs + 1} \tag{3.52}$$

Determination of $V_{oc}(s)$ for Thevenin equivalent circuit

We find the open-circuit voltage using one source at a time, and then use the principle of superposition to determine the overall open-circuit voltage.

(a) **With current source $I(s)$ only**

The circuit with the current source only (short the voltage source) is shown in Figure 3.32c. The source current goes through the two parallel elements only, whose equivalent impedance is $Z_R Z_C / Z_R + Z_C$. Hence the voltage across it, which is also the open-circuit voltage (since no current and hence, no voltage drop along the inductor), is given by

$$V_{oci} = \frac{Z_R Z_C}{(Z_R + Z_C)} I(s) \tag{3.53a}$$

(b) **With voltage source $V(s)$ only**

The circuit with the voltage source only (i.e., open the current source) is shown in Figure 3.32d.

The voltage drop across R should be equal to that across C, and hence the currents in these two elements must be in the same direction. But, the sum of the currents through these parallel elements must be zero, by the node equation (since the open-circuit current is zero). Hence, each current must be zero and the voltages V_R and V_C must be zero. Furthermore, due to the open circuit, the voltage V_L across the inductor must be zero. Then from the loop equation, we have

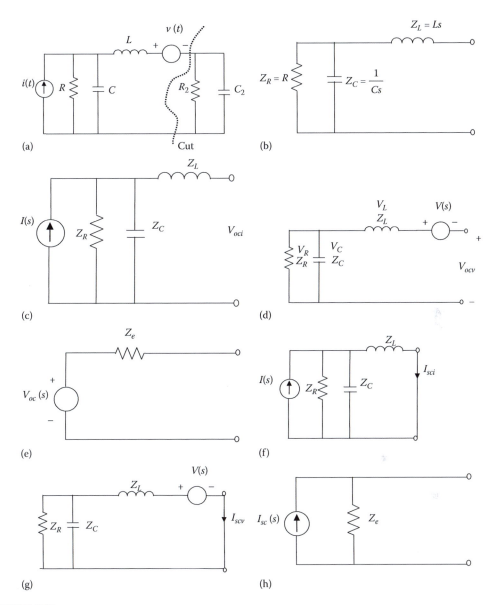

FIGURE 3.32
(a) An electrical impedance circuit; (b) circuit with the sources killed; (c) circuit with current source only; (d) circuit with voltage source only; (e) Thevenin equivalent circuit; (f) circuit with current source only; (g) circuit with voltage source only; (h) Norton equivalent circuit.

$$V_{ocv} + V(s) = 0$$

Or,

$$V_{ocv} = -V(s) \qquad (3.53b)$$

Note the positive direction of potential drop for the open-circuit voltage, as needed for the Thevenin equivalent voltage source.

By superposition, the overall open-circuit voltage is

$$V_{oc}(s) = V_{oci} + V_{ocv} = \frac{Z_R Z_C}{(Z_R + Z_C)} I(s) - V(s) \tag{3.53}$$

The resulting Thevenin equivalent circuit is shown in Figure 3.32e.

Determination of $I_{sc}(s)$ for Norton equivalent circuit

We find the short-circuit current by taking one source at a time, and then using the principle of superposition.

(a) With current source $I(s)$ only

The circuit with the current source only (short the voltage source) is shown in Figure 3.32f.

The source current goes through the three parallel elements, and the currents are divided inversely with the respective impedances. Hence, the current through the inductor is (note the positive direction as marked, for the Norton equivalent current source)

$$I_{sci} = \frac{1/Z_L}{(1/Z_R + 1/Z_C + 1/Z_L)} I(s) \tag{3.54a}$$

(b) With voltage source $V(s)$ only

The circuit with the voltage source only (open the current source) is shown in Figure 3.32g.

Note from the circuit that the short-circuit current is the current that flows through the overall impedance of the circuit (series inductor and a parallel resistor and capacitor combination). According to the polarity of the voltage source, this current is in the opposite direction to the positive direction marked in Figure 3.32g. We have

$$I_{scv}(s) = -\frac{V(s)}{(Z_L + Z_R Z_C/(Z_R + Z_C))} \tag{3.54b}$$

By superposition, the overall short-circuit current is

$$I_{sc}(s) = I_{sci} + I_{scv} = \frac{1/Z_L}{(1/Z_R + 1/Z_C + 1/Z_L)} I(s) - \frac{V(s)}{(Z_L + Z_R Z_C/(Z_R + Z_C))} \tag{3.54}$$

The resulting Norton equivalent circuit is shown in Figure 3.32h.

3.8.2 Mechanical Circuit Analysis Using Linear Graphs

For extending the equivalent-circuit analysis to mechanical systems, we use the force–current analogy where electrical impedance is analogous to mechanical mobility (*A*-type transfer functions) and electrical admittance is analogous to mechanical impedance (*T*-type transfer functions). This analogy is summarized in Table 3.9.

Accordingly, the following two steps do the reduction of a linear graph, in the frequency domain:

1. For each branch of the linear graph mark the mobility function (not mechanical impedance).
2. Carry out linear-graph analysis and reduction as if we are dealing with an electrical circuit, in view of the analogy given in Table 3.9.

TABLE 3.9

Mechanical and Electrical Transfer Function
Analogy

Mechanical Circuit	Electrical Circuit Analogy
Mobility function	Electrical impedance
Force	Current
Voltage	Velocity

In particular, we do the following:

1. For parallel branches, the mobilities are combined by inverse relation $(M = (M_1 M_2)/(M_1 + M_2))$. *Note*: Velocity is common; force is divided inversely to branch mobilities.

2. For series branches, the mobilities add $(M = M_1 + M_2)$. *Note*: Force is common; velocity is divided in proportion to mobility.

3. Killing a force source means open-circuiting it (so, transmitted force = 0).

4. Killing a velocity source means short-circuiting it (so, velocity across = 0).

Example 3.9

Figure 3.33 shows a simplified model of a vehicle with a seat that is occupied by a passenger. The vehicle body (excluding the seat) is represented by a lumped mass m_v, and the vehicle suspension and tires are represented by a spring of stiffness k_v and a viscous damper of damping constant b_v. The mass of the seat and its occupant is m_s. The interface between the seat and the vehicle body is represented by a spring of stiffness k_s and a viscous damper of damping constant b_s. The force generated by the engine, which is exerted on the engine body, is denoted by $f(t)$. This is an input to the system. The road profile applies a velocity $v(t)$ to the vehicle in the vertical direction as it is driven along the road. This is the second input to the system.

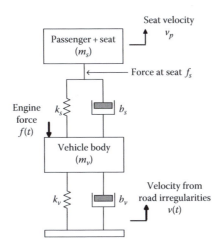

FIGURE 3.33
A model of a moving vehicle with a seat occupied by a passenger for ride quality analysis.

It is desired to determine:

(a) Vertical velocity response v_p of the seat
(b) Vertical force f_s transmitted to the seat through its support interface

in the frequency domain, as a result of the two inputs (f and v) applied simultaneously.
Carry out the following steps in determining these two variables in the frequency domain:

1. Draw the linear graph of the system and mark the mobility functions for all the branches (except the source elements).
2. Simplify the linear graph by combining branches as appropriate (series branches: add mobilities; parallel branches: inverse rule applies for mobilities) and mark on the simplified linear graph the mobilities of the combined branches.
3. Based on the two objectives of the problem (i.e., (a) to determine the vertical velocity response v_p of the seat, and (b) vertical force f_s transmitted to the seat through its support interface) determine which part of the circuit (linear graph) should be cut, for applying the Thevenin's theorem (*Note*: The variable that needs be determined in the particular problem should be associated with the part of the circuit that is cut).
4. Based on the objective of each problem ((a) or (b)) establish whether the Thevenin equivalence or the Norton equivalence is needed (specifically: use Thevenin equivalence if a through variable needs to be determined, because this gives two series elements with a common through variable; use Norton equivalence if an across variable needs to be determined, because this gives two parallel elements with a common across variable).
5. Determine the equivalent sources and mobilities of the equivalent circuits for the two problems.
6. Using the two equivalent circuits determine the two variables of interest.
7. Examine the results to establish ways of reducing the magnitudes of the two variables, as needed in order to improve the ride quality.

Note: Neglect the effects of gravity (i.e., assume that the system is horizontal).

Solution

Part 1:
The linear graph of the system is shown in Figure 3.34a. Note the negative sign for the engine force, according to the positive direction shown in the linear graph.

Part 2:
The reduced linear graph, which is obtained by combining the parallel branches of the spring and damper is shown in Figure 3.34b. Note that the input (source) branches and the output (response to be determined) branches should not be combined in the linear graph reduction.

Part 3:
The branch to be cut, in the application of Thevenin's theorem, is indicated by the broken-line loop in Figure 3.34b. Note that this cut is suitable for the determination of both outputs F_s and V_p.

Part 4:
The Norton equivalent circuit is convenient when determining V_p, which is an across variable.
The Thevenin equivalent circuit is convenient when determining F_s, which is a through variable.

Parts 5 and 6:
(a) **Determination of V_p**
      The necessary Norton equivalent circuit is shown in Figure 3.34c. We need to determine the equivalent mobility M_e of the circuit (with the cut branch removed) after killing the two source elements and the equivalent force source F_{sc}, which is the short-circuit force with the cut terminals shorted.

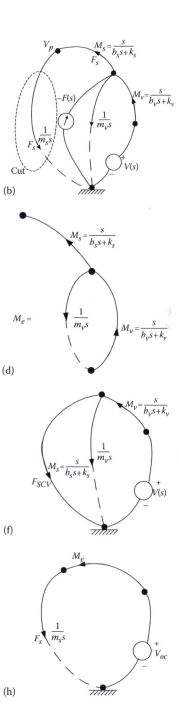

FIGURE 3.34
(a) Linear graph of the system; (b) the reduced linear graph with the branch to be cut indicated; (c) Norton equivalent circuit; (d) determination of equivalent mobility; (e) short-circuit force with the force source alone; (f) short-circuit force with the velocity source alone; (g) Thevenin equivalent circuit of the Norton equivalent circuit; (h) Thevenin equivalent circuit.

(continued)

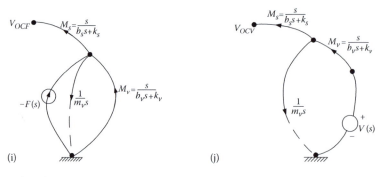

FIGURE 3.34 (continued)
(i) open-circuit velocity with the force source alone; (j) open-circuit velocity with the velocity source alone.

The equivalent mobility M_e is determined using Figure 3.34d.
From Figure 3.34d it is seen that:

$$M_e = \frac{s}{b_s s + k_s} + \frac{1}{m_v s + \dfrac{b_v s + k_v}{s}} = \frac{s}{b_s s + k_s} + \frac{s}{m_v s^2 + b_v s + k_v} \tag{i}$$

Now, the equivalent force source is determined using superposition with the two source elements separately (because the system is linear), as follows:

(i) With force source alone
     The corresponding circuit is shown in Figure 3.34e.
     The force is divided in proportion to the impedances in parallel branches. Hence, from Figure 3.34e we get

$$F_{SCF} = \frac{(b_s s + k_s)/s(-F(s))}{(b_s s + k_s)/s + m_v s + (b_v s + k_v)/s} = \frac{-(b_s s + k_s)F(s)}{m_v s^2 + (b_v + b_s)s + k_v + k_s}$$

(ii) With velocity source alone
      The corresponding circuit is shown in Figure 3.34f.

$$\text{Force out of the velocity source} = \frac{V(s)}{\dfrac{s}{(b_v s + k_v)} + \dfrac{1}{\left(\dfrac{b_s s + k_s}{s} + m_v s\right)}}$$

The force is divided in proportion to the impedances in parallel branches. Hence, from Figure 3.34f we get

$$F_{SCV} = \frac{(b_s s + k_s)/s}{(b_s s + k_s)/s + m_v s} \cdot \frac{V(s)}{\dfrac{s}{(b_v s + k_v)} + \dfrac{1}{\left(\dfrac{b_s s + k_s}{s} + m_v s\right)}} = \frac{(b_s s + k_s)(b_v s + k_v)V(s)}{s[m_v s^2 + (b_v + b_s)s + k_v + k_s]}$$

By superposition,

$$F_{SC} = F_{SCF} + F_{SCV} = \frac{(b_s s + k_s)[-sF(s) + (b_v s + k_v)V(s)]}{s[m_v s^2 + (b_v + b_s)s + k_v + k_s]}$$ (ii)

Now, from Figure 3.34c we get

$$V_p = \frac{\dfrac{1}{m_s s} M_e}{\left(\dfrac{1}{m_s s} + M_e\right)} F_{SC} = \frac{M_e}{(1 + M_e m_s s)} F_{SC}$$ (iii)

At this point, the next result (part (b)) can be obtained simply by using the fact

$$\frac{V_p}{F_s} = \frac{1}{m_s s}$$ (iv)

We get

$$F_s = m_s s V_p = \frac{M_e m_s s}{(1 + M_e m_s s)} F_{SC}$$ (viii*)

Alternatively, the formal approach of using the Thevenin equivalent circuit to determine the through variable F_s is shown next by using a short-cut approach and a more elaborate approach.

(b1) Determination of F_s (short-cut approach)

Here we use the Thevenin equivalent circuit of the previously determined Norton equivalent circuit, as given in Figure 3.34g.

We have

$$F_s = \frac{M_e F_{SC}}{\left(\dfrac{1}{m_s s} + M_e\right)} = \frac{M_e m_s s}{(1 + M_e m_s s)} F_{SC}$$ (viii**)

This is identical to the result obtained from the previous result.
The formal (long) approach of obtaining the same result is given next.

(b2) Determination of F_s (formal approach)

Here, we use the Thevenin equivalent circuit shown in Figure 3.34h by the formal way by starting from the branch cut shown in Figure 3.34b. *Note:* The same cut (see Figure 3.34b) as for the Norton equivalent circuit is appropriate here because the variable to be determined (F_s) is associated with this cut.

Clearly, the equivalent mobility M_e is the same as that for the Norton equivalent circuit. We have (see Equation (i))

$$M_e = \frac{s}{b_s s + k_s} + \frac{s}{m_v s^2 + b_v s + k_v}$$ (i*)

The equivalent velocity source V_{OC} is determined using superposition with the original two source elements (force source and velocity source) separately (because the system is linear), as follows:

(i) With force source alone
     The corresponding circuit is shown in Figure 3.34i.
     From Figure 3.34i, we get

$$V_{OCF} = \frac{-F(s)}{m_v s + (b_v s + k_v)/s} = \frac{-sF(s)}{m_v s^2 + b_v s + k_v}$$

(ii) With velocity source alone
      The corresponding circuit is shown in Figure 3.34j.
      From Figure 3.34j, we get

$$V_{OCV} = \frac{\dfrac{1}{m_v s}}{\left(\dfrac{1}{m_v s} + \dfrac{s}{b_v s + k_v} \right)} V(s) = \frac{(b_v s + k_v)V(s)}{[m_v s^2 + b_v s + k_v]}$$

By superposition,

$$V_{OC} = V_{OCF} + V_{OCV} = \frac{-sF(s)}{m_v s^2 + b_v s + k_v} + \frac{(b_v s + k_v)V(s)}{m_v s^2 + b_v s + k_v} = \frac{-sF(s) + (b_v s + k_v)V(s)}{m_v s^2 + b_v s + k_v} \qquad \text{(v)}$$

Now, from Figure 3.34h we get

$$F_s = \frac{V_{OC}}{\left(\dfrac{1}{m_s s} + M_e \right)} = \frac{m_s s}{(1 + M_e m_s s)} V_{OC} \qquad \text{(vi)}$$

Note from the results (ii) and (v) that

$$F_{SC} = \frac{(b_s s + k_s)(m_v s^2 + b_v s + k_v)}{s[m_v s^2 + (b_v + b_s)s + k_v + k_s]} V_{OC} = \frac{V_{OC}}{M_e} \qquad \text{(vii)}$$

Substitute (vii) into (vi) to get the previously obtained result:

$$F_s = \frac{M_e m_s s}{(1 + M_e m_s s)} F_{SC} \qquad \text{(viii)}$$

Part 7:
Substitute (v) and (i) in (vi), and see (iv). We have

$$F_s = m_s s V_p = \frac{m_s s(b_s s + k_s)[-sF(s) + (b_v s + k_v)V(s)]}{(b_s s + k_s)(m_v s^2 + b_v s + k_v) + m_s s^2[m_v s^2 + (b_v + b_s)s + k_v + k_s]} \qquad \text{(ix)}$$

It is seen that in generating both F_s and V_p, the excitation forces (two source signals) have to pass through the common amplification factor:

$$A = \cfrac{1}{(m_v s^2 + b_v s + k_v) + \cfrac{m_s s^2 [m_v s^2 + (b_v + b_s)s + k_v + k_s]}{(b_s s + k_s)}}$$

This amplification factor A has to be minimized in order to achieve good ride quality. This is a somewhat complex problem. However, by examining (x) it is clear that, as a general guidance, one must try to reduce the suspension parameters b_s and k_s, and increase the vehicle parameters m_v, b_v, and k_v.

Summary of Thevenin approach for mechanical circuits

We now summarize the general steps in applying the Thevenin's theorem to mechanical circuits that are represented by linear graphs, in the Laplace/frequency domain.

General steps

1. Draw the linear graph for the system and mark the mobility functions for all the branches (except the source elements).

2. Simplify the linear graph by combining branches as appropriate (series branches: add mobilities; parallel branches: inverse rule applies for mobilities) and mark the mobilities of the combined branches.

3. Depending on the problem objective (e.g., determine a particular force, velocity, transfer function) determine which part of the circuit (linear graph) should be cut (i.e., the variable or function of interest should be associated with the part that is removed from the circuit) so that the equivalent circuit of the remaining part has to be determined.

4. Depending on the problem objective establish whether the Thevenin equivalence or Norton equivalence is needed (specifically: use the Thevenin equivalence if a through variable needs to be determined, because this gives two series elements with a common through variable. Use the Norton equivalence if an across variable needs to be determined, because this gives two parallel elements with a common across variable).

5. Determine the equivalent source and mobility of the equivalent circuit.

6. Using the equivalent circuit determine the variable or function of interest.

3.9 Block Diagrams

The transfer-function model $G(s)$ of a single-input single-output (SISO) system can be represented by a single block with an input and an output. A disadvantage of this representation is that no information regarding how the various elements or components that are interconnected within the system can be uniquely determined from the transfer function. It contains only a unique input–output description. However, the internal structure of a dynamic system can be indicated by a more elaborate graphical representation. One such representation is provided by linear graphs. Another, detailed representation can be provided by a block diagram with many blocks representing system elements or components that are connected together. Such a detailed block diagram may be used to uniquely indicate the state variables in a particular model.

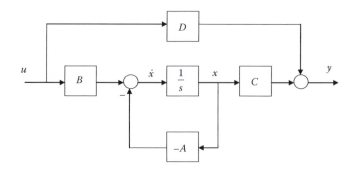

FIGURE 3.35
A block-diagram representation of a state-space model.

For example, consider the state-space model equations (3.5) and (3.6). A block diagram that uniquely possesses this model is shown in Figure 3.35. Note the *feedforward* path corresponding to **D**. The feedback paths (corresponding to **A**) do not necessarily represent a feedback control system (where an external controller generates "active" feedback paths). The internal feedback paths shown in Figure 3.35 are *natural feedback* paths.

In a block diagram, two or more blocks in cascade can be replaced by a single block having the product of individual transfer functions. The circle in Figure 3.35 is a *summing junction*. A negative sign at the arrowhead of an incoming signal corresponds to subtraction of that signal. As mentioned earlier, $1/s$ can be interpreted as integration, and s as differentiation.

In generating and simplifying block diagrams, the rules indicated in Table 3.10 are quite useful. All the entries of the table are quite obvious, and may be verified by inspection. We note the following:

1. Circle (summing junction): Two or more signals are added together forming new signals.

   Note: Negative sign at arrowhead of incoming signal $\rightarrow$ Subtract the signal

2. Two or more blocks in cascade = Single block having product of individual transfer functions.

3. Two or more blocks in parallel = Single block having sum of individual transfer functions.

However, an explanation is appropriate for the last entry of the table, where the equivalent block for a feedback loop is given. The result may be obtained as follows:

The feedback signal at the summing junction $= -Hx_2$

Hence, the signal reaching the block G is $x_1 - Hx_2$

Accordingly, output of the block G is $G(x_1 - Hx_2)$ which is equal to x_2.

We have: $G(x_1 - Hx_2) = x_2$

Straightforward algebra gives

$$x_2 = \frac{G}{1+GH} x_1 \tag{3.55}$$

TABLE 3.10

Basic Relations for Block-Diagram Reduction

Description	Equivalent Representation	

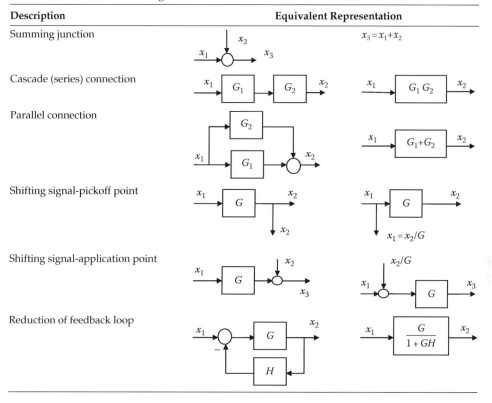

The equivalence of Figure 3.35 and the relations (3.5) and (3.6) should be obvious. Alternatively, the rules for block diagram reduction (given in Table 3.10) can be used to show that the system transfer function is given by

$$\frac{Y(s)}{U(s)} = G(s) = \frac{CB}{(s-A)} + D \tag{3.56}$$

This is the scalar version of the matrix-vector equation 5.15a.

3.9.1 Simulation Block Diagrams

In a simulation block diagram each block contains either an integrator ($1/s$) or a constant gain term. The name originates from classical analog computer applications in which hardware modules of summing amplifiers and integrators (along with other units such as potentiometers and resistors) are interconnected to simulate dynamic systems. Recently, the same type of block diagrams has been in wide use for the purpose of computer simulation of dynamic systems, for example, in software tools such as Simulink.

In summary,
A simulation block diagram consists only of

1. Integration blocks
2. Constant gain blocks
3. Summing junctions

In a simulation block diagram, we have

1. State variables = Outputs of the integrators
2. State equations = Equations for signals going into the integration blocks
3. Algebraic output equation = Equation for the summing junction that generates y (far right)

Also, simulation block diagrams

- Are useful in computer simulation of dynamic systems
- Can be obtained from input–output models (see examples given next) or state-space models (see previous example or converse of the next examples)
- Can be used to develop state-space models
- Are not unique (many forms are possible for a given system)

3.9.2 Principle of Superposition

For a linear system, the *principle of superposition* applies. In particular if, with zero initial conditions, x is the response of a system to an input u, then $d^r x/dt^r$ is the response to the input $d^r u/dt^r$. Consequently, by the principle of superposition, $\alpha_1 x + \alpha_2 d^r x/dt^r$ is the response to the input $\alpha_1 u + \alpha_2 d^r u/dt^r$. This form of the principle of superposition is quite useful in the analytical manipulation of block diagrams.

3.9.3 Causality and Physical Realizability

Consider a dynamic system that is represented by the single input–output differential equation model (3.41), with $n > m$. The physical realizability of the system should dictate the causality (cause–effect) of this system that u should be the input and y should be the output. Its transfer function is given by Equation 3.42. Here, n is the order of the system, $\Delta(s)$ is the characteristic polynomial (of order n), and $N(s)$ is the numerator polynomial (of order m) of the system.

We can prove the above by contradiction. Suppose that $m > n$. Then, if we integrate Equation 3.41 n times, we will have y and its integrals on the left-hand side but the right-hand side will contain at least one derivative of u. Since the derivative of a step function is an impulse, this implies that a finite change in input will result in an infinite change in the output (response). Such a scenario will require infinite power, and is not physically realizable. It follows that a physically realizable system cannot have a numerator order greater than the denominator order, in its transfer function. If in fact $m > n$, then, what it means physically is that y should be the system input and u should be the system output. In other words, the causality should be reversed in this case. For a physically realizable system, a simulation block diagram can be established using integrals ($1/s$) alone, without the need of derivatives (s). Note that pure derivatives are physically not realizable. If $m > n$, the simulation block diagram will need at least one derivative for linking u to y. That will not be physically realizable, again, because it would imply the possibility of producing an infinite response by a finite input. In other words, the simulation block diagram of a physical realizable system will not require feedforward paths containing pure derivatives.

3.10 Response Analysis

An analytical model can provide information regarding how the system responds when excited by an initial condition (i.e., free, natural response) or when a specific excitation (input) is applied (i.e., forced response). Such a study may be carried out by

1. Solution of the differential equations (analytical)
2. Computer simulation (numerical)

A response analysis carried out using either approach, is valuable in many applications such as design, control, testing, validation, and qualification. For large-scale and complex systems, a purely analytical study may not be feasible, and one may have to resort to numerical approaches and computer simulation.

3.10.1 Analytical Solution

The response of a dynamic system may be obtained analytically by solving the associated differential equations, subject to the initial conditions. This may be done by

1. Direct solution (in the time domain)
2. Solution using Laplace transform

Consider a linear time-invariant model given by the input–output differential equation

$$a_n \frac{d^n y}{dt^n} + a_{n-1} \frac{d^{n-1} y}{dt^{n-1}} + \cdots + a_0 y = u \tag{3.57}$$

At the outset, note that it is not necessary to specifically include derivative terms on the right-hand side of Equation 3.57; for example $b_0 u + b_1(du/dt) + \cdots + b_m(d^m u/dt^m)$, because once we have the solution (say, y_s) for (3.57) we can use the *principle of superposition* to obtain the general solution, which is given by $b_0 y_s + b_1(dy_s/dt) + \cdots + b_m(d^m y_s/dt^m)$. Hence, we will consider only the case (3.57).

3.10.1.1 Homogeneous Solution

The natural characteristics of a dynamic system do not depend on the input to the system, and is determined by the homogeneous equation (i.e., with the input $= 0$):

$$a_n \frac{d^n y}{dt^n} + a_{n-1} \frac{d^{n-1} y}{dt^{n-1}} + \cdots + a_0 y = 0 \tag{3.58}$$

Its solution—the homogeneous solution—is denoted by y_h and it depends on the system's initial conditions, and takes the form

$$y_h = ce^{\lambda t} \tag{3.59}$$

where c is an arbitrary constant and, in general, λ can be complex. Substitute 3.59 in 3.58. We get

$$a_n\lambda^n + a_{n-1}\lambda^{n-1} + \cdots + a_0 = 0 \tag{3.60}$$

This is called the *characteristic equation* of the system. *Note*: the polynomial $a_n\lambda^n + a_{n-1}\lambda^{n-1} + \cdots + a_0$ is called the *characteristic polynomial*. Equation 3.60 has n roots $\lambda_1, \lambda_2, \ldots, \lambda_n$. These are called *poles* or *eigenvalues* of the system. Assuming that they are distinct (i.e., unequal), the overall solution to (3.58) becomes

$$y_h = c_1e^{\lambda_1 t} + c_2e^{\lambda_2 t} + \cdots + c_ne^{\lambda_n t} \tag{3.61}$$

The unknown constants $c_1, c_2, \ldots, c_n$ are determined using the necessary n initial conditions $y(0), \dot{y}(0), \ldots, \dfrac{d^{n-1}y(0)}{dt^{n-1}}$.

3.10.1.1.1 Repeated Poles

Suppose that at least two eigenvalues from the solution of (3.60) are equal. Without loss of generality suppose in (3.61) that $\lambda_1 = \lambda_2$. Then the general solution for (3.60) in the case $\lambda_1 = \lambda_2$ is

$$y_h = (c_1 + c_2 t)e^{\lambda_1 t} + c_3e^{\lambda_3 t} + \cdots + c_ne^{\lambda_n t} \tag{3.62}$$

This idea can be easily generalized for the case of three or more repeated poles (by adding terms containing t^2, t^3, and so on).

3.10.1.2 Particular Solution

The effect of the input is incorporated into the *particular solution*, which is defined as one possible function for y that satisfies Equation 3.57. We denote this by y_p. Several important input functions and the corresponding form of y_p that satisfies Equation 3.57 are given in Table 3.11.

The parameters A, B, A_1, A_2, B_1, B_2, and D in Table 3.11 are determined by substituting the pair $u(t)$ and y_p into (3.57) and then equating the like terms. This approach is called the *method of undetermined coefficients*.

The total response of the system (3.57) is given by $y = y_h + y_p$.

Note: It is incorrect to determine $c_1, c_2, \ldots, c_n$ by first substituting the ICs into y_h and then adding y_p to the resulting y_h.

3.10.1.3 Convolution Integral

The system response (output) to a unit-impulse excitation (input) acted at time $t = 0$, is known as the *impulse-response function* and is denoted by $h(t)$. The system output (response) $y(t)$ to an arbitrary input $u(t)$ may be expressed in terms of its impulse-response function using the convolution integral:

TABLE 3.11

Particular Solutions for Useful Input Functions

Input $u(t)$	Particular Solution y_p
c	A
$ct + d$	$At + B$
$\sin ct$	$A_1 \sin ct + A_2 \cos ct$
$\cos ct$	$B_1 \sin ct + B_2 \cos ct$
e^{ct}	De^{ct}

$$y(t) = \int_0^\infty h(t-\tau)u(\tau)d\tau = \int_0^\infty h(\tau)u(t-\tau)d\tau \qquad (3.63)$$

This is in fact the forced response, under zero initial conditions. It is also a particular integral (particular solution) of the system.

Note: The limits of integration in (3.63) can be set in various manners in view of the fact that $u(t)$ and $h(t)$ are zero for $t < 0$ (e.g., the lower limit may be set at τ and the upper limit at t).

3.10.1.4 Stability

A stable system may be defined as one whose natural response (i.e., free, zero-input, initial-condition response) decays to zero. This is in fact the well-known *asymptotic stability*. If the initial-condition response oscillates within finite bounds, we say the system is *marginally stable*. For a linear, time-invariant system of the type (3.57), the free response is of the form (3.61), assuming no repeated poles. Hence, if none of the eigenvalues λ_i have positive real parts, the system is considered stable, because in that case, the response (3.61) does not grow unboundedly. In particular, if the system has a single eigenvalue that is zero, or if the eigenvalues are purely imaginary pairs, the system is marginally stable. If the system has two or more poles that are zero, we will have terms of the form $c_1 + ct$ in (3.61) and hence it will grow polynomially (not exponentially). Then the system will be *unstable*. Even in the presence of repeated poles, however, if the real parts of the eigenvalues are negative, however, the system is stable (because the decay of the exponential terms in the response will be faster than the growth of the polynomial terms—see Equation 3.62 for example).

Note: Since physical systems have real parameters, their eigenvalues must occur as conjugate pairs, if complex.

Since the sign of the real part of the eigenvalues govern stability, it can be represented on the eigenvalue plane (or the pole plane, s-plane, or root plane). This is illustrated in Figure 3.36.

3.10.2 First-Order Systems

Consider the first order dynamic system with time constant τ, input u, and output y, as given by

$$\tau\dot{y} + y = u(t) \qquad (3.64)$$

Suppose that the system starts from $y(0) = y_0$ and a step input of magnitude A is applied at that initial condition. The homogeneous solution is $y_h = ce^{-t/\tau}$. The particular solution (see Table 3.11) is given by $y_p = A$. Hence, the total response is $y = y_h + y_p = ce^{-t/\tau} + A$.
By substituting the IC $y(0) = y_0$ we get

$$y_{step} = \underbrace{(y_0 - A)e^{-t/\tau}}_{\substack{\text{Homogeneous} \\ y_h}} + \underbrace{A}_{\substack{\text{Particular} \\ y_p}} = \underbrace{y_0 e^{-t/\tau}}_{\substack{\text{Free response} \\ y_x}} + \underbrace{A(1 - e^{-t/\tau})}_{\substack{\text{Forced response} \\ y_f}} \qquad (3.65)$$

The steady-state value is given by $t \to \infty$: $y_{ss} = A$. It is seen from (3.65) that the forced response to a unit step input (i.e., $A = 1$) is $(1 - e^{-t/\tau})$. Due to linearity of the system, the forced response to a unit impulse input is

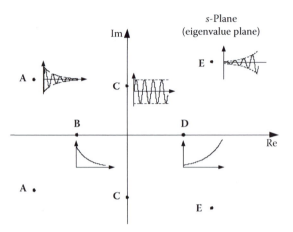

FIGURE 3.36
Dependence of stability on the pole location (**A** and **B** are stable pole locations; **C** is a marginally stable location; **D** and **E** are unstable locations).

$$\frac{d}{dt}(1 - e^{-t/\tau}) = \frac{1}{\tau}e^{-t/\tau}.$$

Hence, the total response to an impulse input of magnitude P is

$$y_{\text{impulse}} = y_0 e^{-t/\tau} + \frac{P}{\tau}e^{-t/\tau} \tag{3.66}$$

This result follows from the fact that

$$\frac{d}{dt}(\text{Unit step}) = \text{Unit impulse} \tag{3.67}$$

and because, due to linearity, when the input is differentiated, the output is correspondingly differentiated.

3.10.3 Second-Order Systems

A general high-order system can be represented by a suitable combination of first-order and second-order models, using the principles of modal analysis. Hence, it is useful to study the response behavior of second-order systems as well. Examples of second-order systems include mass-spring-damper systems and capacitor-inductor-resistor circuits, which we have studied previously. These are called simple oscillators because they exhibit oscillations in the natural response (free, unforced response) when the level of damping is sufficiently low.

3.10.3.1 Free Response of Undamped Oscillator

The equation of free (i.e., no excitation force) motion of an undamped simple oscillator is of the general form

$$\ddot{x} + \omega_n^2 x = 0 \tag{3.68}$$

For a mechanical system of mass m and stiffness k, we have the undamped natural frequency

$$\omega_n = \sqrt{\frac{k}{m}} \tag{3.69a}$$

For an electrical circuit with capacitance C and inductance L we have

$$\omega_n = \sqrt{\frac{1}{LC}} \tag{3.69b}$$

Note: These results can be immediately established from the electromechanical analogy where we use: $m \to C$; $k \to 1/L$; $b \to 1/R$.
The general free response is

$$x = A \sin(\omega_n t + \phi) \tag{3.70}$$

The parameter A is the amplitude and ϕ is the phase angle of the response to be determined by the initial conditions (for x and $\dot{x}$); say, $x(0) = x_o$, $\dot{x}(0) = v_o$.

Note: The system response exactly repeats itself in time-periods of T or at a *cyclic frequency* $f = \frac{1}{T}$ (cycles/s or Hz). The frequency ω (subscript n is dropped for convenience) is the *angular frequency* given by $\omega = 2\pi f$.

3.10.3.2 Free Response of Damped Oscillator

Energy dissipation in a mechanical oscillator may be represented by a damping element. For an electrical circuit, a resistor accounts for energy dissipation. In either case, the equation motion of a damped simple oscillator without an input, may be expressed as

$$\ddot{x} + 2\zeta\omega_n\dot{x} + \omega_n^2 x = 0 \tag{3.71}$$

Here ζ is the *damping ratio*, defined as

$$\zeta = \text{Damping ratio} = \frac{\text{Damping constant}}{\text{Damping constant for critically damped conditions}}$$

The response depends on this, as summarized in Box 3.1.

3.10.3.3 Forced Response of Damped Oscillator

The forced response depends on both the natural characteristics of the system (free response) and the nature of the input. Mathematically, as noted before, the total response is the sum of the homogeneous solution and the particular solution. Consider a damped simple oscillator, with input $u(t)$ scaled such that it has the same units as the response y:

$$\ddot{y} + 2\zeta\omega_n\dot{y} + \omega_n^2 y = \omega_n^2 u(t) \tag{3.72}$$

BOX 3.1   FREE (NATURAL) RESPONSE OF A DAMPED SIMPLE OSCILLATOR

System equation:

$$\ddot{x} + 2\zeta\omega_n\dot{x} + \omega_n^2 x = 0$$

Undamped natural frequency $\omega_n = \sqrt{\dfrac{k}{m}}$ or $\omega_n = \sqrt{\dfrac{1}{LC}}$

Damping ratio $\zeta = \dfrac{b}{2\sqrt{km}}$ or $\zeta = \dfrac{1}{2R}\sqrt{\dfrac{L}{C}}$

Note: Electromechanical analogy $m \to C; k \to 1/L; b \to 1/R$

Characteristic equation: $\lambda^2 + 2\zeta\omega_n\lambda + \omega_n^2 = 0$
   Roots (eigenvalues or poles): λ_1 and $\lambda_2 = -\zeta\omega_n \pm \sqrt{\zeta^2 - 1}\,\omega_n$
Response: $x = C_1 e^{\lambda_1 t} + C_2 e^{\lambda_2 t}$ for unequal roots $(\lambda_1 \neq \lambda_2)$
   $x = (C_1 + C_2 t)e^{\lambda t}$ for equal roots $(\lambda_1 = \lambda_2 = \lambda)$

Initial conditions: $x(0) = x_0$ and $\dot{x}(0) = v_0$
Case 1: Underdamped $(\zeta < 1)$
   Poles are complex conjugates: $-\zeta\omega_n \pm j\omega_d$

   Damped natural frequency $\omega_d = \sqrt{1 - \zeta^2}\,\omega_n$

$$x = e^{-\zeta\omega_n t}\left[C_1 e^{j\omega_d t} + C_2 e^{-j\omega_d t}\right] \qquad A_1 = C_1 + C_2 \text{ and } A_2 = j(C_1 - C_2)$$

$$= e^{-\zeta\omega_n t}\left[A_1 \cos\omega_d t + A_2 \sin\omega_d t\right] \quad C_1 = \frac{1}{2}(A_1 - jA_2) \text{ and } C_2 = \frac{1}{2}(A_1 + jA_2)$$

$$= Ae^{-\zeta\omega_n t}\sin(\omega_d t + \phi) \qquad A = \sqrt{A_1^2 + A_2^2} \text{ and } \tan\phi = \frac{A_1}{A_2}$$

ICs give: $A_1 = x_0$ and $A_2 = \dfrac{v_0 + \zeta\omega_n x_0}{\omega_d}$

Logarithmic decrement per radian: $\alpha = \dfrac{1}{2\pi n}\ln r = \dfrac{\zeta}{\sqrt{1 - \zeta^2}}$
   where $r = \dfrac{x(t)}{x(t + nT)} = $ decay ratio over n complete cycles. For small ζ: $\zeta \cong \alpha$

Case 2: Overdamped $(\zeta > 1)$
   Poles are real and negative: $\lambda_1, \lambda_2 = -\zeta\omega_n \pm \sqrt{\zeta^2 - 1}\,\omega_n$

$$x = C_1 e^{\lambda_1 t} + C_2 e^{\lambda_2 t}$$

$$C_1 = \frac{v_0 - \lambda_2 x_0}{\lambda_1 - \lambda_2} \quad \text{and} \quad C_2 = \frac{v_0 - \lambda_1 x_0}{\lambda_2 - \lambda_1}$$

Case 3: Critically damped $(\zeta = 1)$
   Two identical poles: $\lambda_1 = \lambda_2 = \lambda = -\omega_n$

$$x = (C_1 + C_2 t)e^{-\omega_n t} \quad \text{with } C_1 = x_0 \text{ and } C_2 = v_0 + \omega_n x_0$$

We will consider the response of this system to three types of inputs:

1. Impulse input
2. Step input
3. Harmonic (sinusoidal) input

The nature of the response for the first two types of inputs is presented in Box 3.2. In particular, the step response of a damped oscillator is shown in Figure 3.37.
    The harmonic response is presented in Box 3.3.

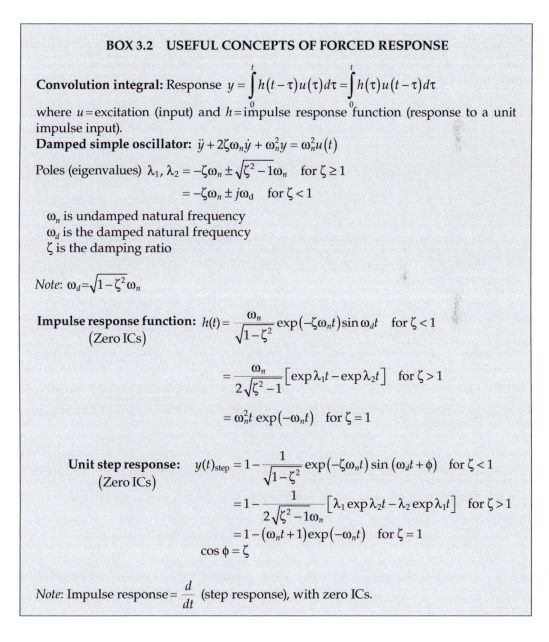

BOX 3.2   USEFUL CONCEPTS OF FORCED RESPONSE

Convolution integral: Response $y = \int_0^t h(t-\tau)u(\tau)d\tau = \int_0^t h(\tau)u(t-\tau)d\tau$

where $u=$ excitation (input) and $h=$ impulse response function (response to a unit impulse input).

Damped simple oscillator: $\ddot{y} + 2\zeta\omega_n\dot{y} + \omega_n^2 y = \omega_n^2 u(t)$

Poles (eigenvalues) $\lambda_1, \lambda_2 = -\zeta\omega_n \pm \sqrt{\zeta^2 - 1}\,\omega_n$ for $\zeta \geq 1$

$$= -\zeta\omega_n \pm j\omega_d \quad \text{for } \zeta < 1$$

ω_n is undamped natural frequency
ω_d is the damped natural frequency
ζ is the damping ratio

Note: $\omega_d = \sqrt{1-\zeta^2}\,\omega_n$

Impulse response function: $h(t) = \dfrac{\omega_n}{\sqrt{1-\zeta^2}} \exp(-\zeta\omega_n t)\sin\omega_d t$ for $\zeta < 1$
    (Zero ICs)

$$= \frac{\omega_n}{2\sqrt{\zeta^2-1}}\left[\exp\lambda_1 t - \exp\lambda_2 t\right] \quad \text{for } \zeta > 1$$

$$= \omega_n^2 t \exp(-\omega_n t) \quad \text{for } \zeta = 1$$

Unit step response: $y(t)_{step} = 1 - \dfrac{1}{\sqrt{1-\zeta^2}}\exp(-\zeta\omega_n t)\sin(\omega_d t + \phi)$ for $\zeta < 1$
    (Zero ICs)

$$= 1 - \frac{1}{2\sqrt{\zeta^2-1}\,\omega_n}\left[\lambda_1 \exp\lambda_2 t - \lambda_2 \exp\lambda_1 t\right] \quad \text{for } \zeta > 1$$

$$= 1 - (\omega_n t + 1)\exp(-\omega_n t) \quad \text{for } \zeta = 1$$

$$\cos\phi = \zeta$$

Note: Impulse response $= \dfrac{d}{dt}$ (step response), with zero ICs.

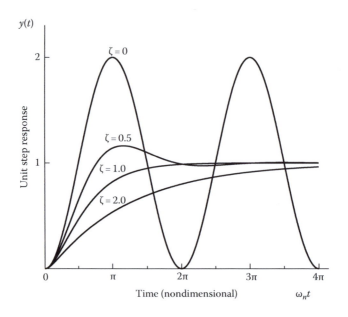

FIGURE 3.37
Step response of a damped oscillator.

3.10.4 Response Using Laplace Transform

Once a transfer function model of a system is available, its response can be determined using the Laplace transform approach. The steps are

1. Using Laplace transform table (Appendix B) determine the Laplace transform ($U(s)$) of the input.
2. Multiply by the transfer function ($G(s)$) to obtain the Laplace transform of the output: $Y(s) = G(s)U(s)$.

   Note: The initial conditions may be introduced in this step by first expressing the system equation in the polynomial form in s and then adding the ICs to each derivative term in the characteristic polynomial.
3. Convert the expression in step 2 into a convenient form (e.g., by partial fractions).
4. Using Laplace transform table, obtain the inverse Laplace transform of $Y(s)$, which gives the response $y(t)$.

Let us illustrate this approach by determining the step response of a simple oscillator.

3.10.4.1 Step Response Using Laplace Transforms

Consider the oscillator system given by (3.72). Since $\mathcal{L}\mathcal{U}(t) = 1/s$, the unit step response of the dynamic system (3.72), with zero ICs, can be obtained by taking the inverse Laplace transform of

$$Y_{\text{step}}(s) = \frac{1}{s} \frac{\omega_n^2}{(s^2 + 2\zeta\omega_n s + \omega_n^2)} = \frac{1}{s} \frac{\omega_n^2}{\Delta(s)} \tag{3.73a}$$

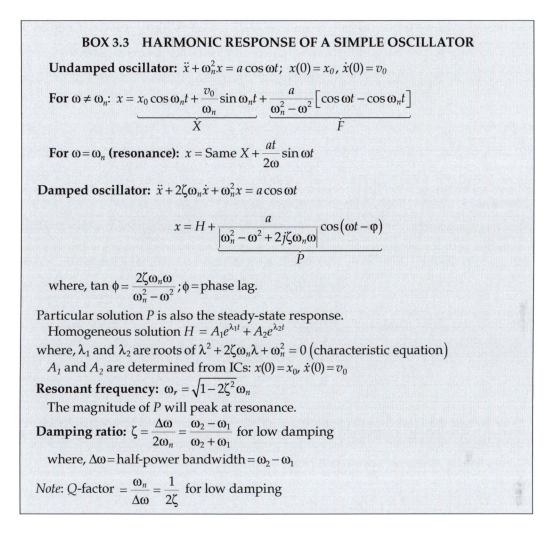

BOX 3.3   HARMONIC RESPONSE OF A SIMPLE OSCILLATOR

Undamped oscillator: $\ddot{x} + \omega_n^2 x = a\cos\omega t$; $x(0) = x_0$, $\dot{x}(0) = v_0$

For $\omega \neq \omega_n$: $x = \underbrace{x_0\cos\omega_n t + \dfrac{v_0}{\omega_n}\sin\omega_n t}_{X} + \underbrace{\dfrac{a}{\omega_n^2 - \omega^2}\Big[\cos\omega t - \cos\omega_n t\Big]}_{F}$

For $\omega = \omega_n$ **(resonance):** $x = \text{Same } X + \dfrac{at}{2\omega}\sin\omega t$

Damped oscillator: $\ddot{x} + 2\zeta\omega_n\dot{x} + \omega_n^2 x = a\cos\omega t$

$$x = H + \underbrace{\dfrac{a}{\left|\omega_n^2 - \omega^2 + 2j\zeta\omega_n\omega\right|}\cos(\omega t - \varphi)}_{P}$$

where, $\tan\phi = \dfrac{2\zeta\omega_n\omega}{\omega_n^2 - \omega^2}$; $\phi = $ phase lag.

Particular solution P is also the steady-state response.
   Homogeneous solution $H = A_1 e^{\lambda_1 t} + A_2 e^{\lambda_2 t}$
where, λ_1 and λ_2 are roots of $\lambda^2 + 2\zeta\omega_n\lambda + \omega_n^2 = 0$ $\big($characteristic equation$\big)$
   A_1 and A_2 are determined from ICs: $x(0) = x_0$, $\dot{x}(0) = v_0$

Resonant frequency: $\omega_r = \sqrt{1 - 2\zeta^2}\,\omega_n$
   The magnitude of P will peak at resonance.

Damping ratio: $\zeta = \dfrac{\Delta\omega}{2\omega_n} = \dfrac{\omega_2 - \omega_1}{\omega_2 + \omega_1}$ for low damping

   where, $\Delta\omega = $ half-power bandwidth $= \omega_2 - \omega_1$

Note: Q-factor $= \dfrac{\omega_n}{\Delta\omega} = \dfrac{1}{2\zeta}$ for low damping

Here the characteristic polynomial of the system is denoted as $\Delta(s) = (s^2 + 2\zeta\omega_n s + \omega_n^2)$. To facilitate using the Laplace transform table, partial fractions of Equation 3.73a are determined as:

$$Y_{\text{step}}(s) = \frac{1}{s} + \frac{s - 2\zeta\omega_n}{(s^2 + 2\zeta\omega_n s + \omega_n^2)} = \frac{1}{s} + \frac{s - 2\zeta\omega_n}{\Delta(s)} \tag{3.73b}$$

Next, using Laplace transform tables, the inverse transform of (3.73b) is obtained, and verified to be identical to what is given in Box 3.2.

3.10.4.2 Incorporation of Initial Conditions

When the initial conditions of the system are not zero, they have to be explicitly incorporated into the derivative terms of the system equation, when converting into the Laplace domain. Except for this, the analysis using the Laplace transform approach is identical to that with zero ICs. In fact, the total solution is equal to the sum of the solution with zero ICs and the solution corresponding to the initial-conditions. We will illustrate the approach using an example.

3.10.4.3 Step Response of a First-Order System

Let us revisit the first order dynamic system with time constant τ, input u, and output y, as given by

$$\tau \dot{y} + y = u(t) \tag{3.64}$$

The initial condition is $y(0) = y_0$. A step input of magnitude A is applied at that initial condition.

From Laplace tables (see Appendix B), convert each term in (3.64) into the Laplace domain as follows:

$$\tau[sY(s) - y_0] + Y(s) = \frac{A}{s} \tag{3.74a}$$

Note how the IC is included in the derivative term, as clear from the Laplace tables. On simplification we get

$$Y(s) = \frac{\tau y_0}{(\tau s + 1)} + \frac{A}{s(\tau s + 1)} = \frac{\tau y_0}{(\tau s + 1)} + \frac{A}{s} - \frac{A\tau}{(\tau s + 1)} \tag{3.74b}$$

Now we use the Laplace tables to determine the inverse Laplace transform each term in (3.74b). We get: $y_{Step} = y_0 e^{-t/\tau} + Ak(1 - e^{-t/\tau})$

This is identical to the previous result (3.65).

3.10.5 Determination of Initial Conditions for Step Response

When a step input is applied to a system, the initial values of the system variables may change instantaneously. However, not all variables will change in this manner since the value of a state variable cannot change instantaneously. We will illustrate some related considerations using an example.

Example 3.10

The circuit shown in Figure 3.38 consists of an inductor L, a capacitor C, and two resistors R and R_o. The input is the voltage $v_i(t)$ and the output is the voltage v_o across the resistor R_o.

(a) Obtain a complete state-space model for the system.
(b) Obtain an input–output differential equation for the system.
(c) Obtain expressions for undamped natural frequency and the damping ratio of the system.
(d) The system starts at steady state with an input of 5 V (for all $t < 0$). Then suddenly, the input is dropped to 1 V (for all $t > 0$), which corresponds to a step input as shown in Figure 3.39. For $R = R_o = 1\,\Omega$, $L = 1\,H$, and $C = 1\,F$, what are the initial conditions of the system and their derivatives at both $t = 0^-$ and $t = 0^+$? What are the final (steady state) values of the state variables and the output variable? Sketch the nature of the system response.

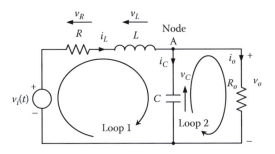

FIGURE 3.38
An electrical circuit.

Solution

(a)
State variables:
Current through independent inductors (i_L); Voltage across independent capacitors (v_C)
Constitutive equations:

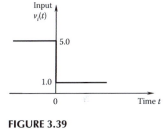

FIGURE 3.39
A step input.

$$v_L = L\frac{di_L}{dt}; \quad i_C = C\frac{dv_C}{dt}; \quad v_R = Ri_L; \quad v_o = Ri_o$$

First two equations are for independent energy storage elements, and they form the state-space shell.
Continuity equation:
Node A (Kirchhoff's current law):

$$i_L - i_C - i_o = 0$$

Compatibility equations:
Loop 1 (Kirchhoff's voltage law): $v_i - v_R - v_L - v_C = 0$
Loop 2 (Kirchhoff's voltage law): $v_C - v_o = 0$
Eliminate auxiliary variables. We have the state equations:

$$L\frac{di_L}{dt} = v_L = v_i - v_R - v_C = v_i - Ri_L - v_C$$

$$C\frac{dv_C}{dt} = i_C = i_L - i_o = i_L - \frac{v_o}{R_o} = i_L - \frac{v_C}{R_o}$$

State equations:

$$\frac{di_L}{dt} = \frac{1}{L}[-Ri_L - v_C + v_i] \tag{i}$$

$$\frac{dv_C}{dt} = \frac{1}{C}\left[i_L - \frac{v_C}{R_o}\right] \tag{ii}$$

Output equation: $v_o = v_C$

Vector-matrix representation: $\dot{x} = Ax + Bu;\ y = Cx$
where

$$\text{System matrix } A = \begin{bmatrix} -R/L & -1/L \\ 1/C & -1/(R_oC) \end{bmatrix}; \quad \text{Input gain matrix } B = \begin{bmatrix} 1/L \\ 0 \end{bmatrix};$$

$$\text{Measurement gain matrix } C = \begin{bmatrix} 0 & 1 \end{bmatrix}; \quad \text{State vector} = x = \begin{bmatrix} i_L \\ v_C \end{bmatrix};$$

$$\text{Input} = u = [v_i]; \quad \text{Output} = y = [v_o]$$

(b)
From (ii):

$$i_L = C\frac{dv_C}{dt} + \frac{v_C}{R_o}$$

Substitute in (i) for i_L:

$$L\frac{d}{dt}\left(C\frac{dv_C}{dt} + \frac{v_C}{R_o}\right) = -R\left(C\frac{dv_C}{dt} + \frac{v_C}{R_o}\right) - v_C + v_i$$

This simplifies to the input–output differential equation (since $v_o = v_C$)

$$LC\frac{d^2v_o}{dt^2} + \left(\frac{L}{R_o} + RC\right)\frac{dv_o}{dt} + \left(\frac{R}{R_o} + 1\right)v_o = v_i \tag{iii}$$

(c)
The input–output differential equation is of the form

$$\frac{d^2v_o}{dt^2} + 2\zeta\omega_n\frac{dv_o}{dt} + \omega_n^2 v_o = \frac{1}{LC}v_i$$

Hence,

$$\text{Natural frequency } \omega_n = \sqrt{\frac{1}{LC}\left(\frac{R}{R_o} + 1\right)} \tag{iv}$$

$$\text{Damping ratio } \zeta = \frac{1}{2\sqrt{LC\left(\frac{R}{R_o} + 1\right)}}\left(\frac{L}{R_o} + RC\right) \tag{v}$$

Note: $1/LC$ has units of (frequency)^2.
RC and L/R_o have units of "time" (i.e., time constant).

(d)
Initial conditions:
For $t < 0$ (initial steady state): $di_L/dt = 0$; $dv_C/dt = 0$

Hence,

(i): $\dfrac{di_L(0^-)}{dt} = 0 = \dfrac{1}{L}\left[-Ri_L(0^-) - v_C(0^-) + v_i(0^-)\right]$

(ii): $\dfrac{dv_C(0^-)}{dt} = 0 = \dfrac{1}{C}\left[i_L(0^-) - \dfrac{v_C(0^-)}{R_o}\right]$

Substitute the given parameter values $R = R_o = 1\,\Omega$, $L = 1\,H$, and $C = 1\,F$, and the input $v_i(0^-) = 5.0$:

$$-i_L(0^-) - v_C(0^-) + 5 = 0$$

$$i_L(0^-) - v_C(0^-) = 0$$

We get

$$i_L(0^-) = 2.5\ A,\ v_C(0^-) = 2.5\ V$$

State variables cannot undergo step changes (because that violates the corresponding physical laws—constitutive equations). Specifically:
Inductor cannot have a step change in current (needs infinite voltage).
Capacitor cannot have a step change in voltage (needs infinite current).
Hence,

$$i_L(0^+) = i_L(0^-) = 2.5\ A$$

$$v_C(0^+) = v_C(0^-) = 2.5\ V$$

Note: Since $v_i(0^+) = 1.0$

(i): $\dfrac{di_L(0^+)}{dt} = -i_L(0^+) - v_C(0^+) + 1.0 = -2.5 - 2.5 + 1.0 = -4.0\ A/s \neq 0$

(ii): $\dfrac{dv_C(0^+)}{dt} = i_L(0^+) - v_C(0^+) = 2.5 - 2.5 = 0.0\ V/s$

Final values:
As $t \to \infty$ (at final steady state)

$$\frac{di_L}{dt} = 0$$

$$\frac{dv_C}{dt} = 0$$

and $v_i = 1.0$
Substitute:

(i): $\dfrac{di_L(\infty)}{dt} = 0 = -i_L(\infty) - v_C(\infty) + 1.0$

(ii): $\dfrac{dv_C(\infty)}{dt} = 0 = i_L(\infty) - v_C(\infty)$

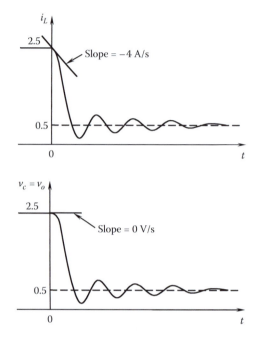

FIGURE 3.40
Responses of the state variables.

Solution: $i_L(\infty)=0.5$ A, $v_C(\infty)=0.5$ V
For the given parameter values,

(iii): $\dfrac{d^2 v_o}{dt^2} + 2\dfrac{dv_o}{dt} + 2v_o = 1$

Hence, $\omega_n = \sqrt{2}$ and $2\zeta\omega_n = 2$, or, $\zeta = 1/\sqrt{2}$.
    This is an under-damped system, producing an oscillatory response as a result. The nature of the responses of the two state variables is shown in Figure 3.40. *Note*: Output $v_o = v_C$.

3.11 Computer Simulation

Simulation of the response of a dynamic system by using a digital computer is perhaps the most convenient and popular approach to response analysis. An important advantage is that any complex, nonlinear, and time variant system may be analyzed in this manner. The main disadvantage is that the solution is not analytic, and is valid only for a specific excitation, under the particular initial conditions, over a limited time interval, and so on. Of course, symbolic approaches of obtaining analytical solutions using a digital computer are available as well. We will consider here numerical simulation only.

    The key operation of digital simulation is integration over time. This typically involves integration of a differential equation of the form

$$\dot{y} = f(y, u, t) \tag{3.75}$$

where u is the input (excitation) and y is the output (response). Note that the function f is nonlinear and time-variant in general. The most straightforward approach to digital integration of this equation is by using the *trapezoidal rule*, which is the Euler's method, as given by

$$y_{n+1} = y_n + f(y_n, u_n, t_n)\Delta t \quad n = 0, 1, \ldots \tag{3.76}$$

Here t_n is the nth time instant, $u_n = u(t_n)$, $y_n = y(t_n)$; and Δt is the integration time step ($\Delta t = t_{n+1} - t_n$). This approach is generally robust. But depending on the nature of the function f, the integration can be ill behaved. Also, Δt has to be chosen sufficiently small.

For complex nonlinearities in f, a better approach of digital integration is the Runge–Kutta method. In this approach, in each time step, first the following four quantities are computed:

$$g_1 = f(y_n, u_n, t_n) \tag{3.77a}$$

$$g_2 = f\left[\left(y_n + g_1\frac{\Delta t}{2}\right), u_{n+\frac{1}{2}}, \left(t_n + \frac{\Delta t}{2}\right)\right] \tag{3.77b}$$

$$g_3 = f\left[\left(y_n + g_2\frac{\Delta t}{2}\right), u_{n+\frac{1}{2}}, \left(t_n + \frac{\Delta t}{2}\right)\right] \tag{3.77c}$$

$$g_4 = f[(y_n + g_3\Delta t), u_{n+1}, t_{n+1}] \tag{3.77d}$$

Then, the integration step is carried out according to

$$y_{n+1} = y_n + (g_1 + 2g_2 + 2g_3 + g_4)\frac{\Delta t}{6} \tag{3.78}$$

Note that $u_{n+\frac{1}{2}} = u\left(t_n + \frac{\Delta t}{2}\right)^*$

Other sophisticated approaches of digital simulation are available as well. Perhaps the most convenient computer-based approach to simulation of a dynamic model is by using a graphic environment that uses block diagrams. Several such environments are commercially available. One that is widely used is Simulink, which is an extension to MATLAB®.

3.11.1 Use of Simulink® in Computer Simulation

Perhaps the most convenient computer-based approach to simulation of a dynamic model is by using a graphic environment that uses block diagrams. Several such environments are commercially available. One that is widely used is Simulink*, and is available as an extension to MATLAB* (also see Appendix D). It provides a graphical environment for

---

* MATLAB and Simulink are properties of Mathworks, Inc.

modeling, simulating, and analyzing dynamic linear and nonlinear systems. Its use is quite convenient. First, a suitable block diagram model of the system is developed on the computer screen, and stored. The Simulink environment provides almost any block that is used in a typical block diagram. These include transfer functions, integrators, gains, summing junctions, inputs (i.e., source blocks) and outputs (i.e., graph blocks or scope blocks). Such a block may be selected and inserted into the workspace as many times as needed, by clicking and dragging using the mouse. These blocks may be converted as required, using directed lines. A block may be opened by clicking on it and the parameter values and text may be inserted or modified as needed. Once the simulation block diagram is generated in this manner, it may be run and the response may be observed through an output block (graph block or scope block). Since Simulink is integrated with MATLAB, data can be easily transferred between programs within various tools and applications.

3.11.1.1 Starting Simulink®

First enter the MATLAB environment. You will have the MATLAB command prompt ≫. To start Simulink, enter the command: simulink. Alternatively, you may click on the "Simulink" button at the top of the MATLAB command window. The Simulink Library Browser window should now appear on the screen. Most of the blocks needed for modeling basic systems can be found in the subfolders of the main Simulink folder.

3.11.1.2 Basic Elements

There are two types of elements in Simulink: **blocks** and **lines**. Blocks are used to generate (or input), modify, combine, output, and display signals. Lines are used to transfer signals from one block to another.

Blocks: The subfolders below the Simulink folder show the general classes of blocks available for use. They are

- Continuous: Linear, continuous-time system elements (integrators, transfer functions, state-space models, etc.)
- Discrete: Linear, discrete-time system elements (integrators, transfer functions, state-space models, etc.)
- Functions and tables: User-defined functions and tables for interpolating function values
- Math: Mathematical operators (sum, gain, dot product, etc.)
- Nonlinear: Nonlinear operators (Coulomb/viscous friction, switches, relays, etc.)
- Signals and systems: Blocks for controlling/monitoring signals and for creating subsystems
- Sinks: For output or display signals (displays, scopes, graphs, etc.)
- Sources: To generate various types of signals (step, ramp, sinusoidal, etc.)

Blocks may have zero or more input terminals and zero or more output terminals.

Lines: A directed line segment transmits signals in the direction indicated by its arrow. Typically, a line must transmit signals from the output terminal of one block to the input terminal of another block. One exception to this is that a line may be used to tap off the signal from another line. In this manner, the tapped original signal can be sent to other (one or more) destination blocks. However, a line can never inject a signal into another line; combining (or, summing) of signals has to be done by using a summing junction. A signal can be either a scalar signal (single signal) or a vector signal (several signals in parallel). The lines used to transmit scalar signals and vector signals are identical. Whether it is a scalar or vector is determined by the blocks connected by the line.

3.11.1.3 Building an Application

To build a system for simulation, first bring up a new model window for creating the block diagram. To do this, click on the "New Model" button in the toolbar of the Simulink Library Browser. Initially the window will be blank. Then, build the system using the following three steps:

1. Gather blocks

   From the Simulink Library Browser, collect the blocks you need in your model. This can be done by simply clicking on a required block and dragging it into your workspace.

2. Modify the blocks

   Simulink allows you to modify the blocks in your model so that they accurately reflect the characteristics of your system. Double-click on the block to be modified. You can modify the parameters of the block in the "Block Parameters" window. Simulink gives a brief explanation of the function of the block in the top portion of this window.

3. Connect the blocks

The block diagram must accurately reflect the system to be modeled. The selected Simulink blocks have be properly connected by lines, to realize the correct block diagram. Draw the necessary lines for signal paths by dragging the mouse from the starting point of a signal (i.e., output terminal of a block) to the terminating point of the signal (i.e., input terminal of another block). Simulink converts the mouse pointer into a crosshair when it is close to an output terminal, to begin drawing a line, and the pointer will become a double crosshair when it is close enough to be snapped to an input terminal. When drawing a line, the path you follow is not important. The lines will route themselves automatically. The terminals points are what matter. Once the blocks are connected, they can be moved around for neater appearance. A block can be simply clicked and dragged to its desired location (the signal lines will remain connected and will reroute themselves).

It may be necessary to branch a signal and transmit it to more than one input terminal. To do this, first placing the mouse cursor at the location where the signal is to be branched (tapped). Then, using either the CTRL key in conjunction with the left mouse button or just the right mouse button, drag the new line to its intended destination.

3.11.1.4 Running a Simulation

Once the model is constructed, you are ready to simulate the system. To do this, go to the **Simulation** menu and click on **Start**, or just click on the "Start/Pause Simulation" button in the model window toolbar (this will look like the "Play" button on a VCR). The simulation will be carried out and the necessary signals will be generated.

General tips:

1. You can save your model by selecting **Save** from the file menu and clicking the **OK** button, (you should give a name to a file).

2. The results of a simulation can be sent to the MATLAB window by the use of the **"to workshop"** icon from the Sinks window.

3. Use the **Demux** (i.e., demultiplexing) icon to convert a vector into several scalar lines. The **Mux** icon takes several scalar inputs and multiplexes them into a vector. (This is useful, for example, when transferring the results from a simulation to the MATLAB workspace).

4. A sign of a **Sum** icon may be changed by double clicking on the icon and changing the sign. The number of inputs to a **Sum** icon may be changed by double clicking on the icon and correctly setting the number of inputs in the window.

5. Be sure to set the integration parameters in the simulation menu. In particular, the default minimum and maximum step sizes must be changed (they should be around 1/100 to 1/10 of the dominant (i.e., slowest) time constant of your system).

Example 3.11

Consider the time-domain model given by

$$\dddot{y} + 13\ddot{y} + 56\dot{y} + 80y = \dddot{u} + 6\ddot{u} + 11\dot{u} + 6u$$

We build the Simulink model, as given in Figure 3.41a.
The system response to an impulse input is shown in Figure 3.41b.

Example 3.12

Consider the model of a robotic sewing system, as considered in Example 3.4 and Figure 3.20. With the state vector $x = [\omega_r\, f_r\, v_h\, f_c\, v_c]^T$; the input vector $u = [T_r\, f_l]^T$; and the output vector $y = [f_c\, \omega_r]^T$, the following state-space model is obtained:

$$\dot{x} = Ax + Bu; \quad y = Cx + Du$$

where

$$A = \begin{bmatrix} -r^2 b_r/J_r & -r/J_r & 0 & 0 & 0 \\ rk_r & 0 & -k_r & 0 & 0 \\ 0 & 1/m_h & -(b_c + b_h)/m_h & 1/m_h & b_c/m_h \\ 0 & 0 & -k_c & 0 & k_c \\ 0 & 0 & b_c/m_c & -1/m_c & -b_c/m_c \end{bmatrix},$$

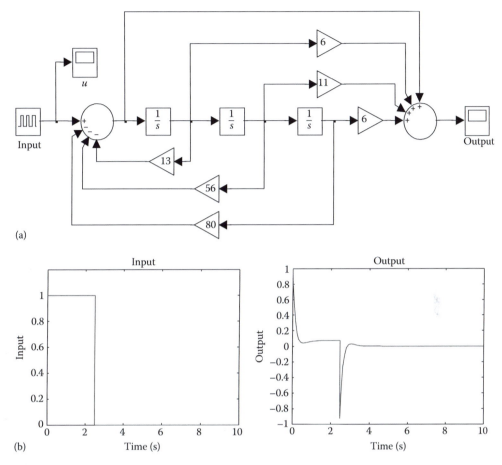

(a)

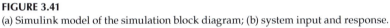

(b)

FIGURE 3.41
(a) Simulink model of the simulation block diagram; (b) system input and response.

$$
\boldsymbol{B} = \begin{bmatrix} 1/J_r & 0 \\ 0 & 0 \\ 0 & 0 \\ 0 & 0 \\ 0 & 1/m_c \end{bmatrix}
$$

$$
\boldsymbol{C} = \begin{bmatrix} 0 & 0 & 0 & 1 & 0 \\ 1 & 0 & 0 & 0 & 0 \end{bmatrix}; \quad \boldsymbol{D} = 0
$$

To carry out a simulation using Simulink, we use the following parameter values:

$m_c = 0.6\,\text{kg}$
$k_c = 100\,\text{N/m}$
$b_c = 0.3\,\text{N/m/s}$
$m_h = 1\,\text{kg}$
$b_h = 1\,\text{N/m/s}$

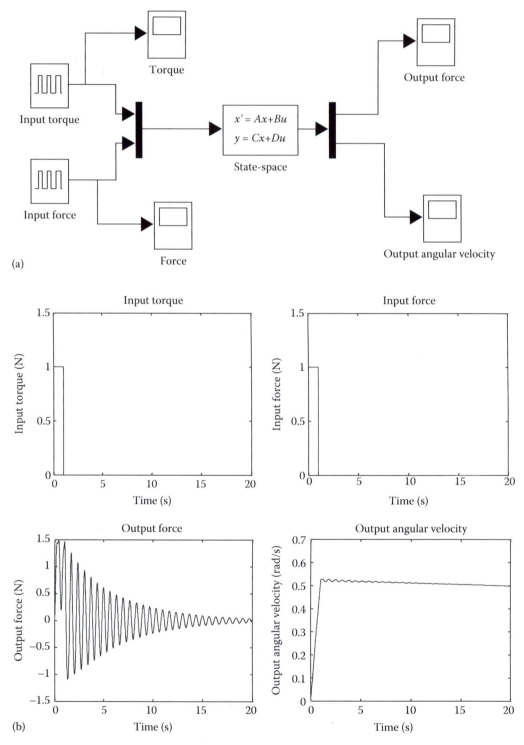

FIGURE 3.42
(a) Simulink model of a robotic sewing machine; (b) simulation results.

$k_r = 200\,\text{N/m}$
$b_r = 1\,\text{N/m/s}$
$J_r = 2\,\text{kg} \cdot \text{m}^2$
$r = 0.05\,\text{m}$

The matrices of the linear model are obtained as

$$
A = \begin{bmatrix}
-0.00125 & -0.025 & 0 & 0 & 0 \\
10 & 0 & -200 & 0 & 0 \\
0 & 1 & -1.3 & 1 & 0.3 \\
0 & 0 & -100 & 0 & 100 \\
0 & 0 & 0.5 & -1.67 & 0.5
\end{bmatrix} ; \quad
B = \begin{bmatrix}
0.5 & 0 \\
0 & 0 \\
0 & 0 \\
0 & 0 \\
0 & 1.67
\end{bmatrix} ;
$$

$$
C = \begin{bmatrix}
0 & 0 & 0 & 1 & 0 \\
1 & 0 & 0 & 0 & 0
\end{bmatrix} ; \quad
D = \begin{bmatrix}
0 & 0 \\
0 & 0
\end{bmatrix}
$$

The Simulink model is built, as shown in Figure 3.42a.
The response of the system to two impulse inputs is shown in Figure 3.42b.

Problems

3.1 What is a "dynamic" system, a special case of any system?
  A typical input variable is identified for each of the following examples of dynamic systems. Give at least one output variable for each system.

(a) Human body: neuroelectric pulses
(b) Company: information
(c) Power plant: fuel rate
(d) Automobile: steering wheel movement
(e) Robot: voltage to joint motor
(f) Highway bridge: vehicle force

3.2 (a) Briefly explain/justify why voltage and not current is the natural state variable for an electrical capacitor; and current and not voltage is the natural state variable for an electrical inductor.

(b) List several advantages of using as state variables, the across variables of independent *A*-type energy storage elements and through variables of independent *T*-type energy storage elements, in the development of a state-space model for an engineering system.

(c) List three things to which the order of an electromechanical dynamic system is equal.

3.3 What are the basic lumped elements of

   (i) A mechanical system?

   (ii) An electrical system?

   Indicate whether a distributed-parameter method is needed or a lumped-parameter model is adequate in the study of following dynamic systems:

   (a) Vehicle suspension system (motion)

   (b) Elevated vehicle guideway (transverse motion)

   (c) Oscillator circuit (electrical signals)

   (d) Environment (weather) system (temperature)

   (e) Aircraft (motion and stresses)

   (f) Large transmission cable (capacitance and inductance)

3.4 (a) Give logical steps of the analytical modeling process for a general physical system.

   (b) Once a dynamic model is derived, what other information would be needed for analyzing its time response (or for computer simulation)?

   (c) A system is divided into two subsystems, and models are developed for these subsystems. What other information would be needed to obtain a model for the overall system?

3.5 Describe two approaches of determining the parameters of a lumped-parameter model that is (approximately) equivalent to a distributed-parameter (i.e., continuous) dynamic system.

   One end of a heavy spring of mass m_s and stiffness k_s is attached to a lumped mass m. The other end is attached to a support that is free to move, as shown in Figure P3.5.

   Using the method of natural frequency equivalence, determine an equivalent lumped-parameter model for the spring where the equivalent lumped mass is located at the free end (support end) of the system. The natural frequencies of a heavy spring with one end fixed and the other end free is given by

$$\omega_n = \frac{\pi}{2}(2n-1)\sqrt{\frac{k_s}{m_s}}$$

   where n is the mode number.

3.6 (a) Why are analogies important in modeling of dynamic systems?

   (b) In the force–current analogy, what mechanical element corresponds to an electrical capacitor?

   (c) In the velocity–pressure analogy, is the fluid inertia element analogous to the mechanical inertia element?

3.7 (a) Briefly explain why a purely thermal system typically does not have an oscillatory response whereas a fluid system can.

   (b) Figure P3.7 shows a pressure-regulated system that can provide a high-speed jet of liquid. The system consists of a pump, a spring-loaded accumulator, and

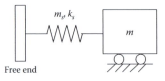

FIGURE P3.5
A mechanical system with a heavy spring and attached mass.

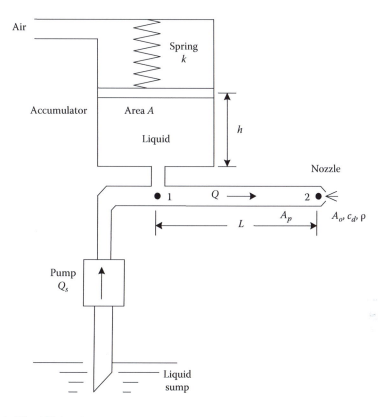

FIGURE P3.7
Pressure regulated liquid jet system.

a fairly long section of piping that ends with a nozzle. The pump is considered as a flow source of value Q_s. The following parameters are important:

A = area of cross section (uniform) of the accumulator cylinder
k = spring stiffness of the accumulator piston
L = length of the section of piping from the accumulator to the nozzle
A_p = area of cross section (uniform, circular) of the piping
A_o = discharge area of the nozzle
C_d = discharge coefficient of the nozzle
Q = mass density of the liquid

Assume that the liquid is incompressible. The following variables are important:

$P_{1r} = P_1 - P_r$ = pressure at the inlet of the accumulator with respect to the ambient reference r
Q = volume flow rate through the nozzle
h = height of the liquid column in the accumulator

Note that the piston (wall) of the accumulator can move against the spring, thereby varying h.

(i) Considering the effects of the movement of the spring loaded wall and also the gravity head of the liquid, obtain an expression for the equivalent fluid capacitance C_a of the accumulator in terms of k, A, ρ, and g. Are the two capacitances that contribute to C_a (i.e., wall stretching and gravity) connected in parallel or in series?

Note: Neglect the effect of bulk modulus of the liquid.

(ii) Considering the capacitance C_a, the inertance I of the fluid volume in the piping (length L and cross section area A^p), and the resistance of the nozzle only, develop a nonlinear state-space model for the system. The state vector $x = [P_{1r}\ Q]^T$, and the input $u = [Qs]$.

For flow in the (circular) pipe with a parabolic velocity profile, the inertance $I = 2\rho L / A_p$ and for the discharge through the nozzle $Q = A_o c_d \sqrt{2P_{2r}/\rho}$

in which

P_{2r} is the pressure inside the nozzle with respect to the outside reference (r)

c_d is the discharge coefficient

3.8 A model for the automatic gage control (AGC) system of a steel rolling mill is shown in Figure P3.8. The rolls are pressed using a single acting hydraulic actuator with a valve

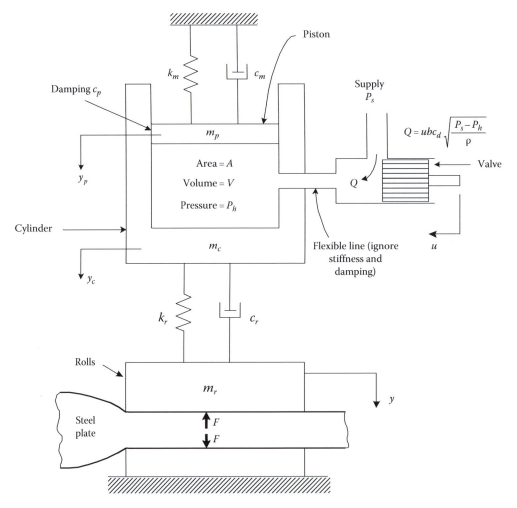

FIGURE P3.8
Automatic gage control (AGC) system of a steel rolling mill.

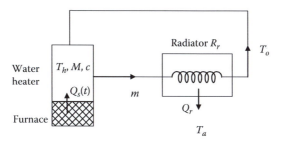

FIGURE P3.9
A household heating system.

displacement of u. The rolls are displaced through y, thereby pressing the steel that is being rolled. The rolling force F is completely known from the steel parameters for a given y.

(i)   Identify the inputs and the controlled variable in this control system.

(ii) In terms of the variables and system parameters indicated in Figure P3.8, write dynamic equations for the system, including valve nonlinearities.

(iii) What is the order of the system? Identify the response variables.

(iv) What variables would you measure (and feedback through suitable controllers) in order to improve the performance of the control system?

3.9 A simplified model of a hot water heating system is shown in Figure P3.9.

Q_s = rate of heat supplied by the furnace to the water heater (1000 kW)

T_a = ambient temperature (°C)

T_h = temperature of water in the water heater is assumed uniform (°C)

T_o = temperature of the water leaving the radiator (°C)

Q_r = rate of heat transfer from the radiator to the ambience (kW)

M = mass of water in the water heater (500 kg)

$\dot{m}$ = mass rate of water flow through the radiator (25 kg/min)

c = specific heat of water (4200 J/kg/°C).

The radiator satisfies the equation

$$T_h - T_a = R_r Q_r$$

where R_r is the thermal resistance of the radiator $(2 \times 10{-}3 \ °C/kW)$

(a) What are the inputs to the system?

(b) Using T_h as a state variable, develop a state–space model for the system.

(c) Give output equations for Q_r and T_o.

3.10 1. In the electrothermal analogy of thermal systems, where voltage is analogous to temperature and current is analogous to heat transfer rate, explain why there exists a thermal capacitor but not a thermal inductor. What is a direct consequence of this fact with regard to the natural (free or unforced) response of a purely thermal system?

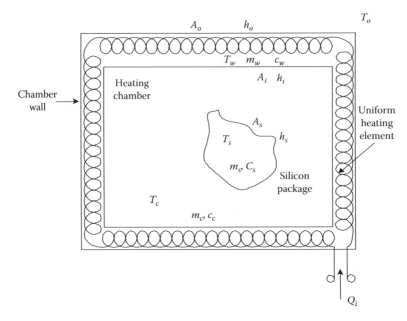

FIGURE P3.10
A model of the heat treatment of a package of silicon.

2. A package of semiconductor material consisting primarily of wafers of crystalline silicon substrate with minute amounts of silicon dioxide is heat treated at high temperature as an intermediate step in the production of transistor elements. An approximate model of the heating process is shown in Figure P3.10.

The package is placed inside a heating chamber whose walls are uniformly heated by a distributed heating element. The associated heat transfer rate into the wall is Q_i. The interior of the chamber contains a gas of mass mc and specific heat c_c, and it is maintained at a uniform temperature T_c. The temperature of silicon is T_s and that of the wall is T_w. The outside environment is maintained at temperature T_o. The specific heats of the silicon package and the wall are denoted by c_s and c_w, respectively, and the corresponding masses are denoted by m_s and m_w as shown. The convective heat transfer coefficient at the interface of silicon and gas inside the chamber is h_s, and the effective surface area is A_s. Similarly, h_i and h_o denote the convective heat transfer coefficients at the inside and outside surfaces of the chamber wall, and the corresponding surface areas are A_i and A_o, respectively.

(a) Using T_s, T_c, and T_w as state variables, write three state equations for the process.

(b) Express these equations in terms of the parameters $C_{hs} = m_s c_s$, $C_{hc} = m_c c_c$, $C_{hw} = m_w c_w$, $R_s = 1/h_s A_s$, $R_i = 1/h_i A_i$, and $R_o = 1/h_o A_o$. Explain the electrical analogy and physical significance of these parameters.

(c) What are the inputs to the process? If T_s is the output of importance, obtain the matrices A, B, C, and D of the state-space model.

(d) Comment on the accuracy of the model in the context of the actual physical process of producing semiconductor elements.

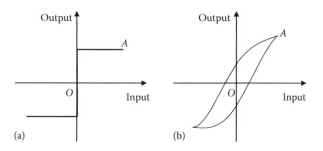

FIGURE P3.11
Two types of nonlinearities: (a) Ideal saturation; (b) hysteresis.

3.11 What precautions may be taken in developing and operating a mechanical system, in order to reduce system nonlinearities?

Read about the following nonlinear phenomena:

(i) Saturation

(ii) Hysteresis

(iii) Jump phenomena

(iv) Frequency creation

(v) Limit cycle

(vi) Deadband

Two types of nonlinearities are shown in Figure P3.11.

In each case, indicate the difficulties of developing an analytical for operation near:

(i) Point O

(ii) Point A

3.12 Characteristic curves of an armature-controlled dc motor are as shown in Figure P3.12. These are torque versus speed curves, measured at a constant armature voltage, at steady state. For the neighborhood of point P, a linear model of the form

$$\hat{\omega} = k_1\hat{v} + k_2\hat{T}$$

needs to be determined, for use in motor control. The following information is given:

The slope of the curve at $P = -a$

The voltage change in the two adjacent curves at point $P = \Delta V$

Corresponding speed change (at constant load torque through P) $= \Delta\omega$.

Estimate the parameters k_1 and k_2.

3.13 An air-circulation fan system of a building is shown in Figure P3.13a, and a simplified model of the system may be developed, as represented in Figure P3.13b.

The induction motor is represented as a torque source $\tau(t)$. The speed ω of the fan, which determines the volume flow rate of air, is of interest. The moment of inertia of

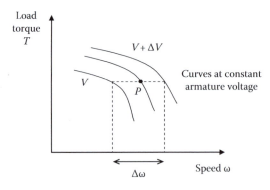

FIGURE P3.12
Characteristic curves of an armature-controlled dc motor.

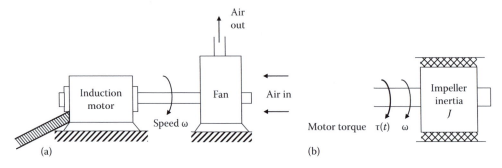

FIGURE P3.13
(a) A motor–fan combination of a building ventilation system; (b) a simplified model of the ventilation fan.

the fan impeller is J. The energy dissipation in the fan is modeled by a linear viscous damping component (of damping constant b) and a quadratic aerodynamic damping component (of coefficient d).

(a) Show that the system equation may be given by

$$J\dot{\omega} + b\omega + d|\omega|\omega = \tau(t)$$

(b) Suppose that the motor torque is given by

$$\tau(t) = \bar{\tau} + \hat{\tau}_a \sin \Omega t$$

in which $\bar{\tau}$ is the steady torque and $\hat{\tau}$ a is a very small amplitude (compared to $\bar{\tau}$) of the torque fluctuations at frequency Ω. Determine the steady-state operating speed $\bar{\omega}$ (which is assumed positive) of the fan.

(c) Linearize the model about the steady-state operating conditions and express it in terms of the speed fluctuations $\hat{\omega}$ (d). From this, estimate the amplitude of the speed fluctuations.

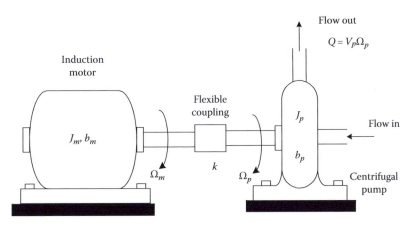

FIGURE P3.14
A centrifugal pump driven by an inductor motor.

3.14 (a) Linearized models of nonlinear systems are commonly used in model-based control of processes. What is the main assumption that is made in using a linearized model to represent a nonlinear system?

(b) A three-phase induction motor is used to drive a centrifugal pump for incompressible fluids. To reduce misalignment and associated problems such as vibration, noise, and wear, a flexible coupling is used for connecting the motor shaft to the pump shaft. A schematic representation of the system is shown in Figure P3.14.

Assume that the motor is a "torque source" of torque T_m, which is being applied to the motor of inertia J_m. Also, the following variables and parameters are defined for the system:

J_p = moment of inertia of the pump impeller assembly

Ω_m = angular speed of the motor rotor/shaft

Ω_p = angular speed of the pump impeller/shaft

k = torsional stiffness of the flexible coupling

T_f = torque transmitted through the flexible coupling

Q = volume flow rate of the pump

b_m = equivalent viscous damping constant of the motor rotor

Also, assume that the net torque required at the ump shaft, to pump fluid steadily at a volume flow rate Q, is given by $b_p \Omega_p$,

where $Q = V_p \Omega_p$ and V_p = volumetric parameter of the pump (assumed constant).

Using T_m as *the* input and Q as the output of the system, develop a complete state-space model for the system. Identify the model matrices **A**, **B**, **C**, and **D** in the usual notation, in this model. What is the order of the system?

(c) In part (a) suppose that the motor torque is given by

$$T_m = \frac{aSV_f^2}{\left[1 + (S/S_b)^2\right]}$$

where the fractional slip S of the motor is defined as

$$S = 1 - \frac{\Omega_m}{\Omega_s}$$

Note that a and S_b are constant parameters of the motor. Also,
  Ω_s is the no-load (i.e., synchronous) speed of the motor
  V_f is the amplitude of the voltage applied to each phase winding (field) of the motor

In *voltage control*, V_f is used as the input, and in *frequency control* Ω_s is used as the input. For combined voltage control and frequency control, derive a linearized state-space model, using the incremental variables $\hat{V}_f$ and $\hat{\Omega}_s$, about the operating values $\bar{V}_f$ and $\bar{\Omega}_s$, as the inputs to the system, and the incremental flow $\bar{Q}$ as the output.

3.15 (a) What are A-type elements and what are T-type elements? Classify mechanical inertia, mechanical spring, fluid inertia and fluid capacitor into these two types. Explain a possible conflict that can arise due to this classification.

(b) A system that is used to pump an incompressible fluid from a reservoir into an open overhead tank is schematically shown in Figure P3.15. The tank has a uniform across section of area A.

  The pump is considered a pressure source of pressure difference $P(t)$. A valve of constant k_v is placed near the pump in the long pipe line, which leads to the overhead tank. The valve equation is $Q = k_v\sqrt{P_1 - P_2}$ in which Q is the volume flow rate of the fluid. The resistance to the fluid flow in the pipe may be modeled as $Q = k_p\sqrt{P_2 - P_3}$ in which k_p is a pipe flow constant. The effect of the accelerating fluid is represented by the linear equation $I(dQ/dt) = P_3 - P_4$ in which I denotes the fluids inertance. Pressures P_1, P_2, P_3, and P_4 are as marked along the pipe length, in Figure P3.10. Also, P_0 denotes the ambient pressure.

  (i) Using Q and P_{40} as the state variables, the pump pressure $P(t)$ as the input variable, and the fluid level H in the tank as the output variable, obtain a complete (nonlinear) state-space model for the system. *Note:* $P_{40} = P_4 - P_0$. Density of the fluid $= \rho$.

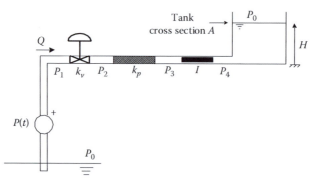

FIGURE P3.15
A pumping system for an overhead tank.

(ii) Linearize the state equations about an operating point given by flow rate $\bar{Q}$. Determine the model matrices A, B, C, and D for the linear model.

(iii) What is the combined linear resistance of the valve and piping? What characteristic of a nonlinear system does this result illustrate?

3.16 An automated wood cutting system contains a cutting unit, which consists of a dc motor and a cutting blade, linked by a flexible shaft and a coupling. The purpose of the flexible shaft is to position the blade unit at any desirable configuration, away from the motor itself. The coupling unit helps with the shaft alignment (compensates for possible misalignment). A simplified, lumped-parameter, dynamic model of the cutting device is shown in Figure P3.16.

The following parameters and variables are shown in the figure:

J_m = axial moment of inertia of the motor rotor

b_m = equivalent viscous damping constant of the motor bearings

k = torsional stiffness of the flexible shaft

J_c = axial moment of inertia of the cutter blade

b_c = equivalent viscous damping constant of the cutter bearings

T_m = magnetic torque of the motor

ω_m = motor speed

T_k = torque transmitted through the flexible shaft

ω_c = cutter speed

T_L = load torque on the cutter from the workpiece (wood)

In comparison with the flexible shaft, the coupling unit is assumed rigid, and is also assumed light. The cutting load is given by $T_L = c|\omega_c|\omega_c$.

The parameter c, which depends on factors such as the depth of cut and the material properties of the workpiece, is assumed constant in the present analysis.

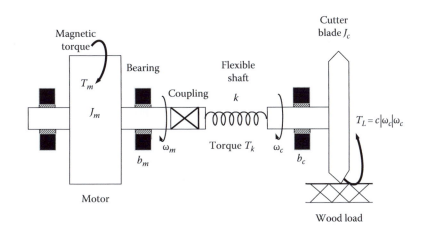

FIGURE P3.16
A wood cutting machine.

(a) Using T_m as the input, T_L as the output, and $[\omega_m \; T_k \; \omega_c]^T$ as the state vector, develop a complete (nonlinear) state model for the system shown in Figure P3.16. What is the order of the system?

(b) Using the state model derived in part (a), obtain a single input–output differential equation for the system, with T_m as the input and ω_c as the output.

(c) Consider the steady operating conditions where $T_m = \bar{T}_m$, $\omega_m = \bar{\omega}_m$, $T_k = \bar{T}_k$, $\omega_c = \bar{\omega}_c$, $T_L = \bar{T}_L$ are all constants. Express the operating point values $\bar{\omega}_m$, $\bar{T}_k$, $\bar{\omega}_c$, and $\bar{T}_L$ in terms of $\bar{T}_m$ and the model parameters only. You must consider both cases $\bar{T}_m > 0$ and $\bar{T}_m < 0$.

(d) Now consider an incremental change $\hat{T}_m$ in the motor torque and the corresponding changes $\hat{\omega}_m$, $\hat{T}_k$, $\hat{\omega}_c$, and $\hat{T}_L$ in the system variables. Determine a linear state model (A, B, C, D) for the incremental dynamics of the system in this case, using $x = [\hat{\omega}_m \; \hat{T}_k \; \hat{\omega}_c]^T$ as the state vector, $u = [\hat{T}_m]$ as the input, and $y = [\hat{T}_L]$ as the output.

(e) In the incremental model (see part (a)), if the twist angle of the flexible shaft (i.e., $\theta_m - \theta_c$) is used as the output what will be a suitable state model? What is the system order then?

(f) In the incremental model, if the angular position θ_c of the cutter blade is used as the output variable, explain how the state model obtained in part (a) should be modified. What is the system order in this case?

Hint for part (b):

$$\frac{d}{dt}(|\omega_c|\omega_c) = 2|\omega_c|\dot{\omega}_c$$

$$\frac{d^2}{dt^2}(|\omega_c|\omega_c) = 2|\omega_c|\ddot{\omega}_c + 2\omega_c^2\, \mathrm{sgn}(\omega_c)$$

Note: These results may be derived as follows: since $|\omega_c| = \omega_c \,\mathrm{sgn}\,\omega_c$ we have

$$\frac{d}{dt}(|\omega_c|\omega_c) = \frac{d}{dt}(\omega_c^2\,\mathrm{sgn}\,\omega_c) = 2\omega_c\dot{\omega}_c\,\mathrm{sgn}\,\omega_c = 2|\omega_c|\dot{\omega}_c$$

and

$$\frac{d^2}{dt^2}(|\omega_c|\omega_c) = 2|\omega_c|\ddot{\omega}_c + 2\dot{\omega}_c^2\,\mathrm{sgn}(\omega_c)$$

Note: Since $\mathrm{sgn}(\omega_c) = +1$ for $\omega_c > 0$; $= -1$ for $\omega_c < 0$; it is a constant and its time derivative is zero (except at $\omega_c = 0$, which is not important here as it corresponds to the static condition).

3.17 A simplified model of an elevator is shown in Figure P3.17. The model parameters are:

J = moment of inertia of the cable pulley

r = radius of the pulley

k = stiffness of the cable

m = mass of the car and its occupants

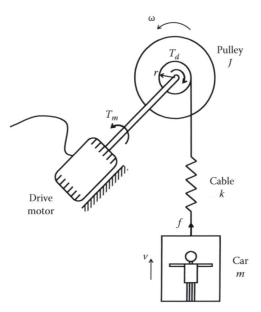

FIGURE P3.17
A simplified model of an elevator.

(a) Which system parameters are variable? Explain.

(b) Suppose that the damping torque $T_d(\omega)$ at the bearing of the pulley is a nonlinear function of the angular speed ω of the pulley. Suppose that:

$$\text{State vector } x = [\omega \quad f \quad v]^T$$

with
  f = tension force in the cable
  v = velocity of the car (taken positive upwards)

Input vector $u = [T_m]^T$
with
  T_m = torque applied by the motor to the pulley (positive in the direction indicated in Figure P3.17)

Output vector as $y = [v]$
Obtain a complete, nonlinear, state-space model for the system.

(c) With T_m as the input and v as the output, convert the state-space model into a nonlinear input–output differential equation model. What is the order of the system?

(d) Give an equation whose solution provides the steady-state operating speed $\bar{v}$ of the elevator car.

(e) Linearize the nonlinear input/output differential-equation model obtained in part (c), for small changes $\hat{T}_m$ of the input and $\hat{v}$ of the output, about an operating point.

Note: $\bar{T}_m$ = steady – stateoperating – point torque of the motor (assumed to be known).

Hint: Denote dT_d/dt as $b(\omega)$.

(f) Linearize the state-space model obtained in part (b) and give the model matrices A, B, C, and D in the usual notation. Obtain the linear input/output differential equation from this state-space model and verify that it is identical to what was obtained in part (e).

3.18 Consider the L-C-R electrical circuit shown in Figure P3.18. Draw a linear graph for this circuit. Identify the primary loops and write equations for them. Sketch a mechanical system that is analogous to the given electrical circuit.

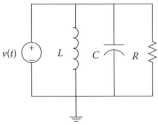

FIGURE P3.18
An *L-C-R* circuit.

3.19 Commercial motion controllers are digitally controlled (microprocessor-controlled) high-torque devices capable of applying a prescribed motion to a system. Such controlled actuators can be considered as velocity sources. Consider an application where a rotary motion controller is used to position an object, which

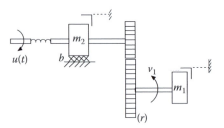

FIGURE P3.19
Rotary-motion system with a gear transmission.

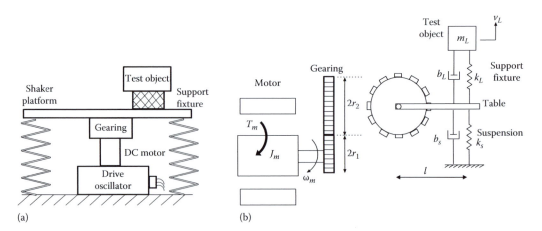

(a)   (b)

FIGURE P3.20
(a) A dynamic-testing system; (b) a model of the dynamic testing system.

is coupled through a gearbox. The system is modeled as in Figure P3.19. Develop a state-space model for this system using the linear graph approach.

3.20 (a) List several advantages of using linear graphs in developing a state-space model of a dynamic system.

(b) Electrodynamic shakers are commonly used in the dynamic testing of products. One possible configuration of a shaker/test-object system is shown in Figure P3.20a. A simple, linear, lumped-parameter model of the mechanical system is shown in Figure P3.20b.

Note: The driving motor is represented by a torque source T_m. The following parameters are indicated:

J_m = equivalent moment of inertia of motor rotor, shaft, coupling, gears, and the shaker platform

r_1 = pitch circle radius of the gear wheel attached to the motor shaft

r_1 = pitch circle radius of the gear wheel rocking the shaker platform

l = lever arm from the center of the rocking gear to the support location of the test object

m_L = equivalent mass of the test object and its support fixture

k_L = stiffness of the support fixture

b_L = equivalent viscous damping constant of the support fixture

k_s = stiffness of the suspension system of the shaker table

b_s = equivalent viscous damping constant of the suspension system

Since the inertia effects are lumped into equivalent elements it may be assumed that the shafts, gearing, platform and the support fixtures are light. The following variables are of interest:

ω_m = angular speed of the drive motor

v_L = vertical speed of motion of the test object

f_L = equivalent dynamic force of the support fixture (force in spring k_L)

f_s = equivalent dynamic force of the suspension system (force in spring k_s)

(i) Obtain an expression for the motion ratio:

$$r = \frac{\text{Vertical movement of the shaker table at the test object support location}}{\text{Angular movement of the drive motor shaft}}$$

(ii) Draw a linear graph to represent the dynamic model.

(iii) Using $x = [\omega_m, f_s, f_L, v_L]^T$ as the state vector, $u = [T_m]$ as the input, and $y = [v_L, f_L]^T$ as the output vector, obtain a complete state-space model for the system. For this purpose, you must use the linear graph drawn in part (ii).

3.21 The circuit shown in Figure P3.21 has an inductor L, a capacitor C, a resistor R, and a voltage source $v(t)$. Considering that L is analogous to a spring, and C is analogous to an inertia, follow the standard steps to obtain the state equations. First sketch the linear graph denoting the currents through and the voltages across the elements L, C, and R by (f_1, v_1), (f_2, v_2) and (f_3, v_3), respectively, and then proceed in the usual manner.

(i) What is the system matrix and what is the input distribution matrix for your choice of state variables?

(ii) What is the order of the system?

(iii) Briefly explain what happens if the voltage source $v(t)$ is replaced by a current source $i(t)$.

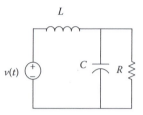

FIGURE P3.21
An electrical circuit.

3.22 Consider an automobile traveling at a constant speed on a rough road, as sketched in Figure P3.22a. The disturbance input due to road irregularities can be considered as a velocity source $u(t)$ at the tires in the vertical direction. An approximate one-dimensional model shown in Figure P3.22b may be used to study the "heave" (up and down) motion of the automobile. Note that v_1 and v_2 are the velocities of the lumped masses m_1 and m_2, respectively.

(a) Briefly state what physical components of the automobile are represented by the model parameters k_1, m_1, k_2, m_2, and b_2. Also, discuss the validity of the assumptions that are made in arriving at this model.

(b) Draw a linear graph for this model, orient it (i.e., mark the directions of the branches), and completely indicate the system variables and parameters.

(c) By following the step-by-step procedure of writing constitutive equations, node equations and loop equations, develop a complete state-space model for this system. The outputs are v_1 and v_2. What is the order of the system?

(d) If instead of the velocity source $u(t)$, a force source $f(t)$ that is applied at the same location, is considered as the system input, draw a linear graph for this modified model. Obtain the state equations for this model. What is the order of the system now?

Note: In this problem, you may assume that the gravitational effects are completely balanced by the initial compression of the springs with reference to which all motions are defined.

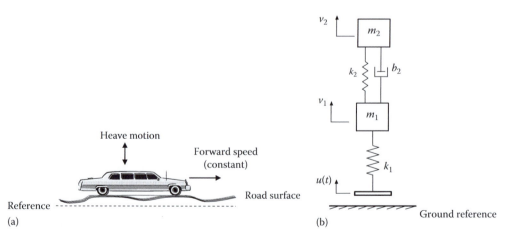

(a)

(b)

FIGURE P3.22
(a) An automobile traveling at constant speed; (b) a crude model of the automobile for the heave motion analysis.

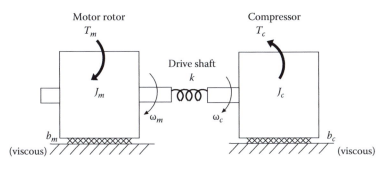

FIGURE P3.23
A model of a motor–compressor unit.

3.23 An approximate model of a motor–compressor combination used in a process control application is shown n Figure P3.23.

Note that T, J, k, b, and ω denote torque, moment of inertia, torsional stiffness, angular viscous damping constant, and angular speed, respectively, and the subscripts m and c denote the motor rotor and compressor impeller, respectively.

(a) Sketch a translatory mechanical model that is analogous to this rotatory mechanical model.
(b) Draw a linear graph for the given model, orient it, and indicate all necessary variables and parameters on the graph.
(c) By following a systematic procedure and using the linear graph, obtain a complete state-space representation of the given model. The outputs of the system are compressor speed ω_c and the torque T transmitted through the drive shaft.

3.24 A model for a single joint of a robotic manipulator is shown in Figure P3.24. The usual notation is used. The gear inertia is neglected and the gear reduction ratio is taken as 1:r (*Note*: $r < 1$).

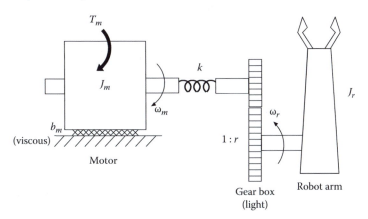

FIGURE P3.24
A model of a single-degree-of-freedom robot.

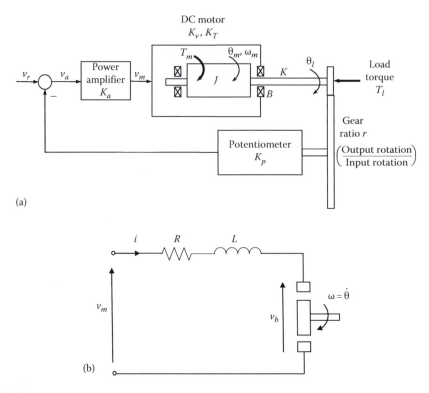

(a)

(b)

FIGURE P3.25
(a) A rotatory electromechanical system; (b) the armature circuit.

(a) Draw a linear graph for the model, assuming that no external (load) torque is present at the robot arm.

(b) Using the linear graph derive a state model for this system. The input is the motor magnetic torque T_m and the output is the angular speed ω_r of the robot arm. What is the order of the system?

(c) Discuss the validity of various assumptions made in arriving at this simplified model for a commercial robotic manipulator.

3.25 Consider the rotatory feedback control system shown schematically by Figure P3.25a. The load has inertia J, stiffness K and equivalent viscous damping B as shown. The armature circuit for the dc fixed field motor is shown in Figure P3.25b. The following relations are known:

The back e.m.f. $v_B = K_v \omega$

The motor torque $T_m = K_T i$

(a) Identify the system inputs.

(b) Write the linear system equations.

3.26 (a) What is the main physical reason for oscillatory behavior in a purely fluid system?

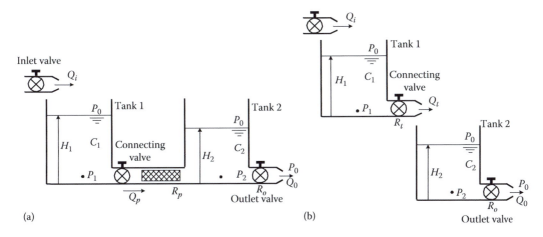

FIGURE P3.26
(a) An interacting two-tank fluid system; (b) a noninteracting two-tank fluid system.

Why do purely fluid systems with large tanks connected by small-diameter pipes rarely exhibit an oscillatory response?

(b) Two large tanks connected by a thin horizontal pipe at the bottom level are shown in Figure P3.26a. Tank 1 receives an inflow of liquid at the volume rate Q_i when the inlet valve is open. Tank 2 has an outlet valve, which has a fluid flow resistance of R_o and a flow rate of Q_0 when opened. The connecting pipe also has a valve, and when opened, the combined fluid flow resistance of the valve and the thin pipe is R_p. The following parameters and variables are defined:

$C_1, C_2 =$ fluid (gravity head) capacitances of tanks 1 and 2

$\rho =$ mass density of the fluid

$g =$ acceleration due to gravity

$P_1, P_2 =$ pressure at the bottom of tanks 1 and 2

$P_0 =$ ambient pressure

Using $P_{10} = P_1 - P_0$ and $P_{20} = P_2 - P_0$ as the state variables and the liquid levels H_1 and H_2 in the two tanks as the output variables, derive a complete, linear, state-space model for the system.

(c) Suppose that the two tanks are as in Figure P3.26b. Here tank 1 has an outlet valve at its bottom whose resistance is R_t and the volume flow rate is Q_t when open. This flow directly enters tank 2, without a connecting pipe. The remaining characteristics of the tanks are the same as in part (b).
Derive a state-space model for the modified system in terms of the same variables as in part (b).

3.27 A common application of dc motors is in accurate positioning of a mechanical load. A schematic diagram of a possible arrangement is shown in Figure P3.27. The actuator of the system is an armature-controlled dc motor. The moment of inertia of its rotor is J_r and the angular speed is ω_r. The mechanical damping of the motor (including that of its bearings) is neglected in comparison to that of the load.

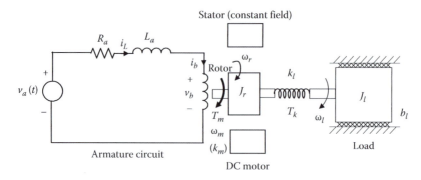

FIGURE P3.27
An electromechanical model of a rotatory positioning system.

The armature circuit is also shown in Figure P3.27, which indicates a back e.m.f. v_b (due to the motor rotation in the stator field), a leakage inductance L_a, and a resistance R_a. The current through the leakage inductor is i_L. The input signal is the armature voltage $v_a(t)$ as shown. The interaction of the rotor magnetic field and the stator magnetic field (*Note*: the rotor field rotates at an angular speed ω_m) generates a "magnetic" torque T_m which is exerted on the motor rotor.

The stator provides a constant magnetic field to the motor, and is not important in the present problem. The dc motor may be considered as an ideal electromechanical transducer that is represented by a linear-graph transformer. The associated equations are

$$\omega_m = \frac{1}{k_m} v_b$$

$$T_m = -k_m i_b$$

where k_m is the torque constant of the motor. *Note*: The negative sign in the second equation arises due to the specific sign convention used for a transformer, in the conventional linear graph representation.

The motor is connected to a rotatory load of moment of inertia J_l using a long flexible shaft of torsional stiffness k_l. The torque transmitted through this shaft is denoted by T_k. The load rotates at an angular speed ω_l and experiences mechanical dissipation, which is modeled by a linear viscous damper of damping constant b_l.

Answer the following questions:

(a) Draw a suitable linear graph for the entire system shown in Figure P3.27, mark the variables and parameters (you may introduce new, auxiliary variables but not new parameters), and orient the graph.

(b) Give the number of branches (b), nodes (n), and the independent loops (l) in the complete linear graph. What relationship do these three parameters satisfy? How many independent node equations, loop equations, and constitutive equations can be written for the system? Verify the sufficiency of these equations to solve the problem.

(c) Take current through the inductor (i_L), speed of rotation of the motor rotor (ω_r), torque transmitted through the load shaft (T_k), and speed of rotation of the load (ω_l) as the four state variables, the armature supply voltage $v_a(t)$ as the input variable, and the shaft torque T_k and the load speed ω_l as the output variables. Write the independent node equations, independent loop equations, and the constitutive equations for the complete linear graph. Clearly show the state-space shell.

(d) Eliminate the auxiliary variables and obtain a complete state-space model for the system, using the equations written in part (c) above. Express the matrices *A, B, C,* and *D* of the state–space model in terms of the system parameters R_a, L_a, k_m, J_r, k_l, b_l, and J_l only.

3.28 Consider the simplified model of a vehicle shown in Figure P3.28, which can be used to study the heave (vertical up and down) and pitch (front–back rotation) motions due to the road profile and other disturbances. For our purposes, let us assume that the road disturbances exciting the front and back suspensions are independent. The equations of motion for heave (y) and pitch (θ) are written about the static equilibrium configuration of the vehicle model (hence, gravity does not enter into the equations) for small motions:

$$m\ddot{y} = k_1(u_1 - y + l_1\theta) + k_2(u_2 - y + l_2\theta) + b_1(\dot{u}_1 - \dot{y} + l_1\dot{\theta}) + b_2(\dot{u}_2 - \dot{y} + l_2\dot{\theta})$$

$$J\ddot{\theta} = -l_1\left[k_1(u_1 - y + l_1\theta) + b_1(\dot{u}_1 - \dot{y} + l_1\dot{\theta})\right] + l_2\left[k_2(u_2 - y + l_2\theta) + b_2(\dot{u}_2 - \dot{y} + l_2\dot{\theta})\right]$$

Determine the transfer functions that relate the responses y and θ to the inputs u_1 and u_2.

3.29 Consider the single-degree-of-freedom systems shown in Figure P3.29. The system is represented by a point mass m, and the suspension system is modeled as a spring of stiffness k and a viscous damper of damping constant b. The model shown in Figure 3.29a is used to study force transmissibility.

Draw its impedance circuit.

The model shown in Figure P3.29b is used in determining the motion transmissibility. Draw its impedance (or, mobility) circuit.

Note: Mobility elements are suitable for motion transmissibility studies.

Show that the force transmissibility of system (a) is equal to the motion transmissibility of system (b).

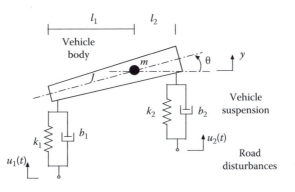

FIGURE P3.28
A model of a vehicle with its suspension system.

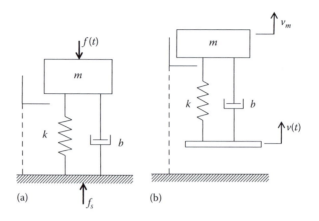

FIGURE P3.29
Single-degree-of-freedom systems: (a) Fixed on ground; (b) with support motion.

Derive an expression for this common transmissibility functions in terms of the given system parameters.

3.30 Consider the two-degree-of-freedom systems shown in Figure P3.30. The main system is represented by two masses linked through a spring and a damper. Mass m_1 is considered the critical mass (It is equally acceptable to consider mass m_2 as the critical mass).

The model shown in Figure 3.30a is used to study force transmissibility.
Draw its impedance circuit.

The model shown in Figure P3.30b is used in determining the motion transmissibility. Draw its impedance (or, mobility) circuit.

Show that the force transmissibility of system (a) is equal to the motion transmissibility of system (b).

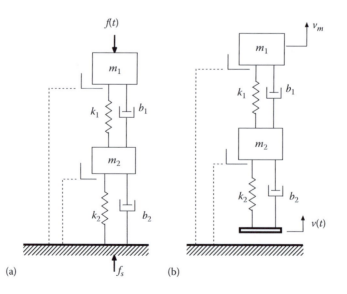

FIGURE P3.30
Systems with two degrees of freedom: (a) Fixed on ground; (b) with support motion.

3.31 A manufacturer of rubber parts uses a conventional process of steam-cured molding of latex. The molded rubber parts are first cooled and buffed (polished) and then sent for inspection and packing. A simple version of a rubber buffing machine is shown in Figure P3.31a. It consists of a large hexagonal drum whose inside surfaces are all coated with a layer of bonded emery. The drum is supported horizontally along its axis on two heavy-duty, self-aligning bearings at the two ends, and is rotated using a three-phase induction motor. The drive shaft of the drum is connected to the motor shaft through a flexible coupling, in order to compensate for possible misalignments of the axes. The buffing process consists of filling the drum with rubber parts, steadily rotating the drum for a specified period of time, and finally vacuum cleaning the drum and its contents. Dynamics of the machine affects the mechanical loading on various parts of the system such as the motor, coupling, bearings, shafts, and the support structure.

In order to study the dynamic behavior, particularly at the startup stage and under disturbances during steady-state operation, an engineer develops a simplified model of the buffing machine. This model is shown in Figure P3.31b. The motor is modeled as a torque source T_m that is applied on the rotor having moment of inertia J_m and resisted by a viscous damping torque of damping constant b_m. The connecting shafts and the coupling unit are represented by a torsional spring of stiffness k_L. The drum and its contents are represented by an equivalent constant moment of inertia J_L. There is a resisting torque on the drum, even at steady operating speed, due to the eccentricity of the contents of the drum. This is represented by a constant torque T_r. Furthermore, energy dissipation due to the buffing action (between the rubber parts and the emery surfaces of the drum) is represented by a nonlinear damping torque T_{NL}, which may be approximated by $T_{NL} = c|\dot{\theta}_L|\dot{\theta}_L$ with $c > 0$.

Note that θ_m and θ_L are the angles of rotation of the motor rotor and the drum, respectively, and these are measured from inertial reference lines that correspond to a relaxed configuration of spring k_L.

(a) Comment on the assumptions made in the modeling process of this system and briefly discuss the validity (or accuracy) of the model.

(b) Show that the model equations are

$$J_m\ddot{\theta}_m = T_m - k_L(\theta_m - \theta_L) - b_m\dot{\theta}_m$$

$$J_L\ddot{\theta}_L = k_L(\theta_m - \theta_L) - c|\dot{\theta}_L|\dot{\theta}_L - T_r$$

What are the inputs of this system?

(c) Using the speeds $\dot{\theta}_m$ and $\dot{\theta}_L$, and the spring torque T_k as the state variables, and the twist of the spring as the output, obtain a complete state-space model for his nonlinear system.

What is the order of the state model?

(d) Suppose that under steady operating conditions, the motor torque is $\bar{T}_m$, which is constant. Determine an expression for the constant speed $\bar{\omega}$ of the drum in terms of $\bar{T}_m$, T_r and appropriate system parameters under these conditions. Show that, as intuitively clear, we must have $\bar{T}_m > T_r$, for this steady operation to be feasible. Also obtain an expression for the spring twist at steady state, in terms of $\bar{\omega}$, T_r and the system parameters.

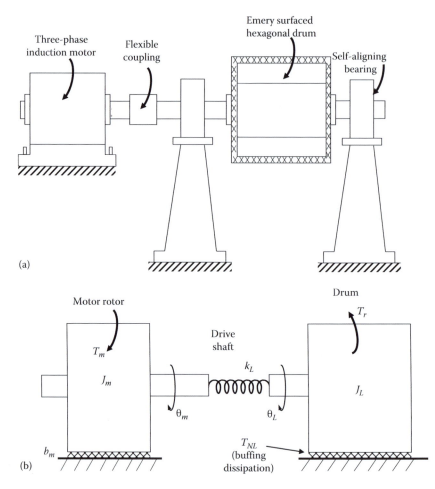

FIGURE P3.31
A rubber buffing machine: (a) Schematic diagram; (b) dynamic model.

(e) Linearize the system equations about the steady operation condition and express the two equations in terms of the following "incremental" variables:

q_1 = variation of θ_m about the steady value

q_2 = variation of θ_L about the steady value

u = disturbance increment of T_m from the steady value $\bar{T}_m$

(f) For the linearized system, obtain the input–output differential equation, first considering q_1 as the output and next considering q_2 as the output. Comment about and justify the nature of the homogeneous (characteristic-equation) parts of the two equations. Discuss, by examining the physical nature of the system, why only the derivatives of q_1 and q_2 and not the variables themselves are present in these input–output equations.

Explain why the derivation of the input–output differential equations will become considerably more difficult if a damper is present between the two inertia elements J_m and J_L.

(g) Consider the input–output differential equation for q_1. By introducing an auxiliary variable, draw a simulation block diagram for this system. (Use integrators, summers, and coefficient blocks only.) Show how this block diagram can be easily modified to represent the following cases:

    (i) q_2 is the output

    (ii) $\dot{q}_1$ is the output

    (iii) $\dot{q}_2$ is the output

    What is the order of the system (or the number of free integrators needed) in each of the four cases of output considered in this example?

(h) Considering the spring twist $(q_1 - q_2)$ as the output, draw a simulation block diagram for the system. What is the order of the system in this case?

    Hint: For this purpose you may use the two linearized second order differential equations obtained in part (e).

(i) Comment on why the "system order" is not the same for the five cases of output considered in parts (g) and (h).

3.32 Consider again the ground-based mechanical oscillator as shown in Figure P3.29a and analyzed in Problem 3.29. Repeat the analysis, this time using the concepts of Thevenin or Norton equivalent circuits and linear graph reduction in the frequency domain, and determine an expression for force transmissibility.

3.33 Consider again the oscillator with support motion as shown in Figure P3.29b and analyzed in Problem 3.29. Repeat the analysis, this time using the concepts of Thevenin or Norton equivalent circuits and linear graph reduction in the frequency domain, and determine an expression for motion transmissibility.

3.34 Consider again the ground-based two-degree-of-freedom oscillator, as shown in Figure P3.30a and analyzed in Problem 3.30. Repeat the analysis, this time using the concepts of Thevenin or Norton equivalent circuits and linear graph reduction in the frequency domain, and determine an expression for force transmissibility.

3.35 Consider again the two-degree-of-freedom oscillator with support motion as shown in Figure P3.30b and analyzed in Problem 3.30. Repeat the analysis, this time using the concepts of Thevenin or Norton equivalent circuits and linear graph reduction in the frequency domain, and determine an expression for motion transmissibility.

3.36 Consider the control system shown in Figure P3.36.

    The back e.m.f. $v_B = K_V \omega$

    The motor torque $T_m = K_T i$

    Draw a simulation block diagram for the system.

3.37 It is required to study the dynamics behavior of an automobile during the very brief period of a sudden start from rest. Specifically, the vehicle acceleration a in the direction of primary motion, as shown in Figure P3.37a, is of interest and should be considered as the system output. The equivalent force $f(t)$ of the engine, applied in the direction of primary motion, is considered as the system input. A simple dynamic model that may be used for the study is shown in Figure P3.37b.

Note: k is the equivalent stiffness, primarily due to tire flexibility, and b is the equivalent viscous damping constant, primarily due to dissipations at the tires and other moving parts of the vehicle, taken in the direction of a. Also, m is the mass of the vehicle.

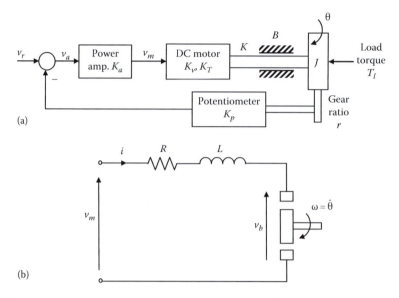

(a)

(b)

FIGURE P3.36
(a) A rotatory electromechanical system; (b) the armature circuit.

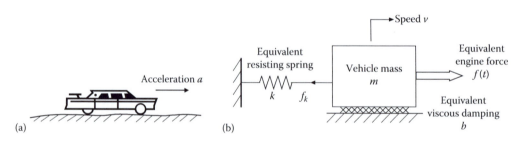

(a)   (b)

FIGURE P3.37
(a) Vehicle suddenly accelerating from rest; (b) a simplified model of the accelerating vehicle.

(a) Discuss advantages and limitations of the proposed model for the particular purpose.

(b) Using force f_k of the spring (stiffness k) and velocity v of the vehicle as the state variables, engine force $f(t)$ as the input and the vehicle acceleration a as the output, develop a complete state-space model for the system.

(*Note*: You must derive the matrices A, B, C, and D for the model).

(c) Draw a simulation block diagram for the model, employing integration and gain blocks, and summation junctions only.

(d) Obtain the input/output differential equation of the system. From this, derive the transfer function (a/f in the Laplace domain).

(e) Discuss the characteristics of this model by observing the nature of matrix D, feed-forwardness of the block diagram, input and output orders of the I/O differential equation, and the numerator and denominator orders of the system transfer function.

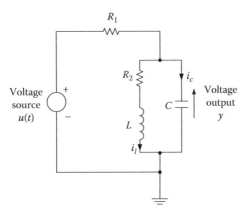

FIGURE P3.38
An *RLC* circuit driven by a voltage source.

3.38 The electrical circuit shown in Figure P3.38 has two resistor R_1 and R_2, an inductor L, a capacitor C, and a voltage source $u(t)$. The voltage across the capacitor is considered the output y of the circuit.

(a) What is the order of the system and why?

(b) Show that the input–output equation of the circuit is given by

$$a_2 \frac{d^2 y}{dt^2} + a_1 \frac{dy}{dt} + a_0 y = b_1 \frac{du}{dt} + b_0 u$$

Express the coefficients a_0, a_1, a_2, b_0 and b_1 in terms of the circuit parameters R_1, R_2, L, and C.

(c) Starting with the auxiliary differential equation:

$$a_2 \ddot{x} + a_1 \dot{x} + a_0 x = u$$

and using $x = [x \ \dot{x}]^T$ as the state vector, obtain a complete state-space model for the system in Figure P3.38.

(d) Clearly explain why, for the system in Figure P3.38, neither the current i_c through the capacitor nor the time derivative of the output ($\dot{y}$) can be chosen as a state variable.

3.39 The movable arm with read/write head of a disk drive unit is modeled as a simple oscillator. The unit has an equivalent bending stiffness $k = 10$ dyne·cm/rad and damping constant b. An equivalent rotation $u(t)$ radians is imparted at the read/write head. This in turn produces a (bending) moment to the read/write arm, which has an equivalent moment of inertia $J = 1 \times 10^{-3}$ gm·cm², and bends the unit at an equivalent angle θ about the centroid.

(a) Write the input–output differential equation of motion for the read/write arm unit.

(b) What is the undamped natural frequency of the unit in rad/s?

(c) Determine the value of b for 5% critical damping.

(d) Write the frequency transfer function of the model.

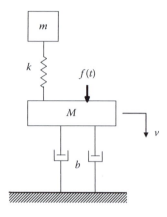

3.40 A rotating machine of mass M is placed on a rigid concrete floor. There is an isolation pad made of elastomeric material between the machine and the floor, and is modeled as a viscous damper of damping constant b. In steady operation there is a predominant harmonic force component $f(t)$, which is acting on the machine in the vertical direction at a frequency equal to the speed of rotation (n rev/s) of the machine. To control the vibrations produced by this force, a dynamic absorber of mass m and stiffness k is mounted on the machine. A model of the system is shown in Figure P3.40.

FIGURE P3.40
A mounted machine with a dynamic absorber.

(a) Determine the frequency transfer function of the system, with force $f(t)$ as the input and the vertical velocity v of mass M as the output.

(b) What is the mass of the dynamic absorber that should be used in order to virtually eliminate the machine vibration (a tuned absorber)?

3.41 The frequency transfer function for a simple oscillator is given by

$$G(\omega) = \frac{\omega_n^2}{[\omega_n^2 - \omega^2 + 2j\zeta\omega_n\omega]}$$

(a) If a harmonic excitation $u(t) = a\cos\omega_n t$ is applied to this system what is the steady-state response?

(b) What is the magnitude of the resonant peak?

(c) Using your answers to parts (a) and (b) suggest a method to measure damping in a mechanical system.

(d) At what excitation frequency is the response amplitude maximum under steady state conditions?

(e) Determine an approximate expression for the half-power (3 dB) bandwidth at low damping. Using this result, suggest an alternative method for the damping measurement.

3.42 (a) An approximate frequency transfer function of a system was determined by Fourier analysis of measured excitation–response data and fitting into an appropriate analytical expression (by curve fitting using the least squares method). This was found to be $G(f) = 5/(10 + j2\pi f)$.
What is its magnitude, phase angle, real part, and imaginary part at $f = 2\,\text{Hz}$? If the reference frequency is taken as $1\,\text{Hz}$, what is the transfer function magnitude at $2\,\text{Hz}$ expressed in dB?

(b) A dynamic test on a structure using a portable shaker revealed the following: The accelerance between two locations (shaker location and accelerometer

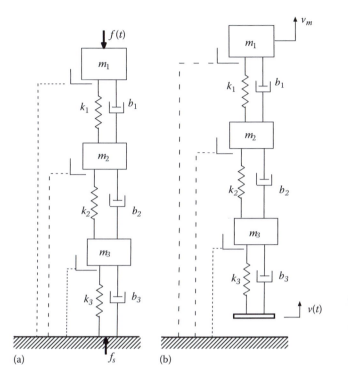

FIGURE P3.43
Two mechanical systems: (a) For determining force transmissibility; (b) for determining motion transmissibility.

location) measured at a frequency ratio of 10 was 35 dB. Determine the corresponding mobility and mechanical impedance at this frequency ratio.

3.43 Figure P3.43 shows two systems (a) and (b), which may be used to study force transmissibility and motion transmissibility, respectively. Clearly discuss whether the force transmissibility F_s/F (in the Laplace domain) in system (a) is equal to the motion transmissibility V_m/V (in the Laplace domain) in system (b), by carrying out the following steps:

1. Draw the linear graphs for the two systems and mark the mobility functions for all the branches (except the source elements).

2. Simplify the two linear graphs by combining branches as appropriate (series branches: add mobilities; parallel branches; inverse rule applies for mobilities) and mark the mobilities of the combined branches.

3. Based on the objectives of the problem, i.e., determination of the force transmissibility of system (a), and motion transmissibility of system (b), for applying Thevenin's theorem, determine which part of the circuit (linear graph) should be cut (*Note*: The variable of interest in the particular transmissibility function should be associated with the part of the circuit that is cut).

4. Based on the objectives of the problem, establish whether Thevenin equivalence or Norton equivalence is needed (specifically: use Thevenin equivalence if a through variable needs to be determined, because this gives two series

elements with a common through variable. Use Norton equivalence if an across variable needs to be determined, because this gives two parallel elements with a common across variable).

5. Determine the equivalent sources and mobilities of the equivalent circuits of the two systems.

6. Using the two equivalent circuits determine the transmissibility functions of interest.

7. By analysis, examine whether the two mobility functions obtained in this manner are equivalent.

Note: Neglect the effects of gravity (i.e., assume that the systems are horizontal, supported on frictionless rollers).

Bonus: Extend your results to an n-degree-of-freedom system (i.e., one with n mass elements), structured as in Figure P3.43a and b.

3.44 The unit step response of a system, with zero initial conditions, was found to be $1.5(1 - e^{-10t})$. What is the input–output differential equation of the system? What is the transfer function?

3.45 Consider the first order system (model)

$$\tau \dot{y} + y = ku \tag{i}$$

Suppose that the unit step response of a first order system with zero ICs, was found to be (say, by curve fitting of experimental data) $y_{step} = 2.25(1 - e^{-5.2t})$
Determine the system parameters: time constant τ and the gain parameter k.

Note: This is a model identification (experimental modeling) example.

3.46 A system at rest is subjected to a unit step input $\mathcal{U}(t)$. Its response is given by

$$y = 2e^{-t}(\cos t - \sin t)U(t).$$

(a) Write the input–output differential equation for the system.

(b) What is its transfer function?

(c) Determine the damped natural frequency, undamped natural frequency, and the damped ratio.

(d) Write the response of the system to a unit impulse and sketch it.

3.47 A system is given by the transfer function $\dfrac{y}{u} = \dfrac{\omega_n^2}{s^2 + 2\zeta\omega_n s + \omega_n^2}$.
where
   u is the input
   y is the output
   s is the Laplace variable
   ζ, ω_n are the system parameters

(a) Write the input–output differential equation of the system.

It is well known that the response of this system to a unit step input with zero initial conditions, $y(0^-) = 0$ and $\dot{y}(0^-) = 0$ is given by

$$y = 1 - \frac{1}{\sqrt{1-\zeta^2}} e^{-\zeta \omega_n t} \sin(\omega_d t + \phi) \quad \text{for } 0 \le \zeta < 1$$

where $\omega_d = \sqrt{1-\zeta^2}\,\omega_n$ and $\cos\phi = \zeta$

(b) Determine $y(0^+)$ and $\dot{y}(0^+)$ for this response.

Now consider the system given by the transfer function

$$\frac{y}{u} = \frac{\omega_n^2(\tau s + 1)}{(s^2 + 2\zeta\omega_n s + \omega_n^2)}$$

where τ is an additional system parameter. The remaining parameters are the same as those given for the previous system.

(c) Write the input–output differential equation for this modified system.

(d) Without using Laplace transform tables, but using the result given for the original system, determine the response of the modified system to a unit step input with zero initial conditions: $y(0^-)=0$ and $\dot{y}(0^-)=0$.

The response must be expressed in terms of the given system parameters (ω_n, ζ, τ).

(e) Determine $y(0^+)$ and $\dot{y}(0^+)$ for this response. Comment on your result, if it is different from the values for $y(0^-)=0$ and $\dot{y}(0^-)=0$.

3.48 An "iron butcher" is a head-cutting machine that is commonly used in the fish-processing industry. Millions of dollars worth salmon is wasted annually due to inaccurate head-cutting using these somewhat outdated machines. The main cause of wastage is the "over-feed problem." This occurs when a salmon is inaccurately positioned with respect to the cutter blade so that the cutting location is beyond the collarbone and into the body of a salmon. An effort has been made to correct this situation by sensing the position of the collarbone and automatically positioning the cutter blade accordingly.

A schematic representation of an electromechanical positioning system of a salmon-head cutter is shown in Figure P3.48a. Positioning of the cutter is achieved

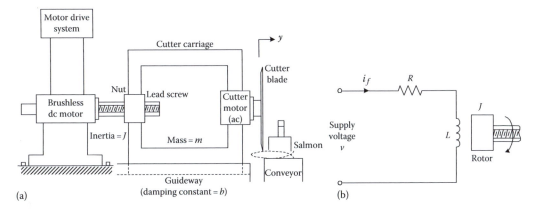

FIGURE P3.48
(a) A positioning system for an automated fish cutting machine; (b) the field circuit of the permanent-magnet rotor dc motor.

through a lead screw and nut arrangement, which is driven by a brushless dc motor. The cutter carriage is integral with the nut of the lead screw and the ac motor that derives the cutter blade, and has an overall mass of m (kg). The carriage slides along a lubricated guideway and provides an equivalent viscous damping force of damping constant b (N/m/s). The overall moment of inertia of the motor rotor and the lead screw is J (N·m²) about the axis of rotation. The motor is driven by a drive system, which provides a voltage v to the stator field windings of the motor. Note that the motor has a permanent magnet rotor. The interaction between the field circuit and, the motor rotor is represented by Figure P3.48b.

The magnetic torque T_m generated by the motor is given by

$$T_m = k_m i_f$$

The force F_L exerted by the lead screw in the y-direction of the cutter carriage is given by $F_L = (e/h)T_L$, in which

$$h = \frac{\text{Translatory motion of the nut}}{\text{Rotatory motion of lead screw}}$$

and e is the mechanical efficiency of the lead screw-nut unit.

The remaining parameters and variables, as indicated in Figure P3.48, should be self-explanatory.

(a) Write the necessary equations to study the displacement y of the cutter in response to an applied voltage v to the motor. What is the order of the system? Obtain the input–output differential equation for the system and from that determine the characteristic equation. What are the roots (poles or eigenvalues) of the characteristic equation?

(b) Using summation junctions, integration blocks, and constant gain blocks only, draw a complete block diagram of the system, with v as the input and y as the output.

(c) Obtain a state-space model for the system, using v as the input and y as the output.

(d) Assume that L/R ratio is very small and can be neglected. Obtain an expression for the response y of the system to a step input with zero initial conditions. Show from this expression that the behavior of the system is unstable in the present form (i.e., without feedback control).

3.49 The circuit shown in Figure P3.49 consists of an inductor L, a capacitor C, and two resistors R and R_o. The input is the source voltage $v_s(t)$ and the output is the voltage v_o across the resistor R_o.

(a) Explain why the current iL through the inductor and the voltage vC across the capacitor are suitable state variables for this circuit.

(b) Using i_L and v_C as the state variables, obtain a complete state-space model for the system. Specifically, express system equations in the vector-matrix form:

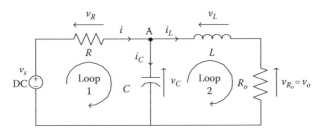

FIGURE P3.49
An electrical circuit with *R-L-C* elements.

$$\dot{x} = Ax + Bu$$

$$y = Cx + Du$$

in the usual notation, where x is the state vector, u is the input vector, and y is the output vector, and determine all the elements of the four matrices A, B, C, and D in terms of the circuit parameters R, R_o, L, and C.

(c) The system starts at steady state with a source voltage of 1 V (for all $t < 0$). Then suddenly, the source voltage (i.e., input) is increased to 10 V (for all $t > 0$), which corresponds to a step input. For $R = R_o = 1\,\Omega$, $L = 1\,H$, and $C = 1\,F$, determine the numerical values of the initial conditions of the following system variables at both $t = 0^-$ and $t = 0^+$:

(i) Voltage v_L across the inductor

(ii) Current i_C through the capacitor

(iii) Current i through the resistor R

(iv) Current i_L

(v) Voltage v_C

(vi) Output voltage v_o

Hint: A state variable cannot change its value instantaneously.

3.50 (a) Answer "true" or "false" for the following:
The order of a system is equal to

(i) The number of states in a state-space model of the system.

(ii) The order of the input–output differential equation of the system.

(iii) The number of initial conditions needed to completely determine the time response of the system.

(iv) The number of independent energy-storage elements in a lumped-parameter model of the system.

(v) The number of independent energy storage elements and energy dissipation elements in a lumped-parameter model of the system.

(b) A fluid pump has an impeller of moment of inertia J and is supported on frictionless bearings. It is driven by a powerful motor at speed ω_m, which may be treated as a velocity source, through a flexible shaft of torsional stiffness K. The fluid load to which the pump impeller is subjected to may be approximated by

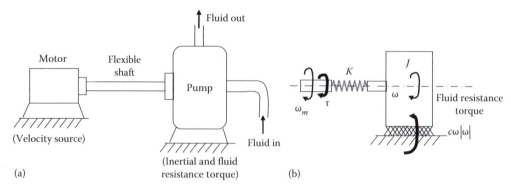

FIGURE P3.50

(a) A pump driven by a powerful motor; (b) a lumped-parameter model.

a load torque $c\omega|\omega|$, where ω is the speed of the pump impeller. A schematic diagram of the system is shown in Figure P3.50a and a lumped-parameter model is shown in Figure P3.50b.

Note that the motor speed ω_m is the input to the system. Treat the speed ω of the pump impeller as the output of the system.

(i) Using the torque τ in the drive shaft and the speed ω of the pump as the state variables develop a complete (nonlinear) state-space model of the system.

(ii) What is the order of the system?

(iii) Under steady operating conditions, with constant input ω_m, (when the rates of changes of the state variables can be neglected) determine expressions for the operating speed ω_o of the pump and the operating torque τ_o of the drive shaft, in terms of the given quantities (e.g., ω_m, K, J, c).

(iv) Linearize the state-space model about the steady operating conditions in part (iii), using the incremental state variables $\hat{\tau}$ and $\hat{\omega}$, and the incremental input variable $\hat{\omega}_m$.

(v) From the linearized state-space model, obtain a linear input–output differential equation (in terms of the incremental input $\hat{\omega}_m$ and incremental output $\hat{\omega}$).

(vi) Obtain expressions for the undamped natural frequency and the damping ratio of the linearized system, in terms of the parameters ω_o, K, J, c.

3.51 Consider the simple oscillator: $\ddot{y} + 2\zeta\omega_n\dot{y} + \omega_n^2 y = \omega_n^2 u(t)$

For the under-damped case $(0 < \zeta < 1)$ determine the response to a unit step input, under general initial conditions $y(0)$ and $\dot{y}(0)$, using Laplace tables.

3.52 Consider the simple oscillator shown in Figure P3.52, with parameters $m = 4\,\text{kg}$, $k = 1.6 \times 10^3\,\text{N/m}$, and the two cases of damping:

1. $b = 80\,\text{N/m/s}$
2. $b = 320\,\text{N/m/s}$

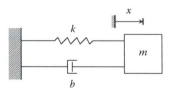

FIGURE P3.52

A damped simple oscillator.

Using MATLAB determine the free response in each case for an initial condition excitation.

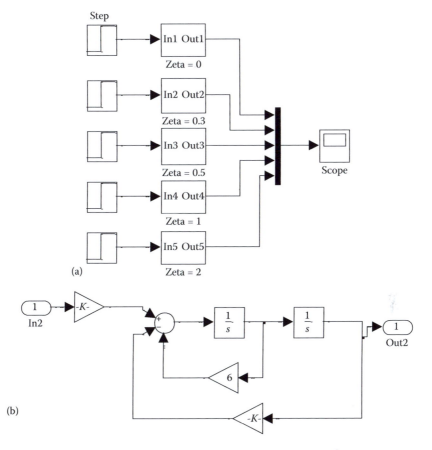

FIGURE P3.53
Use of Simulink to obtain the step response of a simple oscillator: (a) Overall Simulink model; (b) Simulink sub-model for each case of damping.

3.53 Consider the following equation of motion of the single-degree-of-freedom system (damped simple oscillator) shown in Figure P3.52:

$$\ddot{y} + 2\zeta\omega_n\dot{y} + \omega_n^2 y = \omega_n^2 u(t)$$

With an undamped natural frequency of $\omega_n = 10\,\text{rad/s}$, the step responses may be conveniently determined using Simulink for the following cases of damping ratio ζ: 0, 0.3, 0.5, 1.0, and 2.0.

In particular, the block diagram model for the simulation can be formed as shown in Figure P3.53a, where each case of damping is simulated using the sub-model in Figure P3.53b. Obtain the step response for these five cases of damping.

4

Component Interconnection and Signal Conditioning

Study Objectives

- Principles of component interconnection
- Input impedance, output impedance, and impedance matching
- Loading errors (electrical loading and mechanical loading)
- Signal conditioning/conversion
- Amplifiers, filters (analog and digital), modulation, and demodulation
- Ground loop noise and its elimination
- A/D conversion (ADC) and D/A conversion (DAC)
- Resistance bridge circuits and impedance bridge circuits

4.1 Introduction

A mechatronic system is typically a multi-domain (mixed) system, which consists of more than one type of component properly interconnected and integrated to perform the intended functions. In particular, mechanical, electrical, electronic, and computer hardware are integrated to form a mechatronic system. It follows that component interconnection is an important topic in the field of mechatronic engineering. When two components are interconnected, signals will flow through them. The nature and type of the signals that are present at the interface of two components will depend on the nature and type of the components. For example, when a motor is coupled with a load through a gear (transmission) unit, mechanical power flows at the interfaces of these components. Then, we are particularly interested in such signals as angular velocity and torque. In particular, these signals would be modified or "conditioned" as they are transmitted through the gear transmission. Similarly, when a motor is connected to its electronic drive system, command signals of motor control, typically available as voltages, would be converted into appropriate currents for energizing the motor windings so as to generate the necessary torque. Again, signal conditioning or conversion is important here. In general, then, signal conditioning is important in the context of component interconnection and integration and becomes an important subject in the study of mechatronic engineering.

This chapter addresses the interconnection of components such as sensors, signal conditioning circuitry, actuators, and power transmission devices in a mechatronic system. Desirable impedance characteristics for such components are discussed. Impedance and signal modification play crucial roles in component interconnection or interfacing. When two or more components are interconnected, the behavior of the individual components in the integrated system can deviate significantly from their behavior when each component operates independently. The matching of components in a multicomponent system, particularly with respect to their impedance characteristics, should be done carefully in order to improve the system performance and accuracy. In particular, when two devices are interfaced, it is essential to guarantee that a signal leaving one device and entering the other will do so at proper signal levels (the values of voltage, current, speed, force, power, etc.) in the proper form (electrical, mechanical, analog, digital, modulated, demodulated, etc.) and without distortion (where loading problems, nonlinearities, and noise have to be eliminated and where impedance considerations become important). Particularly for transmission, a signal should be properly modified (by amplification, modulation, digitizing, etc.) so that the signal/noise ratio of the transmitted signal is sufficiently large at the receiver. The significance of signal modification is clear from these observations.

The tasks of signal-modification may include *signal conditioning* (e.g., amplification and analog and digital filtering), *signal conversion* (e.g., analog-to-digital conversion, digital-to-analog conversion, voltage-to-frequency conversion, and frequency-to-voltage conversion), *modulation* (e.g., amplitude modulation, frequency modulation, phase modulation, pulse-width modulation (PWM), pulse-frequency modulation, and pulse-code modulation), and *demodulation* (the reverse process of modulation). In addition, many other types of useful signal modification operations can be identified. For example, *sample and hold circuits* are used in digital data acquisition systems. Devices such as *analog and digital multiplexers* and *comparators* are needed in many applications of data acquisition and processing. Phase shifting, curve shaping, offsetting, and linearization can also be classified as signal modification.

This chapter describes signal conditioning and modification operations that are useful in mechatronic systems and applications. First, the basic concepts of impedance and component matching are studied. The operational amplifier (op-amp) is introduced as a basic element in signal conditioning and impedance matching circuitry for electronic systems. Various types of signal conditioning and modification devices such as amplifiers, filters, modulators, demodulators, bridge circuits, analog-to-digital converters, and digital-to-analog converters are discussed. The concepts presented here are applicable to many types of components in a general mechatronic system. Discussions and developments given here can be quite general. Nevertheless, specific hardware components and designs are considered particularly in relation to component interfacing and signal conditioning.

4.2 Impedance Characteristics

When components such as sensors and transducers, control boards, process (plant) equipment, and signal-conditioning hardware are interconnected, it is necessary to *match* impedances properly at each interface in order to realize their rated performance level. One adverse effect of improper impedance matching is the *loading effect*. For example, in a measuring system, the measuring instrument can distort the signal that is being measured. The resulting error can far exceed other types of measurement error. Both electrical and mechanical loading are possible. Electrical loading errors result from connecting an output unit such as a measuring device that has a low input impedance to an input device such as a signal source. Mechanical loading errors can result in an input device due to inertia, friction, and other resistive forces generated by an interconnected output component.

Impedance can be interpreted either in the traditional electrical sense or in the mechanical sense, depending on the type of signals that are involved. For example, a heavy accelerometer mounted on a monitored object can introduce an additional dynamic load on the object, which will modify the actual acceleration at the monitored location. This is *mechanical loading*. A voltmeter can modify the currents (and voltages) in a circuit. This is *electrical loading*. Analogously, a thermocouple junction can modify the temperature that is being measured as a result of the heat transfer into the junction. In mechanical and electrical systems, loading errors can appear as phase distortions as well. Digital hardware also can produce loading errors. For example, an analog-to-digital conversion (ADC) board can load the amplifier output from a strain gage bridge circuit, thereby affecting the digitized data.

Another adverse effect of improper impedance consideration is inadequate output signal levels, which make the output functions such as signal processing and transmission, component driving, and actuation of a final control element or plant very difficult. In the context of sensor-transducer technology, many types of transducers (e.g., piezoelectric accelerometers, impedance heads, and microphones) have high output impedances on the order of a thousand megohms ($1\,M\Omega = 1 \times 10^6\,\Omega$). These devices generate low output signals and they would require conditioning to step up the signal level. *Impedance-matching amplifiers*, which have high input impedances and low output impedances (a few ohms), are used for this purpose (e.g., charge amplifiers are used in conjunction with piezoelectric sensors). A device with a high input impedance has the further advantage that it usually consumes less power (v^2/R is low) for a given input voltage. The fact that a low input impedance device extracts a high level of power from the preceding output device may be interpreted as the reason for a loading error.

4.2.1 Cascade Connection of Devices

Consider a standard two-port electrical device. Some definitions are in order.

4.2.1.1 Output Impedance

The *output impedance* Z_o is defined as the ratio of the open-circuit (i.e., no-load) voltage at the output port to the short-circuit current at the output port. The open-circuit voltage at the output port is the output voltage present when there is no current flowing at the output port. This is the case if the output port is not connected to a load (impedance). As soon as a load is connected at the output of the device, a current will flow through it, and the output voltage will drop to a value less than that of the open-circuit voltage. To measure the open-circuit voltage, the rated input voltage is applied at the input port and maintained at a constant, and the output voltage is measured using a voltmeter that has a very high (input) impedance. To measure the short-circuit current, a very low-impedance ammeter is connected at the output port.

4.2.1.2 Input Impedance

The *input impedance* Z_i is defined as the ratio of the rated input voltage to the corresponding current through the input terminals while the output terminals are maintained as an open circuit.

These definitions are given for electrical devices. As discussed in Chapter 3, generalization is possible by interpreting voltage and velocity as *across variables* and current and force as *through variables*. Then, mechanical *mobility* should be used in place of electrical impedance in the associated analysis.

4.2.1.3 Cascade Connection

According to the given definitions, input impedance Z_i and output impedance Z_o can be represented schematically as in Figure 4.1a. Here, v_o is the open-circuit output voltage. When a load is connected at the output port, the voltage across the load will be different from v_o. This is caused by the presence of a current through Z_o. In the frequency domain, v_i and v_o are represented by their respective *Fourier spectra*. The corresponding transfer relation can be expressed in terms of the complex frequency response (transfer) function $G(j\omega)$ under open-circuit (no-load) conditions:

$$v_o = Gv_i \tag{4.1}$$

Now consider two devices connected in cascade, as shown in Figure 4.1b. It can be easily verified that the following relations apply:

$$v_{o1} = G_1 v_i \tag{4.2}$$

$$v_{i2} = \frac{Z_{i2}}{Z_{o1} + Z_{i2}} v_{o1} \tag{4.3}$$

$$v_o = G_2 v_{i2} \tag{4.4}$$

These relations can be combined to give the overall input/output relation

$$v_o = \frac{Z_{i2}}{Z_{o1} + Z_{i2}} G_2 G_1 v_i \tag{4.5}$$

We see from Equation 4.5 that the overall frequency transfer function differs from the ideally expected product $(G_2 G_1)$ by the factor

$$\frac{Z_{i2}}{Z_{o1} + Z_{i2}} = \frac{1}{Z_{o1}/Z_{i2} + 1} \tag{4.6}$$

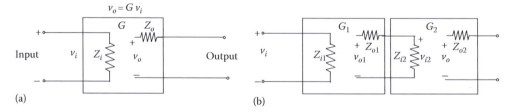

(a)   (b)

FIGURE 4.1
(a) Schematic representation of input impedance and output impedance. (b) Cascade connection of two two-port devices.

Note that cascading has "distorted" the frequency response characteristics of the two devices. If $Z_{o1}/Z_{i2} \ll 1$, this deviation becomes insignificant. From this observation, it can be concluded that when frequency response characteristics (i.e., dynamic characteristics) are important in a cascaded device, cascading should be done so that the output impedance of the first device is much smaller than the input impedance of the second device.

Example 4.1

A lag network used as the compensatory element of a mechatronic system is shown in Figure 4.2a. Show that its transfer function is given by $v_o/v_i = Z_2/(R_1 + Z_2)$, where $Z_2 = R_2 + (1/C_s)$. What is the input impedance and what is the output impedance for this circuit? Also, if two such lag circuits are cascaded as shown in Figure 4.2b, what is the overall transfer function? How would you make this transfer function become close to the ideal result $\{Z_2/(R_1 + Z_2)\}^2$?

Solution
To solve this problem, first note that in Figure 4.2a, the voltage drop across the element $R_2 + 1/(C_s)$

is $v_o = \left(R_2 + \dfrac{1}{C_s} \right) \Big/ \left\{ R_1 + R_2 + \dfrac{1}{C_s} \right\} v_i.$

Hence, $v_o/v_i = Z_2/(R_1 + R_2)$. Now, the input impedance Z_i is derived by using the input current $i = v_i/(R_1 + Z_2)$ as $Z_i = v_i/i = R_1 + Z_2$, and the output impedance Z_o is derived by using the short-circuit current $i_{sc} = v_i/R_1$ as

$$Z_o = \frac{v_o}{i_{sc}} = \frac{Z_2/(R_1 + Z_2)v_i}{v_i/R_1} = \frac{R_1 Z_2}{R_1 + Z_2} \tag{i}$$

Next, consider the equivalent circuit shown in Figure 4.2c. Since Z is formed by connecting Z_2 and $(R_1 + Z_2)$ in parallel, we have

$$\frac{1}{Z} = \frac{1}{Z_2} + \frac{1}{R_1 + Z_2} \tag{ii}$$

The voltage drop across Z is

$$v_o' = \frac{Z}{R_1 + Z} v_i \tag{iii}$$

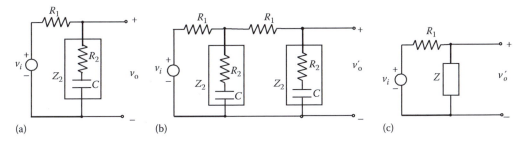

FIGURE 4.2
(a) A single circuit module. (b) Cascade connection of two modules. (c) An equivalent circuit for (b).

Now apply the single-circuit module result (i) to the second circuit stage in Figure 4.2b; thus, $v_o = (Z_2/(R_1 + Z_2))v_o'$. By substituting Equation (iii), we get $v_o = \dfrac{Z_2}{(R_1 + R_2)} \dfrac{Z}{(R_1 + Z)} v_i$. The overall transfer function for the cascaded circuit is

$$G = \frac{v_o}{v_i} = \frac{Z_2}{(R_1 + Z_2)} \frac{Z}{(R_1 + Z)} = \frac{Z_2}{(R_1 + R_2)} \frac{1}{(R_1/Z + 1)}$$

Now, by substituting Equation (ii), we get

$$G = \left[\frac{Z_2}{R_1 + Z_2}\right]^2 \frac{1}{1 + R_1 Z_2/(R_1 + Z_2)^2}$$

We observe that the ideal transfer function is approached by making $R_1 Z_2/(R_1 + Z_2)^2$ small compared with unity.

4.2.2 Impedance Matching

When two electrical components are interconnected, current (and energy) will flow between the two components. This will change the original (unconnected) conditions. This is known as the (electrical) loading effect and it has to be minimized. At the same time, adequate power and current would be needed for signal communication, conditioning, display, etc. Both situations can be accommodated through proper matching of impedances when the two components are connected. Usually, an impedance matching amplifier (impedance transformer) would be needed between the two components.

From the analysis given in the preceding section, it is clear that the signal-conditioning circuitry should have a considerably large input impedance in comparison with the output impedance of the sensor-transducer unit in order to reduce loading errors. The problem is quite serious in measuring devices such as piezoelectric sensors, which have very high output impedances. In such cases, the input impedance of the signal-conditioning unit might be inadequate to reduce loading effects; also, the output signal level of these high-impedance sensors is quite low for signal transmission, processing, actuation, and control. The solution for this problem is to introduce several stages of amplifier circuitry between the output of the first hardware unit (e.g., sensor) and the input of the second hardware unit (e.g., data acquisition unit). The first stage of such an interfacing device is typically an *impedance-matching amplifier* that has very high input impedance, very low output impedance, and almost unity gain. The last stage is typically a stable high-gain amplifier stage to step up the signal level. Impedance-matching amplifiers are, in fact, *op-amps* with feedback.

When connecting a device to a signal source, loading problems can be reduced by making sure that the device has a high input impedance. Unfortunately, this will also reduce the level (amplitude, power) of the signal received by the device. In fact, a high-impedance device may reflect back some harmonics of the source signal. A *termination resistance* that matches the output impedance of the source (e.g., 50 Ω) may be connected in parallel with the device in order to reduce this problem.

In many data acquisition systems, the output impedance of the output amplifier is made equal to the transmission line impedance. When maximum power amplification is desired, *conjugate matching* is recommended. In this case, the input impedance and output impedance of the matching amplifier are made equal to the complex conjugates of the source impedance and the load impedance, respectively.

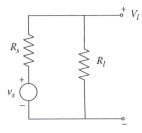

FIGURE 4.3
A load powered by a dc power supply.

Example 4.2

Consider a dc power supply of voltage v_s and an output impedance (resistance) of R_s. It is used to power a load of resistance R_l, as shown in Figure 4.3. What should be the relationship between R_s and R_l if the objective is to maximize the power absorbed by the load?

Solution
Current through the circuit is $i_l = v_s/(R_l + R_s)$.
   Accordingly, the voltage across the load is $v_l = i_l R_l = v_s R_l/(R_l + R_s)$
   The power absorbed by the load is

$$p_l = i_l v_l = \frac{v_s^2 R_l}{[R_l + R_s]^2} \qquad \text{(i)}$$

For maximum power, we need

$$\frac{dp_l}{dR_l} = 0 \qquad \text{(ii)}$$

Differentiate the right-hand side expression of (i) with respect to R_l in order to satisfy (ii). This gives the requirement for the maximum power as

$$R_l = R_S$$

4.2.3 Impedance Matching in Mechanical Systems

The concepts of impedance matching can be extended to mechanical systems (and to multi-domain systems) in a straightforward manner. The procedure follows from the familiar electro-mechanical analogies (see Chapter 3). As a specific application, consider a mechanical load driven by a motor. Often, direct driving is not practical due to the limitations of the speed–torque characteristics of the available motors. By including a suitable gear transmission between the motor and the load, it is possible to modify the speed-torque characteristics of the drive system as felt by the load. This is a process of impedance matching.

Example 4.3

Consider the mechanical system where a torque source (motor) of torque T and moment of inertia J_m is used to drive a purely inertial load of moment of inertia J_l as shown in Figure 4.4a. What is the resulting angular acceleration $\ddot{\theta}$ of the system? Neglect the flexibility of the connecting shaft.
   Now suppose that the load is connected to the same torque source through an ideal (loss free) gear of motor-to-load speed ratio r:1, as shown in Figure 4.4b. What is the resulting acceleration $\ddot{\theta}_g$ of the load?
   Obtain an expression for the normalized load acceleration $a = \ddot{\theta}_g/\ddot{\theta}$ in terms of r and $p = J_l/J_m$. Sketch a versus r for $p = 0.1$, 1.0, and 10.0. Determine the value of r in terms of p that will maximize the load acceleration a.
   Comment on the results obtained in this problem.

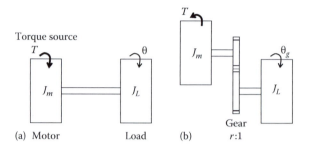

FIGURE 4.4
An inertial load driven by a motor. (a) Without gear transmission. (b) With a gear transmission.

Solution

For the unit without the gear transmission: Newton's second law gives $(J_m+J_L)\ddot{\theta}= T$. Hence,

$$\ddot{\theta} = \frac{T}{J_m + J_L} \tag{i}$$

For the unit with the gear transmission: see the free-body diagram shown in Figure 4.5 in the case of a loss-free (i.e., 100% efficient) gear transmission.

Newton's second law gives

$$J_m r\ddot{\theta}_g = T - \frac{T_g}{r} \tag{ii}$$

and

$$J_L\ddot{\theta}_g = T_g \tag{iii}$$

where T_g is the gear torque on the load inertia. By eliminating T_g in (ii) and (iii), we get

$$\ddot{\theta}_g = \frac{rT}{(r^2 J_m + J_L)} \tag{iv}$$

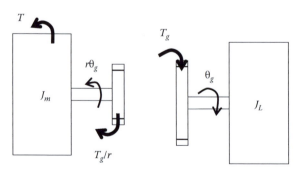

FIGURE 4.5
Free-body diagram.

Divide (iv) by (i)

$$\frac{\ddot{\theta}_g}{\ddot{\theta}} = a = \frac{r(J_m + J_L)}{(r^2 J_m + J_L)} = \frac{r(1 + J_L/J_m)}{(r^2 + J_L/J_m)}$$

or

$$a = \frac{r(1+p)}{(r^2 + p)} \tag{v}$$

where $p = J_L/J_m$.

From (v) note that for $r = 0$, $a = 0$ and for $r \to \infty$, $a \to 0$. The peak value of a is obtained through differentiation $\dfrac{\partial a}{\partial r} = \dfrac{(1+p)[(r^2 + p) - r \times 2r]}{(r^2 + p)^2} = 0$.

By taking the positive root, we get

$$r_p = \sqrt{p} \tag{vi}$$

where r_p is the value of r corresponding to peak a. The peak value of a is obtained by substituting (vi) in (v):

$$a_p = \frac{1+p}{2\sqrt{p}} \tag{vii}$$

Also, note from (v) that when $r = 1$ we have $a = r = 1$. Hence, all curves (v) should pass through the point (1, 1).

The relation (v) is sketched in Figure 4.6 for $p = 0.1$, 1.0, and 10.0. The peak values are tabulated below.

P	r_p	a_p
0.1	0.316	1.74
1.0	1.0	1.0
10.0	3.16	1.74

Note from Figure 4.6 that the transmission speed ratio can be chosen, depending on the inertia ratio, to maximize the load acceleration. In particular, we can state the following:

1. When $J_L = J_m$, pick a direct-drive system (no gear transmission; i.e., $r = 1$).
2. When $J_L < J_m$, pick a speed-up gear at the peak value of r $(= J_L/J_m)$.
3. When $J_L > J_m$, pick a speed-down gear at the peak value of r.

4.3 Amplifiers

An amplifier adjusts the signal level (voltage, current, power, etc.) in a device. Analogous across variables, through variables, and power variables can be defined for nonelectrical signals (e.g., mechanical) as well. Signal levels at various interface locations of components

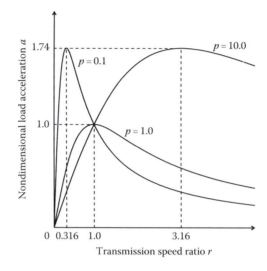

FIGURE 4.6
Normalized acceleration versus speed ratio.

in a mechatronic system have to be properly adjusted for the satisfactory performance of these components and of the overall system. For example, input to an actuator should possess adequate power to drive the actuator. A signal should maintain its signal level above some threshold during transmission so that errors due to signal weakening would not be excessive. Signals applied to digital devices must remain within the specified logic levels. Many types of sensors produce weak signals that have to be upgraded before they can be fed into a monitoring system, data processor, controller, or data logger.

An amplifier is an active device that needs an external power source to operate. Even though various active circuits, amplifiers in particular, are commonly produced in the monolithic form using an original integrated circuit (IC) layout so as to accomplish a particular amplification task, it is convenient to study their performance using discrete circuit models. In fact, high-power amplifiers may be built using discrete components (transistors, resistors, etc.) rather than ICs.

An *op-amp*, which is available as a monolithic IC package, is the basic building block of an amplifier. For this reason, our discussion on amplifiers will evolve from the op-amp.

4.3.1 Operational Amplifier

An *op-amp* got its name due to the fact that originally it was used almost exclusively to perform mathematical operations; for example, in analog computers. In the 1950s, the transistorized op-amp was developed, which used discrete elements such as *bipolar junction transistors* and *resistors*. Still it was too large in size, consumed too much power, and was too expensive for widespread use in general applications. This situation changed in the late 1960s when the IC op-amp was developed in the monolithic form, as a single IC chip. Today, the IC op-amp, which consists of a large number of circuit elements on a *substrate* of typically a single *silicon crystal* (the monolithic form), is a valuable component in almost any signal modification device. Bipolar complementary metal-oxide-semiconductor (CMOS) op-amps in various plastic packages and pin configurations are commonly available.

An op-amp could be manufactured in the discrete-element form using, say, 10 bipolar junction transistors and as many discrete resistors or alternatively (and preferably) in the modern monolithic form as an IC chip that may be equivalent to over 100 discrete elements. In any form, the device has an *input impedance Z_i*, an *output impedance Z_o*, and a gain K. Hence, a schematic model for an op-amp can be given as in Figure 4.7a. Op-amp packages are available in several forms. The 8-pin dual in-line package (DIP) or V package is very common, as shown in Figure 4.7b. The assignment of the pins (pin configuration or *pin-out*) is as shown in the figure, which should be compared with Figure 4.7a. Note the counterclockwise numbering sequence starting with the top left pin next to the semicircular notch (or dot). This convention of numbering is standard for any type of IC package, not just op-amp packages. Other packages include the 8-pin metal-can package or T package, which has a circular shape instead of the rectangular shape of the previous package, and the 14-pin rectangular "quad" package, which contains four op-amps (with a total of eight input pins, four output pins, and two power supply pins). The conventional symbol of an op-amp is shown in Figure 4.7c. Typically, there are five terminals (pins or lead connections) to an op-amp. Specifically, there are two input leads (a positive or noninverting lead with voltage v_{ip} and a negative or inverting lead with voltage v_{in}), an output lead (voltage v_o), and two bipolar power supply leads (+v_s, v_{CC}, or collector supply and −v_s, v_{EE}, or emitter supply). The typical supply voltage is ±15 V. Some of the pins may not be normally connected; for example, pins 1, 5, and 8 in Figure 4.7b.

Note from Figure 4.7a that under the open-loop (no feedback) conditions,

$$v_o = Kv_i \tag{4.7}$$

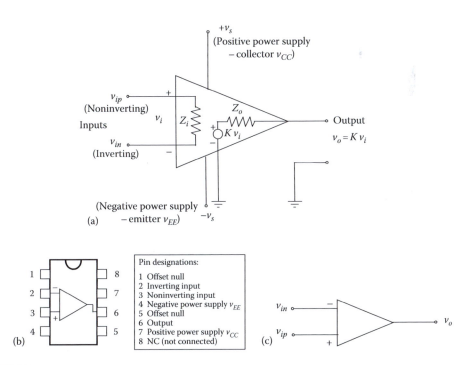

FIGURE 4.7
Operational amplifier. (a) A schematic model. (b) Eight-pin DIP. (c) Conventional circuit symbol.

in which the input voltage v_i is the differential input voltage defined as the algebraic difference between the voltages at the positive and negative lead; thus

$$v_i = v_{ip} - v_{in} \tag{4.8}$$

The open loop voltage gain K is very high (10^4–10^7) for a typical op-amp. Furthermore, the input impedance Z_i could be as high as $10\,\text{M}\Omega$ (typical is $2\,\text{M}\Omega$) and the output impedance is low, of the order of $10\,\Omega$ and may reach about $100\,\Omega$ for some op-amps. Since v_o is typically 1–15 V, from Equation 4.7, it follows that $v_i \cong 0$ since K is very large. Hence, from Equation 4.8 we have $v_{ip} \cong v_{in}$. In other words, the voltages at the two input leads are nearly equal. Now if we apply a large voltage differential v_i (say, 10 V) at the input, then according to Equation 4.7, the output voltage should be extremely high. This never happens in practice, however, since the device saturates quickly beyond moderate output voltages (on the order of 15 V).

From Equations 4.7 and 4.8, it is clear that if the negative input lead is grounded (i.e., $v_{in}=0$) then

$$v_o = K v_{ip} \tag{4.9}$$

and if the positive input lead is grounded (i.e., $v_{ip}=0$), then

$$v_o = -K v_{in} \tag{4.10}$$

This is the reason why v_{ip} is termed *noninverting input* and v_{in} is termed *inverting input*.

Example 4.4

Consider an op-amp having an open loop gain of 1×10^5. If the saturation voltage is 15 V, determine the output voltage in the following cases:

(a) $5\,\mu\text{V}$ at the positive lead and $2\,\mu\text{V}$ at the negative lead
(b) $-5\,\mu\text{V}$ at the positive lead and $2\,\mu\text{V}$ at the negative lead
(c) $5\,\mu\text{V}$ at the positive lead and $-2\,\mu\text{V}$ at the negative lead
(d) $-5\,\mu\text{V}$ at the positive lead and $-2\,\mu\text{V}$ at the negative lead
(e) 1 V at the positive lead and the negative lead is grounded
(f) 1 V at the negative lead and the positive lead is grounded

Solution
This problem can be solved using Equations 4.7 and 4.8. The results are given in Table 4.1. Note that in the last two cases the output will saturate and Equation 4.7 will no longer hold.

Field-effect transistors (FET), for example, metal-oxide-semiconductor field-effect transistors (MOSFET), are commonly used in the IC form of an op-amp (see Chapter 2). The MOSFET type has advantages over many other types; for example, higher input impedance and more stable output (almost equal to the power supply voltage) at saturation, making the MOSFET op-amps preferable over bipolar junction transistor op-amps in many applications.

In analyzing the op-amp circuits under unsaturated conditions, we use the following two characteristics of an op-amp:

1. The voltages of the two input leads should be (almost) equal.
2. The currents through each of the two input leads should be (almost) zero.

TABLE 4.1

Solution to Example 4.4

v_{ip}	v_{in}	v_i	v_o
$5\,\mu V$	$2\,\mu V$	$3\,\mu V$	$0.3\,V$
$-5\,\mu V$	$2\,\mu V$	$-7\,\mu V$	$-0.7\,V$
$5\,\mu V$	$-2\,\mu V$	$7\,\mu V$	$0.7\,V$
$-5\,\mu V$	$-2\,\mu V$	$-3\,\mu V$	$-0.3\,V$
$1\,V$	0	$1\,V$	$15\,V$
0	$1\,V$	$-1\,V$	$-15\,V$

As explained earlier, the first property is credited to high open-loop gain and the second property to high input impedance in an op-amp.

4.3.1.1 Use of Feedback in Op-Amps

An op-amp is a very versatile device, primarily due to its very high input impedance, low output impedance, and very high gain. But, it cannot be used without modification as an amplifier because it is not very stable in the form shown in Figure 4.7. The two main factors that contribute to this problem are frequency response and drift. Stated in another way, op-amp gain K does not remain constant; it can vary with the frequency of the input signal (i.e., the frequency response function is not flat in the operating range); and it can also vary with time (i.e., drift). The frequency response problem arises due to circuit dynamics of an op-amp. This problem is usually not severe unless the device is operated at very high frequencies. The drift problem arises due to the sensitivity of gain K to environmental factors such as temperature, light, humidity, and vibration and also as a result of the variation of K due to aging. Drift in an op-amp can be significant and steps should be taken to eliminate that problem.

It is virtually impossible to avoid the drift in gain and the frequency response error in an op-amp. But an ingenious way has been found to remove the effect of these two problems at the amplifier output. Since gain K is very large, by using feedback we can virtually eliminate its effect at the amplifier output. This *closed loop* form of an op-amp has the advantage that the characteristics and the accuracy of the output of the overall circuit depends on the passive components (e.g., resistors and capacitors) in it, which can be provided at high precision and not the parameters of the op-amp itself. The closed loop form is preferred in almost every application; in particular, the *voltage follower* and *charge amplifier* are devices that use the properties of high Z_i, low Z_o, and high K of an op-amp along with feedback through a high-precision resistor to eliminate errors due to nonconstant K. In summary, the op-amp is not very useful in its open-loop form, particularly because gain K is not steady. But since K is very large, the problem can be removed by using feedback. It is this closed-loop form that is commonly used in the practical applications of an op-amp.

In addition to the unsteady nature of gain, there are other sources of error that contribute to the less-than-ideal performance of an op-amp circuit. Noteworthy sources of error are as follows:

1. The *offset current* present at the input leads due to bias currents that are needed to operate the solid-state circuitry.
2. The *offset voltage* that might be present at the output even when the input leads are open.

3. The unequal gains corresponding to the two input leads (i.e., the *inverting gain* not equal to the *noninverting gain*).

Such problems can produce nonlinear behavior in op-amp circuits, and they can be reduced by proper circuit design and through the use of compensating circuit elements.

4.3.2 Voltage and Current Amplifiers

If an electronic amplifier performs a voltage amplification function, it is termed a *voltage amplifier*. Voltage amplifiers are used to achieve voltage compatibility (or level shifting) in circuits. Current amplifiers are used to achieve current compatibility in electronic circuits. A voltage follower has a unity voltage gain and hence it may be considered as a current amplifier. Besides, it provides impedance compatibility and acts as a buffer between a low-current (high-impedance) output device (signal source or the device that provides the signal) and a high-current (low-impedance) input device (signal receiver or the device that receives the signal) that are interconnected. Hence, the name *buffer amplifier* or *impedance transformer* is sometimes used for a current amplifier with unity voltage gain. If the objective of signal amplification is to upgrade the associated power level, then a *power amplifier* should be used for that purpose. These three types of amplification may be achieved simultaneously from the same amplifier. Furthermore, a current amplifier with unity voltage gain (e.g., a voltage follower) is a power amplifier as well. Usually, voltage amplifiers and current amplifiers are used in the first stages of a signal path (e.g., sensing, data acquisition, and signal generation) where signal levels and power levels are relatively low, while power amplifiers are typically used in the final stages (e.g., final control, actuation, recording, display) where high signal levels and power levels are usually required.

Figure 4.8a gives an op-amp circuit for a voltage amplifier. The feedback resistor R_f serves the purposes of stabilizing the op-amp and providing an accurate voltage gain. The negative lead is grounded through an accurately known resistor R. To determine the voltage gain, recall that the voltages at the two input leads of an op-amp should be equal (in the ideal case). The input voltage v_i is applied to the positive lead of the op-amp. Then, the voltage at point A should also be equal to v_i. Next, recall that the current through the input lead of an op-amp is ideally zero. Hence, by writing the current balance equation for the node point A, we have $(v_o - v_i)/R_f = v_i/R$. This gives the following amplifier equation:

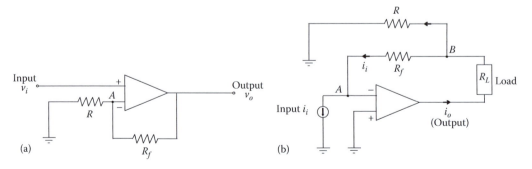

(a)

(b)

FIGURE 4.8
(a) A voltage amplifier. (b) A current amplifier.

$$v_o = \left(1 + \frac{R_f}{R}\right) v_i \qquad (4.11)$$

Here, the voltage gain is given by $K_v = 1 + (R_f/R)$.

Note: The voltage gain can be accurately determined by accurately selecting the "ratio" of the two passive resistor elements R and R_f. However, the magnitude of each resistor is also important. For example, the output short-circuit current of an amplifier is typically in the mA range, and hence R_f may have to be in the kΩ range rather than in the Ω range. Also, note from Equation 4.11 that the output voltage has the same sign as the input voltage. Hence, this is a *noninverting amplifier*. If the voltages are of the opposite sign, we have an *inverting amplifier*.

A current amplifier is shown in Figure 4.8b. The input current i_i is applied to the negative lead of the op-amp as shown, and the positive lead is grounded. There is a feedback resistor R_f that is connected to the negative lead through the load R_L. The resistor R_f provides a path for the input current since the op-amp takes in virtually zero current. There is a second resistor R through which the output is grounded. This resistor is needed for current amplification. To analyze the amplifier, use the fact that the voltage at point A (i.e., at the negative lead) should be zero because the positive lead of the op-amp is grounded (zero voltage). Furthermore, the entire input current i_i passes through the resistor R_f as shown. Hence, the voltage at point B is $R_f i_i$. Consequently, current through the resistor R is $R_f i_i/R$, which is positive in the direction shown. It follows that the output current i_o is given by $i_o = i_i + (R_f/R)i_i$, which is written as

$$i_o = \left(1 + \frac{R_f}{R}\right) i_i \qquad (4.12)$$

The current gain of the amplifier is $K_i = 1 + (R_f/R)$. As before, the amplifier gain can be accurately set using the ratio of the high-precision resistors R and R_f.

4.3.3 Instrumentation Amplifiers

An instrumentation amplifier is typically a special-purpose voltage amplifier dedicated to instrumentation applications. Examples include amplifiers used for producing the output from a bridge circuit (bridge amplifier) and amplifiers used with various sensors and transducers. An important characteristic of an instrumentation amplifier is the adjustable-gain capability. The gain value can be adjusted manually in most instrumentation amplifiers. In more sophisticated instrumentation amplifiers, the gain is *programmable* and can be set by means of digital logic. Instrumentation amplifiers are normally used with low-voltage signals.

4.3.3.1 Differential Amplifier

Usually, an instrumentation amplifier is also a *differential amplifier* (sometimes termed *difference amplifier*). In a differential amplifier, both input leads are used for signal input; whereas in a single-ended amplifier, one of the leads is grounded and only one lead is used for signal input. Ground-loop noise can be a serious problem in single-ended amplifiers. Ground-loop noise can be effectively eliminated using a differential amplifier because noise loops are formed with both inputs of the amplifier and hence these noise signals are subtracted at the amplifier output. Since the noise level is almost the same for both inputs, it is canceled out. Any other noise (e.g., 60 Hz line noise) that might enter both inputs with the same intensity will also be canceled out at the output of a differential amplifier.

A basic differential amplifier that uses a single op-amp is shown in Figure 4.9a. The input–output equation for this amplifier can be obtained in the usual manner. For instance, since current through an op-amp is negligible, the current balance at point B gives $(v_{i2} - v_B)/R = v_B/R_f$ in which v_B is the voltage at B. Similarly, current balance at point A gives $(v_o - v_A)/R_f = (v_A - v_{i1})/R$. Now we use the property $v_A = v_B$ for an op-amp to eliminate v_A and v_B from the first two equations. This gives

$$v_o = \frac{R_f}{R}\left(v_{i2} - v_{i1}\right) \tag{4.13}$$

Two things are clear from Equation 4.13. First, the amplifier output is proportional to the "difference" and not the absolute value of the two inputs v_{i1} and v_{i2}. Second, the voltage gain of the amplifier is R_f/R. This is known as the *differential gain*. It is clear that the differential gain can be accurately set by using high-precision resistors R and R_f.

The basic differential amplifier, shown in Figure 4.9a and discussed above, is an important component of an instrumentation amplifier. An instrumentation amplifier should possess the capability of adjustable gain as well. Furthermore, it is desirable to have a

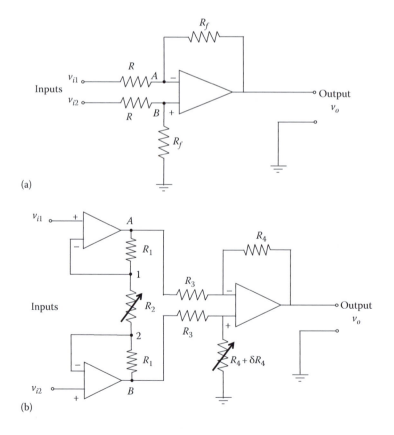

FIGURE 4.9
(a) A basic differential amplifier. (b) A basic instrumentation amplifier.

very high input impedance and a very low output impedance at each input lead. Also, it is desirable for an instrumentation amplifier to possess a high and more stable gain. An instrumentation amplifier that possesses these basic requirements may be built using two amplifier stages, as shown in Figure 4.9b. The amplifier gain can be adjusted using the fine-tunable resistor R_2. Impedance requirements are provided by two voltage-follower type amplifiers, one for each input, as shown. The variable resistance δR_4 is necessary to compensate for errors due to unequal common-mode gain. Let us first consider this aspect and then obtain an equation for the instrumentation amplifier.

4.3.3.2 Common Mode

The voltage that is "common" to both input leads of a differential amplifier is known as the *common-mode voltage*. This is equal to the smaller of the two input voltages. If the two inputs are equal, then the common-mode voltage is obviously equal to each one of the two inputs. When $v_{i1} = v_{i2}$, ideally, the output voltage v_o should be zero. In other words, ideally, any common-mode signals are rejected by a differential amplifier. But, since commercial op-amps are not ideal and since they usually do not have exactly identical gains with respect to the two input leads, the output voltage v_o will not be zero when the two inputs are identical. This *common-mode error* can be compensated for by providing a variable resistor with fine resolution at one of the two input leads of the differential amplifier. Hence, in Figure 4.9b, to compensate for the common-mode error (i.e., to achieve a satisfactory level of common-mode rejection), first the two inputs are made equal and then δR_4 is varied carefully until the output voltage level is sufficiently small (minimum). Usually, δR_4 that is required to achieve this compensation is small compared with the nominal feedback resistance R_4.

Since ideally $\delta R_4 = 0$, we can neglect δR_4 in the derivation of the instrumentation amplifier equation. Note from the basic property of an op-amp with no saturation (specifically, the voltages at the two input leads have to be almost identical) in Figure 4.9b, the voltage at point 2 should be v_{i2} and the voltage at point 1 should be v_{i1}. Next, we use the property that the current through each input lead of an op-amp is negligible. Accordingly, current through the circuit path B $\rightarrow$ 2 $\rightarrow$ 1 $\rightarrow$ A has to be the same. This gives the current continuity equations $(v_B - v_{i2})/R_1 = (v_{i2} - v_{i1})/R_2 = (v_{i1} - v_A)/R_1$ in which v_A and v_B are the voltages at points A and B, respectively. Hence, we get the following two equations:

$$v_B = v_{i2} + \frac{R_1}{R_2}\left(v_{i2} - v_{i1}\right)$$

$$v_A = v_{i1} - \frac{R_1}{R_2}\left(v_{i2} - v_{i1}\right)$$

Now, by subtracting the second equation from the first, we have the equation for the first stage of the amplifier; thus,

$$v_B - v_A = \left(1 + \frac{2R_1}{R_2}\right)\left(v_{i2} - v_{i1}\right) \tag{4.14}$$

Next, from the previous result for a differential amplifier (see Equation 4.13), we have (with $\delta R_4 = 0$)

$$v_o = \frac{R_4}{R_3}\left(v_B - v_A\right) \qquad (4.15)$$

From Equations 4.14 and 4.15, we get the overall equation

$$v_o = \frac{R_4}{R_3}\left(1 + \frac{2R_1}{R_2}\right)\left(v_{i2} - v_{i1}\right) \qquad (4.16)$$

Note that only the resistor R_2 is varied to adjust the gain (differential gain) of the amplifier. In Figure 4.9b, the two input op-amps (the voltage-follower op-amps) do not have to be exactly identical as long as the resistors R_1 and R_2 are chosen to be accurate. This is because the op-amp parameters, such as open-loop gain and input impedance, do not enter into the amplifier equations provided that their values are sufficiently high, as noted earlier.

4.3.4 Amplifier Performance Ratings

The main factors that affect the performance of an amplifier are stability, speed of response (bandwidth and slew rate), and unmodeled signals. We have already discussed the significance of some of these factors.

The level of stability of an amplifier, in the conventional sense, is governed by the dynamics of the amplifier circuitry and may be represented by a *time constant*. But more important consideration for an amplifier is the "parameter variation" due to aging, temperature, and other environmental factors. Parameter variation is also classified as a stability issue, in the context of devices such as amplifiers, because it pertains to the steadiness of the response when the input is maintained steady. Of particular importance is the *temperature drift*. This may be specified as a drift in the output signal per unity change in temperature (e.g., μV/°C).

The speed of response of an amplifier dictates the ability of the amplifier to faithfully respond to transient inputs. Conventional time-domain parameters such as *rise time* may be used to represent this. Alternatively, in the frequency domain, speed of response may be represented by a *bandwidth* parameter. For example, the frequency range over which the frequency response function is considered constant (flat) may be taken as a measure of bandwidth. Since there is some nonlinearity in any amplifier, the bandwidth can depend on the signal level itself. Specifically, *small-signal bandwidth* refers to the bandwidth that is determined using small input signal amplitudes.

Another measure of the speed of response is the *slew rate*, which is defined as the largest possible rate of change of the amplifier output for a particular frequency of operation. Since for a given input amplitude, the output amplitude depends on the amplifier gain, the slew rate is usually defined for unity gain.

Ideally, for a linear device, the frequency response function (transfer function) does not depend on the output amplitude (i.e., the product of the dc gain and the input amplitude). But for a device that has a limited slew rate, the bandwidth (or the maximum operating frequency at which output distortions may be neglected) will depend on the output amplitude. The larger the output amplitude, the smaller the bandwidth for a given slew rate limit. A bandwidth parameter that is usually specified for a commercial op-amp is the *gain-bandwidth product* (GBP). This is the product of the open-loop gain and the bandwidth of the op-amp. For example, for an op-amp with GBP = 15 MHz and an open-loop gain of 100 dB (i.e., 10^5), the bandwidth = $15 \times 10^6/10^5$ Hz = 150 Hz. Clearly, this bandwidth value is rather low. Since the gain of an op-amp with feedback is significantly lower than 100 dB, its effective bandwidth is much higher than that of an open-loop op-amp.

Stability problems and frequency response errors are prevalent in the open loop form of an op-amp. These problems can be eliminated using feedback because the effect of the open loop transfer function on the closed loop transfer function is negligible if the open loop gain is very large, which is the case for an op-amp.

Unmodeled signals can be a major source of amplifier error and these signals include the following:

1. Bias currents
2. Offset signals
3. Common mode output voltage
4. Internal noise

In analyzing op-amps, we assume that the current through the input leads is zero. This is not strictly true because bias currents for the transistors within the amplifier circuit have to flow through these leads. As a result, the output signal of the amplifier will deviate slightly from the ideal value.

Another assumption that we make in analyzing op-amps is that the voltage is equal at the two input leads. In practice, however, offset currents and voltages are present at the input leads, due to minute discrepancies inherent to the internal circuits within an op-amp.

4.3.4.1 Common-Mode Rejection Ratio

Common-mode error in a differential amplifier was discussed earlier. We note that ideally the common mode input voltage (the voltage common to both input leads) should have no effect on the output voltage of a differential amplifier. But, since any practical amplifier has some unbalances in the internal circuitry (e.g., gain with respect to one input lead is not equal to the gain with respect to the other input lead and, furthermore, bias signals are needed for operation of the internal circuitry), there will be an error voltage at the output, which depends on the common-mode input. The common-mode rejection ratio (CMRR) of a differential amplifier is defined as

$$\text{CMRR} = \frac{K v_{cm}}{v_{ocm}} \tag{4.17}$$

in which
  K is the gain of the differential amplifier (i.e., differential gain)
  v_{cm} is the common-mode voltage (i.e., voltage common to both input leads)
  v_{ocm} is the common-mode output voltage (i.e., output voltage due to common-mode input voltage)
  Ideally $v_{ocm} = 0$ and CMRR should be infinity. It follows that the larger the CMRR, the better the differential amplifier performance.

The three types of unmodeled signals mentioned above can be considered as noise. In addition, there are other types of noise signals that degrade the performance of an amplifier. For example, ground-loop noise can enter the output signal. Furthermore, stray capacitances and other types of unmodeled circuit effects can generate internal noise. Usually in amplifier analysis, unmodeled signals (including noise) can be represented by a noise voltage source at one of the input leads. The effects of unmodeled signals can be reduced by using suitably connected compensating circuitry including variable resistors that can be adjusted to eliminate the effect of unmodeled signals at the amplifier output (e.g., see δR_4 in Figure 4.9b). Some useful information about op-amps is summarized in Box 4.1.

BOX 4.1 OPERATIONAL AMPLIFIERS

Ideal op-amp properties:

- Infinite open-loop differential gain
- Infinite input impedance
- Zero output impedance
- Infinite bandwidth
- Zero output for zero differential input

Ideal analysis assumptions:

- Voltages at the two input leads are equal
- Current through either input lead is zero

Definitions:

- Open-loop gain $= \left| \dfrac{\text{Output voltage}}{\text{Voltage difference at input leads}} \right|$ with no feedback

- Input impedance $= \dfrac{\text{Voltage between an input lead and ground}}{\text{Current through that lead}}$ (with other input lead grounded and the output in open circuit)

- Output impedance
  $= \dfrac{\text{Voltage between output lead and ground in open circuit}}{\text{Current through that lead}}$ (with normal input conditions)

- Bandwidth = frequency range in which the frequency response is flat (gain is constant)
- Gain bandwidth product (GBP) = open loop gain × bandwidth at that gain
- Input bias current = average (dc) current through one input lead
- Input offset current = difference in the two input bias currents
- Differential input voltage = voltage at one input lead with the other grounded when the output voltage is zero
- Common-mode gain
  $= \dfrac{\text{Output voltage when input leads are at the same voltage}}{\text{Common input voltage}}$

- Common-mode rejection ratio (CMRR) $= \dfrac{\text{Open loop differential gain}}{\text{Common-mode gain}}$
- Slew rate = rate of change of output of a unity-gain op-amp, for a step input

---

4.3.4.2 AC-Coupled Amplifiers

The dc component of a signal can be blocked off by connecting the signal through a capacitor (*Note*: The impedance of a capacitor is $1/(j\omega C)$ and hence, at zero frequency the impedance will be infinite). If the input lead of a device has a series capacitor, we say that the input is ac-coupled and if the output lead has a series capacitor, then the output is ac-coupled. Typically, an ac-coupled amplifier has a series capacitor both at the input lead and the output lead. Hence, its frequency response function will have a high-pass characteristic; in particular, the dc components will be filtered out. Errors due to bias currents and offset signals are negligible for an ac-coupled amplifier. Furthermore, in an ac-coupled amplifier, stability problems are not very serious.

4.3.5 Ground Loop Noise

In instruments that handle low-level signals (e.g., sensors such as accelerometers, signal conditioning circuitry such as strain gage bridges, and sophisticated and delicate electronic components such as computer disk drives and automobile control modules), electrical noise can cause excessive error unless the proper corrective actions are taken. One form of noise is caused by fluctuating magnetic fields due to nearby ac power lines or electric machinery. This is commonly known as *electromagnetic interference* (EMI). This problem can be avoided by removing the source of EMI so that fluctuating external magnetic fields and currents are not present near the affected instrument. Another solution would be to use *fiber-optic* (optically coupled) signal transmission so that there is no noise conduction along with the transmitted signal from the source to the subject instrument. In the case of hard-wired transmission, if the two signal leads (positive and negative or hot and neutral) are twisted or if shielded cables are used, the induced noise voltages become equal in the two leads, which cancel each other.

Proper grounding practices are important to mitigate unnecessary electrical noise problems and more importantly, to avoid electrical safety hazards. A standard single-phase ac outlet (120 V, 60 Hz) has three terminals: one carrying power (hot), the second being neutral, and the third connected to earth ground (which is maintained at zero potential rather uniformly from point to point in the power network). Correspondingly, the power plug of an instrument should have three prongs. The shorter flat prong is connected to a black wire (hot) and the longer flat prong is connected to a white wire (neutral). The round prong is connected to a green wire (ground), which at the other end is connected to the chassis (or casing) of the instrument (chassis ground). In view of grounding the chassis in this manner, the instrument housing is maintained at zero potential even in the presence of a fault in the power circuit (e.g., a leakage or a short). The power circuitry of an instrument also has a local ground (signal ground), with reference to which its power signal is measured. This is a sufficiently thick conductor within the instrument and it provides a common and uniform reference of 0 V. Consider the sensor signal conditioning example shown in Figure 4.10. The dc power supply can provide both positive (+) and negative (−) outputs. Its zero voltage reference is denoted by common ground (COM), and it is the COM (signal ground) of the device. It should be noted that the COM of the dc power supply is not connected to the chassis ground, the latter being connected to the earth ground through the round prong of the power plug of the power supply. This is necessary to avoid the danger of an electric shock. Note that the COM of the power supply is connected to the signal ground of the signal conditioning module. In this manner, a common 0 V reference is provided for the dc voltage that is supplied to the signal conditioning module.

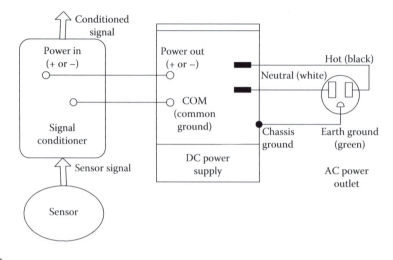

FIGURE 4.10
Instrument grounding—an example.

A main cause of electrical noise is the ground loop, which is created due to improper grounding of instruments. If two interconnected instruments are grounded at two separate locations that are far apart (multiple grounding), ground loop noise can enter the signal leads because of the possible potential difference between the two ground points. The reason is that ground itself is not generally a uniform-potential medium, and a nonzero (and finite) impedance may exist from point to point within this medium. This is, in fact, the case with a typical ground medium such as a common ground wire. An example is shown schematically in Figure 4.11a. In this example, the two leads of a sensor are directly connected to a signal-conditioning device such as an amplifier, one of its input leads (+) being grounded (at point B). The 0 V reference lead of the sensor is grounded through its housing to the earth ground (at point A). Because of nonuniform ground potentials, the two ground points A and B are subjected to a potential difference v_g. This will create a ground loop with the common reference lead, which interconnects the two devices. The solution to this problem is to isolate (i.e., provide an infinite impedance to) either one of the two devices. Figure 4.11b shows the internal isolation of the sensor. The external isolation,

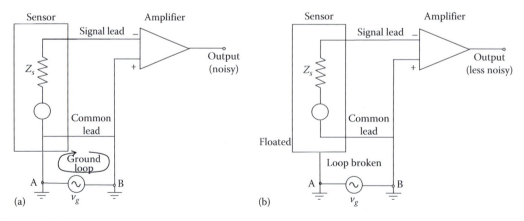

FIGURE 4.11
(a) Illustration of a ground loop. (b) Device isolation to eliminate ground loops (an example of internal isolation).

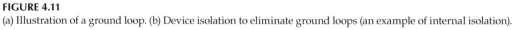

by insulating the housing of the sensor, will also remove the ground loop. Floating off the COM of a power supply (see Figure 4.10) is another approach to eliminating ground loops. Specifically, the COM is not connected to earth ground.

4.4 Filters

A filter is a device that allows through only the desirable part of a signal, rejecting the unwanted part. Unwanted signals can seriously degrade the performance of a mechatronic system. External disturbances, error components in excitations, and noise generated internally within system components and instrumentation are such spurious signals, which may be removed by a filter. As well, a filter is capable of shaping a signal in a desired manner.

In typical applications of acquisition and processing of signals in a mechatronic system, the filtering task would involve the removal of signal components in a specific frequency range. In this context, we can identify the following four broad categories of filters:

1. Low-pass filters
2. High-pass filters
3. Band-pass filters
4. Band-reject (or notch) filters

The ideal frequency-response characteristic (magnitude of the frequency transfer function) of each of these four types of filters is shown in Figure 4.12. The phase distortion of the input signal also should be small within the *pass band* (the allowed frequency range). Practical filters are less than ideal. Their frequency response functions do not exhibit sharp cutoffs as in Figure 4.12 and furthermore some phase distortion will be unavoidable.

A special type of band-pass filter that is widely used in the acquisition and monitoring of response signals is the *tracking filter*. This is simply a band-pass filter with a narrow pass band that is frequency-tunable. The center frequency (mid value) of the pass band is

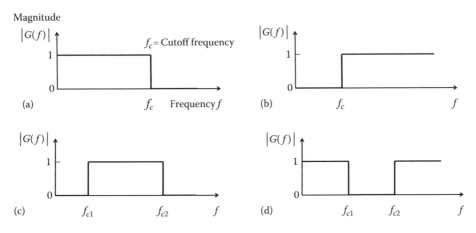

FIGURE 4.12
Ideal filter characteristics: (a) Low-pass filter. (b) High-pass filter. (c) Band-pass filter. (d) Band-reject (notch) filter.

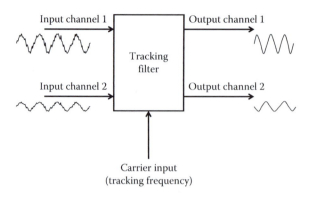

FIGURE 4.13
Schematic representation of a two-channel tracking filter.

variable, usually by coupling it to the frequency of a carrier signal (e.g., drive signal). In this manner, signals whose frequency varies with some basic variable in the system (e.g., rotor speed, frequency of a harmonic excitation signal, frequency of a sweep oscillator) can be accurately tracked in the presence of noise. The inputs to a tracking filter are the signals that are being tracked and the variable *tracking frequency* (*carrier input*). A typical tracking filter that can simultaneously track two signals is schematically shown in Figure 4.13.

Filtering can be achieved by *digital filters* as well as *analog filters*. Before digital signal processing became efficient and economical, analog filters were exclusively used for signal filtering, and are still widely used. In an analog filter, the input signal is passed through an analog circuit. Dynamics of the circuit will determine which (desired) signal components would be allowed through and which (unwanted) signal components would be rejected. An analog filter is typically an *active filter* containing active components such as transistors or op-amps. Earlier versions of analog filters employed discrete circuit elements such as discrete transistors, capacitors, resistors, and even discrete inductors. Since inductors have several shortcomings such as susceptibility to electromagnetic noise, unknown resistance effects, and large size, today they are rarely used in filter circuits. Furthermore, due to the well-known advantages of IC devices, today analog filters in the form of monolithic IC chips are extensively used in modem applications and are preferred over discrete-element filters. Digital filters, which employ digital signal processing to achieve filtering, are also widely used today.

4.4.1 Passive Filters and Active Filters

Passive analog filters employ analog circuits containing passive elements such as resistors and capacitors (and sometimes inductors) only, and an external power source is not needed. Active analog filters employ active elements and components such as transistors and op-amps, which require an external power source, in addition to passive elements. Active filters are widely available in a monolithic IC package and are usually preferred over passive filters. The advantages of active filters include the following:

1. Loading effects and interaction with other components are negligible because active filters can provide a very high input impedance and a very low output impedance.

2. They can be used with low signal levels because both signal amplification and filtering can be provided by the same active circuit.

3. They are widely available in a low-cost and compact IC form.

4. They can be easily integrated with digital devices.

5. They are less susceptible to noise from electromagnetic interference.

The commonly mentioned disadvantages of active filters are the following:

1. They need an external power supply.

2. They are susceptible to "saturation" type nonlinearity at high signal levels.

3. They can introduce many types of internal noise and unmodeled signal errors (offset, bias signals, etc.).

4.4.1.1 Number of Poles

Analog filters are dynamic systems and they can be represented by transfer functions, assuming linear dynamics. The number of poles of a filter is the number of poles in the associated transfer function. This is also equal to the order of the characteristic polynomial of the filter transfer function (i.e., *order* of the filter). Note that poles (or eigenvalues) are the roots of the characteristic equation.

4.4.2 Low-Pass Filters

The purpose of a low-pass filter is to allow all signal components below a certain (cutoff) frequency through and to block off all signal components above that cutoff. Analog low-pass fitters are widely used as *antialiasing filters* to remove the aliasing error in digital signal processing (see Chapter 5). Another typical application would be to eliminate high-frequency noise in a measured system response.

A single-pole passive, active low-pass filter circuit is shown in Figure 4.14a. If two passive filter stages are connected together, the overall transfer function is not equal to the product of the transfer functions of the individual stages. But, if two active filter stages similar to Figure 4.14a are connected together, electrical loading errors will be negligible because the op-amp with feedback (i.e., a voltage follower) introduces a very high input impedance and a very low output impedance. An active filter has the desirable property of very low interaction with any other connected component.

To obtain the filter equation for Figure 4.14a, note that the voltage at node A (v_A) is zero and the currents through two input leads are also zero (the op-amp properties). Hence, the current summation at A (Kirchoff's current law) gives

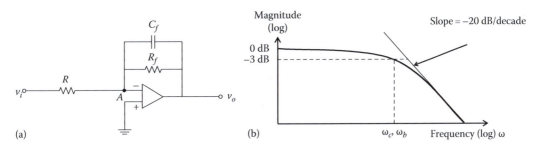

(a)   (b)

FIGURE 4.14
A single-pole low-pass filter. (a) An active filter stage. (b) The frequency response characteristic.

$$\frac{v_i - 0}{R} + \frac{v_o - 0}{R_f} + C_f \frac{d(v_o - 0)}{dt} = 0$$

or

$$\tau_f \frac{dv_o}{dt} + v_o = -kv_i \tag{4.18}$$

where the filter time constant is $\tau_f = R_f C_f$ and the filter gain is $k = R_f/R$. From Equation 4.18, it follows that the filter transfer function is

$$\frac{v_o}{v_i} = G_f(s) = -\frac{k}{(\tau_f s + 1)} \tag{4.19}$$

From this transfer function, it is clear that an analog low-pass filter is essentially a *lag circuit* (i.e., it provides a phase lag).

The frequency response function corresponding to Equation 4.19 is obtained by the setting $s = j\omega$. This gives the response of the filter when a sinusoidal signal of frequency ω is applied. The magnitude $|G(j\omega)|$ of the frequency transfer function gives the signal amplification and phase angle $\angle G(j\omega)$ gives the phase lead of the output signal with respect to the input. The magnitude curve (*Bode magnitude curve*) is shown in Figure 4.14b. Note from Equation 4.19 with $s = j\omega$ that for small frequencies (i.e., $\omega \ll 1/\tau_f$), the magnitude is approximately constant at k. Hence, $1/\tau_f$ can be considered the cutoff frequency ω_c. It can be shown that this is also the *half-power bandwidth* for the low-pass filter. For frequencies much larger than this, the filter transfer function on the Bode magnitude plane (i.e., log magnitude versus log frequency) can be approximated by a straight line with a slope of −20 dB/decade. This slope is known as the *roll-off rate* (see Figure 4.14b). The cutoff frequency and the roll-off rate are the two main design specifications for a low-pass filter. One would prefer a roll-off rate of at least −40 dB/decade and even −60 dB/decade in a practical filter.

4.4.2.1 Low-Pass Butterworth Filter

A low-pass Butterworth filter with 2 poles can provide a roll-off rate of −40 dB/decade and one with 3 poles can provide a roll-off rate of −60 dB/decade. Furthermore, the steeper the roll-off slope, the flatter the filter magnitude curve within the pass band.

A two-pole, low-pass Butterworth filter is shown in Figure 4.15. We could construct a two-pole filter simply by connecting together two single-pole stages of the type shown

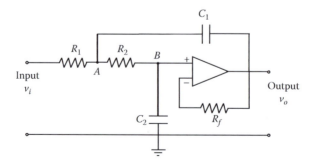

FIGURE 4.15
A two-pole low-pass Butterworth filter.

in Figure 4.14a. Then, we would require two op-amps, whereas the circuit shown in Figure 4.15 achieves the same objective by using only one op-amp (i.e., at a lower cost).

To obtain the filter equation, first we write the two current balance equations (the sum of the currents through R_1 and C_1 passes through R_2 and also through C_2) because the current through the op-amp lead is zero: $\dfrac{v_i - v_A}{R_1} + C_1 \dfrac{d}{dt}(v_o - v_A) = \dfrac{v_A - v_B}{R_2} = C_2 \dfrac{dv_B}{dt}$.

Also, since the op-amp with a feedback resistor R_f is a voltage follower (with unity gain), we have: $v_B = v_o$.

By eliminating the v_A and v_B from these three equations (with time derivative replaced by the Laplace variable s), we get the filter transfer function

$$\frac{v_o}{v_i} = \frac{1}{\left[\tau_1 \tau_2 s^2 + (\tau_2 + \tau_3)s + 1 \right]} = \frac{\omega_n^2}{\left[s^2 + 2\zeta\omega_n^2 + \omega_n^2 \right]} \qquad (4.20)$$

where $\tau_1 = R_1 C_1$, $\tau_2 = R_2 C_2$, and $\tau_3 = R_1 C_2$. This second order transfer function becomes oscillatory if $(\tau_2 + \tau_3)^2 < 4\tau_1\tau_2$. The undamped natural frequency is $\omega_n = 1/(\sqrt{\tau_1\tau_2})$, the damping ratio is $\zeta = (\tau_2 + \tau_3)/(\sqrt{4\tau_1\tau_2})$, and the resonant frequency is $\omega_r = \sqrt{1 - 2\zeta^2}\,\omega_n$. Ideally, we wish to have a zero resonant frequency, which corresponds to a damping ratio value $\zeta = 1/\sqrt{2}$. This corresponds to $(\tau_2 + \tau_3)^2 = 2\tau_1\tau_2$.

The frequency response function of the filter is

$$G(j\omega) = \frac{\omega_n^2}{\left[\omega_n^2 - \omega^2 + 2j\zeta\omega_n\omega \right]} \qquad (4.21)$$

Now for $\omega \ll \omega_n$, the filter frequency response is flat with a unity gain. For $\omega \gg \omega_n$, the filter frequency response can be approximated by

$$G(j\omega) = -\frac{\omega_n^2}{\omega^2}$$

In a log (magnitude) versus log (frequency) scale, this function is a straight line with slope $= -2$. Hence, when the frequency increases by a factor of 10 (i.e., one decade), the $\log_{10}$ (magnitude) drops by 2 units (i.e., 40 dB). In other words, the roll-off rate is -40 dB/decade. Also, ω_n can be taken as the filter cutoff frequency. Hence, $\omega_c = 1/(\sqrt{\tau_1\tau_2})$. It can be easily verified that when $\zeta = 1/\sqrt{2}$, this frequency is identical to the half-power bandwidth (i.e., the frequency at which the transfer function magnitude becomes $1/\sqrt{2}$).

Note: if two single-pole stages (of the type shown in Figure 4.14a) are cascaded, the resulting two-pole filter has an overdamped (nonoscillatory) transfer function, and it is not possible to achieve $\zeta = 1/\sqrt{2}$ as in the present case. Also, note that a three-pole low-pass Butterworth filter can be obtained by cascading the two-pole unit shown in Figure 4.15 with a single-pole unit shown in Figure 4.14a.

4.4.3 High-Pass Filters

Ideally, a high-pass filter allows through it all signal components above a certain (cutoff) frequency and blocks off all signal components below that frequency. A single-pole high-pass filter is shown in Figure 4.16. The filter equation is obtained by considering current

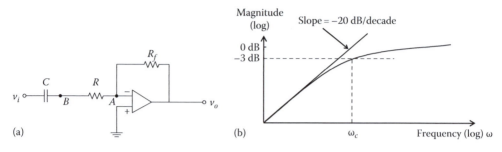

FIGURE 4.16
A single-pole high-pass filter: (a) An active filter stage. (b) Frequency response characteristic.

balance at point A and point B separately in Figure 4.16a: $C\dfrac{d}{dt}(v_i - v_B) = \dfrac{v_B}{R} = -\dfrac{v_o}{R_f}$. By eliminating v_B, we get

$$\tau \frac{dv_o}{dt} + v_o = -\tau_f \frac{dv_i}{dt} \tag{4.22}$$

in which the filter time constant $\tau = RC$ and $\tau_f = R_f C$. Introducing the Laplace variable s, the filter transfer function is

$$\frac{v_o}{v_i} = G_f(s) = -\frac{\tau_f s}{(\tau s + 1)} \tag{4.23}$$

This corresponds to a "lead circuit" (i.e., an overall phase lead is provided by this transfer function). The frequency response function $G(j\omega)$ is obtained by setting $s = j\omega$. Since its magnitude is zero for $\omega \ll (1/\tau, 1/\tau_f)$ and it is unity for $\omega \gg (1/\tau, 1/\tau_f)$, we have the cutoff frequency $\omega_c = \min(1/\tau, 1/\tau_f)$.

Note: Typically, $\tau = \tau_f$.

Signals above this cutoff frequency should be allowed undistorted by an ideal high-pass filter, and signals below the cutoff should be completely blocked off. The actual behavior of the basic high-pass filter discussed above is not that perfect, as observed from the frequency response characteristic shown in Figure 4.16b. It can be easily verified that the half-power bandwidth of the basic high-pass filter is equal to the cutoff frequency, and the roll-up slope is 20 dB/decade. Steeper slopes are desirable. Multiple-pole, high-pass, Butterworth filters can be constructed to give steeper roll-up slopes and reasonably flat pass-band magnitude characteristics.

4.4.4 Band-Pass Filters

An ideal band-pass filter passes all signal components within a finite frequency band and blocks off all signal components outside that band. The lower frequency limit of the pass band is called the *lower cutoff frequency* (ω_{c1}) and the upper frequency limit of the band is called the *upper cutoff frequency* (ω_{c2}). The most straightforward way to form a band-pass filter is to cascade a high-pass filter of cutoff frequency ω_{c1} with a low-pass filter of cutoff frequency ω_{c2}. This will require two op-amps. A simple version that requires only one op-amp is shown in Figure 4.17a.

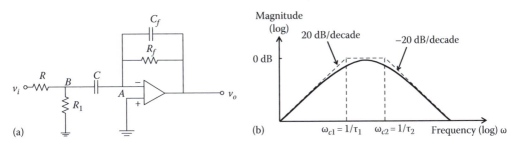

FIGURE 4.17
(a) A band-pass filter. (b) Frequency-response characteristic.

To determine the filter equation, we write the current balance equations node A and node B, separately:

$$\frac{v_i - v_B}{R} = \frac{v_B}{R_1} + C\frac{dv_B}{dt}$$

$$C\frac{dv_B}{dt} + \frac{v_o}{R_f} + C_f\frac{dv_o}{dt} = 0$$

By eliminating v_B in the Laplace domain, we get the filter transfer function:

$$\frac{v_o}{v_i} = G_f(s) = -\frac{k\tau s}{(\tau_f s + 1)(\tau s + k_1 + 1)} \tag{4.24}$$

with $\tau = RC$, $\tau_f = R_f C_f$, $k = R_f/R$, and $k_1 = R/R_1$.

Clearly, this transfer function corresponds a cascaded high-pass filter of time constant $\tau_1 = \tau/(k_1 + 1)$ (or cut-off frequency $\omega_{c1} = 1/\tau_1$) and a low-pass filter of time constant $\tau_2 = \tau_f$ (or cut-off frequency $\omega_{c2} = 1/\tau_2$), albeit with the use of just one op-amp. As indicated in Figure 4.17b, we must have $\tau_1 > \tau_2$, which will give the filter pass band of $(\omega_{c1}, \omega_{c2}) = (1/\tau_1, 1/\tau_2)$. More complex (higher order) band-pass filters with sharper cutoffs and flatter pass bands are commercially available.

4.4.4.1 Resonance-Type Band-Pass Filters

There are many applications where a filter with a very narrow pass band is required. The tracking filter mentioned in the beginning of the section on analog filters is one such application. A filter circuit with a sharp resonance can serve as a narrow-band filter. A cascaded single-pole low-pass filter and a single-pole high-pass filter do not generate an oscillatory response (because the filter poles are all real) and, hence, it does not form a resonance-type filter. A notable shortcoming of a resonance-type filter is that the frequency response within the bandwidth (pass band) is not flat. Hence, quite nonuniform signal attenuation takes place inside the pass band.

4.4.4.2 Band-Reject Filters

Band-reject filters or *notch filters* are commonly used to filter out a narrow band of noise components from a signal. For example, 60 Hz line noise in a signal can be eliminated by using a notch filter with a notch frequency of 60 Hz. While the previous three types of filters achieve their frequency response characteristics through the poles of the filter transfer function, a notch filter achieves its frequency response characteristic through its zeros (roots of the numerator polynomial equation). Some useful information about filters is summarized in Box 4.2.

4.4.5 Digital Filters

Any physical dynamic system can be interpreted as an analog filter, which can be represented by a differential equation with respect to time. It takes an analog input signal $u(t)$,

---

BOX 4.2   FILTERS

Active filters (need external power)
Advantages:

- Smaller loading errors and interaction (have high input impedance and low output impedance, and hence do not affect the input circuit conditions, output signals, and other components).
- Lower cost
- Better accuracy

Passive filters (no external power, use passive elements)
Advantages:

- Useable at very high frequencies (e.g., radio frequency)
- No need of a power supply

Filter types

- Low pass: allows frequency components up to cutoff and rejects the higher frequency components.
- High pass: rejects frequency components up to cutoff and allows the higher frequency components.
- Band pass: allows frequency components within an interval and rejects the rest.
- Notch (or band reject): rejects frequency components within an interval (usually, a narrow band) and allows the rest.

Definitions

- Filter order: the number of poles in the filter circuit or transfer function.
- Anti-aliasing filter: low-pass filter with cutoff at less than half the sampling rate (i.e., at less than the Nyquist frequency), for digital processing.
- Butterworth filter: a high-order filter with a quite flat pass band.
- Chebyshev filter: an optimal filter with uniform ripples in the pass band.
- Sallen–Key filter: an active filter whose output is in phase with input.

which is defined continuously in time t, and generates an analog output $y(t)$. A digital filter is a device that accepts a sequence of discrete input values (say, sampled from an analog signal at sampling period Δt) represented by $\{u_k\} = \{u_0, u_1, u_2, \ldots\}$ and generates a sequence of discrete output values: $\{y_k\} = \{y_0, y_1, y_2, \ldots\}$. It follows that a digital filter is a discrete-time system and it can be represented by a *difference equation*.

An nth order linear difference equation can be written in the form

$$a_0 y_k + a_1 y_{k-1} + \cdots + a_n y_{k-n} = b_0 u_k + b_1 u_{k-1} + \cdots + b_m u_{k-m} \tag{4.25}$$

This is a *recursive algorithm* in the sense that it generates one value of the output sequence using previous values of the output sequence and all values of the input sequence up to the present time point. Digital filters represented in this manner are termed *recursive digital filters*. There are filters that employ digital processing where a block (a collection of samples) of the input sequence is converted by a one-shot computation into a block of the output sequence. They are not recursive filters. Nonrecursive filters usually employ digital Fourier analysis, the *fast Fourier transform* (FFT) algorithm, in particular. We restrict our discussion below to recursive digital filters.

4.4.5.1 Software and Hardware Implementations

In digital filters, signal filtering is accomplished through the digital processing of the input signal. The sequence of input data (usually obtained by sampling and digitizing the corresponding analog signal) is processed according to the recursive algorithm of the particular digital filter. This generates the output sequence. The resulting digital output can be converted into an analog signal using a digital-to-analog converter (DAC).

A recursive digital filter is an implementation of a recursive algorithm that governs the particular filtering scheme (e.g., low-pass, high-pass, band-pass, and band-reject). The filter algorithm can be implemented either by software or by hardware. In software implementation, the filter algorithm is programmed into a digital computer. The processor (e.g., microprocessor or digital signal processor or DSP) of the computer can process an input data sequence according to the run-time filter program stored in the memory (in machine code) to generate the filtered output sequence.

The digital processing of data is accomplished by means of logic circuitry that can perform basic arithmetic operations such as addition. In the software approach, the processor of a digital computer makes use of these basic logic circuits to perform digital processing according to the instructions of a software program stored in the computer memory. Alternatively, a hardware digital processor can be built to perform a somewhat complex, yet fixed, processing operation. In this approach, the program of computation is said to be in hardware. The hardware processor is then available as an IC chip whose processing operation is fixed and cannot be modified (through software). The logic circuitry in the IC chip is designed to accomplish the required processing function. Digital filters implemented by this hardware approach are termed *hardware digital filters*.

The software implementation of digital filters has the advantage of flexibility; specifically, the filter algorithm can be easily modified by changing the software program that is stored in the computer. If, on the other hand, a large number of filters of a particular (fixed) structure is commercially needed then it would be economical to design the filter as an IC chip and replicate the chip in mass production. In this manner, very low-cost digital filters can be produced. A hardware filter can operate at a much faster speed in comparison with a software filter because in the former case, processing takes place automatically through logic circuitry in the filter chip without having to access by the processor, a software program, and various data items stored in the memory. The main disadvantage of a hardware filter is that its algorithm and parameter values cannot be modified, and the filter is dedicated to performing a fixed function.

4.5 Modulators and Demodulators

Sometimes signals are deliberately modified to maintain their accuracy during transmission, conditioning, and processing. In signal modulation, the data signal, known as the *modulating signal*, is employed to vary a property (such as amplitude or frequency) of a *carrier signal*. In this manner, the carrier signal is "modulated" by the data signal. After transmitting or conditioning the modulated signal, typically the data signal has to be recovered by removing the carrier signal. This is known as *demodulation* or *discrimination*.

A variety of modulation techniques exist, and several other types of signal modification (e.g., digitizing) could be classified as signal modulation even though they might not be commonly termed as such. Four types of modulations are illustrated in Figure 4.18. In *amplitude modulation* (AM), the amplitude of a periodic carrier signal is varied according to the amplitude of the data signal (modulating signal) and the frequency of the carrier signal (*carrier frequency*) being kept constant. Suppose that the transient signal shown in Figure 4.18a is the modulating signal and a high-frequency sinusoidal signal is used as the carrier signal. The resulting amplitude-modulated signal is shown in Figure 4.18b.

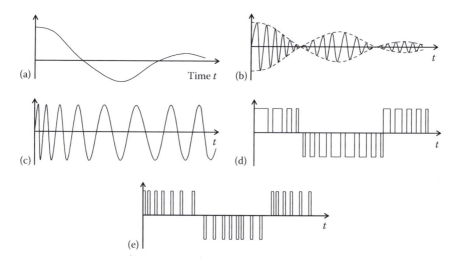

FIGURE 4.18
(a) Modulating signal (data signal). (b) Amplitude-modulated (AM) signal. (c) Frequency-modulated (FM) signal. (d) Pulse-width-modulated (PWM) signal. (e) Pulse-frequency-modulated (PFM) signal.

Amplitude modulation is used in telecommunication, transmission of radio and TV signals, instrumentation, and signal conditioning. The underlying principle is particularly useful in applications such as the sensing and instrumentation of mechatronic systems, fault detection, and diagnosis in rotating machinery.

In *frequency modulation* (FM), the frequency of the carrier signal is varied in proportion to the amplitude of the data signal (modulating signal), while keeping the amplitude of the carrier signal constant. Suppose that the data signal shown in Figure 4.18a is used to frequency-modulate a sinusoidal carrier signal. The modulated result will appear as in Figure 4.18c. Since information is carried as frequency rather than amplitude, any noise that might alter the signal amplitude will have virtually no effect on the transmitted data. Hence, frequency modulation is less susceptible to noise than amplitude modulation. Furthermore, since in frequency modulation the carrier amplitude is kept constant, signal weakening and noise effects that are unavoidable in long-distance data communication will have less effect than in amplitude modulation, particularly if the data signal level is low in the beginning. Besides, the high magnitude of carrier frequency enables the signal to travel through physical barriers. But more sophisticated techniques and hardware are needed for signal recovery (demodulation) in transmission, because frequency modulation demodulation involves frequency discrimination rather than amplitude detection. Frequency modulation is also widely used in radio transmission and in data recording and replay.

In PWM, the carrier signal is a pulse sequence. The pulse width is changed in proportion to the amplitude of the data signal while keeping the pulse spacing constant. This is illustrated in Figure 4.18d. Suppose that the high level of the PWM signal corresponds to the "on" condition of a circuit and the low level corresponds to the "off" condition. Then, as shown in Figure 4.19, the pulse width is equal to the on time ΔT of the circuit within each signal cycle period T. The duty cycle of the PWM is defined as the percentage on time in a pulse period, and is given by

$$\text{Duty cycle} = \frac{\Delta T}{T} \times 100\% \qquad (4.26)$$

PWM signals are extensively used in mechatronic systems for controlling electric motors and other mechanical devices such as valves (hydraulic, pneumatic) and machine tools. Note that in a given (short) time interval, the average value of the PWM signal is an estimate of the average value of the data signal in that period. Hence, PWM signals can be used directly in controlling a process without having to demodulate it. The advantages of PWM include better energy efficiency (less dissipation) and better performance with nonlinear devices. For example, a device may stick at low speeds, due to Coulomb friction. This can be avoided by using a PWM signal having an amplitude that is sufficient to overcome friction, while maintaining the required average control signal, which might be very small.

In *pulse-frequency modulation* (PFM), the carrier signal is a pulse sequence as well. In this method, the frequency of the pulses is changed in proportion to the value of the data signal while keeping the pulse width constant. PFM has the advantages of ordinary frequency modulation. Additional advantages result due to the fact that electronic

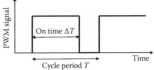

FIGURE 4.19
Duty cycle of a PWM signal.

circuits (digital circuits in particular) can handle pulses very efficiently. Furthermore, pulse detection is not susceptible to noise because it involves distinguishing between the presence and absence of a pulse rather than the accurate determination of the pulse amplitude (or width). PFM may be used in place of PWM in most applications, with better results.

Another type of modulation is *phase modulation* (PM). In this method, the phase angle of the carrier signal is varied in proportion to the amplitude of the data signal.

Conversion of discrete (sampled) data into the digital (binary) form is also considered a form of modulation. In fact, this is termed *pulse-code modulation* (PCM). In this case, each discrete data sample is represented by a binary number containing a fixed number of binary digits (bits). Since each digit in the binary number can take only two values, 0 or 1, it can be represented by the absence or presence of a voltage pulse. Hence, each data sample can be transmitted using a set of pulses. This is known as *encoding*. At the receiver, the pulses have to be interpreted (or decoded) in order to determine the data value. As with any other pulse technique, PCM is quite immune to noise because decoding involves the detection of the presence or absence of a pulse rather than the determination of the exact magnitude of the pulse signal level. Also, since pulse amplitude is constant, long distance signal transmission (of this digital data) can be accomplished without the danger of signal weakening and associated distortion. Of course, there will be some error introduced by the digitization process itself, which is governed by the finite word size (or dynamic range) of the binary data element. This is known as the *quantization* error and is unavoidable in signal digitization.

In any type of signal modulation, it is essential to preserve the algebraic sign of the modulating signal (data). Different types of modulators handle this in different ways. For example, in PCM, an extra *sign bit* is added to represent the sign of the transmitted data sample. In amplitude modulation and frequency modulation, a *phase-sensitive demodulator* is used to extract the original (modulating) signal with the correct algebraic sign. Note that in these two modulation techniques a sign change in the modulating signal can be represented by a 180° phase change in the modulated signal. This is not quite noticeable in Figure 4.18b and c. In PWM and PFM, a sign change in the modulating signal can be represented by changing the sign of the pulses, as shown in Figure 4.18d and e. In phase modulation, a positive range of phase angles (say 0 to π) could be assigned for the positive values of the data signal and a negative range of phase angles (say $-\pi$ to 0) could be assigned for the negative values of the signal.

4.5.1 Amplitude Modulation

Amplitude modulation can naturally enter into many physical phenomena. More important perhaps is the deliberate (artificial) use of amplitude modulation to facilitate data transmission and signal conditioning. Let us first examine the related mathematics.

Amplitude modulation is achieved by multiplying the data signal (modulating signal) $x(t)$ by a high frequency (periodic) carrier signal $x_c(t)$. Hence, the amplitude-modulated signal $x_a(t)$ is given by

$$x_a(t) = x(t)x_c(t) \qquad (4.27)$$

Any periodic signal, such as harmonic (sinusoidal), square wave, or triangular, can serve as the carrier. The main requirement is that the fundamental frequency of the carrier signal (carrier frequency) f_c has to be significantly large (say, by a factor of 5 or 10) than the highest frequency of interest (bandwidth) of the data signal. Analysis can be simplified by assuming a sinusoidal carrier signal:

$$x_c(t) = a_c \cos 2\pi f_c t \tag{4.28}$$

4.5.1.1 Modulation Theorem

The modulation theorem is also known as the *frequency-shifting theorem*. It states that if a signal $x(t)$ is multiplied by a sinusoidal signal, the Fourier spectrum $X_a(f)$ of the product signal (modulated signal) $x_a(t)$ is simply the Fourier spectrum $X(f)$ of the original signal (modulating signal) shifted through the frequency f_c of the sinusoidal signal (carrier signal).

For a mathematical explanation of the modulation theorem, we use the definition of the Fourier integral transform to get $X_a(f) = a_c \int_{-\infty}^{\infty} x(t) \cos 2\pi f_c t \exp(-j2\pi f t) dt$. Next, since $\cos 2\pi f_c t = \dfrac{1}{2} \left[\exp(j2\pi f_c t) + \exp(-j2\pi f_c t) \right]$, we have

$$X_a(f) = \frac{1}{2} a_c \int_{-\infty}^{\infty} x(t) \exp\left[-j2\pi(f - f_c)t \right] dt + \frac{1}{2} a_c \int_{-\infty}^{\infty} x(t) \exp\left[-j2\pi(f + f_c)t \right] dt$$

or

$$X_a(f) = \frac{1}{2} a_c \left[X(f - f_c) + X(f + f_c) \right] \tag{4.29}$$

Equation 4.29 is the mathematical statement of the modulation theorem and is illustrated in Figure 4.20a and b. It should be kept in mind that the magnitude of the spectrum has been multiplied by $a_c/2$ in amplitude modulation. Furthermore, the data signal is assumed to be *band limited*, with bandwidth f_b. Of course, the theorem is not limited to band-limited signals, but for practical reasons, we need to have some upper limit on the useful frequency of the data signal. Also, for practical reasons (not for the theorem itself), the carrier frequency f_c should be several times larger than f_c so that there is a reasonably wide frequency band from 0 to $(f_c - f_b)$ within which the magnitude of the modulated signal is virtually zero. The significance of this should be clear when we discuss applications of amplitude modulation.

Figure 4.20 shows only the magnitude of the frequency spectra. It should be remembered, however, that every Fourier spectrum has a phase angle spectrum as well. This is not shown for the sake of conciseness. But, clearly the phase-angle spectrum is also similarly affected (frequency shifted) by amplitude modulation.

4.5.1.2 Side Frequencies and Side Bands

Clearly, the modulation theorem is quite general and applies to periodic signals (with discrete spectra) as well as transient data signals with associated continuous Fourier spectra. Consider the case of the periodic data signal of frequency f_o as shown in Figure 4.20c, which has two discrete spectral components of amplitude $a/2$ at $\pm f_o$ (strictly, this corresponds to the Fourier series expansion of the data signal). When this signal is multiplied by the carrier signal of frequency f_c, each frequency component at $\pm f_o$ will be shifted by $\pm f_c$

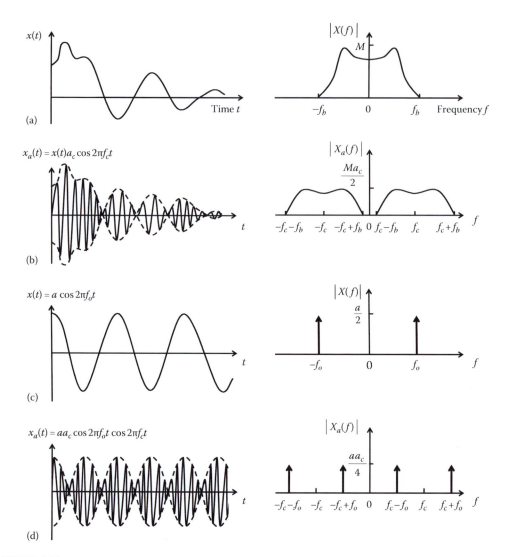

FIGURE 4.20
Illustration of the modulation theorem. (a) A transient data signal and its Fourier spectrum magnitude. (b) Amplitude-modulated signal and its Fourier spectrum magnitude. (c) A sinusoidal data signal. (d) Amplitude modulation by a sinusoidal signal.

to the two new frequency locations $f_c + f_o$ and $-f_c + f_o$ with an associated amplitude $aa_c/4$, as illustrated in Figure 4.20d. Note that the modulated signal does not have a spectral component at the carrier frequency f_c but rather, on each side of it at $f_c \pm f_o$. Hence, these spectral components are termed *side frequencies*. When a band of *side frequencies* is present, it is termed a *side band*. Side frequencies are particularly useful in fault detection and diagnosis of machinery with periodic motions (e.g., gears, bearings, and other rotating devices).

4.5.2 Application of Amplitude Modulation

The main hardware component of an amplitude modulator is an *analog multiplier*. It is commercially available in the monolithic IC form. Alternatively, one can be assembled

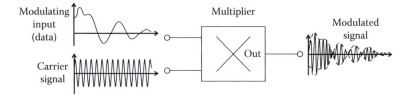

FIGURE 4.21
Representation of an amplitude modulator.

using IC op-amps and various discrete circuit elements. A schematic representation of an amplitude modulator is shown in Figure 4.21. In practice, in order to achieve satisfactory modulation, other components such as signal preamplifiers and filters would be needed.

There are many applications of amplitude modulation. In some applications, modulation is performed intentionally. In others, modulation occurs naturally as a consequence of the physical process, and the resulting signal is used to meet a practical objective. Typical applications of amplitude modulation include the following:

1. Conditioning of general signals (including dc, transient, and low-frequency) by exploiting the advantages of ac signal conditioning hardware
2. Improvement of the immunity of low-frequency signals to low-frequency noise
3. Transmission of general signals (dc, low-frequency, etc.) by exploiting the advantages of ac signals
4. Transmission of low-level signals under noisy conditions
5. Transmission of several signals simultaneously through the same medium (e.g., same telephone line, same transmission antenna, etc.)
6. Fault detection and diagnosis of rotating machinery

The role of amplitude modulation in many of these applications should be obvious if one understands the frequency-shifting property of amplitude modulation. Several other types of applications are also feasible due to the fact that power of the carrier signal can be increased somewhat arbitrarily, irrespective of the power level of the data (modulating) signal. Let us discuss, one by one, the six categories of applications mentioned above.

AC signal conditioning devices such as ac amplifiers are known to be more "stable" than their dc counterparts. In particular, *drift* problems are not as severe and nonlinearity effects are lower in ac signal conditioning devices. Hence, instead of conditioning a dc signal using dc hardware, we can first use the signal to modulate a high-frequency carrier signal. Then, the resulting high-frequency modulated signal may be conditioned more effectively using ac hardware.

The frequency-shifting property of amplitude modulation can be exploited in making low-frequency signals immune to low-frequency noise. Note from Figure 4.20 that using amplitude modulation, the low-frequency spectrum of the modulating signal can be shifted out into a very high frequency region by choosing a carrier frequency f_c that is sufficiently large. Then, any low-frequency noise (within the band 0 to $f_c - f_b$) would not distort the spectrum of the modulated signal. Hence, this noise could be removed by a high-pass filter (with cutoff at $f_c - f_b$) so that it would not affect the data. Finally, the original data signal can be recovered using demodulation. Since the frequency of a noise component can very well be within the bandwidth f_b of the data signal, if amplitude modulation was not employed, noise could directly distort the data signal.

The transmission of ac signals is more efficient than that of dc signals. The advantages of ac transmission include lower energy dissipation problems. As a result, a modulated signal can be transmitted over long distances more effectively than could the original data signal alone. Furthermore, the transmission of low-frequency (large wave-length) signals requires large antennas. Hence, the size of broadcast antenna can be effectively reduced by employing amplitude modulation (with an associated reduction in signal wave length).

Transmission of weak signals over long distances is not desirable because further signal weakening and corruption by noise could produce disastrous results. By increasing the power of the carrier signal to a sufficiently high level, the strength of the modulated signal can be elevated to an adequate level for long-distance transmission.

It is impossible to transmit two or more signals in the same frequency range simultaneously using a single telephone line. This problem can be resolved by using carrier signals with significantly different carrier frequencies to amplitude modulate the data signals. By picking the carrier frequencies sufficiently farther apart, the spectra of the modulated signals can be made to be nonoverlapping, thereby making simultaneous transmission possible. Similarly, with amplitude modulation, simultaneous broadcasting by several radio (AM) broadcast stations in the same broadcast area has become possible.

4.5.2.1 Fault Detection and Diagnosis

In the practice of electromechanical systems, the principle of amplitude modulation is particularly useful in fault detection and the diagnosis of rotating machinery. Here, modulation is not deliberately introduced, but rather results from the dynamics of the machine. Flaws and faults in a rotating machine are known to produce periodic forcing signals at frequencies higher than, and typically at an integer multiple of, the rotating speed of the machine. For example, backlash in a gear pair will generate forces at the tooth-meshing frequency (equal to the product: number of teeth × gear rotating speed). Flaws in roller bearings can generate forcing signals at frequencies proportional to the rotating speed times the number of rollers in the bearing race. Similarly, blade passing in turbines and compressors, and eccentricity and unbalance in the rotor can generate forcing components at frequencies that are integer multiples of the rotating speed. The resulting system response is clearly an amplitude-modulated signal, where the rotating response of the machine modulates the high frequency forcing response. This can be confirmed experimentally through Fourier analysis (FFT) of the resulting response signals. For a gearbox, for example, it will be noticed that, instead of getting a spectral peak at the gear tooth-meshing frequency, two side bands are produced around that frequency. Faults can be detected by monitoring the evolution of these side bands. Furthermore, since side bands are the result of the modulation of a specific forcing phenomenon (e.g., gear-tooth meshing, bearing-roller hammer, turbine-blade passing, unbalance, eccentricity, misalignment, etc.), one can trace the source of a particular fault (i.e., diagnose the fault) by studying the Fourier spectrum of the measured response.

Amplitude modulation is an integral part of many types of sensors. In these sensors, a high-frequency carrier signal (typically the ac excitation in a primary winding) is modulated by the motion. Actual motion can be detected by the demodulation of the output. Examples of sensors that generate modulated outputs are differential transformers (linear variable differential transformer [LVDT], rotary variable differential transformer [RVDT]), magnetic-induction proximity sensors, eddy-current proximity sensors, ac tachometers, and strain-gage devices that use ac bridge circuits. Signal conditioning and transmission would be facilitated by amplitude modulation in these cases. The signal has to be demodulated at the end for most practical purposes such as analysis and recording.

4.5.3 Demodulation

Demodulation, discrimination, or detection is the process of extracting the original data signal from a modulated signal. In general, demodulation has to be phase sensitive in the sense that the algebraic sign of the data signal should be preserved and determined by the demodulation process. In *full-wave demodulation*, an output is generated continuously. In *half-wave demodulation*, no output is generated for every alternate half-period of the carrier signal.

A simple and straightforward method of demodulation is by detection of the envelope of the modulated signal. For this method to be feasible, the carrier signal must be quite powerful (i.e., the signal level has to be high) and the carrier frequency also should be very high. An alternative method of demodulation, which generally provides more reliable results, involves a further step of modulation performed on the already-modulated signal followed by low-pass filtering. This method can be explained by referring to Figure 4.20.

Consider the amplitude-modulated signal $x_a(t)$ shown in Figure 4.20b. If this signal is multiplied by the sinusoidal carrier signal $2/a_c \cos 2\pi f_c t$, we get

$$\tilde{x}(t) = \frac{2}{a_c} x_a(t) \cos 2\pi f_c t \tag{4.30}$$

Now, by applying the modulation theorem (Equation 4.29) to Equation 4.30, we get the Fourier spectrum of $\tilde{x}(t)$ as

$$\tilde{X}(f) = \frac{1}{2} \frac{2}{a_c} \left[\frac{1}{2} a_c \left\{ X(f - 2f_c) + X(f) \right\} + \frac{1}{2} a_c \left\{ X(f) + X(f + 2f_c) \right\} \right]$$

or

$$\tilde{X}(f) = X(f) + \frac{1}{2} X(f - 2f_c) + \frac{1}{2} X(f + 2f_c) \tag{4.31}$$

The magnitude of this spectrum is shown in Figure 4.22a. Observe that we have recovered the spectrum $X(f)$ of the original data signal, except for the two *side bands* that are present at locations far removed (centered at $\pm 2f_c$) from the bandwidth of the original signal. By sending the signal $\tilde{x}(t)$ through a low-pass filter using a filter with a cutoff at f_b, the original data signal can be recovered. A schematic representation of this method of amplitude demodulation is shown in Figure 4.22b.

4.6 Analog-to-Digital Conversion

Inputs to a digital device (typically, a digital computer) and outputs from a digital device are necessarily present in the digital form. Hence, when a digital device is connected to

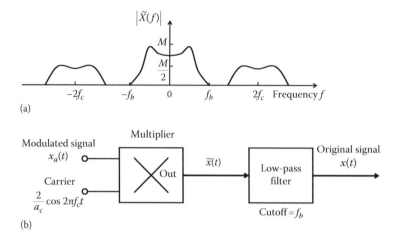

(a)

(b)

FIGURE 4.22
Amplitude demodulation. (a) Spectrum of the signal after the second modulation. (b) Demodulation schematic diagram (modulation + filtering).

an analog device, the interface hardware and associated driver software have to perform several important functions. Two of the most important interface functions are *digital-to-analog conversion* (DAC or D/A) and *analog-to-digital conversion* (ADC or A/D). A digital output from a digital device has to be converted into the analog form for feeding into an analog device such as actuator or analog recording or display unit. Also, an analog signal has to be converted into the digital form, according to an appropriate code, before being read by a digital processor or computer.

Mechatronic systems use digital data acquisition for a variety of purposes such as condition monitoring and performance evaluation, fault detection and diagnosis, product/service quality assessment, dynamic testing, system identification (i.e., experimental modeling), and control. Consider the feedback control system shown in Figure 4.23. Typically, the measured responses (outputs) of a physical system (called process or plant) are available in the analog form as continuous signals (functions of continuous time). Furthermore, typically, the excitation signals (or control inputs) for a physical system have to be provided in the analog form. A digital computer is an integral component of a typical mechatronic system and is commonly incorporated in the form of single-chip microcontrollers and single-board computers together with such single-IC components as DSP.

Both ADC and DAC elements are components in a typical input/output (I/O) board (or data acquisition and control card, or DAQ). Complete I/O cards for mechatronic applications are available from such companies as National Instruments®, Servo to Go, Inc., Precision MicroDynamics, Inc., and Keithly Instruments (Metrabyte), Inc. An I/O board can be plugged into the slot of a personal computer (PC) and automatically linked with the bus of the PC. The main components of an I/O board are shown in Figure 4.24. The multiplexer selects the appropriate input channel. The signal is amplified by a programmable amplifier (prog amp) prior to analog-to-digital conversion. The sample-and-hold (S/H) element samples the analog signal and maintains its value at the sampled level until conversion by the ADC. The first-in-first-out (FIFO) element stores the ADC output until it is accessed by the PC for digital processing. The I/O board can provide an analog output through the DAC. Furthermore, a typical I/O board can provide digital outputs as well. An

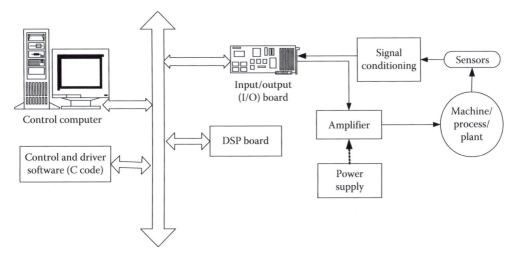

FIGURE 4.23
Components of a data acquisition (DAQ) and control loop.

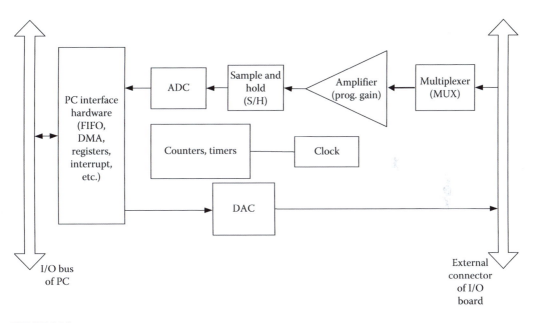

FIGURE 4.24
Main components of an I/O board of a PC.

encoder (i.e., a pulse-generating position sensor) can be directly interfaced to I/O boards that are intended for use in motion control applications. Specifications of a typical I/O board are given in Box 4.3. Particular note should be made about the sampling rate. This is the rate at which an analog input signal is sampled by the ADC. The Nyquist frequency (or the bandwidth limit) of the sampled data would be half this number (50 kHz for the I/O board specified in Box 4.3). When multiplexing is used (i.e., several input channels are read at the same time), the effective sampling rate for each channel will be reduced by a factor equal to the number of channels. For the I/O board specified in Box 4.3, when 16 channels

---

**BOX 4.3   TYPICAL SPECIFICATIONS OF A
PLUG-IN I/O OR DAQ BOARD FOR A PC**

Number of analog input channels = 16 single ended or 8 differential
Analog input ranges = ±5 V; 0–10 V; ±10 V; 0–20 V
Input gain ranges (programmable) = 1, 2, 5, 10, 20, 50, 100
Sampling rate for A/D conversion = 100,000 samples/s (100 kHz)
Word size (resolution) of ADC = 12 bits
Number of D/A output channels = 4
Word size (resolution) of DAC = 12 bits
Ranges of analog output = 0–10 V (unipolar mode); ±10 V (bipolar mode)
Number of digital input lines = 12
Low voltage of input logic = 0.8 V (maximum)
High voltage of input logic = 2.0 V (minimum)
Number of digital output lines = 12
Low voltage of output logic = 0.45 V (maximum)
High voltage of output logic = 2.4 V (minimum)
Number of counters/timers = 3
Resolution of a counter/timer = 16 bits

---

are sampled simultaneously, the effective sampling rate will be 100 kHz/16 = 6.25 kHz, giving a Nyquist frequency of 3.125 kHz.

Commercial DAQ hardware support user-written software (e.g., in C++) for tasks of data acquisition, processing, communication, and control. Also, commercial software tools (e.g., LabVIEW®) may be used for the same purposes. DAC and ADC are discussed now. Digital-to-analog converters are simpler and lower in cost than analog-to-digital converters. Furthermore, some types of analog-to-digital converters employ a digital-to-analog converter to perform their function. For these reasons, we will first discuss DAC.

4.6.1 Digital-to-Analog Conversion

The function of a digital-to-analog converter (DAC) is to convert a sequence of digital words stored in a *data register* (called DAC register), typically in the straight binary form, into an analog signal. The data in the DAC register may be arriving from a data bus of a computer. Each binary digit (bit) of information in the register may be present as a state of a bistable (two-stage) logic device, which can generate a voltage pulse or a voltage level to represent that bit. For example, the "off state" of a bistable logic element, the "absence" of a voltage pulse, the "low level" of a voltage signal, or the "no change" in a voltage level can represent binary 0. Conversely, the "on state" of a bistable device, the "presence" of a voltage pulse, the "high level" of a voltage signal, or the "change" in a voltage level will represent binary 1. The combination of these bits forming the digital word in the DAC register will correspond to some numerical value for the analog output signal. Then, the purpose of the DAC is to generate an output voltage (signal level) that has this numerical value and maintain the value until the next digital word is converted into the analog form. Since a voltage output cannot be arbitrarily large or small for practical reasons, some form of scaling would have to be employed in the DAC process. This scale will depend on the reference voltage v_{ref} used in the particular DAC circuit.

A typical DAC unit is an active circuit in the IC form and may consist of a data register (digital circuits), solid-state switching circuits, resistors, or op-amp powered by an external power supply, which can provide the reference voltage for the DAC. The reference voltage will determine the maximum value of the output (full-scale voltage). As noted before, the IC chip that represents the DAC is usually one of many components mounted on a printed circuit (PC) board, which is the I/O board (or I/O card, interface board, or data acquisition and control board). This board is plugged into a slot of the data acquisition and control PC (see Figures 4.23 and 4.24).

There are many types and forms of DAC circuits. The form will depend mainly on the manufacturer and requirements of the user or of the particular application. Most types of DAC are variations of two basic types: the weighted type (or summer type or adder type) and the ladder type. The latter type of DAC is more desirable even though the former type could be somewhat simpler and less expensive.

4.6.1.1 Weighted Resistor DAC

A schematic representation of a weighted-resistor DAC (or *summer DAC* or *adder DAC*) is shown in Figure 4.25. Note that this is a general *n*-bit DAC and *n* is the number of bits in the output register. The binary word in the register is $w = [b_{n-1}b_{n-2}b_{n-3} \ldots b_1 b_0]$ in which b_i is the bit in the *i*th position and it can take the value 0 or 1 depending on the value of the digital output. The decimal value of this binary word is given by

$$D = 2^{n-1}b_{n-1} + 2^{n-2}b_{n-2} + \cdots + 2^0 b_0 \tag{4.32}$$

Note: The least significant bit (LSB) is b_0 and the most significant bit (MSB) is b_{n-1}. The analog output voltage v of the DAC has to be proportional to D.

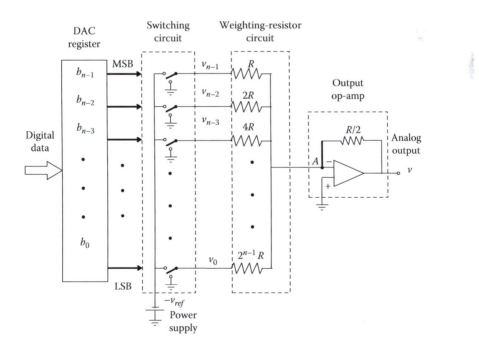

FIGURE 4.25
Weighted-resistor (adder) DAC.

Each bit b_i in the digital word w will activate a solid–state microswitch in the switching circuit, typically by sending a switching voltage pulse. If $b_i = 1$, the circuit lead will be connected to the $-v_{ref}$ supply providing an input voltage $v_i = -v_{ref}$ to the corresponding weighting resistor $2^{n-i-1} R$. If, on the other hand $b_i = 0$, then the circuit lead will be connected to ground, thereby providing an input voltage $v_i = 0$ to the same resistor. Note that the MSB is connected to the smallest resistor (R) and the LSB is connected to the largest resistor ($2^{n-1} R$). By writing the summation of currents at node A of the output op-amp, we get $\frac{v_{n-1}}{R} + \frac{v_{n-2}}{2R} + \cdots + \frac{v_0}{2^{n-1}R} + \frac{v}{R/2} = 0$. In writing this equation, we have used the two principal facts for an op-amp: the voltage is the same at both input leads and the current through each lead is zero. *Note*: The + lead is grounded and hence node A should have zero voltage. Now since $v_i = -b_i v_{ref}$ where $b_i = 0$ or 1 depending on the bit value (state of the corresponding switch), we have

$$v = \left[b_{n-1} + \frac{b_{n-2}}{2} + \cdots + \frac{b_0}{2^{n-1}} \right] \frac{v_{ref}}{2} \tag{4.33}$$

Clearly, as required, the output voltage v is proportional to the value D of the digital word w.

The *full-scale value* (*FSV*) of the analog output occurs when all b_i are equal to 1. Hence, $FSV = \left[1 + \frac{1}{2} + \cdots + \frac{1}{2^{n-1}} \right] \frac{v_{ref}}{2}$. Using the commonly known formula for the sum of a geometric series $1 + r + r^2 + \cdots + r^{n-1} = (1 - r^n)/(1 - r)$, we get

$$FSV = \left(1 - \frac{1}{2^n} \right) v_{ref} \tag{4.34}$$

Note: This value is slightly smaller than the reference voltage v_{ref}.

A major drawback of the weighted-resistor DAC is that the range of the resistance value in the weighting circuit is very wide. This presents a practical difficulty, particularly when the size (number of bits n) of the DAC is large. The use of resistors having widely different magnitudes in the same circuit can create accuracy problems. For example, since the MSB corresponds to the smallest weighting resistor, it follows that the resistors must have very high precision.

4.6.1.2 Ladder DAC

A digital-to-analog converter that uses an R–$2R$ ladder circuit is known as a ladder DAC. This circuit uses only two types of resistors: one with resistance R and the other with $2R$. Hence, the precision of the resistors is not as stringent as what is needed for the weighted-resistor DAC. A schematic representation of an R–$2R$ ladder DAC is shown in Figure 4.26. In this device, the switching circuit can operate just like in a weighted-resistor DAC. To obtain the input–output equation for the ladder DAC, suppose that, as before, the voltage output from the solid–state switch associated with b_i of the digital word is v_i. Furthermore, suppose that $\tilde{v}_i$ is the voltage at node i of the ladder circuit, as shown in Figure 4.26. Now, writing the current summation at node i we get

$$\frac{v_i - \tilde{v}_i}{2R} + \frac{\tilde{v}_{i+1} - \tilde{v}_i}{R} + \frac{\tilde{v}_{i-1} - \tilde{v}_i}{R} = 0,$$

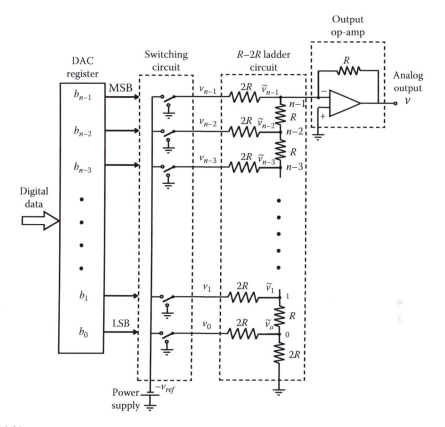

FIGURE 4.26
Ladder DAC.

Hence,

$$\frac{1}{2}v_i = \frac{5}{2}\tilde{v}_i - \tilde{v}_{i-1} - \tilde{v}_{i+1} \quad \text{for } i = 1, 2, \ldots, n-2 \tag{i}$$

Equation (i) is valid for all nodes except node 0 and node $n-1$. It is seen that the current summation for node 0 gives $\dfrac{v_0 - \tilde{v}_0}{2R} + \dfrac{\tilde{v}_1 - \tilde{v}_0}{R} + \dfrac{0 - \tilde{v}_0}{2R} = 0$, hence,

$$\frac{1}{2}v_0 = 2\tilde{v}_0 - \tilde{v}_1 \tag{ii}$$

The current summation for node $n-1$ gives $\dfrac{v_{n-1} - \tilde{v}_{n-1}}{2R} + \dfrac{v - \tilde{v}_{n-1}}{R} + \dfrac{\tilde{v}_{n-2} - \tilde{v}_{n-1}}{R} = 0$. Now, since the + lead of the op-amp is grounded, we have $\tilde{v}_{n-1} = 0$. Hence,

$$\frac{1}{2}v_{n-1} = -\tilde{v}_{n-2} - v \tag{iii}$$

Next, by using Equations (i) through (iii) along with the fact that $\tilde{v}_{n-1}=0$, we can write the following series of equations:

$$\frac{1}{2}v_{n-1} = -\tilde{v}_{n-2} - v$$

$$\frac{1}{2^2}v_{n-2} = \frac{1}{2}\frac{5}{2}\tilde{v}_{n-2} - \frac{1}{2}\tilde{v}_{n-3}$$

$$\frac{1}{2^3}v_{n-3} = \frac{1}{2^2}\frac{5}{2}\tilde{v}_{n-3} - \frac{1}{2^2}\tilde{v}_{n-4} - \frac{1}{2^2}\tilde{v}_{n-2} \qquad\text{(iv)}$$

$$\vdots$$

$$\frac{1}{2^{n-1}}v_1 = \frac{1}{2^{n-2}}\frac{5}{2}\tilde{v}_1 \frac{1}{2^{n-2}}\tilde{v}_0 - \frac{1}{2^{n-2}}\tilde{v}_2$$

$$\frac{1}{2^2}v_0 = \frac{1}{2^{n-1}}2\tilde{v}_0 - \frac{1}{2^{n-1}}\tilde{v}_1$$

If we sum these n equations, first denoting $S = \frac{1}{2^2}\tilde{v}_{n-2} + \frac{1}{2^3}\tilde{v}_{n-3} + \cdots + \frac{1}{2^{n-1}}\tilde{v}_1$, we get $\frac{1}{2}v_{n-1} + \frac{1}{2^2}v_{n-2} + \cdots + \frac{1}{2^n}v_0 = 5S - 4S - S + \frac{1}{2^{n-1}}2\tilde{v}_0 - \frac{1}{2^{n-2}}\tilde{v}_0 - v = -v$. Finally, since $v_i = -b_i v_{ref}$, we have the analog output as

$$v = \left[\frac{1}{2}b_{n-1} + \frac{1}{2^2}b_{n-2} + \cdots + \frac{1}{2^n}b_0\right]v_{ref} \qquad\text{(4.35)}$$

This result is identical to Equation 4.33, which we obtained for the weighted-resistor DAC. Hence, as before, the analog output is proportional to the value D of the digital word and, furthermore, the full-scale value of the ladder DAC is given by the previous Equation 4.34 as well.

4.6.1.3 DAC Error Sources

For a given digital word, the analog output voltage from a DAC would not be exactly equal to what is given by the analytical formulas (e.g., Equation 4.33) that were derived earlier.

DAC error = actual output − ideal output

The error may be normalized with respect to the full-scale value. There are many causes of DAC error. Typical error sources include parametric uncertainties and variations, circuit time constants, switching errors, and variations and noise in the reference voltage. Several types of error sources and representations are discussed below.

4.6.1.3.1 Code Ambiguity

In many digital codes (e.g., in the straight binary code), incrementing a number by an LSB will involve more than 1 bit switching. If the speed of switching from 0 to 1 is different from that for 1 to 0, and if switching pulses are not applied to the switching circuit simultaneously, the switching of the bits will not take place simultaneously. For example, in a 4 bit DAC, incrementing from decimal 2 to decimal 4 will involve changing the digital word from 0011 to 0100. This requires two bit-switchings from 1 to 0 and one bit switching from 0 to 1. If 1 to 0 switching is faster than the 0 to 1 switching, then an intermediate value given by 0000 (decimal zero) will be generated with a corresponding analog output. Hence, there will be a momentary code ambiguity and associated error in the DAC signal. This problem can be reduced (and eliminated in the case of single bit increments) if a *gray code* is used to represent the digital data. Improving the switching circuitry will also help reduce this error.

4.6.1.3.2 Settling Time

The circuit hardware in a DAC unit will have some dynamics with associated time constants and perhaps oscillations (underdamped response). Hence, the output voltage cannot instantaneously settle to its ideal value upon switching. The time required for the analog output to settle within a certain band (say $\pm 2\%$ of the final value or $\pm 1/2$ resolution), following the application of the digital data, is termed settling time. Naturally, the settling time should be smaller for better (faster and more accurate) performance. As a rule of thumb, the settling time should be approximately half the data arrival time. *Note*: The data arrival time is the time interval between the arrival of two successive data values, and is given by the inverse of the data arrival rate.

4.6.1.3.3 Glitches

The switching of a circuit will involve sudden changes in the magnetic flux due to current changes. This will induce the voltages that produce unwanted signal components. In a DAC circuit, these induced voltages due to rapid switching can cause signal spikes, which will appear at the output. The error due to these noise signals is not significant at low conversion rates.

4.6.1.3.4 Parametric Errors

As discussed before, resistor elements in a DAC might not be very precise, particularly when resistors within a wide range of magnitudes are employed, as in the case of weighted-resistor DAC. These errors appear at the analog output. Furthermore, aging and environmental changes (primarily, change in temperature) will change the values of circuit parameters, particularly resistance. This will also result in DAC error. These types of errors due to imprecision of circuit parameters and variations of parameter values are termed parametric errors. The effects of such errors can be reduced by several ways including the use of compensation hardware (and perhaps software), and directly by using precise and robust circuit components and employing good manufacturing practices.

4.6.1.3.5 Reference Voltage Variations

Since the analog output of a DAC is proportional to the reference voltage v_{ref}, any variations in the voltage supply will directly appear as an error. This problem can be overcome by using stabilized voltage sources with sufficiently low output impedance.

4.6.1.3.6 Monotonicity

Clearly, the output of a DAC should change by its resolution ($\delta y = v_{ref}/2^n$) for each step of one LSB increment in the digital value. This ideal behavior might not exist in some practical DACs due to such errors as those mentioned above. At least the analog output should not decrease as the value of the digital input increases. This is known as the monotonicity requirement, and it should be met by a practical digital-to-analog converter.

4.6.1.3.7 Nonlinearity

Suppose that the digital input to a DAC is varied from [0 0 $\cdots$ 0] to [1 1 $\cdots$ 1] in steps of one LSB. As mentioned above, ideally the analog output should increase in constant jumps of $\delta y = v_{ref}/2^n$ giving a staircase-shaped analog output. If we draw the best linear fit for this ideally monotonic staircase response, it will have a slope equal to the resolution/step. This slope is known as the ideal scale factor. The nonlinearity of a DAC is measured by the largest deviation of the DAC output from this best linear fit. Note that in the ideal case, the nonlinearity is limited to half the resolution $1/2(\delta y)$.

One cause of nonlinearity is clearly the faulty bit-transitions. Another cause is circuit nonlinearity in the conventional sense. Specifically, due to nonlinearities in circuit elements such as op-amps and resistors, the analog output will not be proportional to the value of the digital word dictated by the bit switchings (faulty or not). This latter type of nonlinearity can be accounted for by using calibration.

4.6.2 Analog-to-Digital Conversion

Analog signals, which are continuously defined with respect to time, have to be sampled at discrete time points and the sample values have to be represented in the digital form (according to a suitable code) to be read into a digital system such as a microcomputer. An analog-to-digital converter (ADC) is used to accomplish this. For example, since the response measurements of a mechatronic system are usually available as analog signals, these signals have to be converted into the digital form before being passed on to a digital computer for analysis and possibly generating a control command. Hence, the computer interface for the measurement channels should contain one or more ADCs (see Figure 4.23).

DACs and ADCs are usually situated on the same digital interface (DAQ) board (see Figure 4.24). But, the analog-to-digital conversion process is more complex and time consuming than the digital-to-analog conversion process. Furthermore, many types of ADCs use DACs to accomplish the analog-to-digital conversion. Hence, ADCs are usually more costly, and their conversion rate is usually slower in comparison with DACs. Several types of analog-to-digital converters are commercially available. The principle of operation may vary depending on the type. A few commonly known types are discussed here.

4.6.2.1 Successive Approximation ADC

This type of analog-to-digital converter is very fast and is suitable for high-speed applications. The speed of conversion depends on the number of bits in the output register of the ADC but is virtually independent of the nature of the analog input signal. A schematic diagram for a successive approximation ADC is shown in Figure 4.27. Note that a DAC is an integral component of this ADC. The sampled analog signal (from a *sample and hold circuit*) is applied to a *comparator* (typically a *differential amplifier*). Simultaneously, a

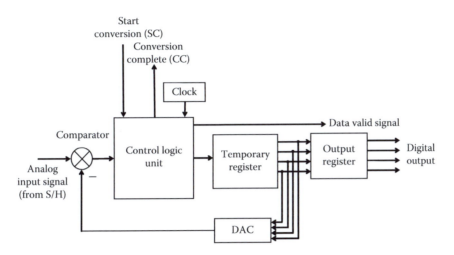

FIGURE 4.27
Successive approximation ADC.

"start conversion" (SC) control pulse is sent into the *control logic unit* by the external device (perhaps a microcomputer) that controls the operation of the ADC. Then, no new data will be accepted by the ADC until a "conversion complete" (CC) pulse is sent out by the control logic unit. Initially, the registers are cleared so that they contain all zero bits. Now, the ADC is ready for its first conversion approximation.

The first approximation begins with a clock pulse. Then, the control logic unit will set the MSB of the temporary register (DAC control register) to 1, all the remaining bits in that register being zero. This digital word in the temporary register is supplied to the DAC. Note that the analog output of the DAC is now equal to half full-scale value. This analog signal is subtracted from the analog input by the comparator. If the output of the comparator is positive, the control logic unit will keep the MSB of the temporary register at binary 1 and will proceed to the next approximation. If the comparator output is negative, the control logic unit will change the MSB to binary 0 before proceeding to the next approximation.

The second approximation will start at another clock pulse. This approximation will consider the second MSB of the temporary register. As before, this bit is set to 1 and the comparison is made. If the comparator output is positive, this bit is left at value 1 and the third MSB is considered. If the comparator output is negative, the bit value will be changed to 0 before proceeding to the third MSB.

In this manner, all the bits in the temporary register are set successively starting from the MSB and ending with the LSB. The contents of the temporary register are then transferred to the output register and a "data valid" signal is sent by the control logic unit, signaling the interfaced device (computer) to read the contents of the output register. The interfaced device will not read the register if a data valid signal is not present. Next, a CC pulse is sent out by the control logic unit, and the temporary register is cleared. The ADC is now ready to accept another data sample for digital conversion. Note that the conversion process is essentially the same for every bit in the temporary register. Hence, the total conversion time is approximately *n* times the conversion time for one bit. Typically, one bit conversion can be completed within one clock period.

It should be clear that if the maximum value of an analog input signal exceeds the full-scale value of a DAC, then the excess signal value cannot be converted by the ADC. The excess value will directly contribute to error in the digital output of the ADC. Hence, this situation should be avoided either by properly scaling the analog input or by properly selecting the reference voltage for the internal DAC unit.

In the foregoing discussion, we have assumed that the value of the analog input signal is always positive. Otherwise, the sign of the signal has to be accounted for by some means. For example, the sign of the signal can be detected from the sign of the comparator output initially, when all bits are zero. If the sign is negative, then the same A/D conversion process as for a positive signal is carried out after switching the polarity of the comparator. Finally, the sign is correctly represented in the digital output (e.g., by the two's complement representation for negative quantities). Another approach to account for signed (bipolar) input signals is to offset the signal by a sufficiently large constant voltage so that the analog input is always positive. After the conversion, the digital number corresponding to this offset is subtracted from the converted data in the output register in order to obtain the correct digital output. In what follows, we shall assume that the analog input signal is positive.

4.6.2.2 Dual Slope ADC

This analog-to-digital converter uses an RC integrating circuit. Hence, it is also known as an *integrating ADC*. This ADC is simple and inexpensive. In particular, an internal DAC is not utilized and hence, DAC errors as mentioned previously will not enter the ADC output. Furthermore, the parameters R and C in the integrating circuit do not enter the ADC output. As a result, the device is self-compensating in terms of circuit-parameter variations due to temperature, aging, etc. A shortcoming of this ADC is its slow conversion rate because, for accurate results, the signal integration has to proceed for a longer time in comparison with the conversion time for a successive approximation ADC.

Analog-to-digital conversion in a dual slope ADC is based on timing (i.e., counting the number of clock pulses during a capacitor-charging process). The principle of operation can be explained with reference to the integrating circuit shown in Figure 4.28a. Note that v_i is a constant input voltage to the circuit and v is the output voltage. Since the "+" lead of the op-amp is grounded, the "−" lead (and node A) also will have zero voltage. Also, the currents through the op-amp leads are negligible. Hence, the current balance at node A gives $v_i/R + C(dv/dt) = 0$. Integrating this equation for constant v_i, we have

$$v(t) = v(0) - \frac{v_i t}{RC} \tag{4.36}$$

Equation 4.36 will be utilized in obtaining a principal result for the dual slope ADC.

A schematic diagram for a dual slope ADC is shown in Figure 4.28b. Initially, the capacitor C in the integrating circuit is discharged (zero voltage). Then, the analog signal v_s is supplied to the switching element and held constant by the S/H circuit. Simultaneously, a "start conversion" (SC) control signal is sent to the control logic unit. This will clear the timer and the output register (i.e., all bits are set to zero) and will send a pulse to the switching element to connect the input v_s to the integrating circuit. Also, a signal is sent to the timer to initiate timing (counting). The capacitor C will begin to charge. Equation 4.36 is now applicable with input $v_i = v_s$ and the initial state $v(0) = 0$. Suppose that the integrator output v becomes $-v_c$ at time $t = t_1$. Hence, from Equation 4.36, we have

$$v_c = \frac{v_s t_1}{RC} \tag{i}$$

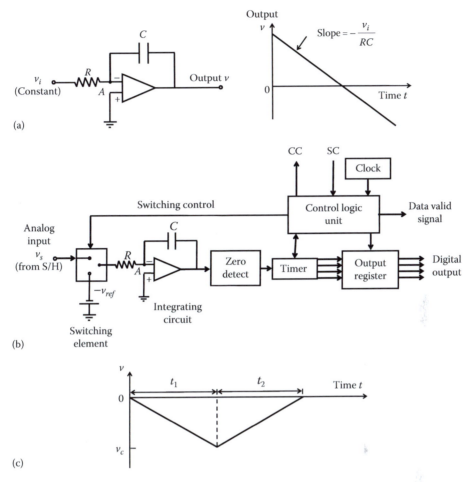

FIGURE 4.28
(a) *RC* integrating circuit. (b) Dual slope ADC. (c) Dual slope charging-discharging curve.

The timer will keep track of the capacitor charging time (as a clock pulse count n) and will inform the control logic unit when the elapsed time is t_1 (i.e., when the count is n_1). Note that t_1 and n_1 are fixed (and known) parameters but voltage v_c depends on the value of v_s and is unknown.

At this point, the control logic unit will send a signal to the switching unit, which will connect the input lead of the integrator to a negative supply voltage $-v_{ref}$. Simultaneously, a signal is sent to the timer to clear its contents and start timing (counting) again. Now the capacitor begins to discharge. The output of the integrating circuit is monitored by the "zero-detect" unit. When this output becomes zero, the zero-detect unit sends a signal to the timer to stop counting. The zero-detect unit could be a comparator (differential amplifier) having one of the two input leads set at zero potential.

Now suppose that the elapsed time is t_2 (with a corresponding count of n_2). It should be clear that Equation 4.36 is valid for the capacitor discharging process as well. Note that $v_i = -v_{ref}$ and $v(0) = -v_c$ in this case. Also, $v(t) = 0$ at $t = t_2$. Hence, from Equation 4.36, we have $0 = -v_c + (v_{ref}t_2)/RC$. Hence,

$$v_c = \frac{v_{ref} t_2}{RC} \tag{ii}$$

When dividing Equation (i) by (ii), we get $v_s = v_{ref}(t_2/t_1)$. But, the timer pulse count is proportional to the elapsed time. Hence, $t_2/t_1 = n_2/n_1$. Now we have

$$v_s = \frac{v_{ref}}{n_1} n_2 \tag{4.37}$$

Since v_{ref} and n_1 are fixed quantities, v_{ref}/n_1 can be interpreted as a scaling factor for the analog input. Then, it follows from Equation 4.37 that the second count n_2 is proportional to the analog signal sample v_s. Note that the timer output is available in the digital form. Accordingly, the count n_2 is used as the digital output of the ADC.

At the end of the capacitor discharge period, the count n_2 in the timer is transferred to the output register of the ADC, and the "data valid" signal is set. The contents of the output register are now ready to be read by the interfaced digital system, and the ADC is ready to convert a new sample.

The charging–discharging curve for the capacitor during the conversion process is shown in Figure 4.28c. The slope of the curve during charging is $-v_s/RC$ and the slope during discharging is $+v_{ref}/RC$. The reason for the use of the term "dual slope" to denote this ADC is therefore clear.

As mentioned before, any variations in R and C do not affect the accuracy of the output. But, it should be clear from the foregoing discussion that the conversion time depends on the capacitor discharging time t_2 (note that t_1 is fixed), which in turn depends on v_c and hence on the input signal value v_s (see Equation (i)). It follows that, unlike the successive approximation ADC, the dual slope ADC has a conversion time that directly depends on the magnitude of the input data sample. This is a disadvantage in a way because in many applications we prefer to have a constant conversion rate.

The above discussion assumed that the input signal is positive. For a negative signal, the polarity of the supply voltage v_{ref} has to be changed. Furthermore, the sign has to be properly represented in the contents of the output register as, for example, in the case of successive approximation ADC.

4.6.2.3 Counter ADC

The counter-type ADC has several aspects in common with the successive approximation ADC. Both are comparison-type (or closed-loop) ADCs. Both use a DAC unit internally to compare the input signal with the converted signal. The main difference is that in a counter ADC the comparison starts with the LSB and proceeds down. It follows that, in a counter ADC, the conversion time depends on the signal level because the counting (comparison) stops when a match is made, resulting in shorter conversion times for smaller signal values.

A schematic diagram for a counter ADC is shown in Figure 4.29. Note that this is quite similar to Figure 4.27. Initially, all registers are cleared (i.e., all bits and counts are set to zero). As an analog data signal (from the sample and hold circuit) arrives at the comparator, an SC pulse is sent to the control logic unit. When the ADC is ready for conversion (i.e., when the "data valid" signal is on), the control logic unit initiates the counter. Now, the counter sets its count to 1, and the LSB of the DAC register is set to 1 as well. The

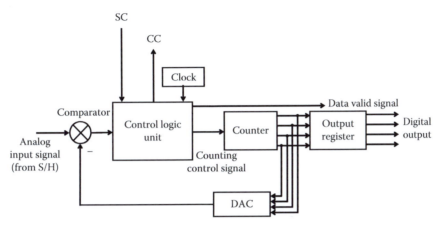

FIGURE 4.29
Counter ADC.

resulting DAC output is subtracted from the analog input by means of the comparator. If the comparator output is positive, the count is incremented by one and this causes the binary number in the DAC register to be incremented by one LSB. The new (increased) output of the DAC is now compared with the input signal. This cycle of count incrementing and comparison is repeated until the comparator output becomes less than or equal to zero. At that point, the control logic unit sends out a CC signal and transfers the contents of the counter to the output register. Finally, the "data valid" signal is turned on, indicating that the ADC is ready for a new conversion cycle, and the contents of the output register (the digital output) is available to be read by the interfaced digital system.

The count of the counter is available in the binary form, which is compatible with the output register as well as the DAC register. Hence, the count can be transferred directly to these registers. The count when the analog signal is equal to (or slightly less than) the output of the DAC is proportional to the analog signal value. Hence, this count represents the digital output. Again, the sign of the input signal has to be properly accounted for in the bipolar operation.

4.6.2.4 ADC Performance Characteristics

For ADCs that use a DAC internally, the same error sources that were discussed previously for DACs will apply. Code ambiguity at the output register will not be a problem because the converted digital quantity is transferred instantaneously to the output register. Code ambiguity in the DAC register can still cause errors in ADCs that use a DAC. Conversion time is a major factor, this being much larger for an ADC. In addition to resolution and dynamic range, quantization error will be applicable to an ADC. These considerations, which govern the performance of an ADC are discussed next.

4.6.2.4.1 Resolution and Quantization Error

The number of bits n in an ADC register determines the resolution and dynamic range of an ADC. For an n-bit ADC, the size of the output register is n bits. Hence, the smallest possible increment of the digital output is one LSB. The change in the analog input that results in a change of one LSB at the output is the resolution of the ADC. For the unipolar (unsigned) case, the available range of the digital outputs is from 0 to $2^n - 1$. This represents the dynamic range. It follows that, as for a DAC, the dynamic range of an n-bit ADC is given by the ratio $DR = 2^n - 1$ or in decibels $DR = 20 \log_{10}(2^n - 1)$ dB. The *full-scale value* of an ADC is the value of the analog input that corresponds to the maximum digital output.

Suppose that an analog signal within the dynamic range of a particular ADC is converted by that ADC. Since the analog input (sample value) has an infinitesimal resolution and the digital representation has a finite resolution (one LSB), an error is introduced in the process of analog-to-digital conversion. This is known as the *quantization error*. A digital number undergoes successive increments in constant steps of 1 LSB. If an analog value falls at an intermediate point within a step of single LSB, a quantization error is caused as a result. Rounding off of the digital output can be accomplished as follows: The magnitude of the error when quantized up is compared with that when quantized down; say, using two hold elements and a differential amplifier. Then, we retain the digital value corresponding to the lower error magnitude. If the analog value is below the 1/2 LSB mark, then the corresponding digital value is represented by the value at the beginning of the step. If the analog value is above the 1/2 LSB mark, then the corresponding digital value is the value at the end of the step. It follows that with this type of rounding off, the quantization error does not exceed 1/2 LSB.

4.6.2.4.2 Monotonicity, Nonlinearity, and Offset Error

Considerations of monotonicity and nonlinearity are important for an ADC as well as for a DAC. In the case of an ADC, the input is an analog signal and the output is digital. Disregarding quantization error, the digital output of an ADC will increase in constant steps in the shape of an ideal staircase function, when the analog input is increased from 0 in steps of the device resolution (δy). This is the ideally monotonic case. The best straight-line fit to this curve has a slope equal to $1/\delta y$ (LSB/Volts). This is the *ideal gain* or *ideal scale factor*. Still there will be an *offset error* of 1/2 LSB because the best linear fit will not pass through the origin. Adjustments can be made for this offset error.

Incorrect bit-transitions can take place in an ADC, due to various errors that might be present and also possibly due to circuit malfunctions. The best linear fit under such faulty conditions will have a slope different from the ideal gain. The difference is the gain error. Nonlinearity is the maximum deviation of the output from the best linear fit. It is clear that with perfect bit transitions, in the ideal case, a nonlinearity of 1/2 LSB would be present. Nonlinearities larger than this would result due to incorrect bit transitions. As in the case of a DAC, another source of nonlinearity in an ADC is circuit nonlinearities, which would deform the analog input signal before being converted into the digital form.

4.6.2.4.3 ADC Rate

It is clear that analog-to-digital conversion is much more time consuming than digital-to-analog conversion. The conversion time is a very important factor because the rate at which conversion can take place governs many aspects of data acquisition, particularly in real-time applications. For example, the data sampling rate has to synchronize with the ADC rate. This, in turn, will determine the Nyquist frequency (half the sampling rate), which corresponds to the bandwidth of the sampled signal and is the maximum value of useful frequency that is retained as a result of sampling. Furthermore, the sampling rate will dictate the requirements of storage and memory. Another important consideration related to the conversion rate of an ADC is the fact that a signal sample has to be maintained at the same value during the entire process of conversion into the digital form. This would require a *hold circuit*, and this circuit should be able to perform accurately at the largest possible conversion time for the particular ADC unit.

The time needed for a sampled analog input to be converted into the digital form will depend on the type of ADC. Usually, in a comparison type ADC (which uses an internal DAC), each bit transition will take place in one clock period Δt. Also, in an integrating (dual slope) ADC, each clock count will need a time of Δt. On this basis, for the three types of ADC that we have discussed, the following figures can be given for their conversion times.

1. Successive-approximation ADC
   In this case, for an n-bit ADC, n comparisons are needed. Hence, the conversion time is given by

$$t_c = n \cdot \Delta t \tag{4.38}$$

   in which Δt is the clock period. Note that for this ADC, t_c does not depend on the signal level (analog input).

2. Dual-slope (integrating) ADC
   In this case, the conversion time is the total time needed to generate the two counts n_1 and n_2 (see Figure 4.28c). Hence,

$$t_c = (n_1 + n_2)\Delta t \tag{4.39}$$

   Note that n_1 is a fixed count. But n_2 is a variable count, which represents the digital output and is proportional to the analog input (signal level). Hence, in this type of ADC, the conversion time depends on the analog input level. The largest output for an n-bit converter is $2n - 1$. Hence, the largest conversion time may be given by

$$t_{c\max} = (n_1 + 2^n - 1)\Delta T \tag{4.40}$$

3. Counter ADC
   For a counter ADC, the conversion time is proportional to the number of bit transitions (1 LSB per step) from zero to the digital output n_o. Hence, the conversion time is given by

$$t_c = n_o\Delta t \tag{4.41}$$

   in which n_o is the digital output value (in decimal).

Note that for this ADC as well, t_c depends on the magnitude of the input data sample. For an n-bit ADC, since the maximum value of n_o is $2^n - 1$, we have the maximum conversion time

$$t_{c\max} = (2^n - 1)\Delta t \tag{4.42}$$

By comparing Equations 4.38, 4.40, and 4.42, it can be concluded that the successive-approximation ADC is the fastest of the three types discussed.

The total time taken to convert an analog signal will depend on other factors besides the time taken for the conversion of sampled data into digital form. For example, in multiple-channel data acquisition (multiplexing), the time taken to select the channels has to be counted in. Furthermore, the time needed to sample the data and the time needed to transfer the converted digital data into the output register have to be included. In fact, the *conversion rate* for an ADC is the inverse of this overall time needed for a conversion cycle. Typically, however, the conversion rate depends primarily on the bit conversion time in the case of a comparison-type ADC and on the integration time in the case of an integration-type ADC.

Example 4.5

A typical time period for a comparison step or counting step in an ADC is $\Delta t = 5\,\mu s$. Hence, for an 8 bit successive approximation ADC, the conversion time is $40\,\mu s$. The corresponding sampling rate would be of the order of (less than) $1/40 \times 10^{-6} = 25 \times 10^3$ samples/s (or 25 kHz). The maximum conversion rate for an 8 bit counter ADC would be about $5 \times (2^8 - 1) = 1275\,\mu s$. The corresponding sampling rate would be on the order of 780 samples/s.

Note: This is considerably slow. The maximum conversion time for a dual slope ADC would likely be larger (i.e., slower rate).

Example 4.6

Consider an 8 bit ADC. For an input signal of range 10 V, the *quantization error* will be $10/(2^8 - 1)$ $V = 39.2\,mV$. Suppose the A/D *conversion time* is 100 ns (i.e., $0.1\,\mu s$). If the input signal has a maximum transient frequency of 0.5 kHz and a corresponding transient rate of 5×10^3 V/s, the signal will change by a maximum amount of $5 \times 10^3 \times 0.1 \times 10^{-6}\,V = 0.5\,mV$. This is the signal uncertainty (*amplitude uncertainty*) due to the transient nature of the signal during conversion. This uncertainty must be less than the quantization error, which is the case in this example.

4.6.3 Sample-and-Hold Circuitry

Typical applications of data acquisition use analog-to-digital conversion. The analog input to an ADC can be very transient, and furthermore, the process of analog-to-digital conversion itself is not instantaneous (ADC time can be much larger than the digital-to-analog conversion time). Specifically, the incoming analog signal might be changing at a rate higher than the ADC rate. Then, the input signal value will vary during the conversion period and there will be an ambiguity as to what analog input value corresponds to a particular digital output value. Hence, it is necessary to sample the analog input signal and maintain the input to the ADC at this sampled value until the analog-to-digital conversion is completed. In other words, since we are typically dealing with analog signals that can vary at a high speed, it would be necessary to S/H the input signal during each analog-to-digital conversion cycle. Each data sample must be generated and captured by the S/H circuit on the issue of the SC control signal, and the captured voltage level has to be maintained constant until a CC control signal is issued by the ADC unit.

The main element in an S/H circuit is the holding capacitor. A schematic diagram of a sample and hold circuit is shown in Figure 4.30. The analog input signal is supplied through a voltage follower to a solid–state switch. The switch typically uses a FET, such as the MOSFET. The switch is closed in response to a "sample pulse" and is opened in response to a "hold pulse." Both control pulses are generated by the control logic unit of the ADC. During the time interval between these two pulses, the holding capacitor is charged to the voltage of the sampled input. This capacitor voltage is then supplied to the ADC through a second voltage follower.

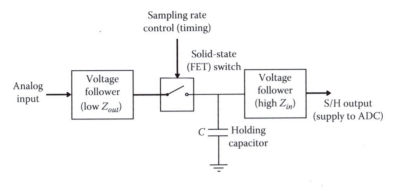

FIGURE 4.30
A sample and hold circuit.

The functions of the two voltage followers are explained in this section. When the FET switch is closed in response to a sample command (pulse), the capacitor has to be charged as quickly as possible. The associated time constant (charging time constant) τ_c is given by $\tau_c = R_s C$ in which R_s is the source resistance and C is the capacitance of the holding capacitor. Since τ_c has to be very small for fast charging, and since C is fixed by the holding requirements (typically C is of the order of 100 pF where $1\ \text{pF} = 1 \times 10^{-12}$ F), we need a very small source resistance. The requirement is met by the input voltage follower (which is known to have a very low output impedance), thereby providing a very small R_s. Furthermore, since a voltage follower has a unity gain, the voltage at the output of this input voltage follower would be equal to the voltage of the analog input signal, as required.

Next, once the FET switch is opened in response to a hold command (pulse), the capacitor should not discharge. This requirement is met due to the presence of the output voltage follower. Since the input impedance of a voltage follower is very high, the current through its leads would be almost zero. Because of this, the holding capacitor will have a virtually zero discharge rate under "hold" conditions. Furthermore, we like the output of this second voltage follower to be equal to the voltage of the capacitor. This condition is also satisfied due to the fact that a voltage follower has a unity gain. Hence, the sampling would be almost instantaneous and the output of the S/H circuit would be maintained (almost) at a constant during the holding period, due to the presence of the two voltage followers. Note that the practical S/H circuits are *zero-order-hold* devices by definition.

4.6.4 Multiplexers

A multiplexer (MUX) is used to select one channel at a time from a bank of signal channels and connect it to a common hardware unit. In this manner, a costly and complex hardware unit can be time-shared among several signal channels. Typically, channel selection is done in a sequential manner at a fixed channel-select rate. There are two types of multiplexers: analog multiplexers and digital multiplexers. An analog multiplexer is used to scan a group of analog channels. Alternatively, a digital multiplexer is used to read one data word at a time sequentially from a set of digital data words.

The process of distributing a single channel of data among several output channels is known as demultiplexing. A demultiplexer (or data distributor) performs the reverse function of a multiplexer (or scanner). A demultiplexer may be used, for example, when the same (processed) signal from a digital computer is needed for several purposes (e.g., digital display, analog reading, digital plotting, or control).

4.7 Bridge Circuits

A full bridge is a circuit having four arms connected in a lattice form. Four nodes are formed in this manner. Two opposite nodes are used for excitation (voltage or current supply) of the bridge and the remaining two opposite nodes provide the bridge output. A bridge circuit is used to make some form of measurement. Typical measurements include change in resistance, change in inductance, change in capacitance, oscillating frequency, or some variable (stimulus) that causes these changes. There are two basic methods of making the measurement: the bridge balance method and the imbalance output method. A bridge is said to be balanced when its output voltage is zero.

In the bridge balance method, we start with a balanced bridge. When making a measurement, the balance of the bridge will be upset due to the associated variation. As a result, a nonzero output voltage will be produced. The bridge can be balanced again by varying one of the arms of the bridge (assuming, of course, that some means are available for fine adjustments that may be required). The "change" that is required to restore the balance is in fact the "measurement." The bridge can be balanced precisely using a servo device, in this method.

In the imbalance output method as well, we usually start with a balanced bridge. As before, the balance of the bridge will be upset as a result of the change in the variable that is being measured. Now, instead of balancing the bridge again, the output voltage of the bridge due to the resulted imbalance is measured and used as the bridge measurement.

There are many types of bridge circuits. If the supply to the bridge is dc, then we have a *dc bridge*. Similarly, an *ac bridge* has an ac excitation. A resistance bridge has only resistance elements in its four arms and it is typically a dc bridge. An impedance bridge has impedance elements consisting of resistors, capacitors, and inductors in one or more of its arms. This is necessarily an ac bridge. If the bridge excitation is a constant voltage supply, we have a constant-voltage bridge. If the bridge supply is a constant current source, we get a constant-current bridge.

4.7.1 Wheatstone Bridge

A Wheatstone bridge is a resistance bridge with a constant dc voltage supply (i.e., it is a constant-voltage resistance bridge). A Wheatstone bridge is particularly useful in strain-gage measurements and consequently in force, torque, and tactile sensors that employ strain-gage techniques. Since a Wheatstone bridge is used primarily in the measurement of small changes in resistance, it can be used in other types of sensing applications as well. For example, in resistance temperature detectors (RTD), the change in resistance in a metallic (e.g., platinum) element, as caused by a change in temperature, is measured using a bridge circuit. *Note*: The temperature coefficient of resistance is positive for a typical metal (i.e., the resistance increases with temperature). For platinum, this value (change in resistance per unit of resistance per unit of change in temperature) is about $0.00385/°C$.

Consider the Wheatstone bridge circuit shown in Figure 4.31a. Assuming that the bridge output is open-circuit (i.e., very high load resistance), the output v_o may be expressed as

$$v_o = v_A - v_B = \frac{R_1 v_{ref}}{(R_1 + R_2)} - \frac{R_3 v_{ref}}{(R_3 + R_4)} = \frac{(R_1 R_4 - R_2 R_3)}{(R_1 + R_2)(R_3 + R_4)} v_{ref} \qquad (4.43)$$

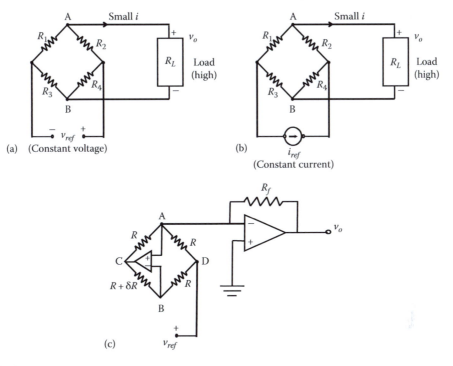

FIGURE 4.31
(a) Wheatstone bridge (constant-voltage resistance bridge). (b) Constant-current resistance bridge. (c) A linearized bridge.

For a balanced bridge, the numerator of the right-hand side expression of Equation 4.43 must vanish. Hence, the condition for bridge balance is

$$\frac{R_1}{R_2} = \frac{R_3}{R_4} \tag{4.44}$$

Suppose that at first $R_1 = R_2 = R_3 = R_4 = R$. Then, according to Equation 4.44, the bridge is balanced. Now increase R_1 by δR. For example, R_1 may represent the only active strain gage while the remaining three elements in the bridge are identical dummy elements. In view of Equation 4.43, the change in the bridge output due to the change δR is given by

$\delta v_o = \dfrac{\left[(R+\delta R)R - R^2\right]}{(R+\delta R+R)(R+R)} v_{ref} - 0$, which can be written as

$$\frac{\delta v_o}{v_{ref}} = \frac{\delta R/R}{(4+2\delta R/R)} \tag{4.45a}$$

Note: The output is nonlinear in $\delta R/R$. If, however, $\delta R/R$ is assumed small in comparison to 2, we have the linearized relationship

$$\frac{\delta v_o}{v_{ref}} = \frac{\delta R}{4R} \tag{4.46}$$

The factor ¼ on the right-hand side of Equation 4.46 representing the *sensitivity* is the bridge, as it gives the change in the bridge output for a given change in the active resistance while the other parameters are kept fixed. Strictly speaking, the bridge sensitivity is given by $\delta v_o / \delta R$, which is equal to $v_{ref}/(4R)$.

The error due to linearization, which is a measure of nonlinearity, may be given as the percentage

$$N_P = 100 \left(1 - \frac{\text{Linearized output}}{\text{Actual output}} \right) \% \tag{4.47}$$

Hence, from Equations 4.45 and 4.46, we have

$$N_P = 50 \frac{\delta R}{R} \% \tag{4.48}$$

Example 4.7

Suppose that in Figure 4.31a, at first $R_1 = R_2 = R_3 = R_4 = R$. Now increase R_1 by δR and decrease R_2 by δR. This will represent two active elements that act in reverse, as in the case of two strain gage elements mounted on the top and bottom surfaces of a beam in bending. Show that the bridge output is linear in δR in this case.

Solution

From Equation 4.43, we get $\delta v_o = \dfrac{\left[(R+\delta R)R - R^2 \right]}{(R+\delta R + R - \delta R)(R+R)} v_{ref} - 0$. This simplifies to $\delta v_o / v_{ref} = \delta R / 4R$, which is linear. Similarly, it can be shown using Equation 4.43 that the pair of changes $R_3 \to R + \delta R$ and $R_4 \to R - \delta R$ will result in a linear relation for the bridge output.

4.7.2 Constant-Current Bridge

When large resistance variations δR are required for a measurement, the Wheatstone bridge may not be satisfactory due to its nonlinearity, as indicated by Equation 4.45. The constant-current bridge is less nonlinear and is preferred in such applications. However, it needs a current-regulated power supply, which is typically more costly than a voltage-regulated power supply.

As shown in Figure 4.31b, the constant-current bridge uses a constant-current excitation i_{ref} instead of a constant voltage supply. The output equation for a constant-current bridge can be determined from Equation 4.43 simply by knowing the voltage at the current source. Suppose that this voltage is v_{ref} with the polarity as shown in Figure 4.31a. Now, since the load current is assumed small (high-impedance load), the current through R_2 is equal to the current through R_1 and is given by $v_{ref}/(R_1 + R_2)$. Similarly, current through R_4 and R_3 is given by $v_{ref}/(R_3 + R_4)$. Accordingly, by current summation, we get $i_{ref} = \dfrac{v_{ref}}{(R_1 + R_2)} + \dfrac{v_{ref}}{(R_3 + R_4)}$, which can be expressed as

$$v_{ref} = \frac{(R_1 + R_2)(R_3 + R_4)}{(R_1 + R_2 + R_3 + R_4)} i_{ref} \tag{4.49}$$

This result may be directly obtained from the equivalent resistance of the bridge, as seen by the current source. By substituting Equation 4.49 in Equation 4.43, we have the output equation for the constant-current bridge:

$$v_o = \frac{(R_1 R_4 - R_2 R_3)}{(R_1 + R_2 + R_3 + R_4)} i_{ref} \qquad (4.50)$$

Note from Equation 4.50 that the bridge-balance requirement (i.e., $v_o = 0$) is again given by Equation 4.44.

To estimate the nonlinearity of a constant-current bridge, we start with the balanced condition $R_1 = R_2 = R_3 = R_4 = R$ and change R_1 by δR while keeping the remaining resistors inactive. Again, R_1 will represent the active element (sensing element) of the bridge and may correspond to an active strain gage. The change in output δv_o is given by $\delta v_o = \dfrac{\left[(R + \delta R)R - R^2\right]}{(R + \delta R + R + R + R)} i_{ref} - 0$, which may be written as

$$\frac{\delta v_o}{R i_{ref}} = \frac{\delta R / R}{(4 + \delta R / R)} \qquad (4.51a)$$

By comparing the denominator on the right-hand side of this equation with Equation 4.45, we observe that the constant-current bridge is less nonlinear. Specifically, using the definition given by Equation 4.47, the percentage nonlinearity may be expressed as

$$N_p = 25 \frac{\delta R}{R} \% \qquad (4.52)$$

It is noted that the nonlinearity is halved by using a constant-current excitation instead of a constant-voltage excitation.

Example 4.8

Suppose that in the constant-current bridge circuit shown in Figure 4.31b, at first $R_1 = R_2 = R_3 = R_4 = R$. Assume that R_1 and R_4 represent strain gages mounted on the same side of a rod in tension. Due to the tension, R_1 increases by δR and R_4 also increases by δR. Derive an expression for the bridge output (normalized) in this case and show that it is linear. What would be the result if R_2 and R_3 represent the active tensile strain gages in this example?

Solution

From Equation 4.50, we get $\delta v_o = \dfrac{\left[(R + \delta R)(R + \delta R) - R^2\right]}{(R + \delta R + R + R + R + \delta R)} i_{ref} - 0$. By simplifying and canceling the common term in the numerator and the denominator, we get the following linear relation:

$$\frac{\delta v_o}{R i_{ref}} = \frac{\delta R / R}{2} \qquad (4.51b)$$

If R_2 and R_3 are the active elements, it is clear from Equation 4.50 that we get the same linear result, except for a sign change, as

$$\frac{\delta v_o}{R i_{ref}} = -\frac{\delta R / R}{2} \qquad (4.51c)$$

4.7.3 Hardware Linearization of Bridge Outputs

From the foregoing developments and as illustrated in the examples, it should be clear that the output of a resistance bridge is not linear in general, with respect to the change in the resistance of the active elements. Particular arrangements of the active elements can result in a linear output. It is seen from Equations 4.43 and 4.50 that when there is only one active element, the bridge output is nonlinear. Such a nonlinear bridge can be linearized using hardware; particularly op-amp elements. To illustrate this approach, consider a constant-voltage resistance bridge. We modify it by connecting two op-amp elements, as shown in Figure 4.31c. The output amplifier has a feedback resistor R_f. The output equation for this circuit can be obtained by using the properties of an op-amp in the usual manner. In particular, the potentials at the two input leads must be equal and the current through these leads must be zero. From the first property, it follows that the potentials at nodes A and B are both zero. Let the potential at node C be denoted by v. Now use the second property and write the current summations at nodes A and B.

$$\text{Node } A: \frac{v}{R} + \frac{v_{ref}}{R} + \frac{v_o}{R_f} = 0 \tag{i}$$

$$\text{Node } B: \frac{v_{ref}}{R} + \frac{v}{R+\delta R} = 0 \tag{ii}$$

Substitute Equation (ii) in (i) to eliminate v, and simplify to get the linear result:

$$\frac{\delta v_o}{v_{ref}} = \frac{R_f}{R}\frac{\delta R}{R} \tag{4.45b}$$

Compare this result with Equation 4.45a for the original bridge with a single active element. Note that when $\delta R = 0$, from (ii) we get, $v = v_{ref}$, and from (i) we get $v_o = 0$. Hence, v_o and δv_o are identical, as used in Equation 4.45b.

4.7.4 Bridge Amplifiers

The output signal from a resistance bridge is usually very small in comparison with the reference signal, and it has to be amplified in order to increase its voltage level to a useful value (e.g., for use in system monitoring, data logging, or control). A bridge amplifier is used for this purpose. This is typically an *instrumentation amplifier*, which is essentially a sophisticated *differential amplifier*. The bridge amplifier is modeled as a simple gain K_a, which multiplies the bridge output.

4.7.5 Half-Bridge Circuits

A half bridge may be used in some applications that require a bridge circuit. A half bridge has only two arms and the output is tapped from the midpoint of these two arms. The ends of the two arms are excited by two voltages, one of which is positive and the other negative. Initially, the two arms have equal resistances so that nominally the bridge output is zero. One of the arms has the active element. Its change in resistance results in a nonzero output voltage. It is noted that the half-bridge circuit is somewhat similar to a potentiometer circuit (a voltage divider).

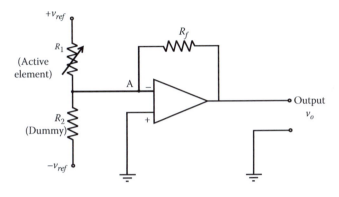

FIGURE 4.32
A half bridge with an output amplifier.

A half-bridge amplifier consisting of a resistance half bridge and an output amplifier is shown in Figure 4.32. The two bridge arms have resistances R_1 and R_2, and the output amplifier uses a feedback resistance R_f. To get the output equation, we use the two basic facts for an unsaturated op-amp; the voltages at the two input leads are equal (due to high gain) and the current in either lead is zero (due to high input impedance). Hence, the voltage at node A is zero and the current balance equation at node A is given by $\dfrac{v_{ref}}{R_1} + \dfrac{(-v_{ref})}{R_2} + \dfrac{v_o}{R_f} = 0$. This gives

$$v_o = R_f \left(\frac{1}{R_2} - \frac{1}{R_1} \right) v_{ref} \qquad (4.53)$$

Now, suppose that initially $R_1 = R_2 = R$ and the active element R_1 changes by δR. The corresponding change in output is $\delta v_o = R_f \left(\dfrac{1}{R} - \dfrac{1}{R + \delta R} \right) v_{ref} - 0$, which may be written as

$$\frac{\delta v_o}{v_{ref}} = \frac{R_f}{R} \frac{\delta R/R}{(1 + \delta R/R)} \qquad (4.54)$$

Note: R_f/R is the amplifier gain. Now in view of Equation 4.47, the percentage nonlinearity of the half-bridge circuit is

$$N_p = 100 \frac{\delta R}{R} \% \qquad (4.55)$$

It follows that the nonlinearity of a half-bridge circuit is worse than that for the Wheatstone bridge.

4.7.6 Impedance Bridges

An impedance bridge is an ac bridge. It contains general impedance elements Z_1, Z_2, Z_3, and Z_4 in its four arms, as shown in Figure 4.33a. The bridge is excited by an ac (supply) voltage v_{ref}. Note that v_{ref} would represent a carrier signal and the output voltage v_o has to

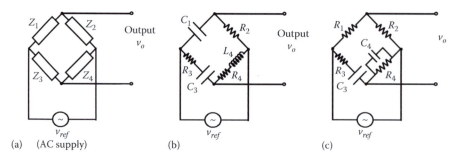

FIGURE 4.33
(a) General impedance bridge. (b) Owen bridge. (c) Wien-bridge oscillator.

be demodulated if a transient signal representative of the variation in one of the bridge elements is needed. Impedance bridges could be used, for example, to measure capacitances in *capacitive sensors* and changes of inductance in *variable-inductance sensors* and *eddy-current sensors*. Also, impedance bridges can be used as *oscillator circuits*. An oscillator circuit could serve as a constant-frequency source of a signal generator (e.g., in product dynamic testing) or it could be used to determine an unknown circuit parameter by measuring the oscillating frequency.

When analyzing using frequency-domain concepts, it is seen that the frequency spectrum of the impedance-bridge output is given by

$$v_o(\omega) = \frac{(Z_1 Z_4 - Z_2 Z_3)}{(Z_1 + Z_2)(Z_3 + Z_4)} v_{ref}(\omega) \tag{4.56}$$

This reduces to Equation 4.43 in the dc case of a Wheatstone bridge. The balanced condition is given by

$$\frac{Z_1}{Z_2} = \frac{Z_3}{Z_4} \tag{4.57}$$

This equation is used to measure an unknown circuit parameter in the bridge. Let us consider two particular impedance bridges.

4.7.6.1 Owen Bridge

The Owen bridge is shown in Figure 4.33b. It may be used, for example, to measure both inductance L_4 and capacitance C_3, by the bridge-balance method. To derive the necessary equation, note that the voltage–current relation for an inductor is $v = L(di/dt)$ and for a capacitor it is $i = C(dv/dt)$. It follows that the voltage/current transfer function (in the Laplace domain) for an inductor is $v(s)/i(s) = Ls$, and that for a capacitor is $v(s)/i(s) = 1/Cs$. Accordingly, the impedance of an inductor element at frequency ω is $Z_L = j\omega L$ and the impedance of a capacitor element at frequency ω is $Z_c = 1/j\omega C$. By applying these results for the Owen bridge, we have $Z_1 = 1/j\omega C_1$; $Z_2 = R_2$; $Z_3 = R_3 + (1/j\omega C_3)$; and $Z_4 = R_4 + j\omega L_4$ in which ω is the excitation frequency. Now, from Equation 4.57, we have

$\dfrac{1}{j\omega C_1}(R_4 + j\omega L_4) = R_2\left(R_3 + \dfrac{1}{j\omega C_3}\right)$. By equating the real parts and the imaginary parts of this equation, we get the following two equations: $L_4/C_1 = R_2 R_3$ and $R_4/C_1 = R_2/C_3$. Hence, we have $L_4 = C_1 R_1 R_2$ and $C_3 = C_1(R_2/R_4)$. It follows that L_4 and C_3 can be determined with the knowledge of C_1, R_2, R_3, and R_4 under balanced conditions. For example, with fixed C_1 and R_2, an adjustable R_3 could be used to measure the variable L_4, and an adjustable R_4 could be used to measure the variable C_3.

4.7.6.2 Wien-Bridge Oscillator

Now consider the Wien-bridge oscillator shown in Figure 4.33c. For this circuit, we have $Z_1 = R_1$; $Z_2 = R_2$; $Z_3 = R_3 + (1/j\omega C_3)$; and $\dfrac{1}{Z_4} = \dfrac{1}{R_4} + J\omega C_4$. Hence, from Equation 4.57, the bridge-balance requirement is $\dfrac{R_1}{R_2} = \left(R_3 + \dfrac{1}{j\omega C_4}\right)\left(\dfrac{1}{R_4} + j\omega C_4\right)$. By equating the real parts, we get $\dfrac{R_1}{R_2} = \dfrac{R_3}{R_4} + \dfrac{C_4}{C_3}$ and by equating the imaginary parts, we get $0 = \omega C_4 R_3 - (1/\omega C_3 R_4)$. Hence,

$$\omega = \frac{1}{\sqrt{C_3 C_4 R_3 R_4}} \qquad (4.58)$$

Equation 4.58 tells us that the circuit is an oscillator whose natural frequency is given by this equation, under balanced conditions. If the frequency of the supply is equal to the natural frequency of the circuit, large-amplitude oscillations will take place. The circuit can be used to measure an unknown resistance (e.g., in strain gage devices) by first measuring the frequency of the bridge signals at resonance (natural frequency). Alternatively, an oscillator that is excited at its natural frequency can be used as an accurate source of periodic signals (signal generator).

Problems

4.1 Define electrical impedance and mechanical impedance. Identify a defect in these definitions in relation to the force–current analogy. What improvements would you suggest? What roles do input impedance and output impedance play in relation to the accuracy of a measuring device?

4.2 What is meant by "loading error" in a signal measurement? Also, suppose that a piezoelectric sensor of output impedance Z_s is connected to a voltage-follower amplifier of input impedance Z_i as shown in Figure P4.2. The sensor signal is v_i volts and the amplifier output is v_o volts. The amplifier output is connected to a device with very high input impedance. Plot to scale the signal ratio v_o/v_i against the impedance ratio Z_i/Z_s for values of the impedance ratio in the range 0.1–10.

4.3 Explain why a voltmeter should have a high resistance and an ammeter should have a very low resistance. What are some of the design implications of these general requirements for the two types of measuring instruments, particularly with respect to instrument sensitivity, speed of response, and robustness? Use a classical moving-coil meter as the model for your discussion.

4.4 Define mechanical loading and electrical loading in the context of motion sensing, and explain how these loading effects can be reduced.

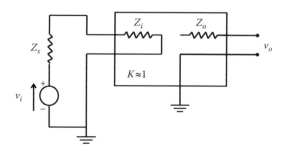

FIGURE P4.2
System with a piezoelectric sensor.

The following table gives ideal values for some parameters of an op-amp. Give typical, practical values for these parameters (e.g., output impedance of 50 Ω).

Parameter	Ideal Value	Typical Value
Input impedance	Infinity	?
Output impedance	Zero	50 Ω
Gain	Infinity	?
Bandwidth	Infinity	?

Also note that under ideal conditions, inverting-lead voltage is equal to the noninverting-lead voltage (i.e., offset voltage is zero).

4.5 Usually, an op-amp circuit is analyzed making use of the following two assumptions:

(i) The potential at the "+" input lead is equal to the potential at the "−" input lead.

(ii) The current through each of the two input leads is zero.

Explain why these assumptions are valid under unsaturated conditions of an op-amp.
An amateur electronics enthusiast connects an op-amp to a circuit without a feedback element. Even when there is no signal applied to the op-amp, the output was found to oscillate between +12 and −12 V once the power supply was turned on. Give a reason for this behavior.
An op-amp has an open-loop gain of 5×10^5 and a saturated output of ±14 V. If the noninverting input is −1 μV and the inverting input is +0.5 μV, what is the output? If the inverting input is 5 μV and the noninverting input is grounded, what is the output?

4.6 Define the following terms in connection with an op-amp:

(a) Offset current

(b) Offset voltage (at input and output)

(c) Unequal gains

(d) Slew rate

Give typical values for these parameters. The open-loop gain and the input impedance of an op-amp are known to vary with frequency and are known to drift (with time) as well. Still, the op-amp circuits are known to behave very accurately. What is the main reason for this?

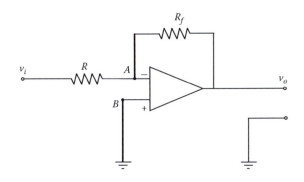

FIGURE P4.7
An amplifier circuit.

4.7 What is a voltage follower? Discuss the validity of the following statements:

(a) A voltage follower is a current amplifier.

(b) A voltage follower is a power amplifier.

(c) A voltage follower is an impedance transformer.

Consider the amplifier circuit shown in Figure P4.7. Determine an expression for the voltage gain K_v of the amplifier in terms of the resistances R and R_f. Is this an inverting amplifier or a noninverting amplifier?

4.8 The speed of response of an amplifier may be represented using three parameters: bandwidth, rise time, and slew rate. For an idealized linear model (transfer function), it can be verified that the rise time and bandwidth are independent of the size of the input and the dc gain of the system. Since the size of the output (under steady conditions) may be expressed as the product of the input size and the dc gain, it is seen that the rise time and the bandwidth are independent of the amplitude of the output, for a linear model.

Discuss how the slew rate is related to the bandwidth and rise time of a practical amplifier. Usually, amplifiers have a limiting slew rate value. Show that the bandwidth decreases with the output amplitude in this case.

A voltage follower has a slew rate of 0.5 V/μs. If a sinusoidal voltage of amplitude 2.5 V is applied to this amplifier, estimate the operating bandwidth. If, instead, a step input of magnitude 5 V is applied, estimate the time required for the output to reach 5 V.

4.9 Define the following terms:

(a) Common-mode voltage

(b) Common-mode gain

(c) CMRR

What is a typical value for the CMRR of an op-amp? Figure P4.9 shows a differential amplifier circuit with a flying capacitor. The switch pairs A and B are turned on and off alternately during operation. For example, first the switches denoted by A are turned on (closed) with the switches B off (open). Next, the switches denoted by A are opened and the switches B are closed. Explain why this arrangement provides good common-mode rejection characteristics.

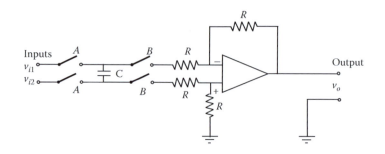

FIGURE P4.9
A differential amplifier with a flying capacitor for common-mode rejection.

4.10 Compare the conventional (textbook) meaning of system stability and the practical interpretation of instrument stability.

An amplifier is known to have a temperature drift of $1\,mV/°C$ and a long-term drift of $25\,\mu V/month$. Define the terms temperature drift and long-term drift. Suggest ways to reduce drift in an instrument.

4.11 Obtain a relationship between the slew rate and the bandwidth for a slew-rate-limited device. An amplifier has a slew rate of $1\,V/\mu s$. Determine the bandwidth of this amplifier when operating at an output amplitude of $5\,V$.

4.12 What are passive filters? List several advantages and disadvantages of passive (analog) filters in comparison with active filters.

A simple way to construct an active filter is to start with a passive filter of the same type and add a voltage follower to the output. What is the purpose of such a voltage follower?

4.13 Give one application each for the following types of analog filters:

   (a) Low-pass filter
   (b) High-pass filter
   (c) Band-pass filter
   (d) Notch filter

Suppose that several single-pole active filter stages are cascaded. Is it possible for the overall (cascaded) filter to possess a resonant peak? Explain.

4.14 A Butterworth filter is said to have a "maximally flat magnitude." Explain what is meant by this. Give another characteristic that is desired from a practical filter.

4.15 Consider an analog low-pass filter given by the transfer function $v_o/v_i = G_f(s) = -k/(\tau_f s + 1)$. Here, $1/\tau_f$ can be considered the cutoff frequency ω_c of the filter. Show that the cutoff frequency is also the *half-power bandwidth* for the low-pass filter. Show that for frequencies much larger than this, the filter transfer function on the Bode magnitude plane (i.e., log magnitude versus log frequency) can be approximated by a straight line with a slope of $-20\,dB/decade$ (the *roll-off rate*).

4.16 What is meant by each of the following terms: modulation, modulating signal, carrier signal, modulated signal, and demodulation? Explain the following types of signal modulation giving an application for each case:

   (a) Amplitude modulation
   (b) Frequency modulation

(c) Phase modulation

(d) PWM

(e) Pulse-frequency modulation

(f) Pulse-code modulation

How could the sign of the modulating signal be accounted for during demodulation in each of these types of modulation?

4.17 Give two situations where amplitude modulation is intentionally introduced and in each situation explain how amplitude modulation is beneficial. Also, describe two devices where amplitude modulation might be naturally present. Could the fact that amplitude modulation is present be exploited to our advantage in these two natural situations as well? Explain.

4.18 The electrical isolation of a device (or circuit) from another device (or circuit) is very useful in the mechatronic practice. An isolation amplifier may be used to achieve this. It provides a transmission link, which is almost "one way," and avoids loading problems. In this manner, damage in one component due to an increase in signal levels in the other components (perhaps due to short-circuits, malfunctions, noise, high common-mode signals, etc.) could be reduced. An isolation amplifier can be constructed from a transformer and a demodulator with other auxiliary components such as filters and amplifiers. Draw a suitable schematic diagram for an isolation amplifier, and explain the operation of this device.

4.19 A LVDT is a displacement sensor, which is commonly used in mechatronic systems. Consider a digital control loop that uses an LVDT measurement for position control of a machine. Typically, the LVDT is energized by a dc power supply. An oscillator provides an excitation signal in the kilohertz range to the primary windings of the LVDT. The secondary winding segments are connected in series opposition. An ac amplifier, demodulator, low-pass filter, amplifier, and ADC are used in the monitoring path. Figure P4.19 shows the various hardware components in the control loop. Indicate the functions of these components.

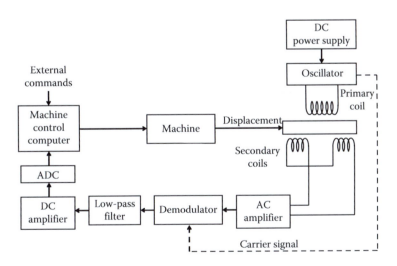

FIGURE P4.19
Components of an LVDT-based machine control loop.

At the null position of the LVDT stroke, there was a residual voltage. A compensating resistor was used to eliminate this voltage. Indicate the connections for this compensating resistor.

4.20 A monitoring system for a ball bearing of a rotating machine is schematically shown in Figure P4.20a. It consists of an accelerometer to measure the bearing vibration and an FFT analyzer to compute the Fourier spectrum of the response signal. This spectrum is examined over a period of one month after the installation of the rotating machine in order to detect any degradation in the bearing performance. An interested segment of the Fourier spectrum can be examined with high resolution by using the "zoom analysis" capability of the FFT analyzer. The magnitude of the original spectrum and that of the spectrum determined 1 month later, in the same zoom region, are shown in Figure P4.20b.

  (a) Estimate the operating speed of the rotating machine and the number of balls in the bearing.

  (b) Do you suspect any bearing problems?

4.21 Explain the following terms:

  (a) Phase sensitive demodulation

  (b) Half-wave demodulation

  (c) Full-wave demodulation

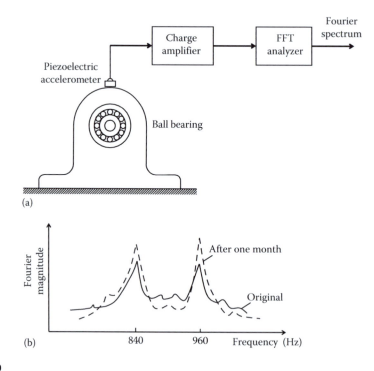

FIGURE P4.20
(a) A monitoring system for a ball bearing. (b) A zoomed Fourier spectrum.

When vibrations in rotating machinery such as gear boxes, bearings, turbines, and compressors are monitored, it is observed that a peak of the spectral magnitude curve does not usually occur at the frequency corresponding to the forcing function (e.g., tooth meshing, ball or roller hammer, blade passing). But, instead, two peaks occur on the two sides of this frequency. Explain the reason for this.

4.22 Define the following terms in relation to an analog-to-digital converter:

(a) Resolution

(b) Dynamic range

(c) Full-scale value

(d) Quantization error

4.23 A sinusoidal voltage signal $v = v_a \sin \omega t$ is read through an A/D converter (ADC) of conversion time t_c. Obtain an expression for the amplitude uncertainty in the digitized data in view of the conversion time of the ADC.

Note: This uncertainty level should not exceed the quantization error of the ADC.

4.24 Single-chip amplifiers with built-in compensation and filtering circuits are becoming popular for signal conditioning tasks in mechatronic systems, particularly those associated with data acquisition, machine monitoring, and control. Signal processing such as integration that would be needed to convert, say, an accelerometer into a velocity sensor, can also be accomplished in the analog form using an IC chip. What are the advantages of such signal modification chips in comparison with the conventional analog signal conditioning hardware that employ discrete circuit elements and separate components to accomplish various signal conditioning tasks?

4.25 Compare the three types of bridge circuits: constant-voltage bridge, constant-current bridge, and half-bridge in terms of nonlinearity, effect of change in temperature, and cost.

Obtain an expression for the percentage error in a half-bridge circuit output due to an error δv_{ref} in the voltage supply v_{ref}. Compute the percentage error in the output if the voltage supply has a 1% error.

4.26 Suppose that in the constant-voltage bridge circuit shown in Figure 4.31a, at first, $R_1 = R_2 = R_3 = R_4 = R$. Assume that R_1 represents a strain gage mounted on the tensile side of a bending beam element and that R_3 represents another strain gage mounted on the compressive side of the bending beam. Due to bending, R_1 increases by δR and R_3 decreases by δR. Derive an expression for the bridge output in this case, and show that it is nonlinear. What would be the result if R_2 represents the tensile strain gage and R_4 represents the compressive strain gage, instead?

4.27 Suppose that in the constant-current bridge circuit shown in Figure 4.31b, at first, $R_1 = R_2 = R_3 = R_4 = R$. Assume that R_1 and R_2 represent strain gages mounted on a rotating shaft, at right angles, and symmetrically about the axis of rotation. Also, in this configuration and in a particular direction of rotation of the shaft, suppose that R_1 increases by δR and R_2 decreases by δR. Derive an expression for the bridge output (normalized) in this case, and show that it is linear. What would be the result if R_4 and R_3 were to represent the active strain gages in this example, the former element being in tension and the latter in compression?

4.28 Consider the constant-voltage bridge shown in Figure 4.31a.

Suppose that the bridge is balanced with the resistors set according to

$$\frac{R_1}{R_2} = \frac{R_3}{R_4} = p$$

Now, if the active element R_1 increases by δR_1, what is the resulting output of the bridge?

Note: This represents the sensitivity of the bridge.

   For what value of the resistance ratio p, would the bridge sensitivity be a maximum? Show that this ratio is almost equal to 1.

4.29 The Maxwell bridge circuit is shown in Figure P4.29. Obtain the conditions for a balanced Maxwell bridge in terms of the circuit parameters R_1, R_2, R_3, R_4, C_1, and L_4. Explain how this circuit could be used to measure variations in both C_1 or L_4.

4.30 The standard LVDT (linear variable differential transducer or transformer) arrangement has a primary coil and two secondary coil segments connected in series opposition. Alternatively, some LVDTs use a bridge circuit to produce their output. An example of a half-bridge circuit for an LVDT is shown in Figure P4.30. Explain the operation of this arrangement. Extend this idea to a full impedance bridge for LVDT measurement.

4.31 The output of a Wheatstone bridge is nonlinear with respect to the variations in a bridge resistance. This nonlinearity is negligible for small changes in resistance. For large variations in resistance, however, some method of calibration or linearization should be employed. One way to linearize the bridge output is to feed back (positively) the output voltage signal into the bridge supply using a feedback op-amp. Consider the Wheatstone bridge circuit shown in Figure 4.31a. Initially, the bridge is balanced with $R_1 = R_2 = R_3 = R_4 = R$. Then, the resistor R_1 is varied to $R + \delta R$. Suppose that the bridge output δv_o is fed back (positive) with a gain of 2 into the bridge supply v_{ref}. Show that this will linearize the bridge equation.

4.32 A furnace used in a chemical process is controlled in the following manner. The furnace is turned on in the beginning of the process. When the temperature within the furnace reaches a certain threshold value T_o, the (temperature) × (time) product is measured in the units of Celsius minutes. When this product reaches a specified value, the furnace is turned off. The available hardware includes a resistance temperature

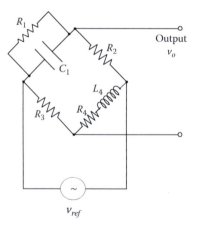

FIGURE P4.29
The Maxwell bridge.

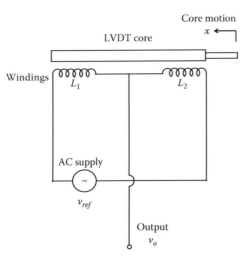

FIGURE P4.30
A half-bridge circuit for an LVDT.

detector (RTD), a differential amplifier, a diode circuit that does not conduct when the input voltage is negative and conducts with a current proportional to the input voltage when the input is positive, a current-to-voltage converter circuit, a voltage-to-frequency converter (VFC), a counter, and an on/off control unit. Draw a block diagram for this control system and explain its operation. Clearly identify the signal-modification operations in this control system, indicating the purpose of each operation.

4.33 Typically, when a digital transducer is employed to generate the feedback signal for an analog controller, a digital-to-analog converter (DAC) would be needed to convert the digital output from the transducer into a continuous (analog) signal. Similarly, when a digital controller is used to drive an analog process, a DAC has to be used to convert the digital output from the controller into the analog drive signal. There are ways, however, to eliminate the need for a DAC in these types of situations.

(a) Show how a shaft encoder and a frequency-to-voltage converter can replace an analog tachometer in an analog speed-control loop.

(b) Show how a digital controller with PWM can be employed to drive a dc motor without the use of a DAC.

4.34 The noise in an electrical circuit can depend on the nature of the coupling mechanism. In particular, the following types of coupling are available:

(a) Conductive coupling
(b) Inductive coupling
(c) Capacitive coupling
(d) Optical coupling

Compare these four types of coupling with respect to the nature and level of noise that is fed through or eliminated in each case. Discuss ways to reduce noise that is fed through in each type of coupling.

The noise due to variations in ambient light can be a major problem in optically coupled systems. Briefly discuss a method that could be used in an optically-coupled device in order to make the device immune to variations in the ambient light level.

4.35 What are the advantages of using optical coupling in electrical circuits? For optical coupling, diodes that emit infrared radiation are often preferred over light emitting diodes (LEDs), which emit visible light. What are the reasons behind this? Discuss why pulse-modulated light (or pulse-modulated radiation) is used in many types of optical systems. List several advantages and disadvantages of laser-based optical systems.

The Young's modulus of a material with known density can be determined by measuring the frequency of the fundamental mode of transverse vibration of a uniform cantilever beam specimen of the material. A photosensor and a timer can be used for this measurement. Describe an experimental setup for this method of determining the modulus of elasticity.

5

Instrument Ratings and Error Analysis

Study Objectives

- Performance of a mechatronic system
- Instrument rating parameters
- Nonlinearities
- Bandwidth; System design using bandwidth considerations
- Signal sampling and aliasing error
- Component error propagation, combination, and analysis
- Absolute error and square-root of sum of squares (SRSS) error
- Statistical process control (SPC)
- Considerations of probability and statistics (also see Appendix C)

5.1 Introduction

A mechatronic system consists of an integration of several components such as sensors, transducers, signal conditioning/modification devices, controllers, and a variety of other electronic and digital hardware. In the design, selection, and prescription of these components, their performance requirements have to be specified or established within the functional needs of the overall mechatronic system. Engineering parameters for performance specification may be defined either in the time domain or in the frequency domain. Instrument ratings of commercial products are often developed on the basis of these engineering parameters. This chapter addresses these and related issues of performance specification.

A sensor detects (feels) the quantity that is being measured (measurand). The transducer converts the detected measurand into a convenient form for subsequent use (recording, control, actuation, etc.). The transducer signal may be filtered, amplified, and suitably modified prior to this. Bandwidth plays an important role in specifying and characterizing these and other components of a mechatronic system. In particular, useful frequency range, operating bandwidth, and control bandwidth are important considerations in mechatronic systems. In this chapter, we will study several important issues related to system bandwidth.

In any multicomponent system, the overall error depends on the component error. A component error degrades the performance of a mechatronic system. This is particularly true for sensors and transducers as their error is directly manifested as incorrectly-known system variables and parameters within the system. Since an error may be separated into a systematic (or deterministic) part and a random (or stochastic) part, statistical considerations are important in error analysis. This chapter also deals with such considerations of error analysis.

5.1.1 Parameters for Performance Specification

All devices that assist in the functions of a mechatronic system can be interpreted as components of the system. The selection of available components for a particular application or the design of new components should rely heavily on the performance specifications for these components. A great majority of instrument ratings provided by manufacturers are in the form of static parameters. In mechatronic applications, dynamic performance specifications are also very important, which will be discussed in a separate chapter on control. In this chapter, we will study instrument ratings and parameters for performance specification of instruments.

5.1.2 Perfect Measurement Device

Consider a measuring device of a mechatronic system, for example. A *perfect measuring device* can be defined as one that possesses the following characteristics:

1. The output of the measuring device instantly reaches the measured value (fast response).
2. The transducer output is sufficiently large (high gain, low output impedance, high sensitivity).
3. The device output remains at the measured value (without drifting or being affected by environmental effects and other undesirable disturbances and noise) unless the measurand (i.e., what is measured) itself changes (stability and robustness).
4. The output signal level of the transducer varies in proportion to the signal level of the measurand (static linearity).
5. The connection of a measuring device does not distort the measurand itself (the loading effects are absent and impedances are matched; see Chapter 4).
6. The power consumption is small (high input impedance; see Chapter 4).

All of these properties are based on dynamic characteristics and therefore can be explained in terms of the dynamic behavior of the measuring device. In particular, items 1 through 4 can be specified in terms of the device response, either in the *time domain* or in the *frequency domain*. Items 2, 5, and 6 can be specified using the *impedance* characteristics of the device.

5.2 Linearity

A device is considered linear if it can be modeled by linear differential equations, with time t as the independent variable. Nonlinear devices are often analyzed using linear techniques by considering small excursions about an operating point. This linearization is accomplished by introducing incremental variables for inputs and outputs. If one increment can cover the entire operating range of a device with sufficient accuracy, it is an indication that the device is linear. If the input/output relations are nonlinear algebraic equations, it represents a *static nonlinearity*. Such a situation can be handled simply by using nonlinear calibration curves that linearize the device without introducing nonlinearity errors. If, on the other hand, the input/output relations are nonlinear differential equations, analysis usually becomes quite complex. This situation represents a *dynamic nonlinearity*.

The transfer-function representation of an instrument implicitly assumes linearity. According to industrial terminology, a linear measuring instrument provides a measured value that varies linearly with the value of the measurand—the variable that is measured. This is consistent with the definition of static linearity. All physical devices are *nonlinear* to some degree. This stems from deviation from the ideal behavior due to causes such as saturation, deviation from Hooke's law in elastic elements, Coulomb friction, creep at joints, aerodynamic damping, backlash in gears and other loose components, and component wearout.

Nonlinearities in devices are often manifested as some peculiar characteristics. In particular, the following properties are important in detecting nonlinear behavior in dynamic systems.

5.2.1 Saturation

Nonlinear devices may exhibit saturation (see Figure 5.1a). This may result from such causes as magnetic saturation, which is common in magnetic-induction devices and transformer-like devices such as differential transformers, plasticity in mechanical components, and nonlinear springs.

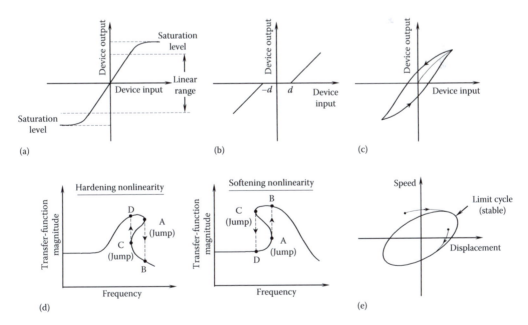

FIGURE 5.1
Common manifestations of nonlinearity in mechatronic-system devices: (a) saturation, (b) dead zone, (c) hysteresis, (d) the jump phenomenon, (e) limit cycle response.

5.2.2 Dead Zone

A dead zone is a region in which a device would not respond to an excitation. Stiction in mechanical devices with Coulomb friction is a good example. Due to stiction, a component would not move until the applied force reaches a certain minimum value. Once the motion is initiated, subsequent behavior can be either linear or nonlinear. A dead zone with subsequent linear behavior is shown in Figure 5.1b.

5.2.3 Hysteresis

Nonlinear devices may produce hysteresis. In hysteresis, the input/output curve changes depending on the direction of motion (as indicated in Figure 5.1c), resulting in a hysteresis loop. This behavior is common in loose components such as gears, which have backlash; in components with nonlinear damping, such as Coulomb friction; and in magnetic devices with ferromagnetic media and various dissipative mechanisms (e.g., eddy current dissipation). For example, consider a coil wrapped around a ferromagnetic core. If a direct current (dc) is passed through the coil, a magnetic field is generated. As the current is increased from zero, the field strength will also increase. Now, if the current is decreased back to zero, the field strength will not return to zero because of residual magnetism in the ferromagnetic core. A negative current has to be applied to demagnetize the core. It follows that the field strength versus current curve looks somewhat like Figure 5.1c. This is magnetic hysteresis. Note that linear viscous damping also exhibits a hysteresis loop in its force–displacement curve. This is a property of any mechanical component that dissipates energy. (The area within the hysteresis loop gives the energy dissipated in one cycle of motion.) In general, if force depends on both displacement (as in the case of a spring) and velocity (as in the case of a damping element), the value of force at a given value of displacement will change with velocity. In particular, the force when the component is moving in one direction (say positive velocity) will be different from the force at the same location when the component is moving in the opposite direction (negative velocity), thereby giving a hysteresis loop in the force–displacement plane. If the relationship of displacement and velocity to force is linear (as in viscous damping), the hysteresis effect is linear. If on the other hand the relationship is nonlinear (as in Coulomb damping and aerodynamic damping), the resulting hysteresis is nonlinear.

5.2.4 The Jump Phenomenon

Some nonlinear devices exhibit an instability known as the jump phenomenon (or *fold catastrophe*) in the frequency response (transfer) function curve. This is shown in Figure 5.1d for both *hardening* devices and *softening* devices. With increasing frequency, jump occurs from A to B, and with decreasing frequency, it occurs from C to D. Furthermore, the transfer function itself may change with the level of input excitation in the case of nonlinear devices.

5.2.5 Limit Cycles

Nonlinear devices may produce limit cycles. An example of the phase plane of velocity versus displacement is given in Figure 5.1e. A limit cycle is a closed trajectory in the state space that corresponds to sustained oscillations at a specific frequency and amplitude without decay or growth. The amplitude of these oscillations is independent of the initial location from which the response started. Also, an external input is not needed to sustain a limit-cycle oscillation. In the case of a stable limit cycle, the response will move onto the limit cycle irrespective of the location in the neighborhood of the limit cycle from which the response was initiated (see Figure 5.1e). In the case of an unstable limit cycle, the response will move away from it with the slightest disturbance.

5.2.6 Frequency Creation

At steady state, nonlinear devices can create frequencies that are not present in the excitation signals. These frequencies might be harmonics (integer multiples of the excitation frequency), subharmonics (integer fractions of the excitation frequency), or nonharmonics (usually rational fractions of the excitation frequency).

Nonlinear systems can be analyzed using the *describing function* approach. When a harmonic input (at a specific frequency) is applied to a nonlinear device, the resulting output at steady state will have a component at this fundamental frequency and will also have components at other frequencies (due to frequency creation by the nonlinear device), typically harmonics. The response may be represented by a Fourier series, which has frequency components that are multiples of the input frequency. The describing function approach neglects all the higher harmonics in the response and retains only the fundamental component. This output component, when divided by the input, produces the describing function of the device. This is similar to the transfer function of the linear device, but unlike for a linear device, the gain and the phase shift will be dependent on the input amplitude. Details of the describing function approach can be found in textbooks on nonlinear control theory.

Several methods are available to reduce or eliminate nonlinear behavior in systems. They include calibration (in the static case); the use of linearizing elements, such as resistors and amplifiers to neutralize the nonlinear effects; and the use of nonlinear feedback. It is also a good practice to take the following precautions:

1. Avoid operating the device over a wide range of signal levels
2. Avoid operation over a wide frequency band
3. Use devices that do not generate large mechanical motions
4. Minimize Coulomb friction and stiction (e.g., using proper lubrication)
5. Avoid loose joints and gear coupling (i.e., use *direct-drive* mechanisms)

5.3 Instrument Ratings

Instrument manufacturers do not usually provide complete dynamic information for their products. In most cases, it is unrealistic to expect complete dynamic models (in the time domain or the frequency domain) and associated parameter values for complex instruments in mechatronic systems. The performance characteristics provided by manufacturers and vendors are primarily static parameters. Known as instrument ratings, these are available as parameter values, tables, charts, calibration curves, and empirical equations. Dynamic characteristics such as transfer functions (e.g., transmissibility curves expressed with respect to excitation frequency) might also be provided for more sophisticated instruments, but the available dynamic information is never complete. Furthermore, the definitions of rating parameters used by manufacturers and vendors of instruments are in some cases not the same as analytical definitions used in textbooks. This is particularly true in relation to the term *linearity*. Nevertheless, instrument ratings provided by manufacturers and vendors are very useful in the selection, installation, operation, and maintenance of components in a mechatronic system. Now, we shall examine some of these performance parameters.

5.3.1 Rating Parameters

The typical rating parameters supplied by instrument manufacturers are as follows:

1. Sensitivity
2. Dynamic range
3. Resolution
4. Linearity
5. Zero drift and full-scale drift
6. Useful frequency range
7. Bandwidth
8. Input and output impedances

We have already discussed the meaning and significance of some of these terms. In this section, we shall look at the conventional definitions given by instrument manufacturers and vendors.

The *sensitivity* of a device (e.g., transducer) is measured by the magnitude (peak, rms value, etc.) of the output signal corresponding to a unit input (e.g., measurand). This may be expressed as the ratio of (incremental output)/(incremental input) or, analytically, as the corresponding partial derivative. In the case of vectorial or tensorial signals (e.g., displacement, velocity, acceleration, strain, force), the direction of sensitivity should be specified.

Cross-sensitivity is the sensitivity along directions that are orthogonal to the primary direction of sensitivity. It is normally expressed as a percentage of direct sensitivity. High sensitivity and low cross-sensitivity are desirable for any input/output device (e.g., measuring instrument). Sensitivity to parameter changes and noise has to be small in any device, however, and this is an indication of its *robustness*. On the other hand, in *adaptive control* and *self-tuning control*, the sensitivity of the system to control parameters has to be sufficiently high. Often, sensitivity and robustness are conflicting requirements.

The *dynamic range* of an instrument is determined by the allowed lower and upper limits of its input or output (response) so as to maintain a required level of output accuracy. This range is usually expressed as a ratio in *decibels*. In many situations, the lower limit of the dynamic range is equal to the resolution of the device. Hence, the dynamic range ratio is usually expressed as (range of operation)/(resolution).

The *resolution* of an input/output instrument is the smallest change in a signal (input) that can be detected and accurately indicated (output) by a transducer, a display unit, or any pertinent instrument. It is usually expressed as a percentage of the maximum range of the instrument or as the inverse of the dynamic range ratio. It follows that dynamic range and resolution are very closely related.

Example 5.1

The meaning of dynamic range (and resolution) can easily be extended to cover digital instruments. For example, consider an instrument that has a 12 bit analog-to-digital (A/D) converter (ADC). Estimate the dynamic range of the instrument.

Solution

In this example, the dynamic range is determined (primarily) by the word size of the ADC. Each bit can take the binary value 0 or 1. Since the resolution is given by the smallest possible increment, i.e., a change by the least significant bit (LSB), it is clear that digital resolution = 1. The largest value represented by a 12 bit word corresponds to the case when all twelve bits are in unity. This value is decimal $2^{12} - 1$. The smallest value (when all twelve bits are zero) is zero. Now, use the definition

$$\text{Dynamic range} = 20 \log_{10} \left[\frac{\text{Range of operation}}{\text{Resolution}} \right] \tag{5.1}$$

The dynamic range of the instrument is given by $20 \log_{10}[(2^{12} - 1)/1] = 72 \text{ dB}$.

Another (perhaps more correct) way of looking at this problem is to consider the resolution to be some value δy, rather than unity, depending on the particular application. For example, δy may represent an output signal increment of 0.0025 V. Next, we note that a 12 bit word can represent a combination of 2^{12} values (i.e., 4096 values), the smallest being y_{min} and the largest being $y_{max} = y_{min} + (2^{12} - 1)\delta y$.

Note that y_{min} can be zero, positive, or negative. The smallest increment between values is δy, which is, by definition, the resolution. There are 2^{12} values with y_{min} and y_{max} (the two end values) inclusive. Then,

$$\text{Dynamic range} = \frac{y_{max} - y_{min}}{\delta y} = \frac{(2^{12} - 1)\delta y}{\delta y} = 12^{12} - 1 = 4095 = 72 \text{ dB}$$

So we end up with the same result for the dynamic range, but the interpretation of resolution is different.

Linearity is determined by the calibration curve of an instrument. The curve of output amplitude (peak or rms value) versus input amplitude under static conditions within the dynamic range of an instrument is known as the *static calibration curve*. Its closeness to a straight line measures the degree of linearity. Manufacturers provide this information either as the maximum deviation of the calibration curve from the least squares straight-line fit of the calibration curve or from some other reference straight line. If the least-squares fit is used as the reference straight line, the maximum deviation is called *independent linearity* (more correctly, independent nonlinearity, because the larger the deviation, the greater the nonlinearity). Nonlinearity may be expressed as a percentage of either the actual reading at an operating point or the full-scale reading.

Zero drift is defined as the drift from the null reading of the instrument when the input is maintained steady for a long period. Note that in this case, the input is kept at zero or any other level that corresponds to the null reading of the instrument. Similarly, *full-scale drift* is defined with respect to the full-scale reading (the input is maintained at the full-scale value). The usual causes of drift include instrument instability (e.g., instability in amplifiers), ambient changes (e.g., changes in temperature, pressure, humidity, and vibration level), changes in power supply (e.g., changes in reference dc voltage or alternating current [ac] line voltage), and parameter changes in an instrument (due to aging, wear and tear, nonlinearities, etc.). Drift due to parameter changes that are caused by instrument nonlinearities is known as *parametric drift, sensitivity drift,* or *scale-factor drift*. For example, a change in spring stiffness or electrical resistance due to changes in ambient temperature results in a parametric drift. Note that parametric drift depends on the input level. Zero drift, however, is assumed to be the same at any input level if the other conditions are kept constant. For example, a change in reading caused by the thermal expansion of the readout mechanism due to changes in ambient temperature is considered a zero drift. Drift in electronic devices can be reduced by using ac circuitry rather than dc circuitry. For example, ac-coupled amplifiers have fewer drift problems than dc amplifiers. Intermittent checking for instrument response levels with zero input is a popular way to calibrate for zero drift. In digital devices, for example, this can be done automatically from time to time between sample points when the input signal can be bypassed without affecting the system operation.

Useful frequency range corresponds to a flat gain curve and a zero phase curve in the frequency response characteristics of an instrument. The maximum frequency in this band is typically less than half (say, one-fifth) of the dominant resonant frequency of the instrument. This is a measure of the instrument bandwidth.

The *bandwidth* of an instrument determines the maximum speed or frequency at which the instrument is capable of operating. A high bandwidth implies faster speed of response. Bandwidth is determined by the dominant natural frequency ω_n or the dominant resonant frequency ω_r of the device. (*Note*: For low damping, ω_r is approximately equal to ω_n). It is inversely proportional to rise time and the dominant time constant. Half-power bandwidth is also a useful parameter. Instrument bandwidth has to be several times greater than the maximum frequency of interest in the input signals. For example, the bandwidth of a measuring device is important particularly when measuring transient signals. Note that bandwidth is directly related to the useful frequency range.

5.4 Bandwidth

Bandwidth plays an important role in specifying and characterizing the components of a mechatronic system. In particular, useful frequency range, operating bandwidth, and control bandwidth are important considerations. In this section, we will study several important issues related to these topics.

Bandwidth has different meanings depending on the particular context and application. For example, when studying the response of a dynamic system, the bandwidth relates to the fundamental resonant frequency and correspondingly to the speed of response for a given excitation. In band-pass filters, the bandwidth refers to the frequency band within which the frequency components of the signal are allowed through the filter; the frequency components outside the band being rejected by it. With respect to measuring instruments, bandwidth refers to the range frequencies within which the instrument measures a signal accurately. In digital communication networks (e.g., the Internet), the bandwidth denotes the capacity of the network in terms of information rate (bits/s). *Note*: These various interpretations of bandwidth are somewhat related. In particular, if a signal passes through a band-pass filter, we know that its frequency content is within the bandwidth of the filter, but we cannot determine the actual frequency content of the signal on the basis of that observation. In this context, the bandwidth appears to represent a frequency uncertainty

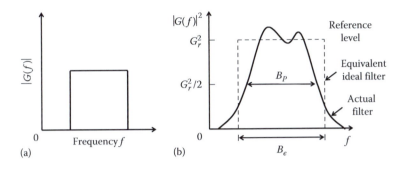

FIGURE 5.2
Characteristics of (a) an ideal band-pass filter and (b) a practical band-pass filter.

in the observation (i.e., the larger the bandwidth of the filter, the less certain you are about the actual frequency content of a signal that passes through the filter).

5.4.1 Transmission Level of a Band-Pass Filter

Practical filters can be interpreted as dynamic systems. In fact, all physical, dynamic systems (e.g., electromechanical systems) are analog filters. It follows that the filter characteristic can be represented by the frequency transfer function $G(f)$ of the filter. A magnitude squared plot of such a filter transfer function is shown in Figure 5.2. In a logarithmic plot, the magnitude-squared curve is obtained by simply doubling the corresponding magnitude curve (in the Bode plot). Note that the actual filter transfer function (Figure 5.2b) is not flat like the ideal filter shown in Figure 5.2a. The reference level G_r is the average value of the transfer function magnitude in the neighborhood of its peak.

5.4.2 Effective Noise Bandwidth

The effective noise bandwidth of a filter is equal to the bandwidth of an ideal filter that has the same reference level and that transmits the same amount of power from a white noise source. Note that white noise has a constant (flat) power spectral density (psd). Hence, for a noise source of unity psd, the power transmitted by the practical filter is given by $\int_0^\infty |G(f)|^2 \, df$, which, by definition, is equal to the power $G_r^2 B_e$ transmitted by the equivalent ideal filter. Hence, the effective noise bandwidth B_e is given by

$$B_e = \int_0^\infty \frac{|G(f)|^2 \, df}{G_r^2} \tag{5.2}$$

5.4.3 Half-Power (or 3 dB) Bandwidth

Half of the power from a unity-psd noise source as transmitted by an ideal filter is $G_r^2 B_r/2$. Hence, $G_r/\sqrt{2}$ is referred to as the *half-power level*. This is also known as a 3 dB level because $20\log_{10}\sqrt{2} = 10\log_{10} 2 = 3$ dB. (*Note*: 3 dB refers to a power ratio of 2 or an amplitude ratio of $\sqrt{2}$. Furthermore, 20 dB corresponds to an amplitude ratio of 10 or a power ratio of 100.) The 3 dB (or half-power) bandwidth corresponds to the width of the filter transfer function at the half-power level. This is denoted by B_p in Figure 5.2b. Note that B_e and B_p are different in general. In an ideal case, where the magnitude-squared filter characteristic has linear rise and fall-off segments, however, these two bandwidths are equal (see Figure 5.3).

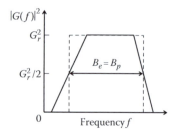

FIGURE 5.3
An idealized filter with linear segments.

5.4.4 Fourier Analysis Bandwidth

In Fourier analysis, bandwidth is interpreted as the *frequency uncertainty* in the spectral results. In analytical Fourier integral transform (FIT) results, which assume that the entire signal is available for analysis, the spectrum is continuously defined over the entire frequency range $[-\infty,\infty]$ and the frequency increment df is infinitesimally small ($df \rightarrow 0$). There is no frequency uncertainty in this case, and the analysis bandwidth is infinitesimally narrow. In digital Fourier transform, the discrete spectral lines are generated at frequency intervals of ΔF. This finite frequency increment ΔF, which is the frequency uncertainty, is therefore the analysis bandwidth B for this analysis. It is known that $\Delta F = 1/T$, where T is the record length of the signal (or window length when a rectangular window is used to select the signal segment for analysis). It also follows that the minimum frequency that has a meaningful accuracy is the bandwidth. This interpretation for analysis bandwidth is confirmed by noting the fact that harmonic components of frequency less than ΔF (or a period greater than T) cannot be studied by observing a signal record of length less than T. An analysis bandwidth carries information regarding distinguishable minimum frequency separation in computed results. In this sense, the bandwidth is directly related to the frequency resolution of analyzed results. The accuracy of the analysis increases by increasing the record length T (or decreasing the analysis bandwidth B).

When a time window other than the rectangular window is used to truncate a signal, then reshaping of the signal segment (data) occurs according to the shape of the window. This reshaping suppresses the sidelobes of the Fourier spectrum of the window and hence reduces the frequency leakage that arises from the truncation of the signal. At the same time, however, an error is introduced due to the information lost through data reshaping. This error is proportional to the bandwidth of the window itself. The effective noise bandwidth of a rectangular window is only slightly less than $1/T$ because the main lobe of its Fourier spectrum is nearly rectangular. Hence, for all practical purposes, the effective noise bandwidth can be taken as the analysis bandwidth. Note that data truncation (multiplication in the time domain) is equivalent to the convolution of the Fourier spectrum (in the frequency domain). The main lobe of the spectrum uniformly affects all spectral lines in the discrete spectrum of the data signal. It follows that a window main lobe having a broader effective-noise bandwidth introduces a larger error into the spectral results. Hence, in digital Fourier analysis, bandwidth is taken as the effective noise bandwidth of the time window that is employed.

5.4.5 Useful Frequency Range

This corresponds to the flat region (static region) in the gain curve and the zero-phase-lead region in the phase curve of a device (with respect to frequency). It is determined by the dominant (i.e., the lowest) resonant frequency f_r of the device. The upper frequency limit f_{max} in the useful frequency range is several times smaller than f_r for a typical input/output device (e.g., $f_{max} = 0.25\, f_r$). A useful frequency range may also be determined by specifying the flatness of the static portion of the frequency response curve. For example, since a single pole or a single zero introduces a slope on the order of $\pm 20\,\text{dB/decade}$ to the Bode magnitude curve of the device, a slope within 5% of this value (i.e., $\pm 1\,\text{dB/decade}$) may be considered flat for most practical purposes. For a measuring instrument, for example, operation in the useful frequency range implies that the significant frequency content of the measured signal is limited to this band. In that case, faithful measurement and fast response are guaranteed because the dynamics of the measuring device will not corrupt the measurement.

5.4.6 Instrument Bandwidth

This is a measure of the useful frequency range of an instrument. Furthermore, the larger the bandwidth of the device, the faster the speed of response. Unfortunately, the larger the bandwidth, the more susceptible the instrument will be to high-frequency noise as well as stability problems. Filtering will be needed to eliminate unwanted noise. The stability can be improved by dynamic compensation. Common definitions of bandwidth include the frequency range over which the transfer-function magnitude is flat, the resonant frequency, and the frequency at which the transfer-function magnitude drops to $1/\sqrt{2}$ (or 70.7%) of the zero-frequency (or static) level. The last definition corresponds to the *half-power bandwidth* because a reduction of amplitude level by a factor of $\sqrt{2}$ corresponds to a power drop by a factor of 2.

5.4.7 Control Bandwidth

This is used to specify the maximum possible speed of control. It is an important specification in both analog control and digital control. In digital control, the data sampling rate (in samples/second) has to be several times higher than the control bandwidth (in hertz) so that sufficient data would be available to compute the control action. Also, from *Shannon's sampling theorem* (see Section 5.5), control bandwidth is given by half the rate at which the control action is computed. The control bandwidth provides the frequency range within which a system can be controlled (assuming that all the devices in the system can operate within this bandwidth).

5.4.8 Static Gain

This is the gain (transfer function magnitude) of a measuring instrument within the useful (flat) range (or at very low frequencies) of the instrument. It is also termed *dc gain*. A high value for static gain results in a high-sensitivity measuring device, which is a desirable characteristic.

Example 5.2

A mechanical device for measuring angular velocity is shown in Figure 5.4. The main element of the tachometer is a rotary viscous damper (damping constant b) consisting of two cylinders. The outer cylinder carries a viscous fluid within which the inner cylinder rotates. The inner cylinder is connected to the shaft whose speed ω_i is to be measured. The outer cylinder is resisted by a linear torsional spring of stiffness k. The rotation θ_o of the outer cylinder is indicated by a pointer on a suitably calibrated scale. Neglecting the inertia of moving parts, perform a bandwidth analysis on this device.

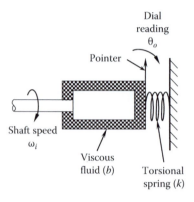

FIGURE 5.4
A mechanical tachometer.

Solution
The damping torque is proportional to the relative velocity of the two cylinders and is resisted by the spring torque. The equation of motion is given by $b(\omega_i - \dot{\theta}_o) = k\theta_o$, or $b\dot{\theta}_o + k\theta_o = b\omega_i$. The transfer function is determined by first replacing the time derivative by the Laplace operator s:

$$\frac{\theta_o}{\omega_i} = \frac{b}{[bs + k]} = \frac{b/k}{[(b/k)s + 1]} = \frac{k_g}{[\tau s + 1]}$$

Note that the static gain or dc gain (transfer-function magnitude at $s = 0$) is $k_g = b/k$ and the time constant is $\tau = b/k$.

We face conflicting design requirements in this case. On the one hand, we like to have a large static gain so that a sufficiently large reading is available. On the other hand, the time constant must be small in order to obtain a quick reading that faithfully follows the measured speed. A compromise must be reached here, depending on the specific design requirements. Alternatively, a signal-conditioning device could be employed to amplify the sensor output.

Now, let us examine the half-power bandwidth of the device. The frequency transfer function is $G(j\omega) = k_g/(\tau j\omega + 1)$. By definition, the half-power bandwidth ω_b is given by $k_g/(|\tau j\omega_b + 1|) = k_g/\sqrt{2}$.
Hence, $(\tau\omega_b)^2 + 1 = 2$. Since both τ and ω_b are positive, we have $\tau\omega_b = 1$ or $\omega_b = 1/\tau$.
Note: The bandwidth is inversely proportional to the time constant. This confirms our earlier statement that bandwidth is a measure of the speed of response.

5.5 Signal Sampling and Aliasing Distortion

Aliasing distortion is an important consideration when dealing with sampled data from a continuous signal. This error may enter into computation in both the time domain and the frequency domain, depending on the domain in which the data are sampled.

5.5.1 Sampling Theorem

If a time signal $x(t)$ is sampled at equal steps of ΔT, no information regarding its frequency spectrum $X(f)$ is obtained for frequencies higher than

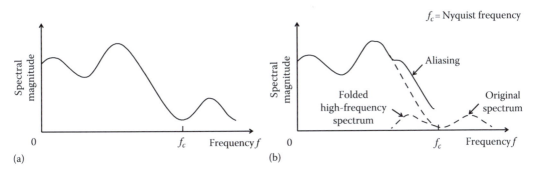

FIGURE 5.5
Aliasing distortion of a frequency spectrum: (a) original spectrum and (b) distorted spectrum due to aliasing.

$$f_c = \frac{1}{2\Delta T} \tag{5.3}$$

This fact is known as *Shannon's sampling theorem*, and the limiting (cut-off) frequency is called the *Nyquist frequency*.

It can be shown that the aliasing error is caused by "folding" of the high-frequency segment of the frequency spectrum beyond the Nyquist frequency into the low-frequency segment. This is illustrated in Figure 5.5. The aliasing error becomes more and more prominent for frequencies of the spectrum closer to the Nyquist frequency. In signal analysis, a sufficiently small sample step ΔT should be chosen in order to reduce the aliasing distortion in the frequency domain, depending on the highest frequency of interest in the analyzed signal. This, however, increases the signal processing time and the computer storage requirements, which is undesirable particularly in real-time analysis. It also can result in stability problems in numerical computations. The Nyquist sampling criterion requires that the sampling rate $(1/\Delta T)$ for a signal should be at least twice the highest frequency of interest. Instead of making the sampling rate very high, a moderate value that satisfies the Nyquist sampling criterion is used in practice, together with an *anti-aliasing filter* to remove the distorted frequency components.

5.5.2 Anti-Aliasing Filter

It should be clear from Figure 5.5 that if the original signal is low-pass filtered at a cut-off frequency equal to the Nyquist frequency, then the aliasing distortion due to sampling would not occur. A filter of this type is called an anti-aliasing filter. Analog hardware filters may be used for this purpose. In practice, it is not possible to achieve perfect filtering. Hence, some aliasing could remain even after using an anti-aliasing filter. Such residual errors may be reduced by using a filter cut-off frequency that is slightly less than the Nyquist frequency. Then the resulting spectrum would only be valid up to this filter cut-off frequency (and not up to the theoretical limit of the Nyquist frequency). Aliasing reduces the valid frequency range in digital Fourier results. Typically, the useful frequency limit is $f_c/1.28$ so that the last 20% of the spectral points near the Nyquist frequency should be neglected. Note that sometimes $f_c/1.28(\cong 0.8f_c)$ is used as the filter cutoff frequency. In this case, the computed spectrum is accurate up to $0.8f_c$ and not up to f_c.

Example 5.3

Consider 1024 data points from a signal sampled at 1 millisecond (ms) intervals.

Sample rate $f_s = 1/0.001$ samples/s $= 1000\,Hz = 1\,kHz$

Nyquist frequency $= 1000/2\,Hz = 500\,Hz$

Due to aliasing, approximately 20% of the spectrum (i.e., the spectrum beyond 400 Hz) will be distorted. Here, we may use an anti-aliasing filter.

Suppose that a digital Fourier transform computation provides 1024 frequency points of data up to 1000 Hz. Half of this number is beyond the Nyquist frequency and will not give any new information about the signal.

Spectral line separation $= 1000/1024\,Hz = 1\,Hz$ (approx.)

Keep only the first 400 spectral lines as the useful spectrum.

Note: Almost 500 spectral lines may be retained if an accurate anti-aliasing filter is used.

Example 5.4

(a) If a sensor signal is sampled at f_s Hz, suggest a suitable cutoff frequency for an anti-aliasing filter to be used in this application.

(b) Suppose that a sinusoidal signal of frequency f_1 Hz is sampled at the rate of f_s samples/s. Another sinusoidal signal of the same amplitude, but of a higher frequency of f_2 Hz was found to yield the same data when sampled at f_s. What is the likely analytical relationship between f_1, f_2, and f_s?

(c) Consider a plant of transfer function $G(s) = k/(1 + \tau s)$.

What is the static gain of this plant? Show that the magnitude of the transfer function reaches $1/\sqrt{2}$ of the static gain when the excitation frequency is $1/\tau$ rad/s. Note that the frequency, $\omega_b = 1/\tau$ rad/s, may be taken as the operating bandwidth of the plant.

(d) Consider a chip refiner that is used in the pulp and paper industry. The machine is used for mechanical pulping of wood chips. It has a fixed plate and a rotating plate driven by an induction motor. The gap between the plates is sensed and is adjustable as well. As the plate rotates, the chips are ground into a pulp within the gap. A block diagram of the plate-positioning control system is shown in Figure 5.6.

Suppose that the torque sensor signal and the gap sensor signal are sampled at 100 and 200 Hz, respectively, into the digital controller, which takes 0.05 s to compute each positional command for the servovalve. The time constant of the servovalve is $0.05/2\pi$ s and that of the mechanical load is $0.2/2\pi$ s. Estimate the control bandwidth and the operating bandwidth of the positioning system.

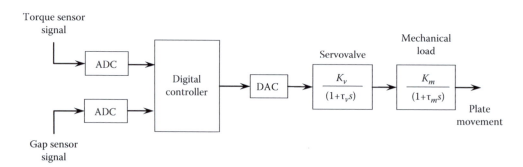

FIGURE 5.6
Block diagram of the plate positioning control system for a chip refiner.

Solution

(a) In theory, the cutoff frequency of the anti-aliasing filter has to be $0.5f_s$, which is the Nyquist frequency. In practice, however, $0.4f_s$ would be desirable, providing a useful spectrum of only up to $0.4f_s$.

(b) $f_2 = f_c + (f_c + f_1) = 2f_c - f_1$ or

$$f_2 = f_s - f_1 \tag{5.4}$$

(c) $G(j\omega) = k/(1 + \tau j\omega) = $ frequency transfer function, where ω is in rad/s.
Static gain is the transfer function magnitude at steady state (i.e., at zero frequency). Hence,
Static gain $= G(0) = k$.
When $\omega = 1/\tau$, we have $G(j\omega) = k/(1 + j)$
Hence, $|G(j\omega)| = k/\sqrt{2}$ at this frequency.
This is the half-power bandwidth.

(d) Due to sampling, the torque signal has a bandwidth of $1/2 \times 100\,\text{Hz} = 50\,\text{Hz}$ and the gap sensor signal has a bandwidth of $1/2 \times 200\,\text{Hz} = 100\,\text{Hz}$. The control cycle time $= 0.05\,\text{s}$, which provides control signals at a rate of $1/0.05\,\text{Hz} = 20\,\text{Hz}$.

Since $20\,\text{Hz} < \min(50/2\,\text{Hz}, 100/2\,\text{Hz})$, we have adequate bandwidth from the sampled sensor signals to compute the control signal. The control bandwidth from the digital controller $= 1/2 \times 20\,\text{Hz}$ (From Shannon's sampling theoreom) $= 10\,\text{Hz}$.
But, the servovalve is also part of the controller. Its bandwidth $= 1/\tau_v$ rad/s $= 1/2\pi\tau_v$ Hz $= 2\pi/(2\pi \times 0.05)$ Hz $= 20\,\text{Hz}$.
Hence, control bandwidth $= \min (10\,\text{Hz}, 20\,\text{Hz}) = 10\,\text{Hz}$.
Bandwidth of the mechanical load $= 1/\tau_m$ rad/s $= 1/2\pi\tau_m$ Hz $= 2\pi/(2\pi \times 0.2)$ Hz $= 5\,\text{Hz}$.
Hence, operating bandwidth of the system $= \min (10\,\text{Hz}, 5\,\text{Hz}) = 5\,\text{Hz}$.

5.5.3 Another Illustration of Aliasing

A simple illustration of aliasing is given in Figure 5.7. Here, two sinusoidal signals of frequency $f_1 = 0.2\,\text{Hz}$ and $f_2 = 0.8\,\text{Hz}$ are shown (Figure 5.7a). Suppose that the two signals are sampled at the rate of $f_s = 1$ sample/s. The corresponding Nyquist frequency is $f_c = 0.5\,\text{Hz}$. It is seen that, at this sampling rate, the data samples from the two signals are identical. In

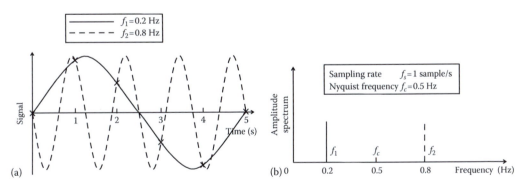

FIGURE 5.7
A simple illustration of aliasing: (a) two harmonic signals with identical sampled data and (b) frequency spectra of the two harmonic signals.

other words, the high-frequency signal cannot be distinguished from the low-frequency signal. Hence, a high-frequency signal component of frequency 0.8 Hz will appear as a low-frequency signal component of frequency 0.2 Hz. This is aliasing, as is clear from the signal spectrum shown in Figure 5.7b. Specifically, the spectral segment of the signal beyond the Nyquist frequency (f_c) cannot be recovered.

Example 5.5

Suppose that the frequency range of interest in a particular signal is 0–200 Hz. We are interested in determining the sampling rate (digitization speed) and the cutoff frequency for the anti-aliasing (low-pass) filter.

The Nyquist frequency f_c is given by $f_c/1.28 = 200$. Hence, $f_c = 256$ Hz.

The sampling rate (or digitization speed) for the time signal that is needed to achieve this range of analysis is $F = 2f_c = 512$ Hz. With this sampling frequency, the cutoff frequency for the anti-aliasing filter could be set at a value between 200 and 256 Hz.

Example 5.6

Consider the digital control system for a mechanical position application, as shown schematically in Figure 5.8. The control computer generates a control signal according to an algorithm, on the basis of the desired position and actual position, as measured by an optical encoder. This digital signal is converted into the analog form using a digital-to-analog converter (DAC) and is supplied to the drive amplifier. Accordingly, the current signals needed to energize the motor windings are generated by the amplifier. The inertial element, which has to be positioned, is directly (and rigidly) linked to the motor rotor and is resisted by a spring and a damper, as shown.

Suppose that the combined transfer function of the drive amplifier and the electromagnetic circuit (torque generator) of the motor is given by $k_e / (s^2 + 2\zeta_e \omega_e s + \omega_e^2)$ and the transfer function of the mechanical system including the inertia of the motor rotor is given by $k_m / (s^2 + 2\zeta_m \omega_m s + \omega_m^2)$.

Here,

k is the equivalent gain

ζ is the damping ratio

ω is the natural frequency with the subscripts $()_e$ and $()_m$ denoting the electrical and mechanical components, respectively

Also, ΔT_c is the time taken to compute each control action and ΔT_p is the pulse of the position sensing encoder. The following numerical values are given:

$$\omega_e = 1000\pi \text{ rad/s}, \ \zeta_e = 0.5, \ \omega_m = 100\pi \text{ rad/s, and } \zeta_m = 0.3$$

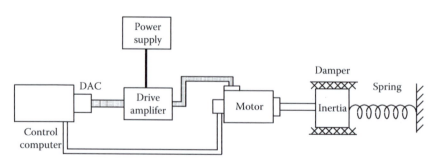

FIGURE 5.8
Digital control system for a mechanical positioning application.

For the purpose of this example, you may neglect loading effects and coupling effects due to component cascading and signal feedback.

(i) Explain why the control bandwidth of this system cannot be much larger than 50 Hz.
(ii) If $\Delta T_c = 0.02$ s, estimate the control bandwidth of the system.
(iii) Explain the significance of ΔT_p in this application. Why should ΔT_p typically not be greater than $0.5\Delta T_c$?
(iv) Estimate the operating bandwidth of the positioning system, assuming that significant plant dynamics are to be avoided.
(v) If $\omega_m = 500\pi$ rad/s and $\Delta T_c = 0.02$ s, with the remaining parameters kept as specified above, estimate the operating bandwidth of the system, again not exciting significant plant dynamics.

Solution

(i) The drive system has a resonant frequency less than 500 Hz. Hence, the flat region of the spectrum of the drive system would be about 1/10th of this; i.e., 50 Hz. This would limit the maximum spectral component of the drive signal to about 50 Hz. Hence, the control bandwidth would be limited by this value.
(ii) The rate at which the digital control signal is generated $= 1/0.02$ Hz $= 50$ Hz. By Shannon's sampling theorem, the effective (useful) spectrum of the control signal is limited to $(1/2)\times50$ Hz $= 25$ Hz. Even though the drive system can accommodate a bandwidth of about 50 Hz, the control bandwidth would be limited to 25 Hz, due to digital control, in this case.
(iii) Note that ΔT_p corresponds to the sampling period of the measurement signal (for feedback). Hence, its useful spectrum would be limited to $1/(2\Delta T_p)$ by Shannon's sampling theorem. Consequently, the feedback signal will not be able to provide any useful information of the process beyond the frequency $1/(2\Delta T_p)$. To generate a control signal at the rate of $1/\Delta T_c$ samples/s, the process information has to be provided at least up to $1/\Delta T_c$ Hz. To provide this information, we must have

$$\frac{1}{2\Delta T_p} \geq \frac{1}{\Delta T_c} \quad \text{or} \quad \Delta T_p \leq 0.5\, \Delta T_c. \tag{5.5}$$

Note: This condition guarantees that at least two points of sampled data from the sensor are used for computing each control action.

(iv) The resonant frequency of the plant (positioning system) is approximately (less than) $100\pi/2\pi$ Hz ≈ 50 Hz. At frequencies near this, the resonance will interfere with control and should be avoided if possible, unless the resonances (or modes) of the plant themselves need to be modified through control. At frequencies much larger than this, the process will not significantly respond to the control action and will not be of much use (the plant will be felt like a rigid wall). Hence, the operating bandwidth has to be sufficiently smaller than 50 Hz, say 25 Hz, in order to avoid plant dynamics.

Note: This is a matter of design judgment based on the nature of the application (e.g., excavator, disk drive). Typically, however, one needs to control the plant dynamics. In that case, it is necessary to use the entire control bandwidth (i.e., maximum possible control speed) as the operating bandwidth. In the present case, even if the entire control bandwidth (i.e., 25 Hz) is used as the operating bandwidth, it still avoids the plant resonance.

(v) The plant resonance in this case is about $500\pi/2\pi$ Hz ≈ 250 Hz. This limits the operating bandwidth to about $250\pi/2$ Hz ≈ 125 Hz, so as to avoid plant dynamics. But, the control bandwidth is about 25 Hz because $\Delta T_c = 0.02$ s. The operating bandwidth cannot be greater than this value and would be ≈ 25 Hz.

5.6 Bandwidth Design of a Control System

Based on the foregoing concepts, it is now possible to give a set of simple steps for designing a control system on the basis of bandwidth considerations.

Step 1: Decide on the maximum frequency of operation (BW_o) of the system based on the requirements of the particular application.

Step 2: Select the process components (electromechanical) that have the capacity to operate at BW_o and perform the required tasks.

Step 3: Select feedback sensors with a flat frequency spectrum (operating frequency range) greater than $4 \times BW_o$.

Step 4: Develop a digital controller with a sampling rate greater than $4 \times BW_o$ for the sensor feedback signals (keeping within the flat spectrum of the sensors) and a direct-digital control cycle time (period) of $1/(2 \times BW_o)$. Note that the digital control actions are generated at a rate of $2 \times BW_o$.

Step 5: Select the control drive systems (interface analog hardware, filters, amplifiers, actuators, etc.) that have a flat frequency spectrum of at least BW_o.

Step 6: Integrate the system and test the performance. If the performance specifications are not satisfied, make the necessary adjustments and test again.

5.6.1 Comment about Control Cycle Time

In the engineering literature, it is often used that $\Delta T_c = \Delta T_p$, where $\Delta T_c =$ control cycle time (period at which the digital control actions are generated) and $\Delta T_p =$ the period at which the feedback sensor signals are sampled (see Figure 5.9a). This is acceptable in systems where the significant frequency range of the plant is sufficiently smaller than $1/\Delta T_p$ (and $1/\Delta T_c$). In that case, the sampling rate $1/\Delta T_p$ of the feedback measurements (and the Nyquist frequency $0.5/\Delta T_p$) will still be sufficiently larger than the significant frequency range of the plant (see Figure 5.9b), hence, the control system will function satisfactorily. But, the bandwidth criterion presented in this section satisfies $\Delta T_p \leq \Delta T_c$. This is a more desirable option. For example, in Figure 5.9c, two measurement samples are used in computing each control action. Here, the Nyquist frequency of the sampled feedback signals is double that of the previous case, and it will cover a larger (double) frequency range of the plant.

5.7 Instrument Error Analysis

The analysis of error in an instrument or a multicomponent mechatronic system is a very challenging task. Difficulties arise for many reasons, particularly the following:

1. The true value is usually unknown.

2. The instrument reading may contain random errors that cannot be determined exactly.

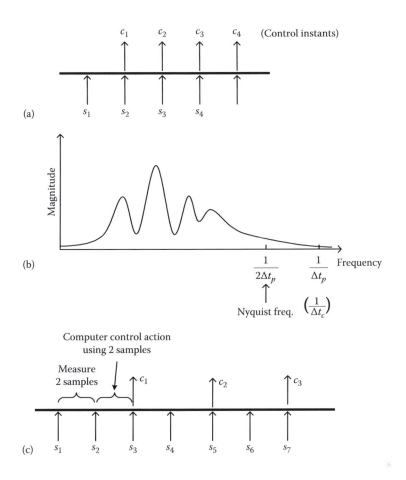

FIGURE 5.9
(a) Conventional sampling of feedback sensor signals for direct digital control, (b) acceptable frequency characteristic of a plant for case (a), (c) improved sampling criterion for feedback signals in direct digital control.

3. The error may be a complex (i.e., not simple) function of many variables (input variables and state variables or response variables).

4. The system/instrument may be made up of many components that have complex interrelations (dynamic coupling, multiple degree-of-freedom responses, nonlinearities, etc.), and each component may contribute to the overall error.

The first item is a philosophical issue that would lead to an argument similar to the chicken-and-egg controversy. For instance, if the true value is known, there is no need to measure it; and if the true value is unknown, it is impossible to determine exactly how inaccurate a particular reading is. In fact, this situation can be addressed to some extent by using statistical representations of error, which takes us to the second item listed. The third and fourth items may be addressed by error combination in multivariable systems and by error propagation in complex multicomponent systems. It is not feasible here to provide a full treatment of all these topics. Only an introduction to simple analytical techniques will be given, using illustrative examples. The concepts discussed here are useful not only in statistical error analysis but also in the field of *statistical process control* (SPC)—the use of statistical signals to improve the performance of a process. Performing the statistical analysis of a response signal and drawing its *control chart*, along with an *upper control line* and a *lower control line*, are key procedures in SPC.

5.7.1 Statistical Representation

In general, error is a random variable. It is defined as

$$\text{Error} = (\text{instrument reading}) - (\text{true value}) \tag{5.6}$$

Randomness associated with a *measurand* can be interpreted in two ways. First, since the true value of the measurand is a fixed quantity, randomness can be interpreted as the randomness in error that is usually originating from the random factors in instrument response. Second, looking at the issue in a more practical manner, error analysis can be interpreted as an "estimation problem" in which the objective is to estimate the true value of a measurand from a known set of readings. In this latter point of view, the "estimated" true value itself becomes a random variable. No matter what approach is used, however, the same statistical concepts may be used in representing error. Appendix C provides an introduction to the analytical basis of the required procedures.

5.7.2 Accuracy and Precision

The instrument ratings, as mentioned before, affect the overall *accuracy* of an instrument. Accuracy can be assigned either to a particular reading or to an instrument. Note that instrument accuracy depends not only on the physical hardware of the instrument but also on the operating conditions (e.g., design conditions that are the normal, steady operating conditions, or extreme transient conditions, such as emergency start-up and shutdown). *Measurement accuracy* determines the closeness of the measured value to the true value. *Instrument accuracy* is related to the worst accuracy obtainable within the dynamic range of the instrument in a specific operating environment. *Measurement error* is defined as

$$\text{Error} = (\text{measured value}) - (\text{true value}) \tag{5.7}$$

Correction, which is the negative of error, is defined as

$$\text{Correction} = (\text{true value}) - (\text{measured value}) \tag{5.8}$$

Each of these can also be expressed as a percentage of the true values. The accuracy of an instrument may be determined by measuring a parameter whose true value is known, near the extremes of the dynamic range of instrument, under certain operating conditions. For this purpose, standard parameters or signals than can be generated at very high levels of accuracy would be needed. The National Institute for Standards and Testing (NIST) is usually responsible for the generation of these standards. Nevertheless, accuracy and error values cannot be determined to 100% exactness in typical applications because the true value is not known to begin with. In a given situation, we can only make estimates for accuracy by using ratings provided by the instrument manufacturer or by analyzing data from previous measurements and models.

The causes of error include instrument instability, external noise (disturbances), poor calibration, inaccurate information (e.g., poor analytical models, inaccurate control laws, and digital control algorithms), parameter changes (e.g., due to environmental changes, aging, and wearout), unknown nonlinearities, and improper use of the instrument.

Errors can be classified as *deterministic* (or *systematic*) and *random* (or *stochastic*). Deterministic errors are those caused by well-defined factors, including nonlinearities and offsets in readings. These usually can be accounted for by proper calibration and analysis practices. Error ratings and calibration charts are used to remove systematic errors from instrument readings. Random errors are caused by uncertain factors entering into instrument response. These include device noise, line noise, and effects of unknown random variations in the operating environment. A statistical analysis using sufficiently large amounts of data is necessary to estimate random errors. The results are usually expressed as a mean error, which is the systematic part of random error, and a standard deviation or confidence interval for instrument response. *Precision* is not synonymous with accuracy. Reproducibility (or repeatability) of an instrument reading determines the precision of an instrument. An instrument that has a high offset error might be able to generate a response at high precision even though this output is clearly inaccurate. For example, consider a timing device (clock) that very accurately indicates time increments (say, up to the nearest nanosecond). If the reference time (starting time) is set incorrectly, the time readings will be in error even though the clock has a very high precision.

Instrument error may be represented by a random variable that has a mean value μ_e and a standard deviation σ_e. If the standard deviation is zero, the variable is considered deterministic. In that case, the error is said to be deterministic or repeatable. Otherwise, the error is said to be random. The precision of an instrument is determined by the standard deviation of error in the instrument response. The readings of an instrument may have a large mean value of error (e.g., large offset), but if the standard deviation is small, the instrument has high precision. Hence, a quantitative definition for precision would be

$$\text{Precision} = \frac{\text{measurement range}}{\sigma_e} \tag{5.9}$$

A lack of precision originates from random causes and poor construction practices. It cannot be compensated for by recalibration, just as the precision of a clock cannot be improved by resetting the time. On the other hand, accuracy can be improved by recalibration. Repeatable (deterministic) accuracy is inversely proportional to the magnitude of the mean error μ_e.

Matching instrument ratings with specifications is very important in selecting instruments for a mechatronic application. Several additional considerations should be looked into as well. These include geometric limitations (size, shape, etc.), environmental conditions (e.g., chemical reactions including corrosion, extreme temperatures, light, dirt accumulation, electromagnetic fields, radioactive environments, shock, and vibration), power requirements, operational simplicity, availability, past record and reputation of the manufacturer and of the particular instrument, and cost-related economic aspects (initial cost, maintenance cost, cost of supplementary components such as signal-conditioning and processing devices, design life and associated frequency of replacement, and cost of disposal and replacement). Often, these considerations become the ultimate deciding factors in the selection process.

5.7.3 Error Combination

Error in a response variable of a device or in an estimated parameter of a system would depend on errors present in measured variables and parameter values that are used to determine the unknown variable or parameter. Knowing how component errors are propagated within a multicomponent system and how individual errors in system variables and parameters contribute toward the overall error in a particular response variable or parameter would be important in estimating error limits in complex mechatronic systems. For example, if the output power of a rotational manipulator is computed by measuring torque and speed at the output shaft, error margins in the two measured "response variables" (torque and speed) would be directly combined into the error in the power computation. Similarly, if the natural frequency of a simple suspension system is determined by measuring the mass and spring stiffness "parameters" of the suspension, the natural frequency estimate would be directly affected by possible errors in mass and stiffness measurements. Extending this idea further, the overall error in a mechatronic system depends on the individual error levels in various components (sensors, actuators, controller hardware, filters, amplifiers, etc.) of the system and on the manner in which these components are physically interconnected and physically interrelated. For example, in a robotic manipulator, the accuracy of the actual trajectory of the end effector will depend on the accuracy of sensors and actuators at the manipulator joints and on the accuracy of the robot controller. Note that we are dealing with a generalized idea of error propagation that considers errors in system variables (e.g., input and output signals, such as velocities, forces, voltages, currents, temperatures, heat transfer rates, pressures, and fluid flow rates), system parameters (e.g., mass, stiffness, damping, capacitance, inductance, resistance, thermal conductivity, and viscosity), and system components (e.g., sensors, actuators, filters, amplifiers, control circuits, thermal conductors, and valves).

For the analytical development of a basic result in error combination, we will start with a functional relationship of the form

$$y = f(x_1, x_2, \ldots x_r) \tag{5.10}$$

Here, x_i are the independent system variables or parameter values whose error is propagated into a dependent variable (or parameter value) y. The determination of this functional relationship is not always simple, and the relationship itself may be in error. Since our intention is to make a reasonable estimate for possible error in y due to the combined effect of errors from x_i, an approximate functional relationship would be adequate in most cases. Let us denote error in a variable by the differential of that variable. Taking the differential of Equation 5.10, we get

$$\delta y = \frac{\partial f}{\partial x_1} \delta x_1 + \frac{\partial f}{\partial x_2} \delta x_2 + \cdots + \frac{\partial f}{\partial x_r} \delta x_r \tag{5.11}$$

for small errors. For those who are not familiar with differential calculus, Equation 5.11 should be interpreted as the first-order terms in a *Taylor series expansion* of Equation 5.10. Now, by rewriting Equation 5.11 in the fractional form, we get

$$\frac{\delta y}{y} = \sum_{i=1}^{r} \left[\frac{x_i}{y} \frac{\partial f}{\partial x_i} \frac{\delta x_i}{x_i} \right] \tag{5.12}$$

Here, $\delta y / y$ represents the overall error and $\delta x_i / x_i$ represents the component error, expressed as fractions. We shall consider two types of estimates for overall error.

5.7.4 Absolute Error

Since error δx_i could be either positive or negative, an upper bound for the overall error is obtained by summing the absolute value of each right-hand-side term in Equation 5.12. This estimate e_{ABS}, which is termed *absolute error*, is given by

$$e_{ABS} = \sum_{i=1}^{r} \left| \frac{x_i}{y} \frac{\partial f}{\partial x_i} \right| e_i \tag{5.13}$$

Note that component error e_i and absolute error e_{ABS} in Equation 5.13 are always positive quantities; when specifying error, however, both positive and negative limits should be indicated or implied (e.g., $\pm e_{ABS}$, $\pm e_i$).

5.7.5 SRSS Error

Equation 5.13 provides a conservative (upper bound) estimate for overall error. Since the estimate itself is not precise, it is often wasteful to introduce such a high conservatism. A nonconservative error estimate that is frequently used in practice is the *SRSS* error. As the name implies, this is given by

$$e_{SRSS} = \left[\sum_{i=1}^{r} \left(\frac{x_i}{y} \frac{\partial f}{\partial x_i} e_i \right)^2 \right]^{1/2} \tag{5.14}$$

This is not an upper bound estimate for error. In particular, $e_{SRSS} < e_{ABS}$ when more than one nonzero error contribution is present. The SRSS error relation is particularly suitable when component error is represented by the standard deviation of the associated variable or parameter value and when the corresponding error sources are independent. Now we will present several examples of error combination.

Example 5.7

Using the absolute value method for error combination, determine the fractional error in each item x_i so that the contribution from each item to the overall error e_{ABS} is the same.

Solution
For equal contribution, we must have

$$\left| \frac{x_1}{y} \frac{\partial f}{\partial x_1} \right| e_1 = \left| \frac{x_2}{y} \frac{\partial f}{\partial x_2} \right| e_2 = \cdots = \left| \frac{x_r}{y} \frac{\partial f}{\partial x_r} \right| e_r$$

Hence, $r \left| \dfrac{x_i}{y} \dfrac{\partial f}{\partial x_i} \right| e_i = e_{ABS}$. Thus,

$$e_i = e_{ABS} \bigg/ \left(r \left| \frac{x_i}{y} \frac{\partial f}{\partial x_i} \right| \right) \tag{5.15}$$

Example 5.8

The result obtained in the previous example is useful in the design of multicomponent systems and in the cost-effective selection of instrumentation for a particular application. Using Equation 5.15, arrange the items x_i in their order of significance.

Solution

Note that Equation 5.15 may be written as

$$e_i = K \Big/ \left| x_i \frac{\partial f}{\partial x_i} \right| \tag{5.16}$$

where K is a quantity that does not vary with x_i. It follows that for equal error contribution from all items, error in x_i should be inversely proportional to $|x_i(\partial f/\partial x_i)|$. In particular, the item with the largest $|x_i(\partial f/\partial x_i)|$ should be made the most accurate. In this manner, allowable relative accuracy for various components can be estimated. Since, in general, the most accurate device is also the most costly one, instrumentation cost can be optimized if components are selected according to the required overall accuracy using a criterion such as that implied by Equation 5.16.

Example 5.9

Figure 5.10 schematically shows an optical device for measuring displacement. This sensor is essentially an optical potentiometer. The potentiometer element is uniform and has a resistance R_c. A photoresistive layer is sandwiched between this element and is a perfect conductor of electricity. A light source, which moves with the object whose displacement is being measured, directs a beam of light whose intensity is I, on to a narrow rectangular region of the photoresistive layer. As a result, this region becomes resistive with resistance R that bridges the potentiometer element and the conductor element, as shown. An empirical relation between R and I was found to be

$$\ln\left(\frac{R}{R_0}\right) = \left(\frac{I_0}{I}\right)^{1/4}$$

in which the resistance R is in kΩ and the light intensity I is expressed in watts per square meter (W/m²). The parameters R_0 and I_0 are empirical constants having the same units as R and I, respectively. These two parameters generally have some experimental errors.

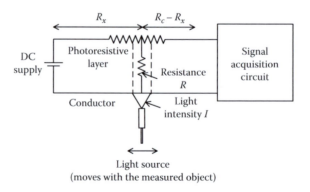

FIGURE 5.10
An optical displacement sensor.

(a) Sketch the curve of R versus I and explain the significance of the parameters R_0 and I_0.

(b) Using the absolute error method, show that the combined fractional error e_R in the bridging resistance R can be expressed as

$$e_R = e_{R0} + \frac{1}{4}\left(\frac{I_0}{I}\right)^{1/4}\left[e_I + e_{I0}\right]$$

in which e_{R0}, e_I, and e_{I0} are the fractional errors in R_0, I, and I_0, respectively.

(c) Suppose that the empirical error in the sensor model can be expressed as $e_{R0} = \pm 0.01$ and $e_{I0} = \pm 0.01$ and due to variations in the supply to the light source and in ambient lighting conditions, the fractional error in I is also ± 0.01. If the error E_R is to be maintained within ± 0.02, at what light intensity level (I) should the light source operate? Assume that the empirical value of I_0 is 2.0 W/m².

(d) Discuss the advantages and disadvantages of this device as a dynamic displacement sensor.

Solution

(a) $\ln\dfrac{R}{R_0} = \left(\dfrac{I_0}{I}\right)^{1/4}$

R_0 represents the minimum resistance provided by the photoresistive bridge (i.e., at very high light intensity levels). When $I = I_0$, the bridge resistance R is about $2.7 R_0$, hence, I_0 represents a lower bound for the intensity for proper operation of the sensor. A suitable upper bound for the intensity would be $10 I_0$ for satisfactory operation. At this value, $R \approx 1.75 R_0$, as shown in Figure 5.11.

(b) $\ln R - \ln R_0 = \left(\dfrac{I_0}{I}\right)^{1/4}$

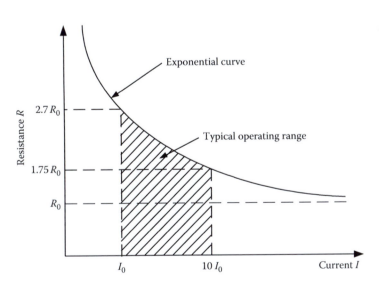

FIGURE 5.11
Characteristic curve of the sensor.

Differentiate

$$\frac{\delta R}{R} - \frac{\delta R_0}{R_0} = \frac{1}{4}\left(\frac{I_0}{I}\right)^{-3/4}\left[\frac{\delta I_0}{I} - \frac{I_0}{I^2}\delta I\right] = \frac{1}{4}\left(\frac{I_0}{I}\right)^{1/4}\left[\frac{\delta I_0}{I_0} - \frac{\delta I}{I}\right]$$

Hence, with the absolute method of error combination,

$$e_R = e_{R0} + \frac{1}{4}\left(\frac{I_0}{I}\right)^{1/4}\left[e_{I0} + e_I\right]$$

(c) With the given numerical values, we have

$$0.02 = 0.01 + \frac{1}{4}\left(\frac{I_0}{I}\right)^{1/4}[0.01 + 0.01] \Rightarrow \left(\frac{I_0}{I}\right)^{1/4} = 2$$

or

$$I = \frac{1}{16}I_0 = \frac{2.0}{16}\text{ W/m}^2 = 0.125\text{ W/m}^2$$

Note: For larger values of I, the absolute error in R_0 would be smaller. For example, for $I = 10I_0$, we have $e_R = 0.01 + \frac{1}{4}\left(\frac{1}{10}\right)^{1/4}[0.01 + 0.01] \approx 0.013$.

(d) *Advantages*
- Noncontacting
- Small moving mass (low inertial loading)
- All advantages of a potentiometer

Disadvantages
- Nonlinear and exponential variation of R
- Effect of ambient lighting
- Possible nonlinear behavior of the device (input–output relation)
- Effect of variations in the supply to the light source
- Effect of aging of the light source

5.8 Statistical Process Control

In SPC, statistical analysis of process responses is used to generate control actions. This method of control is applicable in many situations of process control, including manufacturing quality control, control of chemical process plants, computerized office management systems, inventory control systems, and urban transit control systems. A major step in SPC is to compute control limits (or action lines) on the basis of measured data from the process.

5.8.1 Control Limits or Action Lines

Since a very high percentage of readings from an instrument should lie within $\pm 3\sigma$ about the mean value, according to the normal distribution, these boundaries (-3σ and $+3\sigma$) drawn about the mean value may be considered *control limits* or *action lines* in SPC. If any measurements fall outside the action lines, corrective measures such as recalibration, controller adjustment, or redesign should be carried out.

5.8.2 Steps of SPC

The main steps of SPC are as follows:

1. Collect measurements of appropriate response variables of the process
2. Compute the mean value of the data, the upper control limit, and the lower control limit
3. Plot the measured data and draw the two control limits on a control chart
4. If measurements fall outside the control limits, take corrective action and repeat the control cycle (go to step 1)

If the measurements always fall within the control limits, the process is said to be in statistical control.

Example 5.10

Errors in a satellite tracking system were monitored online for a period of 1 h to determine whether recalibration or gain adjustment of the tracking controller would be necessary. Four measurements of the tracking deviation were taken in a period of 5 min, and 12 such data groups were acquired during the 1 h period. Sample means and sample variance of the 12 groups of data were computed. The results are tabulated as follows:

Period i	1	2	3	4	5	6	7	8	9	10	11	12
Sample mean $\bar{X}_i$	1.34	1.10	1.20	1.15	1.30	1.12	1.26	1.10	1.15	1.32	1.35	1.18
Sample variance S_i^2	0.11	0.02	0.08	0.10	0.09	0.02	0.06	0.05	0.08	0.12	0.03	0.07

Draw a control chart for the error process with control limits (action lines) at $\bar{X} \pm 3\sigma$. Establish whether the tracking controller is in statistical control or needs adjustment.

Solution

The overall mean tracking deviation, $\bar{X} = \frac{1}{12} \sum_{i=1}^{12} \bar{X}_i$, is computed to be $\bar{X} = 1.214$. The average sample variance, $\bar{S}^2 = \frac{1}{12} \sum_{i=1}^{12} S_i^2$, is computed to be $\bar{S}^2 = 0.069$. Since there are four readings within each period, the standard deviation σ of group mean $\bar{X}_i$ can be estimated as (see Appendix C)

$$S = \frac{\bar{S}}{\sqrt{4}} = \frac{\sqrt{0.069}}{\sqrt{4}} = 0.131.$$

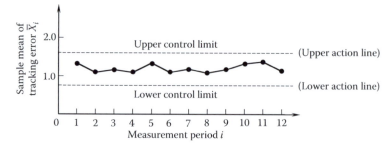

FIGURE 5.12
Control chart for the satellite tracking error example.

The upper control limit (action line) is at (approximately)

$$x = \bar{X} + 3S = 1.214 + 3 \times 0.131 = 1.607$$

The lower control limit (action line) is at

$$x = \bar{X} - 3S = 0.821$$

These two lines are shown on the control chart in Figure 5.12. Since the sample means lie within the two action lines, the process is considered to be in statistical control, and controller adjustments would not be necessary.

Note: If better resolution is required in making this decision, individual readings rather than group means should be plotted in Figure 5.12.

Problems

5.1 What do you consider a perfect measuring device? Suppose that you are asked to develop an analog device for measuring angular position in an application related to the control of a kinematic linkage system (a robotic manipulator, for example). What instrument ratings (or specifications) would you consider crucial in this application? Discuss their significance.

5.2 A tactile (distributed touch) sensor of the gripper of a robotic manipulator consists of a matrix of piezoelectric sensor elements placed at 2 mm apart. Each element generates an electric charge when it is strained by an external load. Sensor elements are multiplexed at very high speeds in order to avoid charge leakage and to read all data channels using a single high-performance charge amplifier. Load distribution on the surface of the tactile sensor is determined from the charge amplifier readings since the multiplexing sequence is known. Each sensor element can read a maximum load of 50 N and can detect load changes on the order of 0.01 N.

(a) What is the spatial resolution of the tactile sensor?

(b) What is the load resolution (in N/m²) of the tactile sensor?

(c) What is the dynamic range?

5.3 A useful rating parameter for a robotic tool is *dexterity*. Though not complete, an appropriate analytical definition for dexterity of a device is

$$\text{dexterity} = \frac{\text{number of degrees of freedom}}{\text{motion resolution}}$$

where the number of degrees of freedom is equal to the number of independent variables that is required to completely define an arbitrary position increment of the tool (i.e., for an arbitrary change in its kinematic configuration).

(a) Explain the physical significance of dexterity and give an example of a device for which the specification of dexterity would be very important.

(b) The power rating of a tool may be defined as the product of maximum force that can be applied by it in a controlled manner and the corresponding maximum speed. Discuss why the power rating of a manipulating device is usually related to the dexterity of the device. Sketch a typical curve of power versus dexterity.

5.4 The resolution of a feedback sensor (or resolution of a response measurement used in feedback) has a direct effect on the accuracy that is achievable in a control system. This is true because the controller cannot correct a deviation of the response from the desired value (set point) unless the response sensor can detect that change. It follows that the resolution of a feedback sensor will govern the minimum (best) possible deviation band (about the desired value) of the system response under feedback control. An angular position servo uses a resolver as its feedback sensor. If peak-to-peak oscillations of the servo load (plant) under steady-state conditions have to be limited to no more than two degrees, what is the worst tolerable resolution of the resolver? Note that, in practice, the feedback sensor should have a resolution better (smaller) than this worst value.

5.5 Consider a mechanical component whose response x is governed by the relationship $f = f(x, \dot{x})$, where f denotes applied (input) force and $\dot{x}$ denotes velocity. Three special cases are:

(a) Linear spring: $f = kx$
(b) Linear spring with a viscous (linear) damper: $f = kx + b\dot{x}$
(c) Linear spring with Coulomb friction: $f = kx + f_c \text{sgn}(\dot{x})$

Suppose that a harmonic excitation of the form $f = f_o \sin \omega t$ is applied in each case. Sketch the force–displacement curves for the three cases at steady state. Which components exhibit hysteresis? Which components are nonlinear? Discuss your answers.

5.6 Discuss how the accuracy of a digital controller may be affected by the following:

(a) Stability and bandwidth of amplifier circuitry
(b) Load impedance of the A/D conversion circuitry

Also, what methods do you suggest to minimize problems associated with these parameters?

5.7 Consider the mechanical tachometer shown in Figure 5.4. Write expressions for sensitivity and bandwidth for the device. Using the example, show that the two performance ratings, sensitivity and bandwidth, generally conflict. Discuss ways to improve the sensitivity of this mechanical tachometer.

5.8 (a) What is an anti-aliasing filter? In a particular application, the sensor signal is sampled at f_s Hz. Suggest a suitable cutoff frequency for an anti-aliasing filter to be used in this application.

5.9 (a) Define the following terms:

- Sensor
- Transducer
- Actuator
- Controller
- Control system
- Operating bandwidth of a control system
- Control bandwidth
- Nyquist frequency

(b) Choose three practical dynamic systems each of which has at least one sensor, one actuator, and a feedback controller.

(i) Briefly describe the purpose and operation of each dynamic system.

(ii) For each system, give a suitable value for the operating bandwidth, control bandwidth, operating frequency range of the sensor, and sampling rate for the sensor signal for feedback control. Clearly justify the values that you have given.

5.10 Discuss and the contrast the following terms:

(a) Measurement accuracy

(b) Instrument accuracy

(c) Measurement error

(d) Precision

Also, for an analog sensor-transducer unit of your choice, identify and discuss various sources of error and ways to minimize or account for their influence.

5.11 (a) Explain why mechanical loading error due to tachometer inertia can be significantly higher when measuring transient speeds than when measuring constant speeds.

(b) A dc tachometer has an equivalent resistance $R_a = 20\ \Omega$ in its rotor windings. In a position plus velocity servo system, the tachometer signal is connected to a feedback control circuit with an equivalent resistance of $2\,k\Omega$. Estimate the percentage error due to electrical loading of the tachometer at steady state.

(c) If the conditions were not steady, how would the electrical loading be affected in this application?

5.12 Briefly explain what is meant by the terms *systematic error* and *random error* of a measuring device. Which statistical parameters may be used to quantify these two types of error? State, giving an example, how *precision* is related to error.

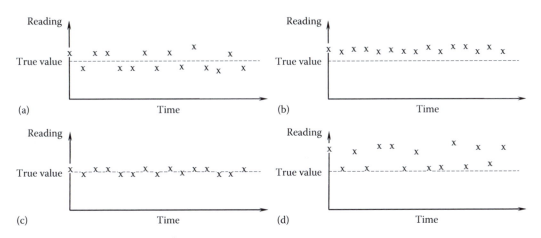

FIGURE P5.13
Four sets of measurements on the same response variable using different sensors.

5.13 Four sets of measurements were taken on the same response variable of a process using four different sensors. The true value of the response was known to be constant. Suppose that the four sets of data are as shown in Figure P5.13a through d. Classify these data sets, and hence the corresponding sensors, with respect to precision and deterministic (repeatable) accuracy.

5.14 (i) Briefly discuss any conflicts that can arise in specifying parameters that can be used to predominantly represent the speed of response and the degree of stability of a process (plant).

(ii) Consider a measuring device that is connected to a plant for feedback control. Explain the significance of the following:

(a) Bandwidth

(b) Resolution

(c) Linearity

(d) Input impedance

(e) Output impedance of the measuring device in the performance of the feedback control system

5.15 Consider the speed control system schematically shown in Figure P5.15. Suppose that the plant and the controller together are approximated by the transfer function $G_p(s) = k/(\tau_p s + 1)$, where τ_p is the plant time constant.

(a) Give an expression for the bandwidth ω_p of the plant without feedback.

(b) If the feedback tachometer is ideal and is represented by a unity (negative) feedback, what is the bandwidth ω_c of the feedback control system?

(c) If the feedback tachometer can be represented by the transfer function $G_s(s) = 1/(\tau_s s + 1)$, where τ_s is the sensor time constant, explain why the bandwidth ω_{cs} of the feedback control system is given by the smaller quantity of $1/\tau_s$ and $(k+1)/(\tau_p + \tau_s)$. Assume that both τ_p and τ_s are sufficiently small.

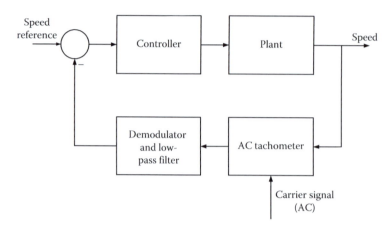

FIGURE P5.15
A speed control system.

Next, suppose that approximately $\tau_p = 0.016$ s. Estimate a sufficient bandwidth in Hz for the tachometer. Also, if $k = 1$, estimate the overall bandwidth of the feedback control system. If $k = 49$, what would be the representative bandwidth of the feedback control system?

For the particular ac tachometer (with the bandwidth value as chosen in the present numerical example), what should be the frequency of the carrier signal? Also, what should be the cutoff frequency of the low-pass filter that is used with its demodulator circuit?

5.16 Using the SRSS method for error combination, determine the fractional error in each component x_i so that the contribution from each component to the overall error e_{SRSS} is the same.

5.17 A single-degree-of-freedom model of a robotic manipulator is shown in Figure P5.17a. The joint motor has rotor inertia J_m. It drives an inertial load that has moment of inertia J_l through a speed reducer of gear ratio 1:r (*Note*: $r < 1$). The control scheme used in this system is the so-called feedforward control (strictly, *computed-torque control*) method. Specifically, the motor torque T_m that is required to accelerate or decelerate the load is computed using a suitable dynamic model and a desired motion trajectory for the manipulator, and the motor windings are excited so as to generate that torque. A typical trajectory would consist of a constant angular acceleration segment followed by a constant angular velocity segment and finally a constant deceleration segment, as shown in Figure P5.17b.

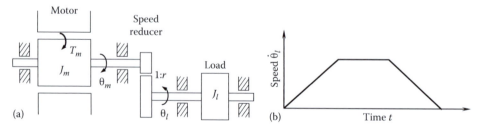

FIGURE P5.17
(a) A single-degree-of-freedom model of a robotic manipulator and (b) a typical reference (desired) speed trajectory for computed-torque control.

(a) Neglecting friction (particularly bearing friction) and inertia of the speed reducer, show that a dynamic model for torque computation during accelerating and decelerating segments of the motion trajectory would be $T_m = (J_m + r^2 J_l) \ddot{\theta}_l / r$, where $\ddot{\theta}_l$ is the angular acceleration of the load, hereafter denoted by α_l. Show that the overall system can be modeled as a single inertia rotating at the motor speed. Using this result, discuss the effect of gearing on a mechanical drive.

(b) Given that $r = 0.1$, $J_m = 0.1\,\text{kg m}^2$, $J_l = 1.0\,\text{kg m}^2$, and $\alpha_l = 5.0\,\text{rad/s}^2$, estimate the allowable error for these four quantities so that the combined error in the computed torque is limited to $\pm 4\%$ and so that each of the four quantities contributes equally toward this error in computed Tm. Use the absolute value method for error combination.

(c) Arrange the four quantities r, J_m, J_l, and α_l in the descending order of required accuracy for the numerical values given in the problem.

(d) Suppose that $J_m = r^2 J_l$. Discuss the effect of error in r on the error in T_m.

5.18 An actuator (e.g., electric motor, hydraulic piston-cylinder) is used to drive a terminal device (e.g., gripper, hand, wrist with active remote center compliance) of a robotic manipulator. The terminal device functions as a force generator. A schematic diagram for the system is shown in Figure P5.18. Show that the displacement error e_x is related to the force error e_f through

$$e_f = \frac{x}{f}\frac{df}{dx} e_x$$

The actuator is known to be 100% accurate for practical purposes, but there is an initial position error δx_o (at $x = x_o$). Obtain a suitable transfer relation $f(x)$ for the terminal device so that the force error e_f remains constant throughout the dynamic range of the device.

5.19 (a) Clearly explain why the SRSS method of error combination is preferred to the "Absolute" method when the error parameters are assumed to be Gaussian and independent.

(b) Hydraulic pulse generators (HPG) may be used in a variety of applications such as rock blasting, projectile driving, and seismic signal generation. In a typical HPG, water at very high pressure is supplied intermittently from an accumulator into the discharge gun through a high-speed control valve. The pulsating water jet is discharged through a shock tube and may be used, for example, for blasting granite. A model for a HPG was found to be

$$E = aV\left(b + \frac{c}{V^{1/3}}\right)$$

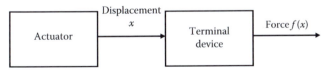

FIGURE P5.18
Block diagram for a terminal device of a robotic manipulator.

where
   E is the hydraulic pulse energy (kJ)
   V is the volume of blast burden (m³)
   a, b, and c are model parameters that may be determined experimentally

Suppose that this model is used to estimate the blast volume of material (V) for a specific amount of pulse energy (E).

(i) Assuming that the estimation error values in the model parameters a, b, and c are independent and may be represented by appropriate standard deviations, obtain an equation relating these fractional errors e_a, e_b, and e_c, to the fractional error e_v of the estimated blast volume.

(ii) Assuming that $a=2175.0$, $b=0.3$, and $c=0.07$ with consistent units, show that a pulse energy of $E=219.0$ kJ can blast a material volume of approximately 0.6^3 m³. If $e_a=e_b=e_c=\pm0.1$, estimate the fractional error e_v of this predicted volume.

5.20 The absolute method of error combination is suitable when the error contributions are additive (same sign). Under what circumstances would the SRSS method be more appropriate than the absolute method?

A simplified block diagram of a dc motor speed control system is shown in Figure P5.20. Show that in the Laplace domain, the fractional error e_y in the motor speed y is given by

$$e_y = -\frac{\tau s}{(\tau s + 1 + k)} e_\tau + \frac{(\tau s + 1)}{(\tau s + 1 + k)} e_k$$

where
   e_τ is the fractional error in the time constant τ
   e_k is the fractional error in the open-loop gain k

The reference speed command u is assumed to be error free. Express the absolute error combination relation for this system in the frequency domain ($s=j\omega$). Using it, show the following:

(a) At low frequencies, the contribution from the error in k will dominate and the error can be reduced by increasing the gain.

(b) At high frequencies, k and τ will make equal contributions toward the speed error, and the error cannot be reduced by increasing the gain.

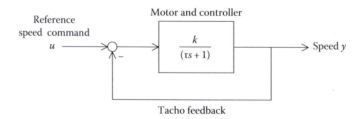

FIGURE P5.20
DC motor speed control system.

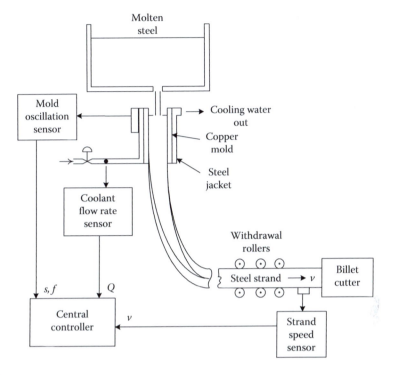

FIGURE P5.21
A steel-billet casting machine.

5.21 (a) Compare and contrast the "absolute error method" with/against the SRSS in analyzing the error combination of multicomponent systems. Indicate situations where one method is preferred over the other.

(b) Figure P5.21 shows a schematic diagram of a machine that is used to produce steel billets. The molten steel in the vessel (called "tundish") is poured into the copper mold having a rectangular cross-section. The mold has a steel jacket with channels to carry cooling water upwards around the copper mold. The mold, which is properly lubricated, is oscillated using a shaker (electromechanical or hydraulic) in order to facilitate stripping of the solidified steel inside it. A set of power-driven friction rollers is used to provide the withdrawal force for delivering the solidified steel strand to the cutting station. A billet cutter (torch or shear type) is used to cut the strand into billets of appropriate length.

The quality of the steel billets produced by this machine is determined on the basis of several factors, which include various types of cracks, deformation problems such as rhomboidity, and oscillation marks. It is known that the quality can be improved through proper control of the following variables:

Q is the coolant (water) flow rate

v is the speed of the steel strand (withdrawal speed)

s is the stroke of the mold oscillations

f is the cyclic frequency of the mold oscillations

Specifically, these variables are measured and transmitted to the central controller of the billet casting machine, which in turn generates proper control commands for the coolant-valve controller, the drive controller of the withdrawal rollers, and the shaker controller.

A nondimensional quality index q has been expressed in terms of the measured variables, as

$$q = \left[1 + \frac{s}{s_0}\sin\frac{\pi}{2}\left(\frac{f}{f_0 + f}\right)\right]\Big/(1 + \beta v/Q)$$

in which s_0, f_0, and β are operating parameters of the control system and are exactly known. Under normal operating conditions, the following conditions are (approximately) satisfied: $Q \approx \beta v$; $f \approx f_0$; $s \approx s_0$.

Note: If the sensor readings are incorrect, the control system will not function properly and the quality of the billets will deteriorate. It is proposed to use the "absolute error method" to determine the influence of the sensor errors on the billet quality.

(i) Obtain an expression for the quality deterioration δq in terms of the fractional errors $\delta v/v$, $\delta Q/Q$, $\delta s/s$, and $\delta f/f$ of the sensor readings.

(ii) If the sensor of the strand speed is known to have an error of $\pm 1\%$, determine the allowable error percentages for the other three sensors so that there is equal contribution of error to the quality index from all four sensors, under normal operating conditions.

5.22 Consider the servo control system that is modeled as in Figure P5.20. Note that k is the equivalent gain and τ is the overall time constant of the motor and its controller.

(a) Obtain an expression for the closed-loop transfer function y/u.

(b) In the frequency domain, show that for equal contribution of parameter error towards the system response, we should have $e_k/e_\tau = \tau\omega/\sqrt{\tau^2\omega^2 + 1}$.

where fractional errors (or variations) are for the gain, $e_k = |\delta k/k|$; and for the time constant, $e_\tau = |\delta\tau/\tau|$.

Using this relation, explain why at low frequencies the control system has a larger tolerance to error in τ than to that in k. Also show that at very high frequencies the two error tolerance levels are almost equal.

5.23 The quality control system in a steel rolling mill uses a proximity sensor to measure the thickness of rolled steel (steel gage) at every 2 ft along the sheet, and the mill controller adjustments are made on the basis of the last 20 measurements. Specifically, the controller is adjusted unless the probability that the mean thickness lies within $\pm 1\%$ of the sample mean exceeds 0.99. A typical set of 20 measurements in millimeters is as follows:

5.10	5.05	4.94	4.98	5.10	5.12	5.07	4.96	4.99	4.95
4.99	4.97	5.00	5.08	5.10	5.11	4.99	4.96	4.90	4.10

Check whether adjustments would be made in the gage controller on the basis of these measurements.

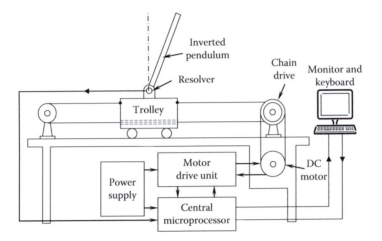

FIGURE P5.24
A microprocessor-controlled inverted pendulum—an application of SPC.

5.24 The dynamics and control of inherently unstable systems, such as rockets, can be studied experimentally using simple scaled-down physical models of the prototype systems. One such study is the classic inverted pendulum problem. An experimental setup for the inverted pendulum is shown in Figure P5.24. The inverted pendulum is supported on a trolley that is driven on a tabletop along a straight line using a chain-and-sprocket transmission operated by a dc motor. The motor is turned by commands from a microprocessor that is interfaced with the drive system of the motor. The angular position of the pendulum rod is measured using a resolver and is transmitted (fed back) to the microprocessor. A strategy of SPC is used to balance the pendulum rod. Specifically, control limits are established from an initial set of measurement samples of the pendulum angle. Subsequently, if the angle exceeds one control limit, the trolley is accelerated in the opposite direction using an automatic command to the motor. The control limits are also updated regularly. Suppose that the following 20 readings of the pendulum angle were measured (in degrees) after the system had operated for a few minutes:

0.5	−0.5	0.4	−0.3	0.3	0.1	−0.3	0.3	4.0	0.0
0.4	−0.4	0.5	−0.5	−5.0	0.4	−0.4	0.3	−0.3	−0.1

Establish whether the system was in statistical control during the period in which the readings were taken. Comment on this method of control.

5.25 (a) You are required to select a sensor for a position control application. List several important considerations that you have to take into account in this selection. Briefly indicate why each of them is important.

(b) A schematic diagram of a chip refiner that is used in the pulp and paper industry is shown in Figure P5.25. This machine is used for mechanical pulping of wood chips. The refiner has one fixed disk and one rotating disk (typical diameter = 2 m). The plate is rotated by an AC induction motor. The plate separation (typical gap = 0.5 mm) is controlled using a hydraulic actuator (piston-cylinder unit with servovalve). Wood chips are supplied to the eye of the refiner by a

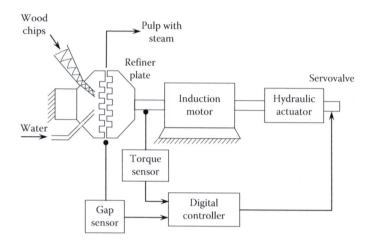

FIGURE P5.25
A single-disk chip refiner.

screw conveyor and are diluted with water. As the refiner plate rotates, the chips are ground into a pulp within the internal grooves of the plates. This is accompanied by the generation of steam due to energy dissipation. The pulp is drawn and further processed for making paper.

An empirical formula relating the plate gap (h) and the motor torque (T) is given by $T = ah/(1 + bh^2)$ with the model parameters a and b known to be positive.

(i) Sketch the curve T versus h. Express the maximum torque T_{max} and the plate gap (h_0) at this torque in terms of a and b only.

(ii) Suppose that the motor torque is measured and the plate gap is adjusted by the hydraulic actuator according to the formula given above. Show that the fractional error in h may be expressed as

$$e_h = \left[e_T + e_a + \frac{bh^2}{(1 + bh^2)} e_b \right] \frac{(1 + bh^2)}{\left| 1 - bh^2 \right|}$$

where e_T, e_a, and e_b are the fractional errors in T, a, and b, respectively, the latter two being representative of model error.

(iii) The normal operating region of the refiner corresponds to $h > h_0$. The interval $0 < h < h_0$ is known as the "pad collapse region" and should be avoided. If the operating value of the plate gap is $h = 2/\sqrt{b}$ and if the error values are given as $e_T = \pm 0.05$, $e_a = \pm 0.02$, and $e_b = \pm 0.025$, compute the corresponding error in the plate gap estimate.

(iv) Discuss why operation at $h = 1/\sqrt{b}$ is not desirable.

6

Sensors and Transducers

Study Objectives

- Terminology of sensors and transducers
- Application situations of sensors in mechatronics
- Modeling, analysis, and performance specification of sensory devices
- Common types of analog sensors for motion measurement
- Variable-inductance, magnetic, capacitive, and piezoelectric sensors
- Use of piezoresistive sensors (strain gages)
- Torque, force, and tactile sensors
- Ultrasonic and optical sensors
- Fluid-thermal sensors
- Digital transducers (optical encoders)
- Camera-based sensors (image processing and computer vision)

6.1 Introduction

Sensors and transducers may be used in a mechatronic system in several ways:

1. To measure the system outputs for feedback control
2. To measure system inputs (desirable inputs, unknown inputs, and disturbances) for feedforward control
3. To measure output signals for system monitoring, diagnosis, evaluation, parameter adjustment, and supervisory control
4. To measure input and output signal pairs for system testing and experimental modeling (i.e., for system identification)

The proper selection and integration of sensors and transducers are crucial in instrumenting a mechatronic system. The characteristics of a *perfect measuring device* were discussed in Chapter 5. Even though real sensors and transducers can behave quite differently in practice, when developing a mechatronic system we should use the ideal behavior as a reference for the design specifications.

Potentiometers, differential transformers, resolvers, synchros, strain gages, tachometers, piezoelectric devices, bellows, diaphragms, flow meters, thermocouples, thermistors,

TABLE 6.1

Sensors and Actuators Used in Some Common Engineering Applications

Process	Typical Sensors	Typical Actuators
Aircraft	Displacement, speed, acceleration, elevation, heading, force pressure, temperature, fluid flow, voltage, current, global positioning system (GPS)	DC motors, stepper motors, relays, valve actuators, pumps, heat sources, jet engines
Automobile	Displacement, speed, force, pressure, temperature, fluid flow, fluid level, voltage, current	DC motors, stepper motors, valve actuators, pumps, heat sources
Home heating system	Temperature, pressure, fluid flow	Motors, pumps, heat sources
Milling machine	Displacement, speed, force, acoustics, temperature, voltage, current	DC motors, ac motors
Robot	Optical image, displacement, speed, force, torque, voltage, current	DC motors, stepper motors, ac motors, hydraulic actuators
Wood drying kiln	Temperature, relative humidity, moisture content, air flow	AC motors, dc motors, pumps, heat sources

and resistance temperature detectors (RTDs) are examples of sensors used in mechatronic systems. Typically, actuators (see Chapter 7) go hand in hand with sensors and transducers. Several engineering/mechatronic applications and their use of sensors and actuators are noted in Table 6.1. In this chapter, the significance of sensors and transducers in a mechatronic system is indicated; important criteria in selecting sensors and transducers for mechatronic applications are presented; and several representative sensors and transducers and their operating principles, characteristics, and applications are described.

6.1.1 Terminology

The variable that is being measured is termed the *measurand*. Examples are acceleration and velocity of a vehicle, torque into robotic joints, the temperature and pressure of a process plant, and current through an electric circuit. A measuring device passes through two stages while measuring a signal. First, the measurand is "felt" or *sensed*. Then, the measured signal is *transduced* (or converted) into the form of the device output. In fact, the sensor, which "senses" the response, automatically converts (i.e., transduces) this "measurement" into the sensor output—the response of the sensor element. For example, a piezoelectric accelerometer senses acceleration and converts it into an electric charge, an electromagnetic tachometer senses velocity and converts it into a voltage, and a shaft encoder senses a rotation and converts it into a sequence of voltage pulses. Hence, the terms sensor and transducer are used interchangeably to denote a sensor–transducer unit. The sensor and transducer stages are functional stages, and sometimes it is not easy or even feasible to separately identify physical elements associated with them. Furthermore, this separation is not very important in using existing devices. The proper separation of sensor and transducer stages (physically as well as functionally) can be crucial, however, when designing new measuring devices.

Typically, the measured signal is *transduced* (or converted) into a form that is particularly suitable for transmitting, recording, conditioning, processing, activating a controller, or driving an actuator. For this reason, the output of a transducer is often an electrical signal. The measurand is usually an analog signal because it represents the output of a dynamic system. For example, the charge signal from a piezoelectric accelerometer has to be converted into a voltage signal of an appropriate level using a charge amplifier. For use in a digital controller, it has to be digitized using an analog-to-digital (A/D) converter (ADC—see Chapter 4). In digital transducers, the transducer output is discrete. This facilitates the direct interface of a transducer with a digital processor.

A complex measuring device can have more than one sensing stage. Often, the measurand goes through several transducer stages before it is available for control and actuating purposes. Furthermore, filtering may be needed to remove measurement noise. Hence, signal conditioning is usually needed between the sensor and the controller as well as the controller and the actuator. Charge amplifiers, lock-in amplifiers, power amplifiers, switching amplifiers, linear amplifiers, tracking filters, low-pass filters, high-pass filters, and notch filters are some of the signal-conditioning devices used in mechatronic systems (see Chapter 4). In some literature, signal-conditioning devices such as electronic amplifiers are also classified as transducers. Since we are treating signal-conditioning and modification devices separately from measuring devices, this unified classification is avoided whenever possible, and the term *transducer* is used primarily in relation to measuring instruments. Note that it is somewhat redundant to consider electrical-to-electrical sensors–transducers as measuring devices because electrical signals need conditioning only before they are used to carry out a useful task. In this sense, electrical-to-electrical transduction should be considered a "conditioning" function rather than a "measuring" function. Additional components, such as power supplies and surge-protection units, are often needed in control systems, but they are only indirectly related to control functions. Relays and other switching devices and modulators and demodulators may also be included.

Pure transducers depend on nondissipative coupling in the transduction stage. *Passive transducers* (sometimes called *self-generating transducers*) depend on their power transfer characteristics for operation and do not need an external power source. It follows that pure transducers are essentially passive devices. Some examples are *electromagnetic, thermoelectric, radioactive, piezoelectric,* and *photovoltaic* transducers. External power is required to operate active sensors/transducers, and they do not depend on power conversion characteristics for their operation. A good example is a *resistive* transducer, such as a potentiometer, which depends on its power dissipation through a resistor to generate the output signal. Note that an active transducer requires a separate power source (power supply) for operation, whereas a passive transducer draws its power from a measured signal (measurand). Since passive transducers derive their energy almost entirely from the measurand, they generally tend to distort (or load) the measured signal to a greater extent than an active transducer would. Precautions can be taken to reduce such loading effects. On the other hand, passive transducers are generally simple in design, more reliable, and less costly. In the present classification of transducers, we are dealing with power in the immediate transducer stage associated with the measurand, not the power used in subsequent signal conditioning. For example, a piezoelectric charge generation is a passive process. But, a charge amplifier, which uses an auxiliary power source, would be needed in order to condition the generated charge.

Next, we will study several analog sensor–transducer devices that are commonly used in mechatronic system instrumentation. We will not attempt to present an exhaustive discussion of all types of sensors; rather, we will consider a representative selection.

Such an approach is reasonable in view of the fact that even though the scientific principles behind various sensors may differ, many other aspects (e.g., performance parameters, signal conditioning, interfacing, modeling procedures) can be common to a large extent.

6.1.2 Motion Sensors and Transducers

By motion, we mean the following four kinematic variables:

* Displacement (including position, distance, proximity, and size or gage)
* Velocity
* Acceleration
* Jerk

Note that each variable is the time derivative of the preceding one. Motion measurements are extremely useful in controlling mechanical responses and interactions in mechatronic systems. Numerous examples can be cited: The rotating speed of a workpiece and the feed rate of a tool are measured in controlling machining operations. Displacements and speeds (both angular and translatory) at joints (revolute and prismatic) of robotic manipulators or kinematic linkages are used in controlling manipulator trajectory. In high-speed ground transit vehicles, acceleration and jerk measurements can be used for active suspension control to obtain improved ride quality. Angular speed is a crucial measurement that is used in the control of rotating machinery, such as turbines, pumps, compressors, motors, and generators in power-generating plants. Proximity sensors (to measure displacement) and accelerometers (to measure acceleration) are the two most common types of measuring devices used in machine protection systems for condition monitoring, fault detection, diagnostic, and online (often real-time) control of large and complex machinery. The accelerometer is often the only measuring device used in controlling dynamic test rigs. Displacement measurements are used for valve control in process applications. Plate thickness (or gage) is continuously monitored by the automatic gage control (AGC) system in steel rolling mills.

A one-to-one relationship may not always exist between a measuring device and a measured variable. For example, although strain gages are devices that measure strains (and hence, stresses and forces), they can be adapted to measure displacements by using a suitable *front-end auxiliary sensor element*, such as a cantilever (or spring). Furthermore, the same measuring device may be used to measure different variables through appropriate data interpretation techniques. For example, piezoelectric accelerometers with built-in microelectronic integrated circuitry are marketed as piezoelectric velocity transducers. Resolver signals, which provide angular displacements, are differentiated to get angular velocities. Pulse-generating (or digital) transducers, such as optical encoders and digital tachometers, can serve as both displacement transducers and velocity transducers, depending on whether the absolute number of pulses are counted or the pulse rate is measured. Note that the pulse rate can be measured either by counting the number of pulses during a unit interval of time or by gating a high-frequency clock signal through the pulse width. Furthermore, in principle, any force sensor can be used as an acceleration sensor, velocity sensor, or displacement sensor depending on whether an inertia element (converting acceleration into force), a damping element (converting velocity into force), or a spring element (converting displacement into force), respectively, is used as the front-end auxiliary sensor.

We might question the need for separate transducers to measure the four kinematic variables—displacement, velocity, acceleration, and jerk—because any one variable is related to any other through simple integration or differentiation. It should be possible, in theory, to measure only one of these four variables and use either analog processing (through analog circuit hardware) or digital processing (through a dedicated processor) to obtain any of the remaining motion variables. The feasibility of this approach is highly limited, however, and it depends crucially on several factors, including the following:

1. The nature of the measured signal (e.g., steady, highly transient, periodic, narrowband, broadband)
2. The required frequency content of the processed signal (or the frequency range of interest)
3. The signal-to-noise ratio (SNR) of the measurement
4. Available processing capabilities (e.g., analog or digital processing, limitations of the digital processor, and interface, such as the speed of processing, sampling rate, and buffer size)
5. Controller requirements and the nature of the plant (e.g., time constants, delays, complexity, hardware limitations)
6. Required accuracy in the end objective (on which processing requirements and hardware costs will depend)

For instance, differentiation of a signal (in the time domain) is often unacceptable for noisy and high-frequency narrowband signals. In any event, costly signal-conditioning hardware might be needed for preprocessing prior to differentiating a signal. As a rule of thumb, in low-frequency applications (on the order of 1 Hz), displacement measurements generally provide good accuracies. In intermediate-frequency applications (less than 1 kHz), velocity measurement is usually favored. In measuring high-frequency motions with high noise levels, acceleration measurement is preferred. Jerk is particularly useful in ground transit (ride quality), manufacturing (forging, rolling, and similar impact-type operations), and shock isolation applications (for delicate and sensitive equipment).

6.2 Potentiometer

The potentiometer, or *pot*, is a displacement transducer. This active transducer consists of a uniform coil of wire or a film of high-resistive material—such as carbon, platinum, or conductive plastic—whose resistance is proportional to its length. A constant voltage v_{ref} is applied across the coil (or film) using an external dc voltage supply. The transducer output signal v_o is the dc voltage between the movable contact (wiper arm) sliding on the coil and one terminal of the coil, as shown schematically in Figure 6.1a. The slider displacement x is proportional to the output voltage:

$$v_o = kx \tag{6.1}$$

This relationship assumes that the output terminals are open-circuit; that is, a load of infinite impedance (or resistance in the present dc case) is present at the output terminal,

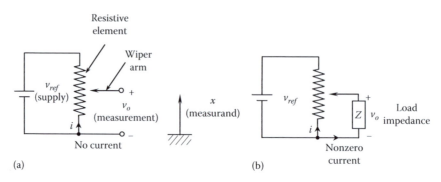

FIGURE 6.1
(a) Schematic diagram of a potentiometer; (b) potentiometer loading.

so that the output current is zero. In actual practice, however, the load (the circuitry into which the pot signal is fed—e.g., conditioning, interfacing, processing, or control circuitry) has a finite impedance. Consequently, the output current (the current through the load) is nonzero, as shown in Figure 6.1b. The output voltage thus drops to $\tilde{v}_o$, even if the reference voltage v_{ref} is assumed to remain constant under load variations (i.e., the output imped-ance of the voltage source is zero); this consequence is known as the *loading effect* of the transducer. Under these conditions, the linear relationship given by Equation 6.1 would no longer be valid, causing an error in the displacement reading. Loading can affect the transducer reading in two ways: by changing the reference voltage (i.e., loading the volt-age source) and by loading the transducer. To reduce these effects, a voltage source that is not seriously affected by load variations (e.g., a regulated or stabilized power supply that has a low output impedance) and data acquisition circuitry (including signal-conditioning circuitry) that has a high input impedance should be used.

The resistance of a potentiometer should be chosen with care. On the one hand, an ele-ment with high resistance is preferred because this results in reduced power dissipation for a given voltage, which has the added benefit of reduced thermal effects. On the other hand, increased resistance increases the output impedance of the potentiometer and results in loading nonlinearity error unless the load resistance is also increased proportionately. Low-resistance pots have resistances less than $10\,\Omega$. High-resistance pots can have resis-tances on the order of $100\,k\Omega$. Conductive plastics can provide high resistances—typically about $100\,\Omega/\text{mm}$—and are increasingly used in potentiometers. Reduced friction (low mechanical loading), reduced wear, reduced weight, and increased resolution are advan-tages of using conductive plastics in potentiometers.

6.2.1 Performance Considerations

The potentiometer is a *resistively coupled transducer*. The force required to move the slider arm comes from the motion source, and the resulting energy is dissipated through fric-tion. This energy conversion, unlike pure mechanical-to-electrical conversions, involves relatively high forces, and the energy is wasted rather than being converted into the out-put signal of the transducer. Furthermore, the electrical energy from the reference source is also dissipated through the resistor element (coil or film), resulting in an undesirable temperature rise and coil degradation. These are two obvious disadvantages of a poten-tiometer. In coil-type pots, there is another disadvantage, which is the finite *resolution*. The selection of a potentiometer involves many considerations. A primary factor is the required resolution for the specific application. Power consumption, loading, and size are also important factors.

A coil, instead of a straight wire, is used to increase the resistance per unit travel of the slider arm. But the slider contact jumps from one turn to the next in this case. Accordingly, the resolution of a coil-type potentiometer is determined by the number of turns in the coil. Resolutions better (smaller) than 0.1% (i.e., 1000 turns) are available with coil potentiometers. Virtually infinitesimal (incorrectly termed infinite) resolutions are now possible with high-quality resistive film potentiometers that use conductive plastics. In this case, the resolution is limited by other factors, such as mechanical limitations and SNR. Nevertheless, resolutions on the order of 0.01 mm are possible with good rectilinear potentiometers.

The *sensitivity* of a potentiometer represents the change (Δv_o) in the output signal associated with a given small change (Δx) in the measurand (the displacement). Hence, the sensitivity S is given by $S = \Delta v_o / \Delta x$ or in the limit

$$S = \frac{\partial v_o}{\partial x} \tag{6.2}$$

This is usually nondimensionalized using the actual value of the output signal (v_o) and the actual value of the displacement (x).

Some limitations and disadvantages of potentiometers as displacement measuring devices are given below:

1. The force needed to move the slider (against friction and arm inertia) is provided by the displacement source. This mechanical loading distorts the measured signal itself.
2. High-frequency (or highly transient) measurements are not feasible because of such factors as slider bounce, friction and inertia resistance, and induced voltages in the wiper arm and primary coil.
3. Variations in the supply voltage cause error.
4. Electrical loading error can be significant when the load resistance is low.
5. Resolution is limited by the number of turns in the coil and by the coil uniformity. This will limit small-displacement measurements.
6. Wearout and heating up (with associated oxidation) in the coil or film and slider contact cause accelerated degradation.

However, there are several advantages associated with potentiometer devices, including the following:

1. They are relatively inexpensive.
2. Potentiometers provide high-voltage (low-impedance) output signals, requiring no amplification in most applications. Transducer impedance can be varied simply by changing the coil resistance and supply voltage.

Although pots are primarily used as displacement transducers, they can be adapted to measure other types of signals, such as pressure and force, using appropriate auxiliary sensor (front-end) elements. For instance, a Bourdon tube or bellows may be used to convert pressure into displacement, and a cantilever element may be used to convert force or moment into displacement.

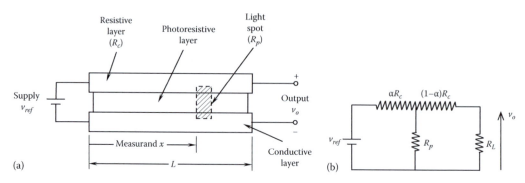

FIGURE 6.2
(a) An optical potentiometer; (b) equivalent circuit ($\alpha = x/L$).

6.2.2 Optical Potentiometer

The optical potentiometer, shown schematically in Figure 6.2a, is a displacement sensor. A layer of photoresistive material is sandwiched between a layer of ordinary resistive material and a layer of conductive material. The layer of resistive material has a total resistance of R_c, and it is uniform (i.e., it has a constant resistance per unit length). This corresponds to the coil resistance of a conventional potentiometer. The photoresistive layer is practically an electrical insulator when no light is projected on it. The displacement of the moving object (whose displacement is being measured) causes a moving light beam to be projected on a rectangular area of the photoresistive layer. This light-activated area attains a resistance of R_p, which links the resistive layer that is above the photoresistive layer and the conductive layer that is below the photoresistive layer. The supply voltage to the potentiometer is v_{ref}, and the length of the resistive layer is L. The light spot is projected at a distance x from one end of the resistive element, as shown in Figure 6.2.

An equivalent circuit for the optical potentiometer is shown in Figure 6.2b. Here it is assumed that a load of resistance R_L is present at the output of the potentiometer, with the voltage across being v_o. Current through the load is v_o/R_L. Hence, the voltage drop across $(1-\alpha)R_c + R_L$, which is also the voltage across R_p, is given by $[(1-\alpha)R_c + R_L]v_o/R_L$. Note that $\alpha = x/L$ is the fractional position of the light spot. The current balance at the junction of the three resistors in Figure 6.2b is

$$\frac{v_{ref} - \left[(1-\alpha)R_c + R_L\right]v_o/R_L}{\alpha R_c} = \frac{v_o}{R_L} + \frac{\left[(1-\alpha)R_c + R_L\right]v_o/R_L}{R_p},$$

which can be written as

$$\frac{v_o}{v_{ref}} \left\{ \frac{R_c}{R_L} + 1 + \frac{x}{L}\frac{R_c}{R_p}\left[\left(1 - \frac{x}{L}\right)\frac{R_c}{R_L} + 1\right]\right\} = 1 \tag{6.3}$$

When the load resistance R_L is quite large in comparison with the element resistance R_c, we have $R_c/R_L \simeq 0$. Hence, Equation 6.3 becomes

$$\frac{v_o}{v_{ref}} = 1 \bigg/ \left[\frac{x}{L}\frac{R_c}{R_p} + 1\right] \tag{6.4}$$

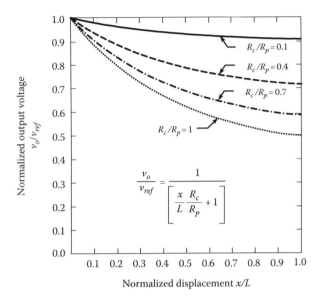

FIGURE 6.3
Behavior of the optical potentiometer at high load resistance.

This relationship is still nonlinear in v_o/v_{ref} vs. x/L. The nonlinearity decreases, however, with decreasing R_c/R_p. This is also seen from Figure 6.3 where Equation 6.4 is plotted for several values of R_c/R_p. Then, for the case of $R_c/R_p = 0.1$, the original equation (6.3) is plotted in Figure 6.4 for several values of load resistance ratio. As expected, the behavior of the optical potentiometer becomes more linear for higher values of load resistance.

$$\frac{v_o}{v_{ref}}\left\{\frac{R_c}{R_L} + 1 + 0.1\frac{x}{L}\left[\left(1-\frac{x}{L}\right)\frac{R_c}{R_L}+1\right]\right\} = 1$$

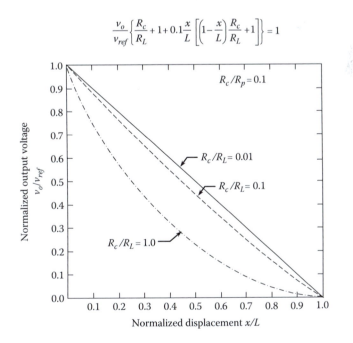

FIGURE 6.4
Behavior of the optical potentiometer for $R_c/R_p = 0.1$.

The potentiometer has disadvantages such as loading problems (both mechanical and electrical), limited speed of operation, considerable time constants, wear, noise, and thermal effects. Many of these problems arise from the fact that it is a "contact" device where its slider has to be in intimate contact with the resistance element of the pot, and it also has to be an integral part of the moving object whose displacements need to be measured. Next, we will consider several noncontact motion sensors.

6.3 Variable-Inductance Transducers

Motion transducers that employ the principle of electromagnetic induction are termed variable-inductance transducers. When the flux linkage (defined as magnetic flux density times the number of turns in the conductor) through an electrical conductor changes, a voltage is induced in the conductor. This, in turn, generates a magnetic field, which opposes the primary field. Hence, a mechanical force is necessary to sustain the change of flux linkage. If the change in flux linkage is brought about by a relative motion, the associated mechanical energy is directly converted (induced) into electrical energy. This is the basis of *electromagnetic induction* and it is the principle of operation of electrical generators and variable-inductance transducers. Note that in these devices, the change of flux linkage is caused by a mechanical motion, and mechanical-to-electrical energy transfer takes place under near-ideal conditions. The induced voltage or change in inductance may be used as a measure of the motion. Variable-inductance transducers are generally electromechanical devices coupled by a magnetic field.

There are many different types of *variable-inductance transducers*. The following three primary types can be identified:

1. Mutual-induction transducers
2. Self-induction transducers
3. Permanent-magnet transducers

Those variable-inductance transducers that use a nonmagnetized ferromagnetic medium to alter the reluctance (magnetic resistance) of the flux path are known as *variable-reluctance transducers*. Some of the mutual-induction transducers and most of the self-induction transducers are of this type. Permanent-magnet transducers are not considered variable-reluctance transducers.

6.3.1 Mutual-Induction Transducers

The basic arrangement of a mutual-induction transducer constitutes two coils: the *primary winding* and the *secondary winding*. One of the coils (primary winding) carries an alternating-current (ac) excitation, which induces a steady ac voltage in the other coil (secondary winding). The level (amplitude, root-mean-square [rms] value, etc.) of the induced voltage depends on the flux linkage between the coils. None of these transducers employ contact sliders or slip-rings and brushes as do resistively coupled transducers (potentiometer). Consequently, they will have an increased design life and low mechanical loading. In mutual-induction transducers, a change in the flux linkage is effected by one of two common techniques. One technique is to move an object made of ferromagnetic material within the flux path. This changes the reluctance of the flux path, with an associated change of the flux linkage in the secondary coil. This is the operating principle of the linear-variable differential transformer (LVDT), the rotatory-variable differential transformer (RVDT), and the mutual-induction proximity probe. All of these are, in fact, variable-reluctance transducers. The other common way to change the flux linkage is to move one coil with respect to the other. This is the operating principle of the resolver, the synchro-transformer, and some types of ac tachometers. However, these are not variable-reluctance transducers.

The motion can be measured by using the secondary signal in several ways. For example, the ac signal in the secondary coil may be "demodulated" by rejecting the *carrier signal* (i.e., the signal component at the excitation frequency) and directly measuring the resulting signal, which represents the motion. This method is particularly suitable for measuring transient motions. Alternatively, the amplitude or the rms value of the secondary (induced) voltage may be measured. Another method is to measure the change of inductance (or reactance, which is equal to $Lj\omega$, since $v = L(di/dt)$) in the secondary circuit directly, by using a device such as an inductance bridge circuit (see Chapter 4).

6.3.2 Linear-Variable Differential Transformer

The differential transformer is a noncontact displacement sensor, which does not possess many of the shortcomings of the potentiometer. It is a variable-inductance transducer and is also a variable-reluctance transducer and a mutual-induction transducer. Unlike the potentiometer, the LVDT is considered to be a passive transducer because the measured displacement provides energy for "changing" the induced voltage, even though an external power supply is used to energize the primary coil, which in turn induces a steady voltage at the carrier frequency in the secondary coil. In its simplest form (see Figure 6.5), the LVDT consists of an insulating, nonmagnetic "form" (a cylindrical structure on which a coil is wound and is integral with the housing), which has a primary coil in the midsegment and a secondary coil symmetrically wound in the two end segments, as depicted schematically in Figure 6.5b. The housing is made of magnetized stainless steel in order to shield the sensor from outside fields. The primary coil is energized by an ac supply of voltage v_{ref}. This will generate, by mutual induction, an ac of the same frequency in the secondary coil. A core made of ferromagnetic material is inserted coaxially through the cylindrical form without actually touching it, as shown. As the core moves, the reluctance of the flux path changes. The degree of flux linkage depends on the axial position of the core. Since the two secondary coils are connected in series opposition (as shown in Figure 6.6), so that the potentials induced in the two secondary coil segments oppose each other, it is seen that the net induced voltage is zero when the core is centered between the two secondary winding segments. This is known as the *null position*. When the core is displaced from this position, a nonzero induced voltage will be generated. At steady state, the amplitude v_o of this induced voltage is proportional to the core displacement x in the linear (operating) region (see Figure 6.5c). Consequently, v_o may be used as a measure of the displacement. Note that because of opposed secondary windings, the LVDT provides the direction as well as the magnitude of displacement. If the output signal is not demodulated, the direction is determined by the phase angle between the primary (reference) voltage and the secondary (output) voltage, which includes the carrier signal.

For an LVDT to measure transient motions accurately, the frequency of the reference voltage (the carrier frequency) has to be at least 10 times larger than the largest significant

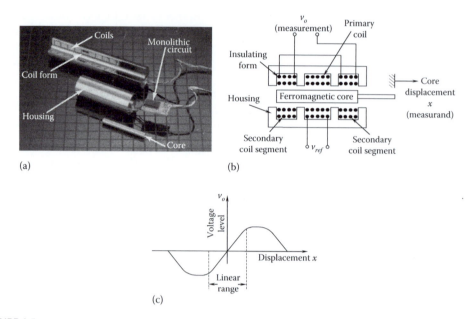

(a)   (b)

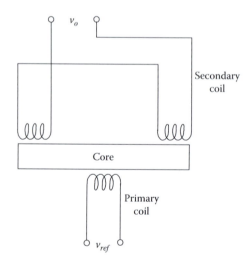

(c)

FIGURE 6.5
LVDT: (a) A commercial unit (Courtesy of Schaevitz Sensors, Measurement Specialties, Inc. Hampton, VA.); (b) schematic diagram; (c) a typical operating curve.

FIGURE 6.6
Series opposition connection of secondary windings.

frequency component in the measured motion, and typically can be as high as 20 kHz. For quasidynamic displacements and slow transients on the order of a few hertz, a standard ac supply (at 60 Hz line frequency) is adequate. The performance (particularly sensitivity and accuracy) is known to improve with the excitation frequency, however. Since the amplitude of the output signal is proportional to the amplitude of the primary signal, the reference voltage should be regulated to get accurate results. In particular, the power source should have a low output impedance.

6.3.2.1 Phase Shift and Null Voltage

An error known as *null voltage* is present in some differential transformers. This manifests itself as a nonzero reading at the null position (i.e., at zero displacement). This is usually 90° out of phase from the main output signal and, hence, is known as *quadrature error*. Nonuniformities in the windings (unequal impedances in the two segments of the secondary winding) are a major reason for this error. The null voltage may also result from harmonic noise components in the primary signal and nonlinearities in the device. Null voltage is usually negligible (typically about 0.1% of the full scale). This error can be eliminated from the measurements by employing appropriate signal-conditioning and calibration practices.

The output signal from a differential transformer is normally not in phase with the reference voltage. The inductance in the primary coil and the leakage inductance in the secondary coil are mainly responsible for this phase shift. Since demodulation involves the extraction of the modulating signal by rejecting the carrier frequency component from the secondary signal, it is important to understand the size of this phase shift. The level of dependence of the phase shift on the load (including the secondary circuit) can be reduced by increasing the load impedance.

6.3.2.2 Signal Conditioning

Signal conditioning associated with differential transformers includes filtering and amplification. Filtering is needed to improve the SNR of the output signal. Amplification is necessary to increase the signal strength for data acquisition and processing. Since the reference frequency (carrier frequency) is induced into (and embedded in) the output signal, it is also necessary to interpret the output signal properly, particularly for transient motions.

The secondary (output) signal of an LVDT is an amplitude-modulated signal where the signal component at the carrier frequency is modulated by the lower-frequency transient signal produced as a result of the core motion (x). Two methods are commonly used to interpret the crude output signal from a differential transformer: rectification and demodulation. Block diagram representations of these two procedures are given in Figure 6.7. In the first method (*rectification*), the ac output from the differential transformer is rectified to obtain a dc signal. This signal is amplified and then low-pass filtered to eliminate any high-frequency noise components. The amplitude of the resulting signal provides the transducer reading. In this method, a phase shift in the LVDT output has to be checked separately to determine the direction of motion. In the second method (*demodulation*), the carrier frequency component is rejected from the output signal by comparing it with a phase-shifted and amplitude-adjusted version of the primary (reference) signal. Note that phase shifting is necessary because, as discussed before, the output signal is not in phase with the reference signal. The result is the modulating signal (proportional to x), which is subsequently amplified and filtered.

As a result of advances in miniature integrated circuit (IC) technology, differential transformers with built-in microelectronics for signal conditioning are commonly available today. A dc differential transformer uses a dc power supply (typically ±15 V) to activate it. A built-in oscillator circuit generates the carrier signal. The rest of the device is identical to an ac differential transformer. The amplified full-scale output voltage can be as high as ±10 V. Let us illustrate the demodulation approach of signal conditioning for an LVDT, using an example.

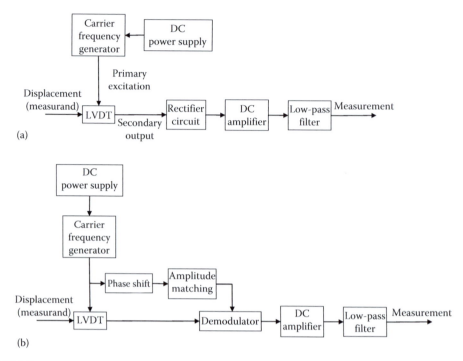

FIGURE 6.7
Signal-conditioning methods for a differential transformer: (a) Rectification; (b) demodulation.

Example 6.1

Figure 6.8 shows a schematic diagram of a simplified signal conditioning system for an LVDT. The system variables and parameters are as indicated in the figure.
In particular,

> $x(t)$ is the displacement of the LVDT core (measurand, to be measured)
> ω_c is the frequency of the carrier voltage
> v_o is the output signal of the system (measurement)

The resistances R_1, R_2, and R, and the capacitance C are as marked. In addition, we may introduce a transformer parameter r for the LVDT, as required.

(i) Explain the functions of the various components of the system shown in Figure 6.8.
(ii) Write equations for the amplifier and filter circuits and, using them, give expressions for the voltage signals v_1, v_2, v_3, and v_o marked in Figure 6.8. Note that the excitation in the primary coil is $v_p \sin \omega_c t$.
(iii) Suppose that the carrier frequency is $\omega_c = 500\,\text{rad/s}$ and the filter resistance $R = 100\,\text{k}\Omega$. If no more than 5% of the carrier component should pass through the filter, estimate the required value of the filter capacitance C. Also, what is the useful frequency range (measurement bandwidth) of the system in rad/s with these parameter values?
(iv) If the displacement $x(t)$ is linearly increasing (i.e., speed is constant), sketch the signals $u(t)$, v_1, v_2, v_3, and v_o as functions of time.

Solution

(i) The LVDT has a primary coil, which is excited by an ac voltage of $v_p \sin \omega_c t$. The ferromagnetic core is attached to the moving object whose displacement $x(t)$ is to be measured. The

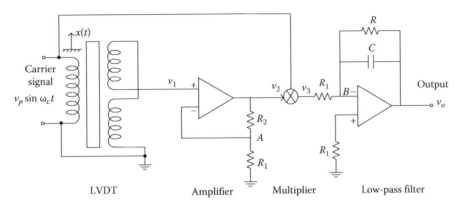

FIGURE 6.8
Signal conditioning system for an LVDT.

two secondary coils are connected in series opposition so that the LVDT output is zero at the null position and that the direction of motion can be detected as well. The amplifier is a noninverting type. It amplifies the output of the LVDT, which is an ac (carrier) signal of frequency ω_c that is modulated by the core displacement $x(t)$.

The multiplier circuit produces the product of the primary (carrier) signal and the secondary (LVDT output) signal. This is an important step in demodulating the LVDT output.

The product signal from the multiplier has a high-frequency ($2\omega_c$) carrier component added to the modulating component ($x(t)$). The low-pass filter removes this unnecessary high-frequency component to obtain the demodulated signal, which is proportional to the core displacement $x(t)$.

(ii) **Noninverting Amplifier**

Note that the potentials at the $+$ and $-$ terminals of the op-amp are nearly equal. Also, currents through these leads are nearly zero. (These are the two common assumptions used for an op-amp; see Chapter 4). Then, the current balance at node A gives $(v_2 - v_1)/R_2 = v_1/R_1$. Hence, $v_2 = kv_1$ with $k = (R_1 + R_2)/R_1 = $ amplifier gain.

Loss-pass filter:

Since the $+$ lead of the op-amp has approximately zero potential (ground), the voltage at point B is also approximately zero. The current balance for node B gives

$$\frac{v_3}{R_1} + \frac{v_o}{R} + C\dot{v}_o = 0.$$

Hence,

$$\tau \frac{dv_o}{dt} + v_o = -\frac{R}{R_1} v_3$$

where $\tau = RC = $ filter time constant. The transfer function of the filter is $v_o/v_3 = -k_o/(1 + \tau s)$ with the filter gain $k_o = R/R_1$. In the frequency domain, $v_o/v_3 = -k_o/(1 + \tau j\omega)$.

Finally, neglecting the phase shift in the LVDT, we have $v_1 = v_p r \, x(t) \sin \omega_c t$; $v_2 = v_p rk \, x(t) \sin \omega_c t$; $v_3 = v_p^2 rk \, x(t) \sin^2 \omega_c t$; or

$$v_3 = \frac{v_p^2 rk}{2} x(t)\left[1 - \cos 2\omega_c t\right].$$

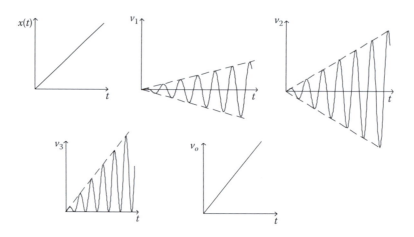

FIGURE 6.9
Nature of the signals at various locations in an LVDT measurement circuit.

The carrier signal will be filtered out by the low-pass filter with an appropriate cut-off frequency. Then, $v_o = ((v_p^2 r k_o)/2)x(t)$

(iii) Filter magnitude $= k_o / \left(\sqrt{1 + \tau^2 \omega^2} \right)$

For no more than 5% of the carrier ($2\omega_c$) component to pass through, we must have $k_o / \left(\sqrt{1 + \tau^2 (2\omega_c)^2} \right) \le (5/100)k_o$; or $\tau\omega_c \ge 10$ (approximately). Pick $\tau\omega_c = 10$. With $R = 100\,\mathrm{k\Omega}$, $\omega_c = 500\,\mathrm{rad/s}$ we have $C \times 100 \times 10^3 \times 500 = 10$. Hence, $C = 0.2\,\mu\mathrm{F}$.

According to the carrier frequency (500 rad/s), we should be able to measure displacements $x(t)$ up to about 50 rad/s. But the flat region of the filter is up to about $\omega\tau = 0.1$, which with the present value of $\tau = 0.02\,\mathrm{s}$ gives a bandwidth of only 5 rad/s.

(iv) See Figure 6.9 for a sketch of various signals in the LVDT measurement system.

The advantages of the LVDT include the following:

1. It is essentially a noncontacting device with no frictional resistance. Near-ideal electromechanical energy conversion and a lightweight core will result in very small resistive forces. Hysteresis (both magnetic hysteresis and mechanical backlash) is negligible.

2. It has low output impedance, typically on the order of $100\,\Omega$. (Signal amplification is usually not needed beyond what is provided by the conditioning circuit.)

3. Directional measurements (positive/negative) are obtained.

4. It is available in small sizes (e.g., 1 cm long with maximum travel of 2 mm).

5. It has a simple and robust construction (inexpensive and durable).

6. Fine resolutions are possible (theoretically, infinitesimal resolution; practically, much better than a coil potentiometer).

In variable-inductance devices, the induced voltage is generated through the rate of change of the magnetic flux linkage. Therefore, displacement readings are distorted by velocity; similarly, velocity readings are affected by acceleration. For the same displacement value, the transducer reading will depend on the velocity at that displacement. This error is known as the *rate error* increases with the ratio (cyclic velocity of the core)/(carrier frequency).

Hence, the rate error can be reduced by increasing the carrier frequency. The reason for this is as follows.

At high frequencies, the induced voltage due to the transformer effect (having a frequency of the primary signal) is greater than the induced voltage due to the rate (velocity) effect of the moving member. Hence, the error will be small. To estimate a lower limit for the carrier frequency in order to reduce rate effects, we may proceed as follows. For an LVDT, let

$$\frac{\text{Maximum speed of operation}}{\text{Stroke of LVDT}} = \omega_o \qquad (6.5)$$

The excitation frequency of the primary coil should be chosen $5\omega_o$ or more.

6.3.3 Resolver

Some mutual-induction displacement transducers use the relative motion between the primary coil and the secondary coil to produce a change in flux linkage. Such a device is the resolver. This is not a variable-reluctance transducer, however, because it does not employ a ferromagnetic moving element.

A resolver is widely used for measuring angular displacements. A simplified schematic diagram of the resolver is shown in Figure 6.10. The *rotor* contains the primary coil. It consists of a single two-pole winding element energized by an ac supply voltage v_{ref}. The rotor is directly attached to the object whose rotation is being measured. The *stator* consists of two sets of windings placed 90° apart. If the angular position of the rotor with respect to one pair of stator windings is denoted by θ, the induced voltage in this pair of windings is given by

$$v_{o1} = a v_{ref} \cos \theta \qquad (6.6)$$

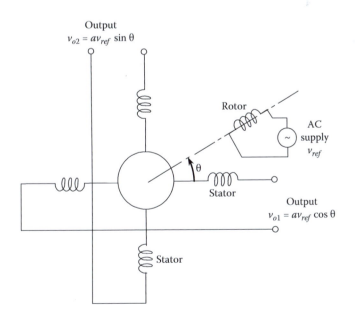

FIGURE 6.10
Schematic diagram of a resolver.

The induced voltage in the other pair of windings is given by

$$v_{o2} = av_{ref} \sin\theta \qquad (6.7)$$

Note that these are amplitude-modulated signals—the carrier signal v_{ref} is modulated by the motion θ. The constant parameter a depends primarily on the geometric and material characteristics of the device, for example, the ratio of the number of turns in the rotor and stator windings.

Either of the two output signals v_{o1} and v_{o2} may be used to determine the angular position in the first quadrant ($0 \le \theta \le 90°$). Both signals are needed, however, to determine the displacement (direction as well as magnitude) in all four quadrants ($0 \le \theta \le 360°$) without causing any ambiguity. For instance, the same sine value is obtained for both $90° + \theta$ and $90° - \theta$ (i.e., a positive rotation and a negative rotation from the $90°$ position), but the corresponding cosine values have opposite signs, thus providing the proper direction.

6.3.3.1 Demodulation

As for differential transformers (e.g., LVDT), the transient displacement signals of a resolver can be extracted by demodulating its modulated outputs. This is accomplished by filtering out the carrier signal, thereby extracting the modulating signal. The two output signals v_{o1} and v_{o2} of a resolver are termed quadrature signals. Suppose that the carrier (primary) signal is

$$v_{ref} = v_a \sin\omega t \qquad (6.8)$$

The induced quadrate signals are $v_{o1} = av_a \cos\theta \sin\omega t$ and $v_{o2} = av_a \sin\theta \sin\omega t$. Multiply each quadrature signal by v_{ref} to get

$$v_{m1} = v_{o1}v_{ref} = av_a^2 \cos\theta \sin^2\omega t = \frac{1}{2}av_a^2 \cos\theta \left[1 - \cos 2\omega t\right]$$

$$v_{m2} = v_{o2}v_{ref} = av_a^2 \sin\theta \sin^2\omega t = \frac{1}{2}av_a^2 \sin\theta \left[1 - \cos 2\omega t\right]$$

Since the carrier frequency ω is about 10 times the maximum frequency content in the angular displacement θ, one can use a low-pass filter with a cut-off set at $\omega/20$ in order to remove the carrier components in v_{m1} and v_{m2}. This gives the following demodulated outputs:

$$v_{f1} = \frac{1}{2}av_a^2 \cos\theta \qquad (6.9)$$

$$v_{f2} = \frac{1}{2}av_a^2 \sin\theta \qquad (6.10)$$

Note that Equations 6.9 and 6.10 provide both $\cos\theta$ and $\sin\theta$, and hence the magnitude and sign of θ.

The output signals of a resolver are nonlinear (trigonometric) functions of the angle of rotation. (Historically, resolvers were used to compute trigonometric functions or to "resolve" a vector into orthogonal components.) In robot control applications, this is sometimes viewed as a blessing. For the computed torque control of robotic manipulators, for example, trigonometric functions of the joint angles are needed in order to compute the required input signals (joint torques). Consequently, when resolvers are used to measure joint angles in manipulators, there is an associated reduction in processing time because the trigonometric functions are available as direct measurements.

The primary advantages of the resolver include the following:

1. Fine resolution and high accuracy
2. Low output impedance (high signal levels)
3. Small size (e.g., 10 mm diameter)

Its main limitations are as follows:

1. Nonlinear output signals (an advantage in some applications where trigonometric functions of the rotations are needed)
2. Bandwidth limited by supply frequency
3. Slip rings and brushes would be needed if complete and multiple rotations have to be measured (which adds mechanical loading and also creates component wear, oxidation, and thermal and noise problems)

6.3.4 Self-Induction Transducers

These transducers are based on the principle of self-induction. Unlike mutual-induction transducers, only a single coil is employed. This coil is activated by an ac supply voltage v_{ref} of sufficiently high frequency. The current produces a magnetic flux, which is linked back with the coil. The level of flux linkage (or self-inductance) can be varied by moving a ferromagnetic object within the magnetic field. This movement changes the reluctance of the flux path and the inductance in the coil. The change in self-inductance, which can be measured using an inductance-measuring circuit (e.g., an inductance bridge; see Chapter 5), represents the measurand (displacement of the object). Note that self-induction transducers are usually variable-reluctance devices.

A typical self-induction transducer is a *self-induction proximity sensor*. A schematic diagram of this device is shown in Figure 6.11. This device can be used as a displacement sensor for transverse displacements. For instance, the distance between the sensor tip and ferromagnetic surface of a moving object, such as a beam or shaft, can be measured. Proximity sensors are used in a wide variety of applications pertaining to noncontacting displacement sensing and dimensional gaging. Some typical applications are as follows:

1. Measurement and control of the gap between a robotic welding torch head and the work surface
2. Gaging the thickness of metal plates in manufacturing operations (e.g., rolling and forming)
3. Detecting surface irregularities in machined parts

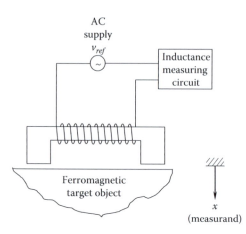

FIGURE 6.11
Schematic diagram of a self-induction proximity sensor.

4. Angular speed measurement at steady state by counting the number of rotations per unit time

5. Measurement of vibration in rotating machinery

6. Level detection (e.g., in the filling, bottling, and chemical process industries)

7. Monitoring of bearing assembly processes

High-speed displacement measurements can result in velocity error (rate error) when variable-inductance displacement sensors (including self-induction transducers) are used. This effect may be reduced, as in other ac-activated variable-inductance sensors, by increasing the carrier frequency.

6.3.5 Eddy Current Transducers

If a conducting (i.e., low-resistivity) medium is subjected to a fluctuating magnetic field, eddy currents are generated in the medium. The strength of the eddy currents increases with the strength of the magnetic field and the frequency of the magnetic flux. This principle is used in eddy current proximity sensors. Eddy current sensors may be used as either dimensional gaging devices or displacement sensors.

A schematic diagram of an eddy current proximity sensor is shown in Figure 6.12a. Unlike variable-inductance proximity sensors, the target object of the eddy current sensor does not have to be made of a ferromagnetic material. A conducting target object is needed, but a thin film of conducting material, such as household aluminum foil glued onto a nonconducting target object would be adequate. The probe head has two identical coils, which will form two arms of an impedance bridge. The coil closer to the probe face is the *active coil*. The other coil is the *compensating coil*. It compensates for ambient changes, particularly thermal effects. The remaining two arms of the bridge will consist of purely resistive elements (see Figure 6.12b). The bridge is excited by a radio-frequency voltage supply. The frequency may range from 1 to 100 MHz. This signal is generated from a radio-frequency converter (an oscillator) that is typically powered by a 20 V dc supply. When the target (sensed) object is absent, the output of the impedance bridge is zero, which corresponds to the balanced condition. When the target object is moved close to the sensor,

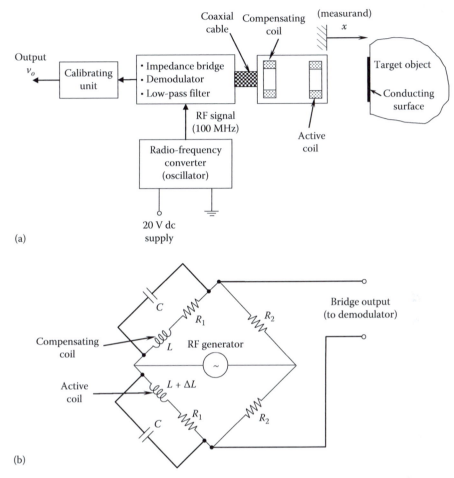

FIGURE 6.12
Eddy current proximity sensor: (a) Schematic diagram; (b) impedance bridge.

eddy currents are generated in the conducting medium because of the radio-frequency magnetic flux from the active coil. The magnetic field of the eddy currents opposes the primary field, which generates these currents. Hence, the inductance of the active coil increases, creating an imbalance in the bridge. The resulting output from the bridge is an amplitude-modulated signal containing the radio-frequency carrier. This signal can be demodulated by removing the carrier. The resulting signal (modulating signal) measures the transient displacement of the target object. Low-pass filtering is used to remove high-frequency leftover noise in the output signal once the carrier is removed. For large displacements, the output is not linearly related to the displacement. Furthermore, the sensitivity of an eddy current probe depends nonlinearly on the nature of the conducting medium, particularly the resistivity. For example, for low resistivities, sensitivity increases with resistivity; for high resistivities, sensitivity decreases with resistivity. A calibrating unit is usually available with commercial eddy current sensors to accommodate various target objects and nonlinearities. The gage factor is usually expressed in volts/millimeter. Note that eddy current probes can also be used to measure resistivity and surface hardness (which affects resistivity) in metals.

The facial area of the conducting medium on the target object has to be slightly larger than the frontal area of the eddy current probe head. If the target object has a curved surface, its radius of curvature has to be at least four times the diameter of the probe. These are not serious restrictions, because the typical diameter of a probe head is about 2 mm. Eddy current sensors are medium-impedance devices; $1000\,\Omega$ output impedance is typical. Sensitivity is on the order of 5 V/mm. Since the carrier frequency is very high, eddy current devices are suitable for highly transient displacement measurements; for example, bandwidths of up to 100 kHz. Another advantage of the eddy current sensor is that it is a non-contacting device; hence, there is no mechanical loading on the moving (target) object.

6.3.6 Permanent-Magnet Tachometers

The third category of variable-inductance transducers represents permanent-magnet transducers, which have a permanent magnet to generate a uniform and steady magnetic field. A relative motion between the magnetic field and an electrical conductor induces a voltage, which is proportional to the speed at which the conductor crosses the magnetic field (i.e., the rate of change of flux linkage). In some designs, a unidirectional magnetic field generated by a dc supply (i.e., an electromagnet) is used in place of a permanent magnet. Nevertheless, they are generally termed permanent-magnet transducers. We will consider a dc tachometer and an ac tachometer in this category. The ac induction tachometer, which does not fall into this category (since it does not use a permanent magnet) is also discussed for completeness.

6.3.7 DC Tachometer

This is a permanent-magnet dc velocity sensor in which the principle of electromagnetic induction between a permanent magnet and a conducting coil is used. Depending on the configuration, either rectilinear speeds or angular speeds can be measured. Schematic diagrams of the two configurations are shown in Figure 6.13. Note that these are passive transducers, because the energy for the output signal v_o is derived from the motion (measured signal) itself. The entire device is usually enclosed in a steel casing to shield (isolate) it from ambient magnetic fields.

In the rectilinear velocity transducer (Figure 6.13a), the conductor coil is wound on a core and placed centrally between two magnetic poles, which produce a cross-magnetic field. The core is attached to the moving object whose velocity v must be measured. This velocity is proportional to the induced voltage v_o. Alternatively, a moving magnet and a fixed coil may be used as a dc tachometer. This arrangement is perhaps more desirable since it

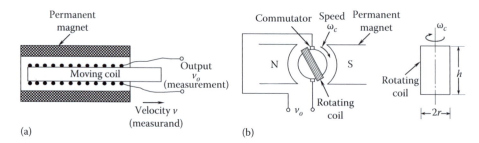

FIGURE 6.13
Permanent-magnet dc transducers: (a) Rectilinear velocity transducer; (b) dc tachometer.

eliminates the need for any sliding contacts (slip rings and brushes) for the output leads, thereby reducing mechanical loading error, wear, and related problems.

The dc tachometer (or tachogenerator) is a common transducer for measuring angular velocities. Its principle of operation is the same as that for a dc generator (or back-driving of a dc motor). This principle of operation is illustrated in Figure 6.13b. The rotor is directly connected to the rotating object. The output signal that is induced in the rotating coil is picked up as dc voltage v_o using a suitable *commutator* device—typically consisting of a pair of low-resistance carbon brushes—that is stationary but makes contact with the rotating coil through split slip rings so as to maintain the direction of the induced voltage throughout each revolution (see commutation in dc motors—Chapter 9). According to Faraday's law, the induced voltage is proportional to the rate of change of magnetic flux linkage. For a coil of height h and width $2r$ that has n turns, moving at an angular speed ω_c in a uniform magnetic field of flux density β, this is given by

$$v_o = (2nhr\beta)\omega_c = k\omega_c \tag{6.11}$$

This proportionality between v_o and ω_c is used to measure the angular speed ω_c. The proportionality constant k is known as the *back-emf constant* or the *voltage constant*.

6.3.7.1 Electronic Commutation

Slip rings, brushes, and associated drawbacks can be eliminated in a dc tachometer by using electronic commutation. In this case, a permanent-magnet rotor together with a set of stator windings are used. The output of the tachometer is drawn from the stationary (stator) coil. It has to be converted to a dc signal using an electronic switching mechanism, which has to be synchronized with the rotation of the tachometer (see Section 7.3.3). As a result of switching and associated changes in the magnetic field of the output signal, induced voltages known as *switching transients* will result. This is a drawback in electronic commutation.

6.3.7.2 Loading Considerations

The torque required to drive a tachometer is proportional to the current generated (in the dc output). The associated proportionality constant is the *torque constant*. With consistent units, in the case of ideal energy conversion, this constant is equal to the voltage constant. Since the tachometer torque acts on the moving object whose speed is measured, high torque corresponds to high mechanical loading, which is not desirable. Hence, it is needed to reduce the tachometer current as much as possible. This can be realized by making the input impedance of the signal-acquisition device (i.e., voltage reading and interface hardware) for the tachometer as large as possible. Furthermore, distortion in the tachometer output signal (voltage) can result because of the reactive (inductive and capacitive) loading of the tachometer. When dc tachometers are used to measure transient velocities, some error will result from the rate (acceleration) effect. This error generally increases with the maximum significant frequency that must be retained in the transient velocity signal, which in turn depends on the maximum speed that has to be measured. All these types of error can be reduced by increasing the load impedance.

Note: A *digital tachometer* is a velocity transducer, which is governed by somewhat different principles. It generates voltage pulses at a frequency proportional to the angular speed. Hence, it is considered a digital transducer.

6.3.8 Permanent-Magnet AC Tachometer

This device has a permanent magnet rotor and two separate sets of stator windings as schematically shown in Figure 6.14a. One set of windings is energized using an ac reference (carrier) voltage. Induced voltage in the other set of windings is the tachometer output. When the rotor is stationary or moving in a quasistatic manner, the output voltage is a constant-amplitude signal much like the reference voltage. As the rotor moves at a finite speed, an additional induced voltage, which is proportional to the rotor speed, is generated in the secondary winding. This is due to the rate of change of flux linkage into the secondary coil from the rotating magnet. The overall output from the secondary coil is an amplitude-modulated signal whose amplitude is proportional to the rotor speed. For transient velocities, it will be necessary to demodulate this signal in order to extract the transient velocity signal (i.e., the modulating signal) from the overall (modulated) output. The direction of velocity is determined from the phase angle of the modulated signal with respect to the carrier signal. Note that in an LVDT, the amplitude of the ac magnetic flux (linkage) is altered by the position of the ferromagnetic core. But in an ac permanent-magnet tachometer, a dc magnetic flux is generated by the magnetic rotor and when the rotor is stationary it does not induce a voltage in the coils. The flux linked with the stator windings changes due to the rotation of the rotor, and the rate of change of the linked flux is proportional to the speed of the rotor.

For low-frequency applications (5 Hz or less), a standard ac supply at line frequency (60 Hz) may be adequate to power an ac tachometer. For moderate-frequency applications, a 400 Hz supply may be used. For high-frequency (high-bandwidth) applications, a high-frequency signal generator (oscillator) may be used as the primary signal. In high-bandwidth applications, carrier frequencies as high as 1.5 kHz are commonly used. The typical sensitivity of an ac permanent-magnet tachometer is on the order of 50–100 mV/rad/s.

6.3.9 AC Induction Tachometer

This tachometer is similar in construction to a two-phase induction motor (see Chapter 7). The stator arrangement is identical to that of the ac permanent-magnet tachometer, as presented before. The rotor has windings, which are shorted and not energized by an external source, as shown in Figure 6.14b. One of the stator windings is powered by an ac supply. This induces a voltage in the rotor windings and it is a modulated signal. The high-frequency (carrier) component of this induced signal is due to the direct transformer action of the primary ac. The other (modulating) component is induced by the speed of the rotation of the rotor, and its magnitude is proportional to the speed of rotation. The nonenergized stator (secondary) windings provide the output of the tachometer. This voltage output is a result of both the stator (primary) windings and the rotor windings. As a

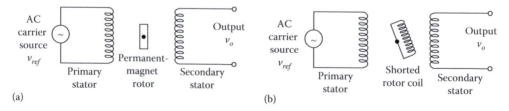

(a)

(b)

FIGURE 6.14
(a) An ac permanent-magnet tachometer; (b) an ac induction tachometer.

result, the tachometer output has a carrier ac component whose frequency is the same as the primary signal frequency and a modulating component, which is proportional to the speed of rotation. Demodulation would be needed to extract the component that is proportional to the angular speed of the rotor.

The main advantage of ac tachometers over their conventional dc counterparts is the absence of slip-ring and brush devices, since the output is obtained from the stator. In particular, the signal from a dc tachometer usually has a voltage ripple, known as the *commutator ripple* or *brush noise*, which are generated as the split ends of the slip ring pass over the brushes and as a result of contact bounce, etc. The frequency of the commutator ripple is proportional to the speed of operation; consequently, filtering it out using a notch filter is difficult (a speed-tracking notch filter would be needed). Also, there are problems with frictional loading and contact bounce in dc tachometers, and these problems are absent in ac tachometers. Note, however, that a dc tachometer with electronic commutation does not use slip rings and brushes. But they produce switching transients, which are undesirable.

As for any sensor, the noise components will dominate at low levels of output signal. In particular, since the output of a tachometer is proportional to the measured speed, at low speeds, the level of noise, as a fraction of the output signal, can be large. Hence, the removal of noise takes an increased importance at low speeds.

It is known that at high speeds the output from an ac tachometer is somewhat nonlinear (primarily due to the saturation effect). Furthermore, signal demodulation is necessary, particularly for measuring transient speeds. Another disadvantage of ac tachometers is that the output signal level depends on the supply voltage; hence, a stabilized voltage source that has a very small output-impedance is necessary for accurate measurements.

6.4 Variable-Capacitance Transducers

Variable-inductance devices and variable capacitance devices are *variable-reactance* devices. (Note that the *reactance* of an inductance L is given by $j\omega L$ and that of a capacitance C is given by $1/(j\omega C)$, since $v = L(di/dt)$ and $i = C(dv/dt)$.) For these reasons, capacitive transducers fall into the category of *reactive* transducers. They are typically high impedance sensors, particularly at low frequencies, as is clear from the impedance (reactance) expression for a capacitor. Also, capacitive sensors are noncontacting devices in the common usage. They require specific signal conditioning hardware. In addition to analog capacitive sensors, digital (pulse-generating) capacitive transducers, such as digital tachometers, are also available.

A capacitor is formed by two plates that can store an electric charge. The charge generates a potential difference, which may be maintained using an external voltage. The capacitance C of a two-plate capacitor is given by

$$C = \frac{kA}{x} \tag{6.12}$$

where
   A is the common (overlapping) area of the two plates
   x is the gap width between the two plates
   k is the dielectric constant (or permittivity $k = \varepsilon = \varepsilon_r \varepsilon_o$; ε_r is the relative permittivity, ε_o is the permittivity in vacuum) which depends on the dielectric properties of the medium between the two plates

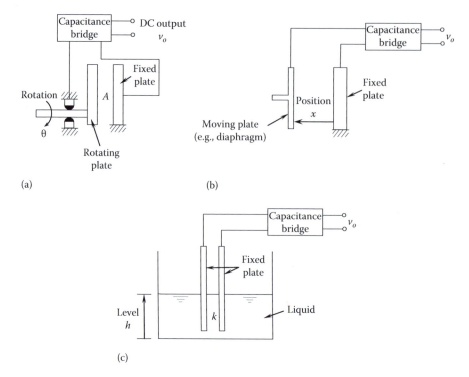

FIGURE 6.15

Schematic diagrams of capacitive sensors: (a) Capacitive rotation sensor; (b) capacitive displacement sensor; (c) capacitive liquid level sensor.

A change in any one of the three parameters in Equation 6.12 may be used in the sensing process; for example, to measure small transverse displacements, large rotations, and fluid levels. Schematic diagrams for measuring devices that use this feature are shown in Figure 6.15. In Figure 6.15a, the angular displacement of one of the plates causes a change in A. In Figure 6.15b, a transverse displacement of one of the plates changes x. Finally, in Figure 6.15c, a change in k is produced as the fluid level between the capacitor plates changes. In all cases, the associated change in capacitance is measured directly or indirectly and is used to estimate the measurand. A popular method is to use a capacitance bridge circuit to measure the change in capacitance, in a manner similar to how an inductance bridge (see Chapter 4) is used to measure changes in inductance. Other methods include measuring a change in such quantities as charge (using a charge amplifier), voltage (using a high input-impedance device in parallel), and current (using a very low impedance device in series) that will result from the change in capacitance in a suitable circuit. An alternative method is to make the capacitor a part of an inductance-capacitance (L-C) oscillator circuit—the natural frequency of the oscillator ($1/\sqrt{LC}$) measures the capacitance. (Incidentally, this method may also be used to measure inductance.)

6.4.1 Capacitive Rotation Sensor

In the arrangement shown in Figure 6.15a, one plate of the capacitor rotates with a rotating object (shaft) and the other plate is kept stationary. Since the common area A is proportional to the angle of rotation θ, Equation 6.12 may be written as

$$C = K\theta \tag{6.13}$$

where K is a sensor constant. This is a linear relationship between C and θ. The capacitance may be measured by any convenient method. The sensor is linearly calibrated to give the angle of rotation.

The sensitivity of this angular displacement sensor is $S = \partial C / \partial \theta = K$, which is constant throughout the measurement. This is expected because the sensor relationship is linear. *Note*: In the nondimensional form, the sensitivity of the sensor is unity, implying "direct" sensitivity.

6.4.2 Capacitive Displacement Sensor

The arrangement shown in Figure 6.15b provides a sensor for measuring the transverse displacements and proximities. One of the capacitor plates is attached to the moving object and the other plate is kept stationary. The sensor relationship is

$$C = \frac{K}{x} \tag{6.14}$$

The constant K has a different meaning here. The corresponding sensitivity is given by $S = \partial C / \partial x = -(K/x^2)$. Again, the sensitivity is unity (negative) in the nondimensional form, which indicates the direct sensitivity of the sensor.

Note that Equation 6.14 is a nonlinear relationship. A simple way to linearize this transverse displacement sensor is to use an inverting amplifier, as shown in Figure 6.16. Note that C_{ref} is a fixed reference capacitance; the value is accurately known. Since the gain of the operational amplifier is very high, the voltage at the negative lead (point A) is zero for most practical purposes (because the positive lead is grounded). Furthermore, since the input impedance of the op-amp is also very high, the current through the input leads is negligible. These are the two common assumptions used in op-amp analysis (see Chapter 4). Accordingly, the charge balance equation for node point A is $v_{ref}C_{ref} + v_oC = 0$. Now, in view of Equation 6.14, we get the following linear relationship for the output voltage v_o in terms of the displacement x:

$$v_o = -\frac{v_{ref}C_{ref}}{K}x \tag{6.15}$$

Hence, the measurement of v_o gives the displacement through a linear relationship. The sensitivity of the device can be increased by increasing v_{ref} and C_{ref}. The reference voltage

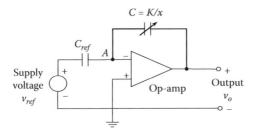

FIGURE 6.16
Linearizing amplifier circuit for a capacitive transverse displacement sensor.

may be either dc or ac with frequency as high as 25 kHz (for high-bandwidth measurements). With an ac reference voltage, the output voltage is a modulated signal, which has to be demodulated to measure transient displacements, as discussed before in the context of variable-inductance sensors.

The arrangement shown in Figure 6.15c can be used as well for displacement sensing. In this case, a solid dielectric element, which is free to move in the longitudinal direction of the capacitor plates, is attached to the moving object whose displacement is to be measured. The dielectric constant of the capacitor changes as the common area between the dielectric element and the capacitor plates varies due to the motion. The same arrangement may be used as a liquid level sensor, in which case the dielectric medium is the measured liquid, as shown in Figure 6.15c.

6.4.3 Capacitance Bridge Circuit

Sensors that are based on the change in capacitance (reactance) will require some means of measuring that change. Furthermore, changing capacitance that is not caused by a change in measurand; for example, due to a change in humidity, temperature, etc., will cause errors for which they should be compensated. Both these goals are accomplished using a capacitance bridge circuit. An example is shown in Figure 6.17.

In this circuit,

$Z_2 = 1/j\omega C_2$ is the reactance (i.e., capacitive impedance) of the capacitive sensor (of capacitance C_2)

$Z_1 = 1/j\omega C_1$ is the reactance of the compensating capacitor C_1

Z_4, Z_3 is the bridge completing impedances (typically, reactances)

$v_{ref} = v_a \sin \omega t$ is the excitation ac voltage

$v_o = v_b \sin(\omega t - \phi)$ is the bridge output

ϕ is the phase lag of the output with respect to the excitation

Using the two assumptions for an op-amp (potentials at the negative and positive leads are equal and the current through these leads is zero; see Chapter 5), we can write the current balance equations

$$\frac{v_{ref} - v}{Z_1} + \frac{v_o - v}{Z_2} = 0; \quad \frac{v_{ref} - v}{Z_3} + \frac{0 - v}{Z_4} = 0$$

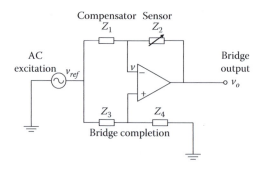

FIGURE 6.17
A bridge circuit for capacitive sensors.

where v is the common voltage at the op-amp leads. Next, eliminate v in these equations to obtain

$$v_o = \frac{(Z_4/Z_3 - Z_2/Z_1)}{1 + Z_4/Z_3} v_{ref} \qquad (6.16)$$

It is noted that when $Z_2/Z_1 = Z_4/Z_3$, the bridge output $v_o = 0$, and the bridge is said to be balanced. Since all capacitors in the bridge are similarly affected by ambient changes, a balanced bridge will maintain that condition even under ambient changes, unless the sensor reactance Z_2 is changed due to the measurand itself. It follows that the ambient effects are compensated (at least up to the first order) by a bridge circuit. From Equation 6.16, it is clear that the bridge output due to a sensor change of δZ, starting from a balanced state, is given by

$$\delta v_o = -\frac{v_{ref}}{Z_1(1 + Z_4/Z_3)} \delta Z \qquad (6.17)$$

The amplitude and phase angle of δv_o with respect to v_{ref} will determine δZ, assuming that Z_1 and Z_4/Z_3 are known.

6.5 Piezoelectric Sensors

Some substances, such as barium titanate, single-crystal quartz, and lead zirconate-titanate (PZT) can generate an electrical charge and an associated potential difference when they are subjected to mechanical stress or strain. This piezoelectric effect is used in piezoelectric transducers. Direct application of the piezoelectric effect is found in pressure and strain measuring devices, touch screens of computer monitors, and a variety of microsensors. Many indirect applications also exist. They include piezoelectric accelerometers and velocity sensors and piezoelectric torque sensors and force sensors. It is also interesting to note that piezoelectric materials deform when subjected to a potential difference (or charge or electric field). Some delicate test equipment (e.g., in vibration testing) use piezoelectric actuating elements (reverse piezoelectric action) to create fine motions. Also, piezoelectric valves (e.g., flapper valves), with direct actuation using voltage signals, are used in pneumatic and hydraulic control applications and in ink-jet printers. Miniature stepper motors based on the reverse piezoelectric action are available. Microactuators based on the piezoelectric effect are found in a number of applications including hard-disk drives (HDD). Modern piezoelectric materials include lanthanum-modified PZT (or PLZT) and piezoelectric polymeric polyvinylidene fluoride (PVDF).

The piezoelectric effect arises as a result of charge polarization in an anisotropic material (having nonsymmetric molecular structure), as a result of an applied strain. This is a reversible effect. In particular, when an electric field is applied to the material so as to change the ionic polarization, the material will regain its original shape. Natural piezoelectric materials are by and large crystalline whereas synthetic piezoelectric materials tend to be ceramics. When the direction of the electric field and the direction of strain (or stress) are the same, we have direct sensitivity. Other cross sensitivities can be defined in a 6×6 matrix with reference to three orthogonal direct axes and three rotations about these axes.

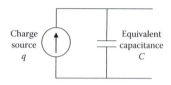

Consider a piezoelectric crystal in the form of a disc with two electrodes plated on the two opposite faces. Since the crystal is a dielectric medium, this device is essentially a capacitor, which may be modeled by a capacitance C, as in Equation 6.12. Accordingly, a piezoelectric sensor may be

FIGURE 6.18
Equivalent circuit representation of a piezoelectric sensor.

represented as a *charge source* with a capacitive impedance in parallel (Figure 6.18). An equivalent circuit (Thevenin equivalent representation) can be given as well, where the capacitor is in series with an equivalent voltage source. The impedance from the capacitor is given by

$$Z = \frac{1}{j\omega C} \tag{6.18}$$

As is clear from Equation 6.18, the output impedance of piezoelectric sensors is very high, particularly at low frequencies. For example, a quartz crystal may present an impedance of several megohms at 100 Hz, increasing hyperbolically with decreasing frequencies. This is one reason why piezoelectric sensors have a limitation on the useful lower frequency. The other reason is the charge leakage.

6.5.1 Sensitivity

The sensitivity of a piezoelectric crystal may be represented either by its *charge sensitivity* or by its *voltage sensitivity*. Charge sensitivity is defined as $S_q = \partial q / \partial F$ where q denotes the generated charge and F denotes the applied force. For a crystal with surface area A, this equation may be expressed as

$$S_q = \frac{1}{A} \frac{\partial q}{\partial p} \tag{6.19}$$

where p is the stress (normal or shear) or pressure applied to the crystal surface.

Voltage sensitivity S_v is given by the change in voltage due to a unit increment in pressure (or stress) per unit thickness of the crystal. Thus, in the limit, we have

$$S_v = \frac{1}{d} \frac{\partial v}{\partial p} \tag{6.20}$$

where d denotes the crystal thickness. Now, since $\delta q = C \delta v$, by using Equation 6.12 for a capacitor element, the following relationship between charge sensitivity and voltage sensitivity is obtained:

TABLE 6.2

Sensitivities of Several Piezoelectric Material

Material	Charge Sensitivity S_q (pC/N)	Voltage Sensitivity S_v (mV·m/N)
PZT	110	10
Barium titanate	140	6
Quartz	2.5	50
Rochelle salt	275	90

$$S_q = kS_v \qquad (6.21)$$

Note that k is the dielectric constant (permittivity) of the crystal capacitor, as defined in Equation 6.12. The overall sensitivity of a piezoelectric device can be increased through the use of properly designed multielement structures (bimorphs).

The sensitivity of a piezoelectric element is dependent on the direction of loading. This is because the sensitivity depends on the molecular structure (e.g., crystal axis). Direct sensitivities of several piezoelectric materials along their most sensitive crystal axis are listed in Table 6.2.

6.5.2 Types of Accelerometers

It is known from Newton's second law that a force (f) is necessary to accelerate a mass (or inertia element) and its magnitude is given by the product of mass (M) and acceleration (a). This product (Ma) is commonly termed *inertia force*. The rationale for this terminology is that if a force of magnitude Ma were applied to the accelerating mass in the direction opposing the acceleration, then the system could be analyzed using static equilibrium considerations. This is known as *d'Alembert's principle*. The force that causes acceleration is itself a measure of the acceleration (mass is kept constant). Accordingly, mass can serve as a front-end element to convert acceleration into a force. This is the principle of operation of common accelerometers. There are many different types of accelerometers, ranging from strain gage devices to those that use electromagnetic induction. For example, the force which causes acceleration may be converted into a proportional displacement using a spring element, and this displacement may be measured using a convenient displacement sensor. Examples of this type are differential-transformer accelerometers, potentiometer accelerometers, and variable-capacitance accelerometers. Alternatively, the strain at a suitable location of a member that was deflected due to inertia force may be determined using a strain gage. This method is used in strain gage accelerometers. Vibrating-wire accelerometers use the accelerating force to tension a wire. The force is measured by detecting the natural frequency of vibration of the wire (which is proportional to the square root of tension). In servo force-balance (or null-balance) accelerometers, the inertia element is restrained from accelerating by detecting its motion and feeding back a force (or torque) to exactly cancel out the accelerating force (torque). This feedback force is determined, for instance, by knowing the motor current and it is a measure of the acceleration.

6.5.3 Piezoelectric Accelerometer

The piezoelectric accelerometer (or *crystal accelerometer*) is an acceleration sensor that uses a piezoelectric element to measure the inertia force caused by acceleration. A piezoelectric velocity transducer is simply a piezoelectric accelerometer with a built-in integrating amplifier in the form of a miniature IC.

The advantages of piezoelectric accelerometers over other types of accelerometers are their lightweight and high-frequency response (up to about 1 MHz). However, piezoelectric transducers are inherently high output impedance devices, which generate small voltages (on the order of 1 mV). For this reason, special impedance-transforming amplifiers (e.g., charge amplifiers) have to be employed to condition the output signal and to reduce loading error.

A schematic diagram for a compression-type piezoelectric accelerometer is shown in Figure 6.19. The crystal and the inertia mass are restrained by a spring of very high stiffness. Consequently, the fundamental natural frequency or resonant frequency of the device becomes high (typically 20 kHz). This gives a reasonably wide useful range (typically up to 5 kHz). The lower limit of the useful range (typically 1 Hz) is set by factors such as the limitations of the signal-conditioning system, the mounting methods, the charge leakage in the piezoelectric element, the time constant of the charge-generating dynamics, and the SNR. A typical frequency response curve of a piezoelectric accelerometer is shown in Figure 6.20.

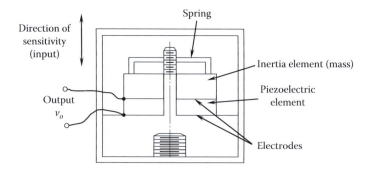

FIGURE 6.19
A compression-type piezoelectric accelerometer.

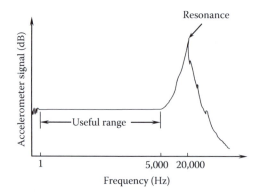

FIGURE 6.20
A typical frequency response curve for a piezoelectric accelerometer.

In a compression-type crystal accelerometer, the inertia force is sensed as a compressive normal stress in the piezoelectric element. There are also piezoelectric accelerometers where the inertia force is applied to the piezoelectric element as a shear strain or as a tensile strain.

For an accelerometer, acceleration is the signal that is being measured (the measurand). Hence, accelerometer sensitivity is commonly expressed in terms of electrical charge per unit of acceleration or voltage per unit of acceleration (compare this with Equations 6.19 and 6.20). Acceleration is measured in units of acceleration due to gravity (g), and charge is measured in picocoulombs (pC), which are units of 10^{-12} coulombs (C). Typical accelerometer sensitivities are 10 pC/g and 5 mV/g. The sensitivity depends on the piezoelectric properties, the way in which the inertia force is applied to the piezoelectric element (e.g., compressive, tensile, shear) and the mass of the inertia element. If a large mass is used, the reaction inertia force on the crystal will be large for a given acceleration, thus generating a relatively large output signal. However, large accelerometer mass results in several disadvantages, particularly,

1. The accelerometer mass distorts the measured motion variable (mechanical loading effect).
2. A heavy accelerometer has a lower resonant frequency and hence a lower useful frequency range (Figure 6.20).

For a given accelerometer size, improved sensitivity can be obtained by using the shear-strain configuration. In this configuration, several shear layers can be used (e.g., in a *delta arrangement*) within the accelerometer housing, thereby increasing the effective shear area and hence the sensitivity in proportion to the shear area. Another factor that should be considered in selecting an accelerometer is its *cross sensitivity* or transverse sensitivity. Cross sensitivity is present because a piezoelectric element can generate a charge in response to forces and moments (or torques) in orthogonal directions as well. The problem can be aggravated due to manufacturing irregularities of the piezoelectric element, including material unevenness and incorrect orientation of the sensing element, and due to poor design. Cross sensitivity should be less than the maximum error (percentage) that is allowed for the device (typically 1%).

The technique employed to mount the accelerometer on an object can significantly affect the useful frequency range of the accelerometer. Some common mounting techniques are as follows:

1. Screw-in base
2. Glue, cement, or wax
3. Magnetic base
4. Spring-base mount
5. Hand-held probe

Drilling holes in the object can be avoided by using the second through fifth methods, but the useful range can decrease significantly when spring-base mounts or handheld probes are used (typical upper limit of 500 Hz). The first two methods usually maintain the full useful range (e.g., 5 kHz), whereas the magnetic attachment method reduces the upper frequency limit to some extent (typically 3 kHz).

6.5.4 Charge Amplifier

Piezoelectric signals cannot be read using low-impedance devices. The two primary reasons for this are (1) high output impedance in the sensor results in small output signal levels and large loading errors and (2) the charge can quickly leak out through the load.

In order to overcome these problems to a great extent, a charge amplifier is commonly used as the signal-conditioning device for piezoelectric sensors (see Chapter 4). Because of impedance transformation, the impedance at the output of the charge amplifier becomes much smaller than the output impedance of the piezoelectric sensor. This virtually eliminates loading error and provides a low-impedance output for purposes such as signal communication, acquisition, recording, processing, and control. Also, by using a charge amplifier circuit with a relatively large time constant, the speed of charge leakage can be decreased. For example, consider a piezoelectric sensor and charge amplifier combination, as represented by the circuit in Figure 6.21. Let us examine how the rate of charge leakage is reduced by using this arrangement. Sensor capacitance, feedback capacitance of the charge amplifier, and feedback resistance of the charge amplifier are denoted by C, C_f, and R_f, respectively. The capacitance of the cable, which connects the sensor to the charge amplifier, is denoted by C_c.

For an op-amp of gain K, the voltage at its inverting (negative) input is $-v_o/K$, where v_o is the voltage at the amplifier output. Note that the noninverting (positive) input of the op-amp is grounded (i.e., maintained at zero potential). Due to very high input impedance of the op-amp, the currents through its input leads will be negligible. The current balance at point A gives

$$\dot{q} + C\frac{\dot{v}_o}{K} + C_c\frac{\dot{v}_o}{K} + C_f\left(\dot{v}_o + \frac{\dot{v}_o}{K}\right) + \frac{v_o + v_o/K}{R_f} = 0.$$

Since gain K is very large (typically 10^5–10^9) compared with unity, this differential equation may be approximated as

$$R_f C_f \frac{dv_o}{dt} + v_o = -R_f \frac{dq}{dt}.$$

Alternatively, it is possible to directly obtain this result from the two common assumptions (equal inverting and noninverting lead potentials and zero lead currents; see Chapter 4)

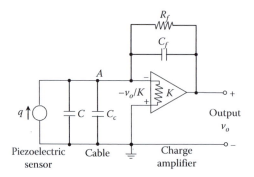

FIGURE 6.21
A piezoelectric sensor and charge amplifier combination.

for an op-amp. Then the potential at the negative (inverting) lead would be zero, as the positive lead is grounded. Also, as a result, the voltage across C_c would be zero. Hence, the current balance at point A gives

$$\dot{q} + \frac{v_o}{R_f} + C_f \dot{v}_o = 0,$$

which is identical to the previous result. The corresponding transfer function is $v_o(s)/q(s) = -R_f s/[R_f C_f s + 1]$, where s is the Laplace variable. Now, in the frequency domain $(s = j\omega)$, we have $v_o(j\omega)/q(j\omega) = -R_f j\omega/[R_f C_f j\omega + 1]$. Note that the output is zero at zero frequency $(\omega = 0)$. Hence, a piezoelectric sensor cannot be used for measuring constant (dc) signals. At very high frequencies, on the other hand, the transfer function approaches the constant value $-1/C_f$, which is the calibration constant for the device.

The transfer function of the sensor–amplifier unit as obtained above represents a first-order system with a time constant of $\tau_c = R_f C_f$. Suppose that the charge amplifier is properly calibrated (by the factor $-1/C_f$) so that the frequency transfer function can be written as

$$G(j\omega) = \frac{j\tau_c \omega}{[j\tau_c \omega + 1]} \tag{6.22}$$

The magnitude M of this transfer function is given by $M = \tau_c \omega \big/ \left(\sqrt{\tau_c^2 \omega^2 + 1}\right)$. As $\omega \to \infty$, note that $M \to 1$. Hence, at infinite frequency there is no sensor error. Measurement accuracy depends on the closeness of M to 1. Suppose that we want the accuracy to be better than a specified value M_o. Accordingly, we must have

$$\frac{\tau_c \omega}{\sqrt{\tau_c^2 \omega^2 + 1}} > M_o,$$

or

$$\tau_c \omega > \frac{M_o}{\sqrt{1 - M_0^2}}.$$

If the required lower frequency limit is ω_{min}, the time constant requirement is

$$\tau_c > \frac{M_o}{\omega_{min} \sqrt{1 - M_0^2}},$$

or

$$R_f C_f > \frac{M_o}{\omega_{min} \sqrt{1 - M_0^2}} \tag{6.23}$$

It follows that, for a specified level of accuracy, a specified lower limit on the frequency of operation may be achieved by increasing the time constant (i.e., by increasing R_f, C_f, or both). The feasible lower limit on the frequency of operation (ω_{min}) can be set by adjusting the time constant.

In theory, it is possible to measure velocity by first converting velocity into a force using a viscous damping element and measuring the resulting force using a piezoelectric sensor. This principle may be used to develop a piezoelectric velocity transducer. The practical implementation of an ideal velocity–force transducer is quite difficult. Hence, commercial piezoelectric velocity transducers use a piezoelectric accelerometer and a built-in (miniature) integrating amplifier. The overall size of such a unit can be as small as 1 cm. With double integration hardware, a piezoelectric displacement transducer is obtained. Alternatively, an ideal spring element (or cantilever), which converts displacement into a force (or bending moment or strain), may be employed to stress the piezoelectric element, resulting in a displacement transducer. Such devices are usually not practical for low-frequency (few hertz) applications because of the poor low-frequency characteristics of piezoelectric elements.

6.6 Strain Gages

Many types of force/torque sensors are based on strain gage measurements. Although strain gages measure strain, the measurements can be directly related to stress and force. Hence, it is appropriate to discuss strain gages under force/torque sensors. Note, however, that strain gages may be used in a somewhat indirect manner (using auxiliary front-end elements) to measure other types of variables, including displacement, acceleration, pressure, and temperature. Two common types of resistance strain gages will be discussed next. Specific types of force/torque sensors will be studied in the subsequent sections.

6.6.1 Equations for Strain Gage Measurements

The change of electrical resistance in material when mechanically deformed is the property used in resistance-type strain gages. The resistance R of a conductor that has length l and the area of cross section A is given by

$$R = \frac{\rho l}{A} \tag{6.24}$$

where ρ denotes the *resistivity* of the material. Taking the logarithm of Equation 6.24, we have $\log R = \log \rho + \log(l/A)$. Now, taking the differential, we obtain

$$\frac{dR}{R} = \frac{d\rho}{\rho} + \frac{d(l/A)}{l/A} \tag{6.25}$$

The first term on the right-hand side of Equation 6.25 depends on the change in resistivity, and the second term represents deformation. It follows that the change in resistance comes from the change in shape as well as from the change in resistivity of the material. For linear deformations, the two terms on the right-hand side of Equation 6.25 are linear functions of strain ε; the proportionality constant of the second term, in particular, depends on Poisson's ratio of the material. Hence, the following relationship can be written for a strain gage element:

$$\frac{\delta R}{R} = S_s \varepsilon \tag{6.26}$$

The constant S_s is known as the *gage factor* or *sensitivity* of the strain gage element. The numerical value of this constant ranges from 2 to 6 for most *metallic strain gage* elements and from 40 to 200 for *semiconductor strain gages*. These two types of strain gages will be discussed later.

The change in resistance of a strain gage element, which determines the associated strain (Equation 6.26), is measured using a suitable electrical circuit. Many variables, including displacement, acceleration, pressure, temperature, liquid level, stress, force, and torque can be determined using strain measurements. Some variables (e.g., stress, force, and torque) can be determined by measuring the strain of the dynamic object itself at suitable locations. In other situations, an auxiliary front-end device may be required to convert the measurand into a proportional strain. For instance, pressure or displacement may be measured by converting them to a measurable strain using a diaphragm, bellows, or bending element. Acceleration may be measured by first converting it into an inertia force of a suitable mass (seismic) element, then subjecting a cantilever (strain member) to that inertia force and, finally, measuring the strain at a high-sensitivity location of the cantilever element (see Figure 6.22). Temperature may be measured by measuring the thermal expansion or deformation in a bimetallic element. *Thermistors* are temperature sensors made of semiconductor material whose resistance changes with temperature. RTDs operate by the same principle, except that they are made of metals, not of semiconductor material. These temperature sensors, and the piezoelectric sensors discussed previously, should not be confused with strain gages. Resistance strain gages are based on resistance change due to strain or the *piezoresistive* property of materials.

Modern strain gages are manufactured primarily as metallic foil (for example, using the copper-nickel alloy known as constantan) or semiconductor elements (e.g., silicon with trace impurity boron). They are manufactured by first forming a thin film (foil) of metal or a single crystal of semiconductor material and then cutting it into a suitable grid pattern, either mechanically or by using photoetching (opto-chemical) techniques. This process is much more economical and is more precise than making strain gages with metal filaments. The strain gage element is formed on a backing film of electrically insulated material (e.g., polymide plastic). Using epoxy, this element is cemented or bonded onto the

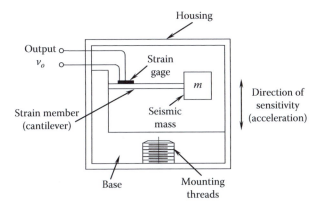

FIGURE 6.22
A strain gage accelerometer.

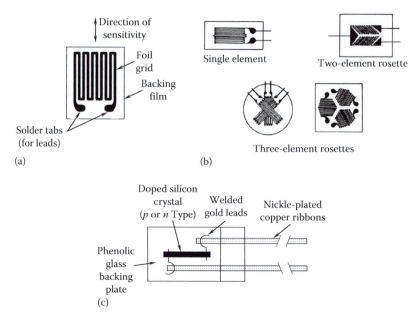

FIGURE 6.23
(a) Strain gage nomenclature; (b) typical foil-type strain gages; (c) a semiconductor strain gage.

member whose strain is to be measured. Alternatively, a thin film of insulating ceramic substrate is melted onto the measurement surface, on which the strain gage is mounted directly. The direction of sensitivity is the major direction of the elongation of the strain gage element (Figure 6.23a). To measure strains in more than one direction, multiple strain gages (e.g., various rosette configurations) are available as single units. These units have more than one direction of sensitivity. Principal strains in a given plane (the surface of the object on which the strain gage is mounted) can be determined by using these multiple strain gage units (see Appendix A). Typical foil-type gages are shown in Figure 6.23b, and a semiconductor strain gage is shown in Figure 6.23c.

A direct way to obtain strain gage measurement is to apply a constant dc voltage across a series-connected pair of strain gage elements (of resistance R) and a suitable resistor R_c and to measure the output voltage v_o across the strain gage under open-circuit conditions (using a voltmeter with high input impedance). It is known as a *potentiometer circuit* or *ballast circuit*. This arrangement has several weaknesses. Any ambient tempera-ture variation will directly introduce some error because of associated change in the strain gage resistance and the resistance of the connecting circuitry. Also, measurement accuracy will be affected by possible variations in the supply voltage v_{ref}. Furthermore, the electrical loading error will be significant unless the load impedance is very high. Perhaps the most serious disadvantage of this circuit is that the change in signal due to strain is usually a very small percentage of the total signal level in the circuit output. This problem can be reduced to some extent by decreasing v_o, which may be accom-plished by increasing the resistance R_c. This, however, reduces the sensitivity of the circuit. Any changes in the strain gage resistance due to ambient changes will directly enter the strain gage reading unless R and R_c have identical coefficients with respect to ambient changes.

A more favorable circuit for use in strain gage measurements is the *Wheatstone bridge*, as discussed in Chapter 4. One or more of the four resistors R_1, R_2, R_3, and R_4 in the bridge may represent strain gages.

6.6.1.1 Bridge Sensitivity

Strain gage measurements are calibrated with respect to a balanced bridge. When the strain gages in the bridge deform, the balance is upset. If one of the arms of the bridge has a variable resistor, it can be changed to restore balance. The amount of this change measures the amount by which the resistance of the strain gages changed, thereby measuring the applied strain. This is known as the *null-balance method* of strain measurement. This method is inherently slow because of the time required to balance the bridge each time a reading is taken. A more common method, which is particularly suitable for making dynamic readings from a strain gage bridge, is to measure the output voltage resulting from the imbalance caused by the deformation of active strain gages in the bridge. To determine the *calibration constant* of a strain gage bridge, the sensitivity of the bridge output to changes in the four resistors in the bridge should be known. For small changes in resistance, using straightforward calculus, this may be determined as

$$\frac{\delta v_o}{v_{ref}} = \frac{(R_2 \delta R_1 - R_1 \delta R_2)}{(R_1 + R_2)^2} - \frac{(R_4 \delta R_3 - R_3 \delta R_4)}{(R_3 + R_4)^2} \tag{6.27}$$

This result is subject to the bridge balance condition

$$\frac{R_1}{R_2} = \frac{R_3}{R_4} \tag{6.28}$$

because changes are measured from the balanced condition. Note from Equation 6.27 that if all four resistors are identical (in value and material), resistance changes due to ambient effects cancel out among the first-order terms (δR_1, δR_2, δR_3, δR_4), producing no net effect on the output voltage from the bridge. A closer examination of Equation 6.27 will reveal that only the adjacent pairs of resistors (e.g., R_1 with R_2 and R_3 with R_4) have to be identical in order to achieve this environmental compensation. Even this requirement can be relaxed. In fact, compensation is achieved if R_1 and R_2 have the same temperature coefficient and if R_3 and R_4 have the same temperature coefficient.

6.6.1.2 The Bridge Constant

Any of the four resistors in a bridge circuit may represent active strain gages; for example, tension in R_1 and compression in R_2, as in the case of two strain gages mounted symmetrically at 45° about the axis of a shaft in torsion. In this manner, the overall sensitivity of a strain gage bridge can be increased. It is clear from Equation 6.27 that if all four resistors in the bridge are active, the best sensitivity is obtained; for example, if R_1 and R_4 are in tension and R_2 and R_3 are in compression so that all four differential terms have the same sign. If more than one strain gage is active, the bridge output may be expressed as

$$\frac{\delta v_o}{v_{ref}} = k \frac{\delta R}{4R} \qquad (6.29)$$

where $k = \dfrac{\text{bridge output in the general case}}{\text{bridge output if only one stain gage is active}}$.

This constant is known as the *bridge constant*. The larger the bridge constant, the better the sensitivity of the bridge.

Example 6.2

A strain gage load cell (force sensor) consists of four identical strain gages, forming a Wheatstone bridge, that are mounted on a rod that has square cross section. One opposite pair of strain gages is mounted axially and the other pair is mounted in the transverse direction, as shown in Figure 6.24a. To maximize the bridge sensitivity, the strain gages are connected to the bridge as shown in Figure 6.24b. Determine the bridge constant k in terms of *Poisson's ratio v* of the rod material.

Solution

Suppose that $\delta R_1 = \delta R$. Then, for the given configuration, we have $\delta R_2 = -v\delta R$, $\delta R_3 = -v\delta R$, and $\delta R_4 = \delta R$. Note that from the definition of Poisson's ratio v,

Transverse strain $= (-v) \times$ longitudinal strain

Now, it follows from Equation 6.27 that

$$\frac{\delta v_o}{v_{ref}} = 2(1+v)\frac{\delta R}{4R},$$

according to which the bridge constant is given by $k = 2(1 + v)$.

6.6.1.3 Calibration Constant

The calibration constant C of a strain gage bridge relates the strain that is measured to the output of the bridge. Specifically,

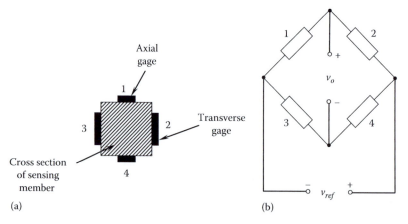

(a)   (b)

FIGURE 6.24
An example of four active strain gages: (a) Mounting configuration on the load cell; (b) bridge circuit.

$$\frac{\delta v_o}{v_{ref}} = C_\varepsilon \tag{6.30}$$

Now, in view of Equations 6.26 and 6.29, the calibration constant may be expressed as

$$C = \frac{k}{4} S_s \tag{6.31}$$

where
k is the *bridge constant*
S_s is the *sensitivity* or *gage factor* of the strain gage

Ideally, the calibration constant should remain constant over the measurement range of the bridge (i.e., independent of strain ε and time t) and should be stable with respect to ambient conditions. In particular, there should not be any creep, nonlinearities such as hysteresis, or thermal effects.

6.6.1.4 Data Acquisition

For measuring dynamic strains, either the servo null-balance method or the imbalance output method should be employed (see Chapter 4). In the imbalance output method, the output from the active bridge is directly measured as a voltage signal and calibrated to provide the measured strain. An ac bridge (powered by an ac voltage) may be used. The supply frequency should be about 10 times the maximum frequency of interest in the dynamic strain signal (bandwidth). A supply frequency on the order of 1 kHz is typical. This signal is generated by an oscillator and is fed into the bridge. The transient component of the output from the bridge is very small (typically less than 1 mV and possibly a few microvolts). This signal has to be amplified, demodulated (especially if the signals are transient), and filtered to provide the strain reading. The calibration constant of the bridge should be known in order to convert the output voltage to strain. Strain gage bridges powered by dc voltages are common, however. They have the advantage of simplicity with regard to the necessary circuitry and portability. The advantages of ac bridges include improved stability (reduced drift) and accuracy and reduced power consumption.

6.6.1.5 Accuracy Considerations

Foil gages are available with resistances as low as 50 Ω and as high as several kilohms. The power consumption of a bridge circuit decreases with increased resistance. This has the added advantage of decreased heat generation. Bridges with a high range of measurement (e.g., a maximum strain of 0.01 m/m) are available. The accuracy depends on the linearity of the bridge, environmental effects (particularly temperature), and mounting techniques. For example, zero shift, due to strain produced when the cement or epoxy that is used to mount the strain gage dries, will result in calibration error. Creep will introduce errors during static and low-frequency measurements. Flexibility and hysteresis of the bonding cement (or epoxy) will bring about errors during high-frequency strain measurements. Resolutions on the order of 1 μm (i.e., one *microstrain*) are common.

The cross sensitivity of a strain gage is the sensitivity to strain that is orthogonal to the measured strain. This cross sensitivity should be small (say, less than 1% of the direct sensitivity). Manufacturers usually provide cross sensitivity factors for their strain gages. This factor, when multiplied by the cross strain present in a given application, gives the error in the strain reading due to cross sensitivity.

Often, strains in moving members are sensed for control purposes. Examples include real-time monitoring and failure detection in machine tools, measurement of power, measurement of force and torque for feedforward and feedback control in dynamic systems, biomechanical devices, and tactile sensing using instrumented hands in industrial robots. If the motion is small or the device has a limited stroke, strain gages mounted on the moving member can be connected to the signal-conditioning circuitry and power source using coiled flexible cables. For large motions, particularly in rotating shafts, some form of commutating arrangement has to be used. Slip rings and brushes are commonly used for this purpose. When ac bridges are used, a mutual-induction device (rotary transformer) may be used with one coil located on the moving member and the other coil stationary. To accommodate and compensate for errors (e.g., losses and glitches in the output signal) caused by commutation, it is desirable to place all four arms of the bridge, rather than just the active arms, on the moving member.

6.6.2 Semiconductor Strain Gages

In some low-strain applications (e.g., dynamic torque measurement), the sensitivity of foil gages is not adequate to produce an acceptable strain gage signal. Semiconductor (SC) strain gages are particularly useful in such situations. The strain element of an SC strain gage is made of a single crystal of *piezoresistive* material such as silicon, doped with a trace impurity such as boron. A typical construction is shown in Figure 6.25. The gage factor (sensitivity) of an SC strain gage is about two orders of magnitude higher than that of a metallic foil gage (typically 40–200), as seen for silicon from the data given in Table 6.3. The resistivity is also higher, providing reduced power consumption and heat generation. Another advantage of SC strain gages is that they deform elastically to fracture. In particular, mechanical hysteresis is negligible. Furthermore, they are smaller and lighter, providing less cross sensitivity, reduced distribution error (i.e., improved spatial resolution), and negligible error due to mechanical loading. The maximum strain that is measurable using

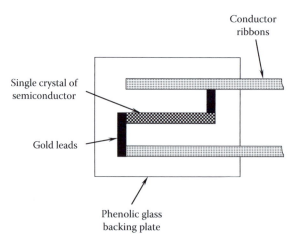

FIGURE 6.25
Component details of a semiconductor strain gage.

TABLE 6.3

Properties of Common Strain Gage Material

Material	Composition	Gage Factor (Sensitivity)	Temperature Coefficient of Resistance (10^{-6}/°C)
Constantan	45% Ni, 55% Cu	2.0	15
Isoelastic	36% Ni, 52% Fe, 8% Cr, 4% (Mn, Si, Mo)	3.5	200
Karma	74% Ni, 20% Cr, 3% Fe, 3% Al	2.3	20
Monel	67% Ni, 33% Cu	1.9	2000
Silicon	*p*-Type	100–170	70–700
Silicon	*n*-Type	−140 to −100	70–700

a semiconductor strain gage is typically 0.003 m/m (i.e., 3000 $\mu\varepsilon$). Strain gage resistance can be several hundred ohms (typically 120 Ω or 350 Ω).

There are several disadvantages associated with semiconductor strain gages, however, which can be interpreted as advantages of foil gages. The undesirable characteristics of SC gages include the following:

1. The strain–resistance relationship is more nonlinear.

2. They are brittle and difficult to mount on curved surfaces.

3. The maximum strain that can be measured is an order of magnitude smaller (typically, less than 0.01 m/m).

4. They are more costly.

5. They have a much larger temperature sensitivity.

The first disadvantage is illustrated in Figure 6.26. There are two types of semiconductor strain gages: the *p*-type, which are made of a semiconductor (e.g., silicon) doped with an acceptor impurity (e.g., boron) and the *n*-type, which are made of a semiconductor doped with a donor impurity (e.g., arsenic). In *p*-type strain gages, the direction of sensitivity is along the (1, 1, 1) crystal axis, and the element produces a "positive" (*p*) change in resistance

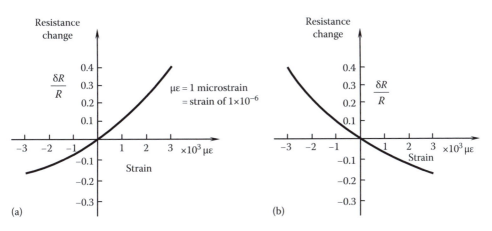

FIGURE 6.26

Nonlinear behavior of a semiconductor (silicon–boron) strain gage: (a) A *p*-type gage; (b) an *n*-type gage.

in response to a positive strain. In n-type strain gages, the direction of sensitivity is along the $(1, 0, 0)$ crystal axis and the element responds with a "negative" (n) change in resistance to a positive strain. In both types, the response is nonlinear and can be approximated by the quadratic relationship

$$\frac{\delta R}{R} = S_1 \varepsilon + S_2 \varepsilon^2 \tag{6.32}$$

The parameter S_1 represents the *linear gage factor* (*linear sensitivity*), which is positive for p-type gages and negative for n-type gages. Its magnitude is usually somewhat larger for p-type gages, corresponding to better sensitivity. The parameter S_2 represents the degree of nonlinearity, which is usually positive for both types of gages. Its magnitude, however, is typically a little smaller for p-type gages. It follows that p-type gages are less nonlinear and have higher strain sensitivities. The nonlinear relationship given by Equation 6.32 or the nonlinear characteristic curve (Figure 6.26) should be used when measuring moderate to large strains with semiconductor strain gages. Otherwise, the nonlinearity error would be excessive.

Example 6.3

For a semiconductor strain gage characterized by the quadratic strain–resistance relationship, Equation 6.32, obtain an expression for the equivalent gage factor (sensitivity) S_s, using least squares error linear approximation and assuming that strains in the range $\pm\varepsilon_{max}$ have to be measured. Derive an expression for the percentage nonlinearity. Taking $S_1 = 117$, $S_2 = 3600$, and $\varepsilon_{max} = 1 \times 10^{-2}$, calculate S_s and the percentage nonlinearity.

Solution

The linear approximation of Equation 6.32 may be expressed as $[\delta R/R]_L = S_s \varepsilon$.

The error is given by $e = \dfrac{\delta R}{R} - \left[\dfrac{\delta R}{R} \right]_L = S_1 \varepsilon + S_2 \varepsilon^2 - S_s \varepsilon = \left(S_1 - S_s \right) \varepsilon + S_2 \varepsilon^2$

The quadratic integral error is $J = \displaystyle\int_{-\varepsilon_{max}}^{\varepsilon_{max}} e^2 d\varepsilon = \int_{-\varepsilon_{max}}^{\varepsilon_{max}} \left[\left(S_1 - S_s \right) \varepsilon + S_2 \varepsilon^2 \right]^2 d\varepsilon$

We have to determine S_s that will result in minimum J. Hence, we use $\partial J/\partial S_s = 0$. This gives

$$\int_{-\varepsilon_{max}}^{\varepsilon_{max}} (-2\varepsilon) \left[\left(S_1 - S_s \right) \varepsilon + S_2 \varepsilon^2 \right]^2 d\varepsilon = 0$$

On performing the integration, we get

$$S_s = S_1 \tag{6.33}$$

The maximum error is at $\varepsilon = \pm\varepsilon_{max}$. The maximum error value is obtained (by substituting $S_s = S_1$ and $\varepsilon = \pm\varepsilon_{max}$) as $e_{max} = S_2 \varepsilon_{max}^2$.

The true change in resistance (nondimensional) from $-\varepsilon_{max}$ to $+\varepsilon_{max}$ is obtained using Equation 6.32

$$\frac{\Delta R}{R} = \left(S_1 \varepsilon_{max} + S_2 \varepsilon_{max}^2 \right) - \left(-S_1 \varepsilon_{max} + S_2 \varepsilon_{max}^2 \right) = 2 S_1 \varepsilon_{max}.$$

Hence, the percentage nonlinearity is given by $N_p = (\text{max error/range}) \times 100\% = (S_2\varepsilon_{max}^2/2S_1\varepsilon_{max}) \times 100\%$ or

$$N_p = 50S_2\varepsilon_{max}/S_1\%$$ (6.34)

Now, with the given numerical values, we have $S_s = 117$ and $N_p = 50 \times 3600 \times 1 \times 10^{-2}/117\% = 15.4\%$. *Note*: We obtained this high value for nonlinearity because the given strain limits were high. Usually, the linear approximation is adequate for strains up to $\pm 1 \times 10^{-3}$.

The higher temperature sensitivity listed as a disadvantage of semiconductor strain gages may be considered an advantage in some situations. For instance, it is this property of high temperature sensitivity that is used in piezoresistive temperature sensors. Furthermore, using the fact that the temperature sensitivity of a semiconductor strain gage can be determined very accurately, accurate methods can be employed for temperature compensation in strain gage circuitry, and temperature calibration can also be done accurately. In particular, a passive SC strain gage may be used as an accurate temperature sensor for compensation purposes.

6.7 Torque Sensors

The sensing of torque and force is useful in many applications, including the following:

1. In robotic tactile and manufacturing applications—such as gripping, surface gaging, and material forming—where exerting an adequate load on an object is the primary purpose of the task

2. In the control of fine motions (e.g., fine manipulation and micromanipulation) and in assembly tasks, where a small motion error can cause large damaging forces or performance degradation

3. In control systems that are not fast enough when motion feedback alone is employed, where force feedback and feedforward force control can be used to improve accuracy and bandwidth

4. In process testing, monitoring, and diagnostic applications, where torque sensing can detect, predict, and identify abnormal operation, malfunction, component failure, or excessive wear (e.g., in monitoring machine tools such as milling machines and drills)

5. In the measurement of power transmitted through a rotating device, where power is given by the product of torque and angular velocity in the same direction

6. In controlling complex nonlinear mechanical systems, where the measurement of force and acceleration can be used to estimate unknown nonlinear terms and an appropriate nonlinear feedback can linearize or simplify the system (nonlinear feedback control)

In most applications, sensing is done by detecting an effect of torque or the cause of torque. As well, there are methods for measuring torque directly. Common methods of torque sensing include the following:

1. Measuring strain in a sensing member between the drive element and the driven load, using a strain gage bridge
2. Measuring displacement in a sensing member (as in the first method)—either directly, using a displacement sensor, or indirectly, by measuring a variable, such as magnetic inductance or capacitance, that varies with displacement
3. Measuring reaction in support structure or housing (by measuring a force) and the associated lever arm length
4. In electric motors, measuring the field or armature current that produces motor torque; in hydraulic or pneumatic actuators, measuring actuator pressure
5. Measuring torque directly, using piezoelectric sensors, for example
6. Employing a servo method—balancing the unknown torque with a feedback torque generated by an active device (say, a servomotor) whose torque characteristics are precisely known
7. Measuring the angular acceleration caused by the unknown torque in a known inertia element

Note that force sensing may be accomplished by essentially the same techniques.

6.7.1 Strain Gage Torque Sensors

The most straightforward method of torque sensing is to connect a torsion member between the drive unit and the load in series and to measure the torque in the torsion member. If a circular shaft (solid or hollow) is used as the torsion member, the torque–strain relationship becomes relatively simple (see Appendix A) and is given by

$$\varepsilon = \frac{r}{2GJ}T \tag{6.35}$$

where
  T is the torque transmitted through the member
  ε is the principal strain (45° to axis) at radius r of the member
  J is the polar moment of area of cross section of the member
  G is the shear modulus of the material

Also, the shear stress τ at a radius r of the shaft is given by

$$\tau = \frac{Tr}{J} \tag{6.36}$$

It follows from Equation 6.35 that torque T can be determined by measuring the direct strain ε on the shaft surface along a principal stress direction (i.e., at 45° to the shaft axis). This is the basis of torque sensing using strain measurements. Using the general bridge equation (6.30) along with (6.31) in Equation 6.35, we can obtain torque T from bridge output δv_o:

$$T = \frac{8GJ}{kS_s r} \frac{\delta v_o}{v_{ref}} \tag{6.37}$$

where S_s is the gage factor (or sensitivity) of the strain gages. The bridge constant k depends on the number of active strain gages used. Strain gages are assumed to be mounted along a principal direction. Three possible configurations are shown in Figure 6.27. In configurations (a) and (b), only two strain gages are used and the bridge constant $k=2$. Note that both axial and bending loads are compensated with the given configurations because resistance in both gages will be changed by the same amount (same sign and same magnitude), which cancels out up to the first order for the bridge circuit connection shown in Figure 6.27. Configuration (c) has two pairs of gages mounted on the two opposite surfaces of the shaft. The bridge constant is doubled in this configuration, and here again, the sensor self-compensates for axial and bending loads up to the first order $[O(\delta R)]$.

The design of a torsion element for torque sensing can be viewed as the selection of the polar moment of area J of the element to meet the following four requirements:

1. The strain capacity limit specified by the strain gage manufacturer is not exceeded.

2. A specified upper limit on nonlinearity for the strain gage is not exceeded for linear operation.

3. Sensor sensitivity is acceptable in terms of the output signal level of the differential amplifier (see Chapter 4) in the bridge circuit.

4. The overall stiffness (bandwidth, steady-state error, etc.) of the system is acceptable.

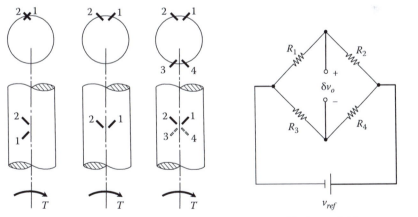

Strain gage bridge

Bridge constant (k)	2	2	4
Axial loads compensated	Yes	Yes	Yes
Bending loads compensated	Yes	Yes	Yes
Configuration	(a)	(b)	(c)

FIGURE 6.27
Strain gage configurations for a circular shaft torque sensor.

TABLE 6.4

Design Criteria for a Strain Gage Torque-Sensing Element

Criterion	Specification	Governing Formula for Polar Moment of Area (J)
Strain capacity of strain gage element	ε_{max} and T_{max}	$> \dfrac{r}{2G} \cdot \dfrac{T_{max}}{\varepsilon_{max}}$
Strain gage nonlinearity	N_p and T_{max}	$> \dfrac{25rS_2}{GS_1} \cdot \dfrac{T_{max}}{N_p}$
Sensor sensitivity	v_o and T_{max}	$\leq \dfrac{K_a k S_s r v_{ref}}{8G} \cdot \dfrac{T_{max}}{v_o}$
Sensor stiffness (system bandwidth and gain)	K	$\geq \dfrac{L}{G} \cdot K$

The governing formulas for the polar moment of area J of the torque sensor, based on the four criteria discussed earlier, are summarized in Table 6.4.

Example 6.4

A joint of a direct-drive robotic arm is sketched in Figure 6.28. Note that the rotor of the drive motor is an integral part of the driven link, without the use of gears or any other speed reducers. Also, the motor stator is an integral part of the drive link. A tachometer measures the joint speed (relative), and a resolver measures the joint rotation (relative). Gearing is used to improve

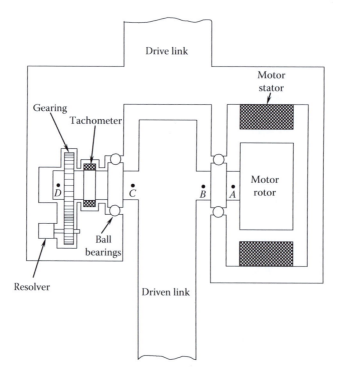

FIGURE 6.28
A joint of a direct-drive robotic arm.

the performance of the resolver. Neglecting mechanical loading from sensors and gearing, but including bearing friction, sketch the torque distribution along the joint axis. Suggest a location (or locations) for measuring the net torque transmitted to the driven link using a strain gage torque sensor.

Solution

For simplicity, assume point torques. By denoting the motor torque by T_m, the total rotor inertia torque and frictional torque in the motor by T_I, and the frictional torques at the two bearings by T_{f1} and T_{f2}, the torque distribution can be sketched as shown in Figure 6.29. The net torque transmitted to the driven link is T_L. The locations available for installing strain gages include A, B, C, and D. Note that T_L is given by the difference between the torques at B and C. Hence, strain gage torque sensors should be mounted at B and C and the difference of the readings should be taken for accurate measurement of T_L. Since bearing friction is small for most practical purposes, a single torque sensor located at B will provide reasonably accurate results. The motor torque T_m is also approximately equal to the transmitted torque when bearing friction and motor loading effects (inertia and friction) are negligible. This is the reason behind using motor current (field or armature) to measure joint torque in some robotic applications.

Although the manner in which strain gages are configured on a torque sensor can be exploited to compensate for cross-sensitivity effects arising from factors such as tensile and bending loads, it is advisable to use a torque-sensing element that inherently possesses low sensitivity to these factors, which cause error in a torque measurement. The tubular torsion element discussed in this section is convenient for analytical purposes because of the simplicity of the associated expressions for design parameters. Its mechanical design and integration into a practical system are convenient as well. Unfortunately, this member is not optimal with respect to rigidity (stiffness) for both bending and tensile loads. Alternative shapes and structural arrangements have to be considered when inherent rigidity (insensitivity) to cross loads is needed. Furthermore, a tubular element has the same principal strain at all locations on the element surface. This does not give us a choice with respect to mounting locations of strain gages in order to maximize the torque sensor sensitivity. Another disadvantage of the basic tubular torsion member is that, due to the curved surface, much care is needed in mounting fragile semiconductor gages, which

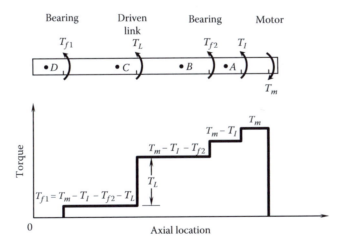

FIGURE 6.29
Torque distribution along the axis of a direct-drive manipulator joint.

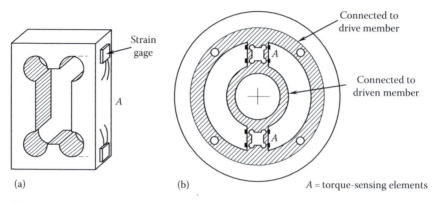

FIGURE 6.30
A bending element for torque sensing: (a) Shape of the sensing element; (b) element location.

could be easily damaged even with slight bending. Hence, a sensor element that has flat surfaces to mount the strain gages would be desirable.

A torque-sensing element that has the foregoing desirable characteristics (i.e., inherent insensitivity to cross loading, nonuniform strain distribution on the surface, and availability of flat surfaces to mount strain gages) is shown in Figure 6.30. Notice that two sensing elements are connected radially between the drive unit and the driven member. The sensing elements undergo bending to transmit a torque between the driver and the driven member. Bending strains are measured at locations of high sensitivity and are taken to be proportional to the transmitted torque. An analytical determination of the calibration constant is not easy for such complex sensing elements, but experimental determination is straightforward. A finite element analysis may be used as well for this purpose. Note that the strain gage torque sensors measure the direction as well as the magnitude of the torque transmitted through it.

6.7.2 Force Sensors

Force sensors are useful in numerous applications. For example, cutting forces generated by a machine tool may be monitored to detect tool wear and an impending failure; to diagnose the causes of failure; to control the machine tool, with feedback; and to evaluate the product quality. In vehicle testing, force sensors are used to monitor impact forces on the vehicles and crash-test dummies. Robotic handling and assembly tasks are controlled by measuring the forces generated at the end effector. The measurement of excitation forces and corresponding responses is employed in experimental modeling (model identification) of mechanical systems. The direct measurement of forces is useful in the nonlinear feedback control of mechanical systems.

Force sensors that employ strain gage elements or piezoelectric (quartz) crystals with built-in microelectronics are common. For example, thin-film and foil sensors that employ the strain gage principle for measuring forces and pressures are commercially available. A sketch of an industrial load cell, which uses the strain-gage method is shown in Figure 6.31. Both impulsive forces and slowly varying forces can be monitored using this sensor. Some types of force sensors are based on measuring a deflection caused by the force. Relatively high deflections (fraction of a mm) would be necessary for this technique to be feasible. Commercially available sensors range from sensitive devices, which can detect forces on the order of thousandth of a newton (1 N) to heavy-duty load

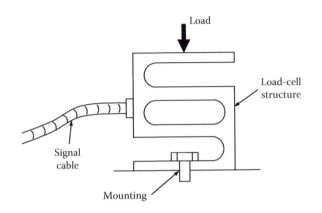

FIGURE 6.31
An industrial force sensor (load cell).

cells, which can handle very large forces (e.g., 10,000 N). Since the techniques of torque sensing can be extended in a straightforward manner to force sensing, further discussion of the topic is not undertaken here. Typical rating parameters for several types of sensors are given in Table 6.5.

6.8 Tactile Sensing

Tactile sensing is usually interpreted as *touch sensing*, but tactile sensing is different from a simple "clamping" where very few discrete force measurements are made. In tactile sensing, a force "distribution" is measured using a closely spaced array of force sensors and usually exploiting the skin-like properties of the sensor array.

Tactile sensing is particularly important in two types of operations: (1) grasping and (2) object identification. In grasping, the object has to be held in a stable manner without being allowed to slip and without being damaged. Object identification includes recognizing or determining the shape, location, and orientation of an object as well as detecting or identifying surface properties (e.g., density, hardness, texture, flexibility) and defects. Ideally, these tasks would require two types of sensing: continuous spatial sensing of time-variable contact forces and sensing of surface deformation profiles (time-variable).

These two types of data are generally related through the constitutive relations (e.g., stress–strain relations) of the touch surface of the tactile sensor or of the object that is being grasped. As a result, either the almost-continuous-spatial sensing of tactile forces or the sensing of a tactile deflection profile, separately, is often termed tactile sensing. Note that the learning experience is also an important part of tactile sensing. For example, picking up a fragile object such as an egg and picking up an object that has the same shape but is made of a flexible material are not identical processes; they require some learning through touch, particularly when vision capability is not available.

6.8.1 Tactile Sensor Requirements

Significant advances in tactile sensing have taken place in the robotics area. Applications, which are very general and numerous, include the automated inspection of surface

TABLE 6.5

Rating Parameters of Several Sensors and Transducers

Transducer	Measurand	Measurand Frequency (Max/Min)	Output Impedance	Typical Resolution	Accuracy	Sensitivity
Potentiometer	Displacement	5 Hz/dc	Low	0.1 mm	0.1%	200 mV/mm
LVDT	Displacement	2500 Hz/dc	Moderate	0.001 mm or less	0.3%	50 mV/mm
Resolver	Angular displacement	500 Hz/dc (limited by excitation frequency)	Low	2 min	0.2%	10 mV/deg
Tachometer	Velocity	700 Hz/dc	Moderate (50 Ω)	0.2 mm/s	0.5%	5 mV/mm/s 75 mV/rad/s
Eddy current proximity sensor	Displacement	100 kHz/dc	Moderate	0.001 mm 0.05% full scale	0.5%	5 V/mm
Piezoelectric accelerometer	Acceleration (velocity, etc.)	25 kHz/1 Hz	High	1 mm/s^2	1%	0.5 mV/m/s^2
Semiconductor strain gage	Strain (displacement, acceleration, etc.)	1 kHz/dc (limited by fatigue)	200 Ω	1–10 με (1 με = 10^{-6} unity strain)	1%	1 V/ε, 2000 με max
Loadcell	Force (10–1000 N)	500 Hz/dc	Moderate	0.01 N	0.05%	1 mV/N
Laser	Displacement/shape	1 kHz/dc	100 Ω	1.0 μm	0.5%	1 V/mm
Optical encoder	Motion	100 kHz/dc	500 Ω	10 bit	±½ bit	10^4/rev

profiles and joints (e.g., welded or glued parts) for defects, material handling or parts transfer (e.g., pick and place), parts assembly (e.g., parts mating), parts identification and gaging in manufacturing applications (e.g., determining the size and shape of a turbine blade picked from a bin), and fine-manipulation tasks (e.g., production of arts and craft, robotic engraving, and robotic microsurgery). Note that some of these applications might need only simple touch (force–torque) sensing if the parts being grasped are properly oriented and if adequate information about the process and the objects is already available.

Naturally, the frequently expressed design objective for tactile sensing devices has been to mimic the capabilities of human fingers. Specifically, the tactile sensor should have a compliant covering with skin-like properties, along with enough degrees of freedom for flexibility and dexterity, adequate sensitivity and resolution for information acquisition, adequate robustness and stability to accomplish various tasks, and some local intelligence for identification and learning purposes. Although the spatial resolution of a human fingertip is about 2 mm, still finer spatial resolutions (less than 1 mm) can be realized if information through other senses (e.g., vision), prior experience, and intelligence are used simultaneously during the touch. The force resolution (or sensitivity) of a human fingertip is on the order of 1 g. Also, human fingers can predict "impending slip" during grasping so that corrective actions can be taken before the object actually slips. At an elementary level, this requires the knowledge of shear stress distribution and friction properties at the common surface between the object and the hand. Additional information and an "intelligent" processing capability are also needed to predict slip accurately and to take corrective actions to prevent slipping. These are, of course, ideal goals for a tactile sensor, but they are not unrealistic in the long run. The typical specifications for an industrial tactile sensor are as follows:

1. Spatial resolution of about 2 mm
2. Force resolution (sensitivity) of about 2 g
3. Force capacity (maximum touch force) of about 1 kg
4. Response time of 5 ms or less
5. Low hysteresis (low energy dissipation)
6. Durability under harsh working conditions
7. Robustness and insensitivity to change in environmental conditions (temperature, dust, humidity, vibration, etc.)
8. Capability to detect and even predict slip

Although the technology of tactile sensing has not peaked yet, and the widespread use of tactile sensors in industrial applications is still to come, several types of tactile sensors that meet and even exceed the foregoing specifications are commercially available. In the future developments of these sensors, two separate groups of issues need to be addressed:

1. Ways to improve the mechanical characteristics and design of a tactile sensor so that accurate data with high resolution can be acquired quickly using the sensor
2. Ways to improve signal analysis and processing capabilities so that useful information can be extracted accurately and quickly from the data acquired through tactile sensing

Under the second category, we also have to consider techniques for using tactile information in the feedback control of dynamic processes. In this context, the development of control algorithms, rules, and inference techniques for intelligent controllers that use tactile information has to be addressed.

6.8.2 Construction and Operation of Tactile Sensors

The touch surface of a tactile sensor is usually made of an elastomeric pad or flexible membrane. Starting from this common basis, the principle of operation of a tactile sensor differs primarily depending on whether the distributed force is sensed or the deflection of the tactile surface is measured. The common methods of tactile sensing include the following:

1. Use a closely spaced set of strain gages or other types of force sensors to sense the distributed force.
2. Use a conductive elastomer as the tactile surface. The change in its resistance as it deforms will determine the distributed force.
3. Use a closely spaced array of deflection sensors or proximity sensors (e.g., optical sensors) to determine the deflection profile of the tactile surface.

Note that since force and deflection are related through a constitutive law for the tactile sensor (touch pad), only one type of measurement, not both force and deflection, is needed in tactile sensing. A force distribution profile or a deflection profile obtained in this manner may be treated as a two-dimensional (2D) array or an "image" and may be processed (filtered, function-fitted, etc.) and displayed as a tactile image or used in applications (object identification, manipulation control, etc.).

The contact force distribution in a tactile sensor is commonly measured using an array of force sensors located under the flexible membrane. Arrays of piezoelectric sensors and metallic or semiconductor strain gages (piezoresistive sensors) in sufficient density (number of elements per unit area) may be used for the measurement of the tactile force distribution. In particular, semiconductor elements are poor in mechanical strength but have good sensitivity. Alternatively, the skin-like membrane itself can be made from a conductive elastomer (e.g., graphite-leaded neoprene rubber) whose resistance changes can be sensed and used in determining the force and deflection distribution. In particular, as the tactile pressure increases, the resistance of the particular elastomer segment decreases and the current conducted through it (due to an applied constant voltage) will increase. Conductors can be etched underneath the elastomeric pad to detect the current distribution in the pad through proper signal acquisition circuitry. Some common problems with conductive elastomers are electrical noise, nonlinearity, hysteresis, low sensitivity, drift, low bandwidth, and poor material strength.

The deflection profile of a tactile surface may be determined using a matrix of proximity sensors or deflection sensors. Electromagnetic and capacitive sensors may be used in obtaining this information. The principles of the operation of these types of sensors have been discussed previously in this chapter. Optical tactile sensors use light-sensitive elements (photosensors) to sense the intensity of light (or laser beams) reflected from the tactile surface. In one approach (extrinsic), the proximity of a light-reflecting surface, which is attached to the back of a transparent tactile pad, is measured. Since the light intensity depends on the distance from the light-reflecting surface to the photosensor, the deflection profile can be determined. In another approach (intrinsic), the deformation of the tactile pad alters the light transmission characteristics of the pad. As a result, the intensity distribution of the transmitted light, as detected by an array of photosensors, determines the deflection profile. Optical methods have the advantages of being free from electromagnetic noise and safe in explosive environments, but they can have errors due to stray light reaching the sensor, variation in the intensity of the light source, and changes in environmental conditions (e.g., dirt, humidity, and smoke).

Example 6.5

A tactile sensor pad consists of a matrix of conductive elastomer elements. The resistance R_t in each tactile element is given by $R_t = a/F_t$, where F_t is the tactile force applied to the element and a is a constant. The circuit shown in Figure 6.32 is used to acquire the tactile sensor signal v_o, which measures the local tactile force F_t. The entire matrix of tactile elements may be scanned by addressing the corresponding elements through an appropriate switching arrangement.

For the signal acquisition circuit shown in Figure 6.32, obtain a relationship for the output voltage v_o in terms of the parameters a, R_o, and others if necessary and the variable F_t. Show that $v_o = 0$ when the tactile element is not addressed (i.e., when the circuit is switched to the reference voltage 2.5 V).

Solution
Define v_i = input to the circuit (2.5 or 0.0 V); v_{o1} = output of the first op-amp.
We use the following properties of an op-amp (see Chapter 4):

1. Voltages at the two input leads are equal
2. Currents through the two input leads are zero

Hence, note the same v_i at both input leads of the first op-amp (and at node A) and the same zero voltage at both input leads of the second op-amp (and at node B), because one of the leads is grounded.

The current balance at A:

$$\frac{5.0 - v_i}{R} = \frac{v_i - v_{o1}}{R} \Rightarrow v_{o1} = 2v_i - 5.0 \tag{i}$$

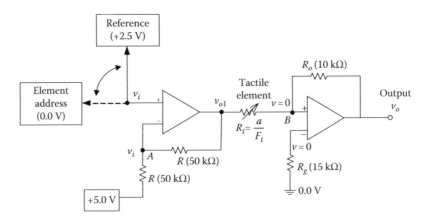

FIGURE 6.32
A signal acquisition circuit for a conductive-elastomer tactile sensor.

The current balance at B:

$$\frac{v_{o1}-0}{R_t} = \frac{0-v_o}{R_o} \quad \Rightarrow \quad v_o = -v_{o1}\frac{R_o}{R_t} \tag{ii}$$

By substituting (i) into (ii) and also substituting the given expression for R_t, we get

$$v_o = \frac{R_o}{a}F_t(5.0 - 2v_i).$$

By substituting the two switching values for v_i, we have

$$v_o = \frac{5R_o}{a}F_t \quad \text{when addressed.}$$

$$= 0 \quad \text{when reference.}$$

6.8.3 Optical Tactile Sensors

A schematic representation of an optical tactile sensor (built at the Man-Machine Systems Laboratory at the Massachusetts Institute of Technology [MIT]) is shown in Figure 6.33. If a beam of light (or laser) is projected onto a reflecting surface, the intensity of light reflected back and received by a light receiver depends on the distance (proximity) of the reflecting surface. For example, in Figure 6.33a, more light is received by the light receiver when the reflecting surface is at position 2 than when it is at position 1. But if the reflecting surface actually touches the light source, the light will be completely blocked off and no light will reach the receiver. Hence, in general, the proximity–intensity curve for an optical proximity sensor will be nonlinear and will have the shape shown in Figure 6.33a. Using this (calibration) curve, we can determine the position (x) once the intensity of the light received at the photosensor is known. This is the principle of operation of many optical tactile sensors. In the system shown in Figure 6.33b, the flexible tactile element consists of a thin, light-reflecting surface embedded within an outer layer (touch pad) of high-strength

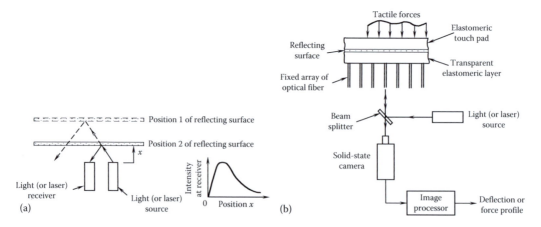

FIGURE 6.33
(a) The principle of an optical proximity sensor; (b) schematic representation of a fiber-optic tactile sensor.

rubber and an inner layer of transparent rubber. Optical fibers are uniformly and rigidly mounted across this inner layer of rubber so that light can be projected directly onto the reflecting surface.

The light source, the beam splitter, and the solid-state (charge-coupled device, or CCD) camera form an integral unit, which can be moved laterally in known steps to scan the entire array of optical fiber if a single image frame of the camera does not cover the entire array. The splitter plate reflects part of the light from the light source onto a bundle of optical fiber. This light is reflected by the reflecting surface and is received by the solid-state camera. Since the intensity of the light received by the camera depends on the proximity of the reflecting surface, the gray-scale intensity image detected by the camera will determine the deflection profile of the tactile surface. Using appropriate constitutive relations for the tactile sensor pad, the tactile force distribution can be determined as well. The image processor conditions (filtering, segmenting, etc.) the successive image frames received by the frame grabber and computes the deflection profile and the associated tactile force distribution in this manner. The image resolution will depend on the pixel (picture-element) size of each image frame (e.g., 512×512 pixels, 1024×1024 pixels, etc.) as well as the spacing of the fiber-optic matrix. Note that the force resolution or sensitivity of the tactile sensor can be improved at the expense of the thickness of the elastomeric layer, which determines the robustness of the sensor.

In the described fiber-optic tactile sensor (Figure 6.33), the optical fibers serve as the medium through which light or laser rays are transmitted to the tactile site. This is an "extrinsic" use of fiber optics for sensing. Alternatively, an "intrinsic" application can be developed where an optical fiber serves as the sensing element itself. Specifically, the tactile pressure is directly applied to a mesh of optical fibers. Since the amount of light transmitted through a fiber will decrease due to deformation caused by the tactile pressure, the light intensity at a receiver can be used to determine the tactile pressure distribution.

Yet another alternative of an optical tactile sensor is available. In this design, the light source and the receiver are located at the tactile site itself; optical fibers are not used. The principle of operation of this type of tactile sensor is shown in Figure 6.34. When the elastomeric touch pad is pressed at a particular location, a pin attached to the pad at that point moves (in the x direction), thereby obstructing the light received by the photodiode from the light-emitting diode (LED). The output signal of the photodiode measures the pin movement.

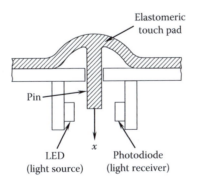

FIGURE 6.34
An optical tactile sensor with localized light sources and photosensors.

6.8.4 Piezoresistive Tactile Sensors

One type of piezoresistive tactile sensor uses an array of semiconductor strain gages mounted under the touch pad on a rigid base. In this manner, the force distribution on the touch pad is measured directly.

Example 6.6

When is tactile sensing preferred over the sensing of a few point forces? A piezoelectric tactile sensor has 25 force-sensing elements per square centimeter. Each sensor element in the sensor can withstand a maximum load of 40 N and can detect load changes on the order of 0.01 N. What is the force resolution of the tactile sensor? What is the spatial resolution of the sensor? What is the dynamic range of the sensor in decibels?

Solution
Tactile sensing is preferred when it is not a simple-touch application. Shape, surface characteristics, and flexibility characteristics of a manipulated (handled or grasped) object can be determined using tactile sensing.

Force resolution = 0.01 N

$$\text{Spatial resolution} = \frac{\sqrt{1}}{\sqrt{25}} \text{ cm} = 2 \text{ mm}$$

$$\text{Dynamic range} = 20 \log_{10}\left(\frac{40}{0.01}\right) = 72 \text{ dB}$$

6.8.5 Dexterity

Dexterity is an important consideration in the sophisticated manipulators and robotic hands that employ tactile sensing. The dexterity of a device is conventionally defined as the ratio (number of degrees of freedom in the device)/(motion resolution of the device). We will call this *motion dexterity*.

We can define another type of dexterity called *force dexterity*, as follows:

$$\text{Force dexterity} = \frac{\text{number of degrees of freedom}}{\text{force resolution}} \tag{6.38}$$

Both types of dexterity are useful in mechanical manipulation where tactile sensing is used.

6.9 Gyroscopic Sensors

Gyroscopic sensors are used for measuring angular orientations and angular speeds of aircraft, ships, vehicles, and various mechanical devices. These sensors are commonly used in control systems for stabilizing vehicle systems. Since a spinning body (a gyroscope) requires an external torque to turn (precess) its axis of spin, it is clear that if this gyro is mounted on a rigid vehicle so that there are a sufficient number of degrees of freedom (at most three) between the gyro and the vehicle, the spin axis will remain unchanged in space, regardless of the motion of the vehicle. Hence, the axis of spin of the gyro will provide a reference with respect to which the vehicle orientation (e.g., azimuth or yaw, pitch, and roll angles) and angular speed can be measured. The orientation can be measured by using angular sensors at the pivots of the structure, which mounts the gyro on the vehicle. The angular speed about an orthogonal axis can be determined; for example, by measuring the precession torque (which is proportional to the angular speed) using a strain-gage sensor; or by measuring using a resolver, the deflection of a torsional spring that restrains the precession. The angular deflection in the latter case is proportional to the precession torque and hence to the angular speed.

6.9.1 Rate Gyro

A rate gyro is used to measure angular speeds. The arrangement shown in Figure 6.35a may be used to explain its principle of operation.

A rigid disk (gyroscopic disk) of a polar moment of inertia J is spun at angular speed ω about frictionless bearings using a constant-speed motor spinning about an axis. The angular momentum H about the same axis is given by

$$H = J\omega \qquad (6.39)$$

This vector is shown by the solid line in Figure 6.35b. Due to the angular speed (rate) Ω, which is the quantity to be measured (measurand or sensor input), the vector H will turn through angle $\Omega \cdot \Delta t$ in an infinitesimal time Δt, as shown. The magnitude of the resulting

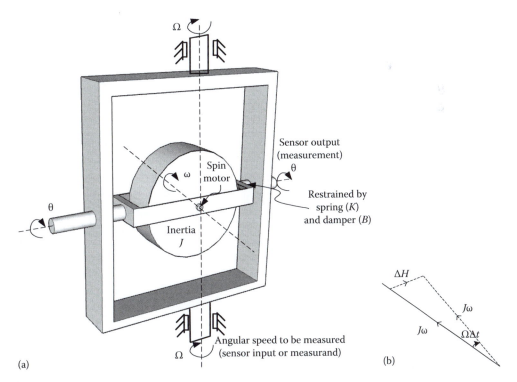

(a)
(b)

FIGURE 6.35
(a) A rate gyro; (b) gyroscopic torque needed to change the direction of an angular momentum vector.

change in angular momentum is $\Delta H = J\omega \cdot \Omega \cdot \Delta t$; or the rate of change of angular momentum is $dH/dt = J\omega \cdot \Omega$. To perform this rotation (precession), a torque has to be applied in the orthogonal direction shown by ΔH in Figure 6.35b, which is the same as the direction of rotation θ in Figure 6.35a. If this direction is restrained by a torsional spring of stiffness K and a damper with rotational damping constant B, the corresponding resistive torque is $K\theta + B\dot{\theta}$. Newton's second law (torque = rate of change of angular momentum) gives $J\omega\Omega = K\theta + B\dot{\theta}$ or

$$\Omega = \frac{K\theta + B\dot{\theta}}{J\omega} \tag{6.40}$$

From this result, it is seen that when B is very small, angular rotation θ at the gimbal bearings (measured, for example, by a resolver) will be proportional to the angular speed to be measured (Ω).

6.9.2 Coriolis Force Devices

Consider a mass m moving at velocity v relative to a rigid frame. If the frame itself rotates at an angular velocity ω, it is known that the acceleration of m has a term given by $2\omega \times v$. This is known as the *Coriolis acceleration*. The associated force $2m\omega \times v$ is the *Coriolis force*. This force can be sensed either directly using a force sensor or by measuring a resulting deflection in a flexible element, and may be used to determine the variables (ω or v) in the Coriolis force. Note that Coriolis force is somewhat similar to gyroscopic force even though the concepts are different. For this reason, devices based on the Coriolis effect are also commonly termed gyroscopes. Coriolis concepts are gaining popularity in microelectromechanical systems (MEMS)-based sensors, which use MEMS technologies.

6.10 Optical Sensors and Lasers

The laser (light amplification by stimulated emission of radiation) produces electromagnetic radiation in the ultraviolet, visible, or infrared bands of the spectrum. A laser can provide a single-frequency (*monochromatic*) light source. Furthermore, the electromagnetic radiation in a laser is *coherent* in the sense that all waves generated have constant phase angles. The laser uses oscillations of atoms or molecules of various elements. The laser is useful in fiber optics. But it can also be used directly in sensing and gaging applications. The helium–neon (He–Ne) laser and the semiconductor laser are commonly used in optical sensor applications.

The characteristic component in a fiber-optic sensor is a bundle of glass fibers (typically a few hundred) that can carry light. Each optical fiber may have a diameter on the order of a few μm to about 0.01 mm. There are two basic types of fiber-optic sensors. In one type—the "indirect" or the *extrinsic* type—the optical fiber acts only as the medium in which the sensor light is transmitted. In this type, the sensing element itself does not consist of optical fibers. In the second type—the "direct" or the *intrinsic* type—the optical fiber itself acts as the sensing element. When the conditions of the sensed medium change, the light-propagation properties of the optical fibers change (for example, due to the micro-bending of a straight fiber as a result of an applied force), providing a measurement of the change in conditions. Examples of the first (extrinsic) type of sensor include fiber-optic position sensors, proximity sensors, and tactile sensors. The second (intrinsic) type of sensor is found, for example, in fiber-optic gyroscopes, fiber-optic hydrophones, and some types of micro-displacement or force sensors MEMS devices.

6.10.1 Fiber-Optic Position Sensor

A schematic representation of a fiber-optic position sensor (or proximity sensor or displacement sensor) is shown in Figure 6.36.

The optical fiber bundle is divided into two groups: transmitting fibers and receiving fibers. Light from the light source is transmitted along the first bundle of fibers to the target object whose position is being measured. Light reflected (or diffused) onto the receiving fibers by the surface of the target object is carried to a photodetector. The intensity of the light received by the photodetector will depend on position x of the target object. In particular, if $x=0$, the transmitting bundle will be completely blocked off and the light intensity at the receiver will be zero. As x is increased, the intensity of the received light will increase because more and more light will be reflected onto the tip of the receiving bundle. This will reach a peak at some value of x. When x is increased beyond that value, more and more light will be reflected outside the receiving bundle; hence, the intensity of the received light will drop. In general then, the proximity–intensity curve for an optical proximity sensor will be nonlinear and will have the shape shown in Figure 6.37. Using this (calibration) curve, we can determine the position (x) once the intensity of the light

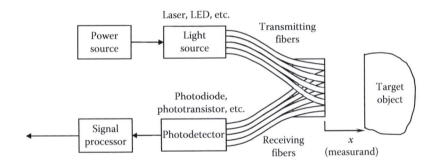

FIGURE 6.36
A fiber-optic position sensor.

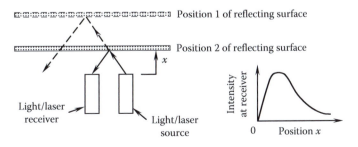

FIGURE 6.37
The principle of a fiber-optic proximity sensor.

received at the photosensor is known. The light source could be a laser (structured light), infrared light-source, or some other type, such as an LED. The light sensor (photodetector) could be some device such as a photodiode or a photo-field-effect transistor (photo-FET). This type of fiber-optic sensor can be used with a suitable front-end device (such as bellows, springs, etc.) to measure pressure, force, etc. as well.

6.10.2 Laser Interferometer

This sensor is useful in the accurate measurement of small displacements. In this fiber-optic position sensor, the same bundle of fibers is used for sending and receiving a monochromatic beam of light (typically, a laser). Alternatively, monomode fibers, which transmit only monochromatic light (of a specific wavelength) may be used for this purpose. In either case, as shown in Figure 6.38, a beam splitter (*A*) is used so that part of the light is directly reflected back to the bundle tip and the other part reaches the target object (as in Figure 6.36) and is reflected back from it (using a reflector mounted on the object) onto the bundle tip. In this manner, part of the light returning through the bundle had not traveled beyond the beam splitter while the other part had traveled between the beam splitter (*A*) and the object (through an extra distance equal to twice the separation between the beam splitter and the object). As a result, the two components of light will have a phase difference ϕ, which is given by

$$\phi = \frac{2x}{\lambda} \times 2\pi \tag{6.41}$$

where
  x is the distance of the target object from the beam splitter
  λ is the wavelength of monochromatic light

The returning light is directed to a light sensor using a beam splitter (*B*). The sensed signal is processed using principles of interferometry to determine ϕ and from Equation 6.41, the distance x. Very fine resolutions better than a fraction of a micrometer (μm) can be obtained using this type of fiber-optic position sensor.

The advantages of fiber optics include insensitivity to electrical and magnetic noise (due to optical coupling); safe operation in explosive, high-temperature, corrosive, and hazardous environments; and high sensitivity. Furthermore, mechanical loading and wear problems do not exist because fiber-optic position sensors are noncontact devices

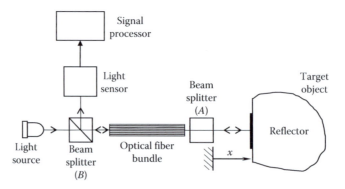

FIGURE 6.38
Laser interferometer position sensor.

with no moving parts. The disadvantages include direct sensitivity to variations in the intensity of the light source and dependence on ambient conditions (temperature, dirt, moisture, smoke, etc.). Compensation can be made, however, with respect to temperature. An *optical encoder* is a digital (or pulse-generating) motion transducer, which will be discussed later in this chapter. Here, a light beam is intercepted by a moving disk that has a pattern of transparent windows. The light that passes through, as detected by a photosensor, provides the transducer output. These sensors may also be considered in the extrinsic category.

As an *intrinsic* application of fiber optics in sensing, consider a straight optical fiber element that is supported at the two ends. In this configuration, almost 100% of the light at the source end will transmit through the optical fiber and reach the detector (receiver) end. Now, suppose that a slight load is applied to the optical fiber segment at its mid span. It will deflect slightly due to the load, and as a result the amount of light received at the detector can drop significantly. For example, a microdeflection of just 50 µm can result in a drop in intensity at the detector by a factor of 25. Such an arrangement may be used in deflection, force, and tactile sensing. Another intrinsic application is the fiber-optic gyroscope, as described next.

6.10.3 Fiber-Optic Gyroscope

This is an angular speed sensor that uses fiber optics. Contrary to the implication of its name, however, it is not a gyroscope in the conventional sense. Two loops of optical fibers wrapped around a cylinder are used in this sensor and they rotate with the cylinder at the same angular speed, which needs to be sensed. One loop carries a monochromatic light (or laser) beam in the clockwise (cw) direction; the other loop carries a beam from the same light (laser) source in the counterclockwise (ccw) direction (see Figure 6.39). Since the laser beam traveling in the direction of the rotation of the cylinder attains a higher frequency than that of the other beam, the difference in the frequencies (known as the Sagnac effect) of the two laser beams received at a common location will measure the angular speed of the cylinder. This may be accomplished through interferometry because the combined signal is a sine beat. As a result, light and dark patterns (fringes) will be present in the detected light, and they will measure the frequency difference and hence the rotating speed of the optical fibers.

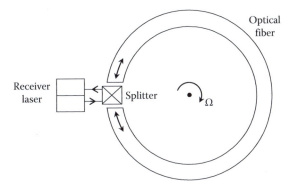

FIGURE 6.39
A fiber-optic, laser gyroscope.

In a laser (ring) gyroscope, it is not necessary to have a circular path for the laser. Triangular and square paths are commonly used as well. In general, the beat frequency $\Delta\omega$ of the combined light from two laser beams traveling in opposite directions is given by

$$\Delta\omega = \frac{4A}{p\lambda}\Omega \qquad (6.42)$$

where
 A is the area enclosed by the travel path (πr^2 for a cylinder of radius r)
 p is the length (perimeter) of the traveled path ($2\pi r$ for a cylinder)
 λ is the wavelength of the laser
 Ω is the angular speed of the object (or optical fiber)

The length of the optical fiber wound around the rotating object can exceed 100 m and can be about 1 km. The angular displacements can be measured with a laser gyro simply by counting the number of cycles and clocking fractions of cycles. Acceleration can be determined by digitally determining the rate of the change in speed. In a laser gyro, there is an alternative to use two separate loops of optical fiber, wound in opposite directions. The same loop can be used to transmit light from the same laser from the opposite ends of the fiber. A beam splitter has to be used in this case, as shown in Figure 6.39.

6.10.4 Laser Doppler Interferometer

The laser Doppler interferometer is used for the accurate measurement of speed. To understand the operation of this device, we should explain two phenomena: the Doppler effect and light wave interference. The latter phenomenon is used in the laser interferometer position sensor, which was discussed before. Consider a wave source (e.g., a light source or sound source) that is moving with respect to a receiver (observer). If the source moves toward the receiver, the frequency of the received wave appears to have increased; if the source moves away from the receiver, the frequency of the received wave appears to have decreased. The change in frequency is proportional to the velocity of the source relative to the receiver. This phenomenon is known as the *Doppler effect*. Now consider a monochromatic (single-frequency) light wave of frequency f (say, 5×10^{14} Hz) emitted by a laser source. If this ray is reflected by a target object and received by a light detector, the frequency of the received wave would be $f_2 = f + \Delta f$. The frequency increase Δf will be proportional to the velocity v of the target object, which is assumed to be positive when moving toward the light source. Specifically,

$$\Delta f = \frac{2f}{c}v = kv \qquad (6.43)$$

where c is the speed of light in the particular medium (typically, air). Now by comparing the frequency f_2 of the reflected wave with the frequency $f_1 = f$ of the original wave, we can determine Δf and, hence, the velocity v of the target object.

The change in frequency Δf due to the Doppler effect can be determined by observing the fringe pattern due to light wave interference. To understand this, consider the two waves $v_1 = a\sin 2\pi f_1 t$ and $v_2 = a\sin 2\pi f_2 t$. If we add these two waves, the resulting wave would be $v = v_1 + v_2 = a(\sin 2\pi f_1 t + \sin 2\pi f_2 t)$, which can be expressed as

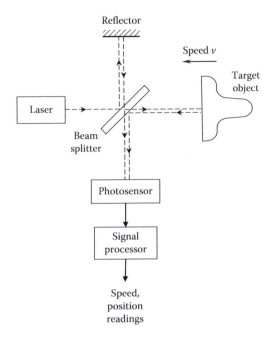

FIGURE 6.40
A laser-Doppler interferometer for measuring velocity and displacement.

$$v = 2a \sin \pi \left(f_2 + f_1 \right) t \cos \pi \left(f_2 - f_1 \right) t \qquad (6.44)$$

It follows that the combined signal beats at the beat frequency $\Delta f/2$. Since f_2 is very close to f_1 (because Δf is small compared with f), these beats will appear as dark and light lines (fringes) in the resulting light wave. This is known as *wave interference*. Note that Δf can be determined by two methods: by measuring the spacing of the fringes and by counting the beats in a given time interval or by timing successive beats using a high-frequency clock signal.

The velocity of the target object is determined in this manner. Displacement can be obtained simply by digital integration (or by accumulating the count). A schematic diagram for the laser Doppler interferometer is shown in Figure 6.40. Industrial interferometers usually employ a He–Ne laser, which has waves of two frequencies close together. In that case, the arrangement shown in Figure 6.40 has to be modified to take into account the two frequency components.

Note that the laser interferometer discussed before (Figure 6.38) directly measures displacement rather than speed. It is based on measuring the phase difference between the direct and returning laser beams, not the Doppler effect (frequency difference).

6.11 Ultrasonic Sensors

Audible sound waves have frequencies in the range of 20 Hz–20 kHz. Ultrasound waves are pressure waves, just like sound waves, but their frequencies are higher ("ultra") than the audible frequencies. Ultrasonic sensors are used in many applications, including medical imaging, ranging systems for cameras with autofocusing capability, level sensing, and speed sensing. For example, in medical applications, ultrasound probes of frequencies 40 kHz, 75 kHz, 7.5 MHz, and 10 MHz are commonly used. Ultrasound can be generated according to several principles. For example, high-frequency (gigahertz) oscillations in a piezoelectric crystal subjected to an electrical potential is used to generate very high-frequency ultrasound. Another method is to use the *magnetostrictive* property of ferromagnetic material. Ferromagnetic materials deform when subjected to magnetic fields. Respondent oscillations generated by this principle can produce ultrasonic waves. Another method of generating ultrasound is to apply a high-frequency voltage to a metal-film capacitor. A microphone can serve as an ultrasound detector (receiver).

Analogous to fiber-optic sensing, there are two common ways of employing ultrasound in a sensor. In one approach—the *intrinsic* method—the ultrasound signal undergoes changes as it passes through an object, due to acoustic impedance and the absorption characteristics of the object. The resulting signal (image) may be interpreted to determine properties of the object, such as texture, firmness, and deformation. This approach has been utilized, for example, in an innovative firmness sensor for herring roe. It is also the principle used in medical ultrasonic imaging. In the other approach—the *extrinsic* method—the time of flight of an ultrasound burst from its source to an object and then back to a receiver is measured. This approach is used in distance and position measurement and in dimensional gaging. For example, an ultrasound sensor of this category has been used in the thickness measurement of fish. This is also the method used in camera autofocusing.

In distance (range, proximity, displacement) measurement using ultrasound, a burst of ultrasound is projected at the target object and the time taken for the echo to be received is clocked. A signal processor computes the position of the target object, possibly compensating for environmental conditions. This configuration is shown in Figure 6.41. The applicable relation is

$$x = \frac{ct}{2} \qquad (6.45)$$

where
t is the time of flight of the ultrasound pulse (from generator to receiver)
x is the distance between the ultrasound generator/receiver and the target object
c is the speed of sound in the medium (typically, air)

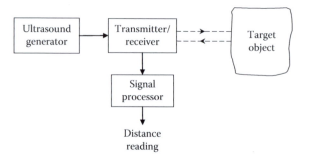

FIGURE 6.41
An ultrasonic position sensor.

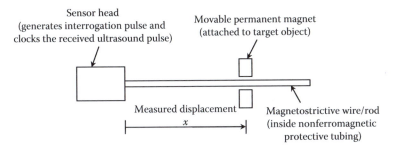

FIGURE 6.42
A magnetostrictive ultrasound displacement sensor.

Distances as small as a few cm to several meters may be accurately measured by using this approach, with fine resolution (e.g., a millimeter or less). Since the speed of ultrasonic wave propagation depends on the medium and the temperature of the medium (typically air), errors will enter into the ultrasonic readings unless the sensor is compensated for the variations in the medium; particularly for temperature.

Alternatively, the velocity of the target object can be measured using the Doppler effect by measuring (clocking) the change in frequency between the transmitted wave and the received wave. The "beat" phenomenon is employed here. The applicable relation is Equation 6.43, except, now f is the frequency of the ultrasound signal and c is the speed of sound.

6.11.1 Magnetostrictive Displacement Sensors

The ultrasound-based time of flight method is used somewhat differently in a magnetostrictive displacement sensor (e.g., the sensor manufactured as Temposonics by MTS Systems Corp., Cary, NC). The principle behind this method is illustrated in Figure 6.42. The sensor head generates an interrogation current pulse, which travels along the magnetostrictive wire. This pulse interacts with the magnetic field of the permanent magnet and generates an ultrasound pulse (by magnetostrictive action in the wire). This pulse is received (and timed) at the sensor head. The time of flight is proportional to the distance of the magnet from the sensor head. If the target object is attached to the magnet of the sensor, its position (x) can be determined using the time of flight as usual.

Strokes (maximum displacement) ranging from a few cm to 1 or 2 m at resolutions better than 50 μm are possible with these sensors. With a 15 V dc power supply, the sensor can provide a dc output in the range ±5 V. Since the sensor uses a magnetostrictive medium with protective nonferromagnetic tubing, some of the common sources of error in ultrasonic sensors that use air as the medium of propagation can be avoided.

6.12 Thermo-Fluid Sensors

Common thermo-fluid (mechanical engineering) sensors include those measuring pressure, fluid flow rate, temperature, and the heat transfer rate. Such sensors are useful in mechatronic applications as well in view of the fact that the plant (e.g., automobile, machine tool, aircraft, etc.) may involve these measurands. Several common types of sensors in this category are presented next.

6.12.1 Pressure Sensors

Some common methods of pressure sensing are the following:

1. Balance the pressure with an opposing force (or head) and measure this force. Examples are liquid manometers and pistons.
2. Subject the pressure to a flexible front-end (auxiliary) member and measure the resulting deflection. Examples are Bourdon tubes, bellows, and helical tubes.
3. Subject the pressure to a front-end auxiliary member and measure the resulting strain (or stress). Examples are diaphragms and capsules.

Some of these devices are illustrated in Figure 6.43.

In the manometer shown in Figure 6.43a, the liquid column of height h and density ρ provides a counterbalancing pressure head to support the measured pressure p with respect to the reference (ambient) pressure p_{ref}. Accordingly, this device measures the gage pressure given by

$$p - p_{ref} = \rho g h \tag{6.46}$$

where g is the acceleration due to gravity. In the pressure sensor shown in Figure 6.43b, a frictionless piston of area A supports the pressure load with an external force F. The governing equation is

$$p = \frac{F}{A} \tag{6.47}$$

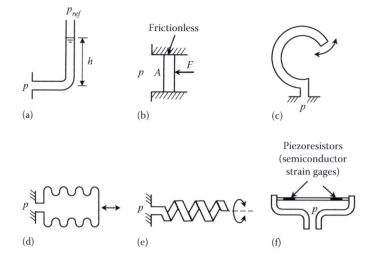

FIGURE 6.43
Typical pressure sensors: (a) Manometer; (b) counterbalance piston; (c) Bourdon tube; (d) bellows; (e) helical tube; (f) diaphragm.

The pressure is determined by measuring F using a force sensor. The Bourdon tube shown in Figure 6.43c deflects with a straightening motion as a result of internal pressure. This deflection can be measured using a displacement sensor (typically, a rotatory sensor) or can be indicated by a moving pointer. The bellows deflect with internal pressure causing a linear motion, as shown in Figure 6.43d. The deflection can be measured using a sensor such as LVDT or a capacitive sensor and can be calibrated to indicate pressure. The helical tube shown in Figure 6.43e undergoes a twisting (rotational) motion when deflected by internal pressure. This deflection can be measured by an angular displacement sensor (RVDT, resolver, potentiometer, etc.) to provide pressure reading through proper calibration. Figure 6.43f illustrates the use of a diaphragm to measure pressure. The membrane (typically metal) will be strained due to pressure. The pressure can be measured by means of strain gages (piezoresistive sensors) mounted on the diaphragm. MEMS pressure sensors that use this principle are available. In one such device, the diaphragm has a silicon wafer substrate integral with it. Through proper doping (using boron, phosphorous, etc.) a microminiature semiconductor strain gage can be formed. In fact, more than one piezoresistive sensor can be etched on the diaphragm and can be used in a bridge circuit to provide the pressure reading, through proper calibration. The most sensitive locations for the piezoresistive sensors are closer to the edge of the diaphragm, where the strains reach the maximum.

6.12.2 Flow Sensors

The volume flow rate Q of a fluid is related to the mass flow rate Q_m through

$$Q_m = \rho Q \tag{6.48}$$

where ρ is the mass density of the fluid. Also, for a flow across an area A at average velocity v, we have

$$Q = Av \tag{6.49}$$

If the flow is not uniform and if a local velocity of the maximum velocity is used, a suitable correction factor has to be included in Equation 6.49.

According to *Bernoulli's equation* for incompressible, ideal flow (no energy dissipation), we have

$$p + \frac{1}{2}\rho v^2 = \text{constant} \tag{6.50}$$

This theorem may be interpreted as conservation of energy. Also, note that the pressure p due to fluid head of height h is given by (gravitational potential energy) $\rho g h$. Using Equation 6.50 and allowing for dissipation (friction), the flow across a constriction (i.e., a fluid resistance element such as an orifice, nozzle, valve, etc.; see Chapter 3) of area A can be shown to obey the relation

$$Q = c_d A \sqrt{\frac{2\Delta p}{\rho}} \tag{6.51}$$

where
  Δp is the pressure drop across the constriction
  c_d is the discharge coefficient for the constriction

The common methods of measuring fluid flow may be classified as follows:

1. Measure pressure across a known constriction or opening; examples include nozzles, Venturi meters, and orifice plates.
2. Measure the pressure head, which will bring the flow to static conditions; a pitot tube, liquid level sensing using floats, etc. are examples.
3. Measure the flow rate (volume or mass) directly; the turbine flow meter and angular-momentum flow meter are examples.
4. Measure the flow velocity; the Coriolis meter, laser-Doppler velocimeter, and ultrasonic flow meter are examples.
5. Measure an effect of the flow and estimate the flow rate using that information; a hot-wire (or hot-film) anemometer and magnetic induction flow meter are examples.

Several examples of flow meters are shown in Figure 6.44.

For the orifice meter shown in Figure 6.44a, Equation 6.51 is applied to measure the volume flow rate. The pressure drop is measured using the techniques outlined earlier. For the pitot tube shown in Figure 6.44b, Bernoulli's Equation 6.50 is applicable, noting that the fluid velocity at the free surface of the tube is zero. This gives the flow velocity

$$v = \sqrt{2gh} \tag{6.52}$$

Note that a correction factor is needed when determining the flow rate because the velocity is not uniform across the flow section. In the angular momentum method shown in Figure 6.44c, the tube bundle through which the fluid flows is rotated by a motor. The motor torque τ and the angular speed ω are measured. As the fluid mass passes through the tube bundle, it is imparted at an angular momentum at a rate governed by the mass flow rate Q_m of the fluid. The motor torque provides the torque needed for this rate of change of angular momentum. Neglecting losses, the governing equation is

$$\tau = \omega r^2 Q_m \tag{6.53}$$

where r is the radius of the centroid of the rotating fluid mass. In a turbine flow meter, the rotation of the turbine wheel located in the flow can be calibrated to directly give the flow rate. In the Coriolis method shown in Figure 6.44d, the fluid is made to flow through a "U" segment, which is hinged to oscillate out of plane (at angular velocity ω) and restrained by springs (with known stiffness) in the lateral direction. If the fluid velocity is v, the resulting Coriolis force (due to Coriolis acceleration $2\omega \times v$) is supported by the springs. The out-of-plane angular speed is measured by a motion sensor. Also, the spring force is measured using a suitable sensor (e.g., displacement sensor). This information will determine the Coriolis acceleration of the fluid particles and hence their velocity.

In the laser-Doppler velocimeter, a laser beam is projected on the fluid flow (through a window) and its frequency shift due to the Doppler effect is measured (see under optical sensors, as described before). This is a measure of the speed of the fluid particles. As

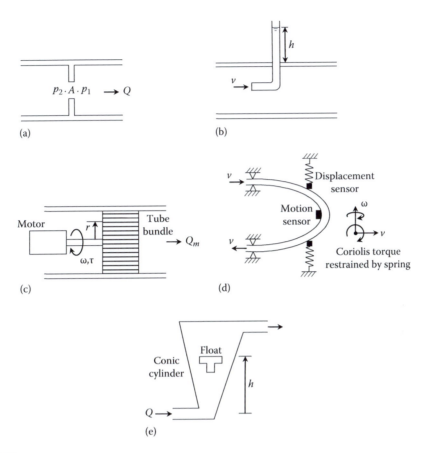

FIGURE 6.44
Several flow meters: (a) Orifice flow meter; (b) pitot tube; (c) angular-momentum flow meter; (d) Coriolis velocity meter; (e) rotameter.

another method of sensing velocity of a fluid, an ultrasonic burst is sent in the direction of flow and the time of flight is measured. An increase in the speed of propagation is due to the fluid velocity and may be determined as usual (see Section 6.11).

In the hot-wire anemometer, a conductor carrying current (i) is placed in the fluid flow. The temperatures of the wire (T) and the surrounding fluid (T_f) are measured along with the current. The coefficient of heat transfer (forced convection) at the boundary of the wire and the moving fluid is known to vary with $\sqrt{v}$ where v is the fluid velocity. Under steady conditions, the heat loss from the wire into the fluid is exactly balanced by the heat generated by the wire due to its resistance (R). The heat balance equation gives

$$i^2 R = c(a + \sqrt{v})(T - T_f) \tag{6.54}$$

This relation can be used to determine v. Instead of a wire, a metal film (e.g., platinum-plated glass tube) may be used.

There are other indirect methods of measuring fluid flow rate. In one method, the drag force on an object suspended in the flow using a cantilever arm is measured (using a strain-gage sensor at the clamped end of the cantilever). This force is known to vary quadratically with the fluid speed. A rotameter (see Figure 6.44e) is another device for measuring fluid flow. This device consists of a conic tube with a uniformly increasing cross-sectional area, which is vertically oriented. A cylindrical object is floated in the conic tube, which is connected to the fluid flow. The weight of the object is balanced by the pressure differential on the object. When the flow speed increases, the object rises within the conic tube, thereby allowing more area between the object and the tube for the fluid to pass. The pressure differential, however, still balances the weight of the object and is constant. Equation 6.51 is used to measure the fluid flow rate, since A increases quadratically with the height of the object. Consequently, the level of the object can be calibrated to give the flow rate.

6.12.3 Temperature Sensors

In most (if not all) temperature measuring devices, the temperature is sensed through "heat transfer" from the source to the measuring device. The physical (or chemical) change in the device caused by this heat transfer is the transducer stage. Several temperature sensors are outlined below.

6.12.3.1 Thermocouple

When the temperature at the junction formed by joining two unlike conductors is changed, its electron configuration changes due to heat transfer. This electron reconfiguration produces a voltage (emf or electromotive force) and is known as the Seebeck effect. Two junctions (or more) of a thermocouple are made by two unlike conductors such as iron and constant, copper and constantan, chrome and alumel, and so on. One junction is placed in a reference source (cold junction) and the other is placed in the temperature source (hot junction), as shown in Figure 6.45. The voltage across the two junctions is measured to give the temperature of the hot junction with respect to the cold junction. Note that the presence of other junctions (such as the ones formed by the wires to the voltage sensor) does not affect the reading as long as these leads are maintained at the same temperature. Very low temperatures (e.g., −250°C) as well as very high temperatures (e.g., 3000°C) can be measured using a thermocouple. Since the temperature–voltage relationship is nonlinear, a correction has to be made when measuring changes in temperature; usually by making polynomial relations. Sensitivity is quite reasonable (e.g., 10 mV/°C), but signal conditioning may be needed in some applications. Fast measurements are possible; small thermocouples having low time constants (e.g., 1 ms) should be used.

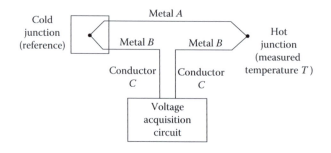

FIGURE 6.45
A thermocouple.

6.12.3.2 Resistance Temperature Detector

An RTD is simply a metal element (in a ceramic tube) whose resistance typically increases with temperature, according to a known function. A linear approximation, as given by Equation 6.65, is adequate when the temperature change is not too large.

$$R = R_0(1 + \alpha T) \qquad (6.55)$$

Temperature is measured using an RTD simply by measuring the change in resistance (say, using a bridge circuit). Metals used in RTDs include platinum, nickel, copper, and various alloys. The temperature coefficient of the resistance (α) of several metals, which can be used in RTDs, are given in Table 6.6.

The useful temperature range of an RTD is about −200°C to +800°C. At high temperatures, these may tend to be less accurate than thermocouple. The speed of response can be lower as well (e.g., a fraction of a second). A commercial RTD unit is shown in Figure 6.46.

6.12.3.3 Thermistor

Unlike an RTD, a thermistor is made of a semiconductor material (e.g., metal oxides such as chromium, cobalt, copper, iron, manganese, and nickel), which usually has a negative change in resistance with temperature (i.e., negative α). The resistance change is detected through a bridge circuit or a voltage circuit. Even though the accuracy provided by a thermistor is usually better than that of an RTD, the temperature–resistance relation is far more nonlinear, as given by

$$R = R_0 \exp\beta\left(\frac{1}{T} - \frac{1}{T_0}\right) \qquad (6.56)$$

TABLE 6.6

Temperature Coefficients of Resistance of Some RTD Metals

Metal	Temperature Coefficient of Resistance α (K)
Copper	0.0043
Nickel	0.0068
Platinum	0.0039

FIGURE 6.46
A commercial RTD unit. (Courtesy of RdF Corp., Hudson, NH.)

The characteristic temperature β (about 4000 K) itself is temperature dependent, thereby adding to the nonlinearity of the device. Hence, proper calibration is essential when operating in a wide temperature range (say, greater than 50°C). Thermistors are quite robust and they provide a fast response and high sensitivity (compared with RTDs).

6.12.3.4 Bimetal Strip Thermometer

The unequal thermal expansion of different materials is used in this device. If strips of the two materials (typically metals) are firmly bonded, thermal expansion causes this element to bend toward the material with lower expansion. This motion can be measured using a displacement sensor or can be indicated using a needle and scale. Household thermostats commonly use this principle for temperature sensing and control (on–off).

6.13 Digital Transducers

A digital transducer is a measuring device that produces a digital output. Any measuring device that presents information as discrete samples and that does not introduce a *quantization error* when the reading is represented in the digital form may be classified as a digital transducer. A transducer whose output is a pulse signal may be considered in this category since the pulses can be counted and presented in the digital form using a counter. Similarly, a transducer whose output is a frequency falls into the same category since it can use a frequency counter to generate a digital output. According to this definition, for example, an analog sensor such as a thermocouple along with an ADC is not a digital transducer. This is because a quantization error is introduced by the ADC process (see Chapter 4).

When the output of a digital transducer is a pulse signal, a common way of reading the signal is by using a counter, either to count the pulses (for high-frequency pulses) or to count clock cycles over one pulse duration (for low-frequency pulses). The count is placed as a digital word in a buffer, which can be accessed by the host (control) computer, typically at a constant frequency (sampling rate). On the other hand, if the output of a digital transducer is automatically available in a coded form (e.g., natural binary code or gray code), it can be directly read by a computer. In the latter case, the coded signal is normally generated by a parallel set of pulse signals; each pulse transition generates one bit of the digital word, and the numerical value of the word is determined by the pattern of the generated pulses. This, for example, is the case with absolute encoders. Data acquisition from (i.e., computer interfacing) a digital transducer is commonly done using a general-purpose input/output (I/O) card, for example, a motion control (servo) card, which may be able to accommodate multiple transducers (e.g., 8 channels of encoder inputs with 24 bit counters) or is done by using a data acquisition card specific to the particular transducer.

There are digital measuring devices that incorporate microprocessors to perform numerical manipulations and conditioning locally and provide output signals in either digital form or analog form. These measuring systems are particularly useful when the required variable is not directly measurable but could be computed using one or more measured outputs (e.g., power = force × speed). Although a microprocessor is an integral part of the measuring device in this case, it performs not a measuring task but, rather, a conditioning task.

Our discussion will be limited primarily to motion transducers. Note, however, that by using a suitable auxiliary front-end sensor, other *measurands*, such as force, torque, temperature, and pressure, may be converted into a motion and subsequently measured using a motion transducer. For example, altitude (or pressure) measurements in aircraft and aerospace applications are made using a pressure-sensing front end, such as a bellows or diaphragm device, in conjunction with an optical encoder (which is a digital transducer) to measure the resulting displacement. Similarly, a bimetallic element may be used to convert temperature into a displacement, which may be measured using a displacement sensor.

As we have done, it is acceptable to call an analog sensor an "analog transducer," since both the sensor stage and the transducer stage are analog in this case. The sensor stage of a digital transducer is typically analog as well. For example, motion, as manifested in physical systems, is continuous in time. Therefore, we cannot generally speak of digital motion sensors. Actually, it is the transducer stage that generates the discrete output signal in a digital motion-measuring device. Hence, we have chosen to term the present category of devices as digital transducers rather than digital sensors.

6.13.1 Advantages of Digital Transducers

There are several advantages of digital signals (or digital representation of information) in comparison with analog signals. Notably,

1. Digital signals are less susceptible to noise, disturbances, or parameter variation in instruments because data can be generated, represented, transmitted, and processed as binary words consisting of bits, which possess two identifiable states.

2. Complex signal processing with very high accuracy and speed are possible through digital means (hardware implementation is faster than software implementation).

3. High reliability in a system can be achieved by minimizing analog hardware components.

4. Large amounts of data can be stored using compact, high-density, data storage methods.

5. Data can be stored or maintained for very long periods of time without any drift or without being affected by adverse environmental conditions.

6. Fast data transmission is possible over long distances without introducing significant dynamic delays, as in analog systems.

7. Digital signals use low voltages (e.g., 0–12 V dc) and low power.

8. Digital devices typically have low overall cost.

6.13.2 Shaft Encoders

Any transducer that generates a coded (digital) reading of a measurement can be termed an encoder. Shaft encoders are digital transducers that are used for measuring angular displacements and angular velocities. Applications of these devices include motion measurement in performance monitoring and the control of robotic manipulators, machine tools, industrial processes (e.g., food processing and packaging, pulp and paper), digital data storage devices, positioning tables, satellite mirror positioning systems, and rotating machinery such as motors, pumps, compressors, turbines, and generators. High resolution (depending on the word size of the encoder output and the number of pulses generated per revolution of the encoder), high accuracy (particularly due to noise immunity and the reliability of digital signals and superior construction), and the relative ease of adoption in digital control systems (because transducer output can be read as a digital word), with an associated reduction in system cost and improvement of system reliability, are some of the relative advantages of digital transducers in general and shaft encoders in particular, in comparison with their analog counterparts.

6.13.2.1 Encoder Types

Shaft encoders can be classified into two categories, depending on the nature and the method of interpretation of the transducer output:

1. Incremental encoders
2. Absolute encoders

The output of an incremental encoder is a pulse signal, which is generated when the transducer disk rotates as a result of the motion that is being measured. By counting the pulses or by timing the pulse width using a clock signal, both angular displacement and angular velocity can be determined. With an incremental encoder, displacement is obtained with respect to some reference point. The reference point can be the home position of the moving component (say, determined by a limit switch) or a reference point on the encoder disk, as indicated by a reference pulse (index pulse) generated at that location on the disk. Furthermore, the index pulse count determines the number of full revolutions.

An absolute encoder (or whole-word encoder) has many pulse tracks on its transducer disk. When the disk of an absolute encoder rotates, several pulse trains—equal in number to the tracks on the disk—are generated simultaneously. At a given instant, the magnitude of each pulse signal will have one of two signal levels (i.e., a binary state), as determined by a level detector (or edge detector). This signal level corresponds to a binary digit (0 or 1). Hence, the set of pulse trains gives an encoded binary number at any instant. The pulse windows on the tracks can be organized into some pattern (code) so that the generated binary number at a particular instant corresponds to the specific angular position of the encoder disk at that time. The pulse voltage can be made compatible with some digital interface logic (e.g., transistor-to-transistor logic, or TTL). Consequently, the direct digital readout of an angular position is possible with an absolute encoder, thereby expediting digital data acquisition and processing. Absolute encoders are commonly used to measure fractions of a revolution. However, complete revolutions can be measured using an additional track, which generates an index pulse, as in the case of an incremental encoder.

The same signal generation (and pick-off) mechanism may be used in both types of transducers. Four techniques of transducer signal generation can be identified:

1. The optical (photosensor) method
2. The sliding contact (electrical conducting) method
3. The magnetic saturation (reluctance) method
4. The proximity sensor method

By far, the optical encoder is most popular and cost effective. The other three approaches may be used in special circumstances where an optical may not be suitable (e.g., under extreme temperatures) or may be redundant (e.g., where a code disk such as a toothed wheel is already available as an integral part of the moving member). For a given type of encoder (incremental or absolute), the method of signal interpretation is identical for all four types of signal generation listed above.

6.13.3 Incremental Optical Encoder

The optical encoder uses an opaque disk (code disk) that has one or more circular tracks with some arrangement of identical transparent windows (slits) in each track. A parallel beam of light (e.g., from a set of LEDs) is projected to all tracks from one side of the disk. The transmitted light is picked off using a bank of photosensors on the other side of the disk, which typically has one sensor for each track. This arrangement is shown in Figure 6.47a, which indicates just one track and one pick-off sensor. The light sensor could be a silicon photodiode or a phototransistor. Since the light from the source is interrupted by the opaque regions of the track, the output signal from the photosensor is a series of voltage pulses. This signal can be interpreted (e.g., through edge

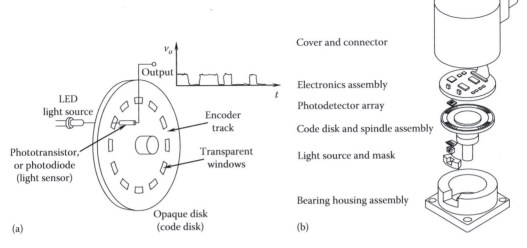

FIGURE 6.47
(a) Schematic representation of an (incremental) optical encoder; (b) components of a commercial incremental encoder. (Courtesy of BEI Electronics, Inc., Goleta, CA. With permission.)

detection or level detection) to obtain the increments in the angular position and also in the angular velocity of the disk. Note that in the standard terminology, the sensor element of such a measuring device is the encoder disk, which is coupled to the rotating object (directly or through a gear mechanism). The transducer stage is the conversion of disk motion (analog) into the pulse signals (which can be coded into a digital word). The opaque background of transparent windows (the window pattern) on an encoder disk may be produced by contact printing techniques. The precision of this production procedure is a major factor that determines the accuracy of optical encoders. Note that a transparent disk with a track of opaque spots will work equally well as the encoder disk of an optical encoder. In either form, the track has a 50% duty cycle (i.e., length of the transparent region = length of the opaque region). A commercially available optical encoder is shown in Figure 6.47b.

There are two possible configurations for an incremental encoder disk: an offset sensor configuration and an offset track configuration.

The first configuration is schematically shown in Figure 6.48. The disk has a single circular track with identical and equally spaced transparent windows. The area of the opaque region between adjacent windows is equal to the window area. Two photodiode sensors (pick-offs 1 and 2 in Figure 6.48) are positioned facing the track at a quarter-pitch (half the window length) apart. The forms of their output signals (v_1 and v_2), after passing them through pulse-shaping circuitry (idealized), are shown in Figure 6.49a and b for the two directions of rotation.

In the second configuration of incremental encoders, two identical tracks are used; one is offset from the other by a quarter-pitch. Each track has its own pick-off sensor, oriented normally facing the track. The two pick-off sensors are positioned on a radial line facing the disk without any circumferential offset unlike the previous configuration. However, the output signals from the two sensors are the same as before (Figure 6.49). Note that an output pulse signal is on half the time and off half the time, giving a 50% duty cycle.

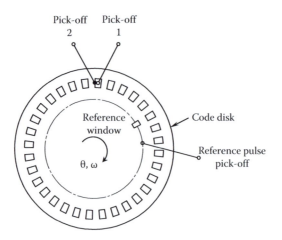

FIGURE 6.48
An incremental encoder disk (offset sensor configuration).

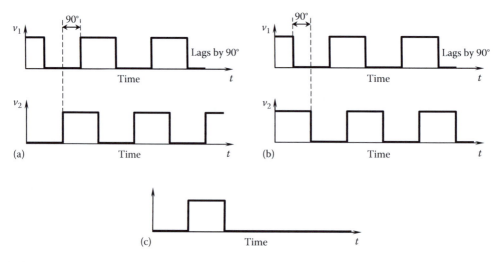

FIGURE 6.49
Shaped pulse signals from an incremental encoder (a) for cw rotation; (b) for ccw rotation; (c) reference pulse signal.

In both configurations, an additional track with a lone window and associated sensor is also usually available. This track generates a reference pulse (index pulse) per revolution of the disk (see Figure 6.49c). This pulse is used to initiate the counting operation. Furthermore, the index pulse count gives the number of complete revolutions, which is required in measuring absolute angular rotations. Note that when the disk rotates at constant angular speed, the pulse width and pulse-to-pulse period (encoder cycle) are constant (with respect to time) in each sensor output. When the disk accelerates, the pulse width decreases continuously; when the disk decelerates, the pulse width increases.

6.13.3.1 Direction of Rotation

The quarter-pitch offset in sensor location (or in track placement) is used to determine the direction of the rotation of the disk. For example, Figure 6.49a shows the shaped (idealized) sensor outputs (v_1 and v_2) when the disk rotates in the cw direction; and Figure 6.49b shows the outputs when the disk rotates in the ccw direction. It is clear from these two figures that in cw rotation, v_1 lags v_2 by a quarter of a cycle (i.e., a phase lag of 90°); and in ccw rotation, v_1 leads v_2 by a quarter of a cycle. Hence, the direction of rotation is obtained by determining the phase difference of the two output signals, using phase-detecting circuitry.

One method for determining the phase difference is to time the pulses using a high-frequency clock signal. Suppose that the counting (timing) begins when the v_1 signal begins to rise (i.e., when a rising edge is detected). Let $n_1 =$ the number of clock cycles (time) up to the time when v_2 begins to rise; and let $n_2 =$ the number of clock cycles up to the time when v_1 begins to rise again. Then, the following logic applies:

if $n_1 > n_2 - n_1 \Rightarrow$ cw rotation

if $n_1 < n_2 - n_1 \Rightarrow$ ccw rotation

This logic for direction detection should be clear from Figure 6.49a and b.

Another scheme can be given for direction. In this case, we first detect a high level (logic high or binary 1) in signal v_2 and then check whether the edge in signal v_1 rises or falls during this period. From Figure 6.49a and b, the following logic applies:

if there is a falling edge in v_1 when v_2 is logic high $\Rightarrow$ cw rotation

if there is a rising edge in v_1 when v_2 is logic high $\Rightarrow$ ccw rotation

6.13.3.2 Hardware Features

The actual hardware of commercial encoders is not as simple as what is suggested by Figure 6.48 (see Figure 6.47b). A more detailed schematic diagram of the signal generation mechanism of an optical incremental encoder is shown in Figure 6.50a. The light generated by the LED is collimated (forming parallel rays) using a lens. This pencil of parallel light passes through a window of the rotating code disk. The masking (grating) disk is stationary and has a track of windows identical to that in the code disk. Because of the presence of the masking disk, light from the LED will pass through more than one window of the code disk, thereby improving the intensity of light received by the photosensor but not introducing any error due to the diameter of the pencil of light being larger than the window length. When the windows of the code disk face the opaque areas of the masking disk, virtually no light is received by the photosensor. When the windows of the code disk face the transparent areas of the masking disk, the maximum amount of light reaches the photosensor. Hence, as the code disk moves, a sequence of triangular (and positive) pulses of light is received by the photosensor. Pulse width, in this case, is a full cycle (i.e., it corresponds to the window pitch) and not a half cycle.

A fluctuation in the supply voltage to the encoder light source also directly influences the light level received by the photosensor. If the sensitivity of the photosensor is not high enough, a low light level might be interpreted as no light, which would result in

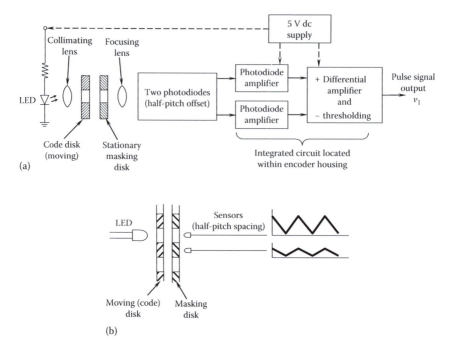

FIGURE 6.50
(a) Internal hardware of an optical incremental encoder; (b) use of two sensors at 180° spacing to generate a shaped pulse.

measurement error. Such errors due to instabilities and changes in the supply voltage can be eliminated by using two photosensors, one placed half a pitch away from the other along the window track, as shown in Figure 6.50b. This arrangement is for contrast detection and it should not be confused with the quarter-of-a-pitch offset arrangement that is required for direction detection. The sensor facing the opaque region of the masking disk will always read a low signal. The other sensor will read a triangular signal whose peak occurs when a moving window overlaps with a window of the masking disk, and whose valley occurs when a moving window faces an opaque region of the masking disk. The two signals from these two sensors are amplified separately and fed into a differential amplifier. The result is a high-intensity triangular pulse signal. A shaped (or binary) pulse signal can be generated by subtracting a threshold value from this signal and identifying the resulting positive (or binary 1) and negative (or binary 0) regions. This procedure will produce a more distinct (or binary) pulse signal that is immune to noise.

The signal amplifiers are IC devices and are housed within the encoder itself. Additional pulse-shaping circuitry may also be present. The power supply has to be provided separately as an external component. The voltage level and pulse width of the output pulse signal are logic-compatible (e.g., TTL) so that they may be read directly using a digital board. Note that if the output level v_1 is positive high, we have a logic high (or binary 1) state. Otherwise, we have a logic low (or binary 0) state. In this manner, a stable and accurate digital output can be obtained even under unstable voltage supply conditions. The schematic diagram in Figure 6.50 shows the generation of only one (v_1) of the two quadrature pulse signals. The other pulse signal (v_2) is generated using identical hardware but at a quarter of a pitch offset. The index pulse (reference pulse) signal is also generated in a similar manner. The cable of the encoder (usually a ribbon cable) has a multi-pin connector. Three of the pins provide the three output pulse signals. Another pin carries the dc supply voltage (typically 5 V) from the power supply into the encoder. Typically, a ground line (ground pin) is included as well. Note that the only moving part in the system shown in Figure 6.50 is the code disk.

6.13.3.3 Displacement Measurement

An incremental encoder measures displacement as a pulse count and it measures velocity as a pulse frequency. A digital processor is able to express these readings in engineering units (rad, degrees, rad/s, etc.) using pertinent parameter values of the physical system. Suppose that the maximum count possible is M pulses and the range of the encoder is $\pm\theta_{max}$. The angular position θ corresponding to a count of n pulses is computed as

$$\theta = \frac{n}{M}\theta_{max} \tag{6.57}$$

6.13.3.4 Digital Resolution

The resolution of an encoder represents the smallest change in measurement that can be measured realistically. Since an encoder can be used to measure both displacement and velocity, we can identify a resolution for each case. First, we will consider displacement resolution, which is governed by the number of windows N in the code disk and the digital size (number of bits) of the buffer (counter output). To begin, we will discuss digital resolution. Suppose that the encoder count is stored as digital data of r bits. Allowing for a *sign bit*, we have $M = 2^{r-1}$. The displacement resolution of an incremental encoder is given by the change in displacement corresponding to a unit change in the count (n). It follows from Equation 6.57 that the *displacement resolution* is given by $\Delta\theta = \theta_{max}/M$. In particular, the *digital resolution* corresponds to a unit change in the bit value. By substituting for M, we have the digital resolution

$$\Delta\theta_d = \frac{\theta_{max}}{2^{r-1}} \tag{6.58a}$$

Typically, $\theta_{max} = \pm 180°$ or $360°$. Then,

$$\Delta\theta_d = \frac{180°}{2^{r-1}} = \frac{360°}{2^r} \tag{6.58b}$$

Note that the minimum count corresponds to the case where all the bits are zero and the maximum count corresponds to the case where all the bits are unity. Suppose that these two readings represent the angular displacements θ_{min} and θ_{max}. We have $\theta_{max} = \theta_{min} + (M-1)\Delta\theta$. By substituting for M, we get $\theta_{max} = \theta_{min} + (2^{r-1} - 1)\Delta\theta_d$, which leads to the conventional definition for digital resolution $\Delta\theta_d = (\theta_{max} - \theta_{min})/(2^{r-1} - 1)$. This result is exactly the same as what is given by Equation 6.58. If θ_{max} is 2π and $\theta_{min} = 0$, then θ_{max} and θ_{min} will correspond to the same position of the code disk. To avoid this ambiguity, we use $\theta_{min} = \theta_{max}/2^{r-1}$. Then, the digital resolution is given by $(360° - 360°/2^r)/(2^r - 1)$, which is identical to Equation 6.58.

6.13.3.5 Physical Resolution

The physical resolution of an encoder is governed by the number of windows N in the code disk. If only one pulse signal is used (i.e., no direction sensing), and if only the rising edges of the pulses are detected (i.e., full cycles of the encoder signal are counted), the physical resolution is given by the pitch angle of the track (i.e., angular separation between adjacent windows), which is $(360/N)°$. But if quadrature signals (i.e., two pulse signals, one out of phase with the other by $90°$ or a quarter of a pitch angle) are available and the capability to detect both rising and falling edges of a pulse is also present, four counts can be made per encoder cycle, thereby improving the resolution by a factor of four. Under these conditions, the physical resolution of an encoder is given by

$$\Delta\theta_p = \frac{360°}{4N} \tag{6.59}$$

To understand this, note in Figure 6.49a (or Figure 6.49b) that when the two signals v_1 and v_2 are added, the resulting signal has a transition at every quarter of the encoder cycle. By detecting each transition (through edge detection or level detection), four pulses can be counted within every main cycle. It should be mentioned that each signal (v_1 or v_2) separately has a resolution of half a pitch, provided that all transitions (rising edges and falling edges) are detected and counted instead of pulses (or high signal levels) being counted. Accordingly, a disk with 10,000 windows has a resolution of $0.018°$ if only one pulse signal is used (and both transitions, rise and fall, are detected). When two signals (with a phase shift of a quarter of a cycle) are used, the resolution improves to $0.009°$. This resolution is achieved directly from the mechanics of the transducer; no interpolation is involved. It assumes, however, that the pulses are nearly ideal and, in particular, that the transitions are perfect. In practice, this cannot be realized if the pulse signals are noisy. Then, pulse shaping will be necessary as mentioned before.

The larger of the two resolutions given by Equations 6.58 and 6.59 governs the displacement resolution of the encoder.

6.13.3.6 Step-Up Gearing

The physical resolution of an encoder can be improved by using step-up gearing so that one rotation of the moving object that is being monitored corresponds to several rotations of the code disk of the encoder. This improvement is directly proportional to the step-up gear ratio (p). Specifically, we have

$$\Delta\theta_p = \frac{360°}{4pN} \tag{6.60}$$

Backlash in the gearing introduces a new error, however. For best results, this backlash error should be several times smaller than the resolution with no backlash.

The digital resolution will not improve by gearing if the maximum angle of rotation of the moving object (say, $360°$) still corresponds to the buffer/register size. Then, the change in the least significant bit (LSB) of the buffer corresponds to the same change in the angle of rotation of the moving object. In fact, the overall displacement resolution can be harmed in this case if excessive backlash is present. But, if the buffer/register size corresponds to a full rotation of the code disk (i.e., a rotation of $360°/p$ in the object) and if the output register (or buffer) is cleared at the end of each such rotation and a separate count of full rotations of the code disk is kept, then the digital resolution will also improve by a factor of p. Specifically, from Equation 6.58 we get

$$\Delta\theta_d = \frac{180°}{p2^{r-1}} = \frac{360°}{p2^r} \tag{6.61}$$

Example 6.7

By using high-precision techniques to imprint window tracks on the code disk, it is possible to attain a window density of 500 windows/cm of diameter. Consider a 3000 window disk. Suppose that step-up gearing is used to improve resolution and the gear ratio is 10. If the word size of the output register is 16 bits, examine the displacement resolution of this device for the two cases where the register size corresponds to (1) a full rotation of the object and (2) a full rotation of the code disk.

Solution

First, consider the case in which gearing is not present. With quadrature signals, the physical resolution is $\Delta\theta_p = 360°/(4 \times 3000) = 0.03°$.

For a range of measurement given by $\pm180°$, a 16 bit output provides a digital resolution of $\Delta\theta_d = 180°/2^{15} = 0.005°$.

Hence, in the absence of gearing, the overall displacement resolution is 0.03°.

Next consider a geared encoder with a gear ratio of 10 and neglect gear backlash. The physical resolution improves to 0.003°. But, in case (1), the digital resolution remains unchanged at best. Hence, the overall displacement resolution improves to 0.005° as a result of gearing. In case (2), the digital resolution improves to 0.0005°. Hence, the overall displacement resolution becomes 0.03°.

In summary, the displacement resolution of an incremental encoder depends on the following factors:

1. The number of windows on the code track (or disk diameter)
2. Gear ratio
3. The word size of the measurement buffer

6.13.3.7 Interpolation

The output resolution of an encoder can be further enhanced by interpolation. This is accomplished by adding equally spaced pulses in between every pair of pulses generated by the encoder circuit. These auxiliary pulses are not true measurements, and they can be interpreted as a linear interpolation scheme between true pulses. One method of accomplishing this interpolation is by using the two pick-off signals that are generated by the encoder (quadrature signals). These signals are nearly sinusoidal (or triangular) prior to shaping (say, by level detection). They can be filtered to obtain two sine signals that are 90° out of phase (i.e., a sine signal and a cosine signal). By weighted combination of these two signals, a series of sine signals can be generated such that each signal lags the preceding signal by any integer fraction of 360°. By level detection or edge detection (rising and falling edges), these sine signals can be converted into square wave signals. Then, by logical combination of the square waves, an integer number of pulses can be generated within each encoder cycle. These are the interpolation pulses that are added to improve the encoder resolution. In practice, about twenty interpolation pulses can be added between a pair of adjacent main pulses.

6.13.3.8 Velocity Measurement

Two methods are available for determining velocities using an incremental encoder:

1. Pulse-counting method
2. Pulse-timing method.

In the first method, the pulse count over a fixed time period (the successive time period at which the data buffer is read) is used to calculate the angular velocity. For a given period of data reading, there is a lower speed limit below which this method is not very accurate. To compute the angular velocity ω using this method, suppose that the count during a time period T is n pulses. Hence, the average time for one pulse is T/n. If there are N windows on the disk, assuming that quadrature signals are not used, the angle moved during one pulse is $2\pi/N$. Hence,

$$\text{Speed } \omega = \frac{2\pi/N}{T/n} = \frac{2\pi n}{NT} \tag{6.62}$$

If quadrature signals are used, replace N by $4N$ in Equation 6.62.

In the second method, the time for one encoder cycle is measured using a high-frequency clock signal. This method is particularly suitable for accurately measuring low speeds. In this method, suppose that the clock frequency is f Hz. If m cycles of the clock signal are counted during an encoder period (interval between two adjacent windows, assuming that quadrature signals are not used), the time for that encoder cycle (i.e., the time to rotate through one encoder pitch) is given by m/f. With a total of N windows on the track, the angle of rotation during this period is $2\pi/N$ as before. Hence,

$$\text{Speed } \omega = \frac{2\pi/N}{m/f} = \frac{2\pi f}{Nm} \tag{6.63}$$

If quadrature signals are used, replace N by $4N$ in Equation 6.63.

A single incremental encoder can serve as both position sensor and speed sensor. Hence, a position loop and a speed loop in a control system can be closed using a single encoder, without having to use a conventional (analog) speed sensor such as a tachometer. The speed resolution of the encoder (depending on the method of speed computation—pulse counting or pulse timing) can be chosen to meet the accuracy requirements for the speed control loop. A further advantage of using an encoder rather than a conventional (analog) motion sensor is that an ADC would be unnecessary. For example, the pulses generated by the encoder may be used as *interrupts* for the control computer. These interrupts are then directly counted (by an up/down counter or indexer) or timed (by a clock in the data acquisition computer) within the control computer, thereby providing position and velocity readings.

6.13.3.9 Velocity Resolution

The velocity resolution of an incremental encoder depends on the method that is employed to determine velocity. Since the pulse-counting method and the pulse-timing method are both based on counting, the velocity resolution is given by the change in angular velocity that corresponds to a change (increment or decrement) in the count by one.

For the pulse-counting method, it is clear from Equation 6.62 that a unity change in the count n corresponds to a speed change of

$$\Delta\omega_c = \frac{2\pi}{NT} \tag{6.64}$$

where
N is the number of windows in the code track
T is the time period over which a pulse count is read

Equation 6.64 gives the velocity resolution by this method. Note that the engineering value (in rad/s) of this resolution is independent of the angular velocity itself, but when expressed as percentage of the speed, the resolution becomes better (smaller) at higher speeds. Furthermore, note from Equation 6.64 that the resolution improves with the number of windows and the count reading (sampling) period. Under transient conditions, the accuracy of a velocity reading decreases with increasing T (because, according to Shannon's sampling theorem—see Chapter 5—the sampling frequency has to be at least double the highest frequency of interest in the velocity signal). Hence, the sampling period should not be increased indiscriminately. As usual, if quadrature signals are used, N in Equation 6.64 has to be replaced by $4N$ (i.e., the resolution improves by ×4).

In the pulse-timing method, the velocity resolution is given by (see Equation 6.63)

$$\Delta\omega_t = \frac{2\pi f}{Nm} - \frac{2\pi f}{N(m+1)} = \frac{2\pi f}{Nm(m+1)} \tag{6.65a}$$

where f is the clock frequency. For large m, $(m+1)$ can be approximated by m. Then, by substituting Equation 6.63 in 6.65a, we get

$$\Delta\omega_t \approx \frac{2\pi f}{Nm^2} = \frac{N\omega^2}{2\pi f} \tag{6.65b}$$

Note that in this case, the resolution degrades quadratically with speed. This resolution degrades with the speed even when it is considered as a fraction of the measured speed:

$$\frac{\Delta\omega_t}{\omega} = \frac{N\omega}{2\pi f} \tag{6.66}$$

This observation confirms the previous suggestion that the pulse-timing method is appropriate for low speeds. For a given speed and clock frequency, the resolution further degrades with increasing N. This is true because, when N is increased, the pulse period shortens and hence the number of clock cycles per pulse period also decreases. The resolution can be improved, however, by increasing the clock frequency.

Example 6.8

An incremental encoder with 500 windows in its track is used for speed measurement. Suppose that:

a. In the pulse-counting method, the count (buffer) is read at the rate of 10 Hz
b. In the pulse-timing method, a clock of frequency 10 MHz is used

Determine the percentage resolution for each of these two methods when measuring a speed of

  i. 1 rev/s
  ii. 100 rev/s

Solution

  i. Speed = 1 rev/s
     With 500 windows, we have 500 pulses/s
     a. Pulse counting method

     $$\text{Counting period} = \frac{1}{10 \text{ Hz}} = 0.1 \text{ s}$$

     Pulse count (in 0.1 s) = $500 \times 0.1 = 50$

     $$\text{Percentage resolution} = \frac{1}{50} \times 100\% = 2\%$$

     b. Pulse timing method

     At 500 pulses/s, pulse period = $\frac{1}{500}$ s = 2×10^{-3} s

     With a 10 MHz clock, the clock count = $10 \times 10^6 \times 2 \times 10^{-3} = 20 \times 10^3$

     $$\text{Percentage resolution} = \frac{1}{20 \times 10^3} \times 100\% = 0.005\%$$

  ii. Speed = 100 rev/s
      With 500 windows, we have 50,000 pulses/s
      a. Pulse counting method
         Pulse count (in 0.1 s) = $50,000 \times 0.1 = 5000$

     $$\text{Percentage resolution} = \frac{1}{5000} \times 100\% = 0.02\%$$

      b. Pulse timing method

      At 50,000 pulses/s, the pulse period = $\frac{1}{50,000}$ s = 20×10^{-6} s

      With a 10 MHz clock, clock count = $10 \times 10^6 \times 20 \times 10^{-6} = 200$

     $$\text{Percentage resolution} = \frac{1}{200} \times 100\% = 0.5\%$$

      The results are summarized in Table 6.7.

Results given in Table 6.7 confirm that in the pulse-counting method the resolution improves with speed, and hence it is more suitable for measuring high speeds. Furthermore, in the pulse-timing method, the resolution degrades with speed, and hence it is more suitable for measuring low speeds.

TABLE 6.7

Comparison of Speed Resolution from an
Incremental Encoder

	Percentage Resolution	
Speed (Rev/s)	Pulse-Counting Method (%)	Pulse-Timing Method (%)
1.0	2	0.005
100.0	0.02	0.5

6.13.3.10 Step-Up Gearing

Consider an incremental encoder that has an N window per track, and is connected to a rotating shaft through a gear unit with step-up gear ratio p. Formulas for computing the angular velocity of the shaft by

    a. Pulse-counting method

    b. Pulse-timing method

can be easily determined by using Equations 6.62 and 6.63. Specifically, the angle of rotation of the shaft corresponding to the window spacing (pitch) of the encoder disk now is $2\pi/(pN)$. Hence, the corresponding formulas for speed can be obtained by replacing N by pN in Equations 6.62 and 6.63. We have,

for the pulse-counting method

$$\omega = \frac{2\pi n}{pNT} \tag{6.67}$$

for the pulse timing method

$$\omega = \frac{2\pi f}{pNm} \tag{6.68}$$

Note that these relations can also be obtained simply by dividing the encoder disk speed by the gear ratio, which gives the object speed.

As before, the speed resolution is given by the change in speed corresponding to a unity change in the court. Hence,

for the pulse-counting method

$$\Delta\omega_c = \frac{2\pi(n+1)}{pNT} - \frac{2\pi n}{pNT} = \frac{2\pi}{pNT} \tag{6.69}$$

It follows that in the pulse count method, step-up gearing causes an improvement in the resolution.

For the pulse timing method,

$$\Delta\omega_t = \frac{2\pi f}{pNm} - \frac{2\pi f}{pN(m+1)} = \frac{2\pi f}{pNm(m+1)} \cong \frac{pN}{2\pi f}\omega^2 \tag{6.70}$$

Note that in the pulse time approach, for a given speed, the resolution degrades with increasing p.

In summary, the speed resolution of an incremental encoder depends on the following factors:

    1. Number of windows N

    2. Count reading (sampling) period T

3. Clock frequency f

4. Speed ω

5. Gear ratio

In particular, gearing up has a detrimental effect on the speed resolution in the pulse-timing method, but it has a favorable effect in the pulse-counting method.

6.13.3.11 Data Acquisition Hardware

A method for interfacing an incremental encoder to a digital processor (digital controller) is shown schematically in Figure 6.51. In practice, a suitable interface card (e.g., servo card, encoder card, etc.) in the control computer will possess the necessary functional capabilities indicated in the figure. The pulse signals from the encoder are fed into an up/down counter, which has circuitry to detect pulses (for example, by rising-edge detection, falling-edge detection, or level detection) and logic circuitry to determine the direction of motion (i.e., sign of the reading) and to code the count. A pulse in one direction (say, cw) will increment the count by one (an upcount), and a pulse in the opposite direction will decrement the count by one (a downcount). The coded count may be directly read by the host computer through its I/O board without the need for an ADC. The count is transferred to a latch buffer so that the measurement is read from the buffer rather than from the counter itself. This arrangement provides an efficient means of data acquisition because the counting process can continue without interruption while the computer reads the count from the latch buffer. The digital processor (computer) identifies various components in the measurement system using addresses, and this information is communicated to the individual components through the address bus. The start, end, and nature of an action (e.g., data read, clear the counter, clear the buffer) are communicated to various devices by the computer through its control bus. The computer can command an action to a component in one direction of the bus, and the component can respond with a message (e.g., job completed) in the opposite direction. The data (e.g., the count) are transmitted through the data bus. While the computer reads (samples) data from the buffer, the control

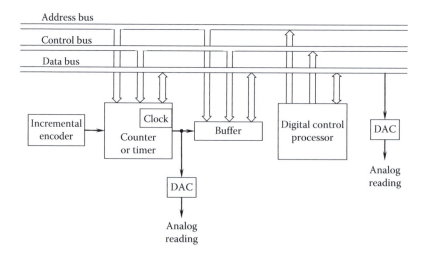

FIGURE 6.51
Computer interface for an incremental encoder.

signals guarantee that no data are transferred to that buffer from the counter. It is clear that the data acquisition consists of handshake operations between the main processor of the computer and auxiliary components. More than one encoder may be addressed, controlled, and read by the same three buses of the computer. The buses are conductors; for example, multicore cables carrying signals in parallel logic. Communication in serial logic is also common but is slower.

An incremental optical encoder generates two pulse signals—one 1/4 of a pitch out of phase with the other. The internal electronics of the encoder may be powered by a 5 V dc supply. The two pulse signals determine the direction of rotation of the motor by one of various means (e.g., sign of the phase difference, timing of the consecutive rising edges). The encoder pulse count is stored in a buffer within the controller and is read at fixed intervals (say, 5 ms). The net count gives the joint position and the difference in count at a fixed time increment gives joint speed.

While measuring a displacement (position) of an object using an incremental encoder, the counter may be continuously monitored as an analog signal through a digital-to-analog converter (DAC; see Figure 6.51). On the other hand, the pulse count is read by the computer only at finite time intervals. Since a cumulative count is required in displacement measurement, the buffer is not cleared in this case once the count is read in by the computer.

In velocity measurement by the pulse-counting method, the buffer is read at fixed time intervals of T, which is also the counting-cycle time. The counter is cleared every time a count is transferred to the buffer so that a new count can begin. With this method, a new reading is available at every sampling instant.

In the pulse-timing method of velocity computation, the counter is actually a timer. The encoder cycle is timed using a clock (internal or external) and the count is passed on to the buffer. The counter is then cleared and the next timing cycle is started. The buffer is periodically read by the computer. With this method, a new reading is available at every encoder cycle. Note that under transient velocities, the encoder-cycle time is variable and is not directly related to the data sampling period. In the pulse-timing method, it is desirable to make the sampling period slightly smaller than the encoder-cycle time, so that no count is missed by the processor.

More efficient use of the digital processor may be achieved by using an interrupt routine. With this method, the counter (or buffer) sends an interrupt request to the processor when a new count is ready. The processor then temporarily suspends the current operation and reads in the new data. Note that in this case the processor does not continuously wait for a reading.

6.13.4 Absolute Optical Encoders

An absolute encoder directly generates a coded digital word to represent each discrete angular position (sector) of its code disk. This is accomplished by producing a set of pulse signals (data channels) equal in number to the word size (number of bits) of the reading. Unlike with an incremental encoder, no pulse counting is involved. An absolute encoder may use various techniques (e.g., optical, sliding contact, magnetic saturation, proximity sensor) to generate the sensor signal, as for an incremental encoder. The optical method, which uses a code disk with transparent and opaque regions and pairs of light sources and photosensors, is the most common technique.

A simplified code pattern on the disk of an absolute encoder that utilizes the direct binary code is shown in Figure 6.52a. The number of tracks (n) in this case is 4, but in practice n is

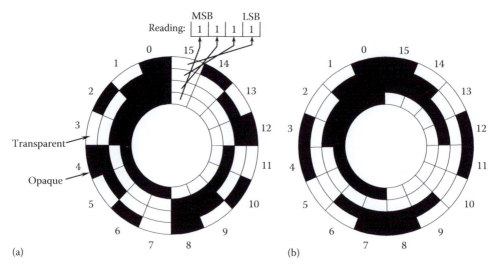

FIGURE 6.52
Illustration of the code pattern of an absolute encoder disk: (a) Binary code; (b) a gray code.

on the order of 14, but may be as high as 22. The disk is divided into 2^n sectors. Each partitioned area of the matrix thus formed corresponds to a bit of data. For example, a transparent area will correspond to binary 1 and an opaque area will correspond to binary 0. Each track has a pick-off sensor similar to what is used in incremental encoders. The set of n pick-off sensors is arranged along a radial line and is facing the tracks on one side of the disk. A light source (e.g., LED) illuminates the other side of the disk. As the disk rotates, the bank of pick-off sensors generates pulse signals that are sent to n parallel data channels (or pins). At a given instant, the particular combination of signal levels in the data channels will provide a coded data word that uniquely determines the position of the disk at that time.

6.13.4.1 Gray Coding

There is a data interpretation problem associated with the straight binary code in absolute encoders. Notice in Table 6.8 that with the straight binary code, the transition from one sector to an adjacent sector may require more than one switching of bits in the binary data. For example, the transition from 0011 to 0100 or from 1011 to 1100 requires three bit switchings, and the transition from 0111 to 1000 or from 1111 to 0000 requires four bit switchings. If the pick-off sensors are not properly aligned along a radius of the encoder disk, or if the manufacturing error tolerances for imprinting the code pattern on the disk were high, or if environmental effects have resulted in large irregularities in the sector matrix, then bit switching from one reading to the next will not take place simultaneously. This results in ambiguous readings during the transition period. For example, in changing from 0011 to 0100, if the LSB switches first, the reading becomes 0010. In decimal form, this incorrectly indicates that the rotation was from angle 3 to angle 2, whereas, it was actually a rotation from angle 3 to angle 4. Such ambiguities in data interpretation can be avoided by using a gray code, as shown in Figure 6.52b for this example. The coded representation of the sectors is given in Table 6.8. Note that in the case of gray code, each adjacent transition involves only one bit switching.

TABLE 6.8

Sector Coding for a 4 Bit Absolute Encoder

Sector Number	Straight Binary Code (MSB→LSB)	A Gray Code (MSB→LSB)
0	0 0 0 0	0 0 0 0
1	0 0 0 1	0 0 0 1
2	0 0 1 0	0 0 1 1
3	0 0 1 1	0 0 1 0
4	0 1 0 0	0 1 1 0
5	0 1 0 1	0 1 1 1
6	0 1 1 0	0 1 0 1
7	0 1 1 1	0 1 0 0
8	1 0 0 0	1 1 0 0
9	1 0 0 1	1 1 0 1
10	1 0 1 0	1 1 1 1
11	1 0 1 1	1 1 1 0
12	1 1 0 0	1 0 1 0
13	1 1 0 1	1 0 1 1
14	1 1 1 0	1 0 0 1
15	1 1 1 1	1 0 0 0

For an absolute encoder, a gray code is not essential for removing the ambiguity in bit switchings of binary code. For example, for a given absolute reading, the two adjacent absolute readings are automatically known. A reading can be checked against these two valid possibilities (or a single possibility if the direction of rotation is known) to see whether the reading is correct. Another approach is to introduce a delay (e.g., Schmitt trigger) to reading the output. In this manner, a reading will be taken only after all the bit switchings have taken place, thereby eliminating the possibility of an ambiguous reading.

6.13.4.2 Code Conversion Logic

A disadvantage of utilizing a gray code is that it requires additional logic to convert the gray-coded number to the corresponding binary number. This logic may be provided in hardware or software. In particular, an "Exclusive-Or" gate can implement the necessary logic, as given by

$$B_{n-1} = G_{n-1}$$

$$B_{k-1} = B_k \oplus G_{k-1} \quad k = n-1, \dots, 1 \tag{6.71}$$

This converts an n bit gray-coded word $[G_{n-1}G_{n-2}\dots G_0]$ into an n bit binary coded word $[B_{n-1}B_{n-2}\dots B_0]$ where the subscript $n-1$ denotes the most significant bit (MSB) and 0 denotes the LSB. For small word sizes, the code may be given as a look-up table (see Table 6.8). Note that the gray code is not unique. Other gray codes, which provide single bit switching between adjacent numbers can be developed.

6.13.4.3 Advantages and Drawbacks

The main advantage of an absolute encoder is its ability to provide absolute angle readings (with a full 360° rotation). Hence, if a reading is missed, it will not affect the next reading. Specifically, the digital output uniquely corresponds to a physical rotation of the code disk, and hence a particular reading is not dependent on the accuracy of a previous reading. This provides immunity to data failure. A missed pulse (or a data failure of some sort) in an incremental encoder would carry an error into the subsequent readings until the counter is cleared.

An incremental encoder has to be powered throughout the operation of the device. Thus, a power failure can introduce an error unless the reading is reinitialized (or calibrated). An absolute encoder has the advantage that it needs to be powered and monitored only when a reading is taken.

Because the code matrix on the disk is more complex in an absolute encoder, and because more light sensors are required, an absolute encoder can be nearly twice as expensive as an incremental encoder. Also, since the resolution depends on the number of tracks present, it is more costly to obtain finer resolutions. An absolute encoder does not require digital counters and buffers, however, unless resolution enhancement is done using an auxiliary track or pulse-timing is used for velocity calculation.

6.13.5 Encoder Error

Errors in shaft encoder readings can come from several factors. The primary sources of these errors are as follows:

1. Quantization error (due to digital word size limitations)
2. Assembly error (eccentricity of rotation, etc.)
3. Coupling error (gear backlash, belt slippage, loose fit, etc.)
4. Structural limitations (disk deformation and shaft deformation due to loading)
5. Manufacturing tolerances (errors from inaccurately imprinted code patterns, inexact positioning of the pick-off sensors, limitations and irregularities in signal generation, and sensing hardware, etc.)
6. Ambient effects (vibration, temperature, light noise, humidity, dirt, smoke, etc.)

These factors can result in inexact readings of displacement and velocity and erroneous detection of the direction of motion.

One form of error in an encoder reading is the hysteresis. For a given position of the moving object, if the encoder reading depends on the direction of motion, the measurement has a hysteresis error. In that case, if the object rotates from position A to position B and back to position A, for example, the initial and the final readings of the encoder will not match. The causes of hysteresis include backlash in gear couplings, loose fits, mechanical deformation in the code disk and shaft, delays in electronic circuitry and components (electrical time constants, nonlinearities, etc.), and noisy pulse signals that make the detection of pulses (say, by level detection or edge detection) less accurate.

The raw pulse signal from an optical encoder is somewhat irregular and does not consist of perfect pulses, primarily because of the variation (somewhat triangular) of the intensity of light received by the optical sensor as the code disk moves through a window, and because of noise in the signal generation circuitry, including the noise created by imperfect light sources and photosensors. Noisy pulses have imperfect edges. As a result, pulse detection through edge detection can result in errors such as multiple triggering for the same edge of a pulse. This can be avoided by including a Schmitt trigger (a logic circuit with electronic hysteresis) in the edge-detection circuit so that slight irregularities in the pulse edges will not cause erroneous triggering, provided that the noise level is within the hysteresis band of the trigger. A disadvantage of this method, however, is that hysteresis will be present even when the encoder itself is perfect. Virtually noise-free pulses can be generated if two photosensors are used to detect adjacent transparent and opaque areas on a track simultaneously and a separate circuit (a comparator) is used to create a pulse that depends on the sign of the voltage difference of the two sensor signals. This method of pulse shaping has been described earlier, with reference to Figure 6.50.

6.13.5.1 Eccentricity Error

The eccentricity (denoted by e) of an encoder is defined as the distance between the center of rotation C of the code disk and the geometric center G of the circular code track. Nonzero eccentricity causes a measurement error known as the *eccentricity error*. The primary contributions to eccentricity are the following:

1. Shaft eccentricity (e_s)
2. Assembly eccentricity (e_t)
3. Track eccentricity (e_1)
4. Radial play (e_p)

Shaft eccentricity results if the rotating shaft on which the code disk is mounted is imperfect, so that its axis of rotation does not coincide with its geometric axis. Assembly eccentricity is caused if the code disk is improperly mounted on the shaft, so that the center of the code disk does not fall on the shaft axis. Track eccentricity comes from irregularities in the imprinting process of the code track, so that the center of the track circle does not coincide with the nominal geometric center of the disk. Radial play is caused by any looseness in the assembly in the radial direction. All four of these parameters are random variables. Let their mean values be μ_s, μ_a, μ_t, and μ_p, and let the standard deviations be σ_s, σ_a, σ_t, and σ_p, respectively. A very conservative upper bound for the mean value of the overall eccentricity is given by the sum of the individual mean values, each value being considered positive. A more reasonable estimate is provided by the rms value, as given by

$$\mu = \sqrt{\mu_s^2 + \mu_a^2 + \mu_t^2 + \mu_p^2}.$$

Furthermore, assuming that the individual eccentricities are independent random variables, the standard deviation of the overall eccentricity is given by $\sigma = \sqrt{\sigma_s^2 + \sigma_a^2 + \sigma_t^2 + \sigma_p^2}$.

Knowing the mean value μ and the standard deviation σ of the overall eccentricity, it is possible to obtain a reasonable estimate for the maximum eccentricity that can occur. *Note:* It is reasonable to assume that the eccentricity has a Gaussian (or normal) distribution.

The eccentricity of an incremental encoder also affects the phase angle between the quadrature signals, if a single track and two pick-off sensors (with circumferential offset) are used. This error can be reduced using the two-track arrangement, with the two sensors positioned along a radial line so that eccentricity will equally affect the two outputs.

6.14 Miscellaneous Digital Transducers

Now several other types of digital transducers that are useful in mechatronics are described. In particular, digital rectilinear transducers are described. Typical applications include x–y positioning tables, machine tools, valve actuators, read–write heads in disk drive systems, robotic manipulators (e.g., at prismatic joints), and robot hands. The principles used in angular motion transducers described in this chapter can be used in measuring rectilinear motions as well. Techniques of signal acquisition, interpretation, conditioning, etc. may find similarities in the devices described below with those presented thus far.

6.14.1 Digital Resolvers

Digital resolvers, or mutual induction encoders, operate somewhat like analog resolvers using the principle of mutual induction. They are known commercially as *Inductosyns*. A digital resolver has two disks facing each other (but not in contact), one (the stator) stationary and the other (the rotor) coupled to the rotating object whose motion is measured. The rotor has a fine electric conductor foil imprinted on it, as is shown schematically in Figure 6.53. The printed pattern is "pulse" shaped, closely spaced, and connected to a high-frequency ac supply (carrier) of voltage v_{ref}. The stator disk has two separate printed patterns that are identical to the rotor pattern, but one pattern on the stator is shifted by a quarter-pitch from the other pattern. The primary voltage in the rotor circuit induces voltages in the two secondary (stator) foils at the same frequency, i.e., the rotor and the stator are *inductively coupled*. These induced voltages are "quadrature" signals. As the rotor turns, the level of the induced voltage changes, depending on the relative position of the foil patterns on the two disks. When the foil pulse patterns coincide, the induced voltage is a maximum (positive or negative), and when the rotor foil pattern has a half-pitch offset from the stator foil pattern, the induced voltage in the adjacent segments cancel each other, producing a zero output. The output (induced) voltages v_1 and v_2 in the two foils

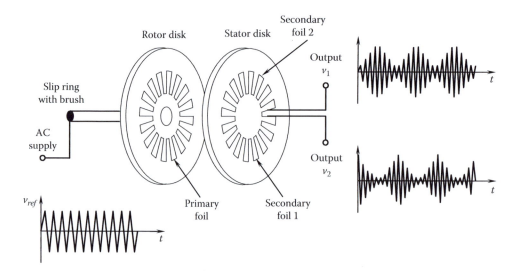

FIGURE 6.53
Schematic diagram of a digital resolver.

of the stator have a carrier component at the supply frequency and a modulating component corresponding to the rotation of the disk. The latter (modulating component) can be extracted through demodulation and converted into a proper pulse signal using pulse-shaping circuitry, as for an incremental encoder. When the rotating speed is constant, the two modulating components are periodic and nearly sinusoidal with a phase shift of 90° (i.e., in quadrature). When the speed is not constant, the pulse width will vary with time.

Angular displacement and angular velocity are determined as in the case of an incremental encoder by counting the pulses. Very fine resolutions (e.g., 0.0005°) may be obtained from a digital resolver, and it is usually not necessary to use step-up gearing or other techniques to improve the resolution. These transducers are usually more expensive than optical encoders. The use of a slip ring and brush to supply the carrier signal may be viewed as a disadvantage.

6.14.2 Digital Tachometers

A pulse-generating transducer whose pulse train is synchronized with a mechanical motion may be treated as a digital transducer for motion measurement. Pulse counting may be used for displacement measurement and the pulse rate (or pulse timing) may be used for velocity measurement. According to this terminology, a shaft encoder (particularly, an incremental optical encoder) may be considered as a digital tachometer. According to the popular terminology, however, a digital tachometer is a device that employs a toothed wheel to measure angular velocities.

A schematic diagram of a digital tachometer is shown in Figure 6.54. This is a magnetic induction, *pulse tachometer* of the variable-reluctance type. The teeth on the wheel are made of a ferromagnetic material. The two magnetic-induction (and variable-reluctance) proximity probes are placed radially facing the teeth, at quarter-pitch apart (pitch = tooth-to-tooth spacing). When the toothed wheel rotates, the two probes generate output signals that are 90° out of phase (i.e., quadrature signals). One signal leads the other in one direction of rotation and lags the other in the opposite direction of rotation. In this manner, a directional reading (i.e., velocity rather than speed) is obtained. The speed is computed as in the case of an incremental encoder.

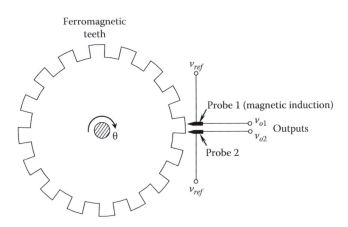

FIGURE 6.54
Schematic diagram of a pulse tachometer.

Alternate types of digital tachometers use eddy current proximity probes or capacitive proximity probes. In the case of an eddy current tachometer, the teeth of the pulsing wheel are made of (or plated with) electricity-conducting material. The probe consists of an active coil connected to an ac bridge circuit excited by a radio-frequency (i.e., in the range 1–100 MHz) signal. The resulting magnetic field (at radio frequency) is modulated by the tooth-passing action. The bridge output may be demodulated and shaped to generate the pulse signal.

In the case of a capacitive tachometer, the toothed wheel forms one plate of the capacitor; the other plate is the probe and is kept stationary. As the wheel turns, the gap width of the capacitor fluctuates. If the capacitor is excited by an ac voltage of high frequency (typically 1 MHz), a nearly pulse-modulated signal at that carrier frequency is obtained. This can be detected through a bridge circuit as before (but using a capacitance bridge rather than an inductance bridge).

The advantages of digital (pulse) tachometers over optical encoders include simplicity, robustness, immunity to environmental effects, other common fouling mechanisms (except magnetic effects), and low cost. Both are noncontacting devices. The disadvantages of a pulse tachometer include poor resolution (determined by the number of teeth and size [bigger and heavier than optical encoders]) and mechanical errors due to loading, hysteresis (i.e., output is not symmetric and depends on the direction of motion), and manufacturing irregularities. Mechanical loading will not be a factor if the toothed wheel already exists as an integral part of the original system that is sensed. The resolution (digital resolution) depends on the word size used for data acquisition.

6.14.3 Hall-Effect Sensors

Consider a semiconductor element subject to a dc voltage v_{ref}. If a magnetic field is applied perpendicular to the direction of this voltage, a voltage v_o will be generated in the third orthogonal direction within the semiconductor element. This is known as the Hall effect (observed by E. H. Hall in 1879). A schematic representation of a Hall-effect sensor is shown in Figure 6.55.

A Hall-effect sensor may be used for motion sensing in many ways—for example, as an analog proximity sensor, a limit switch (digital), or a shaft encoder. Since the output voltage v_o increases as the distance from the magnetic source to the semiconductor element decreases, the output signal v_o can be used as a measure of proximity. This is the principle behind an analog proximity sensor. Alternatively, a certain threshold level of the output voltage v_o can be used to generate a binary output, which represents the presence/absence of an object. This principle is used in a digital limit switch. The use of a toothed ferromagnetic wheel (as for a digital tachometer) to alter the magnetic flux will result in a shaft encoder. The sensitivity of a practical sensor element is of the order of 10 V/T. For a Hall-effect device, the temperature coefficient of resistance is positive and the temperature coefficient of sensitivity is negative. In view of these properties, auto-compensation for temperature may be achieved, as for a semiconductor strain gage.

The longitudinal arrangement of a proximity sensor, in which the moving element approaches head-on toward the sensor, is not suitable when there is a danger of overshooting the target, since it will damage the sensor. A more desirable configuration is the lateral arrangement in which the moving member slides by the sensing face of the sensor. The sensitivity will be lower, however, with this lateral arrangement. The relationship between the output voltage v_o and the distance x of a Hall-effect sensor, measured from the moving member, is nonlinear. Linear Hall-effect sensors use calibration to linearize their output.

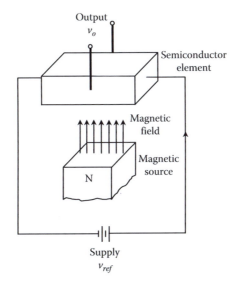

FIGURE 6.55
Schematic representation of a Hall-effect sensor.

A practical arrangement for a motion sensor based on the Hall effect would be to have the semiconductor element and the magnetic source fixed relative to one another in a single package. As a ferromagnetic member is moved into the air gap between the magnetic source and the semiconductor element, the flux linkage is varied. The output voltage v_o is changed accordingly. This arrangement is suitable for both an analog proximity sensor and a limit switch. By using a toothed ferromagnetic wheel as in Figure 6.56 to change v_o, and then by shaping the resulting signal, it is possible to generate a pulse train in proportion to the wheel rotation. This provides a shaft encoder or a digital tachometer. Apart from the familiar applications of motion sensing, Hall-effect sensors are used for *electronic commutation* of brushless dc motors (see Chapter 7) where the field circuit of the motor is appropriately switched depending on the angular position of the rotor with respect to the stator.

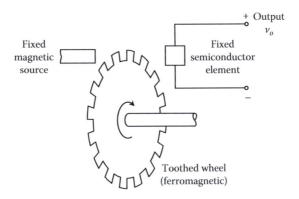

FIGURE 6.56
A Hall effect shaft encoder or digital tachometer.

Hall effect motion transducers are rugged devices and have many advantages. They are not affected by "rate effects" (specifically, the generated voltage is not affected by the rate of change of the magnetic field). Also, their performance is not severely affected by common environmental factors, except magnetic fields. They are noncontacting sensors with associated advantages as mentioned before. Some hysteresis will be present, but it is not a serious drawback in digital transducers. Miniature Hall-effect devices (mm scale) are available.

6.14.4 Linear Encoders

In rectilinear encoders (popularly called linear encoders, where "linear" does not imply linearity but refers to rectilinear motion), rectangular flat plates moving rectilinearly, instead of rotating disks, are used with the same types of signal generation and interpretation mechanisms as for shaft (rotary) encoders. A transparent plate with a series of opaque lines arranged in parallel in the transverse direction forms the stationary plate (grating plate or phase plate) of the transducer. This is called the mask plate. A second transparent plate, with an identical set of ruled lines, forms the moving plate (or the code plate). The lines on both plates are evenly spaced, and the line width is equal to the spacing between adjacent lines. A light source is placed on the moving plate side, and the light transmitted through the common area of the two plates is detected on the other side using one or more photosensors. When the lines on the two plates coincide, the maximum amount of light will pass through the common area of the two plates. When the lines on one plate fall on the transparent spaces of the other plate, virtually no light will pass through the plates. Accordingly, as one plate moves relative to the other, a pulse train is generated by the photosensor and it can be used to determine rectilinear displacement and velocity, as in the case of an incremental encoder.

A suitable arrangement is shown in Figure 6.57. The code plate is attached to the moving object whose rectilinear motion is to be measured. An LED light source and a phototransistor light sensor are used to detect the motion pulses, which can be interpreted just like in the case of a rotary encoder. The phase plate is used, as with a shaft encoder, to enhance the intensity and the discrimination of the detected signal. Two tracks of windows in quadrature (i.e., 1/4 pitch offset) would be needed to determine the direction of motion, as shown in Figure 6.57. Another track of windows at 1/2 pitch offset with the main track

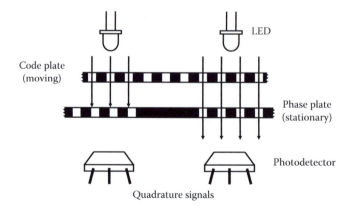

FIGURE 6.57
A rectilinear optical encoder.

(not shown in Figure 6.57) may be used as well on the phase plate to further enhance the discrimination of the detected pulses. Specifically, when the sensor at the main track reads a high intensity (i.e., when the windows on the code plate and the phase plate are aligned) the sensor at the track that is 1/2 pitch away will read a low intensity (because the corresponding windows of the phase plate are blocked by the solid regions of the code plate).

6.14.5 Moiré Fringe Displacement Sensors

Suppose that a piece of transparent fabric is placed on another. If one piece is moved or deformed with respect to the other, we will notice various designs of light and dark patterns (lines) in motion. Dark lines of this type are called Moiré fringes. In fact, the French term Moiré refers to a silk-like fabric, which produces Moiré fringe patterns. An example of a Moiré fringe pattern is shown in Figure 6.58. Consider the rectilinear encoder, which was described above. When a window slits of one plate overlap with the window slits of the other plate, we get an alternating light and dark pattern. This is a special case of Moiré fringes. A Moiré device of this type may be used to measure rigid-body movements of one plate of the sensor with respect to the other.

The application of the Moiré fringe technique is not limited to sensing rectilinear motions. This technology can be used to sense angular motions (rotations) and more generally, distributed deformations (e.g., elastic deformations) of one plate with respect to the other. Consider two plates with gratings (optical lines) of identical pitch (spacing) p. Suppose that initially the gratings of the two plates exactly coincide. Now if one plate is deformed in the direction of the grating lines, the transmission of light through the two plates will not be altered. But, if a plate is deformed in the perpendicular direction to the grating lines, then the window width of that plate will be deformed accordingly. In this case, depending on the nature of the plate deformation, some transparent lines of one plate will be completely covered by the opaque lines of the other plate and some other transparent lines of the first plate will have coinciding transparent lines on the second plate. Thus, the observed image will have dark lines (Moiré fringes) corresponding to the regions with

FIGURE 6.58
A Moiré fringe pattern.

clear/opaque overlaps of the two plates and bright lines corresponding to the regions with clear/clear overlaps of the two plates. The resulting Moiré fringe pattern will provide the deformation pattern of one plate with respect to the other. Such 2D fringe patterns can be detected and observed by arrays of optical sensors (e.g., using a CCD) and by photographic means. In particular, since the "presence" of a fringe is binary information, binary optical sensing techniques (as for optical encoders) and digital imaging techniques may be used with these transducers. Accordingly, these devices may be classified as digital transducers. With the Moiré fringe technique, very small resolutions (e.g., 0.002 mm) can be realized because finer line spacing (in conjunction with wider light sensors) can be used.

To further understand and analyze the fundamentals of Moiré fringe technology, consider two grating plates with an identical line pitch (spacing between the windows) p. Suppose that one plate is kept stationary. This is the plate of master gratings (or reference gratings or main gratings). The other plate (which is the plate containing index gratings or model gratings) is placed over the fixed plated and rotated so that the index gratings form an angle α with the master gratings, as shown in Figure 6.59. The lines shown are in fact the opaque regions, which are identical in size and spacing to the windows in between the opaque regions. A uniform light source is placed on one side of the overlapping pair of plates and the light transmitted through them is observed on the other side. Dark bands called Moiré fringes are seen as a result, as is shown in Figure 6.59.

A Moiré fringe corresponds to the line joining a series of points of intersection of the opaque lines of the two plates, because no light can pass through such points. This is further shown in Figure 6.60. Note that in the present arrangement, the line pitch of the two plates is identical and equal to p. A fringe line formed is shown as the broken line in Figure 6.60. Since the line pattern in the two plates is identical, by symmetry of the arrangement,

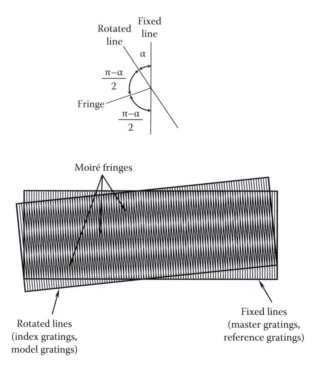

FIGURE 6.59
Formation of Moiré fringes.

the fringe line should bisect the obtuse angle $(\pi - \alpha)$ formed by the intersecting opaque lines. In other words, a fringe line makes an angle of $(\pi-\alpha)/2$ with the fixed gratings. Furthermore, the vertical separation (or the separation in the direction of the fixed gratings) of the Moiré fringes is seen to be $p/\tan\alpha$.

In summary then, the rotation of the index plate with respect to the reference plate can be measured by sensing the orientation of the fringe lines with respect to the fixed (master or reference) gratings. Furthermore, the period of the fringe lines in the direction of the reference gratings is $p/\tan\alpha$ and when the index plate is moved rectilinearly by a distance of one grating pitch, the fringes also shift vertically by its period of $p/\tan\alpha$ (see Figure 6.60). It is clear then that the rectilinear displacement of the index plate can be measured by sensing the fringe spacing. In a 2D pattern of Moiré fringes, these facts can be used as local information in order to sense full-field motions and deformations.

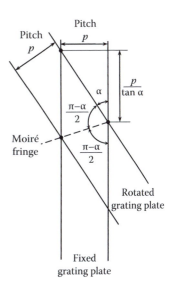

FIGURE 6.60
The orientation of Moiré fringes.

Example 6.9

Suppose that each plate of a Moiré fringe deformation sensor has a line pitch of 0.01 mm. A tensile load is applied to one plate in the direction perpendicular to the lines. Five Moiré fringes are observed in 10 cm of the Moiré image under tension. What is the tensile strain in the plate?

Solution
There is one Moiré fringe in every $10/5 = 2$ cm of the plate. Hence, extension of a 2 cm portion of the plate $= 0.01$ mm, and

$$\text{tensile strain} = \frac{0.01 \text{ mm}}{2 \times 10 \text{ mm}} = 0.0005 \, \varepsilon = 500 \, \mu\varepsilon$$

In this example, we have assumed that the strain distribution (or deformation) of the plate is uniform. Under nonuniform strain distributions, the observed Moiré fringe pattern generally will not be parallel straight lines but rather complex shapes.

6.14.6 Binary Transducers

Digital binary transducers are two-state sensors. The information provided by such a device takes only two states (on/off, present/absent, go/no-go, etc.); it can be represented by one bit. For example, limit switches are sensors used for detecting whether an object has reached its limit (or destination) of mechanical motion and are useful in sensing presence/absence and in object counting. In this sense, a limit switch is considered a digital transducer. Additional logic is needed if the direction of contact is also needed. Limit switches are available for both rectilinear and angular motions. A limit of a movement can be detected by mechanical means using a simple contact mechanism to close a circuit or trigger a pulse. Although a purely mechanical device consisting of linkages, gears, ratchet wheels, pawls, and so forth can serve as a limit switch, electronic and solid-state switches are usually preferred for such reasons as accuracy, durability, a low activating force (practically zero) requirement, low cost, and small size. Any proximity sensor may serve as the sensing element of a limit switch to detect the presence of an object. The proximity sensor signal is then used in a desired manner, for example, to activate a counter, a mechanical switch, or a relay circuit or simply as an input to a digital controller. A microswitch is a solid-state switch that can be used as a limit switch. Microswitches are commonly used in counting operations, for example, to keep a count of completed products in a factory warehouse.

There are many types of binary transducers that are applicable in the detection and counting of objects. They include the following:

1. Electromechanical switches
2. Photoelectric devices
3. Magnetic (Hall-effect, eddy current) devices
4. Capacitive devices
5. Ultrasonic devices

An electromechanical switch is a mechanically activated electric switch. The contact with an arriving object turns on the switch, thereby completing a circuit and providing an electrical signal. This signal provides the "present" state of the object. When the object is removed, the contact is lost and the switch is turned off. This corresponds to the "absent" state.

In the other four types of binary transducers listed above, a signal (light beam, magnetic field, electric field, or ultrasonic wave) is generated by a source (emitter) and is received by a receiver. A passing object interrupts the signal. This event can be detected by usual means using the signal received at the receiver. In particular, the signal level, a rising edge, or a falling edge may be used to detect the event. The following three arrangements of the emitter–receiver pair are common:

1. Through (opposed) configuration
2. Reflective (reflex) configuration
3. Diffuse (proximity, interceptive) configuration

In the through configuration (Figure 6.61a), the receiver is placed directly facing the emitter. In the reflective configuration, the emitter–source pair is located in a single package. The emitted signal is reflected by a reflector, which is placed facing the emitter–receiver package (Figure 6.61b). In the diffuse configuration as well, the emitter–reflector pair is in a single package. In this case, a conventional proximity sensor can serve the purpose of detecting the presence of an object (Figure 6.61c) by using the signal diffused from

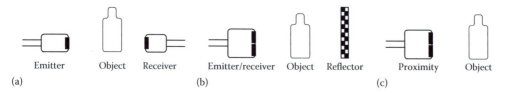

FIGURE 6.61
Two-state transducer configurations: (a) Through (opposed); (b) reflective (reflex); (c) interceptive (proximity).

the intercepting object. When the photoelectric method is used, an LED may serve as the emitter and a phototransistor may serve as the receiver. Infrared LEDs are preferred emitters for phototransistors because their peak spectral responses match.

Many factors govern the performance of a digital transducer for object detection. They include the following:

1. Sensing range (operating distance between the sensor and the object)
2. Response time
3. Sensitivity
4. Linearity
5. Size and shape of the object
6. Material of the object (e.g., color, reflectance, permeability, permittivity)
7. Orientation and alignment (optical axis, reflector, object)
8. Ambient conditions (light, dust, moisture, magnetic field, etc.)
9. Signal conditioning considerations (modulation, demodulation, shaping, etc.)
10. Reliability, robustness, and design life

Example 6.10

The response time of a binary transducer for object counting is the fastest (shortest) time the transducer needs to detect an absent-to-present condition or a present-to-absent condition and generate the counting signal (say, a pulse). Consider the counting process of packages on a conveyor. Suppose that, typically, packages of length 20 cm are placed on the conveyor at 15 cm spacing. A transducer of response time 10 ms is used for counting the packages. Estimate the allowable maximum operating speed of the conveyor.

Solution
If the conveyor speed is v cm/ms, then

$$\text{Package-present time} = \frac{20.0}{v} \text{ ms}$$

$$\text{Package-absent time} = \frac{15.0}{v} \text{ ms}$$

We must have a transducer response time of at least $15.0/v$ ms. Hence,

$$10.0 \leq \frac{15.0}{v}$$

or

$$v \leq 1.5 \text{ cm/ms}$$

The maximum allowable operating speed is 1.5 cm/ms or 15.0 m/s. This corresponds to a counting rate of $1.5/(20.0+15.0)$ packages/ms or about 43 packages/s.

6.14.7 Other Types of Sensors

There are many other types of sensors and transducers, which cannot be discussed here due to space limitation. But, the principles and techniques presented in this chapter may be extended to many of these devices. One area in which a great variety of sensors are used is *factory automation*. Here, in applications of automated manufacturing and robotics, it is important to use proper sensors for specific operations and needs. For example, chemical sensors, camera-based sensors and vision systems, and ultrasonic motion detectors may be used for product quality assessment, control, and human safety requirements. Motion and force, power-line, debris, sound, vibration, temperature, pressure, flow, and liquid-level sensing may be used in machine monitoring and diagnosis. Motion, force, torque, current, voltage, flow, and pressure sensing are important in machine control. Vision, motion, proximity, tactile, force, torque, pressure sensing, and dimensional gaging are useful in task monitoring and control.

Several areas can be identified where new developments and innovations are being made in sensor technology:

1. Microminiature (MEMS and nano) sensors: (IC-based with built-in signal processing)
2. Intelligent sensors: (built-in reasoning or information preprocessing to provide high-level knowledge and decision-making capability)
3. Integrated and distributed sensors: (sensors are integral with the components and agents, which communicate with each other in an overall multi-agent system)
4. Hierarchical sensory architectures: (low-level sensory information is preprocessed to match higher level requirements)

These four areas of activity are also representative of future trends in sensor technology development. In the concluding section, we will give an introduction to camera-based image sensors.

6.15 Image Sensors

An image of an object is indeed a valuable source of information about that object. In this context, the imaging device is the sensor and the image is the sensed data. Depending on the imaging device, an image can be many varieties such as optical, thermal or infrared, x-ray, ultraviolet, acoustic, ultrasound, and so on. Since the image processing methods are rather similar among these imaging devices, we will consider only the digital camera as a sensor. This is a very popular optical imaging device, which is used in a variety of mechatronic applications such as vision guided robotics.

6.15.1 Image Processing and Computer Vision

An image may be processed (analyzed) to obtain a more refined image from which useful information such as edges, contours, areas, and other geometrical information can be determined. This is called image processing. Computer vision goes beyond image processing and performs such operations as object recognition, pattern recognition and classification, abstraction, and knowledge-based decision making using information extracted through image processing. It follows that computer vision involves higher level operations than image processing and is akin to what humans infer based on what they see.

6.15.2 Image-Based Sensory System

A complete image-based sensory system consists of a camera (e.g., CCD camera), data acquisition system (e.g., frame grabber), a computer, and associated software. Such a system is schematically shown in Figure 6.62. Not included in the figure are other useful components such as a structured lighting source, which may be needed to capture good quality and clear images, without shadows and so on.

6.15.2.1 Camera

The camera has an array or matrix of semiconductor elements that are sensitive to the brightness of light coming from an object (through the camera lens). A CCD camera has a CCD as its sensing element. There other types of cameras whose sensing element may be a charge injection device (CID) or a complementary metal-oxide-semiconductor (CMOS) device. Most common and relatively inexpensive are the CCD cameras.

Suppose that a 2D beam of light coming from the sensed object is directed by the camera lens on to the CCD matrix (e.g., 4000×4000) located on the focal plane of the lens in the back of the camera. Each element generates a charge that is proportional to the brightness of the light. A printed circuit board in the camera reads these charge levels row by row (from the bottom to the top row, sequentially) through a row-shifting operation controlled and synchronized by a clock and other hardware. The analog signal from each CCD element is digitized and represented as a "picture element" or *pixel*. The number of bits in a pixel is representative of the number of gray levels it can store. For example, a 4 bit pixel can represent $2^4 = 16$ gray levels from 0 to 15 (black to white). This procedure of generating the pixels of a 2D image is represented in Figure 6.63.

6.15.2.2 Image Frame Acquisition

Image pixels from the buffer of the CCD camera are arranged into an image frame of digital data and are provided to the image processing computer. This data acquisition device (hardware) may be called a "frame grabber board," which may placed in the card cage of the image processing computer itself. The associated driver software is also located in the computer.

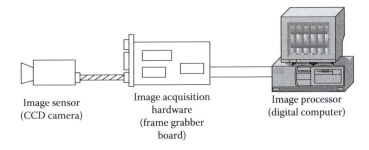

Image sensor
(CCD camera)

Image acquisition
hardware
(frame grabber
board)

Image processor
(digital computer)

FIGURE 6.62
A camera-based sensory system.

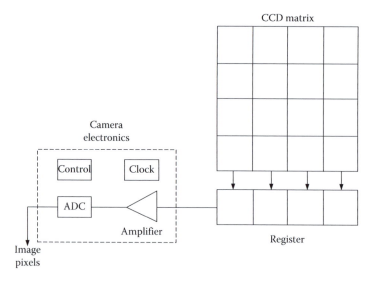

FIGURE 6.63
Digital image generation by a CCD camera.

6.15.2.3 Color Images

A gray-scale image can be represented by a single image frame. At least three image frames are needed to represent a color image. For example in the RGB model, a red (R) image, a green (G) image, and a blue (B) image are formed using red, green, and blue filters. The resulting three separate image frames can be combined for the original color image. Even though the human eye is quite sensitive to the colors R, G, and B, humans cannot typically perceive/describe visual images in terms of their RGB components. In terms of human perception/description of a visual image, a more appropriate model is the HIS model. In this model hue (H) is representative of the dominant color in the image, saturation (S) represents the degree of white light mixed with a dominant color in the image, and intensity (I) represents the level of brightness of the image. There are analytical relationships that convert an RGB model into an HIS model.

6.15.3 Image Processing

There is some element of "analog" image processing carried out by the electronics in the camera (e.g., analog filtering and amplification), the focus now is the "digital" image processing done by the computer. The objective of digital image processing is to remove unwanted elements and noise in the image, enhance the important features, and extract the needed geometric information from the processed data. Several useful operations of image processing are listed below.

1. Filtering (to remove noise and enhance the image) including directional filtering (to enhance edges, for edge detection)
2. Segmentation (to subdivide an enhanced image, identify geometric shapes/objects, and capture properties such as the area and dimensions of the identified geometric entities)
3. Thresholding (to generate a two-level black-and-white image where the gray levels above a set threshold are assigned white and those below the threshold are

assigned black. Thresholding is usually the step prior to the segmentation of an image)

4. Morphological processing (sequential shrinking, filtering, stretching, etc. to prune out unwanted image components and extract those that are important)

5. Subtraction (e.g., subtract the background from the image)

6. Template matching (to match a processed image to a template—useful in object detection)

7. Compression (to reduce the quantity of data that is needed to represent the useful information of an image)

6.15.4 Some Applications

The applications of image-based sensors are numerous. Some of these are discussed in this book under mechatronic applications and case studies. Several are listed below.

1. Measurement of a location of an object for cutting, grasping, manipulating, etc.

2. Measurement/estimation of size, shape, weight, color, texture, firmness, etc. for quality assessment or grading of a product.

3. Visual servoing. Here the actual position of an object is measured (using camera images) and is compared with the position of a robotic end effector (gripper, hand, tool, etc.). The difference (error) is used to generate a motion command for the robot so that the end effector would reach the object.

4. Object recognition in various applications of security, safety, and automated processing.

Problems

6.1 In each of the following examples, indicate at least one (unknown) input, which should be measured and used for feedforward control to improve the accuracy of the system.

(a) A servo system for positioning a mechanical load. The servo motor is a field-controlled dc motor with position feedback using a potentiometer and velocity feedback using a tachometer.

(b) An electric heating system for a pipeline carrying a liquid. The exit temperature of the liquid is measured using a thermocouple and is used to adjust the power of the heater.

(c) A room heating system. Room temperature is measured and compared with the set point. If it is low, a valve of a steam radiator is opened; if it is high, the valve is shut.

(d) An assembly robot, which grips a delicate part to pick it up without damaging the part.

(e) A welding robot, which tracks the seam of a part to be welded.

6.2 Giving examples, discuss situations in which the measurement of more than one type of kinematic variable using the same measuring device is

(a) An advantage

(b) A disadvantage

6.3 Write the expression for electrical-loading nonlinearity error (percentage) in a rotatory potentiometer in terms of the angular displacement, maximum displacement (stroke), potentiometer element resistance, and load resistance. Plot the percentage error as a function of the fractional displacement for the three cases $R_L/R_c = 0.1$, 1.0, and 10.0.

6.4 At the null position, the impedances of the two secondary winding segments of an LVDT were found to be equal in magnitude but slightly unequal in phase. Show that the quadrature error (null voltage) is about 90° out of phase with reference to the predominant component of the output signal under open-circuit conditions. *Hint*: This may be proved either analytically or graphically by considering the difference between two rotating directed lines (phasors) that are separated by a very small angle.

6.5 Standard rectilinear displacement sensors such as the LVDT and the potentiometer are used to measure displacements up to 25 cm; within this limit, accuracies as high as $\pm 0.2\%$ can be obtained. For measuring large displacements on the order of 3 m, cable extension displacement sensors, which have an angular displacement sensor as the basic sensing unit, may be used. One type of rectilinear displacement sensor has a rotatory potentiometer and a light cable, which wraps around a spool that rotates with the wiper arm of the pot. In using this sensor, the free end of the cable is connected to the moving member whose displacement is to be measured. The sensor housing is mounted on a stationary platform, such as the support structure of the system being monitored. A spring motor winds the cable back as the cable retracts. Using suitable sketches, describe the operation of this displacement sensor. Discuss the shortcomings of this device.

6.6 It is known that some of the factors that should be considered in selecting an LVDT for a particular application are linearity, sensitivity, response time, size and weight of the core, size of the housing, primary excitation frequency, output impedance, phase change between primary and secondary voltages, null voltage, stroke, and environmental effects (temperature compensation, magnetic shielding, etc.). Explain why and how each of these factors is an important consideration.

6.7 The signal-conditioning system for an LVDT has the following components: power supply, oscillator, synchronous demodulator, filter, and voltage amplifier. Using a block diagram, show how these components are connected to the LVDT. Describe the purpose of each component. A high-performance LVDT has a linearity rating of 0.01% in its output range of 0.1–1.0 V ac. The response time of the LVDT is known to be 10 ms. What should be the frequency of the primary excitation?

6.8 For directional sensing using an LVDT, it is necessary to determine the phase angle of the induced signal. In other words, *phase-sensitive demodulation* would be needed.

(a) First, consider a linear core displacement starting from a positive value, moving to zero, and then returning to the same position in an equal time period. Sketch the output of the LVDT for this "triangular" core displacement.

(b) Next, sketch the output if the core continued to move to the negative side at the same speed.

By comparing the two outputs, show that phase-sensitive demodulation would be needed to distinguish between the two cases of displacement.

6.9 Joint angles and angular speeds are the two basic measurements used in the direct (low-level) control of robotic manipulators. One type of robot arm uses resolvers to measure angles and differentiate these signals (digitally) to obtain angular speeds. A gear system is used to step up the measurement (typical gear ratio, 1:8). Since the gear wheels are ferromagnetic, an alternative measuring device would be a self-induction or mutual-induction proximity sensor located at a gear wheel. This arrangement, known as a pulse tachometer, generates a pulse (or near-sine) signal, which can be used to determine both angular displacement and angular speed. Discuss the advantages and disadvantages of these two arrangements (resolver and pulse tachometer) in this particular application.

6.10 Compare and contrast the principles of operation of a dc tachometer and an ac tachometer (both permanent-magnet and induction types). What are the advantages and disadvantages of these two types of tachometers?

6.11 Describe three different types of proximity sensors. In some applications, it may be required to sense only two-state positions (e.g., presence or absence, go or no-go). Proximity sensors can be used in such applications, and in that context they are termed proximity switches (or limit switches). For example, consider a parts-handling application in automated manufacturing in which a robot end effector grips a part and picks it up to move it from a conveyor to a machine tool. We can identify four separate steps in the gripping process. Explain how proximity switches can be used for sensing in each of these four tasks:

(a) Make sure that the part is at the expected location on the conveyor.

(b) Make sure that the gripper is open.

(c) Make sure that the end effector has moved to the correct location so that the part is in between the gripper fingers.

(d) Make sure that the part did not slip when the gripper was closed.

Note: A similar use of limit switches is found in lumber mills where tree logs are cut (bucked) into smaller logs, bark removed (debarked), cut into a square/rectangular log using a chip'n saw operation, and sawed into smaller dimensions (e.g., two by four cross sections) for marketing.

6.12 In some industrial processes, it is necessary to sense the condition of a system at one location and, depending on that condition, activate an operation at a location far from that location. For example, in a manufacturing environment, when the count of the finished parts exceeds some value, as sensed in the storage area, a milling machine could be shut down or started. A proximity switch could be used for sensing and a networked (e.g., Ethernet-based) control system could be used for process control. Since activation of the remote process usually requires a current that is larger than the rated load of a proximity switch, one would have to use a relay circuit, which is operated by the proximity switch. One such arrangement is shown in Figure P6.12. Note that the relay circuit can be used to operate a device such as a valve, a motor, a pump, or a heavy-duty switch. Discuss an application of the arrangement shown in Figure P6.12 in the food-packaging industry. A mutual-induction proximity sensor with the following ratings is used in this application:

Sensor diameter = 1 cm

Sensing distance (proximity) = 1 mm

Supply to primary winding = 110 ac at 60 Hz

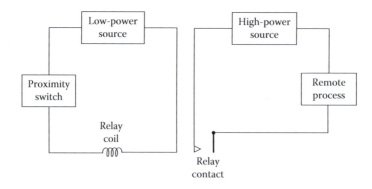

FIGURE P6.12
Proximity switch-operated relay circuit.

Load current rating (in secondary) = 200 mA

Discuss the limitations of this proximity sensor

6.13 Compression molding is used in making parts of complex shapes and varying sizes. Typically, the mold consists of two platens, the bottom platen being attached to the press table and the top platen operated by a hydraulic press. Metal or plastic sheets—for example, for the automotive industry—can be compression-molded in this manner. The main requirement in controlling the press is to position the top platen accurately with respect to the bottom platen (say, with a 0.001 in. or 0.025 mm tolerance), and it has to be done quickly (say, in a few seconds). How many degrees of freedom have to be sensed (how many position sensors are needed) in controlling the mold? Suggest typical displacement measurements that would be made in this application and the types of sensors that could be employed. Indicate sources of error that cannot be perfectly compensated for in this application.

6.14 Seam tracking in robotic arc welding needs accurate position control under dynamic conditions. The welding seam has to be accurately followed (tracked) by the welding torch. Typically, the position error should not exceed 0.2 mm. A proximity sensor could be used for sensing the gap between the welding torch and the welded part. It is necessary to install the sensor on the robot end effector so that it tracks the seam at some distance (typically 1 in. or 2.5 cm) ahead of the welding torch. Explain why this is important. If the speed of welding is not constant and the distance between the torch and the proximity sensor is fixed, what kind of compensation would be necessary in controlling the end effector position? The sensor sensitivity of several volts per millimeter is required in this position control application. What type of proximity sensor would you recommend?

6.15 An angular motion sensor, which operates somewhat like a conventional resolver, has been developed at Wright State University. The rotor of this resolver is a permanent magnet. A 2 cm diameter Alnico-2 disk magnet, diametrically magnetized as a two-pole rotor, has been used. Instead of the two sets of stationary windings placed at 90° in a conventional resolver, two Hall-effect sensors placed at 90° around the permanent-magnet rotor are used for detecting quadrature signals. Note that Hall-effect sensors can detect moving magnetic sources. Describe the operation of this modified resolver and explain how this device could be used to measure angular motions continuously. Compare this device with a conventional resolver, giving the advantages and disadvantages.

6.16 An active suspension system is proposed for a high-speed ground transit vehicle in order to achieve improved ride quality. The system senses jerk (rate of change of acceleration) due to road disturbances and adjusts system parameters accordingly.

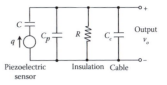

FIGURE P6.17
Equivalent circuit for a quartz crystal (piezoelectric) accelerometer.

(a) Draw a suitable schematic diagram for the proposed control system and describe the appropriate measuring devices.

(b) Suggest a way to specify the "desired" ride quality for a given type of vehicle. (Would you specify one value of jerk, a jerk range, or a jerk curve with respect to time or frequency?)

(c) Discuss the drawbacks and limitations of the proposed control system with respect to such factors as reliability, cost, feasibility, and accuracy.

6.17 A design objective in most mechatronic applications is to achieve small time constants. An exception is the time constant requirements for a piezoelectric sensor. Explain why a large time constant, on the order of 10 s, is desirable for a piezoelectric sensor in combination with its signal conditioning system.

An equivalent circuit for a piezoelectric accelerometer, which uses a quartz crystal as the sensing element, is shown in Figure P6.17. The charge generated is denoted by q, and the voltage output at the end of the accelerometer cable is v_o. The piezoelectric sensor capacitance is modeled by C_p, and the overall capacitance experienced at the sensor output, whose primary contribution is due to cable capacitance, is denoted by C_c. The resistance of the electric insulation in the accelerometer is denoted by R. Write a differential equation relating v_o to q. What is the corresponding transfer function? Using this result, show that the accuracy of the accelerometer improves when the sensor time constant is large and when the frequency of the measured acceleration is high. For a quartz crystal sensor with $R = 1 \times 10^{11}\,\Omega$ and $C_s = 300\,pF$ and a circuit with $C_c = 700\,pF$, compute the time constant.

6.18 Applications of accelerometers are found in the following areas:

(a) Transit vehicles (automobiles—microsensors for airbag sensing in particular, aircraft, ships, etc.), (b) power cable monitoring, (c) robotic manipulator control, (d) building structures, (e) shock and vibration testing, and (f) position and velocity sensing.

Describe one direct use of acceleration measurement in each application area.

6.19 A strain gage accelerometer uses a semiconductor strain gage mounted at the root of a cantilever element with the seismic mass mounted at the free end of the cantilever. Suppose that the cantilever element has a square cross section with a dimension of $1.5 \times 1.5\,mm^2$. The equivalent length of the cantilever element is 25 mm, and the equivalent seismic mass is 0.2 gm. If the cantilever is made of an aluminum alloy with Young's modulus $E = 69 \times 10^9\,N/m^2$, estimate the useful frequency range of the accelerometer in hertz. *Hint*: When force F is applied to the free end of a cantilever, the deflection y at that location may be approximated by the formula

$$y = \frac{Fl^3}{3EI}$$

where

    l is the cantilever length

    I is the second moment area of the cantilever cross section about the bending axis $= bh^3/12$

    b is the cross section width

    h is the cross section height

6.20 Applications of piezoelectric sensors are numerous; push-button devices and switches, airbag MEMS sensors in vehicles, pressure and force sensing, robotic tactile sensing, accelerometers, glide testing of computer disk-drive heads, excitation sensing in dynamic testing, respiration sensing in medical diagnostics, and graphics input devices for computers. Discuss the advantages and disadvantages of piezoelectric sensors.

    What is the cross sensitivity of a sensor? Indicate how the anisotropy of piezoelectric crystals (i.e., charge sensitivity quite large along one particular crystal axis) is useful in reducing cross-sensitivity problems in a piezoelectric sensor.

6.21 As a result of advances in microelectronics, piezoelectric sensors (such as accelerometers and impedance heads) are now available in miniature form with built-in charge amplifiers in a single integral package. When such units are employed, additional signal conditioning is usually not necessary. An external power supply unit is needed, however, to provide power for the amplifier circuitry. Discuss the advantages and disadvantages of a piezoelectric sensor with built-in microelectronics for signal conditioning.

    A piezoelectric accelerometer is connected to a charge amplifier. An equivalent circuit for this arrangement is shown in Figure 6.21.

(a) Obtain a differential equation for the output vo of the charge amplifier with acceleration a as the input, in terms of the following parameters: $S_a =$ the charge sensitivity of the accelerometer (charge/acceleration); $R_f =$ the feedback resistance of the charge amplifier; $\tau_c =$ the time constant of the system (charge amplifier).

(b) If an acceleration pulse of magnitude a_o and duration T is applied to the accelerometer, sketch the time response of the amplifier output v_o. Show how this response varies with τ_c. Using this result, show that the larger the τ_c the more accurate the measurement.

6.22 Give the typical values for the output impedance and the time constant of the following measuring devices:

(a) Potentiometer

(b) Differential transformer

(c) Resolver

(d) Piezoelectric accelerometer

    A RTD has an output impedance of $500\,\Omega$. If the loading error has to be maintained near 5%, estimate a suitable value for the load impedance.

6.23 A signature verification pen has been developed by IBM Corporation. The purpose of the pen is to authenticate the person who provides the signature by detecting whether the user is forging someone else's signature. The instrumented pen has analog sensors. Sensor signals are conditioned using microcircuitry built into the pen and sampled into a digital computer at the rate of 80 samples/s using an ADC. Typically, about 1000 data samples are collected per signature. Prior to the pen's use, authentic signatures are collected off-line and stored in a reference database. When a signature and the corresponding identification code are supplied to the computer for verification, a program in the processor retrieves the authentic signature from the database, by referring to the identification code, and then compares the two sets of data for authenticity. This process takes about 3 s. Discuss the types of sensors that could be used in the pen. Estimate the total time required for a signal verification. What are the advantages and disadvantages of this method in comparison to having the user punch in an identification code alone or provide the signature without the identification code?

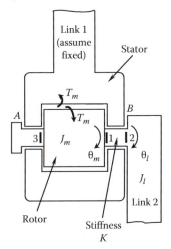

FIGURE P6.24
Torque sensing locations for a manipulator joint.

6.24 Consider the joint of a robotic manipulator, shown schematically in Figure P6.24. Torque sensors are mounted at locations 1, 2, and 3. If the electromagnetic torque generated at the motor rotor is T_m, write equations for the torque transmitted to link 2, the frictional torque at bearing A, the frictional torque at bearing B, and the reaction torque on link 1 in terms of the measured torques, the inertia torque of the rotor, and T_m.

6.25 A strain gage sensor to measure the torque T_m generated by a motor is shown schematically in Figure P6.25. The motor is floated on frictionless bearings. A uniform rectangular lever arm is rigidly attached to the motor housing and its projected end is restrained by a pin joint. Four identical strain gages are mounted on the lever arm, as shown. Three of the strain gages are at point A, which is located at a distance a from the motor shaft and the fourth strain gage is at point B, which is located at a distance $3a$ from the motor shaft. The pin joint is at a distance l from the motor shaft. Strain

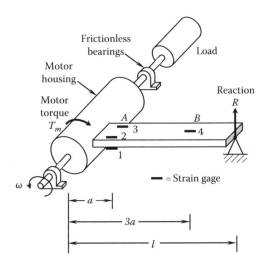

FIGURE P6.25
A strain gage sensor for measuring motor torque.

gages 2, 3, and 4 are on the top surface of the lever arm and gage 1 is on the bottom surface. Obtain an expression for T_m in terms of the bridge output δv_o and the following additional parameters:

S_s is the gage factor (strain gage sensitivity)
v_{ref} is the supply voltage to the bridge
b is the width of the lever arm cross section
h is the height of the lever arm cross section
E is the Young's modulus of the lever arm.

Verify that the bridge sensitivity does not depend on a and l. Describe the means to improve the bridge sensitivity. Explain why the sensor reading is only an approximation to the torque transmitted to the load. Give a relation to determine the net normal reaction force at the bearings, using the bridge output.

6.26 Discuss the advantages and disadvantages of the following techniques in the context of measuring transient signals.

(a) DC bridge circuits versus ac bridge circuits

(b) Slip ring and brush commutators versus ac transformer commutators

(c) Strain gage torque sensors versus variable-inductance torque sensors

(d) Piezoelectric accelerometers versus strain gage accelerometers

(e) Tachometer velocity transducers versus piezoelectric velocity transducers

6.27 Briefly describe how strain gages may be used to measure

(a) Force   (d) Pressure

(b) Displacement   (e) Temperature.

(c) Acceleration

Show that if a compensating resistance Rc is connected in series with the supply voltage vref to a strain gage bridge that has four identical members, each with resistance R, the output equation is given by

$$\frac{\delta v_o}{v_{ref}} = \frac{R}{(R+R_c)} \frac{kS_s}{4} \varepsilon$$

in the usual rotation.

A foil-gage load cell uses a simple (one-dimensional) tensile member to measure force. Suppose that k and S_s are insensitive to temperature change. If the temperature coefficient of R is α_1, that of the series compensating resistance R_c is α_2, and that of the Young's modulus of the tensile member is $(-\beta)$, determine an expression for R_c that would result in automatic (self) compensation for temperature effects. Under what conditions is this arrangement realizable?

6.28 The read–write head in a disk drive of a digital computer should float at a constant but small height (say, a fraction of a μm) above the disk surface. Because of the aerodynamics resulting from the surface roughness and the surface deformations of the disk, the head can be excited into vibrations that could cause head–disk contacts. These contacts, which are called head–disk interferences (HDIs), are clearly undesirable. They can occur at very high frequencies (say, 1 MHz). The purpose of a glide test is to detect HDIs and to determine the nature of these interferences. Glide testing can be used to determine the effect of parameters such as the flying height of the head and the speed of the disk and to qualify (certify the quality of) disk drive units. Indicate the basic instrumentation needed in glide testing. In particular, suggest the types of sensors that could be used and their advantages and disadvantages.

6.29 What are the typical requirements for an industrial tactile sensor? Explain how a tactile sensor differs from a simple touch sensor. Define the spatial resolution and force resolution (or sensitivity) of a tactile sensor.

The spatial resolution of your fingertip can be determined by a simple experiment using two pins and a helper. Close your eyes. Instruct the helper to apply one pin or both pins randomly to your fingertip so that you will feel the pressure of the tip of the pins. You should respond by telling the helper whether you feel both pins or just one pin. If you feel both pins, the helper should decrease the spacing of the two pins in the next round of tests. The test should be repeated in this manner by successively decreasing the spacing between the pins until you feel only one pin when both pins are actually applied. Then measure the distance between the two pins in millimeters. The largest spacing between the two pins that will result in this incorrect sensation corresponds to the spatial resolution of your fingertip. Repeat this experiment for all your fingers, repeating the test several times on each finger. Compute the average and the standard deviation. Then perform the test on other subjects. Discuss your results. Do you notice large variations in the results?

6.30 The *motion dexterity* of a device is defined as the ratio (the number of degrees of freedom in the device)/(the motion resolution of the device). The *force dexterity* may be defined as (the number of degrees of freedom in the device)/(the force resolution of the device). Given a situation where both types of dexterity mean the same thing and a situation where the two terms mean different things, outline how force dexterity of a device (say, an end effector) can be improved by using tactile sensors. Provide the dexterity requirements for the following tasks by indicating whether motion dexterity or force dexterity is preferred in each case:

(a) Gripping a hammer and driving a nail with it

(b) Threading a needle

(c) Seam tracking of a complex part in robotic arc welding

(d) Finishing the surface of a complex metal part using robotic grinding

6.31 Using the usual equation for a dc strain-gage bridge, show that if the resistance elements R_1 and R_2 have the same temperature coefficient of resistance and if R_3 and R_4 have the same temperature coefficient of resistance, the temperature effects are compensated for up to the first order.

A microminiature (MEMS) strain-gage accelerometer uses two semiconductor strain gages, one integral with the cantilever element near the fixed end (root) and the other mounted at an unstrained location of the accelerometer. The entire unit including the cantilever and the strain gages has a silicon IC construction and measures smaller than 1 mm in size. Outline the operation of the accelerometer. What is the purpose of the second strain gage?

6.32 A simple rate gyro, which may be used to measure angular speeds, is shown in Figure P6.32. The angular speed of spin is ω and is kept constant at a known value. The angle of the rotation of the gyro about the gimbal axis (or the angle of twist of the torsional spring) is θ and is measured using a displacement sensor. The angular speed of the gyro about the axis that is orthogonal to both gimbal axis and spin axis is Ω. This is the angular speed of the supporting structure (vehicle), which needs to be measured. Obtain a relationship between Ω and θ in terms of parameters such as the following:

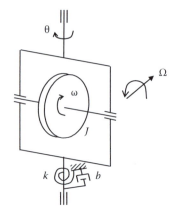

J is the moment of inertia of the spinning wheel
k is the torsional stiffness of the spring restraint at the gimbal bearings
b is the damping constant of rotational motion about the gimbal axis and the spinning speed

How would you improve the sensitivity of this device?

FIGURE P6.32
A rate gyro speed sensor.

6.33 Level sensors are used in a wide variety of applications, including soft drink bottling, food packaging, monitoring of storage vessels, mixing tanks, and pipelines. Consider the following types of level sensors, and briefly explain the principle of operation of each type in level sensing. Also, what are the limitations of each type?

(a) Capacitive sensors

(b) Inductive sensors

(c) Ultrasonic sensors

(d) Vibration sensors

6.34 Consider the following types of position sensors: inductive, capacitive, eddy current, fiber-optic, and ultrasonic. For the following conditions, indicate which of these types are not suitable and explain why:

(a) Environment with variable humidity

(b) Target object made of aluminum

(c) Target object made of steel

(d) Target object made of plastic

(e) Target object several feet away from the sensor location

(f) Environment with significant temperature fluctuations

(g) Smoke-filled environment

6.35 Discuss advantages and disadvantages of fiber-optic sensors. Consider the fiber-optic position sensor. In the curve of light intensity received versus *x*, in which region would you prefer to operate the sensor, and what are the corresponding limitations?

6.36 The manufacturer of an ultrasonic gage states that the device has applications in measuring cold roll steel thickness, determining parts positions in robotic assembly, lumber sorting, measurement of particle board and plywood thickness, ceramic tile dimensional inspection, sensing the fill level of food in a jar, pipe diameter gaging, rubber tire positioning during fabrication, gaging of fabricated automotive components, edge detection, location of flaws in products, and parts identification. Discuss whether the following types of sensors are also equally suitable for some or all of the foregoing applications. In each case where you think that a particular sensor is not suitable for a given application, give reasons to support your claim.

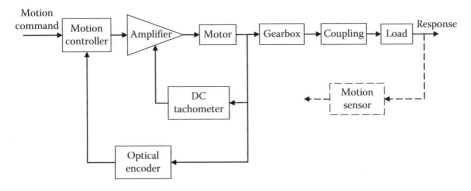

FIGURE P6.37
Block diagram of a motion control system.

(a) Fiber-optic position sensors

(b) Self-induction proximity sensors

(c) Eddy current proximity sensors

(d) Capacitive gages

(e) Potentiometers

(f) Differential transformers

6.37 (a) Consider the motion control system that is shown by the block diagram in Figure P6.37.

    (i) Giving examples of typical situations, explain the meaning of the block represented as "Load" in this system.

    (ii) Indicate the advantages and shortcomings of moving the motion sensors from the motor shaft to the load response point, as indicated by the broken lines in the figure.

  (b) Indicate, giving reasons, what type of sensors you will recommend for the following applications:

    (i) In a soft drink bottling line, for online detection of improperly fitted metal caps on glass bottles

    (ii) In a paper processing plant, to simultaneously measure both the diameter and eccentricity of rolls of newsprint

    (iii) To measure the dynamic force transmitted from a robot to its support structure, during operation

    (iv) In a plywood manufacturing machine, for online measurement of the thickness of plywood

    (v) In a food canning plant, to detect defective cans (with damage to flange, side seam, etc.)

    (vi) To read codes on food packages

6.38 Consider the two quadrature pulse signals (say, A and B) from an incremental encoder. Using sketches of these signals, show that in one direction of rotation, signal B is at a high level during the up-transition of signal A and in the opposite direction of rotation, signal B is at a low level during the up-transition of signal A. Note that the direction of motion can be determined in this manner, by using level detection of one signal during the up-transition of the other signal.

6.39 Explain why the speed resolution of a shaft encoder depends on the speed itself. What are some of the other factors that affect speed resolution? The speed of a dc motor was increased from 50 to 500 rpm. How would the speed resolution change if the speed were measured using an incremental encoder by the

(a) Pulse-counting method?

(b) Pulse-timing method?

6.40 Describe methods of improving the displacement resolution and the velocity resolution in an encoder. An incremental encoder disk has 5000 windows. The word size of the output data is 12 bits. What is the angular (displacement) resolution of the device? Assume that quadrature signals are available but that no interpolation is used.

6.41 An incremental optical encoder that has N windows per track is connected to a shaft through a gear system with gear ratio p. Derive formulas for calculating the angular velocity of the shaft by the

(a) Pulse-counting method

(b) Pulse-timing method

What is the speed resolution in each case? What effect does step-up gearing have on the speed resolution?

6.42 What is the main advantage of using a gray code instead of straight binary code in an encoder? Give a table corresponding to a gray code for a 4 bit absolute encoder. What is the corresponding code pattern on the encoder disk?

6.43 Discuss the construction features and operation of an optical encoder for measuring *rectilinear* displacements and velocities.

6.44 A centrifuge is a device that is used to separate components in a mixture. In an industrial centrifugation process, the mixture to be separated is placed in the centrifuge and rotated at high speed. The centrifugal force on a particle depends on the mass, radial location, and the angular speed of the particle. This force is responsible for separating the particles in the mixture.

The angular motion and temperature of the container are the two key variables that have to be controlled in a centrifuge. In particular, a specific centrifugation curve is used that consists of an acceleration segment, a constant-speed segment, and a braking (deceleration) segment and this corresponds to a trapezoidal speed profile. An optical encoder may be used as the sensor for microprocessor-based speed control in the centrifuge. Discuss whether an absolute encoder is preferred for this purpose. Give the advantages and possible drawbacks of using an optical encoder in this application.

6.45 Suppose that a feedback control system (Figure P6.45) is expected to provide an accuracy within $\pm\Delta y$ for a response variable y. Explain why the sensor that measures y should have a resolution of $\pm(\Delta y/2)$ or better for this accuracy to be possible. An x–y table has a travel of 2 m. The feedback control system is expected to provide an accuracy of ± 1 mm. An optical encoder is used to measure the position for feedback

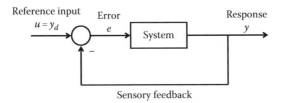

FIGURE P6.45
A feedback control loop.

in each direction (x and y). What is the minimum bit size that is required for each encoder output buffer? If the motion sensor used is an absolute encoder, how many tracks and how many sectors should be present on the encoder disk?

6.46 The pulses generated by the coding disk of an incremental optical encoder are approximately triangular (actually, upward shifted sinusoidal) in shape. Explain the reason for this. Describe a method for converting these triangular (or shifted sinusoidal) pulses into sharp rectangular pulses.

6.47 Explain how the resolution of a shaft encoder could be improved by pulse interpolation. Specifically, consider the arrangement shown in Figure P6.47. When the masking windows are completely covered by the opaque regions of the moving disk, no light is received by the photosensor. The peak level of light is received when the windows of the moving disk coincide with the windows of the masking disk. The variation of the light intensity from the minimum level to the peak level is approximately linear (generating a triangular pulse), but more accurately sinusoidal, and may be given by

$$v = v_o\left(1 - \cos\frac{2\pi\theta}{\Delta\theta}\right)$$

where θ denotes the angular position of the encoder window with respect to the masking window, as shown, $\Delta\theta$ is the window pitch angle. Note that, in the sense of rectangular pulses, the pulse corresponds to the motion in the interval $\Delta\theta/4 \leq \theta \leq 3\Delta\theta/4$. By using this sinusoidal approximation for a pulse, as given above, show that one can improve the resolution of an encoder indefinitely simply by measuring the shape of each pulse at clock cycle intervals using a high-frequency clock signal.

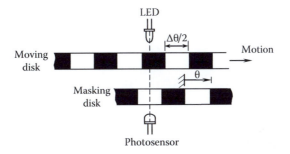

FIGURE P6.47
An encoder with a masking disk.

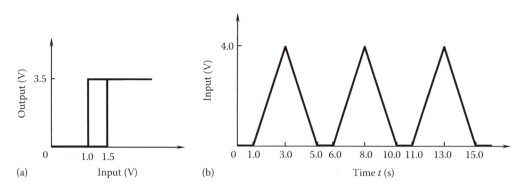

FIGURE P6.48
(a) The I/O characteristic of a Schmitt trigger; (b) a triangular input signal.

6.48 A Schmitt trigger is a semiconductor device that can function as a level detector or a switching element with hysteresis. The presence of hysteresis can be used, for example, to eliminate chattering during switching caused by noise in the switching signal. In an optical encoder, a noisy signal detected by the photosensor may be converted into a clean signal of rectangular pulses by this means. The I/O characteristic of a Schmitt trigger is shown in Figure P6.48a. If the input signal is as shown in Figure P6.48b, determine the output signal.

6.49 Compare and contrast an optical incremental encoder against a potentiometer by giving the advantages and disadvantages for an application involving the sensing of a rotatory motion.

A schematic diagram for the servo control loop of one joint of a robotic manipulator is given in Figure P6.49.

The motion command for each joint of the robot is generated by the robot controller, in accordance with the required trajectory. An optical incremental encoder is used for both position and velocity feedback in each servo loop. Note that for a six-degree-of-freedom robot, there will be six such servo loops. Describe the function of each hardware component shown in the figure and explain the operation of the servo loop.

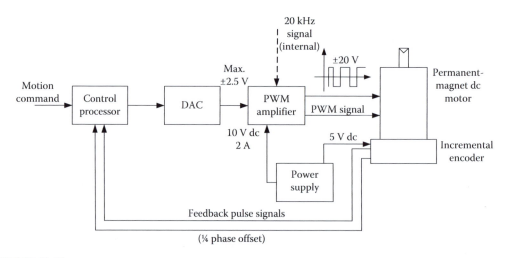

FIGURE P6.49
A servo loop of a robot.

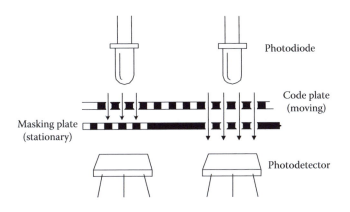

FIGURE P6.50
Photodiode–detector arrangement of a linear optical encoder.

After several months of operation, the motor of one joint of the robot was found to be faulty. An enthusiastic engineer quickly replaced the motor with an identical one without realizing that the encoder of the new motor was different. In particular, the original encoder generated 200 pulses/rev whereas the new encoder generated 720 pulses/rev. When the robot was operated, the engineer noticed an erratic and unstable behavior at the repaired joint. Discuss reasons for this malfunction and suggest a way to correct the situation.

6.50 (a) A position sensor is used in a microprocessor-based feedback control system for accurately moving the cutter blades of an automated meat-cutting machine. The machine is an integral part of the production line of a meat processing plant. What are the primary considerations in selecting the position sensor for this application? Discuss the advantages and disadvantages of using an optical encoder in comparison with an LVDT in this context.

(b) Figure P6.50 illustrates one arrangement of the optical components in a linear incremental encoder.

The moving code plate has uniformly spaced windows as usual, and the fixed masking plate has two groups of identical windows, one above each of the two photodetectors. These two groups of fixed windows are positioned in a half-pitch out of phase so that when one detector receives light from its source directly through the aligned windows of the two plates, the other detector has the light from its source virtually obstructed by the masking plate.

Explain the purpose of the two sets of photodiode–detector units, giving a schematic diagram of the necessary electronics. Can the direction of motion be determined with the arrangement shown in Figure P6.50? If so, explain how this could be done. If not, describe a suitable arrangement for detecting the direction of motion.

6.51 (a) What features and advantages of a digital transducer will distinguish it from a purely analog sensor?

(b) Consider a "linear incremental encoder," which is used to measure rectilinear positions and speeds. The moving element is a nonmagnetic plate containing a series of identically magnetized areas uniformly distributed along its length. The pickoff transponder is a mutual-induction-type proximity sensor

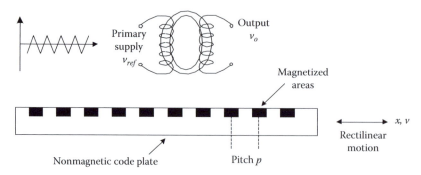

FIGURE P6.51
A linear incremental encoder of the magnetic induction type.

(i.e., a transformer) consisting of a toroidal core with a primary winding and a secondary winding. A schematic diagram of the encoder is shown in Figure P6.51. The primary excitation v_{ref} is a high-frequency sine wave.

Explain the operation of this position encoder, clearly indicating what types of signal conditioning would be needed to obtain a pure pulse signal. Also, sketch the output v_o of the proximity sensor as the code plate moves very slowly. Which position of the code plate does a high value of the pulse signal represent and which position does a low value represent?

(c) Suppose that the "pulse period timing" method is used to measure speed (v) using this encoder. The pitch distance of the magnetic spots on the plate is p, as shown in Figure P6.51. If the clock frequency of the pulse period timer is f, give an expression for the speed v in terms of the clock cycle count m.

Show that the speed resolution Δv for this method may be approximated by

$$\Delta v = \frac{v^2}{pf}$$

It follows that the dynamic range $v/\Delta v = pf/v$.

If the clock frequency is 20 MHz, the code pitch is 0.1 mm, and the required dynamic range is 100 (i.e., 40 dB), what is the maximum speed in m/s that can be measured by this method?

6.52 What is a Hall-effect tachometer? Discuss the advantages and disadvantages of a Hall effect motion sensor in comparison with an optical motion sensor (e.g., an optical encoder).

6.53 Discuss the advantages of solid-state limit switches over mechanical limit switches. Solid-state limit switches are used in many applications, particularly in the aircraft and aerospace industries. One such application is in landing gear control to detect the up, down, and locked conditions of the landing gear. High reliability is of utmost importance in such applications. The mean time between failure (MTBF) of over 100,000 h is possible with solid-state limit switches. Using your engineering judgment, give an MTBF value for a mechanical limit switch.

6.54 Mechanical force switches are used in applications where only a force limit, rather than a continuous force signal, has to be detected. Examples include detecting closure force (torque) in valve closing, detecting fit in parts assembly, automated clamping devices, robotic grippers and hands, overload protection devices in process/machine monitoring, and product filling in containers by weight. Expensive and sophisticated force sensors are not needed in such applications because a continuous history of a force signal is not needed. Furthermore, they are robust and reliable and can safely operate in hazardous environments. Using a sketch, describe the construction of a simple spring-loaded force switch.

6.55 Consider the following three types of photoelectric object counters (or object detectors or limit switches):

1. Through (opposed) type
2. Reflective (reflex) type
3. Diffuse (proximity, interceptive) type

Classify these devices into long-range (up to several meters), intermediate range (up to 1 m), and short-range (up to a fraction of a meter) detection.

6.56 A brand of autofocusing camera uses a microprocessor-based feedback control system consisting of a CCD imaging system, a microprocessor, a drive motor, and an optical encoder. The purpose of the control system is to focus the camera automatically based on the image of the subject as sensed by a matrix of CCDs (a set of metal-oxide-semiconductor field-effect transistors, or MOSFETs). The light rays from the subject that pass through the lens will fall onto the CCD matrix. This will generate a matrix (image frame) of charge signals, which are shifted one at a time, row by row, into an output buffer (or frame grabber) and passed on to the microprocessor after conditioning the resulting video signal. The CCD image obtained by sampling the video signal is analyzed by the microprocessor to determine whether the camera is focused properly. If not, the lens is moved by the motor so as to achieve focusing. Draw a schematic diagram for the autofocusing control system and explain the function of each component in the control system, including the encoder.

6.57 Today, image processing and machine vision are used in many industrial tasks including process control, monitoring, pattern classification, and object recognition. In an industrial system based on image processing, an imaging device such as a CCD camera is used as the sensing element. The camera provides an image (picture) to an image processor of a scene related to the industrial process (the measurement). The computed results from the image processor are used to determine the necessary information about the process (plant).

A digital camera has an image plate consisting of a matrix of MOSFET elements. The electrical charge that is held by each MOSFET element is proportional to the intensity of light falling on the element. The output circuit of the camera has a charge-amplifier-like device (capacitor-coupled), which is supplied by each MOSFET element. The MOSFET element that is to be connected to the output circuit at a given instant is determined by the control logic, which systematically scans the matrix of MOSFET elements. The capacitor circuit provides a voltage that is proportional to the charge in each MOSFET element.

(a) Draw a schematic diagram for a process monitoring system based on machine vision, which uses a CCD camera. Indicate the necessary signal modification operations at various stages in the monitoring loop, showing whether analog filters, amplifiers, ADC, and DAC are needed and if so, at which locations.

An image may be divided into *pixels* (or picture elements) for representation and subsequent processing. A pixel has a well-defined coordinate location in the picture frame, relative to some reference coordinate frame. In a CCD camera, the number of pixels per image frame is equal to the number of CCD elements in the image plate. The information carried by a pixel (in addition to its location) is the photointensity (or *gray level*) at the image location. This number has to be expressed in the digital form (using a certain number of bits) for digital image processing. The need for very large data-handling rates is a serious limitation on a real-time controller that uses machine vision.

(b) Consider a CCD image frame of the size 488×380 pixels. The refresh rate of the picture frame is 30 frames/s. If 8 bits are needed to represent the gray level of each pixel, what is the associated data (baud) rate?

(c) Discuss whether you prefer hardware processing or programmable software-based processing in a process monitoring system based on machine vision.

7

Actuators

Study Objectives

- The purpose of actuators in a mechatronic system
- Types of actuators
- Stepper motors and dc motors (including brushless dc motors)
- AC motors (induction motors and synchronous motors)
- Linear actuators
- Hydraulic and pneumatic actuators
- Modeling and analysis of actuators
- Practical performance and parameters of actuators
- Sizing and selection of actuators for practical applications
- Instrumentation, drive hardware, and control of actuators

7.1 Introduction

This chapter introduces the subject of actuators, as related to mechatronics. The actuator is the device that mechanically drives a mechatronic system. Joint motors in a robotic manipulator are good examples of such actuators. Actuators may be used as well to operate controller components (final control elements), such as servovalves, as well. Actuators in this category are termed *control actuators*. Actuators that automatically use response error signals from a process in feedback to correct the operation of the process (i.e., to drive the process to achieve a desired response) are termed *servoactuators*. In particular, the motors that use measurements of position, speed, and perhaps load torque and armature current or field current in feedback to drive a load to realize a specified motion are termed *servomotors*.

One broad classification separates actuators into two types: incremental-drive actuators and continuous-drive actuators. Stepper motors, which are driven in fixed angular steps, represent the class of incremental-drive actuators. They can be considered to be digital actuators, which are pulse-driven devices. Each pulse received at the driver of a digital actuator causes the actuator to move by a predetermined, fixed increment of displacement. Continuous-drive devices are very popular in mechatronic applications. Examples are direct current (dc) torque motors, induction motors, hydraulic and pneumatic motors, and piston-cylinder drives (rams). Microactuators are actuators that are able to generate very small (microscale) actuating forces/torques and motions. In general, they can be neither developed nor analyzed as scaled-down versions of regular actuators. Separate and more innovative procedures of design, construction, and analysis are necessary for microactuators. Micromachined, millimeter-size micromotors with submicron accuracy are useful in modern information storage systems. Distributed or multilayer actuators constructed using piezoelectric, electrostrictive, magnetostrictive, or photostrictive materials are used in advanced and complex applications such as adaptive structures. An actuator may be directly connected to the driven load and this is known as the "direct-drive" arrangement. More commonly, however, a transmission device may be needed to convert the actuator motion into a desired load motion for the proper matching of the actuator with the driven load. The stepper motor, dc motor, alternating current (ac) induction motor, and hydraulic actuator are particularly studied in this chapter. The modeling, selection, drive system, and control of various actuators are discussed and the procedures of actuator selection are also addressed.

7.2 Stepper Motors

Stepper motors are a popular type of actuator. They are driven in fixed angular steps (increments). Each step of rotation is the response of the motor rotor to an input pulse (or a digital command). In this manner, the stepwise rotation of the rotor can be synchronized with pulses in a command-pulse train, assuming of course that no steps are missed, thereby making the motor respond faithfully to the input signal (pulse sequence) in an open-loop manner. Like a conventional continuous-drive motor, a stepper motor is also an electromagnetic actuator in that it converts electromagnetic energy into mechanical energy to perform mechanical work. The terms *stepper motor, stepping motor,* and *step motor* are synonymous and are often used interchangeably.

One common feature in any stepper motor is that the stator of the motor contains several pairs of field windings (or phase windings) that can be switched on to produce electromagnetic pole pairs (N and S). These pole pairs effectively pull the motor rotor in sequence so as to generate the torque for motor rotation. By switching the currents in the phases in an appropriate sequence, either a clockwise (CW) rotation or a counterclockwise (CCW) rotation can be produced. The polarities of a stator pole may have to be reversed in some types of stepper motors in order to carry out a stepping sequence. Although the commands that generate the switching sequence for a phase winding could be supplied by a microprocessor or a personal computer (a software approach), it is customary to generate it through hardware logic in a device called a *translator* or an *indexer*. This approach is more effective because the switching logic for a stepper motor is fixed, as noted in the foregoing discussion. *Microstepping* provides much smaller step angles. This is achieved by changing the phase currents by small increments (rather than on, off, and reversal) so that the detent (equilibrium) position of the rotor shifts in correspondingly small angular increments.

7.2.1 Stepper Motor Classification

Most classifications of stepper motors are based on the nature of the motor rotor. One such classification considers the magnetic character of the rotor. Specifically, a variable-reluctance (VR) stepper motor has a soft-iron rotor while a permanent-magnet (PM) stepper motor has a magnetized rotor. The two types of motors operate in a somewhat similar manner. Specifically, the stator magnetic field (polarity) is stepped so as to change the minimum reluctance (or detent) position of the rotor in increments. Hence, both types of motors undergo similar changes in reluctance (magnetic resistance) during operation. A disadvantage of VR stepper motors is that since the rotor is not magnetized, the holding torque is zero when the stator windings are not energized (power off). Hence, there is no capability to hold the load at a given position under power-off conditions unless mechanical brakes are employed. A hybrid stepper motor possesses characteristics of both VR steppers and PM steppers. The rotor of a hybrid stepper motor consists of two rotor segments connected by a shaft. Each rotor segment is a toothed wheel and is called a *stack*. The two rotor stacks form the two poles of a permanent magnet located along the rotor axis. Hence, an entire stack of rotor teeth is magnetized to be a single pole (which is different from the case of a PM stepper where the rotor has multiple poles). The rotor polarity of a hybrid stepper can be provided either by a permanent magnet or by an electromagnet using a coil activated by a unidirectional dc source and placed on the stator to generate a magnetic field along the rotor axis.

Another practical classification that is used in this book is based on the number of "stacks" of teeth (or rotor segments) present on the rotor shaft. In particular, a hybrid stepper motor has two stacks of teeth. Further sub-classifications are possible, depending on the tooth pitch (angle between adjacent teeth) of the stator and tooth pitch of the rotor. In a *single-stack stepper motor*, the rotor tooth pitch and the stator tooth pitch generally have to be unequal so that not all teeth in the stator are ever aligned with the rotor teeth at any instant. It is the misaligned teeth that exert the magnetic pull, generating the driving torque. In each motion increment, the rotor turns to the minimum reluctance (stable equilibrium) position corresponding to that particular polarity distribution of the stator. In *multiple-stack stepper motors*, operation is possible even when the rotor tooth pitch is equal to the stator tooth pitch, provided that at least one stack of rotor teeth is rotationally shifted (misaligned) from the other stacks by a fraction of the rotor tooth pitch. In this design, it is this *inter-stack misalignment* that generates the drive torque for each motion step. It should be obvious that unequal-pitch multiple stack steppers are also a practical possibility. In this design, each rotor stack operates as a separate single-stack stepper motor. A photograph of the internal components of a two-stack stepper motor is given in Figure 7.1.

7.2.2 Hybrid Stepper Motor

Hybrid steppers are arguably the most common variety of stepping motors in engineering applications. A hybrid stepper motor has two stacks of rotor teeth on its shaft. The two rotor stacks are magnetized to have opposite polarities, as shown in Figure 7.2. There are two stator segments surrounding the two rotor stacks. Both rotor and stator have teeth and their pitch angles are equal. Each stator segment is wound to a single phase, and accordingly, the number of phases is two. It follows that a hybrid stepper is similar in mechanical design and stator winding to a multi-stack, equal-pitch, VR stepper. There are some dissimilarities, however. First, the rotor stacks are magnetized. Second, the inter-stack misalignment is ¼ of a tooth pitch (see Figure 7.3).

A full cycle of the switching sequence for the two phases is given by [0, 1], [−1, 0], [0, −1], [1, 0], [0 1] for one direction of rotation. In fact, this sequence produces a downward movement (CW rotation, looking from the left end) in the arrangement shown in Figure 7.3, starting from the state of [0, 1] shown in the figure (phase 1 off and phase 2 on with N polarity). For the opposite direction, the sequence is simply reversed; thus, [0, 1], [1, 0], [0, −1], [−1, 0], [0, 1]. Clearly, the step angle is given by

FIGURE 7.1
A commercial two-stack stepper motor. (Courtesy of Danaher Motion, Rockford, IL. With permission.)

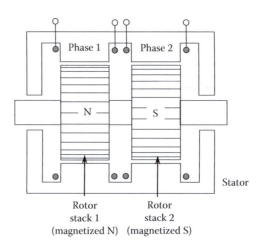

FIGURE 7.2
A hybrid stepper motor.

$$\Delta\theta = \frac{\theta}{4} \tag{7.1}$$

where $\theta = \theta_r = \theta_s$ = tooth pitch angle.

Just like in the case of a PM stepper motor, a hybrid stepper has the advantage providing a holding torque (detent torque) even under power-off conditions. Furthermore, a hybrid stepper can provide very small step angles, high stepping rates, and generally good torque–speed characteristics.

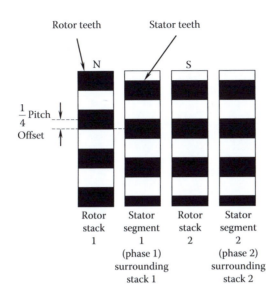

Rotor teeth   Stator teeth

N   S

$\frac{1}{4}$ Pitch

Offset

Rotor stack 1   Stator segment 1 (phase 1) surrounding stack 1   Rotor stack 2   Stator segment 2 (phase 2) surrounding stack 2

FIGURE 7.3
Rotor stack misalignment (1/4 pitch) in a hybrid stepper motor (schematically shows the state where phase 1 is off and phase 2 is on with N polarity).

Example 7.1

The half-stepping sequence for the motor represented in Figures 7.2 and 7.3 may be determined quite conveniently. Starting from the state [0, 1], if phase 1 is turned on to state "−1" without turning off phase 2, then phase 1 will oppose the pull of phase 2, resulting in a detent position halfway between the full stepping detent position. Next, if phase 2 is turned off while keeping phase 1 in "−1," the remaining half step of the original full step will be completed. In this manner, the half-stepping sequence for CW rotation is obtained as: [0, 1], [−1, 1], [−1, 0], [−1, −1], [0, −1], [1, −1], [1, 0], [1, 1], [0, 1]. For CCW rotation, this sequence is simply reversed. Note that, as expected, in half-stepping, both phases remain on during every other half step.

7.2.3 Microstepping

Full-stepping or half-stepping can be achieved simply by using an appropriate switching scheme of the phases (stator poles) of a stepper motor. For example, half-stepping occurs when phase switchings alternate between one-phase-on and two-phase-on states. Full-stepping occurs when either one-phase-on switching or two-phase-on switching is used exclusively for every step. In both these cases, the current level (or state) of a phase is either 0 (off) or 1 (on). Rather than using two current levels (the binary case), it is possible to apply several levels of phase current between these two extremes, thereby achieving much smaller step angles. This is the principle behind microstepping.

Microstepping is achieved by properly changing the phase currents in small steps instead of switching them on and off (as in the case of full-stepping and half-stepping). The principle behind this can be understood by considering two identical stator poles (wound with identical windings), as shown in Figure 7.4. When the currents through the windings are identical (in magnitude and direction), the resultant magnetic field will lie symmetrically between the two poles. If the current in one pole is decreased while the other current is kept unchanged, the resultant magnetic field will move closer to the pole with the larger

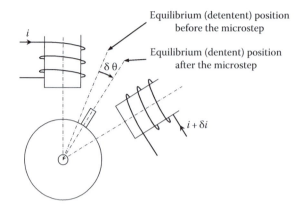

FIGURE 7.4
The principle of microstepping.

current. Since the detent position (equilibrium position) depends on the position of the resultant magnetic field, it follows that very small step angles can be achieved simply by controlling (varying the relative magnitudes and directions of) the phase currents.

Step angles of 1/125 of a full step or smaller could be obtained through microstepping. For example, 10,000 steps/revolution may be achieved. Note that the step size in a sequence of microsteps is not identical. This is because stepping is done through the microsteps of the phase current (and the magnetic field generated by it), which has a nonlinear relation with the step angle.

Motor drive units with the microstepping capability are more costly, but microstepping provides the advantages of accurate motion capabilities, including finer resolution, overshoot suppression, and smoother operation (reduced jitter and less noise) even in the neighborhood of a resonance in the motor-load combination. A disadvantage is that usually there is a reduction in the motor torque as a result of microstepping.

7.2.4 Driver and Controller

In principle, the stepper motor is an open-loop actuator. In its normal operating mode, the stepwise rotation of the motor is synchronized with the command pulse train. Under highly transient conditions near rated torque, "pulse missing" can be a problem.

A stepper needs a "control computer" or at least a hardware "indexer" to generate the pulse commands and a "driver" to interpret the commands and correspondingly generate the proper currents for the phase windings of the motor. This basic arrangement is shown in Figure 7.5a. For feedback control, the response of the motor has to be sensed (say, using an optical encoder) and fed back into the controller (see the dotted line in Figure 7.5a) to take the necessary corrective action to the pulse command when an error is present. The basic components of the driver for a stepper motor are identified in Figure 7.5b. It consists of a logic circuit called a "translator" to interpret the command pulses and switch the appropriate analog circuits to generate the phase currents. Since sufficiently high current levels are needed for the phase windings, depending on the motor capacity, the drive system includes amplifiers powered by a power supply.

The command pulses are generated either by a control computer (a desktop computer or a microprocessor), the software approach, or by a variable-frequency oscillator (or

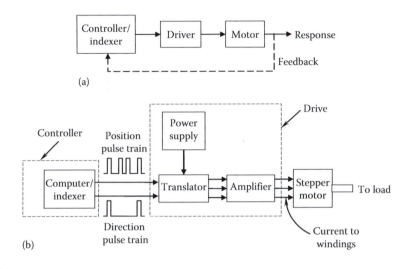

FIGURE 7.5
(a) The basic control system of a stepper motor; (b) The basic components of a driver.

an indexer), the basic hardware approach. For bidirectional motion, two pulse trains are necessary: the position-pulse train and the direction-pulse train, which are determined by the required motion trajectory. The position pulses identify the exact times at which angular steps should be initiated. The direction pulses identify the instants at which the direction of rotation should be reversed. Only a position pulse train is needed for unidirectional operation. The generation of the position pulse train for steady-state operation at a constant speed is a relatively simple task. In this case, a single command identifying the stepping rate (pulse rate), corresponding to the specified speed, would suffice. The logic circuitry within the translator will latch onto a constant-frequency oscillator with the frequency determined by the required speed (stepping rate) and continuously cycle the switching sequence at this frequency. This is a hardware approach to open-loop control of a stepping motor. For steady-state operation, the stepping rate can be set by manually adjusting the knob of a potentiometer connected to the translator. For simple motions (e.g., starting from rest and stopping after reaching a certain angular position), the commands that generate the pulse train (commands to the oscillator) can be set manually. Under the more complex and transient operating conditions that are present when following intricate motion trajectories, however, a computer-based (or microprocessor-based) generation of the pulse commands, using programmed logic, would be necessary. This is a software approach, which is usually slower than the hardware approach. Sophisticated feedback control schemes can be implemented as well through such a computer-based controller.

The *translator* module has logic circuitry to interpret a pulse train and "translate" it into the corresponding switching sequence for stator field windings (on/off/reverse state for each phase of the stator). The translator also has solid-state switching circuitry (using gates, latches, triggers, etc.) to direct the field currents to the appropriate phase windings according to the particular switching state. A "packaged" system typically includes both indexer (or controller) functions and driver functions. As a minimum, it possesses the capability to generate command pulses at a steady rate, thus assuming the role of the pulse generator (or indexer) as well as the translator and switching amplifier functions. The stepping rate or direction may be changed manually using knobs or through a user interface.

The translator may not have the capability to keep track of the number of steps taken by the motor (i.e., a step counter). A packaged device that has all these capabilities, including pulse generation, the standard translator functions, and drive amplifiers, is termed a *preset indexer*. It usually consists of an oscillator, digital microcircuitry (integrated-circuit [IC] chips) for counting and for various control functions, a translator, and drive circuitry in a single package. The required angle of rotation, stepping rate, and direction are set manually, by turning the corresponding knobs. With a more sophisticated programmable preset indexer, these settings can be programmed through computer commands from a standard interface. An external pulse source is not needed in this case. A programmable indexer—consisting of a microprocessor and microelectronic circuitry for the control of position and speed and for other programmable functions, memory, a pulse source (an oscillator), a translator, drive amplifiers with switching circuitry, and a power supply—represents a "programmable" controller for a stepping motor. A programmable indexer can be programmed using a personal computer or a hand-held programmer (provided with the indexer) through a standard interface (e.g., RS232 serial interface). Control signals within the translator are on the order of 10 mA, whereas the phase windings of a stepper motor require large currents on the order of several amperes. Control signals from the translator have to be properly amplified and directed to the motor windings by means of "switching amplifiers" for activating the required phase sequence.

Power to operate the translator (for logic circuitry, switching circuitry, etc.) and to operate phase excitation amplifiers comes from a dc *power supply* (typically 24 V dc). A regulated (i.e., the voltage is maintained constant irrespective of the load) power supply is preferred. A packaged unit that consists of the translator (or preset indexer), the switching amplifiers, and the power supply is what is normally termed a *motor-drive system*. The leads of the output amplifiers of the drive system carry currents to the phase windings on the stator (and to the rotor magnetizing coils located on the stator in the case of an electromagnetic rotor) of the stepping motor. The *load* may be connected to the motor shaft directly or through some form of mechanical coupling device (e.g., harmonic drive, tooth-timing belt drive, hydraulic amplifier, rack, or pinion).

7.2.5 Driver Hardware

The driver hardware consists of the following basic components:

1. Digital (logic) hardware to interpret the information carried by the stepping pulse signal and the direction pulse signal (i.e., step instants and the direction of motion) and to provide appropriate signals to the switches (switching transistors) that actuate the phase windings. This is the "translator" component of the drive hardware.
2. The drive circuit for phase windings with switching transistors to actuate the phases (on, off, and reverse in the unifilar case; on and off in the bifilar case).
3. Power supply to power the phase windings.

These three components are commercially available as a single package to operate a corresponding class of stepper motors. Since there is considerable heat generation in a drive module, an integrated *heat sink* (or some means of heat removal) is needed as well. Consider the

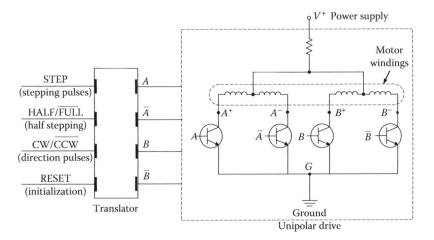

FIGURE 7.6
Basic drive hardware for a two-phase bifilar-wound stepper motor.

drive hardware for a two-phase stepper motor. The phases are denoted by A and B. A schematic representation of the drive system, which is commercially available as a single package, is shown in Figure 7.6. What is indicated is a unipolar drive (no current reversal in a phase winding). As a result, a stepper motor with bifilar windings (two coil segments for each phase) has to be used. The motor has five leads, one of which is the "motor common" or ground (G) and the other four are the terminals of the two bifilar coil segments (A^+, A^-, B^+, B^-).

There are several pins in the drive module, some of which are connected to the motor controller/computer (driver inputs) and some are connected to the motor leads (driver outputs). There are other pins, which correspond to the dc power supply, common ground, various control signals, etc. The pin denoted by STEP (or PULSE) receives the stepping pulse signal (from the motor controller). This corresponds to the required stepping sequence of the motor. A transition from a low level to a high level (or rising edge) of a pulse will cause the motor to move by one step. The direction in which the motor moves is determined by the state of the pin denoted by CW/ $\overline{\text{CCW}}$. A logical high state at this pin (or open connection) will generate switching logic for the motor to move in the CW direction, and a logical low state (or logic common) will generate switching logic for the motor to move in the CCW direction. The pin denoted by HALF/$\overline{\text{FULL}}$ determines whether half stepping or full stepping is carried out. Specifically, a logical low at this pin will generate switching logic for full stepping, and the logical high will generate switching logic for half stepping. The pin denoted by RESET receives the signal for initialization of a stepping sequence. There are several other pins, which are not necessary for the present discussion. The translator interprets the logical states at the STEP, HALF/$\overline{\text{FULL}}$, and CW/ $\overline{\text{CCW}}$ pins and generates the proper logic to activate the switches in the unipolar drive. Specifically, four active logic signals are generated corresponding to A (Phase A on), $\overline{A}$ (Phase A reversed), B (Phase B on), and $\overline{B}$ (Phase B reversed). These logic signals activate the four switches in the bipolar drive, thereby sending current through the corresponding winding segments/leads (A^+, A^-, B^+, B^-) of the motor.

The logic hardware is commonly available as compact chips in the monolithic form. If the motor is unifilar-wound (for a two-phase stepper there should be three leads—a ground wire and two power leads for the two phases), a bipolar drive will be necessary in order to change the direction of the current in a phase winding. A schematic representation of a bipolar drive for a single phase of a stepper is shown in Figure 7.7. Note that when the two transistors marked A are on, the current flows in one direction through the phase winding and when the two transistors marked $\bar{A}$ are on, the current flows in the opposite direction through the same phase winding. What is shown is an H-bridge circuit.

7.2.6 Stepper Motor Selection

The selection of a stepper motor for a specific application cannot be made on the basis of geometric parameters alone. Torque and speed considerations are often more crucial in the selection process. For example, a faster speed of response is possible if a motor with a larger torque-to-inertia ratio is used.

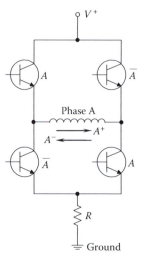

FIGURE 7.7
A bipolar drive for a single phase of a stepper motor (unifilar-wound).

7.2.6.1 Torque Characteristics and Terminology

The torque that can be provided to a load by a stepper motor depends on several factors. For example, the motor torque at a constant speed is not the same as that when the motor "passes through" that speed (i.e., under acceleration, deceleration, or general transient conditions). In particular, at a constant speed, there is no inertia torque. Also, the torque losses due to magnetic induction are lower at constant stepping rates in comparison with the variable stepping rates. It follows that the available torque is larger under steady (constant-speed) conditions. Another factor of influence is the magnitude of the speed. At low speeds (i.e., when the step period is considerably larger than the electrical time constant), the time taken for the phase current to build up or be turned off is insignificant compared with the step time. Then the phase current waveform can be assumed to be rectangular. At high stepping rates, the induction effects dominate and as a result a phase may not reach its rated current during the duration of a step. As a result, the generated torque will be degraded. Furthermore, since the power provided by the power supply is limited, the torque×speed product of the motor is limited as well. Consequently, as the motor speed increases, the available torque must decrease in general. These two are the primary reasons for the characteristic shape of a speed–torque curve of a stepper motor where the peak torque occurs at a very low (typically zero) speed, and as the speed increases, the available torque decreases. Eventually, at a particular limiting speed (known as the no-load speed), the available torque becomes zero.

The characteristic shape of the speed–torque curve of a stepper motor is shown in Figure 7.8. Some terminology is given as well. What is given may be interpreted as experimental data measured under steady operating conditions (and averaged and interpolated). The given torque is called the "pull-out torque" and the corresponding speed is the "pull-out speed." In industry, this curve is known as the "pull-out curve."

Holding torque is the maximum static torque and is different from the maximum (pull-out) torque defined in Figure 7.8. In particular, the holding torque can be about 40% greater than the maximum pull-out torque, which is typically equal to the starting torque (or stand-still torque). Furthermore, the static torque becomes higher if the motor has more than one stator pole per phase and if all these poles are excited at a time. The *residual torque*

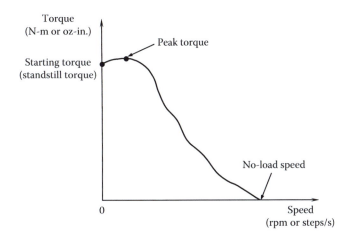

FIGURE 7.8
The speed–torque characteristics of a stepper motor (pull-out curve).

is the maximum static torque that is present when the motor phases are not energized. This torque is practically zero for a VR motor, but is not negligible for a PM motor. In some industrial literature, *detent torque* takes the same meaning as the residual torque. In this context, detent torque is defined as the torque ripple that is present under power-off conditions. A more appropriate definition for detent torque is the static torque at the present detention position (equilibrium position) of the motor, when the next phase is energized. According to this definition, detent toque is equal to $T_{max} \sin 2\pi/p$, where T_{max} is the holding torque and p is the number of phases.

Some further definitions of speed–torque characteristics of a stepper motor are given in Figure 7.9. The pull-out curve or the *slew curve* here takes the same meaning as what is given in Figure 7.8. Another curve known as the *start-stop curve* or *pull-in curve* is given as well.

The pull-out curve (or slew curve) gives the speed at which the motor can run under steady (constant-speed) conditions, under rated current, and using appropriate drive

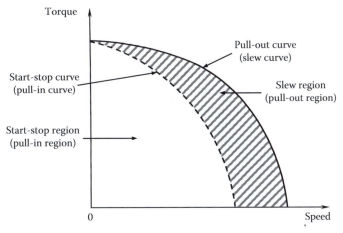

FIGURE 7.9
Further speed–torque characteristics and terminology.

circuitry. But, the motor is unable to steadily accelerate to the slew speed, starting from rest and applying a pulse sequence at constant rate corresponding to the slew speed. Instead, it should be accelerated first up to the pull-in speed by applying a pulse sequence corresponding to this speed. After reaching the start-stop region (pull-in region) in this manner, the motor can be accelerated to the pull-out speed (or to a speed lower than this, within the slew region). Similarly, when stopping the motor from a slew speed, it should be first decelerated (by down-ramping) to a speed in the start-stop region (pull-in region) and only when this region is reached satisfactorily should the stepping sequence be turned off.

Since the drive system determines the current and the switching sequence of the motor phases and the rate at which the switching pulses are applied, it directly affects the speed–torque curve of a motor. Accordingly, what is given in a product data sheet should be interpreted as the speed–torque curve of the particular motor when used with a specified drive system and a matching power supply and when it is operating at rated values.

7.2.6.2 Stepper Motor Selection Process

The effort required in selecting a stepper motor for a particular application can be reduced if the selection is done in a systematic manner. The following steps provide some guidelines for the selection process:

Step 1: List the main requirements for the particular application, according to the conditions and specifications for the particular application. These include operational requirements such as speeds, accelerations, required accuracy and resolution, and load characteristics, such as size, inertia, fundamental natural frequencies, and resistance torques.

Step 2: Compute the operating torque and stepping rate requirements for the particular application.

Newton's second law is the basic equation employed in this step. Specifically, the required torque rating is given by

$$T = T_R + J_{eq} \frac{\omega_{max}}{\Delta t} \tag{7.2}$$

where
T_R is the net resistance torque
J_{eq} is the equivalent moment of inertia (including rotor, load, gearing, dampers, etc.)
ω_{max} is the maximum operating speed
Δt is the time taken to accelerate the load to the maximum speed, starting from rest

Step 3: Using the torque versus stepping rate curves (pull-out curves) for a group of commercially available stepper motors, select a suitable stepper motor.

The torque and speed requirements determined in Step 2 and the accuracy and resolution requirements specified in Step 1 should be used in this step.

Step 4: If a stepper motor that meets the requirements is not available, modify the basic design.

This may be accomplished by changing the speed and torque requirements by adding devices such as gear systems (e.g., harmonic drive) and amplifiers (e.g., hydraulic amplifiers).

Step 5: Select a drive system that is compatible with the motor and that meets the operational requirements in Step 1.

Motors and appropriate drive systems are prescribed in product manuals and catalogs available from the vendors. For relatively simple applications, a manually controlled preset indexer or an open-loop system consisting of a pulse source (oscillator) and a translator could be used to generate the pulse signal to the translator in the drive unit. For more complex transient tasks, a software controller (a microprocessor or a personal computer) or a customized hardware controller may be used to generate the desired pulse command in open-loop operation. Further sophistication may be incorporated by using digital processor-based closed-loop control with encoder feedback, for tasks that require very high accuracy under transient conditions and for operation near the rated capacity of the motor.

The single most useful piece of information in selecting a stepper motor is the torque versus stepping rate curve (the pull-out curve). Other parameters that are valuable in the selection process include the following:

1. The step angle or the number of steps per revolution
2. The static holding torque (the maximum static torque of the motor when powered at rated voltage)
3. The maximum slew rate (maximum steady-state stepping rate possible at the rated load)
4. The motor torque at the required slew rate (pull-out torque, available from the pull-out curve)
5. The maximum ramping slope (maximum acceleration and deceleration possible at the rated load)
6. The motor time constants (no-load electrical time constant and mechanical time constant)
7. The motor natural frequency (without an external load and near detent position)
8. The motor size (dimensions of poles, stator and rotor teeth, air gap and housing, weight, rotor moment of inertia)
9. The power supply ratings (voltage, current, and power)

There are many parameters that determine the ratings of a stepper motor. For example, the static holding torque increases with the number of poles per phase that are energized, decreases with the air gap width and tooth width, and increases with the rotor diameter and stack length. Furthermore, the minimum allowable air gap width should exceed the combined maximum lateral (flexural) deflection of the rotor shaft caused by thermal deformations and the flexural loading, such as magnetic pull, static, and dynamic mechanical loads. In this respect, the flexural stiffness of the shaft, the bearing characteristics, and the thermal expansion characteristics of the entire assembly become important. Field winding parameters (diameter, length, resistivity, etc.) are chosen by giving due consideration to the required torque, power, electrical time constant, heat generation rate, and motor dimensions. Note that a majority of these are design parameters that cannot be modified in a cost-effective manner during the motor selection stage.

7.2.6.3 Positioning (x–y) Tables

A common application of stepper motors is in positioning tables (see Figure 7.10a). Note that a two-axis (x–y) table requires two stepper motors of nearly equal capacity. The values of the following parameters are assumed to be known:

- Maximum positioning resolution (displacement/step)
- Maximum operating velocity, to be attained in less than a specified time
- Weight of the x–y table
- Maximum resistance force (primarily friction) against table motion

A schematic diagram of the mechanical arrangement for one of the two axes of the table is shown in Figure 7.10b. A lead screw is used to convert the rotary motion of the motor into rectilinear motion. Free-body diagrams for the motor rotor and the table are shown in Figure 7.11.

Now we will derive a somewhat generalized relation for this type of application. The equations of motion (from Newton's second law) are

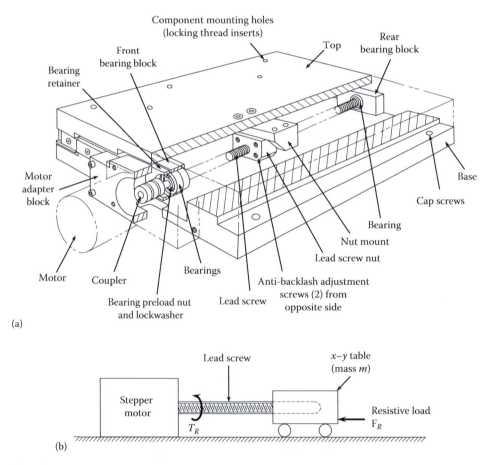

(a)

(b)

FIGURE 7.10
(a) A single axis of a positioning table; (b) an equivalent model.

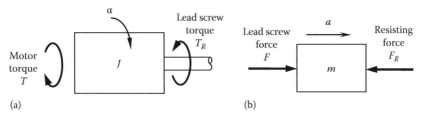

FIGURE 7.11
Free-body diagrams: (a) Motor rotor; (b) table.

$$\text{For the rotor: } T - T_R = J\alpha \tag{7.3}$$

$$\text{For the table: } F - F_R = ma \tag{7.4}$$

where
T is the motor torque T_R
T_R is the resistance torque from the lead screw
J is the equivalent moment of inertia of the rotor
α is the angular acceleration of the rotor
F is the driving force from the lead screw
F_R is the external resistance force on the table
M is the equivalent mass of the table
a is the acceleration of the table

Assuming a rigid lead screw without backlash, the compatibility condition is written as

$$a = r\alpha \tag{7.5}$$

where r denotes the *transmission ratio* (rectilinear motion/angular motion) of the lead screw. The load transmission equation for the lead screw is

$$F = \frac{e}{r}T_R \tag{7.6}$$

where e denotes the *fractional efficiency* of the lead screw. Finally, Equations 7.3 through 7.6 can be combined to give

$$T = \left(J + \frac{mr^2}{e}\right)\frac{a}{r} + \frac{r}{e}F_R \tag{7.7}$$

Example 7.2

A schematic diagram of an industrial conveyor unit is shown in Figure 7.12. In this application, the conveyor moves intermittently at a fixed rate, thereby indexing the objects on the conveyor through a fixed distance d in each time period T. A triangular speed profile is used for each motion interval, having an acceleration and a deceleration that are equal in magnitude (see Figure 7.13). The conveyor is driven by a stepper motor. A gear unit with step-down speed ratio p:1, where $p > 1$, may be used if necessary, as shown in Figure 7.12.

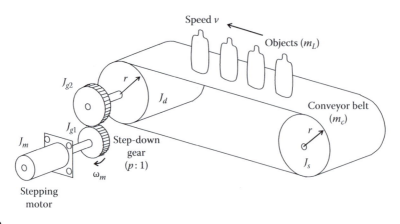

FIGURE 7.12
Conveyor unit with intermittent motion.

(a) Explain why the equivalent moment of inertia J_e at the motor shaft for the overall system is given by

$$J_e = J_m + J_{g1} + \frac{1}{p^2}(J_{g2} + J_d + J_s) + \frac{r^2}{p^2}(m_c + m_L)$$

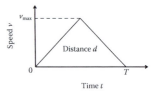

FIGURE 7.13
Speed profile for a motion period of the conveyor.

where J_m, J_{g1}, J_{g2}, J_d, and J_s are the moments of inertia of the motor rotor, drive gear, driven gear, drive cylinder of the conveyor, and the driven cylinder of the conveyor, respectively; m_c and m_L are the overall masses of the conveyor belt and the moved objects (load), respectively; and r is the radius of each of the two conveyor cylinders.

(b) Four models of stepping motor are available for the application. Their specifications are given in Table 7.1 and the corresponding performance curves are given in Figure 7.14. The following values are known for the system:

$d = 10\,\text{cm}$, $T = 0.2\,\text{s}$, $r = 10\,\text{cm}$, $m_c = 5\,\text{kg}$, $m_L = 5\,\text{kg}$, $J_d = J_s = 2.0 \times 10^{-3}\,\text{kg-m}^2$.

Also, two gear units with $p = 2$ and 3 are available, and for each unit $J_{g1} = 50 \times 10^{-6}\,\text{kg-m}^2$ and $J_{g2} = 200 \times 10^{-6}\,\text{kg-m}^2$.

Indicating all calculations and procedures, select a suitable motor unit for this application. You must not use a gear unit unless it is necessary to have one with the available motors.

What is the positioning resolution of the conveyor (rectilinear) for the final system?

Note: Assume an overall system efficiency of 80% regardless of whether a gear unit is used.

Solution

(a) The angular speed of the motor and drive gear $= \omega_m$.
The angular speed of the driven gear and conveyor cylinders $= \omega_m/p$.
The rectilinear speed of the conveyor and objects $v = r\omega_m/p$.

Determination of the equivalent inertia

The determination of the equivalent moment of inertia of the system, referred to as the motor rotor, is an important step of the motor selection. This is done by determining the

TABLE 7.1

Stepper Motor Data

Model		50SM	101SM	310SM	1010SM
NEMA motor frame size			23	34	42
Full step angle	degrees		1.8		
Accuracy	%		±3 (noncumulative)		
Holding torque	oz-in.	38	90	370	1050
	N-m	0.27	0.64	2.61	7.42
Detent torque	oz-in	6	18	25	20
	N-m	0.04	0.13	0.18	0.14
Rated phase current	A	1	5	6	8.6
Rotor inertia	$\times 10^{-3}$ oz-in.-s^2	1.66	5	26.5	114
	$\times 10^{-6}$ kg-m^2	11.8	35	187	805
Maximum radial load	lb		15	35	40
	N		67	156	178
Maximum thrust load	lb		25	60	125
	N		111	267	556
Weight	lb	1.4	2.8	7.8	20
	kg	0.6	1.3	3.5	9.1
Operating temperature	°C		−55 to +50		
Storage temperature	°C		−55 to +130		

Source: Courtesy of Aerotech Inc., Pittsburgh, PA. With permission.

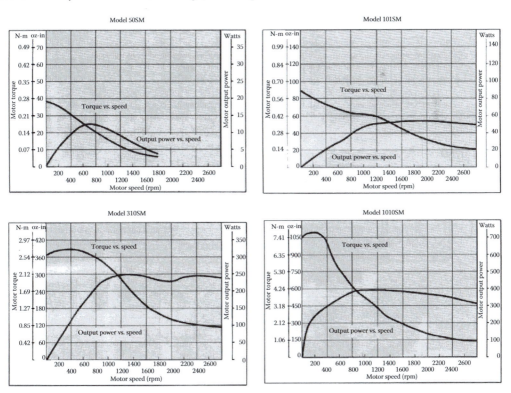

FIGURE 7.14
Stepper motor performance curves. (Courtesy of Aerotech, Inc., Pittsburgh, PA. With permission.)

kinetic energy of the overall system and equating it to the kinetic energy of the equivalent system as follows:

$$
\begin{aligned}
KE &= \frac{1}{2}(J_m + J_{g1})\omega_m^2 + \frac{1}{2}(J_{g2} + J_d + J_s)\left(\frac{\omega_p}{p}\right)^2 + \frac{1}{2}(m_c + m_L)\left(\frac{r\omega_m}{p}\right)^2 \\
&= \frac{1}{2}\left[J_m + J_{g1} + \frac{1}{p^2}(J_{g2} + J_d + J_s) + \frac{r^2}{p^2}(m_c + m_L)\right]\omega_m^2 \\
&= \frac{1}{2}J_e\omega_m^2
\end{aligned}
$$

Hence, the equivalent moment of inertia as felt at the motor rotor is

$$
J_e = J_m + J_{g1} + \frac{1}{p^2}(J_{g2} + J_d + J_s) + \frac{r^2}{p^2}(m_c + m_L) \tag{i}
$$

(a) The area of the speed profile is equal to the distance traveled. Hence,

$$
d = \frac{1}{2}v_{max}T \tag{ii}
$$

Substitute numerical values $0.1 = \frac{1}{2}v_{max}0.2 \Rightarrow v_{max} = 1.0$ m/s.

The acceleration/deceleration of the system

$$
a = \frac{v_{max}}{T/2} = \frac{1.0}{0.2/2} \text{ m/s}^2 = 10.0 \text{ m/s}^2.
$$

The corresponding angular acceleration/deceleration of the motor is

$$
\alpha = \frac{pa}{r} \tag{iii}
$$

With an overall system efficiency of η, the motor torque T_m that is needed to accelerate/decelerate the system is given by

$$
\eta T_m = J_e\alpha = J_e\frac{pa}{r} = \left[J_m + J_{g1} + \frac{1}{p^2}(J_{g2} + J_d + J_s) + \frac{r^2}{p^2}(m_c + m_L)\right]\frac{pa}{r} \tag{iv}
$$

Note: An alternative way to include energy dissipation into this equation is by using two separate terms: frictional torque referred to as the motor rotor and gear efficiency. In the present problem, for simplicity, we use a single efficiency term whether a gear is present or not. In practice, however, it should be clear that the overall efficiency drops when a gear transmission is added.

The maximum speed of the motor is

$$\omega_{max} = \frac{pv_{max}}{r} \tag{v}$$

Without gears ($p=1$), we have from (iv)

$$\eta T_m = \left[J_m + J_d + J_s + r^2(m_c + m_L) \right] \frac{a}{r} \tag{vi}$$

From (v),

$$\omega_{max} = \frac{v_{max}}{r} \tag{vii}$$

Substitute numerical values.

Case 1: Without gears
For an efficiency value $\eta = 0.8$ (i.e., 80% efficient), we have from (vi)

$$0.8T_m = \left[J_m + 2 \times 10^{-3} + 2 \times 10^{-3} + 0.1^2(5+5) \right] \frac{10}{0.1} \text{ N-m}$$

$$\text{Or: } T_m = 125.0[J_m + 0.104] \text{ N-m}$$

$$\text{From (vii): } \quad \omega_{max} = \frac{1.0}{0.1} \text{ rad/s} = 10 \times \frac{60}{2\pi} \text{ rpm} = 95.5 \text{ rpm}$$

The operating speed range is 0 to 95.5 rpm.

Note: The torque at 95.5 rpm is less than the starting torque for the first two motor models and is not so for the second two models (see the speed–torque curves in Figure 7.14). We must use the weakest point (i.e., lowest torque) from the operating speed range in the motor selection process. Allowing for this requirement, Table 7.2 is formed for the four motor models.

It is seen that without a gear unit, the available motors cannot meet the system requirements.

TABLE 7.2

Data for Selecting a Motor Without a Gear Unit

Motor Model	Available Torque at ω_{max} (N-m)	Motor Rotor Inertia ($\times 10^{-6}$ kg-m^2)	Required Torque (N-m)
50SM	0.26	11.8	13.0
101SM	0.60	35.0	13.0
310SM	2.58	187.0	13.0
1010SM	7.41	805.0	13.1

TABLE 7.3

Data for Selecting a Motor with a Gear Unit

Motor Model	Available Torque at ω_{max} (N-m)	Motor Rotor Inertia ($\times 10^{-6}$ kg-m^2)	Required Torque (N-m)
50SM	0.25	11.8	6.53
101SM	0.58	35.0	6.53
310SM	2.63	187.0	6.57
1010SM	7.41	805.0	6.73

Case 2: With gears

Note: Usually the system efficiency drops when a gear unit is introduced. In this exercise, we use the same efficiency for reasons of simplicity.

With an efficiency of 80%, we have $\eta = 0.8$. Then, from (iv)

$$0.8 T_m = \left[J_m + 50 \times 10^{-6} + \frac{1}{p^2}\left(200 \times 10^{-6} + 2 \times 10^{-3} + 2 \times 10^{-3}\right) + \frac{0.1^2}{p^2}(5+5) \right] p \times \frac{10}{0.1} \text{ N-m}$$

$$\rightarrow T_m = 125.0 \left[J_m + 50 \times 10^{-6} + \frac{1}{p^2} \times 104.2 \times 10^{-3} \right] p \text{ N-m}$$

From (v): $\omega_{max} = \dfrac{1.0p}{0.1}$ rad/s $= 10p \times \dfrac{60}{2\pi}$ rpm $\square$ $\omega_{max} = 95.5p$ rpm

First, try the case of $p = 2$; we have $\omega_{max} = 191.0$ rpm. Table 7.3 is formed for the present case.

It is seen that with a gear of speed ratio $p = 2$, motor model 1010SM satisfies the requirement.

With full stepping, the step angle of the rotor $= 1.8°$. The corresponding step in the conveyor motion is the positioning resolution.

With $p = 2$ and $r = 0.1$ m, the position resolution is $\dfrac{1.8°}{2} \times \dfrac{\pi}{180°} \times 0.1 = 1.57 \times 10^{-3}$ m.

7.2.7 Stepper Motor Applications

More than one type of actuator may be suitable for a given application. In this discussion, we indicate situations where a stepper motor is a suitable choice as an actuator. It does not, however, rule out the use of other types of actuators for the same application.

Stepper motors are particularly suitable for positioning, ramping (constant acceleration and deceleration), and slewing (constant speed) applications at relatively low speeds. Typically, they are suitable for short and repetitive motions at speeds lower than 2000 rpm. They are not the best choice for servoing or trajectory following applications because of jitter and step (pulse) missing problems (dc and ac servo motors are better for such applications). Encoder feedback will make the situation better, but at a higher cost and controller complexity. Generally, however, the stepper motor provides a low-cost option in a variety of applications.

The stepper motor is a low-speed actuator that may be used in applications that require torques as high as 15 N-m (2121 oz-in.). For heavy-duty applications, torque amplification may be necessary. One way to accomplish this is by using a hydraulic actuator in cascade with the motor. The hydraulic valve (typically a rectilinear spool valve), which controls the hydraulic actuator (typically a piston-cylinder device), is driven by a stepper motor through suitable gearing for speed reduction as well as for rotary-rectilinear motion conversion.

Torque amplification by an order of magnitude is possible with such an arrangement. Of course, the time constant will increase and the operating bandwidth will decrease because of the sluggishness of hydraulic components. Also, a certain amount of backlash will be introduced by the gear system. Feedback control will be necessary to reduce the position error, which is usually present in open-loop hydraulic actuators.

Stepper motors are incremental actuators. As such, they are ideally suited for digital control applications. High-precision open-loop operation is possible as well, provided that the operating conditions are well within the motor capacity. Early applications of stepper motor were limited to low-speed, low-torque drives. With rapid developments in solid-state drives and microprocessor-based pulse generators and controllers, however, reasonably high-speed operation under transient conditions at high torques and closed-loop control has become feasible. Since brushes are not used in stepper motors, there is no danger in spark generation. Hence, they are suitable in hazardous environments. But, heat generation and associated thermal problems can be significant at high speeds.

There are numerous applications of stepper motors. For example, a stepper motor is particularly suitable in printing applications (including graphic printers, plotters, and electronic typewriters) because the print characters are changed in steps and the printed lines (or paper feed) are also advanced in steps. Stepper motors are commonly used in x–y tables. In automated manufacturing applications, stepper motors are found as joint actuators, end effector (gripper) actuators of robotic manipulators, and as drive units in programmable dies, parts-positioning tables, and tool holders of machine tools (milling machines, lathes, etc.). In automotive applications, pulse windshield wipers, power window drives, power seat mechanisms, automatic carburetor control, process control applications, valve actuators, and parts-handling systems use stepper motors. Other applications of stepper motors include source and object positioning in medical and metallurgical radiography, lens drives in auto-focus cameras, camera movement in computer vision systems, and paper feed mechanisms in photocopying machines.

The advantages of stepper motors include the following:

1. Position error is noncumulative. A high accuracy of motion is possible, even under open-loop control.
2. The cost is relatively low. Furthermore, considerable savings in sensor (measuring system) and controller costs are possible when the open-loop mode is used.
3. Because of the incremental nature of command and motion, stepper motors are easily adoptable to digital control applications.
4. No serious stability problems exist, even under open-loop control.
5. Torque capacity and power requirements can be optimized and the response can be controlled by electronic switching.
6. Brushless construction has obvious advantages.

The disadvantages of stepper motors include the following:

1. They are low speed actuators. The torque capacity is typically less than 15 N-m, which may be low compared with torque motors.
2. They have limited speed (limited by torque capacity and by pulse-missing problems due to faulty switching systems and drive circuits).
3. They have high vibration levels due to stepwise motion.

4. Large errors and oscillations can result when a pulse is missed under open-loop control.
5. Thermal problems can be significant when operating at high speeds.

In most applications, the merits of stepper motors outweigh the drawbacks.

7.3 DC Motors

A dc motor converts direct current electrical energy into rotational mechanical energy. A major part of the torque generated in the rotor (armature) of the motor is available to drive an external load. The dc motor is probably the earliest form of electric motor. Because of features such as high torque, speed controllability over a wide range, portability, well-behaved speed–torque characteristics, easier and accurate modeling, and adaptability to various types of control methods, dc motors are still widely used in numerous mechatronic applications including robotic manipulators, transport mechanisms, disk drives, positioning tables, machine tools, and servovalve actuators.

The principle of operation of a dc motor is illustrated in Figure 7.15. Consider an electric conductor placed in a steady magnetic field at right angles to the direction of the field. Flux density B is assumed to be constant. If a dc current is passed through the conductor, the magnetic flux due to the current will loop around the conductor, as shown in the figure. Consider a plane through the conductor parallel to the direction of flux of the magnet. On one side of this plane, the current flux and the field flux are additive; on the opposite side, the two magnetic fluxes oppose each other. As a result, an imbalance magnetic force F is generated on the conductor, normal to the plane. This force is given by (*Lorentz's law*):

$$F = Bil \tag{7.8}$$

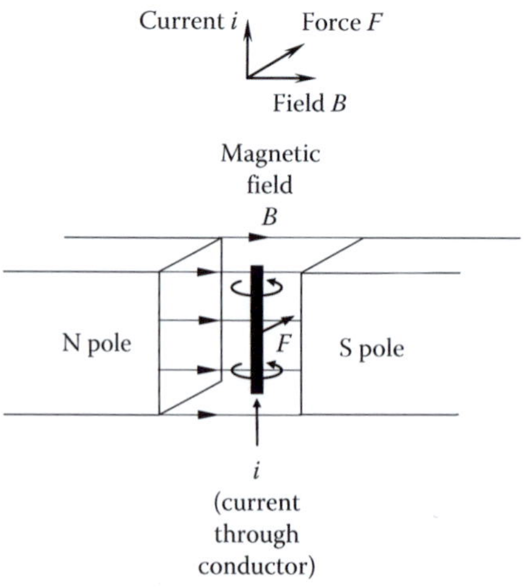

FIGURE 7.15
Operating principle of a dc motor.

where
  B is the flux density of the original field
  i is the current through the conductor
  l is the length of the conductor

Note: If the field flux is not perpendicular to the length of the conductor, it can be resolved into a perpendicular component that generates the force and to a parallel component that has no effect.

The active components of i, B, and F are mutually perpendicular and form a right-hand triad, as shown in Figure 7.15. Alternatively, in the vector representation of these three quantities, the vector F can be interpreted as the **cross product** of the vectors i and B. Specifically, $F = i \times B$.

If the conductor is free to move, the generated force will move it at some velocity v in the direction of the force. As a result of this motion in the magnetic field B, a voltage is induced in the conductor. This is known as the back electromotive force, or *back emf*, and is given by

$$v_b = Blv \tag{7.9}$$

According to *Lenz's law*, the flux due to the back emf v_b will be opposing the flux due to the original current through the conductor, thereby trying to stop the motion. This is the cause of *electrical damping* in motors. Equation 7.8 determines the armature torque (motor torque), and Equation 7.9 governs the motor speed.

7.3.1 Rotor and Stator

A dc motor has a rotating element called the rotor or armature. The rotor shaft is supported on two bearings in the motor housing. The rotor has many closely spaced slots on its periphery. These slots carry the rotor windings, as shown in Figure 7.16a. Assuming the field flux is in the radial direction of the rotor, the force generated in each conductor will be in the tangential direction, thereby generating a torque (force × radius), which drives the rotor. The rotor is typically a laminated cylinder made from a ferromagnetic

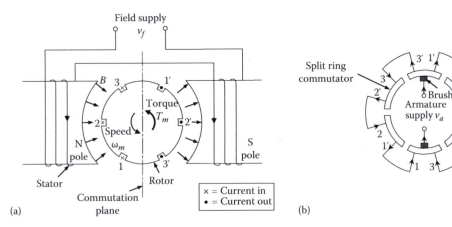

FIGURE 7.16
(a) Schematic diagram of a dc motor; (b) commutator wiring.

material. A ferromagnetic core helps concentrate the magnetic flux toward the rotor. The lamination reduces the problem of magnetic hysteresis and limits the generation of eddy currents and associated dissipation (energy loss by heat generation) within the ferromagnetic material. More advanced dc motors use powdered-iron-core rotors rather than the laminated-iron-core variety, thereby further restricting the generation and conduction/dissipation of eddy currents and reducing various nonlinearities such as hysteresis. The rotor windings (armature windings) are powered by the supply voltage v_a.

The fixed magnetic field (which interacts with the rotor coil and generates the motor torque) is provided by a set of fixed magnetic poles around the rotor. These poles form the stator of the motor. The stator may consist of two opposing poles of a permanent magnet. In industrial dc motors, however, the field flux is usually generated not by a permanent magnet but electrically in the stator windings by an electromagnet, as is shown schematically in Figure 7.16a. Stator poles are constructed from ferromagnetic sheets (i.e., a *laminated construction*). The stator windings are powered by supply voltage v_f, as shown in Figure 7.16a. Furthermore, note that in Figure 7.16a, the net stator magnetic field is perpendicular to the net rotor magnetic field, which is along the commutation plane. The resulting forces that attempt to pull the rotor field toward the stator field may be interpreted as the cause of the motor torque (which is maximum when the two fields are at right angles).

7.3.2 Commutation

A plane known as the "commutation plane" symmetrically divides two adjacent stator poles of opposite polarity. In the two-pole stator shown in Figure 7.16a, the commutation plane is at right angles to the common axis of the two stator poles, which is the direction of the stator magnetic field. It is noted that on one side of the plane, the field is directed toward the rotor, while on the other side the field is directed away from the rotor. Accordingly, when a rotor conductor rotates from one side of the plane to the other side, the direction of the generated torque will be reversed. Such a scenario is not useful since the average torque will be zero in that case.

In order to maintain the direction of torque in each conductor group (one group is numbered 1, 2, 3, and the other group is numbered 1′, 2′, and 3′) in Figure 7.16a, the direction of current in a conductor has to change as the conductor crosses the commutation plane. Physically, this may be accomplished by using a split ring and brush commutator, shown schematically in Figure 7.16b. The armature voltage is applied to the rotor windings through a pair of stationary conducting blocks made of graphite (conducting soft carbon), which maintain sliding contact with the split ring. These contacts are called "brushes" because historically, they were made of bristles of copper wire in the form of a brush. The graphite contacts are cheaper, more durable primarily due to reduced sparking (arcing) problems, and provide more contact area (less electrical contact resistance). Also the contact friction is lower. The split ring segments, equal in number to the conductor slots in the rotor, are electrically insulated from one another, but the adjacent segments are connected by the armature windings in each opposite pair of rotor slots, as shown in Figure 7.16b. For the rotor position shown in Figure 7.16, note that when the split ring rotates in the CCW direction through 30°, the current paths in conductors 1 and 1′ reverse but the remaining current paths are unchanged, thus achieving the required commutation.

7.3.3 Brushless DC Motors

There are several shortcomings of the slip ring and brush mechanisms that are used for current transmission through moving members, even with the advances from the historical copper brushes to modern graphite contacts. The main disadvantages include rapid wearout, mechanical loading, wear and heating due to sliding friction, contact bounce, excessive noise, and electrical sparks (arcing) with the associated dangers in hazardous (e.g., chemical) environments, problems of oxidation, problems in applications that require wash-down (e.g., in food processing), and voltage ripples at switching points. Conventional remedies for these problems, such as the use of improved brush designs and modified brush positions to reduce arcing, are inadequate in sophisticated applications. Also, the required maintenance (to replace brushes and resurface the split-ring commutator) can be rather costly.

Brushless dc motors have permanent-magnet rotors. Since in this case the polarities of the rotor cannot be switched as the rotor crosses a commutation plane, commutation is accomplished by electronically switching the current in the stator winding segments. Note that this is the reverse of what is done in brushed commutation, where the stator polarities are fixed and the rotor polarities are switched when crossing a commutation plane. The stator windings of a brushless dc motor can be considered to be the armature windings, whereas for a brushed dc motor, the rotor is the armature.

Permanent-magnet motors are less nonlinear than the electro-magnet motors because the field strength generated by a permanent magnet is rather constant and independent of the current through a coil. This is true whether the permanent magnet is in the stator (i.e., a brushed motor) or in the rotor (i.e., a brushless dc motor or a PM stepper motor).

7.3.4 DC Motor Equations

Consider a dc motor with separate windings in the stator and the rotor. Each coil has a resistance (R) and an inductance (L). When a voltage (v) is applied to the coil, a current (i) flows through the circuit, thereby generating a magnetic field. As discussed before, a force is produced in the rotor windings and an associated torque (T_m), which turns the rotor. The rotor speed (ω_m) causes the flux linkage of the rotor coil with the stator field to change at a corresponding rate, thereby generating a voltage (*back emf*) in the rotor coil.

Equivalent circuits for the stator and the rotor of a conventional dc motor are shown in Figure 7.17a. Since the field flux is proportional to field current i_f, we can express the magnetic torque of the motor as

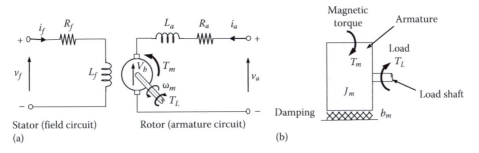

Stator (field circuit)
(a)

Rotor (armature circuit)

(b)

FIGURE 7.17
(a) The equivalent circuit of a conventional dc motor (separately excited); (b) armature mechanical loading diagram.

$$T_m = k i_f i_a \tag{7.10}$$

This directly follows from Equation 7.8. Next, in view of Equation 7.9, the back emf generated in the armature of the motor is given by

$$v_b = k' i_f \omega_m \tag{7.11}$$

The following notation has been used:

i_f = field current

i_a = armature current

ω_m = angular speed of the motor

and k and k' are motor constants, which depend on factors such as the rotor dimensions, the number of turns in the armature winding, and the permeability (inverse of reluctance) of the magnetic medium. In the case of ideal electrical-to-mechanical energy conversion at the rotor (where the rotor coil links with the stator field), we have $T_m \omega_m = v_b i_a$ with consistent units (e.g., torque in Newton-meters, speed in radians per second, voltage in volts, and current in amperes). Then we observe that

$$k = k' \tag{7.12}$$

The field circuit equation is obtained by assuming that the stator magnetic field is not affected by the rotor magnetic field (i.e., the stator inductance is not affected by the rotor) and that there are no eddy current effects in the stator. Then, from Figure 7.17a

$$v_f = R_f i_f + L_f \frac{di_f}{dt} \tag{7.13}$$

where
v_f is the supply voltage to the stator
R_f is the resistance of the field winding
L_f is the inductance of the field winding

The equation for the armature rotor circuit is written as (see Figure 7.17a)

$$v_a = R_a i_a + L_a \frac{di_a}{dt} + v_b \tag{7.14}$$

where
v_a is the supply voltage to the armature
R_a is the resistance of the armature winding
L_a is the leakage inductance in the armature winding

It should be emphasized here that the primary inductance or *mutual inductance* in the armature winding is represented in the back emf term v_b. The leakage inductance, which is usually neglected, represents the fraction of the armature flux that is not linked with the stator and is not used in the generation of useful torque. This includes self-inductance in the armature.

The mechanical equation of the motor is obtained by applying Newton's second law to the rotor. Assuming that the motor drives some load, which requires a load torque T_L to operate, and that the frictional resistance in the armature can be modeled by a linear viscous term, we have (see Figure 7.17b)

$$J_m \frac{d\omega_m}{dt} = T_m - T_L - b_m \omega_m \tag{7.15}$$

where
J_m is the moment of inertia of the rotor
b_m is the equivalent (mechanical) damping constant for the rotor

Note: The load torque may be due, in part, to the inertia of the external load that is coupled to the motor shaft. If the coupling flexibility is neglected, the load inertia may be directly added to (i.e., lumped with) the rotor inertia after accounting for the possible existence of a speed reducer (gear, harmonic drive, etc.). In general, a separate set of equations is necessary to represent the dynamics of the external load. Equations 7.10 through 7.15 form the dynamic model for a dc motor.

7.3.4.1 Steady-State Characteristics

In selecting a motor for a given application, its steady-state characteristics are a major determining factor. In particular, steady-state torque–speed curves are employed for this purpose. The rationale is that, if the motor is able to meet the steady-state operating requirements, with some design conservatism, it should be able to tolerate small deviations under transient conditions of short duration. In the separately excited case shown in Figure 7.17a, where the armature circuit and field circuit are excited by separate and independent voltage sources, it can be shown that the steady-state torque–speed curve is a straight line.

The shape of the steady-state speed–torque curve will be modified if a common voltage supply is used to excite both the field winding and the armature winding. Here, the two windings have to be connected together. There are three common ways the windings of the rotor and the stator are connected. They are known as: shunt-wound motor, series-wound motor, and compound-wound motor. In a shunt-wound motor, the armature windings and the field windings are connected in parallel. In the series-wound motor, they connected in a series. In the compound-wound motor, part of the field windings are connected with the armature windings in the series and the other part is connected in parallel. In a shunt-wound motor at steady-state, the back emf v_b depends directly on the supply voltage. Since the back emf is proportional to the speed, it follows that speed controllability is good with the shunt-wound configuration. In a series-wound motor, the relation between v_b and the supply voltage is coupled through both the armature windings and the field windings. Hence, its speed controllability is relatively poor. But in this case, a relatively large current flows through both windings at low speeds of the motor, giving a higher starting torque. Also, the operation is approximately at constant power in this case. Since both speed controllability and higher starting torque are desirable characteristics, compound-wound motors are used to obtain a performance in between the two extremes. The torque–speed characteristics for the three types of winding connections are shown in Figure 7.18.

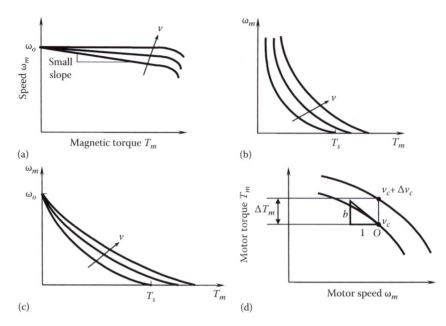

FIGURE 7.18
Torque–speed characteristic curves for dc motors: (a) Shunt-wound; (b) series-wound; (c) compound-wound; (d) general case.

7.3.5 Experimental Model for DC Motor

In general, the speed–torque characteristic of a dc motor is nonlinear. A linearized dynamic model can be extracted from the speed–torque curves. One of the parameters of the model is the damping constant. First, we will examine this.

7.3.5.1 Electrical Damping Constant

Newton's second law governs the dynamic response of a motor. In Equation 7.15, for example, b_m denotes the mechanical (viscous) damping constant and represents the mechanical dissipation of energy. As is intuitively clear, mechanical damping torque opposes motion—hence the negative sign in the $b_m\omega_m$ term in Equation 7.15. Furthermore, note that the magnetic torque T_m of the motor is also dependent on speed ω_m. In particular, the back emf, which is governed by ω_m, produces a magnetic field, which tends to oppose the motion of the motor rotor. This acts as a damper and the corresponding damping constant is given by

$$b_e = -\frac{\partial T_m}{\partial \omega_m} \tag{7.16}$$

This parameter is termed the *electrical damping constant*. Caution should be exercised when experimentally measuring b_e. Note that in constant speed tests, the inertia torque of the rotor will be zero; there is no torque loss due to inertia. The torque measured at the motor shaft includes, as well, the torque reduction due to mechanical dissipation (mechanical damping) within the rotor. Hence, the magnitude b of the slope of the speed–torque curve as obtained by a steady-state test is equal to $b_e + b_m$, where b_m is the equivalent viscous damping constant representing mechanical dissipation at the rotor.

7.3.5.2 Linearized Experimental Model

To extract a linearized experimental model for a dc motor, consider the speed–torque curves shown in Figure 7.18d. For each curve, the excitation voltage v_c is maintained constant. This is the voltage that is used in controlling the motor, and is termed control voltage. It can be, for example, the armature voltage, the field voltage, or the voltage that excites both armature and field windings in the case of combined excitation (e.g., shunt-wound motor). One curve in Figure 7.18d is obtained at control voltage v_c and the other curve is obtained at $v_c + \Delta v_c$. Note also that a tangent can be drawn at a selected point (operating point O) of a speed–torque curve. The magnitude b of the slope (which is negative) corresponds to a damping constant, which includes both electrical and mechanical damping effects. The mechanical damping effects that are included in this parameter depend entirely on the nature of mechanical damping that was present during the test (primarily bearing friction). We have the *damping constant* as the magnitude of the slope at the operating point:

$$b = -\frac{\partial T_m}{\partial \omega_m}\bigg|_{v_c = \text{constant}} \tag{7.17}$$

Next, draw a vertical line through the operating point O. The torque intercept ΔT_m between the two curves can be determined in this manner. Since a vertical line is a constant speed line, we have the *voltage gain*

$$k_v = \frac{\partial T_m}{\partial v_c}\bigg|_{\omega_m = \text{constant}} = \frac{\Delta T_m}{\Delta v_c} \tag{7.18}$$

Now, using the well-known relation for total differential, we have

$$\delta T_m = \frac{\partial T_m}{\partial \omega_m}\bigg|_{v_c} \delta \omega_m + \frac{\partial T_m}{\partial v_c}\bigg|_{\omega_m} \delta v_c = -b\delta \omega_m + k_v \delta v_c \tag{7.19}$$

Equation 7.19 is the linearized model of the motor. This may be used in conjunction with the mechanical equation of the motor rotor, for the incremental motion about the operating point:

$$J_m \frac{d\delta \omega_m}{dt} = \delta T_m - \delta T_L \tag{7.20}$$

Note that Equation 7.20 is the incremental version of Equation 7.15 except that the overall damping constant of the motor (including mechanical damping) is included in Equation 7.19. The torque needed to drive the rotor inertia, however, is not included in Equation 7.19 because the steady-state curves are used in determining the parameters for this equation. The inertia term is explicitly present in Equation 7.20.

7.3.6 Control of DC Motors

Both the speed and torque of a dc motor may have to be controlled for proper performance in a given application of a dc motor. By using proper winding arrangements, dc motors can be operated over a wide range of speeds and torques. Because of this adaptability, dc motors are particularly suitable as variable-drive actuators. Historically, ac motors were employed almost exclusively in constant-speed applications, but their use in variable-speed applications was greatly limited because speed control of ac motors was found to be quite difficult by conventional means. Since variable-speed control of a dc motor is quite convenient and straightforward, dc motors have dominated in industrial control applications for many decades.

Following a specified motion trajectory is called servoing and servomotors (or servoactuators) are used for this purpose. The vast majority of servomotors are dc motors with feedback control of motion. Servo control is essentially a motion control problem, which involves the control of position and speed. There are applications, however, that require torque control, directly or indirectly, but they usually require more sophisticated sensing and control techniques. Control of a dc motor is accomplished by controlling either the stator field flux or the armature flux. If the armature and field windings are connected through the same circuit, both techniques are incorporated simultaneously. Specifically, the two methods of control are the armature control and field control.

7.3.6.1 Armature Control

In an armature-controlled dc motor, the armature voltage v_a is used as the control input, while keeping the conditions in the field circuit constant. In particular, the field current i_f is assumed constant. Consequently, Equations 7.10 and 7.11 can be written as

$$T_m = k_m i_a \tag{7.21}$$

$$v_b = k'_m \omega_m \tag{7.22}$$

The parameters k_m and k'_m are termed the *torque constant* and the *back emf constant*, respectively.

Note: With consistent units, $k_m = k'_m$ in the case of ideal electrical-to-mechanical energy conversion at the motor rotor.

In the Laplace domain, Equation 7.14 becomes

$$v_a - v_b = (L_a s + R_a)i_a \tag{7.23}$$

Note: For convenience, time domain variables (functions of *t*) are used to denote their Laplace transforms (functions of *s*). It is understood, however, that the time functions are not identical to the Laplace functions.

In the Laplace domain, the mechanical Equation 7.15 becomes

$$T_m - T_L = (J_m s + b_m)\omega_m \tag{7.24}$$

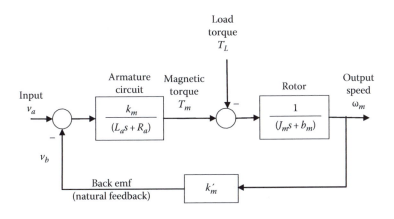

FIGURE 7.19
Open-loop block diagram for an armature-controlled dc motor.

where J_m and b_m denote the moment of inertia and the rotary viscous damping constant, respectively of the motor rotor. Equations 7.22 through 7.24 are represented in the block diagram form in Figure 7.19. Note that the speed ω_m is taken as the motor output. If the motor position θ_m is considered the output, it is obtained by passing ω_m through an integration block $1/s$. Note, further, that the load torque T_L, which is the useful (effective) torque transmitted to the load that is being driven, is an (unknown) input to the system. Usually, T_L increases with ω_m because a larger torque is necessary to drive a load at a higher speed. If a linear (and dynamic) relationship exists between T_L and ω_m at the load, a feedback path can be completed from the output speed to the input load torque through a proper load transfer function (load block). The system shown in Figure 7.19 is not a feedback control system. The feedback path, which represents the back emf, is a "natural feedback" and is characteristic of the process (dc motor); it is not an external control feedback loop.

The overall transfer relation for the system is obtained by first determining the output for one of the inputs with the other input removed, and then adding the two output components obtained in this manner, in view of the *principle of superposition*, which holds for a linear system. We get

$$\omega_m = \frac{k_m}{\Delta(s)} v_a - \frac{(L_a s + R_a)}{\Delta(s)} T_L \tag{7.25}$$

where $\Delta(s)$ is the *characteristic polynomial* of the system, given by

$$\Delta(s) = (L_a s + R_a)(J_m s + b_m) + k_m k'_m \tag{7.26}$$

This is a second-order polynomial in the Laplace variable s.

7.3.6.2 Motor Time Constants

The *electrical time constant* of the armature is

$$\tau_a = \frac{L_a}{R_a} \tag{7.27}$$

which is obtained from Equation 7.14 or 7.23. The mechanical response of the rotor is governed by the *mechanical time constant*

$$\tau_m = \frac{J_m}{b_m} \tag{7.28}$$

which is obtained from Equation 7.15 or 7.24. Usually, τ_m is several times larger than τ_a, because the leakage inductance L_a is quite small (leakage of the flux linkage is negligible for high-quality dc motors). Hence, τ_a can be neglected in comparison with τ_m for most practical purposes. In that case, the transfer functions in Equation 7.25 become first order.

Note that the characteristic polynomial is the same for both transfer functions in Equation 7.25, regardless of the input (v_a or T_L). This should be the case because $\Delta(s)$ determines the natural response of the system and does not depend on the system input. True time constants of the motor are obtained by first solving the characteristic equation $\Delta(s) = 0$ to determine the two roots (poles or eigenvalues) and then taking the reciprocal of the magnitudes (*Note*: only the real part of the two roots is used if the roots are complex). For an armature-controlled dc motor, these true time constants are not the same as τ_a and τ_m because of the presence of the coupling term $k_m k'_m$ in $\Delta(s)$ (see Equation 7.26). This also follows from the presence of the natural feedback path (back emf) in Figure 7.19.

Example 7.3

Determine an expression for the dominant time constant of an armature-controlled dc motor.

Solution
By neglecting the electrical time constant in Equation 7.26, we have the approximate characteristic polynomial $\Delta(s) = R_a(J_m s + b_m) + k_m k'_m \rightarrow \Delta(s) = k'(\tau s + 1)$
  where τ = the overall dominant time constant of the system
  It follows that the dominant time constant is given by

$$\tau = \frac{R_a J_m}{(R_a b_m + k_m k'_m)} \tag{7.29}$$

7.3.6.3 Field Control

In field-controlled dc motors, the armature current is assumed to be kept constant and the field voltage is used as the control input. Since i_a is assumed constant, Equation 7.10 can be written as

$$T_m = k_a i_f \tag{7.30}$$

where k_a is the electromechanical torque constant for the motor. The back emf relation and the armature circuit equation are not used in this case. Equations 7.13 and 7.15 are written in the Laplace form as

$$v_f = (L_f s + R_f) i_f \tag{7.31}$$

$$T_m - T_L = (J_m s + b_m) \omega_m \tag{7.32}$$

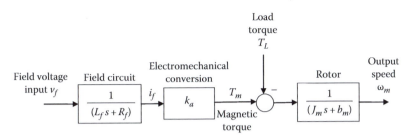

FIGURE 7.20
Open-loop block diagram for a field-controlled dc motor.

Equations 7.30 through 7.32 can be represented by the open-loop block diagram given in Figure 7.20.

Note that even though i_a is assumed constant, this is not strictly true. This should be clear from the armature circuit Equation 7.14. It is the armature supply voltage v_a that is kept constant. Even though L_a can be neglected, i_a depends on the back emf v_b, which changes with the motor speed as well as the field current i_f. Under these conditions, the block representing k_a in Figure 7.20 is not a constant gain, and in fact it is not linear. At least, feedback will be needed into this block from output speed. This will also add another electrical time constant, which depends on the dynamics of the armature circuit. It will also introduce a coupling effect between the mechanical dynamics (of the rotor) and the armature circuit electronics. For the present purposes, however, we assume that k_a is a constant gain.

Now, we return to Figure 7.20. Since the system is linear, the principle of superposition holds. According to this, the overall output ω_m is equal to the sum of the individual outputs due to the two inputs v_f and T_L, taken separately. It follows that the transfer relationship is given by

$$\omega_m = \frac{k_a}{(L_f s + R_f)(J_m s + b_m)} v_f - \frac{1}{(J_m s + b_m)} T_L \tag{7.33}$$

In this case, the electrical time constant originates from the field circuit and is given by

$$\tau_f = \frac{L_f}{R_f} \tag{7.34}$$

The mechanical time constant τ_m of the field-controlled motor is the same as that for the armature-controlled motor and can be defined by Equation 7.28:

$$\tau_m = \frac{J_m}{b_m} \tag{7.28}$$

The characteristic polynomial of the open-loop field-controlled motor is

$$\Delta(s) = (L_f s + R_f)(J_m s + b_m) \tag{7.35}$$

It follows that τ_f and τ_m are the true time constants of the system, unlike in an armature-controlled motor. As in an armature-controlled dc motor, however, the electrical time constant is several times smaller and can be neglected in comparison with the mechanical time constant. Furthermore, as for an armature-controlled motor, the speed and the angular position of a field-controlled motor have to be measured and fed back for accurate motion control.

7.3.7 Feedback Control of DC Motors

The open-loop operation of a dc motor, as represented by Figures 7.19 (armature control) and 7.20 (field control), can lead to excessive error and even instability, particularly because of the unknown load input and also due to the integration effect when position (not speed) is the desired output (as in positioning applications). Feedback control is necessary in these circumstances.

In feedback control, the motor response (position, speed, or both) is measured using an appropriate sensor and fed back into the controller, which generates the control signal for the drive hardware of the motor. An optical encoder can be used to sense both position and speed and a tachometer may be used to measure the speed alone. The following three types of feedback control are important:

1. Velocity feedback
2. Position plus velocity feedback
3. Position feedback with a multi-term controller

7.3.7.1 Velocity Feedback Control

Velocity feedback is particularly useful in controlling the motor speed. In velocity feedback, motor speed is sensed using a device such as a tachometer or an optical encoder, and is fed back to the controller, which compares it with the desired speed and the error is used to correct the deviation. Additional dynamic compensation (e.g., lead or lag compensation) may be needed to improve the accuracy and the effectiveness of the controller and can be provided using either analog circuits or digital processing. The error signal is passed through the compensator in order to improve the performance of the control system.

7.3.7.2 Position Plus Velocity Feedback Control

In position control, the motor angle θ_m is the output. In this case, the open-loop system has a free integrator and the characteristic polynomial is $s(\tau s + 1)$. This is a marginally stable system. In particular, if a slight disturbance or model error is present, it will be integrated out, which can lead to a diverging error in the motor angle. In particular, the load torque T_L is an input to the system and is not completely known. In control systems terminology, this is a disturbance (an unknown input), which can cause unstable behavior in the open-loop system. In view of the free integrator at the position output, the resulting unstable behavior cannot be corrected using velocity feedback alone. Position feedback is needed to remedy the problem. Both position and velocity feedback are needed. The feedback gains for the position and velocity signals can be chosen so as to obtain the desired response (speed of response, overshoot limit, steady-state accuracy, etc.). A block diagram of a position plus velocity feedback control system for a dc motor is shown in Figure 7.21. The motor block is shown in Figure 7.19 for an armature-controlled motor and in Figure 7.20 for a field-control

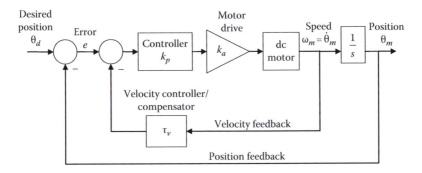

FIGURE 7.21
Position plus velocity feedback control of a dc motor.

motor (*Note*: load torque input is integral in each of these two models). The drive unit of the motor is represented by an amplifier of gain k_a. Control system design involves the selection of proper parameter values for sensors and other components in the control system.

7.3.7.3 Position Feedback with PID Control

A popular method of controlling a dc motor is to use just position feedback and then compensate for the error using a three-term controller having the proportional, integral, and derivative (PID) actions. A block diagram for this control system is shown in Figure 7.22.

In the control system of a dc motor (Figure 7.21 or 7.22), the desired position command may be provided by a potentiometer as a voltage signal. The measurements of position and speed also are provided as voltage signals. Specifically, in the case of an optical encoder, the pulses are detected by a digital pulse counter and read into the digital controller. This reading has to be calibrated to be consistent with the desired position command. In the case of a tachometer, the velocity reading is generated as a voltage, which has to be calibrated to be consistent with the desired position signal.

It is noted that proportional plus derivative control (PPD control or PD control) with position feedback has a similar effect as position plus velocity (speed) feedback control. But, the two are not identical because the latter adds a zero to the system transfer function, requiring further considerations in the controller design and affecting the motor response. In particular, the zero modifies the sign and the ratio in which the two response components corresponding to the two poles contribute to the overall response.

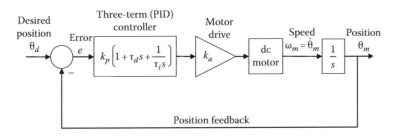

FIGURE 7.22
PID control of the position response of a dc motor.

7.3.8 Motor Driver

The driver of a dc motor is a hardware unit, which generates the necessary current to energize the windings of the motor. By controlling the current generated by the driver, the motor torque can be controlled. By receiving feedback from a motion sensor (encoder, tachometer, etc.), the angular position and the speed of the motor can be controlled.

Note: When an optical encoder is provided with the motor—a typical situation—it is not necessary to use a tachometer as well because the encoder can generate both position and speed measurements.

The drive unit primarily consists of a drive amplifier, with additional circuitry, and a dc power supply. In typical applications of motion control and servoing, the drive unit is a *servoamplifier* with auxiliary hardware. The driver is commanded by a control input provided by a host computer (personal computer or PC) through an interface (input/output [I/O]) card. A suitable arrangement is shown in Figure 7.23. Also, the driver parameters (e.g., amplifier gains) are software programmable and can be set by the host computer.

The control computer receives a feedback signal of the motor motion, through the interface board, and generates a control signal, which is provided to the drive amplifier, again through the interface board. Any control scheme can be programmed (say, in C language) and implemented in the control computer. In addition to typical servo control schemes such as PID and position-plus-velocity feedback, other advanced control algorithms (e.g., optimal control techniques, such as linear quadratic regulator [LQR] and linear quadratic Gaussian [LQG]; adaptive control techniques, such as model-referenced adaptive control; switching control techniques, such as sliding-mode control; nonlinear control schemes, such as the feedback linearization technique [FLT]; and intelligent control techniques, such as fuzzy logic control) may be applied in this manner. If the computer does not have the processing power to carry out the control computations at the required speed (i.e., control bandwidth), a digital signal processor (DSP) may be incorporated into the computer. But, with modern computers, which can provide substantial computing power at low cost, DSPs are not needed in most applications.

7.3.8.1 Interface Board

The I/O card is a hardware module with associated driver software based in a host computer (PC) and connected through its bus (ISA bus). It forms the input–output link between

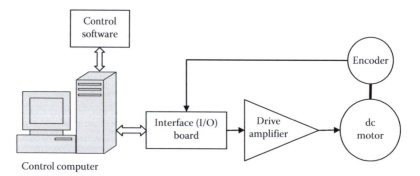

FIGURE 7.23
Components of a dc motor control system.

the motor and the controller. It can provide many (say, eight) analog signals to drive many (eight) motors, and hence termed a *multi-axis* card. It follows that the digital-to-analog conversion (DAC) capability is built into the I/O card (e.g., a 16-bit DAC including a sign bit, ±10 V output voltage range). Similarly, the analog-to-digital conversion (ADC) function is included in the I/O card (e.g., eight analog input channels with a 16-bit ADC including a sign bit, ±10 V output voltage range). These input channels can be used for analog sensors such as tachometers, potentiometers, and strain gauges. Equally important are the encoder channels to read the pulse signals from the optical encoders mounted on the dc servomotors. Typically, the encoder input channels are equal in number to the analog output channels (and the number of axes; e.g., eight). The position pulses are read using counters (e.g., 24-bit counters), and the speed is determined by the pulse rate. The rate at which the encoder pulses are counted can be quite high (e.g., 10 MHz). In addition, a number of bits (e.g., 32) of digital input and output may be available through the I/O card for use in simple digital sensing, control, and switching functions.

7.3.8.2 Drive Unit

The primary hardware component of the motor drive system is the drive amplifier. In typical motion control applications, these amplifiers are called servo amplifiers. Two types of drive amplifiers are commercially available:

1. Linear amplifier
2. Pulse-width modulation (PWM) amplifier

A linear amplifier generates a voltage output, which is proportional to the control input provided to it. Since the output voltage is proportioned by dissipative means (using resistor circuitry), this is a wasteful and in efficient approach. Furthermore, fans and heat sinks have to be provided to remove the generated heat, particularly in continuous operation. To understand the inefficiency associated with a linear amplifier, suppose that the operating output range of the amplifier is 0–20 V and that the amplifier is powered by a 20 V power supply. Under a particular operating condition, suppose that the motor is applied 10 V and draws a current of 4 A. The power used by the motor then is $10 \times 4\,\text{W} = 40\,\text{W}$. Still, the power supply provides 20 V at 5 A, thereby consuming 100 W. This means, 60 W of power is dissipated and the efficiency is only 40%. The efficiency can be made close to 100% using modern PWM amplifiers, which are nondissipative devices depending on high-speed switching at a constant voltage to control the power supplied to the motor, as discussed next.

Modern servo amplifiers use PWM to drive servomotors efficiently under variable-speed conditions without incurring excessive power losses. Integrated microelectronic design makes them compact, accurate, and inexpensive. The components of a typical PWM drive system are shown in Figure 7.24. Other signal conditioning hardware (e.g., filters) and auxiliary components such as isolation hardware, safety devices including tripping hardware, and cooling fan are not shown in the figure. In particular, note the following components connected in a series:

1. A velocity amplifier (a differential amplifier)
2. A torque amplifier
3. A PWM amplifier

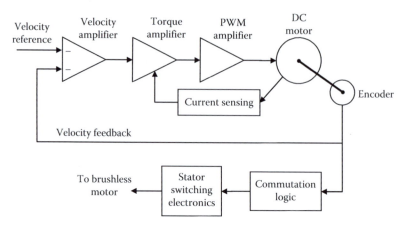

FIGURE 7.24
The main components of a PWM drive system for a dc motor.

The power can come from an ac line supply, which is rectified and regulated in the drive unit to provide the necessary dc power for the electronics. Alternatively, leads may be provided for an external power supply (e.g., 15 V dc). The reference velocity signal and the feedback signal (from an encoder or a tachometer) are connected to the input leads of the velocity amplifier. The resulting difference (error signal) is conditioned and amplified by the torque amplifier to generate a current corresponding to the required torque (corresponding to the driving speed). The motor current is sensed and fed back to this amplifier to improve the torque performance of the motor. The output from the torque amplifier is used as the modulating signal to the PWM amplifier. The reference switching frequency of a PWM amplifier is high (on the order of 25 kHz). PWM is accomplished by varying the duty cycle of the generated pulse signal, through switching control, as explained next. The PWM signal from the amplifier (e.g., at 10 V) is used to energize the field windings of a dc motor. A brushless dc motor needs electronic commutation. This may be accomplished using the encoder signal to time the switching of the current through the stator windings.

Consider the voltage pulse signal shown in Figure 7.25. The following notation is used:

T = pulse period (i.e., the interval between the successive on times)

T_o = on period (i.e., the interval between the on time to the next off time)

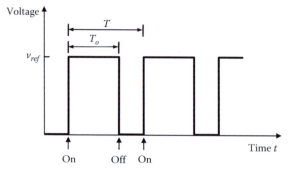

FIGURE 7.25
Duty cycle of a PWM signal.

Then, the *duty cycle* is given by the percentage

$$d = \frac{T_o}{T} \times 100\% \tag{7.36}$$

Note: In PWM, the voltage level v_{ref} and the pulse frequency $1/T$ are kept fixed and T_o is varied.

PWM is achieved by "chopping" the reference voltage so that the average voltage is varied. It is easy to see that, with respect to an output pulse signal, the duty cycle is given by the ratio of the average output to the peak output:

$$\text{Duty cycle} = \frac{\text{Average output}}{\text{Peak output}} \times 100\% \tag{7.37}$$

Equation 7.36 or 7.37 also verifies that the average level of a PWM signal is proportional to the duty cycle (or the on time period T_o) of the signal. It follows that the output level (i.e., the average value) of a PWM signal can be varied simply by changing the signal-on time period (in the range 0 to T) or equivalently by changing the duty cycle (in the range 0%–100%). This relationship between the average output and the duty cycle is linear. Hence, a digital or software means of generating a PWM signal would be to use a straight line from 0 to the maximum signal level, spanning the period (T) of the signal. For a given output level, the straight line segment at this height, when projected on the time axis, gives the required on-time interval (T_o).

Historically, and even today in laboratory projects, for example, the PWM type power amplifiers for motors have been constructed using discrete power-electronic components such as bipolar transistors or field effect transistors (FETs). The H-bridge structure is common. More convenient and cost effective are the PWM drives with monolithic power electronics, which are commercially available in the IC form from such companies as National Semiconductor, Texas Instruments, and Agilent Technologies. A typical motor drive arrangement of this type is shown in Figure 7.26.

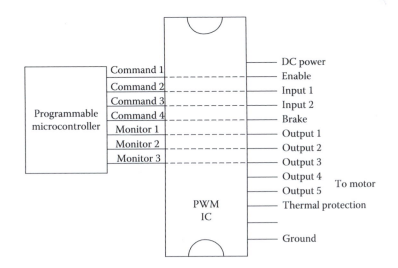

FIGURE 7.26
A motor drive arrangement with commercial IC hardware.

In this arrangement, there is a programmable microcontroller together with the PWM drive IC. The microcontroller provides command for such requirements as the position, speed, and direction of motion to the drive IC. The motor receives the PWM signals for its windings from the drive IC. In addition, the drive IC possesses various capabilities as current sensing, thermal (overheat) protection, and braking. Such a drive system with commercial hardware may be used not just for dc motors but also for a variety of other motors such as steppers.

7.3.9 DC Motor Selection

DC motors, dc servomotors in particular, are suitable for applications requiring continuous operation (continuous duty) at high levels of torque and speed. Brushless permanent-magnet motors with advanced magnetic material provide high torque/mass ratio and are preferred for continuous operation at high throughput (e.g., component insertion machines in the manufacture of printed-circuit boards, portioning and packaging machines, printing machines) and high speeds (e.g., conveyors, robotic arms) in hazardous environments (where spark generation from brushes would be dangerous) and in applications that need minimal maintenance and regular wash-down (e.g., in food processing applications). For applications that call for high torques and low speeds at high precision (e.g., inspection, sensing, product assembly), torque motors or regular motors with suitable speed reducers (e.g., harmonic drive, gear unit commonly using worm gears, etc.) may be employed.

A typical application involves a "rotation stage" producing rotary motion for the load. If an application requires linear (rectilinear) motions, a "linear stage" has to be used. One option is to use a rotary motor with a rotatory-to-linear motion transmission device such as a lead screw or ball screw and nut, rack, and pinion or a conveyor belt. This approach introduces some degree of nonlinearity and other errors (e.g., friction, backlash). For high-precision applications, a linear motor provides a better alternative. The operating principle of a linear motor is similar to that of a rotary motor, except linearly moving armatures on linear bearings or guideways are used instead of rotors mounted on rotary bearings.

When selecting a dc motor for a particular application, a matching drive unit has to be chosen as well. Due consideration must be given to the requirements (specifications) of power, speed, accuracy, resolution, size, weight, and cost when selecting a motor and a drive system. In fact, vendor catalogs give the necessary information for motors and matching drive units, thereby making the selection far more convenient. Also, a suitable speed transmission device (harmonic drive, gear unit, lead screw and nut, etc.) may have to be chosen as well, depending on the application.

7.3.9.1 Motor Data and Specifications

Torque and speed are the two primary considerations in choosing a motor for a particular application. Speed–torque curves are available, in particular. The torques given in these curves are typically the maximum torques (known as peak torques), which the motor can generate at the indicated speeds. A motor should not be operated continuously at these torques (and current levels) because of the dangers of overloading, wear, and malfunction. The peak values have to be reduced (say, by 50%) in selecting a motor to match the torque requirement for continuous operation. Alternatively, the continuous torque values as given by the manufacturer should be used in the motor selection.

Motor manufacturers' data that are usually available to users include the following:

1. Mechanical data
   - Peak torque (e.g., 65 N-m)
   - Continuous torque at zero speed or continuous stall torque (e.g., 25 N-m)
   - Frictional torque (e.g., 0.4 N-m)
   - Maximum acceleration at peak torque (e.g., 33×10^3 rad/s^2)
   - Maximum speed or no-load speed (e.g., 3000 rpm)
   - Rated speed or speed at rated load (e.g., 2400 rpm)
   - Rated output power (e.g., 5100 W)
   - Rotor moment of inertia (e.g., 0.002 kg-m^2)
   - Dimensions and weight (e.g., 14 cm diameter, 30 cm length, 20 kg)
   - Allowable axial load or thrust (e.g., 230 N)
   - Allowable radial load (e.g., 700 N)
   - Mechanical (viscous) damping constant (e.g., 0.12 N-m/k rpm)
   - Mechanical time constant (e.g., 10 ms)
2. Electrical data
   - Electrical time constant (e.g., 2 ms)
   - Torque constant (e.g., 0.9 N-m/A for peak current or 1.2 N-m/A rms current)
   - Back emf constant (e.g., 0.95 V/rad/s for peak voltage)
   - Armature/field resistance and inductance (e.g., 1.0 Ω, 2 mH)
   - Compatible drive unit data (voltage, current, etc.)
3. General data
   - Brush life and motor life (e.g., 5×10^8 revolutions at maximum speed)
   - Operating temperature and other environmental conditions (e.g., 0°C–40°C)
   - Thermal resistance (e.g., 1.5°C/W)
   - Thermal time constant (e.g., 70 min)
   - Mounting configuration

Quite commonly, motors and drive systems are chosen from what is commercially available. Customized production may be required, however, in highly specialized applications and in research and development applications where the cost may not be a primary consideration. The selection process involves matching the engineering specifications for a given application with the data of commercially available motor systems.

7.3.9.2 Selection Considerations

When a specific application calls for large speed variations (e.g., speed tracking over a range of 10 dB or more), armature control is preferred. Note, however, that at low speeds (typically, half the rated speed), poor ventilation and associated temperature buildup can cause problems. At very high speeds, mechanical limitations and heating due to frictional

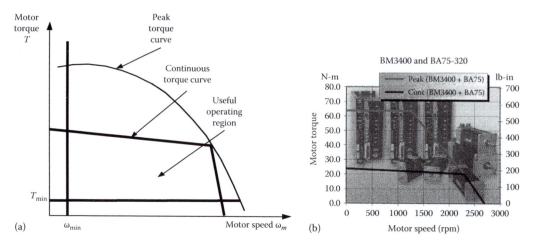

FIGURE 7.27
(a) Representation of the useful operating region for a dc motor; (b) speed–torque characteristics of a commercial brushless dc servomotor with a matching amplifier. (Courtesy of Aerotech, Inc., Pittsburgh, PA.)

dissipation become determining factors. For constant-speed applications, shunt-wound motors are preferred. Finer speed regulation may be achieved using a servo system with encoder or tachometer feedback or with phase-locked operation. For constant power applications, the series-wound or compound-wound motors are preferable over shunt-wound units. If the shortcomings of mechanical commutation and limited brush life are critical, brushless dc motors should be used.

A simple way to determine the operating conditions of a motor is by using its torque–speed curve, as illustrated in Figure 7.27. What is normally provided by the manufacturer is the peak torque curve, which gives the maximum torque the motor (with a matching drive system) can provide at a given speed for short periods (say, 30% duty cycle). The actual selection of a motor should be based on its continuous torque, which is the torque that the motor is able to provide continuously at a given speed for long periods without overheating or damaging the unit. If the continuous torque curve is not provided by the manufacturer, the peak torque curve should be reduced by about 50% (or even by 70%) for matching with the specified operating requirements. The minimum operating torque T_{min} is limited mainly by loading considerations. The minimum speed ω_{min} is determined primarily by the operating temperature. These boundaries along with the continuous torque curve define the useful operating region of the particular motor (and its drive system), as indicated in Figure 7.27a. The optimal operating points are those that fall within this segment on the continuous torque–speed curve. The upper limit on speed may be imposed by taking into account the transmission limitations in addition to the continuous torque–speed capability of the motor system.

7.3.9.3 Motor Sizing Procedure

Motor sizing is the term used to denote the procedure of matching a motor (and its drive system) to a load (demand of the specific application). The load may be given by a load curve, which is the speed–torque curve representing the torque requirements for operating the load at various speeds (see Figure 7.28). Clearly, greater torques are needed to drive

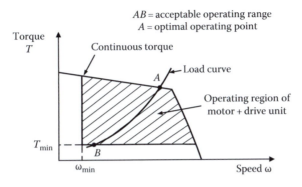

FIGURE 7.28
Sizing a motor for a given load.

a load at higher speeds. For a motor and a load, the acceptable operating range is the interval where the load curve overlaps with the operating region of the motor (segment AB in Figure 7.28). The optimal operating point is the point where the load curve intersects with the speed–torque curve of the motor (point A in Figure 7.28).

Sizing a dc motor is similar to sizing a stepper motor, as studied before. The same equations may be used for computing the load torque (demand). The motor characteristic (speed–torque curve) gives the available torque, as in the case of a stepper motor. The main difference is that a stepper motor is not suitable for continuous operation for long periods and at high speeds, whereas a dc motor can perform well in such situations. In this context, a dc motor can provide high torques, as given by its "peak torque curve," for short periods, and reduced torques, as given by its "continuous torque curve" can provide high torques for long periods of operation. In the motor sizing procedure, then, the peak torque curve may be used for short periods of acceleration and deceleration, but the continuous torque curve (or the peak torque curve reduced by about 50%) must be used for continuous operation for long periods.

7.3.9.4 Inertia Matching

The motor rotor inertia (J_m) should not be very small compared with the load inertia (J_L). This is particularly critical in high speed and highly repetitive (high throughput) applications. Typically, for high-speed applications, the value of J_L/J_m may be in the range of 5–20. For low-speed applications, J_L/J_m can be as high as 100. This assumes direct drive applications.

A gear transmission may be needed between the motor and the load in order to amplify the torque available from the motor, which also reduces the speed at which the load is driven. Then, further considerations have to be made in inertia matching. In particular, by neglecting the inertial and frictional loads due to gear transmission, it can be shown that the best acceleration conditions for the load are possible if

$$\frac{J_L}{J_m} = r^2 \qquad (7.38)$$

where r is the step-down gear ratio (i.e., motor speed/load speed). Since J_L/r^2 is the load inertia as felt at the motor rotor, the optimal condition (Equation 7.38) is when this equivalent inertia (which moves at the same acceleration as the rotor) is equal to the rotor inertia (J_m).

7.3.9.5 Drive Amplifier Selection

Usually, the commercial motors come with matching drive systems. If this is not the case, some useful sizing computations can be done to assist the process of selecting a drive amplifier. As noted before, even though the control procedure becomes linear and convenient when linear amplifiers are used, it is desirable to use PWM amplifiers in view of their high efficiency (and associated low thermal dissipation).

The required current and voltage ratings of the amplifier, for a given motor and a load, may be computed rather conveniently. The required motor torque is given by

$$T_m = J_m \alpha + T_L + T_f \tag{7.39a}$$

where

α is the highest angular acceleration needed from the motor
T_L is the worst-case load torque
T_f is the frictional torque on the motor

If the load is pure inertia (J_L), Equation 7.39a becomes

$$T_m = (J_m + J_L)\alpha + T_f \tag{7.39b}$$

The current required to generate this torque in the motor is given by

$$i = \frac{T_m}{k_m} \tag{7.40}$$

where k_m is the torque constant of the motor.

The voltage required to drive the motor is given by

$$v = k_m' \omega_m + Ri \tag{7.41}$$

where

$k_m' = k_m$ is the back emf constant
R is the winding resistance
ω_m is the highest operating speed of the motor in driving the load

For a PWM amplifier, the supply voltage (from a dc power supply) is computed by dividing the voltage in Equation 7.41 by the lowest duty cycle of operation.

Note 1: For "peak curve" operation, pick an amplifier and power supply with these ratings (voltage, current, power).

Note 2: For "continuous curve" operation, increase the current rating proportionately.

Note 3: If several amplifiers use the same single power supply, increase the power rating of the power supply in proportion.

Summary of motor selection:

- It is a component matching problem
- Components: load, motor, sensors, drive systems, transmission (gear), etc.
- Typically, the motor and sensor (e.g., encoder) come together; the motor may include a harmonic drive (transmission); the drive system (PWM amplifier, power supply, etc.) comes commercially matched to the motor
- Typically: match the motor to the load; select a gear unit if necessary
- Continuous operation has more stringent performance requirements (due to thermal problems) than peak (intermittent) operation

Example 7.4

A load of moment of inertia $J_l = 0.5\,\text{kg-m}^2$ is ramped up from rest to a steady speed of 200 rpm in 0.5 s using a dc motor and a gear unit of step down speed ratio $r = 5$. A schematic representation of the system is shown in Figure 7.29a and the speed profile of the load is shown in Figure 7.29b. The load exerts a constant resistance of $T_R = 55$ N-m throughout the operation. The efficiency of the gear unit is $e = 0.7$. Check whether the commercial brushless dc motor with its drive unit, whose characteristics are shown in Figure 7.27b, is suitable for this application. The moment of inertia of the motor rotor is $J_m = 0.002\,\text{kg-m}^2$.

Solution

The load equation to compute the torque required from the motor is given by

$$T_m = \left(J_m + \frac{J_l}{er^2} \right) r\alpha + \frac{T_R}{er} \tag{7.42}$$

where α is the load acceleration, and the remaining parameters are as defined in the example. The derivation of Equation 7.42 is straightforward. From the given speed profile, we have

Maximum load speed = 200 rpm = 20.94 rad/s

Load acceleration = $\dfrac{20.94}{0.5}$ rad/s^2 = 42 rad/s^2.

Substitute the numerical values in Equation 7.41, under worst-case conditions, to compute the required torque from the motor. We have

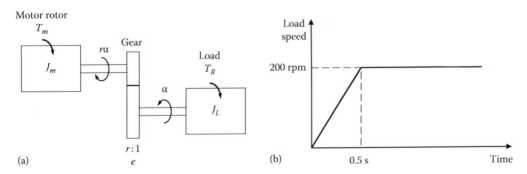

(a)

(b)

FIGURE 7.29
(a) A load driven by a dc motor through a gear transmission; (b) speed profile of the load.

$$T_m = \left(0.002 + \frac{0.05}{0.7 \times 5^2}\right)5 \times 42 + \frac{55.0}{0.7 \times 5} \text{ N-m} = 1.02 + 15.71 \text{ N-m} = 16.73 \text{ N-m}$$

Under worst-case conditions, at least this much of torque would be required from the motor, operating at a speed of $200 \times 5 = 1000$ rpm. Note from Figure 7.27b that the load point (1000 rpm, 16.73 N-m) is sufficiently below even the continuous torque curve of the given motor (with its drive unit). Hence, this motor is adequate for the task.

7.4 Induction Motors

With the widespread availability of ac as an economical form of power for operating industrial machinery and household appliances, much attention has been given to the development of ac motors. Because of the rapid progress made in this area, ac motors have managed to replace dc motors in many industrial applications until the revival of the dc motor, particularly as a servomotor in control system applications. However, ac motors are generally more attractive than conventional dc motors because of their robustness, lower cost, simplicity of construction, and easier maintenance, especially in heavy-duty (high-power) applications (e.g., rolling mills, presses, elevators, cranes, material handlers, and operations in paper, metal, petrochemical, cement, and other industrial plants) and in continuous constant-speed operations (e.g., conveyors, mixers, agitators, extruders, pulping machines, and household and industrial appliances such as refrigerators, heating-ventilation-and-air-conditioning [HVAC] devices such as pumps, compressors and fans). Many industrial applications using ac motors may involve continuous operation throughout the day for over 6 days a week. Also, with advances in control hardware and software and the low cost of power electronics have led to advance controllers for ac motors, which can emulate the performance of variable-speed drives of dc motors; for example, ac servomotors that rival their dc counterpart. Some advantages of ac motors are as follows:

1. Cost-effectiveness
2. Use of convenient power source (standard power grid supplying single-phase and three-phase ac supplies)
3. No commutator and brush mechanisms needed in many types of ac motors
4. Low power dissipation, low rotor inertia, and lightweight in some designs
5. Virtually no electric spark generation or arcing (less hazardous in chemical environments)
6. Capability of accurate constant-speed operation without needing servo control (with synchronous ac motors)
7. No drift problems in ac amplifiers in drive circuits (unlike linear dc amplifiers)
8. High reliability, robustness, easy maintenance, and long life

The primary disadvantages include the following:

1. Low starting torque (zero starting torque in synchronous motors)
2. Need of auxiliary starting devices for ac motors with zero starting torque

3. Difficulty of variable-speed control or servo control (unless modern sold-state and variable-frequency drives with devices with field feedback compensation are employed)

In this section, we will study induction motors (asynchronous motors). Later, synchronous motors will be discussed.

7.4.1 Rotating Magnetic Field

The operation of an ac motor can be explained using the concept of a rotating magnetic field. A rotating field is generated by a set of windings uniformly distributed around a circular stator and is excited by ac signals with uniform phase differences. To illustrate this, consider a standard three-phase supply. The voltage in each phase is 120° out of phase with the voltage in the next phase. The phase voltages can be represented by

$$v_1 = a \cos \omega_p t; \quad v_2 = a \cos \left(\omega_p t - \frac{2\pi}{3} \right); \quad v_3 = a \cos \left(\omega_p t - \frac{4\pi}{3} \right) \qquad (7.43)$$

where ω_p is the frequency of each phase of the ac signal (i.e., the line frequency). Note that v_1 leads v_2 by $2\pi/3$ radians and v_2 leads v_3 by the same angle. Furthermore, since v_1 leads v_3 by $4\pi/3$ radians, it is correct to say that v_1 lags v_3 by $2\pi/3$ radians. In other words, v_1 leads $-v_3$ by $(\pi - 2\pi/3)$, which is equal to $\pi/3$. Now consider a group of three windings, each of which has two segments (a positive segment and a negative segment) uniformly arranged around a circle (stator), as shown in Figure 7.30, in the order $v_1, -v_3, v_2, -v_1, v_3, -v_2$. Note that each winding segment has a phase difference of $\pi/3$ (or 60°) from the adjacent segment. The physical (geometric) spacing of adjacent winding segments is also 60°. Now,

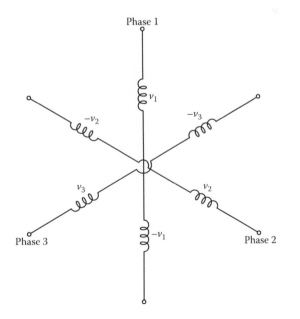

FIGURE 7.30
The generation of a rotating magnetic field using a three-phase supply and two winding sets per phase.

consider the time interval $\Delta t = \pi/(3\omega_p)$. The status of $-v_3$ at the end of a time interval of Δt is identical to the status of v_1 in the beginning of the time interval. Similarly, the status of v_2 after a time Δt becomes that of $-v_3$ in the beginning, and so on. In other words, the voltage status (and hence the magnetic field status) of one segment becomes identical to that of the adjacent segment in a time interval of Δt. This means that the magnetic field generated by the winding segments appears to rotate physically around the circle (stator) at an angular velocity of ω_p.

It is not necessary for the three sets of three-phase windings to be distributed over the entire 360° angle of the circle. Suppose, instead, that these three sets (six segments) of windings are distributed within the first 180° of the circle, at 30° apart and a second three sets are distributed similarly within the remaining 180°. Then, the field would appear to rotate at half the speed ($\omega_p/2$), because in this case, Δt is the time taken for the field to rotate through 30°, not 60°. It follows that the general formula for the angular speed ω_p of the rotating magnetic field generated by a set of winding segments uniformly distributed on a stator and excited by an ac supply, is

$$\omega_f = \frac{\omega_p}{n} \tag{7.44}$$

where
   ω_p is the frequency of the ac signal in each phase (i.e., line frequency)
   n is the number of pairs of winding sets used per phase (i.e., number of pole pairs per phase).

Note that when $n = 1$, there are two coils (+ve and −ve) for each phase (i.e., there are two poles per phase). Similarly, when $n = 2$, there are four coils for each phase. Hence, n denotes the number of "pole pairs" per phase in a stator. In this manner, the speed of the rotating magnetic field can be reduced to a fraction of the line frequency simply by adding more sets of windings. These windings occupy the stator of an ac motor. The number of phases and the number of segments wound to each phase determine the angular separation of the winding segments around the stator. For example, for the three-phase, one pole pair per phase arrangement shown in Figure 7.30, the physical separation of the winding segments is 60°. For a two-phase supply with one pole pair per phase, the physical separation is 90° and the separation is halved to 45° if two pole pairs are used per phase. It is the rotating magnetic field, produced in this manner, which generates the driving torque by interacting with the rotor windings. The nature of this interaction determines whether a particular motor is an induction motor or a synchronous motor.

7.4.2 Induction Motor Characteristics

The stator windings of an induction motor generate a rotating magnetic field, as explained in the previous section. The rotor windings are purely secondary windings, which are not energized by an external voltage. For this reason, no commutator-brush devices are needed in induction motors (see Figure 7.31). The core of the rotor is made of ferromagnetic laminations in order to concentrate the magnetic flux and to minimize dissipation (eddy currents). The rotor windings are embedded in the axial direction on the surface of the rotor and are interconnected in groups. The rotor windings may consist of uninsulated copper or aluminum (or any other conductor) bars (*a cage rotor*), which are fitted

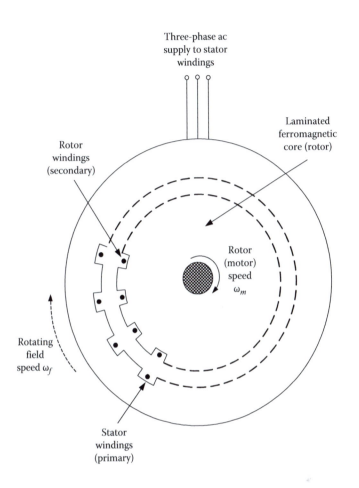

FIGURE 7.31
Schematic diagram of an induction motor.

into slots in the end rings at the two ends of the rotor. These end rings complete the paths for electrical conduction through the rods. Alternatively, wire with one or more turns in each slot (*a wound rotor*) may be used. First, consider a stationary rotor. The rotating field in the stator intercepts the rotor windings, thereby generating an induced current due to mutual induction or transformer action (hence the name induction motor). The resulting secondary magnetic flux interacts with the primary, rotating magnetic flux, thereby producing a torque in the direction of rotation of the stator field. This torque drives the rotor. As the rotor speed increases, initially the motor torque also increases (rather moderately) because of secondary interactions between the stator circuit and the rotor circuit, even though the relative speed of the rotating field with respect to the rotor decreases, which reduces the rate of change of flux linkage and hence the direct transformer action. (*Note:* The relative speed is termed the *slip rate*). Quite soon, the maximum torque (called "breakdown torque") will be reached.

A further increase in rotor speed (i.e., a decrease in slip rate) will sharply decrease the motor torque until at synchronous speed (zero slip rate), the motor torque becomes zero. This behavior of an induction motor is illustrated by the typical characteristic curve given in Figure 7.32. From the starting torque T_s to the maximum torque (breakdown torque) T_{max},

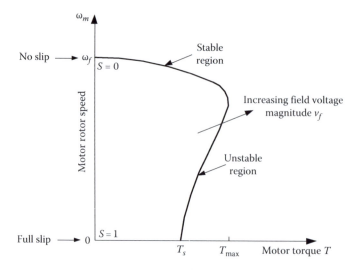

FIGURE 7.32
Torque–speed characteristic curve of an induction motor.

the motor behavior is unstable. This can be explained as follows. An incremental increase in speed will cause an increase in torque, which will further increase the speed. Similarly, an incremental reduction in speed will bring about a reduction in torque that will further reduce the speed. The portion of the curve from T_{max} to the zero-torque (or no-load or synchronous) condition represents the region of stable operation. Under normal operating conditions, an induction motor should operate in this region.

The fractional slip S for an induction motor is given by

$$S = \frac{\omega_f - \omega_m}{\omega_f} \tag{7.45}$$

Even when there is no external load, the synchronous operating condition (i.e., $S=0$) is not achieved at steady-state because of the presence of frictional torque, which opposes the rotor motion. When an external torque (load torque) T_L is present, under normal operating conditions, the slip rate will further increase so as to increase the motor torque to support this load torque. As is clear from Figure 7.32, in the stable region of the characteristic curve, the induction motor is quite insensitive to torque changes; a small change in speed would require a very large change in torque (in comparison with an equivalent dc motor). For this reason, an induction motor is relatively insensitive to load variations and can be regarded as a constant speed machine. Note that if the rotor speed is increased beyond the synchronous speed (i.e., $S<0$), the motor becomes a generator.

7.4.3 Torque–Speed Relationship

It is instructive to determine the torque–speed relationship for an induction motor. This relationship provides insight into possible control methods for induction motors. The equivalent circuits of the stator and the rotor for one phase of an induction motor are shown in Figure 7.33a. The circuit parameters are as follows:

R_f = stator coil resistance

L_f = stator leakage inductance

R_c = stator core iron loss resistance (eddy current effects, etc.)

L_c = stator core (magnetizing) inductance

L_r = rotor leakage inductance

R_r = rotor coil resistance

The magnitude of the ac supply voltage for each phase of the stator windings is v_f at the line frequency ω_p. The rotor current generated by the induced emf is i_r. After allowing for the voltage drop due to stator resistance and stator leakage inductance, the voltage available for mutual induction is denoted by v. This is also the induced voltage in the secondary (rotor) windings at standstill, assuming the same number of turns. This induced voltage changes linearly with slip S, because the induced voltage is proportional to the relative velocity of the rotating field with respect to the rotor (i.e., $\omega_f - \omega_m$). Hence, the induced voltage in the rotor windings (secondary windings) is Sv. Note further that at standstill (when $S = 1$), the frequency of the induced voltage in the rotor is ω_p. At the synchronous speed of rotation (when $S = 0$), this frequency is zero because the magnetic field is fixed and constant relative to the rotor in this case. Now, assuming a linear variation of frequency of the induced voltage between these two extremes, we note that the

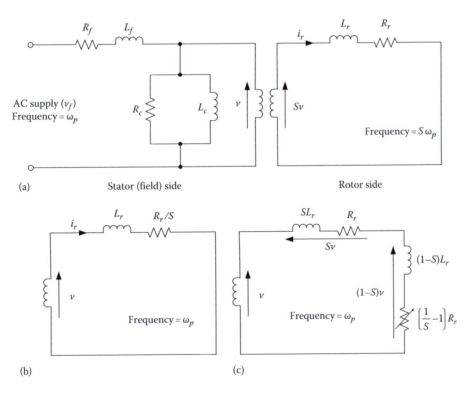

FIGURE 7.33
(a) Stator and rotor circuits for an induction motor; (b) rotor circuit referred to the stator side; (c) representation of available mechanical power using the rotor circuit.

frequency of the induced voltage in the rotor circuit is $S\omega_p$. These observations are indicated in Figure 7.33a.

Using the frequency domain (complex) representation for the out-of-phase currents and voltages, the rotor current i_r in the complex form is given by

$$i_r = \frac{Sv}{(R_r + jS\omega_p L_r)} = \frac{v}{(R_r/S + j\omega_p L_r)} \tag{7.46}$$

From Equation 7.46, it is clear that the rotor circuit can be represented by a resistance R_r/S and an inductance L_r in series and excited by voltage v at frequency ω_p. This is in fact the rotor circuit referred to the stator side, as shown in Figure 7.33b. This circuit can be grouped into two parts, as shown in Figure 7.33c. The inductance SL_r and resistance R_r in a series, with a voltage drop Sv, are identical to the rotor circuit in Figure 7.33a. Note that SL_r has to be used as the inductance in the new equivalent circuit segment, instead of L_r in the original rotor circuit, for the sake of circuit equivalence. The reason is simple. The new equivalent circuit operates at frequency ω_p while the original rotor circuit operates at frequency $S\omega_p$. (*Note*: impedance of an inductor is equal to the product of inductance and frequency of excitation). The second voltage drop $(1 - S)v$ in Figure 7.33c represents the back emf due to rotor–stator field interaction; it generates the capacity to drive an external load (mechanical power). Note here that the back emf governs the current in the rotor circuit and hence the generated torque. It follows that the available mechanical power, per phase, of an induction motor is given by $i_r^2(1/S - 1)R_r$. Hence,

$$T_m\omega_m = pi_r^2\left(\frac{1}{S} - 1\right)R_r \tag{7.47}$$

where
   T_m is the motor torque generated in the rotor
   ω_m is the rotor speed of the motor
   p is the number of supply phases
   i_r is the magnitude of the current in the rotor

The magnitude of the current in the rotor circuit is obtained from Equation 7.46 as

$$i_r = \frac{v}{\sqrt{R_r^2/S^2 + \omega_p^2 L_r^2}} \tag{7.48}$$

By substituting Equation 7.48 in 7.47, we get

$$T_m = pv^2\frac{S(1-S)}{\omega_m}\frac{R_r}{(R_r^2 + S'\omega_p^2 L_r^2)} \tag{7.49}$$

From Equations 7.44 and 7.45, we can express the number of pole pairs per phase of stator winding as

$$n = \frac{\omega_p}{\omega_m}(1 - S) \tag{7.50}$$

Equation 7.50 is substituted in (7.49) to get

$$T_m = \frac{pnv^2 S R_r}{\omega_p(R_r^2 + S^2\omega_p^2 L_r^2)} \tag{7.51}$$

If the resistance and the leakage inductance in the stator are neglected, v is approximately equal to the stator excitation voltage v_f. This gives the torque–slip relationship

$$T_m = \frac{pnv_f^2 S R_r}{\omega_p(R_r^2 + S^2\omega_p^2 L_r^2)} \tag{7.52}$$

Note that by using Equation 7.50, it is possible to express S in Equation 7.52 in terms of the rotor speed ω_m. This results in a torque–speed relationship, which gives the characteristic curve shown in Figure 7.32. Specifically, we employ the fact that the motor speed ω_m is related to slip through

$$S = \frac{\omega_p - n\omega_m}{\omega_p} \tag{7.53}$$

Note, further, from Equation 7.52 that the motor torque is proportional to the square of the supply voltage v_f.

Example 7.5

In the derivation of Equation 7.52, we assumed that the number of effective turns per phase in the rotor is equal to that in the stator. This assumption is generally not valid, however. Determine how the equation should be modified in the general case. Suppose that

$$r = \frac{\text{Number of effective turns per phase in the rotor}}{\text{Number of effective turns per phase in the stator}}$$

Solution
At standstill ($S = 1$), the induced voltage in the rotor is rv and the induced current is i_r/r. Hence, the impedance in the rotor circuit is given by

$$Z_r = \frac{rv}{i_r/r} = r^2 \frac{v}{i_r} \tag{7.54}$$

or

$$Z_r = r^2 Z_{req}$$

It follows that the true rotor impedance (or resistance and inductance) simply has to be divided by r^2 to obtain the equivalent impedance. In this general case of $r \neq 1$, the resistance R_r and the inductance L_r should be replaced by $R_{req} = R_r/r^2$ and $L_{req} = L_r/r^2$ in Equation 7.52.

7.4.4 Induction Motor Control

DC motors are widely used in servo control applications because of their simplicity and flexible speed–torque capabilities. In particular, dc motors are easy to control and they operate accurately and efficiently over a wide range of speeds. The initial cost and the maintenance cost of a dc motor, however, are generally higher than those for a comparable ac motor. AC motors are rugged and are most common in medium- to high-power applications involving fairly constant speed operation. Of late, much effort has been invested in developing improved control methods for ac motors and significant progress has been seen in this area. Today's ac motors with advanced drive systems with frequency control and field feedback compensation can provide speed control that is comparable with the capabilities of dc servomotors (e.g., 1:20 or 26 dB range of speed variation).

Since fractional slip S determines motor speed ω_m, Equation 7.52 suggests several possibilities for controlling an induction motor. Four possible methods for induction motor control are as follows:

1. Excitation frequency control (ω_p)
2. Supply voltage control (v_f)
3. Rotor resistance control (R_r)
4. Pole changing (n)

What is given in parentheses is the parameter that is adjusted in each method of control. The first two approaches of control are popular and desirable and are described in the following section.

7.4.4.1 Excitation Frequency Control

Excitation frequency control can be accomplished using power electronics. For example, a *thyristor* circuit, as in a dc motor drive system, may be used that possesses very effective, efficient, and nondissipative switching characteristics at very high frequencies. Furthermore, thyristors can handle high voltages and power levels.

By using an *inverter* circuit, a variable-frequency ac output can be generated from a dc supply. A single-phase inverter circuit is shown in Figure 7.34. Thyristors 1 and 2 are gated by their firing circuits according to the required frequency of the output voltage v_o. The primary winding of the output transformer is center-tapped. A dc supply voltage v_{ref} is applied to the circuit as shown. If both thyristors are not conducting, the voltage across the capacitor C is zero. Now, if thyristor 1 is gated (i.e., fired), the current in the upper half of the primary winding will build to its maximum and the voltage across that half will reach v_{ref} (since the voltage drop across thyristor 1 is very small). As a result of the corresponding change in the magnetic flux, a voltage v_{ref} (approximately) will be induced in the lower half of the primary winding, complementing the voltage in the upper half. Accordingly, the voltage across the primary winding (or across the capacitor) is approximately $2v_{ref}$. Now, if thyristor 2 is fired, the voltage at point A becomes v_{ref}. Since the capacitor is already charged to $2v_{ref}$, the voltage at point B becomes $3v_{ref}$. This means that a voltage of $2v_{ref}$ is applied across thyristor 1 in the nonconducting direction. As a result, thyristor 1 will be turned off. Then, as before, a voltage $2v_{ref}$ is generated in the primary winding, but in the opposite direction because it is thyristor 2 that is conducting now. In this manner, an approximately rectangular pulse sequence of ac voltage v_o is generated at the circuit output. The frequency of the voltage is equal to the inverse of the firing interval between

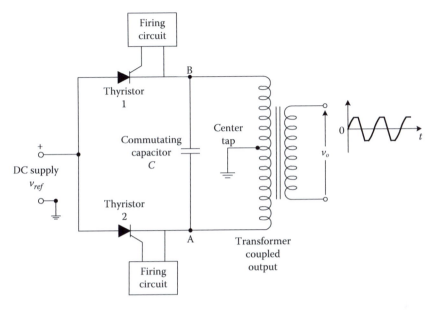

FIGURE 7.34
A single-phase inverter circuit for frequency control.

the two thyristors. A three-phase inverter can be formed by triplicating the single-phase inverter and by phasing the firing times appropriately.

Modern drive units for induction motors use PWM and advanced microelectronic circuitry incorporating a single monolithic IC chip with more than 30,000 circuit elements, rather than discrete semiconductor elements. The block diagram in Figure 7.35a shows a frequency control system for an induction motor. A standard ac supply (three-phase or single-phase) is rectified and filtered to provide the dc supply to the three-phase PWM inverter circuit. This device generates a nearly sinusoidal three-phase output at a specified frequency. Firing of the switching circuitry, for varying the frequency of the ac output, is commanded by a hardware controller. If the control requirements are simple, a variable-frequency oscillator or a voltage-to-frequency converter may be used instead.

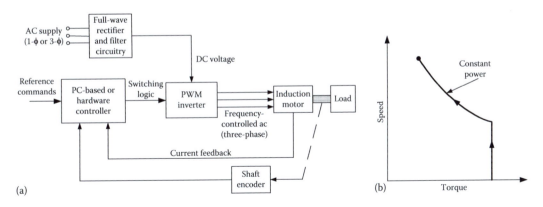

FIGURE 7.35
(a) Variable-frequency control of an induction motor; (b) a typical control strategy.

Alternatively, a digital computer (PC)-based controller may be used to vary the drive frequency and to adjust other control parameters in a more flexible manner, using software. The controller may use hardware logic or software to generate the switching signal (reference frequency), while taking into account external (human-operator) commands and sensor feedback signals. A variable-frequency drive for an ac motor can effectively operate in the open-loop mode. Sensor feedback may be employed, however, for more accurate performance. Feedback signals may include shaft encoder readings (motor angle) for speed control and current (stator current, rotor current in wound rotors, dc current to PWM inverter, etc.) particularly for motor torque control. A typical control strategy is shown in Figure 7.35b. In this case, the control processor provides a two-mode control scheme. In the initial mode, the torque is kept constant while accelerating the motor. In the next mode, the power is kept constant while further increasing the speed. Both modes of operation can be achieved through frequency control. Strategies of specified torque profiles (torque control) or specified speed profiles (speed control) can be implemented in a similar manner.

Programmable microcontrollers and variable-frequency drives for ac motors are commercially available (as for dc motors). One such drive is able to control the excitation frequency in the range of 0.1–400 Hz with a resolution of 0.01 Hz. A three-phase ac voltage in the range of 200–230 V or 380–460 V is generated by the drive, depending on the input ac voltage. AC motors with frequency control are employed in many applications, including the variable-flow control of pumps, fans and blowers, industrial manipulators (robots, hoists, etc.), conveyors, elevators, process plant and factory instrumentation, and flexible operation of production machinery for flexible (variable-output) production. In particular, ac motors with frequency control and sensor feedback are able to function as servomotors.

7.4.4.2 Voltage Control

From Equation 7.51, it is seen that the torque of an induction motor is proportional to the square of the supply voltage. It follows that an induction motor can be controlled by varying its supply voltage. This may be done in several ways. For example, amplitude modulation of the ac supply, using a ramp generator, will directly accomplish this objective by varying the supply amplitude. Alternatively, by introducing zero-voltage regions (i.e., blanking out) periodically (at high frequency) in the ac supply, for example using a thyristor circuit with firing delays as in PWM, will accomplish voltage control by varying the root-mean-square (rms) value of the supply voltage. Voltage control methods are appropriate for small induction motors, but they provide poor efficiency when control over a wide speed range is required. Frequency control methods are recommended in low-power applications. An advantage of voltage control methods over frequency control methods is the lower stator copper loss.

Example 7.6

Show that the fractional slip versus motor torque characteristic of an induction motor, at steady-state, may be expressed by

$$T_m = \frac{aSv_f^2}{\left[1 + \left(S/S_b\right)^2\right]} \tag{7.55}$$

Identify the parameters a and S_b. Show that S_b is the slip corresponding to the breakdown torque (maximum torque) T_{max}. Obtain an expression for T_{max}.

An induction motor with parameter values $a=4\times10^{-3}$ N-m/V^2 and $S_b=0.2$ is driven by an ac supply that has a line frequency of 60 Hz. Stator windings have two pole pairs per phase. Initially, the line voltage is 500 V. The motor drives a mechanical load, which can be represented by an equivalent viscous damper with a damping constant $b=0.265$ N-m/rad/s. Determine the operating point (i.e., the values of torque and speed) for the system. Suppose that the supply voltage is dropped by 50% (to 250 V) using a voltage control scheme. What is the new operating point? Is this a stable operating point? In view of your answer, comment on the use of voltage control in induction motors.

Solution

First, we note that Equation 7.52 can be expressed as Equation 7.55 with

$$a = \frac{pn}{\omega_p R_r} \tag{7.56}$$

and

$$S_b = \frac{R_r}{\omega_p L_r} \tag{7.57}$$

The breakdown torque is the peak torque and is defined by $\partial T_m / \partial \omega_m = 0$.

We express

$$\frac{\partial T_m}{\partial \omega_m} = \frac{\partial T_m}{\partial S}\frac{\partial S}{\partial \omega_m} = -\frac{1}{\omega_f}\frac{\partial T_m}{\partial S}$$

where we have differentiated Equation 7.45 with respect to ω_m and substituted the result. It follows that the breakdown torque is given by $\partial T_m / \partial S = 0$. Now, differentiate Equation 7.55 with respect to S and equate to zero. We get

$$\left[1+\left(\frac{S}{S_b}\right)^2\right]-S\left[2\frac{S}{S_b^2}\right]=0$$

or

$$1-\left(\frac{S}{S_b}\right)^2=0$$

It follows that $S=S_b$ corresponds to the breakdown torque. Substituting in Equation 7.55, we have

$$T_{max} = \frac{1}{2}aS_b v_f^2 \tag{7.58}$$

Next, the speed–torque curve is computed using the given parameter values in Equation 7.55 and plotted as shown in Figure 7.36 for the two cases $v_f=500$ V and $v_f=250$ V. Note that with $S_b=0.2$,

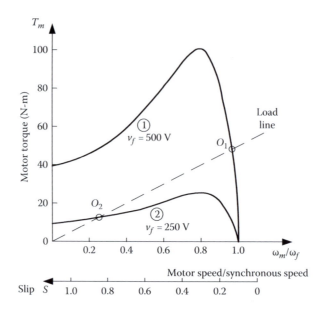

FIGURE 7.36
Speed–torque curves for induction motor voltage control.

we have, from Equation 7.58, $(T_{max})_1 = 100$ N-m and $(T_{max})_2 = 25$ N-m. These values are confirmed from the curves in Figure 7.36.

The load curve is given by $T_m = b\omega_m$ or $T_m = b\omega_f(\omega_m/\omega_f)$. Now, from Equation 7.44, the synchronous speed is computed as

$$\omega_f = \frac{60 \times 2\pi}{2} \text{ rad/s} = 188.5 \text{ rad/s.}$$

Hence, $b\omega_f = 0.265 \times 188.5 = 50$ N-m

This is the slope of the load line shown in Figure 7.36. The points of intersection of the load line and the motor characteristic curve are the steady-state operating points. They are as follows

for case 1 ($v_f = 500$ V):

Operating torque = 48 N-m
Operating slip = 4%
Operating speed = 1728 rpm

for case 2 ($v_f = 250$ V):

Operating torque = 12 N-m
Operating slip = 77%
Operating speed = 414 rpm

Note that when the supply voltage is halved, the torque drops by a factor of four and the speed drops by about 76%. But, what is worse is that the new operating point (O_2) is in the unstable region (i.e., from $S = S_b$ to $S = 1$) of the motor characteristic curve. It follows that large drops in supply voltage are not feasible, and the efficiency of the motor can degrade significantly with voltage control.

7.4.4.3 Field Feedback Control (Flux Vector Drive)

An innovative method for controlling ac motors is through field feedback (or *flux vector*) compensation. This approach can be explained using the equivalent circuit shown in Figure 7.33c. Note that this circuit separates the rotor-equivalent impedance into two parts—a nonproductive part and a torque-producing part—as discussed previously. There are magnetic field vectors (or complex numbers) that correspond to these two parts of circuit impedance. As is clear from Figure 7.33c, these magnetic flux components depend on the slip *S* and hence the rotor speed and also the current. In the present method of control, the magnetic field vector associated with the first part of impedance is sensed using speed measurement (from an encoder) and motor current measurement (from a current-voltage transducer), and is compensated for (i.e., removed through feedback) in the stator circuit. As a result, only the second part of impedance (and magnetic field vector), corresponding to the back emf, will remain. Hence, the ac motor will behave quite like a dc motor that has an equivalent torque-producing back emf. More sophisticated schemes of control may use a model of the motor. Flux-vector control has been commercially implemented in ac motors using customized digital signal processing (DSP) chips and IC hardware. The feedback of rotor current can further improve the performance of a flux-vector drive. A flux-vector drive tends to be more complex and costly than a variable-frequency drive (which is a "scalar" drive). The need for sensory feedback introduces a further burden in this regard.

7.4.5 A Transfer-Function Model for an Induction Motor

The true dynamic behavior of an induction motor is generally nonlinear and time-varying. For small variations about an operating point, however, linear relations can be written. On this basis, a transfer-function model can be established for an induction motor, as we have done for a dc motor. The procedure described in this section uses the steady-state speed–torque relationship for an induction motor to determine the transfer-function model. The basic assumption here is that this steady-state relationship, if the inertia effects are modeled by some other means, can represent the dynamic behavior of the motor for small changes about an operating point (steady-state) with reasonable accuracy.

Suppose that a motor rotor that has moment of inertia J_m and mechanical damping constant b_m (mainly from the bearings) is subjected to a variation δT_m in the motor torque and an associated change $\delta \omega_m$ in the rotor speed, as shown in Figure 7.37. In general, these changes may arise from a change δT_L in the load torque and a change δv_f in supply voltage.

Newton's second law gives

$$\delta T_m - \delta T_L = J_m \delta \dot{\omega}_m + b_m \delta \omega_m \tag{7.59}$$

Now, use a linear steady-state relationship to represent the variation in motor torque as a function of the incremental change $\delta \omega_m$ in speed and a variation δv_f in the supply voltage. We get

$$\delta T_m = -b_e \delta \omega_m + k_v \delta v_f \tag{7.60}$$

By substituting Equation 7.60 in (7.59) and using the Laplace variable *s*, we have

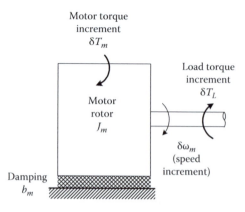

FIGURE 7.37
Incremental load model for an induction motor.

$$\delta\omega_m = \frac{k_v}{\left[J_m s + b_m + b_e\right]}\delta v_f - \frac{1}{\left[J_m s + b_m + b_e\right]}\delta T_L \tag{7.61}$$

In the transfer-function Equation 7.61, note that $\delta\omega_m$ is the output, δv_f is the control input, and δT_L is an unknown (disturbance) input. The motor transfer function $\delta\omega_m/\delta v_f$ is given by

$$G_m(s) = \frac{k_v}{\left[J_m s + b_m + b_e\right]} \tag{7.62}$$

The motor time constant τ is

$$\tau = \frac{J_m}{b_m + b_e} \tag{7.63}$$

Now it remains to identify the parameters b_e (analogous to electrical damping in a dc motor) and k_v (a voltage gain parameter, as for a dc motor). To accomplish this, we use Equation 7.55, which can be written in the form

$$T_m = k(S)v_f^2 \tag{7.64}$$

where

$$k(S) = \frac{aS}{1 + \left(S/S_b\right)^2} \tag{7.65}$$

Now, using the well-known relation in differential calculus

$$\delta T_m = \frac{\partial T_m}{\partial \omega_m}\delta\omega_m + \frac{\partial T_m}{\partial v_f}\delta v_f$$

we have $b_e = -\dfrac{\partial T_m}{\partial \omega_m}$ and $k_v = \dfrac{\partial T_m}{\partial v_f}$. But $\dfrac{\partial T_m}{\partial \omega_m} = \dfrac{\partial T_m}{\partial S}\dfrac{\partial S}{\partial \omega_m} = -\dfrac{1}{\omega_f}\dfrac{\partial T_m}{\partial S}$.

Thus,

$$b_e = \frac{1}{\omega_f}\frac{\partial T_m}{\partial S} \tag{7.66}$$

where ω_f is the synchronous speed of the motor. Now, by differentiating Equation 7.65 with respect to S, we have

$$\frac{\partial k}{\partial S} = a\frac{1-\left(S/S_b\right)^2}{\left[1+\left(S/S_b\right)^2\right]^2} \tag{7.67}$$

Hence,

$$b_e = \frac{av_f^2}{\omega_f}\frac{1-\left(S/S_b\right)^2}{\left[1+\left(S/S_b\right)^2\right]^2} \tag{7.68}$$

Next, by differentiating Equation 7.64 with respect to v_f, we have $\dfrac{\partial T_m}{\partial v_f} = 2k(S)v_f$.

Accordingly, we get

$$k_v = \frac{2aSv_f}{1+\left(S/S_b\right)^2} \tag{7.69}$$

Here, S_b is the fractional slip at the breakdown (maximum) torque and a is a motor torque parameter defined by Equation 7.56. If we wish to include the effects of the electrical time constant τ_e of the motor, we may include the factor $\tau_e s + 1$ in the denominator (characteristic polynomial) on the right-hand side of Equation 7.61. Since τ_e is usually an order of magnitude smaller than τ as given by Equation 7.63, no significant improvement in accuracy results through this modification. Finally, note that the constants b_e and k_v can be obtained graphically using experimentally determined speed–torque curves for an induction motor for several values of the line voltage v_f, using a procedure similar to what we have described for a dc motor.

7.4.6 Single-Phase AC Motors

The multiphase (polyphase) ac motors are normally employed in moderate- to high-power applications (e.g., more than 5 hp). In low-power applications (e.g., motors used in household appliances such as refrigerators, dishwashers, food processors, and hair dryers and in tools such as saws, lawn mowers, and drills), single-phase ac motors are commonly used for they have the advantages of simplicity and low cost.

The stator of a single-phase motor has only one set of drive windings (with two or more stator poles) excited by a single-phase ac supply. If the rotor is running close to the frequency of the line ac, this single phase can maintain the motor torque, operating as an induction motor. But a single phase is obviously not capable of starting the motor. To overcome this problem, a second coil that is out of phase from the first coil is used during the starting period and is turned off automatically once the operating speed is attained. The phase difference is obtained either through a difference in inductance for a given resistance in the two coils or by including a capacitor in the second coil circuit.

7.5 Miscellaneous Actuators

In this section, we will consider several other electrical actors that find application in mechatronics. Briefly discussed here are ac synchronous motors and linear actuators (solenoids and linear motors).

7.5.1 Synchronous Motors

Phase-locked servos and stepper motors can be considered synchronous motors because they run in synchronism with an external command signal (a pulse train) under normal operating conditions. The rotor of a synchronous ac motor rotates in synchronism with the rotating magnetic field generated by the stator windings. The generation principle of this rotating field is identical to that in an induction motor. Unlike an induction motor, however, the rotor windings of a synchronous motor are energized by an external dc source. The rotor magnetic poles generated in this manner will lock themselves with the rotating magnetic field generated by the stator and will rotate at the same speed (synchronous speed). For this reason, synchronous motors are particularly suited for constant-speed applications under variable-load conditions. Synchronous motors with permanent magnet (e.g., samarium-cobalt) rotors are also commercially available. The dc voltage, which is required to energize the rotor windings of a synchronous motor, may come from several sources. An independent dc supply, an external ac supply, and a rectifier (or a dc generator that is driven by the synchronous motor itself) are three ways of generating the dc signal.

One major drawback of the synchronous ac motor is that an auxiliary "starter" is required to start the motor and bring its speed close to the synchronous speed. The reason for this is that in synchronous motors, the starting torque is virtually zero. To understand this, consider the starting conditions. The rotor is at rest and the stator field is rotating (at the synchronous speed). Consequently, there is 100% slip ($S = 1$). When, for example, an N pole of the rotating field in the stator is approaching an S pole in the rotor, the magnetic force will tend to turn the rotor in the direction opposite to the rotating field. When the same N pole of the rotating field has just passed the rotor S pole, the magnetic force will tend to pull the rotor in the same direction as the rotating field. These opposite interactions balance out, producing a zero net torque on the rotor. One method of starting a synchronous motor is by using a small dc motor. Once the synchronous motor reaches the synchronous speed, the dc motor is operated as a dc generator to supply power to the rotor windings. Alternatively, a small induction motor may be used to start the synchronous motor. A more desirable arrangement, which employs the principle, is to include several sets of induction-motor-type rotor windings (cage-type or wound-type) in the synchronous motor rotor itself. In all these cases, the supply to the rotor windings of the synchronous motor is disconnected during the starting conditions and is turned on only when the motor speed comes close to the synchronous speed.

7.5.1.1 Control of a Synchronous Motor

Under normal operating conditions, the speed of a synchronous motor is completely determined by the frequency of the ac supply to the stator windings, because the motor speed is equal to the speed ω_f of the rotating field (see Equation 7.44). Hence, speed control can be achieved by the variable-frequency control method as described for an induction motor. In some applications of ac motors (both induction and synchronous types), clutch devices that link the motor to the driven load are used to achieve variable-speed control (e.g., using an eddy current clutch system that produces a variable coupling force through the eddy currents generated in the clutch). These dissipative techniques are quite wasteful and can considerably degrade the motor efficiency. Furthermore, heat removal methods would be needed to avoid thermal problems. Hence, they are not recommended for high-power applications where motor efficiency is a prime consideration.

Note that unless a permanent magnet rotor is used, a synchronous motor would require a slip ring and brush mechanism to supply the dc voltage to its rotor windings. This is a drawback that is not present in an induction motor.

The steady-state speed–torque curve of a synchronous motor is a straight line parallel to the torque axis. But with proper control (e.g., frequency control), an ac motor can function as a servomotor. Conventionally, a servomotor has a linear torque–speed relationship, which can be approached by an ac servomotor with a suitable drive system. Applications of synchronous ac motors include steel rolling mills, rotary cement kilns, conveyors, hoists, process compressors, recirculation pumps in hydroelectric power plants, and, more recently, servomotors and robotics. Synchronous motors are particularly suitable in high-speed, high-power applications where dc motors might not be appropriate. A synchronous motor can operate with a larger air gap between the rotor and the stator in comparison with an induction motor. This is an advantage for synchronous motors from the mechanical design point of view (e.g., bearing tolerances and rotor deflections due to thermal, static, and dynamic loads). Furthermore, rotor losses are smaller for synchronous motors than for induction motors.

7.5.2 Linear Actuators

Linear actuator stages are common in industrial mechatronic applications. They may employ the same principles as the rotary actuators, but employ linear arrangements for the stator and the moving element, or a rotary motor with a rotary/linear motion transmission unit. Solenoids are typically on-off (or push-pull) type linear actuators, commonly used in relays, valve actuators, switches, and a variety of other applications. Some useful types of linear actuators are presented in the following section.

7.5.2.1 Solenoid

The solenoid is a common rectilinear actuator, which consists of a coil and a soft iron core. When the coil is activated by a dc signal, the soft iron core becomes magnetized. This electromagnet can serve as an on-off (push-pull) actuator, for example to move a ferromagnetic element (moving pole or plunger). The moving element is the load, which is typically restrained by a light spring and a damping element.

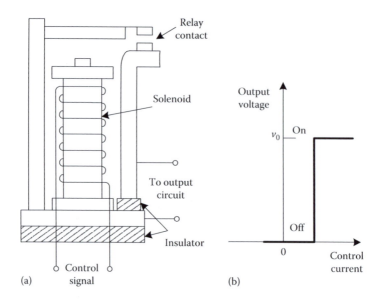

FIGURE 7.38
A solenoid operated relay: (a) Physical components; (b) characteristic curve.

Solenoids are rugged and inexpensive devices. Common applications of solenoids include valve actuators, mechanical switches, relays, and other two-state positioning systems. An example of a relay is shown in Figure 7.38.

A relay of this type may be used to turn on and off such devices as motors, heaters, and valves in industrial systems. They may be controlled by a programmable logic controller (PLC). A time-delay relay provides a delayed on-off action with an adjustable time delay, as is necessary in some process applications.

The percentage on time with respect to the total on-off period is the *duty cycle* of a solenoid. A solenoid will need a sufficiently large current to move a load. There is a limit to the resulting magnetic force because the coil will saturate. In order to avoid this, the ratings of the solenoid should match the needs of the load. There is another performance consideration. For long duty cycles, it is necessary to maintain a current through the solenoid coil for a correspondingly long period. If the initial activating current of the solenoid is maintained over a long period, it will heat up the coil and create thermal problems. Apart from the loss of energy, this situation is undesirable because of safety issues, reduction in the coil life, and the need to have special means for cooling. A common solution is to incorporate a hold-in circuit, which will reduce the current through the solenoid coil shortly after it is activated. A simple hold-in circuit is shown in Figure 7.39.

The resistance R_h is sufficiently large and comparable with the resistance R_s of the solenoid coil. Initially, the capacitor C is fully discharged. Then the transistor is "on" (i.e., forward biased) and is able to conduct from the emitter (E) to the collector (C). When the switch that is normally open (denoted by NO) is turned on (i.e., closed), the dc supply sends a current through the solenoid coil (R_s) and the circuit is completed through the transistor. Since the transistor offers only a low resistance, the resulting current is large enough to actuate the solenoid. As the current flows through the circuit (while the switch is closed), the capacitor C becomes fully charged. The transistor becomes reverse-biased due to the resulting voltage of the capacitor. This turns off the transistor. Then the circuit is completed not through the transistor but through the hod-in resistor R_h. As a result,

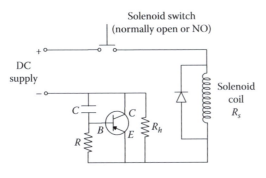

FIGURE 7.39
A hold-in circuit for a solenoid.

the current through the solenoid drops by a factor of $R_s/(R_s + R_h)$. This lower current is adequate to maintain the state of the solenoid without overheating it.

A *rotary solenoid* provides a rotary push-pull motion. Its principle of operation is the same as that of a linear solenoid. Another type of solenoid is the *proportional solenoid*. It is able to produce a rectilinear motion in proportion to the current through the coil. Accordingly, it acts as a linear motor. Proportional solenoids are particularly useful as valve actuators in fluid power systems; for example, as actuators for spool valves in hydraulic piston-cylinder devices (rectilinear actuators) and valve actuators for hydraulic motors (rotary actuators).

7.5.2.2 Linear Motors

It is possible to obtain a rectilinear motion from a rotary electromechanical actuator (motor) by employing an auxiliary kinematic mechanism (motion transmission), such as a cam and follower, a belt and pulley, a rack and pinion, or a lead screw and nut. These devices inherently have problems of friction and backlash. Furthermore, they add inertia and flexibility to the driven load, thereby generating undesirable resonances and motion errors. Proper matching of the transmission inertia and the load inertia is essential. Particularly, the transmission inertia should be less than the load inertia, when referred to one side of the transmission mechanism. Furthermore, extra energy is needed to operate the system against the inertia of the transmission mechanism.

For improved performance, direct rectilinear electromechanical actuators are desirable. These actuators operate according to the same principle as their rotary counterparts, except that flat stators and rectilinearly moving elements (in place of rotors) are employed. They come in the following different types:

1. Stepper linear actuators
2. DC linear actuators
3. AC linear actuators
4. Fluid (hydraulic and pneumatic) pistons and cylinders

We have already indicated the principle of operation of a linear stepper motor. Fluids, pistons, and cylinders are discussed later in this chapter. Linear electric motors are also termed electric cylinders and are suitable as high-precision linear stages of motion applications.

For example, a dc brushless linear motor operates similarly to a rotary brushless motor and uses a similar drive amplifier. Advanced rare-earth magnets are used for the moving member, providing high force/mass ratio. The stator takes the form of a U-channel within which the moving member slides. Linear (sliding) bearings are standard. Since magnetic bearings can interfere with the force generating magnetic flux, air bearings are used in more sophisticated applications. The stator has the "forcer" coil for generating the drive magnetic field and Hall-effect sensors for commutation. Since conductive material will create eddy-current problems, reinforced ceramic epoxy structures are used for the stator channel by leading manufacturers of linear motors. Applications of linear motors include traction devices, liquid-metal pumps, multi-axis tables, Cartesian robots, conveyor mechanisms, and servovalve actuators.

7.6 Hydraulic Actuators

The ferromagnetic material in an electric motor saturates at some level of magnetic flux density (and the electric current, which generates the magnetic field). This limits the torque/mass ratio obtainable from an electric motor. Hydraulic actuators use the hydraulic power of a pressurized liquid. Since high pressures (on the order of 5000 psi) can be used, hydraulic actuators are capable of providing very high forces (and torques) at very high power levels simultaneously to several actuating locations in a flexible manner. The force limit of a hydraulic actuator can be an order of magnitude larger than that of an electromagnetic actuator. This results in higher torque/mass ratios than those available from electric motors, particularly at high levels of torque and power. This is a principal advantage of hydraulic actuators. Note that the actuator mass considered here is the mass of the final actuating element, not including auxiliary devices such as those needed to pressurize and store the fluid. Another advantage of a hydraulic actuator is that it is quite stiff when viewed from the side of the load. This is because a hydraulic medium is mechanically stiffer than an electromagnetic medium. Consequently, the control gains required in a high-power hydraulic control system would be significantly less than the gains required in a comparable electromagnetic (motor) control system. Note that the stiffness of an actuator may be measured by the slope of the speed–torque (force) curve and is representative of the speed of response (or bandwidth).

There are other advantages of *fluid power systems*. Electric motors generate heat. In continuous operation, then, the thermal problems can be serious and special means of heat removal will be necessary. In a fluid power system, however, any heat generated at the load can be quickly transferred to another location away from the load, by the hydraulic fluid itself, and effectively removed by means of a heat exchanger. Another advantage of fluid power systems is that they are self-lubricating and as a result, the friction in valves, cylinders, pumps, hydraulic motors, and other system components will be low and will not require external lubrication. Safety considerations will be less as well because; for example, there is no possibility of spark generation as in motors with brush mechanisms. There are several disadvantages as well. Fluid power systems are more nonlinear than electrical actuator systems. Reasons for this include valve nonlinearities, fluid friction, compressibility, thermal effects, and generally nonlinear constitutive relations. Leakage can create problems. Fluid power systems tend to be noisier than electric motors. Synchronization of multi-actuator operations may be more difficult as well. Also, when the necessary accessories are included, fluid power systems are by and large more expensive and less portable than electrical actuator systems.

Fluid power systems with analog control devices have been in use in engineering applications since the 1940s. Smaller, more sophisticated, and less costly control hardware and microprocessor-based controllers were developed in the 1980s making fluid power control systems as sophisticated, precise, cost-effective, and versatile as electromechanical control systems. Today, miniature fluid power systems with advanced digital control and electronics are used in numerous applications of mechatronics, directly competing with advanced dc and ac motion control systems. Also, logic devices based on "fluidics" or fluid logic devices are preferred over digital electronics in some types of industrial applications. Applications of fluid power systems include vehicle steering and braking systems, active suspension systems, material handling devices, and industrial mechanical manipulators such as hoists, industrial robots, rolling mills, heavy-duty presses, actuators for aircraft control surfaces (ailerons, rudder, and elevators), excavators, actuators for opening and closing of bridge spans, tunnel boring machines, food processing machines, reaction injection molding (RIM) machines, dynamic testing machines, heavy-duty shakers for structures and components, machine tools, ship building, and dynamic props, stage backgrounds, and structures in theatres and auditoriums.

7.6.1 Components of a Hydraulic Control System

A schematic diagram of a basic hydraulic control system is shown in Figure 7.40a. A view of a practical fluid power system is shown in Figure 7.40b. The hydraulic fluid (oil) is pressurized using a pump, which is driven by an ac motor. The typical fluids used are mineral oils or oil in water emulsions. These fluids have the desirable properties of self-lubrication, corrosion resistance, good thermal properties, fire resistance, environmental friendliness, and low compressibility (high stiffness for good bandwidth). Note that the motor converts electrical power into mechanical power, and the pump converts mechanical power into fluid power. In terms of through and across variable pairs, these power conversions can be expressed as

$$(i, v) \xrightarrow{\eta_m} (T, \omega) \xrightarrow{\eta_h} (Q, P)$$

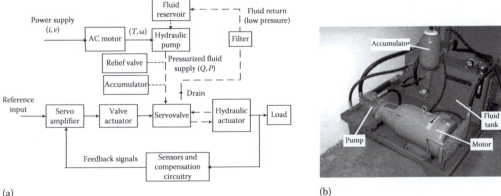

(a) (b)

FIGURE 7.40
(a) Schematic diagram of a hydraulic control system; (b) an industrial fluid-power system.

in the usual notation. The conversion efficiency η_m of a motor is typically very high (over 90%), whereas the efficiency η_h of a hydraulic pump is not as good (about 60%), mainly because of dissipation, leakage, and compressibility effects. Depending on the pump capacity, flow rates in the range of 1,000 to 50,000 gallons per minute (*Note:* 1 gal/min = 3.8 L/min) and pressures from 500 to 5,000 psi (*Note:* 1 kPa = 0.145 psi) can be obtained. The pressure of the fluid from the pump is regulated and stabilized by a relief valve and an accumulator. A hydraulic valve provides a controlled supply of fluid into the actuator, controlling both the flow rate (including direction) and the pressure. In feedback control, this valve uses response signals (motion) sensed from the load to achieve the desired response, hence the name *servovalve*. Usually, the servovalve is driven by an electric *valve actuator*, such as a *torque motor* or a *proportional solenoid*, which in turn is driven by the output from a *servo amplifier*. The servo amplifier receives a reference input command (corresponding to the desired position of the load) as well as a measured response of the load (in feedback). Compensation circuitry may be used in both feedback and forward paths to modify the signals so as to obtain the desired control action. The hydraulic actuator (typically a piston-cylinder device for rectilinear motions or a hydraulic motor for rotary motions) converts fluid power back into mechanical power, which is available to perform useful tasks (i.e., to drive a load). Note that some power in the fluid is lost at this stage. The low-pressure fluid at the drain of the hydraulic servovalve is filtered and returned to the reservoir and is available to the pump.

One might argue that since the power that is required to drive the load is mechanical, it would be much more efficient to use a motor directly to drive that load. There are good reasons for using hydraulic power, however. For example, ac motors are usually difficult to control, particularly under variable-load conditions. Their efficiency can drop rapidly when the speed deviates from the rated speed, particularly when voltage control is used. They need gear mechanisms for low-speed operation with associated problems such as backlash, friction, vibration, and mechanical loading effects. Special coupling devices are also needed. Hydraulic devices usually filter out high-frequency noise, which is not the case with ac motors. Thus, hydraulic systems are ideal for high-power, high-force control applications. In high-power applications, a single high-capacity pump or several pumps can be employed to pressurize the fluid. Furthermore, in low-power applications, several servovalve and actuator systems can be operated to perform different control tasks in a distributed control environment using the same pressurized fluid supply. In this sense, hydraulic systems are very flexible. Hydraulic systems provide excellent speed–force (or torque) capability, variable over a wide range of speeds, without significantly affecting the power-conversion efficiency because the excess high-pressure fluid is diverted to the return line. Consequently, hydraulic actuators are far more controllable than ac motors. As noted before, hydraulic actuators also have an advantage over electromagnetic actuators from the point of view of heat transfer characteristics. Specifically, the hydraulic fluid promptly carries away any heat that is generated locally and releases it through a heat exchanger at a location away from the actuator.

7.6.2 Hydraulic Pumps and Motors

The objective of a hydraulic pump is to provide pressurized oil to a hydraulic actuator. Three common types of hydraulic pumps are

1. Vane pump
2. Gear pump
3. Axial piston pump

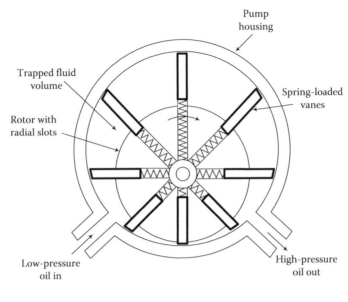

Pump housing

Trapped fluid volume

Spring-loaded vanes

Rotor with radial slots

Low-pressure oil in

High-pressure oil out

FIGURE 7.41
A hydraulic vane pump.

The pump type used in a hydraulic control system is not very significant, except for the pump capacity, in terms of the control functions of the system. But since hydraulic motors can be interpreted as pumps operating in the reverse direction, it is instructional to outline the operation of these three types of pumps.

A sliding-type *vane pump is* shown schematically in Figure 7.41. The vanes slide in the interior of the housing as they rotate with the rotor of the pump. They can move within radial slots on the rotor, thereby maintaining full contact between the vanes and the housing. Springs or the pressurized hydraulic fluid itself may be used for maintaining this contact. The rotor is eccentrically mounted inside the housing. In the first half of the rotation cycle, the fluid is drawn in at the inlet port as a result of the increasing volume between vane pairs as they rotate. In the second half of the rotation cycle, the oil volume trapped between two vanes is eventually compressed because of the decreasing volume of the vane compartment. Note that a pressure rise will result from pushing the liquid volume into the high-pressure side and not allowing it to return to the low-pressure side of the pump, even when there is no significant compressibility in the liquid when it moves from the low-pressure side to the high-pressure side. The typical operating pressure (at the outlet port) of these devices is about 2000 psi (13.8 MPa). The output pressure can be varied by adjusting the rotor eccentricity because this alters the change in the compartment volume during a cycle. A disadvantage of any rotating device with eccentricity is the centrifugal forces that are generated even while rotating at constant speed. Dynamic balancing is needed to reduce this problem.

The operation of an external-gear hydraulic pump (or simply a *gear pump*) is illustrated in Figure 7.42. The two identical gears are externally meshed. The inlet port is facing the gear enmeshing (retracting) region. Fluid is drawn in and trapped between the pairs of teeth in each gear in rotation. This volume of fluid is transported around by the two gear wheels into the gear meshing region at the pump outlet. Here it undergoes an increase in pressure, as in the vane pump, as a result of forcing the fluid into the high-pressure side. Only moderate to low pressure can be realized by gear pumps (about 1000 psi or 7 MPa,

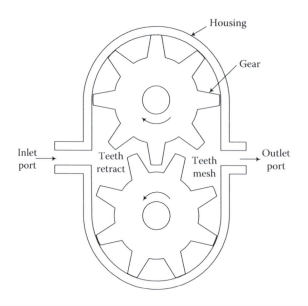

FIGURE 7.42
A hydraulic gear pump.

maximum) because the volume changes that take place in the enmeshing and meshing regions are small (unlike in the vane pumps) and because fluid leakage between teeth and housing can be considerable. However, gear pumps are robust and low-cost devices and they are probably the most commonly used hydraulic pumps.

A schematic diagram of an axial piston hydraulic pump is shown in Figure 7.43. The chamber barrel is rigidly attached to the drive shaft. The two pistons themselves rotate with the chamber barrel, but since the end shoes of the pistons slide inside a slanted (skewed) slot, which is stationary, the pistons simultaneously undergo a reciprocating motion as well in the axial direction. As a chamber opening reaches the inlet port of the pump housing, fluid is drawn in because of the increasing volume between the piston head and the chamber. This fluid is trapped and transported to the outlet port while undergoing compression as a result of the decreasing volume inside the chamber due to the axial motion of the piston. Fluid pressure increases in this process. High outlet pressures (4000 psi, 27.6 MPa, or more) can be achieved using piston pumps. As shown in Figure 7.43, the piston stroke can be increased by increasing the inclination angle of the stroke plate (slot). This, in turn, increases the pressure ratio of the pump. A lever mechanism is usually available to adjust the piston stroke. Piston pumps are relatively expensive.

The efficiency of a hydraulic pump is given by the ratio of the output fluid power to the motor mechanical power; thus,

$$\eta_p = \frac{PQ}{\omega T} \tag{7.70}$$

where
    P is the pressure increase in the fluid
    Q is the fluid flow rate
    ω is the rotating speed of the pump
    T is the drive torque to the pump

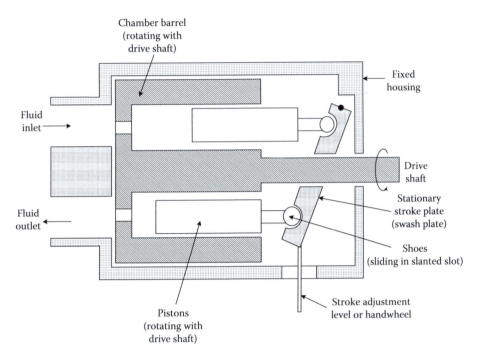

FIGURE 7.43
An axial piston hydraulic pump.

7.6.3 Hydraulic Valves

Fluid valves can perform three basic functions:

1. Change the flow direction
2. Change the flow rate
3. Change the fluid pressure

The valves that accomplish the first two functions are termed *flow-control valves*. The valves that regulate the fluid pressure are termed *pressure-control valves*. A simple relief valve regulates pressure, whereas the poppet valve, gate valve, and globe valve are on/off flow-control valves. Some examples are shown in Figure 7.44. The directional valve

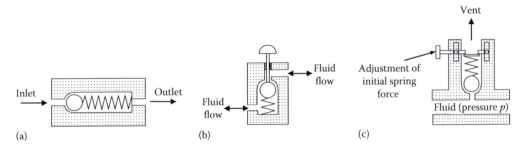

FIGURE 7.44
(a) A check valve (directional valve); (b) a poppet valve (an on/off valve); (c) a relief valve (a pressure regulating valve).

(or check valve) shown in Figure 7.44a allows the fluid flow in one direction and blocks it in the opposite direction. The spring provides sufficient force for the ball to return to the seat when there is no fluid flow. It does not need to sustain any fluid pressure, and hence its stiffness is relatively low. A check valve falls into the category of flow control valves. Figure 7.44b shows a poppet valve. It is normally in the closed position with the ball completely seated to block the flow. When the plunger is pushed down, the ball moves with it, allowing fluid to flow through the seat opening. This on/off valve is bi-directional and may be used to permit fluid flow in either direction. The relief valve shown in Figure 7.44c is in the closed condition under normal conditions. The spring force, which closes the valve (by seating the ball) is adjustable. When the fluid pressure (in a container or a pipe to which the valve is connected) rises above a certain value, as governed by the spring force, the valve opens thereby letting the fluid out through a vent (which may be recirculated in the system). In this manner, the pressure of the system is maintained at a nearly constant level. Typically, an accumulator is used in conjunction with a relief valve to take up undesirable pressure fluctuations and to stabilize the system. Valves are classified by the number of flow paths present under operating conditions. For example, a four-way valve has four ways in which the flow can enter and leave the valve. In high-power fluid systems, two valve stages consisting of a *pilot valve* and a main valve may be used. Here, the pilot valve is a low-capacity, low-power valve, which operates the higher-capacity main valve.

7.6.3.1 Spool Valve

Spool valves are used extensively in hydraulic servo systems. A schematic diagram of a four-way spool valve is shown in Figure 7.45a. This is commonly called a *servovalve* because motion feedback is used by it to control the motion of a hydraulic actuator. The moving unit of the valve is called the *spool*. It consists of a spool rod and one or more expanded regions (or lobes), which are called *lands*. Input displacement (U) applied to the spool rod, using an actuator (torque motor or proportional solenoid), regulates the flow rate (Q) to the main hydraulic actuator as well as the corresponding pressure difference (P) available to the actuator. If the land length is larger than the port width (Figure 7.45b), it is an *overlapped land*. This introduces a dead zone in the neighborhood of the central position of the spool, resulting in decreased sensitivity and increased stability problems. Since it is virtually impossible to exactly match the land size with the port width, the *underlapped land* configuration (Figure 7.45c) is commonly employed. In this case, there is a leakage flow, even in the fully closed position, which decreases the efficiency and increases the steady-state error of the hydraulic control system. For the accurate operation of the valve, the leakage should not be excessive. The direct flow at various ports of the valve and the leakage flows between the lands and the valve housing should be included in a realistic analysis of a spool valve. For small displacements δU about an operating point, the following linearized equations can be written. Since the flow rate Q_2 into the actuator increases as U increases and it decreases as P_2 increases, we have

$$\delta Q_2 = k_q \delta U - k'_c \delta P_2 \tag{7.71}$$

Similarly, since the flow rate Q_1 from the actuator increases with both U and P_1, we have

$$\delta Q_1 = k_q \delta U + k'_c \delta P_1 \tag{7.72}$$

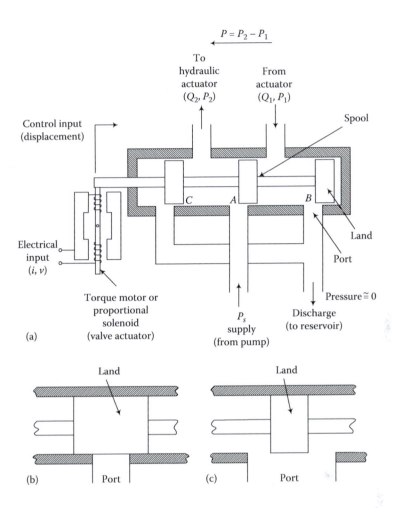

FIGURE 7.45
(a) A four-way spool valve; (b) an overlapped land; (c) an underlapped land.

The gains k_q and k'_c will be defined later.

In fact, if we disregard the compressibility of the fluid, we have $\delta Q_1 = \delta Q_2$, assuming that the hydraulic piston (actuator) is double-acting with equal piston areas on the two sides of the actuator piston. We consider the general case where $Q_1 \neq Q_2$. Note, however, that the inlet port and the outlet port are assumed to have identical characteristics. By adding Equations 7.71 and 7.72 and defining an average flow rate, we have

$$Q = \frac{Q_1 + Q_2}{2} \tag{7.73}$$

and with an equivalent flow-pressure coefficient

$$k_c = \frac{k'_c}{2} \tag{7.74}$$

we get

$$\delta Q = k_q \delta U - k_c \delta P \tag{7.75}$$

where the *flow gain* is

$$k_q = \left(\frac{\partial Q}{\partial U}\right)_P \tag{7.76}$$

and the *flow-pressure coefficient* is

$$k_c = -\left(\frac{\partial Q}{\partial P}\right)_U \tag{7.77}$$

Note further that the *pressure sensitivity* is

$$k_p = \left(\frac{\partial P}{\partial U}\right)_Q = \frac{k_q}{k_c} \tag{7.78}$$

To obtain Equation 7.78, we use the following well-known result from calculus: $\delta Q = \left(\frac{\partial Q}{\partial U}\right)_P \delta U + \left(\frac{\partial Q}{\partial P}\right)_U \delta P$. Since $\delta P/\delta U \to \partial P/\partial U$ as $\delta Q \to 0$, we have

$$\left(\frac{\partial P}{\partial U}\right)_Q = -\left(\frac{\partial Q}{\partial U}\right)_P \bigg/ \left(\frac{\partial Q}{\partial P}\right)_U \tag{7.79}$$

Equation 7.78 directly follows from Equation 7.79.

A valve can be actuated by several methods; for example, manual operation, the use of mechanical linkages connected to the drive load, and the use of electromechanical actuators such as solenoids and torque motors (or force motors). Regular solenoids are suitable for on/off control applications, and proportional solenoids and torque motors are used in continuous control. For precise control applications, electromechanical actuation of the valve (with feedback for servo operation) is preferred.

Large valve displacements can saturate a valve because of the nonlinear nature of the flow relations at the valve ports. Several valve stages may be used to overcome this saturation problem when controlling heavy loads. In this case, the spool motion of the first stage (pilot stage) is the input motion. It actuates the spool of the second stage, which acts as a hydraulic amplifier. The fluid supply to the main hydraulic actuator, which drives the load, is regulated by the final stage of a multistage valve.

7.6.3.2 Steady-State Valve Characteristics

Although the linearized valve Equation 7.75 is used in the analysis of hydraulic control systems, it should be noted that the flow equations of a valve are quite nonlinear. Consequently, the valve constants k_q and k_c change with the operating point. Valve constants can be determined either by experimental measurements or by using an accurate nonlinear model. Now we establish a reasonably accurate nonlinear relationship relating the (average) flow rate Q through the main hydraulic actuator and the pressure difference (load pressure) P provided to the hydraulic actuator.

Assume identical rectangular ports at the supply and discharge points in Figure 7.45a. When the valve lands are in the neutral (central) position, we set $U=0$. We assume that the lands perfectly match the ports (i.e., no dead zone or leakage flows due to clearances). The positive direction of U is taken as shown in Figure 7.45a. For this positive configuration, the flow directions are also indicated in the figure. The flow equations at ports A and B are

$$Q_2 = Ubc_d\sqrt{\frac{2(P_s - P_2)}{\rho}} \tag{7.80}$$

$$Q_1 = Ubc_d\sqrt{\frac{2P_1}{\rho}} \tag{7.81}$$

where
  b is the land width
  c_d is the discharge coefficient at each port
  ρ is the density of the hydraulic fluid
  P_s is the supply pressure of the hydraulic fluid

Note that in Equation 7.81 the pressure at the discharge end is taken to be zero. For steady-state operation, we use

$$Q_1 = Q_2 = Q \tag{7.82}$$

Now, squaring Equations 7.80 and 7.81 and adding, we get

$$2Q^2 = 2(Ubc_d)^2 \frac{(P_s - P)}{\rho}$$

where the pressure difference supplied to the hydraulic actuator is denoted by

$$P = P_2 - P_1 \tag{7.83}$$

Consequently,

$$Q = Ubc_d\sqrt{\frac{P_s - P}{\rho}} \quad \text{for } U > 0 \tag{7.84}$$

When $U<0$, the flow direction reverses; furthermore, port A is now associated with P_1 (not P_2) and port C is associated with P_2. It follows that Equation 7.84 still holds, except that $P_2 - P_1$ is replaced by $P_1 - P_2$. Hence,

$$Q = Ubc_d\sqrt{\frac{P_s + P}{\rho}} \quad \text{for } U < 0 \qquad (7.85)$$

Combining Equations 7.84 and 7.85, we have

$$Q = Ubc_d\sqrt{\frac{P_s - P\,\text{sgn}(U)}{\rho}} \qquad (7.86)$$

This can be written in the nondimensional form

$$\frac{Q}{Q_{max}} = \frac{U}{U_{max}}\sqrt{1 - \frac{P}{P_s}\text{sgn}\left(\frac{U}{U_{max}}\right)} \qquad (7.87)$$

where U_{max} = the maximum valve opening (>0) and

$$Q_{max} = U_{max}bc_d\sqrt{\frac{P_s}{\rho}} \qquad (7.88)$$

Equation 7.87 is plotted in Figure 7.46. As with the speed–torque curve for a motor, it is possible to obtain the valve constants k_q and k_c, defined by Equations 7.76 and 7.77, from the curves given in Figure 7.46 for various operating points. For better accuracy, however, experimentally determined valve characteristic curves should be used.

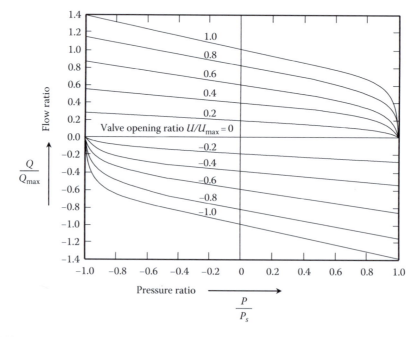

FIGURE 7.46
Steady-state characteristics of a four-way spool valve.

7.6.4 Hydraulic Primary Actuators

Rotary hydraulic actuators (hydraulic motors) operate much like the hydraulic pumps discussed earlier, except that the direction flow is reversed and the mechanical power is delivered by the shaft, rather than taken in. High-pressure fluid enters the actuator. As it passes through the hydraulic motor, the fluid power is used up in turning the rotor and the pressure is dropped. The low-pressure fluid leaves the motor en route to the reservoir. One of the more efficient rotary hydraulic actuators is the axial piston motor, quite similar in construction to the axial piston pump shown in Figure 7.43.

The most common type of rectilinear hydraulic actuator, however, is the hydraulic ram or piston-cylinder actuator. A schematic diagram of such a device is shown in Figure 7.47. This is a *double-acting actuator* because the fluid pressure acts on both sides of the piston. If the fluid pressure is present only on one side of the piston, it is termed a *single-acting actuator*. Single-acting piston-cylinder (ram) actuators are also commonly used for their simplicity and the simplicity of the other control components, such as servovalves, that are needed; although they have the disadvantage of asymmetry. The fluid flow at the ports of a hydraulic actuator is regulated typically by a spool valve. This valve may be operated by a pilot valve (e.g., a flapper valve).

To obtain the equations for the actuator shown in Figure 7.47, we note that the flow rate Q into a chamber depends primarily on two factors:

1. Increase in chamber volume
2. Increase in pressure (compressibility effect of the fluid)

When a piston of area A moves through a distance Y, the flow rate due to the increase in the chamber volume is $\pm A\dot{Y}$. Now, with an increase in pressure δP, the volume of a given fluid mass would decrease by the amount $[-(\partial V/\partial P)\delta P]$. As a result, an equal volume of new fluid would enter the chamber. The corresponding rate of flow is $[-(\partial V/\partial P)(dP/dt)]$. The *bulk modulus* (isothermal or at constant temperature) is given by

$$\beta = -V\frac{\partial P}{\partial V} \tag{7.89}$$

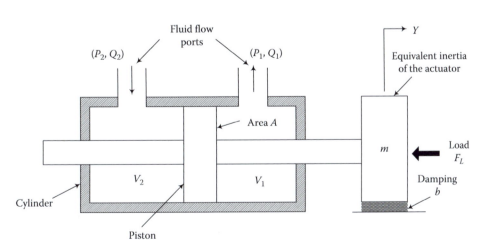

FIGURE 7.47
Double-acting piston–cylinder hydraulic actuator.

Hence, the rate of flow due to the rate of pressure change is given by $[(V/\beta)(dP/dt)]$. Using these facts, the fluid conservation (i.e., flow continuity) equations for the two sides of the actuator chamber in Figure 7.47 can be written as

$$Q_2 = A\frac{dY}{dt} + \frac{V_2}{\beta}\frac{dP_2}{dt} \tag{7.90}$$

$$Q_1 = A\frac{dY}{dt} - \frac{V_1}{\beta}\frac{dP_1}{dt} \tag{7.91}$$

For a realistic analysis, leakage flow rate terms (for leakage between a piston and cylinder and between a piston rod and cylinder) should be included in Equations 7.90 and 7.91. For a linear analysis, these leakage flow rates can be taken as proportional to the pressure difference across the leakage path. Note, further, that V_1 and V_2 can be expressed in terms of Y as follows:

$$V_1 + V_2 = V_o \tag{7.92}$$

$$V_1 - V_2 = V_o' + 2AY \tag{7.93}$$

where V_o and V_o' are constant volumes that depend on the cylinder capacity and on the piston position when $Y = 0$, respectively. Now, for incremental changes about the operating point $V_1 = V_2 = V$, Equations 7.90 and 7.91 can be written as

$$\delta Q_2 = A\frac{d\delta Y}{dt} + \frac{V}{\beta}\frac{d\delta P_2}{dt} \tag{7.94}$$

$$\delta Q_1 = A\frac{d\delta Y}{dt} - \frac{V}{\beta}\frac{d\delta P_1}{dt} \tag{7.95}$$

The "total" Equations 7.90 and 7.91 are already linear for constant V. But, since the valve equation is nonlinear, and since V is not a constant, we should use the "incremental" Equations 7.94 and 7.95 instead of the total equations in a linear model. Adding Equations 7.94 and 7.95 and dividing by 2, we get the hydraulic actuator equation

$$\delta Q = A\frac{d\delta Y}{dt} + \frac{V}{2\beta}\frac{d\delta P}{dt} \tag{7.96}$$

where

$Q = \dfrac{Q_1 + Q_2}{2}$ is the average flow into the actuator

$P = P_2 - P_1$ is the pressure difference on the piston of the actuator

7.6.5 The Load Equation

So far, we have obtained the linearized valve Equation 7.75 and the linearized actuator actuation in (7.96). Determining the load equation remains, which depends on the nature of the load that is driven by the hydraulic actuator. We may represent it by a load force F_L, as shown in Figure 7.47. Note that F_L is a dynamic term, which may represent such effects as flexibility, inertia, and the dissipative effects of the load. In addition, the inertia of the moving parts of the actuator is modeled as a mass m, and the energy dissipation effects associated with these moving parts are represented by an equivalent viscous damping constant b. Accordingly, Newton's second law gives

$$m\frac{d^2Y}{dt^2} + b\frac{dY}{dt} = A(P_2 - P_1) - F_L \tag{7.97}$$

This equation is also linear already. Again, since the valve equation is nonlinear, to be consistent, we should consider incremental motions δY about an operating point. Consequently, we have

$$m\frac{d^2\delta Y}{dt^2} + b\frac{d\delta Y}{dt} = A\delta P - \delta F_L \tag{7.98}$$

where, as before $P = P_2 - P_1$. If the active areas on the two sides of the piston are not equal, a net imbalance force would exist. This could lead to an unstable response under some conditions.

7.6.6 Hydraulic Control Systems

The main components of a hydraulic control system are the following:

1. A servovalve
2. A hydraulic actuator
3. A load
4. Feedback control elements

We have obtained linear equations for the first three components as Equations 7.75, 7.96, and 7.98. Now we rewrite these equations, denoting the incremental variables about an operating point by lowercase letters.

Valve:

$$q = k_q u - k_c p \tag{7.99}$$

Hydraulic actuator:

$$q = A\frac{dy}{dt} + \frac{V}{2\beta}\frac{dp}{dt} \tag{7.100}$$

Load:

$$m\frac{d^2y}{dt^2} + b\frac{dy}{dt} = Ap - f_L \tag{7.101}$$

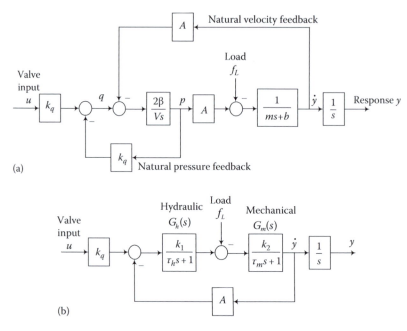

FIGURE 7.48
(a) Block diagram for an open-loop hydraulic control system; (b) an equivalent block diagram.

The feedback elements will depend on the specific feedback control method that is employed. We will revisit this aspect of a hydraulic control system later. Equations 7.99 through 7.101 can be represented by the block diagram shown in Figure 7.48a. This is an open-loop control system because no external feedback elements have been used. Note, however, the presence of a "natural" *pressure feedback* path and a "natural" *velocity feedback* path, which are inherent to the dynamics of the open-loop system.

The block diagram can be reduced to the equivalent form shown in Figure 7.48b. To obtain this equivalent representation, combine the first two summing junctions and then obtain the equivalent transfer function for the pressure feedback loop. This equivalent transfer function can be obtained using the relationship for reducing a feedback control system:

$$G_h = \frac{G}{1 + GH} \tag{7.102}$$

where
   G is the forward transfer function
   H is the feedback transfer function

In the present problem, $G = 2\beta/V_S$ and $H = k_c$. Hence,

$$G_h = \frac{k_1}{\tau_h s + 1} \tag{7.103}$$

where the *pressure gain* parameter is

$$k_1 = \frac{1}{k_c} \tag{7.104}$$

and the *hydraulic time constant* is

$$\tau_h = \frac{V}{2\beta k_c} \tag{7.105}$$

The pressure gain k_1 is a measure of the load pressure p generated for a given flow rate q into the hydraulic actuator. The smaller the pressure coefficient k_c, the larger the pressure gain, as is clear from Equation 7.77. The hydraulic time constant increases with the volume of the actuator fluid chamber and decreases with the bulk modulus of the hydraulic fluid. This is to be expected because the hydraulic time constant depends on the compressibility of the hydraulic fluid.

The mechanical transfer function of the hydraulic actuator is represented by

$$G_m = \frac{k_2}{\tau_m s + 1} \tag{7.106}$$

where the *mechanical time constant* is given by

$$\tau_m = \frac{m}{b} \tag{7.107}$$

and $k_2 = 1/b$. Typically, the mechanical time constant is the dominant time constant, since it is usually larger than the hydraulic time constant.

Example 7.7

A model of the automatic gage control (AGC) system of a steel rolling mill is shown in Figure 7.49. The rollers are pressed using a single-acting hydraulic actuator with valve displacement u. The rollers are displaced through y, thereby pressing the steel that is being rolled. For a given y, the rolling force F is completely known from the steel parameters.

1. Identify the inputs and the controlled variable in this control system.
2. In terms of the variables and system parameters indicated in Figure 7.49, write dynamic equations for the system, including valve nonlinearities.
3. What is the order of the system? Identify the response variables.
4. Draw a block diagram for the system, clearly indicating the hydraulic actuator with valve, the mechanical structure of the mill, inputs, and the controlled variable.
5. What variables would you measure (and feed back through suitable controllers) in order to improve the performance of the control system?

Solution
Part 1: Valve displacement u and rolling force F are inputs. Roll displacement y is the controlled variable.

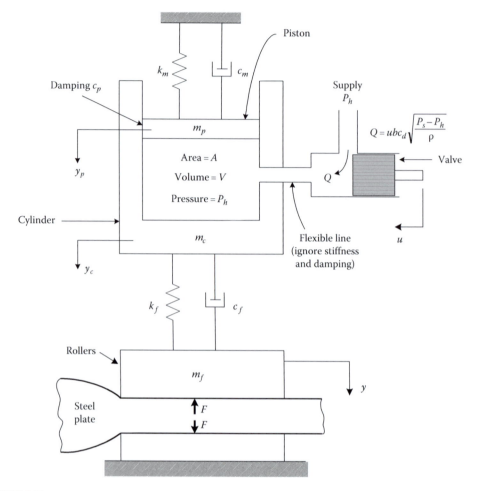

FIGURE 7.49
AGC system of a steel rolling mill.

Part 2: The mechanical-dynamic equations are

$$m_p\ddot{y}_p = -k_m y_p - c_m \dot{y}_p - c_p(\dot{y}_p - \dot{y}_c) - AP_h \tag{i}$$

$$m_c\ddot{y}_c = -k_r(y_c - y) - c_r(\dot{y}_c - \dot{y}) - c_p(\dot{y}_c - \dot{y}_p) + AP_h \tag{ii}$$

$$m_r\ddot{y} = -k_r(y - y_c) - c_r(\dot{y} - \dot{y}_c) - F \tag{iii}$$

Note that the static forces balance and the displacements are measured from the corresponding equilibrium configuration so that gravity terms do not enter into the equations.

The *hydraulic actuator equation* is derived as follows. For the valve, with the usual notation, the flow rate is given by

$$Q = buc_d\sqrt{\frac{P_s - P_h}{\rho}}$$

For the piston–cylinder,

$$Q = A(\dot{y}_c - \dot{y}_p) + \frac{V}{\beta}\frac{dP_h}{dt}$$

Hence,

$$\frac{V}{\beta}\frac{dP_h}{dt} = A(\dot{y}_c - \dot{y}_p) + buc_d\sqrt{\frac{P_s - P_h}{\rho}} \qquad \text{(iv)}$$

Part 3: There are three second-order differential equations (i), (ii), (iii) and one first-order differential equation (iv). Hence, the system is seventh-order. The response variables are the displacements y_p, y_c, y, and the pressure P_h.

Part 4: A block diagram for the hydraulic control system of the steel rolling mill is shown in Figure 7.50.

Part 5: The hydraulic pressure P_h and the roller displacement y are the two response variables that can be conveniently measured and used in feedback control. The rolling force F may be measured and fed forward, but this is somewhat difficult in practice.

Example 7.8

A single-stage pressure control valve is shown in Figure 7.51. The purpose of the valve is to keep the load pressure P_L constant. Volume rates of flow, pressures, and the volumes of fluid subjected to those pressures are indicated in the figure. The mass of the spool and appurtenances is m, the damping constant of the damping force acting on the moving parts is b, and the effective bulk modulus of oil is β. The accumulator volume is V_a. The flow into the valve chamber (volume V_c) is through an orifice. This flow may be taken as proportional to the pressure drop across the orifice, the constant of proportionality being k_o. A compressive spring of stiffness k restricts the spool motion. The initial spring force is set by adjusting the initial compression y_o of the spring.

1. Identify the reference input, the primary output, and a disturbance input for the valve system.
2. By making linearization assumptions and introducing any additional parameters that might be necessary, write equations to describe the system dynamics.
3. Set up a block diagram for the system, showing various transfer functions.

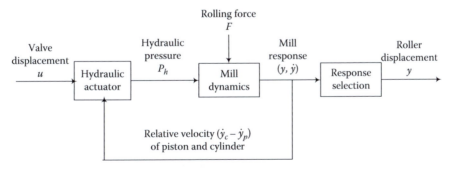

FIGURE 7.50
Block diagram for the hydraulic control system of a steel rolling mill.

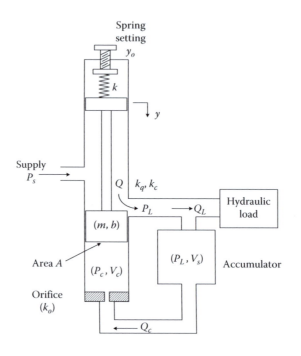

FIGURE 7.51
A single-stage pressure control valve.

Solution
Part 1:
   Input setting $= y_o$
   Primary response (controlled variable) $= P_L$
   Disturbance input $= Q_L$

Part 2:
Suppose that the valve displacement y is measured from the static equilibrium position of the system. The equation of motion for the valve spool device is

$$m\ddot{y} = -b\dot{y} - k(y - y_o) + A(P_s - P_c) \tag{i}$$

The flow through the chamber orifice is given by

$$Q_c = k_o(P_L - P_c) = -A\frac{dy}{dt} + \frac{V_c}{\beta}\frac{dP_c}{dt} \tag{ii}$$

The outflow Q from the spool port increases with y and decreases with the pressure drop $(P_L - P_s)$. Hence, the linearized flow equation is $Q = k_q y - k_c(P_L - P_s)$. Note that k_q and k_c are positive constants, defined previously by Equations 7.76 and 7.77.
   The accumulator equation is

$$Q - Q_c - Q_L = \frac{V_a}{\beta}\frac{dP_L}{dt}$$

Substituting for Q and Q_c, we have

$$k_q y - k_c(P_L - P_s) - k_o(P_L - P_c) - Q_L = \frac{V_a}{\beta}\frac{dP_L}{dt}$$

or

$$k_q y - (k_c + k_o)P_L + (k_c P_s + k_o P_c) - Q_L = \frac{V_a}{\beta}\frac{dP_L}{dt} \qquad \text{(iii)}$$

The equations of motion are (i), (ii), and (iii).

Part 3:
Using Equations (i) through (iii), the block diagram shown in Figure 7.52 can be obtained. Note in particular the feedback path of load pressure P_L. This feedback is responsible for the pressure control characteristic of the valve.

7.6.6.1 Feedback Control

In Figure 7.48a, we have identified two "natural" feedback paths that are inherent in the dynamics of the open-loop hydraulic control system. In Figure 7.48b, we have shown the time constants associated with these natural feedback modules. Specifically, we observe the following:

1. A *pressure feedback path* and an associated *hydraulic time constant* τ_h
2. A *velocity feedback path* and an associated *mechanical time constant* τ_m

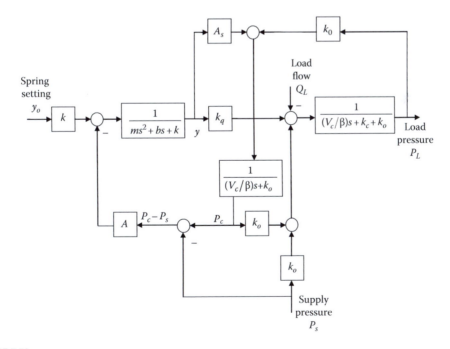

FIGURE 7.52
Block diagram for the single-stage pressure control valve.

The hydraulic time constant is determined by the compressibility of the fluid. The larger the bulk modulus of the fluid, the smaller the compressibility. This results in a smaller hydraulic time constant. Furthermore, τ_h increases with the volume of the fluid in the actuator chamber; hence, this time constant is related to the capacitance of the fluid as well. The mechanical time constant has its origin in the inertia and the energy dissipation (damping) in the moving parts of the actuator. As expected, the actuator becomes more sluggish as the inertia of the moving parts increases, resulting in an increased mechanical time constant.

These natural feedback paths usually provide a stabilizing effect to a hydraulic control system, but they are not adequate for the satisfactory operation of the system. In particular, the position of the actuator is provided by an integrator (see Figure 7.48). In an open-loop operation, the position response will steadily grow and will display an unstable behavior in the presence of the slightest disturbance. Furthermore, the speed of response, which usually conflicts with stability, has to be adequate for proper performance. Consequently, it is necessary to include feedback control into the system. This is accomplished by measuring the response variables and by modifying the system inputs using them, according to some control law.

A schematic representation of a computer-controlled hydraulic system is shown in Figure 7.53. In addition to the motion (both position and speed) of the mechanical load, it is desirable to sense the pressures on the two sides of the piston of the hydraulic actuator for feedback control. There are numerous laws of feedback control, which may be programmed into the control computer. Many of the conventional methods implement a combination of the following three basic control actions:

1. Proportional control (P)
2. Derivative control (D)
3. Integral control (I)

In proportional control, the measured response (or response error) is used directly in the control action. In derivative control, the measured response (or the response error)

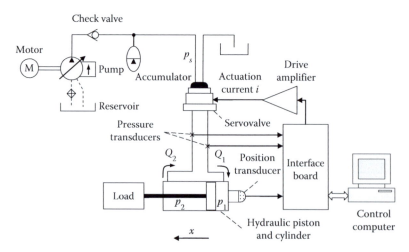

FIGURE 7.53
A computer-controlled hydraulic system.

is differentiated before it is used in the control action. Similarly, in integral control, the response error is integrated and used in the control action. Modification of the measured responses to obtain the control signal is done in many ways, including by electronic, digital, and mechanical means. For example, an analog hardware unit (termed a compensator or controller), which consists of electronic circuitry, may be employed for this purpose. Alternatively, the measured signals, if they are analog, may be digitized and subsequently modified in a required manner through digital processing (multiplication, differentiation, integration, addition, etc.). This is the method used in digital control; either hardware control or software control may be used. The software approach is represented in Figure 7.53.

Consider the feedback (closed-loop) hydraulic control system shown by the block diagram in Figure 7.54. In this case, a general controller is located in the feedback path. Then, a control law may be written as

$$u = u_{ref} - f(y) \tag{7.108a}$$

where $f(y)$ denotes the modifications made to the measured output y in order to form the control (error) signal u. The reference input u_{ref} is specified. Alternatively, if the controller is located in the forward path, as usual, the control law may be given by

$$u = f(u_{ref} - y) \tag{7.108b}$$

Mechanical components may be employed as well to obtain a robust control action.

Fluid power systems in general and hydraulic systems in particular are nonlinear. Nonlinearities have such origins as nonlinear physical relations of the fluid flow, compressibility, nonlinear valve characteristics, friction in the actuator (at the piston rings, which slide inside the cylinder) and the valves, unequal piston areas on the two sides of the actuator piston, and leakage. As a result, accurate modeling of a fluid power system will be difficult, and a linear model will not represent the correct situation except near a small operating region. This situation may be addressed by using an accurate nonlinear model or a series of linear models for different operating regions. In either case, linear control laws (e.g., proportional, integral, and derivative actions) may not be adequate. This situation can be further exacerbated by factors such as parameter variations, unknown disturbances, and noise.

Many advanced control techniques have been applied to fluid power systems, in view of the limitations of such classical control techniques as PID. In one approach, an observer

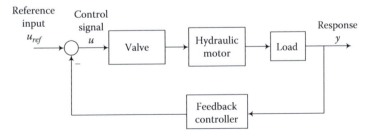

FIGURE 7.54
A closed-loop hydraulic control system.

is used to estimate the velocity and friction in the actuator, and a controller is designed to compensate for friction. Adaptive control is another advanced approach used in hydraulic control systems. In model-referenced adaptive control, the controller pushes the behavior of the hydraulic system towards a reference model. The reference model is designed to display the desired behavior of the physical system. Frequency-domain control techniques, such as H-infinity control (H_∞ control) and quantitative feedback theory (QFT) where the system transfer function is shaped to realize a desired performance, have been studied. They are linear control techniques, which may not work perfectly when applied to a nonlinear system. Impedance control has been studied as well, with respect to hydraulic control systems. In impedance control, the objective is to realize a desired impedance function (*Note*: impedance = force/velocity in the frequency domain) at the output of the control system by manipulating the controller. These advanced techniques are beyond the scope of the present introductory treatment.

7.6.7 Constant-Flow Systems

So far, we have discussed only *valve-controlled* hydraulic actuators. There are two types of valve-controlled systems:

1. Constant pressure systems
2. Constant flow systems

Since there are four flow paths for a four-way spool valve, an analogy can be drawn between a spool-valve controlled hydraulic actuator with a Wheatstone bridge circuit, as shown in Figure 7.55. Each arm of the bridge corresponds to a flow path. As usual, P denotes pressure, which is an across variable analogous to voltage; and Q denotes the

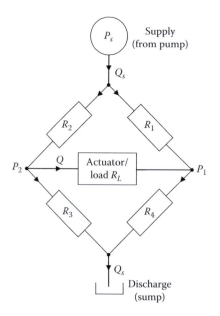

FIGURE 7.55
The bridge circuit representation of a four-way valve and an actuator load.

volume flow rate, which is a through variable analogous to current. The four fluid resistors R_i represent the resistances experienced by the fluid flow in the four paths of the valve. Note that these are variable resistors whose variation is governed by the spool movement (and hence the current of the valve actuator). When the spool moves to one side of the neutral (center) position, two of the resistors (say, R_2 and R_4) change due to the port opening and the remaining two resistors represent the leakage resistances (see Figure 7.45). The reverse is true when the spool moves in the opposite direction from the neutral position. The flow through the actuator is represented by a load resistance R_L, which is connected across the bridge.

In our discussion thus far, we have considered only the *constant pressure system*, in which the supply pressure P_s to the servovalve is maintained constant, but the corresponding supply flow rate Q_s is variable. This system is analogous to a constant-voltage bridge. In a *constant flow system*, the supply flow Q_s is kept constant, but the corresponding pressure P_s is variable. This system is analogous to a constant-current Wheatstone bridge. A constant flow operation requires a constant flow pump, which may be more economical than a variable flow pump. But, it is easier to maintain a constant pressure level by using a pressure regulator and an accumulator. As a result, constant pressure systems are more commonly used in practical applications.

Valve-controlled hydraulic actuators are the most common type used in industrial applications. They are particularly useful when more than one actuator is powered by the same hydraulic supply. Pump-controlled actuators are gaining popularity and are outlined next.

7.6.8 Pump-Controlled Hydraulic Actuators

Pump-controlled hydraulic drives are suitable when only one actuator is needed to drive a process. A typical configuration of a pump-controlled hydraulic drive system is shown in Figure 7.56. A variable flow pump is driven by an electric motor (typically, an ac motor). The pump feeds a hydraulic motor, which in turn drives the load. Control is provided by the flow control of the pump. This may be accomplished in several ways, for example, by controlling the pump stroke (see Figure 7.43) or by controlling the pump speed using a frequency-controlled ac motor. Typical hydraulic drives of this type can provide positioning errors less than $1°$ at torques in the range of 25–250 N-m.

7.6.9 Hydraulic Accumulators

Since hydraulic fluids are quite incompressible, one way to increase the hydraulic time constant is to use an accumulator. An accumulator is a tank that can hold excessive fluid during pressure surges and release this fluid to the system when the pressure slacks. In

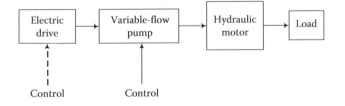

FIGURE 7.56
Configuration of a pump-controlled hydraulic drive system.

this manner, pressure fluctuations can be filtered out from the hydraulic system and the pressure can be stabilized. There are two common types of hydraulic accumulators:

1. Gas-charged accumulators
2. Spring-loaded accumulators

In a gas-charged accumulator, the top half of the tank is filled with air. When high-pressure liquid enters the tank, the air compresses, making room for the incoming liquid. In a spring-loaded accumulator, a movable piston, restrained from the top of the tank by a spring, is used in place of air. The operation of these two types of accumulators is quite similar.

7.6.10 Pneumatic Control Systems

Pneumatic control systems operate in a manner similar to hydraulic control systems. Pneumatic pumps, servovalves, and actuators are quite similar in design to their hydraulic counterparts. The basic differences include the following:

1. The working "fluid" is air, which is far more compressible than hydraulic oils. Hence, thermal effects and compressibility should be included in any meaningful analysis.
2. The outlet of the actuator and the inlet of the pump are open to the atmosphere (no reservoir tank is needed for the working fluid).

By connecting the pump (hydraulic or pneumatic) to an accumulator, the flow into the servovalve can be stabilized and the excess energy can be stored for later use. This minimizes undesirable pressure pulses, vibration, and fatigue loading. Hydraulic systems are stiffer and usually employed in heavy-duty control tasks, whereas pneumatic systems are particularly suitable for medium to low-duty tasks (supply pressures in the range of 500 kPa to 1 MPa). Pneumatic systems are more nonlinear and less accurate than hydraulic systems. Since the working fluid is air and since regulated high-pressure air lines are available in most industrial facilities and laboratories, pneumatic systems tend to be more economical than hydraulic systems. Also, pneumatic systems are more environmentally friendly and cleaner, and the fluid leakage does not cause a hazardous condition. But, they lack the self-lubricating property of hydraulic fluids. Furthermore, atmospheric air has to be filtered and any excess moisture has to be removed before compressing. Heat generated in the compressor has to be removed as well.

Both hydraulic and pneumatic control loops might be present in the same control system. For example, in a manufacturing workcell, hydraulic control can be used for parts transfer, positioning, and machining operations and pneumatic control can be used for tool change, parts grasping, switching, ejecting, and single-action cutting operations. In a fish processing machine, servo-controlled hydraulic actuators have been used for accurately positioning the cutter while pneumatic devices have been used for the grasping and chopping of fish. We will not extend our analysis of hydraulic systems to include air as the working fluid. The reader may consult a book on pneumatic control for information on pneumatic actuators and valves.

7.6.10.1 Flapper Valves

Flapper valves, which are relatively inexpensive and operate at low-power levels, are commonly used in pneumatic control systems. This does not rule them out for hydraulic

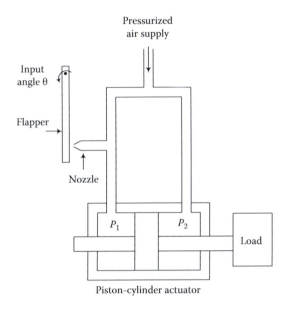

Pressurized air supply

Input angle θ

Flapper

Nozzle

P_1 P_2

Load

Piston-cylinder actuator

FIGURE 7.57
A pneumatic flapper valve system.

control applications, however, where they are popular in pilot valve stages. A schematic diagram of a single-jet flapper valve used in a piston-cylinder actuator is shown in Figure 7.57. If the nozzle is completely blocked by the flapper, the two pressures P_1 and P_2 will be equal, balancing the piston. As the clearance between the flapper and the nozzle increases, the pressure P_1 drops, thus creating an imbalance force on the piston of the actuator. For small displacements, a linear relationship between the flapper clearance and the imbalance force can be assumed.

The operation of a flapper valve requires fluid leakage at the nozzle. This does not create problems in a pneumatic system. In a hydraulic system, however, this not only wastes power but also wastes hydraulic oil and creates a possible hazard, unless a collecting tank and a return line to the oil reservoir are employed. For more stable operation, double-jet flapper valves should be employed. In this case, the flapper is mounted symmetrically between two jets. The pressure drop is still highly sensitive to flapper motion, potentially leading to instability. To reduce instability problems, pressure feedback, using a bellows unit, can be employed.

A two-stage servovalve with a flapper stage and a spool stage is shown in Figure 7.58. Actuation of the torque motor moves the flapper. This changes the pressure in the two nozzles of the flapper, in opposite directions. The resulting pressure difference is applied across the spool, which is moved as a result, which in turn moves the actuator as in the case of a single-stage spool valve. In the system shown in Figure 7.58, there is a feedback mechanism as well between the two stages of valve. Specifically, as the spool moves due to the flapper movement caused by the torque motor, the spool carries the flexible end of the flapper in the opposite direction to the original movement. This creates a back pressure in the opposite direction. Hence, this valve system is said to possess *force feedback* (more accurately, *pressure feedback*).

In general, a multistage servovalve uses several servovalves in a series to drive a hydraulic actuator. The output of the first stage becomes the input to the second stage. As noted

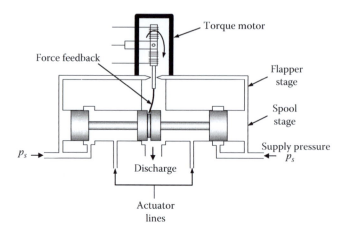

FIGURE 7.58
A two-stage servovalve with pressure feedback.

before, a common combination is a hydraulic flapper valve and a hydraulic spool valve, operating in a series. A multistage servovalve is analogous to a multistage amplifier. The advantages of multistage servovalves are

1. A single-stage servovalve will saturate under large displacements (loads). This may be overcome by using several stages, with each stage being operated in its linear region. Hence, a large operating range (load variations) is possible without introducing excessive nonlinearities, particularly saturation.
2. Each stage will filter out high-frequency noise, giving a lower overall noise-to-signal ratio.

The disadvantages are

1. They cost more and are more complex than single-stage servovalves.
2. Because of the series connection of several stages, the failure of one stage will bring about the failure of the overall system (a reliability problem).
3. Multiple stages will decrease the overall bandwidth of the system (i.e., lower the speed of response).

Example 7.9

Draw a schematic diagram to illustrate the incorporation of pressure feedback, using a bellows, in a flapper-valve pneumatic control system. Describe the operation of this feedback control scheme, giving the advantages and disadvantages of this method of control.

Solution
One possible arrangement for external pressure feedback in a flapper valve is shown in Figure 7.59. Its operation can be explained as follows: if pressure P_1 drops, the bellows will contract, thereby moving the flapper closer to the nozzle, thus increasing P_1. Hence, the bellows acts as a *mechanical feedback* device, which tends to regulate pressure disturbances. The advantages of such a device are

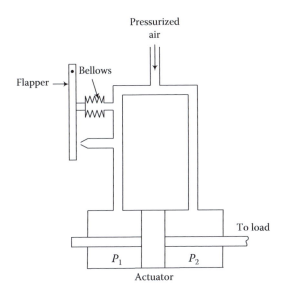

FIGURE 7.59
External pressure feedback for a flapper valve, using a bellows.

1. It is a simple, robust, low-cost mechanical device.
2. It provides mechanical feedback control of pressure variations.

The disadvantages are

1. It can result in a slow (i.e., low-bandwidth) system, if the inertia of the bellows is excessive.
2. It introduces a time delay, which can have a destabilizing effect, particularly at high frequencies.

7.6.11 Hydraulic Circuits

A typical hydraulic control system consists of several components such as pumps, motors, valves, piston-cylinder actuators, and accumulators that are interconnected through piping. It is convenient to represent each component with a standard graphic symbol. The overall system can be represented by a circuit diagram where the symbols for various components are joined by lines to denote flow paths. Circuit representations of some of the many hydraulic components are shown in Figure 7.60. A few explanatory comments would be appropriate. The inward solid pointers in the motor symbols indicate that a hydraulic motor receives hydraulic energy. Similarly, the pointers in the pump symbols show that a hydraulic pump gives out hydraulic energy. In general, the arrows inside a symbol show fluid flow paths. The external spring and arrow in the relief valve symbol shows that the unit is adjustable and spring restrained. There are three basic types of hydraulic line symbols. A solid line indicates a primary hydraulic flow. A broken line with long dashes is a *pilot line*, which indicates the control of a component. For example, the broken line in the relief valve symbol indicates that the valve is controlled by pressure. A broken line with short dashes represents a drain line or leakage flow. In the spool valve symbols, P denotes the supply port (with pressure P_s) and T denotes the discharge port to the reservoir (with gage zero pressure). Finally, note that ports A and B of a four-way spool valve are connected to the two ports of a double-acting hydraulic cylinder (see Figure 7.45a).

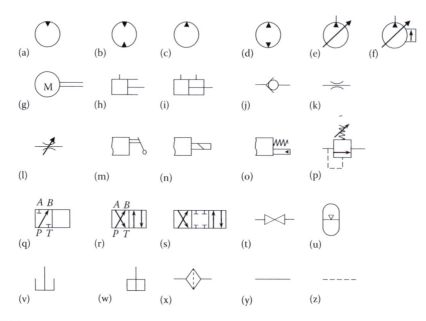

FIGURE 7.60
Typical graphic symbols used in hydraulic circuit diagrams: (a) Motor; (b) reversible motor; (c) pump; (d) reversible pump; (e) variable displacement pump; (f) pressure-compensated variable displacement pump; (g) electric motor; (h) single-acting cylinder; (i) double-acting cylinder; (j) ball-and-seat check valve; (k) fixed orifice; (l) variable flow orifice; (m) manual valve; (n) solenoid-actuated valve; (o) spring centered pilot-controlled valve; (p) relief valve (adjustable and pressure-operated); (q) two-way spool valve; (r) four-way spool valve; (s) three-position four-way valve; (t) manual shut-off valve; (u) accumulator; (v) vented reservoir; (w) pressurized reservoir; (x) filter; (y) main fluid line; (z) pilot line.

Problems

7.1 Figure P7.1 shows a schematic diagram of a stepper motor. What type of stepper is this? Describe the operation of this motor. In particular, discuss whether four separate phases are needed or whether the phases of the opposite stator poles may be connected together, giving a two-phase stepper. What is the step angle of the motor

(a) In full-stepping?
(b) In half-stepping?

7.2 In connection with the phase windings of a stepper motor, explain the following terms:

(a) Unifilar (or monofilar) winding
(b) Bifilar winding
(c) Bipolar winding

Discuss why the torque characteristics of a bifilar-wound motor are better than those of a unifilar-wound motor at high stepping rates.

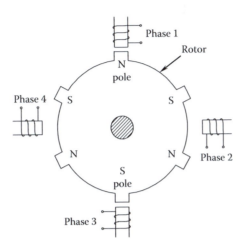

FIGURE P7.1
Schematic diagram of a stepper motor.

7.3 The torque of a stepping motor can be increased by increasing its diameter, for a given coil density (the number of turns per unit area) of the stator poles, for a given current rating. Alternatively, the motor torque can be increased by introducing multiple stacks (i.e., a longer motor) for a given diameter, coil density, and current rating. Giving reasons, indicate which design is generally preferred.

7.4 The principle of operation of a (hybrid) linear stepper motor is indicated in the schematic diagram of Figure P7.4. The toothed platen is a stationary member made of ferromagnetic material, which is not magnetized. The moving member is termed the "forcer," which has four groups of teeth (only one tooth per group is shown in the figure, for convenience). A permanent magnet has its N pole located at the first two groups of teeth and the S pole located at the next two groups of teeth, as shown. Accordingly, the first two groups are magnetized to take the N polarity and the next two groups take the S polarity. The motor has two phases, denoted by A and B. Phase A is wound between the first two groups of teeth and Phase B is wound between the second two groups of teeth of the forcer, as shown. In this manner, when Phase A is energized, it will create an electromagnet with opposite polarities located at the first two groups of teeth. Hence, one of this first two groups of teeth will have its magnetic polarity

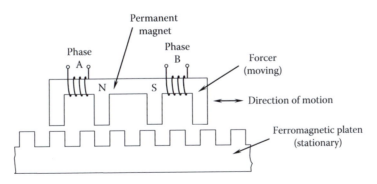

FIGURE P7.4
Schematic representation of a linear hybrid stepping motor.

reinforced while the other group will have its polarity neutralized. Similarly, Phase B, when energized, will strengthen one of the next two groups of teeth while neutralizing the other group. The teeth in the four groups of the forcer have quadrature offsets as follows. The second group has an offset of a ½ tooth pitch with respect to the first group. The third group of teeth has an offset of a ¼ tooth pitch with respect to the first group in one direction, and the fourth group has an offset of a ¼ pitch with respect to the first group in the opposite direction (hence, the fourth group has an offset of a ½ pitch with respect to the third group of teeth). The phase windings are bipolar (i.e., the current can be reversed).

(a) Describe the full-stepping cycle of this motor for motion to the right and for motion to the left.
(b) Give the half-stepping cycle of this motor for motion to the right and for motion to the left.

7.5 When a phase winding of a stepper motor is switched on, ideally the current in the winding should instantly reach the full value (hence providing the full magnetic field instantly). Similarly, when a phase is switched off, its current should become zero immediately. It follows that the ideal shape of phase current history is a rectangular pulse sequence, as shown in Figure P7.5. In actual motors, however, the current curves deviate from the ideal rectangular shape, primarily because of the magnetic induction in the phase windings. Using sketches, indicate how the phase current waveform would deviate from this ideal shape under the following conditions:

(a) Very slow stepping
(b) Very fast stepping at a constant stepping rate
(c) Very fast stepping at a variable (transient) stepping rate

A stepper motor has a phase inductance of 10 mH and a phase resistance of 5 Ω. What is the electrical time constant of each-phase in a stepper motor? Estimate the stepping rate below which magnetic induction effects can be neglected so that the phase current waveform is almost a rectangular pulse sequence.

7.6 Consider a stepper motor that has two poles per phase. The pole windings in each phase may be connected either in parallel or in series, as shown in Figure P7.6. In each case, determine the required ratings for phase power supply (rated current, rated

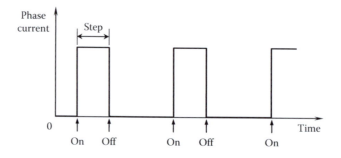

FIGURE P7.5
Ideal phase current waveform for a stepper motor.

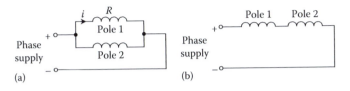

FIGURE P7.6
Pole windings in a phase of a stepper motor that has two poles per phase: (a) Parallel connection; (b) series connection.

voltage, rated power) in terms of current i and resistance R, as indicated in Figure P7.6a. Note that the power rating should be the same for both cases, as is intuitively clear.

7.7 Some industrial applications of stepper motors call for very high stepping rates under variable load (variable motor torque) conditions. Since the motor torque depends directly on the current in the phase windings (typically 5 A per phase), one method of obtaining a variable-torque drive is to use an adjustable resistor in the drive circuit. An alternative method is to use a *chopper drive*. Switching transistors, diodes, or thyristors are used in a chopper circuit to periodically bypass (chop) the current through a phase winding. The chopped current passes through a free-wheeling diode back to the power supply. The chopping interval and chopping frequency are adjustable. Discuss the advantages of chopper drives compared with the resistance drive method.

7.8 Define and compare the following pairs of terms in the context of electromagnetic stepper motors:

(a) Pulses and steps

(b) Step angle and resolution

(c) Residual torque and static holding torque

(d) Translator and drive system

(e) PM stepper motor and VR stepper motor

(f) Single-stack stepper and multiple-stack stepper

(g) Stator poles and stator phases

(h) Pulse rate and slew rate

7.9 Compare the VR stepper motor with the PM stepper motor with respect to the following considerations:

(a) Torque capacity for a given motor size

(b) Holding torque

(c) Complexity of switching circuitry

(d) Step size

(e) Rotor inertia

The hybrid stepper motor possesses characteristics of both the VR and the PM types of stepper motors. Consider a typical construction of a hybrid stepper motor, as shown schematically in Figure P7.9. The rotor has two stacks of teeth made of ferromagnetic material, joined together by a permanent magnet, that assign opposite polarities to the two rotor stacks. The tooth pitch is the same for both stacks, but

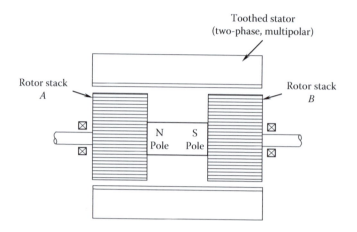

FIGURE P7.9
Schematic diagram of a hybrid stepper motor.

the two stacks have a tooth misalignment of half a tooth pitch ($\theta_r/2$). The stator may consist of a common tooth stack for both rotor stacks, or it may consist of two tooth stack segments that are in complete alignment, one for each rotor stack. The number of teeth in the stator is not equal to the number of teeth in each rotor stack. The stator is made up of several toothed poles that are equally spaced around the rotor. Half the poles are connected to one phase and the other half are connected to the second phase. The current in each phase may be turned on and off or reversed using switching amplifiers. The switching sequence for rotation in one direction (say, CW) would be A^+, B^+, A^-, B^-; for rotation in the opposite direction (CCW), it would be A^+, B^-, A^-, B^+, where A and B denote the two phases and the superscripts $+$ and $-$ denote the direction of current in each phase. This may also be denoted by [1, 0], [0, 1], [−1, 0], [0, −1] for CW rotation and [1, 0], [0, −1], [−1, 0], [0, 1] for CCW rotation.

Consider a motor that has eighteen teeth in each rotor stack and eight poles in the stator, with two teeth per stator pole. The stator poles are wound to the two phases as follows: two radially opposite poles are wound to the same phase with identical polarity. The two radially opposite poles that are at 90° from this pair of poles are also wound to the same phase, but with the field in the opposite direction (i.e., opposite polarity) to the previous pair.

(a) Using suitable sketches of the rotor and stator configurations at the two stacks, describe the operation of this hybrid stepper motor.

(b) What is the step size of the motor?

7.10 A relatively convenient method of electronic damping uses simultaneous multiphase energization, where more than one phase is energized simultaneously and some of the simultaneous phases are excited with a fraction of the normal operating (rated) voltage. A simultaneous two-phase energization technique has been suggested for a three-phase, single-stack stepper motor. If the standard sequence of switching of the phases for forward motion is given by 1–2–3–1, what is the corresponding simultaneous two-phase energization sequence?

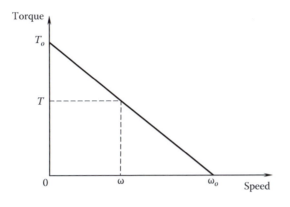

FIGURE P7.11
An approximate speed–torque curve for a stepper motor.

7.11 The torque versus speed curve of a stepper motor is approximated by a straight line, as shown in Figure P7.11. The following two parameters are given:

T_o = torque at zero speed (starting torque or stand-still torque)
ω_o = speed at zero torque (no-load speed)

Suppose that the load resistance is approximated by a rotary viscous damper with a damping constant b. Assuming that the motor directly drives the load, without any speed reducers, determine the steady-state speed of the load and the corresponding drive torque of the stepper motor.

7.12 The speed–torque curve of a stepper motor is shown in Figure P7.12. Explain the shape, particularly the two dips, of this curve.

Suppose that with one phase on, the torque of a stepper motor in the neighborhood of the detent position of the rotor is given by the linear relationship

$$T = -K_m\theta$$

where
θ is the rotor displacement measured from the detent position
K_m is the motor torque constant, magnetic stiffness, or torque gradient

The motor is directly coupled to an inertial load. The combined moment of inertia of the motor rotor and the inertial load is $J = 0.01$ kg-m^2. If $K_m = 628.3$ N-m/rad, at what stepping rates would you expect dips in the speed–torque curve of the motor–load combination?

7.13 Briefly discuss the operation of a microprocessor-controlled stepper motor. How would it differ from the standard setup in which a "preset indexer" is employed? Compare and contrast table lookup, programmed stepping, and hardware stepping methods for stepper motor translation.

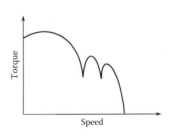

FIGURE P7.12
Typical speed–torque curve of a stepper motor.

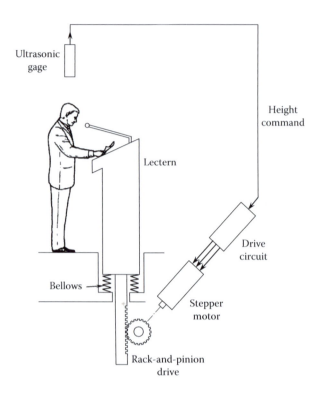

FIGURE P7.14
An automated lectern.

7.14 A lectern (or podium) in an auditorium is designed to adjust its height automatically, depending on the height of the speaker. An ultrasonic gage measures the height of the speaker and sends a command to the logic hardware controller of a stepper motor, which adjusts the lectern vertically through a rack-and-pinion drive. The dead load of the moving parts is supported by a bellow device. A schematic diagram of this arrangement is shown in Figure P7.14. The following design requirements have been specified:

Time to adjust a maximum stroke of $1\,m = 5\,s$
Mass of the lectern $= 50\,kg$
Maximum resistance to vertical motion $= 5\,kg$
Displacement resolution $= 0.5\,cm/step$

  Select a suit*able* stepper motor system for this application. You may use the ratings of the four commercial stepper motors as given in Table 7.1 and Figure 7.14.

7.15 (a) In theory, a stepper motor does not require a feedback sensor for its control. But, in practice, a feedback encoder is needed for accurate control, particularly under transient and dynamic loading conditions. Explain the reasons for this.

   (b) A material transfer unit in an automated factory is sketched in Figure P7.15. The unit consists of a conveyor, which moves objects onto a platform. When an object reaches the platform, the conveyor is stopped and the height of the object is measured using a laser triangulation unit. Then the stepper motor of the platform is activated to raise the object through a distance that is determined on the basis of the object height, for further processing of the object.

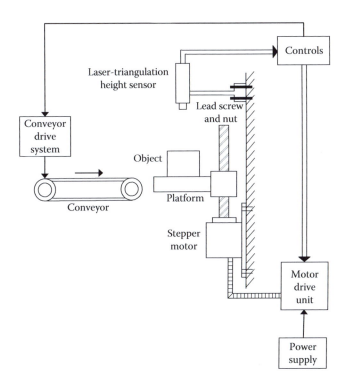

FIGURE P7.15
A material transfer unit in an automated factory.

The following parameters are given:
Mass of the heaviest object that is raised = 3.0 lb (1.36 kg)
Mass of the platform and nut = 3.0 lb
Inertia of the lead screw and coupling = 0.001 oz-in-s^2 (0.07 kg-m^2)
Maximum travel of the platform = 1.0 in (2.54 cm)
Positioning time = 200 ms
Assume a 4-pitch lead screw of 80% efficiency. Also, neglect any external resistance to the vertical motion of the object, apart from gravity.
Out of the four choices of stepper motor that are given in Table 7.1 and Figure 7.14, which one would you pick to drive the platform? Justify your selection by giving all the computational details of the approach.

7.16 (a) Consider a stepper motor of moment of inertia J_m, which drives a purely inertial load of moment of inertia J_L, through a gearbox of speed reduction r:1, as shown in Figure P7.16a.

Note that $\omega_L = \omega_m/r$
where
ω_m is the motor speed
ω_L is the load speed

(i) Show that the motor torque T_m may be expressed as

$$T_m = \left(rJ_m + \frac{J_L}{er} \right) \dot{\omega}_L$$

in which e is the gear efficiency.

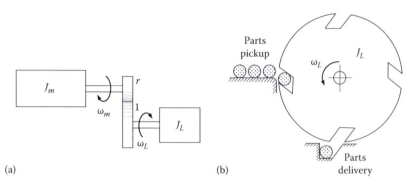

FIGURE P7.16
(a) Stepper motor driving an inertial load; (b) a parts transfer mechanism: an example of inertial load.

  (ii) For optimal conditions of load acceleration, express the required gear ratio
       r in terms of J_L, J_m, and e.

(b) An example of a rotary load that is driven by a stepper motor is shown in Figure
    P7.16b. Here, in each quarter revolution of the load rotor, a part is transferred
    from the pickup position to the delivery position. The equivalent moment of
    inertia of the rotor, which carries a part, is denoted by J_L.
    Suppose that $JL = 12.0 \times 10^{-3}$ kg-m². The required rate of parts transfer is 7 parts/s.
    A stepper motor is used to drive the load. A gearbox may be employed as well.
    Four motor models are available, whose parameters are given in Table P7.16.
       The speed–torque characteristics of the motors are given in Figure 7.14.
    Assume that the step angle of each motor is 1.8°. The gearbox efficiency may
    be taken as 0.8.
  (i) Prepare a table giving the optimal gear ratio, the operating speed of motor,
      the available torque, and the required torque for each of the four models
      of motor, assuming that a gearbox with optimal gear ratio is employed in
      each case. On this basis, which motor would you choose for the present
      application?

  (ii) Now consider the motor chosen in Part (i). Suppose that three gearboxes of
       speed reduction 5, 8, and 10 respectively, may be available to you. Is a gear-
       box required in the present application with the chosen motor? If so, which
       gearbox would you choose? Make your decision by computing the available
       torque and the required torque (with the motor chosen in part (i)), for the
       four values of r given by 1, 5, 8, and 10.

  (iii) What is the positioning resolution of the parts transfer system? What fac-
        tors can affect this value?

7.17 Piezoelectric stepper motors are actuators that convert
vibrations in a piezoelectric element (e.g., PZT) generated by an ac voltage (reverse piezoelectric effect) into
rotary motion. Step angles on the order of 0.001° can be
obtained by this method. In one design, as the piezoelectric PZT rings vibrate due to an applied ac voltage,
radial bending vibrations are produced in a conical
aluminum disk. These vibrations impart twisting

TABLE P7.16

Motor Parameter Values

Motor Model	Motor Inertia J_m (×10⁻⁶ kg-m²)
50SM	11.8×10^{-6}
101SM	35
310SM	187
1010SM	805

(torsional) vibrations onto a beam element. The twisting motion is subsequently converted into a rotary motion of a frictional disk, which is frictionally coupled with the top surface of the beam. Essentially, because of the twisting motion, the two top edges of the beam push the frictional disk tangentially in a stepwise manner. This forms the output member of the piezoelectric stepper motor. List several advantages and disadvantages of this motor. Describe an application in which a miniature stepper motor of this type could be used.

7.18 What factors generally govern (a) the electrical time constant and (b) the mechanical time constant of a motor? Compare the typical values for these parameters and discuss how they affect the motor response.

7.19 Explain the operation of a brushless dc motor. How does it compare with the principle of the operation of a stepper motor?

7.20 Give the steady-state torque–speed relations for a dc motor with the following three types of connections for the armature and filed windings:

(a) A shunt-wound motor

(b) A series-wound motor

(c) A compound-wound motor with $R_{f1} = R_{f2} = 10\,\Omega$

The following parameter values are given: Ra = 5 Ω, Rf = 20 Ω, k = 1 N-m/A², and for a compound-wound motor, $R_{f1} = R_{f2} = 10\,\Omega$.

Note: $T_m = ki_fi_a$

Assume that the supply voltage is 115 V. Plot the steady-state torque–speed curves for these types of winding arrangements.

Using these curves, compare the steady-state performance of the three types of motors.

7.21 What is the electrical damping constant of a dc motor? Determine the expressions for this constant for the three types of dc motor winding arrangements mentioned in Problem 7.20. In which case is this a constant value? Explain how the electrical damping constant could be experimentally determined. How is the dominant time constant of a dc motor influenced by the electrical damping constant? Discuss ways to decrease the motor time constant.

7.22 Explain why the transfer-function representation for a separately excited and armature-controlled dc motor is more accurate than that of a field-controlled motor and still more accurate than those of shunt-wound, series-wound, or compound-wound dc motors. Give a transfer function relation (using Laplace variable s) for a dc motor where the incremental speed $\delta\omega_m$ is the output, the incremental winding excitation voltage δv_c is the control input, and the incremental load torque δT_L on the motor is a disturbance input. Assume that the parameters of the motor model are determined from the experimental speed–torque curves for a constant excitation voltage.

7.23 Using sketches, describe how PWM effectively varies the average value of the modulated signal. Explain how one could obtain the following by PWM:

(a) A zero average

(b) A positive average

(c) A negative average

Indicate how PWM is useful in the control of dc motors. List the advantages and disadvantages of PWM.

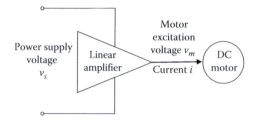

FIGURE P7.24
A linear amplifier for a dc motor.

7.24 Figure P7.24 shows a schematic arrangement for driving a dc motor using a linear amplifier. The amplifier is powered by a dc power supply of regulated voltage v_s. Under a particular condition, suppose that the linear amplifier drives the motor at voltage v_m and current i. Assume that the current drawn from the power supply is also i. Give an expression for the efficiency at which the linear amplifier is operating under these conditions. If $v_s = 50\,V$, $v_m = 20\,V$, and $i = 5$ A, estimate the efficiency of the operation of the linear amplifier.

7.25 For a dc motor, the starting torque and the no-load speed are known, which are denoted by T_s and ω_o, respectively. The rotor inertia is J. Determine an expression for the dominant time constant of the motor.

7.26 A schematic diagram for the servo control loop of one joint of a robotic manipulator is given in Figure P7.26.

The motion command for each joint of the robot is generated by the controller of the robot in accordance with the required trajectory. An optical (incremental) encoder is used for both position and velocity feedback in each servo loop. Note that for a six-degree-of-freedom robot there will be six such servo loops. Describe the function of each hardware component shown in the figure and explain the operation of the servo loop.

After several months of operation, the motor of one joint of the robot was found to be faulty. An enthusiastic engineer quickly replaced the motor with an identical one

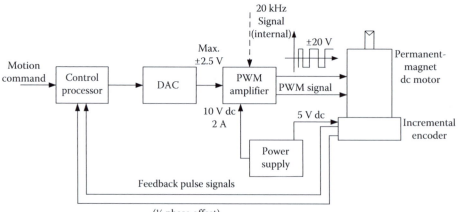

FIGURE P7.26
A servo loop of a robot.

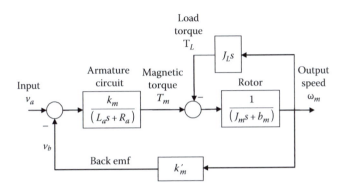

FIGURE P7.27
An armature-controlled dc motor with an inertial load.

without realizing that the encoder of the new motor was different. In particular, the original encoder generated 200 pulses/rev whereas the new encoder generated 720 pulses/rev. When the robot was operated, the engineer noticed an erratic and unstable behavior at the repaired joint. Discuss reasons for this malfunction and suggest a way to correct the situation.

7.27 Consider the block diagram in Figure 7.19, which represents a dc motor, for armature control with the usual notation. Suppose that the load driven by the motor is a pure inertia element (e.g., a wheel or a robot arm) of moment of inertia J_L, which is directly and rigidly attached to the motor rotor.

(a) Show that, in this case, the motor block diagram may be given as in Figure P7.27. Obtain an expression for the transfer function $\omega_m/v_a = G_m(s)$ for the motor with the inertial load in terms of the parameters given in Figure P7.27a.

(b) Now, neglect the leakage inductance L_a. Then, show that the transfer function in Part (a) can be expressed as $G_m(s) = k/(\tau s + 1)$. Give expressions for τ and k in terms of the given system parameters.

(c) Suppose that the motor (with the inertial load) is to be controlled using position plus velocity feedback. The block diagram of the corresponding control system is given in Figure 7.21, where the motor transfer function $G_m(s) = k/(\tau s + 1)$. Determine the transfer function of the (closed-loop) control system $G_{CL}(s) = \theta_m/\theta_d$ in terms of the given system parameters (k, k_p, τ, τ_v). Note that θ_m is the angle of rotation of the motor with an inertial load and θ_d is the desired angle of rotation.

7.28 The moment of inertia of the rotor of a motor (or any other rotating machine) can be determined by a run-down test. With this method, the motor is first brought up to an acceptable speed and then quickly turned off. The motor speed versus time curve is obtained during the run-down period that follows. A typical run-down curve is shown in Figure P7.28. Note that the motor decelerates because of its resisting torque T_r during this period. The slope of the run-down curve is determined at a suitable (operating) value of speed ($\bar{\omega}_m$) in Figure P7.28. Next, the motor is brought up to this speed ($\bar{\omega}_m$) and the torque ($\bar{T}_r$) that is needed to maintain the motor steady at this speed is obtained (either by direct measurement of torque or by computing using the field current measurement and the known value for the torque constant, which

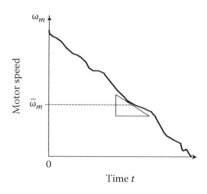

FIGURE P7.28
Data from a run-down test on an electric motor.

is available in the data sheet of the motor). Explain how the rotor inertia J_m may be determined from this information.

7.29 In some types of (indirect-drive) robotic manipulators, joint motors are located away from the joints and torques are transmitted to the joints through transmission devices such as gears, chains, cables, and timing belts. In some other types of manipulators (i.e., direct-drive), joint motors are located at the joints themselves, the rotor being integral with one link and the stator being integral with the joining link. Discuss the advantages and disadvantages of these two designs.

7.30 In brushless motors, commutation is achieved by switching the stator phases at the correct rotor positions (e.g., at the points of intersection of the static torque curves corresponding to the phases, for achieving maximum average static torque). We have noted that the switching points can be determined by measuring the rotor position using an incremental encoder. Incremental encoders are delicate, costly, cannot operate at high temperatures, and will increase the size and cost of the motor package. Also, precise mounting is required for proper operation. The generated signal may be subjected to electromagnetic interference (EMI) depending on the means of signal transmission. Since we only need to know the switching points (i.e., continuous measurement of rotor position is not necessary) and since these points are uniquely determined by the stator magnetic field distribution, a simpler and cost effective alternative to an encoder for detecting the switching points would be to use Hall-effect sensors. Specifically, Hall-effect sensors are located at switching points around the stator (a sensor ring) and a magnet assembly is located around the rotor (in fact, the magnetic poles of the rotor can serve this purpose without needing an additional set of poles). As the rotor rotates, a magnetic pole on the rotor will trigger an appropriate Hall-effect sensor, thereby generating a switching signal (pulse) for commutation at the proper rotor position. A microelectronic switching circuit (or switching transistor) is actuated by the corresponding pulse. Since Hall-effect sensors have several disadvantages, such as hysteresis (and associated nonsymmetry of the sensor signal), low operating temperature ratings (e.g., 125°C), thermal drift problems, and noise due to stray magnetic fields and EMI, it may be more desirable to use fiber optic sensors for brushless commutation. Describe how the fiber-optic method of motor commutation works.

7.31 A brushless dc motor and a suitable drive unit are to be chosen for a continuous drive application. The load has a moment of inertia 0.016 kg-m^2 and faces a constant resisting torque of 35.0 N-m (excluding the inertia torque) throughout the operation. The application involves accelerating the load from rest to a speed of 250 rpm in 0.2 s, maintaining it at this period for extended periods, and then decelerating to rest in 0.2 s. A gear unit with a step-down gear ratio 4 is to be used with the motor. Estimate a suitable value for the moment of inertia of the motor rotor for a fairly optimal design. The gear efficiency is known to be 0.8. Determine a value for continuous torque and a corresponding value for operating speed on which a selection of a motor and a drive unit can be made.

7.32 Compare dc motors with ac motors in general terms. In particular, consider the mechanical robustness, cost, size, maintainability, speed control capability, and implementation of complex control schemes.

7.33 Compare the frequency control with the voltage control in induction motor control, giving the advantages and disadvantages. The steady-state slip-torque relationship of an induction motor is given by

$$T_m = \frac{aSv_f^2}{[1+(S/S_b)^2]}$$

with the parameter values $a = 1 \times 10^{-3}$ N-m/V^2 and $S_b = 0.25$. If the line voltage $v_f = 241$ V, calculate the breakdown torque. If the motor has two pole pairs per phase and if the line frequency is 60 Hz, what is the synchronous speed (in rpm)? What is the speed corresponding to the breakdown torque? If the motor drives an external load, which is modeled as a viscous damper of damping constant $b = 0.03$ N-m/rad/s, determine the operating point of the system. Now, if the supply voltage is dropped to 163 V through voltage control, what is the new operating point? Is this a stable operating point?

7.34 Consider the induction motor in Problem 7.33. Suppose that the line voltage is $v_f = 200$ V and the line frequency is 60 Hz. The motor is rigidly connected to an inertial load. The combined moment of inertia of the rotor and load is $J_{eq} = 5$ kg-m^2. The combined damping constant is $b_{eq} = 0.1$ N-m/rad/s. If the system starts from rest, determine, by computer simulation, the speed time history $\omega_L(t)$ of the load (and motor rotor). (*Hint:* Assume that the motor is a torque source, with torque represented by the steady-state speed–torque relationship.)

7.35 (a) The equation of the rotor circuit of an induction motor (per phase) is given by (see Figure 7.33a)

$$i_r = \frac{Sv}{(R_r + jS\omega_p L_r)} = \frac{v}{(R_r/S + j\omega_p L_r)}.$$

This corresponds to an impedance (i.e., voltage/current in the frequency domain) of $Z = R_r/S + j\omega_p L_r$.

Show that this may be expressed as the sum of two impedance components:

$$Z = [R_r + j\omega_p SL_r] + \left[\left(\frac{1}{S-1}\right)R_r + j\omega_p(1-S)L_r\right].$$

For a line frequency of ω_p, this result is equivalent to the circuit shown in Figure 7.33c. Note that the first component of impedance corresponds to the rotor electrical loss and the second component corresponds to the useful mechanical power.

(b) Consider the characteristic shape of the speed versus torque curve of an induction motor. Typically, the starting torque Ts is less than the maximum torque T_{max}, which occurs at a nonzero speed. Explain the main reason for this.

7.36 Prepare a table to compare and contrast the following types of motors:

(a) Conventional dc motor with brushes

(b) Brushless torque motor (dc)

(c) Stepper motor

(d) Induction motor

(e) ac synchronous motor

In your table, include terms such as power capability, speed controllability, speed regulation, linearity, operating bandwidth, starting torque, power supply requirements, commutation requirements, and power dissipation. Discuss a practical method for reversing the direction of rotation in each of these types of motors.

7.37 Show that the rms value of a rectangular wave can be changed by phase-shifting it and adding to the original signal. What is its applicability in the control of induction motors?

7.38 The direction of the rotating magnetic field in an induction motor (or any other type of ac motor) can be reversed by changing the supply sequence of the phases to the stator poles. This is termed phase-switching. An induction motor can be decelerated quickly in this manner. This is known as "plugging" an induction motor. The slip versus torque relationship of an induction motor may be expressed as $T_m = k(S)v_f^2$.

Show that the same relationship holds under plugged conditions, except that $k(S)$ has to be replaced by $-k(2-S)$. Sketch the curves $k(S)$, $k(2-S)$, and $-k(2-S)$ from $S=0$ to $S=2$. Using these curves, indicate the nature of the torque acting on the rotor during plugging. (*Hint:* $k(S)=(aS)/[1+(S/S_b)^2]$).

7.39 What is a servomotor? AC servomotors that can provide torques on the order of 100 N-m at 3000 rpm are commercially available (*Note:* 1 N-m = 141.6 oz-in.). Describe the operation of an ac servomotor that uses a two-phase induction motor. A block diagram for an ac servo motor is shown in Figure P7.39. Describe the purpose of each

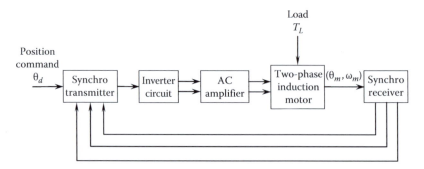

FIGURE P7.39
AC servomotor using a two-phase induction motor and a synchro transformer.

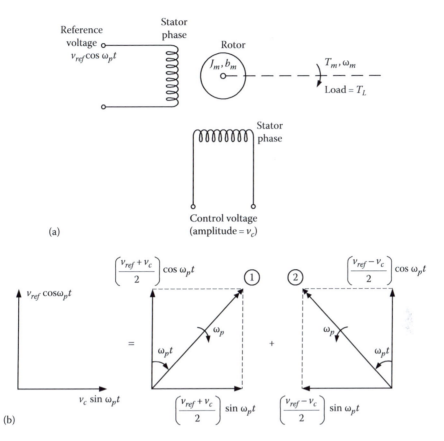

FIGURE P7.40

(a) A two-phase induction motor functioning as an ac servomotor; (b) equivalent representation of the magnetic field vector in the stator.

component in the system and explain the operation of the overall system. What are the advantages of using an ac amplifier after the inverter circuit in comparison with using a dc amplifier before the inverter circuit?

7.40 A two-phase induction motor can serve as an ac servomotor. The field windings are identical and are placed in the stator with a geometric separation of 90°, as shown in Figure P7.40a. One of the phases is excited by a fixed reference ac voltage $v_{ref} \cos \omega_p t$. The other phase is 90° out of phase from the reference phase; it is the control phase with voltage amplitude v_c. The motor is controlled by varying the voltage v_c.

1. With the usual notation, obtain an expression for the motor torque Tm in terms of the rotor speed ω_m and the input voltage vc.

2. Indicate how a transfer function model may be obtained for this ac servo
   (a) Graphically, using the characteristic curves of the motor
   (b) Analytically, using the relationship obtained in part 1

7.41 Consider the two-phase induction motor discussed in Problem 7.40. Show that the motor torque T_m is a linear function of the control voltage v_c when $k(2-S) = k(S)$. How many values of speed (or slip) satisfy this condition? Determine these values.

7.42 A magnetically levitated rail vehicle uses the principle of induction motor for traction. Magnetic levitation is used for the suspension of the vehicle slightly above the emergency guide rails. Explain the operation of the traction system of this vehicle, particularly identifying the stator location and the rotor location. What kinds of sensors would be needed for the control systems for traction and levitation? What type of control strategy would you recommend for the vehicle control?

7.43 What are the common techniques for controlling

(a) DC motors?

(b) AC motors?

Compare these methods with respect to speed controllability.

7.44 Describe the operation of a single-phase ac motor. List several applications of this common actuator. Is it possible to realize a three-phase operation using a single-phase ac supply? Explain your answer.

7.45 Speed control of motors (ac motors as well as dc motors) can be accomplished by using solid-state switching circuitry. In one such method, a solid-state relay is activated using a switching signal generated by a microcomputer so as to turn on and turn off the power into the motor drive circuit at high speed. The speed of the motor can be measured using a sensor such as a tachometer or optical encoder. This signal is read by the microcomputer and is used to modify the switching signal so as to correct the motor speed. Using a schematic diagram, describe the hardware needed to implement this control scheme. Explain the operation of the control system.

7.46 In some applications, it is necessary to apply a force without creating a motion. Discuss one such application. Discuss how an induction motor could be used in such an application. What are the possible problems?

7.47 The harmonic drive principle can be integrated with an electric motor in a particular manner in order to generate a high-torque "gear motor." Suppose that the flexispline of the harmonic drive is made of an electromagnetic material, such as the rotor of a motor. Instead of the mechanical wave generator, suppose that a rotating magnetic field is generated around the fixed spline. The magnetic attraction will cause the tooth engagement between the flexispline and the fixed spline. What type of motor principle may be used in the design of this actuator? Give an expression for the motor speed. How would one control the motor speed in this case?

7.48 List three types of hydraulic pumps and compare their performance specifications. A position servo system uses a hydraulic servo along with a synchro transformer as the feedback sensor. Draw a schematic diagram and describe the operation of the control system.

7.49 Giving typical applications and performance characteristics (bandwidth, load capacity, controllability, etc.), compare and contrast dc servos, ac servos, hydraulic servos, and pneumatic servos.

7.50 What is a multistage servovalve? Describe its operation. What are the advantages of using several valve stages?

7.51 Discuss the origins of the hydraulic time constant in a hydraulic control system that consists of a four-way spool valve and a double-acting cylinder actuator. Indicate the significance of this time constant. Show that the dimensions of the right-hand-side expression in the following equation: $\tau_h = \dfrac{V}{2\beta k_c}$ are [time].

7.52 Sometimes either a PWM ac signal or a dc signal with a superimposed constant-frequency ac signal (or dither) is used to drive the valve actuator (torque motor) of a hydraulic actuator. What is the main reason for this? Discuss the advantages and disadvantages of this approach.

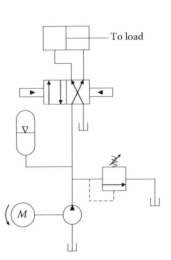

To load

7.53 Compare and contrast valve-controlled hydraulic systems with pump-controlled hydraulic systems. Using a schematic diagram, explain the operation of a pump-controlled hydraulic motor. What are its advantages and disadvantages over a frequency-controlled ac servo?

7.54 Explain why accumulators are used in hydraulic systems. Sketch two types of hydraulic accumulators and describe their operation.

7.55 Identify and explain the components of the hydraulic system given by the circuit diagram in Figure P7.55. Describe the operation of the overall system.

FIGURE P7.55
A hydraulic circuit diagram.

7.56 The sketch in Figure P7.56 shows a half-sectional view of a flow control valve, which is intended to keep the flow to a hydraulic load constant regardless of the variations of the load pressure P_3 (disturbance input).

(a) Briefly discuss the physical operation of the valve, noting that the flow will be constant if the pressure drop across the fixed area orifice is constant.

(b) Write the equations that govern the dynamics of the unit. The mass, the damping constant, and the spring constant of the valve are denoted by m, b, and k, respectively. The volume of oil under pressure P_2 is V, and the bulk modulus of the oil is β. Make the usual linearizing assumptions.

(c) Set up a block diagram for the system from which the dynamics and stability of the valve could be studied.

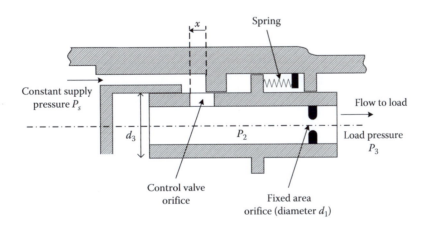

FIGURE P7.56
A flow control valve.

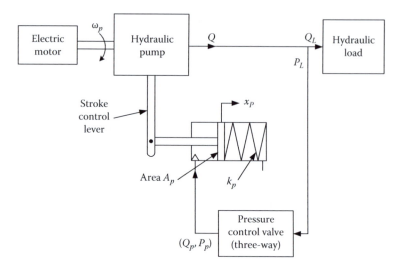

FIGURE P7.57
A pump stroke-regulated hydraulic power supply.

7.57 A schematic diagram of a pump stroke-regulated hydraulic power supply is shown in Figure P7.57. The system uses a three-way pressure control valve of the type described in the book. This valve controls a spring-loaded piston, which in turn regulates the pump stroke by adjusting the swash plate angle of the pump. The load pressure P_L is to be regulated. This pressure can be set by adjusting the preload x_o of the spring in the pressure control valve. The load flow Q_L enters into the hydraulic system as a disturbance input.

(a) Briefly describe the operation of the control system.

(b) Write the equations for the system dynamics, assuming that the pump stroke mechanism and the piston inertia can be represented by an equivalent mass m_p moving through x_p. The corresponding spring constant and damping constant are k_p and b_p, respectively. The piston area is A_p. The mass, spring constant, and damping constant of the valve are m, k, and b, respectively. The valve area is A_v and the valve spool movement is x_v. The volume of oil under pressure P_L is V_l, and the volume of oil under pressure P_p is V_o (volume of oil in the cylinder chamber). The bulk modulus of the oil is β.

(c) Draw a block diagram for the system from which the behavior of the system could be investigated. Indicate the inputs and outputs.

(d) If Q_p is relatively negligible, indicate which control loops can be omitted from the block diagram. Hence, derive an expression for the transfer function $x_p(s)/x_v(s)$ in terms of the system parameters.

7.58 A schematic diagram of a solenoid-actuated flow control valve is shown in Figure P7.58a. The downward motion x of the valve rod is resisted by a spring of stiffness k. The mass of the valve rod assembly (all moving parts) is m and the associated equivalent viscous damping constant is b. The voltage supply to the valve actuator (proportional solenoid) is denoted by v_i. For a given voltage v_i, the solenoid force is a nonlinear (decreasing) function of the valve position x. This steady-state variation of the solenoid force (downward) and the resistive spring force (upward) are shown in

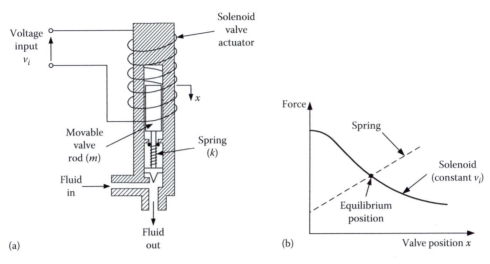

FIGURE P7.58
(a) A solenoid-actuated flow control valve; (b) steady-state characteristics of the valve.

Figure P7.58b. Assuming that the inlet pressure and the outlet pressure of the fluid flow are constant, the flow rate will be determined by the valve position x. Hence, the objective of the valve actuator would be to set x using v_i.

(a) Show that for a given input voltage v_i, the resulting equilibrium position (x) of the valve is always stable

(b) Describe how the relationship between v_i and x could be obtained

    (i) Under quasi-static conditions

    (ii) Under dynamic conditions

7.59 What are the advantages and disadvantages of pneumatic actuators in comparison with electric motors in process control applications? A pneumatic rack-and-pinion actuator is an on/off device that is used as a rotary valve actuator. A piston or diaphragm in the actuator is moved by allowing compressed air into the valve chamber. This rectilinear motion is converted into rotary motion through a rack-and-pinion device in the actuator. Single-acting types with springs return and double-acting types are commercially available. Using a sketch, explain the operation of a piston-type single-acting rack-and-pinion actuator with a spring-restrained piston. Could the force rating, sensitivity, and robustness of the device be improved by using two pistons and racks coupled with the same pinion? Explain.

7.60 A two-axis hydraulic positioning mechanism is used to position the cutter of an industrial fish-cutting machine. The cutter blade is pneumatically operated. The hydraulic circuit of the positioning mechanism is given in Figure P7.60.

Since the two hydraulic axes are independent, the governing equations are similar. State the nonlinear servovalve equations, hydraulic cylinder (actuator) equations, and the mechanical load (cutter assembly) equations for the system. Use the following notation:

x_v = servovalve displacement
K = valve gain (nonlinear)

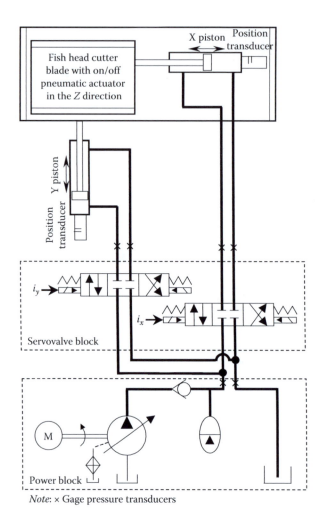

Note: × Gage pressure transducers

FIGURE P7.60
Two-axis hydraulic positioning system of an industrial fish cutter.

P_s = supply pressure
P_1 = head-side pressure of the cylinder with area A_1 and flow Q_1
P_2 = rod-side pressure of the cylinder with area A_2 and flow Q_2
V_h = hydraulic volume in the cylinder chamber
β = bulk modulus of the hydraulic oil
x = actuator displacement
M = mass of the cutter assembly
F_f = frictional force against the motion of the cutter assembly

8

Digital Hardware and Microcontrollers

Study Objectives

- Hardware implementation and software implementation
- Number systems (e.g., binary), logic (two-state), and Boolean algebra
- Codes
- Combinational logic hardware
- Sequential logic devices
- Microprocessors and microcontrollers

8.1 Introduction

Digital hardware can perform a variety of functions in a mechatronic system. Signal acquisition and processing, system monitoring and control, switching, and information display are such functions. Simple functions, actions, and tasks may be implemented using *digital logic hardware* without programmability. Let us call this the "hardware implementation." More complex and variable mechatronic tasks may require programmable digital devices, or in a broader sense, embedded digital computers that are known as *microcontrollers*. Let us call this the "software implementation." Of course, software implementation also requires digital logic hardware and aspects of hardware implementation. Hardware implementation has the advantages of high speed, simplicity, smaller size, and low cost (in mass production). However, its functions are fixed and cannot be changed without difficult hardware changes. Hence, it lacks "flexibility." The software implementation provides flexibility that comes with programmability and the capability of implementing very complex tasks. However, it has the disadvantages of relatively low speed, larger size, and higher cost.

Digital devices, digital computers in particular, use digits (according to some code) to represent information and some logic to process such information. In binary (or, two-state) logic, a variable can take one of two discrete states: *true* (T) or *false* (F). In the *binary* number system, each digit can assume one of only two values: 0 or 1. A digital device may have to process both logical quantities and numerical data. The purpose of a digital circuit might be to turn on or off a device depending on some logical conditions; for example, turn the light on or off in a room depending on whether there are people in it or not. Such an implementation can be done using logic circuitry, which performs logical operations. In some other application, a digital circuit might have to perform numerical computations on a measured signal (available in digital form) and then generate a control signal (in digital form). A digital device can perform such numerical functions as well, using the binary number system where each digit can assume one of only two values: 0 or 1. Because the logic state of each output line of a digital circuit corresponds to the binary 0 or 1, a combination of several output lines can form a numerical output as well (in binary form). It is convenient to use the binary number system in digital devices, not only because logical variables and numerical variables can be processed in the same manner, but also because components that can assume one of two states can be used as the building blocks for digital circuitry. Such 2-state devices are much easier to develop than, for instance, 10-state devices based on the decimal system.

A digital circuit converts digital inputs into digital outputs. There are two types of logic devices, classified as either *combinational logic devices* or *sequential logic devices*. Combinational logic devices are static devices where the present inputs completely (and uniquely) determine the present outputs without using any past information (history) or memory. In contrast, the output of a sequential logic device depends on the past values of the inputs as well as the present values. In other words, they depend on the time sequence of the input data, and hence some form of *memory* would be needed.

Microcontrollers are miniature digital computers of somewhat limited functionality that can be embedded in various locations of a mechatronic system. The digital processor of a microcontroller is a microprocessor. For example, in digital control of a mechatronic system, one or more microcontrollers may serve as the controller. They will use external input commands along with measured responses of the system to generate suitable control signals. These control signals are used to operate the mechatronic system. One or more microprocessors, additional hardware (memory, storage, input/output, etc.), and software capability are used as the controller. This chapter introduces digital logic hardware and microcontrollers.

8.2 Number Systems and Codes

The *base* or the *radix* (denoted by R) of a number system is the maximum number of discrete values each digit of a number can assume. This is also equal to the maximum number of different characters (symbols) that are needed to represent any number in that system. For the decimal system $R = 10$, for the binary system $R = 2$, for the octal system $R = 8$, and for the hexadecimal system $R = 16$. We are quite familiar with the decimal system. The origin of this system is perhaps linked to the fact that a human has 10 fingers in his or her hand. Also, 10 is a convenient and moderate number, which is neither too large nor too small. However, the binary number system is what is natural for digital logic devices and digital computers.

8.2.1 Binary Representation

Since digital devices internally make use of high/low voltage levels, the presence/absence of voltage pulses, and the on/off state of bistable elements (microswitches) for data representation and processing, it is the binary representation that is natural for their internal operations. For example, this is the reason why, instructions, stored data, and addresses of memory locations in a computer are all present in the form of "binary numbers," even though not all such types of information are numerical.

A binary number consists of binary digits (*bits*). Each bit can take the value 0 or 1. Typically, information is arranged in 8 bit groups called *bytes*, internally in a digital computer. A digital computer operates on one data *word* at a time, however, and the word size can be several bytes (e.g., 16 bits in microcontrollers).

If a string of bits represents a binary number, then it can be converted into a decimal number in a straightforward manner. This simply amounts to a conversion from the base-2 (or radix-2) representation into the base-10 representation.

Example 8.1

$$(10101.11)_2 = 1 \times 2^4 + 0 \times 2^3 + 1 \times 2^2 + 0 \times 2^1 + 1 \times 2^0 + 1 \times 2^{-1} + 1 \times 2^{-2}$$

By evaluating the decimal value on the right-hand side, we have

$$(10101.11)_2 = (21.75)_{10}$$

Note: Usually, the subscript is omitted when the decimal system (base-10) is used or when the base is understood by the user without any ambiguity.

Example 8.2

To convert the decimal number 50.578125 into the binary form, treat the integer part (50) and the fractional part (0.578125) separately. The binary number is obtained by repeated division by 2 of the integer part, and the repeated multiplication of 2 of the fractional part. Accordingly,

$$(50)_{10} = (110010)_2$$

$$(0.578125)_{10} = (0.100101)_2$$

Since the decimal point should also be represented as binary information internally, some form of code or convention has to be used in representing the *fractional* and *mixed* (integral + fractional) numbers within a computer. For example, the integer part may be stored in one byte and the fractional part in the adjacent byte. Alternatively, the *floating-point* representation may be used, in which the number is represented by a *mantissa* and an *exponent*. The floating-point representation is particularly suitable when the range of values that are needed to be handled by a digital computer (i.e., its *dynamic range*) is large, the advantage being the word size that is needed to store any number in a given dynamic range would be smaller in the floating-point representation in comparison to the *fixed-point* representation.

Because writing a number in the binary form is lengthy, and conversion between binary and decimal forms is time consuming, it is convenient to employ a base larger than 2. Such a base should also have the advantage of ease of conversion between the binary and that of the base. Two such representations are *octal* (base-8) and *hexadecimal* or *hex* (base-16) number systems. Since the largest octal digit (7) requires 3 bits in the binary representation, conversion from the binary to the octal form can be done simply by grouping the integer part and the fractional part into 3 bit groups, starting from the decimal point and, then, converting each group into an octal digit. Similarly, conversion from the binary to the hex representation can be accomplished by using 4 bit groups.

8.2.2 Negative Numbers

Consider an n bit binary integer number. The most significant bit (MSB) is the leftmost bit, and a digit of 1 there has the value 2^{n-1}. The least significant bit (LSB) is the rightmost bit, and a digit of 1 there has the value 1. When all the n bits are assigned 1s, the value of the number is $2^n - 1$. This is the largest value of an n bit binary number. Similarly, when all the n bits are assigned 0s, the value of the number is 0. This is the smallest value of an n bit number. Hence, an n bit digital storage space can occupy n positive integers ranging from 0 to $2^n - 1$. The above discussion concerns positive numbers. When the number is negative, a special representation is needed to take into account the sign. Several methods are available, as presented below.

8.2.2.1 Signed Magnitude Representation

When negative numbers are represented, we need an extra bit to represent its sign. So, to represent $-(2^n - 1)$ we need $n + 1$ bits. The MSB is the sign bit. In the *signed magnitude* (sm) representation, for a negative number, MSB = 1; and for a positive number, MSB = 0. The remaining bits represent the magnitude of the number. For example, in this representation $(11110)_{sm}$ represents decimal −14 and $(01110)_{sm}$ represents decimal +14. In general, in the sm representation with $n + 1$ bits, we can represent a number ranging from $-(2^n - 1)$ to $+(2^n - 1)$. Also, zero is not uniquely represented in this form, allowing for both a negative zero (e.g., 10000) and a positive zero (e.g., 00000).

8.2.2.2 Two's Complement Representation

In this representation, the MSB is 1 and it has a negative value attached to it. The remaining bits are considered to have positive values. For a positive number, the MSB is 0 and the remaining bits are the same as the binary representation of the number.

Specifically, consider $n + 1$ bits that represent a number in two's complement. Suppose that all $n + 1$ bits are 1s. Then the value of the MSB is -2^n. The value of the remaining n bits is $2^n - 1$. Hence, the overall value of the number is $-2^n + (2^n - 1) = -1$. If the MSB is 1 and the remaining n bits are all 0s, we have the largest (in magnitude) negative number that can be represented by $n + 1$ bits, namely, -2^n. If the MSB is 0 and the remaining n bits are all 1s, we have the largest positive number that can be represented by $n + 1$ bits in the 2's complement form, namely, $2^n - 1$. Hence, in the two's complement form, $n + 1$ bits can represent any number from -2^n to $+(2^n - 1)$. For example, with 5 bits, the smallest number will be $(10000)_{2's} = -16$ and the largest number will be $(01111)_{2's} = +15$. Also, in the two's complement method, zero is uniquely represented (there is no negative zero).

Given a negative binary number, its 2's complement representation can be formed by the following method:

Step 1: Switch the bit values (i.e., change 1s to 0s and 0s to 1s)
Step 2: Append a "1" as the new MSB
Step 3: Add 1 to the resulting binary number

The two's complement representation has an interesting property called *sign extension*. Specifically, any number of leading 1s can be included in a negative 2's complement number (assuming that an adequate number of bits are available for storage), and this does not change its value. Similarly, any number of leading zeros can be included in a positive 2's complement number, and this does not change its value. The latter is more obvious than the former. As examples, $(1011)_{2's}$, $(11011)_{2's}$, $(111011)_{2's}$, $(1111011)_{2's}$, etc., all represent the same negative number −5. Similarly, $(0110)_{2's}$, $(00110)_{2's}$, $(000110)_{2's}$, $(0000110)_{2's}$, etc., all represent the same positive number +6.

8.2.2.3 One's Complement

In the 1's complement form, the MSB is the sign bit; 0 represents a positive number and 1 represents a negative number as usual. The remaining bits give the magnitude of the number. For a positive number, the magnitude is represented in the binary form. For a negative number, the magnitude is represented by the complement of the binary form (i.e., 0s are changed to 1s and 1s are changed to 0s).

Note: For positive numbers the 1's complement form, 2's complement form, and the signed magnitude form are identical. But for negative numbers, these three representations are different. The property of sign extension holds for the 1's complement representation as well.

For example, consider decimal 9, which has the binary form 1001. Then

$$9 = (01001)_{1's} = (001001)_{1's} = (0001001)_{1's}, \text{etc.}$$

$$-9 = (10110)_{1's} = (110110)_{1's} = (1110110)_{1's}, \text{etc.}$$

With $n+1$ bits, in the 1's complement form, we can represent an integer from $-(2^n-1)$ to $+(2^n-1)$. *Note*: Unlike in the two's complement method, zero is not uniquely represented; there is a negative zero (e.g., 10000) as well as a positive zero (e.g., 00000).

8.2.3 Binary Multiplication and Division

Binary arithmetic is quite analogous to decimal arithmetic. The main difference is the number of distinct digits that are available (2 for binary and 10 for decimal). The operations and the rules are the same. In binary addition, only 0 and 1 are used as the primary digits. When the value of a summation becomes 2, it is taken to the next higher order place and added as 1 there (i.e., a carry of 1). Binary subtraction is done by sign reversal and addition.

Binary multiplication too is straightforward, as for decimal multiplication. It should be clear that as long as the rules for the arithmetical operations are provided and sufficient memory (storage space) is available, a digital computer can follow the instructions very fast to produce accurate results.

8.2.4 Binary Gray Codes

The binary representation of numbers, as discussed above, is known as the *natural binary code* or *straight binary code* because it comes directly from the standard representation of numbers with respect to an arbitrary base. A practical drawback of this natural binary code is that more than one bit switching would be needed when advancing from one number to the next number consecutively. Since bit switching might not occur simultaneously in actual hardware, there would be some ambiguity concerning the actual value during switching, depending on which bit is switched first. In absolute encoders used in motion sensing (Chapter 6), a binary gray code is used. Many such binary gray codes could be developed depending on the bit pattern used to represent the starting number (typically zero).

8.2.5 Binary Coded Decimal

In the *binary coded decimal* (BCD) representation, instead of representing the entire decimal number by its binary form, we represent each decimal digit separately by its binary form. Since the binary representation of the largest decimal digit (9) is 1001, in BCD, each decimal digit will require 4 bits. The disadvantage of BCD, however, is that the codes 1010 through 1111 are not used (wasted) because they do not represent separate decimal digits. The advantage is the ease of conversion, similar to the conversion between binary and hex.

Example 8.3

Since $(7)_{10} = (0111)_2$ and $(9)_{10} = (1001)_2$, we have $(79)_{10} = (01111001)_{BCD}$
This result is quite different from the binary representation of 79, which is found to be $(01001111)_2$.

From the input/output point of view, it is convenient to use BCD for arithmetic manipulation because in view of its familiarity, we usually prefer the decimal representation in the real world. But, within a computer, arithmetic manipulation of BCD numbers will require a different (and more complex) set of rules from those used for binary numbers. Even with this disadvantage and inefficient use of memory, the BCD might be preferred in input/output oriented applications such as point-of-sale terminals where processing itself is not complex, but the ease of conversion from and to the decimal form is a major advantage. BCD is also commonly used in digital displays such as the seven-segment numeric light emitting diode (LED) display (see Figure 8.1), where each decimal digit is illuminated separately by energizing a suitable set of the seven LED segments.

8.2.6 ASCII (Askey) Code

A number computed by a microcontroller would be in binary representation, but it has to be converted into a proper form according to some code, depending on whether it is displayed as a decimal number on an LED display, or printed on paper, or used as a control signal to control a process. Information received by a digital computer from a peripheral device such as a keyboard (input device) may include nonnumeric data (e.g., letters in the alphabet and special characters and control commands) as well as numeric data. Furthermore, numbers that appear in a text file (e.g., in a letter or report) take a different meaning internally within a computer from numerical data that will be processed (e.g., multiply, subtract) by the computer. Similarly, information transmitted from a computer to an output device (e.g., display) can also take these different forms. Usually, information enters and leaves a keyboard as a sequence of pulses (in bit serial manner) and a computer handles the information in a bit parallel manner (as bytes or words). The standard code used for the information transfer between a computer and a keyboard is the *United States of America Standard Code for Information Exchange* (U.S.-ASCII), which is usually abbreviated as the ASCII code. Decimal numbers, lowercase (simple) and uppercase (capital) letters

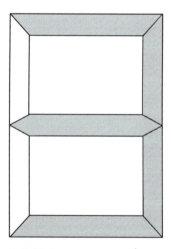

FIGURE 8.1
A seven-segment LED display for a decimal digit (digit 3 is illustrated).

of the alphabet, special keyboard characters, and control commands can be transmitted as binary information using ASCII. Seven bits are needed for ASCII. But one byte (8 bits) is usually employed, with the eighth bit being used as a *parity bit* for checking whether errors have entered the transmitted data (*odd parity* or *even parity* check).

8.3 Logic and Boolean Algebra

Digital circuits can perform logical and binary operations at very high speeds and very accurately. Similarly, they can perform numerical operations, particularly when implemented in digital computers. Logical and numerical operations are directly useful in all types of mechatronic systems. An understanding of the concepts of binary logic is important in the analysis and design of digital logic circuits. Crisp sets and binary logic are analogous. Furthermore, Boolean algebra is useful in the representation and analysis of sets and binary logic.

8.3.1 Logic

Conventional logic deals with statements called "propositions." In binary (or *two-valued*) logic, a proposition can assume one of only two truth-values: true (T), false (F). An example of a proposition would be "John is over 50 years old." Now consider the following propositions: (1) charcoal is white, (2) snow is cold, (3) temperature is above 60°C. Here, proposition 1 has the truth value F, and proposition 2 has the truth value T. But, for proposition 3, the truth value depends on the actual value of the temperature: if it is above 60°C, the truth value is T and otherwise it is F.

Propositions may be connected/modified by logical connectives such as *AND, OR, NOT*, and *IMPLIES*. These basic logical operations are defined below.

8.3.1.1 Negation

The negation of a proposition A is "*NOT A*" and may be denoted as $\sim A$ (also $\bar{A}$). It is clear that when A is *TRUE*, then *NOT A* is *FALSE* and *vice versa*. These properties of negation may be expressed by a *truth table* as shown in Table 8.1a. A truth table gives the truth values of a combined proposition in terms of the truth values of its individual components.

8.3.1.2 Disjunction

The disjunction of the two propositions A and B is "*A OR B*" and is denoted by the symbol $A \vee B$. Its truth table is given in Table 8.1b. In this case, the combined proposition is true if at least one of its constituents is true. This is not the "Exclusive *OR*" where "*A OR B*" is false also when both A and B are true, and is true only when either A is true or B is true. Disjunction in logic corresponds to the "union" in sets.

8.3.1.3 Conjunction

The conjunction of two propositions A and B is "*A AND B*" and is denoted by a symbol $A \wedge B$. Its truth table is given in Table 8.1c. In this case, the combined proposition is true

TABLE 8.1

Truth Tables of Some Logical Connectives

(a) Negation (*NOT*)

A	~A	
T	F	
F	T	

(b) Disjunction (*OR*)

A	B	A ∨ B
T	T	T
T	F	T
F	T	T
F	F	F

(c) Conjunction (*AND*)

A	B	A ∧ B
T	T	T
T	F	F
F	T	F
F	F	F

(d) Implication (*IF-THEN*)

A	B	A → B
T	T	T
T	F	F
F	T	T
F	F	T

if and only if both constituents are true. The conjunction in logic is analogous to the intersection of sets.

8.3.1.4 Implication

Consider two propositions A and B. The statement "A implies B" is the same as "*IF A THEN B*." This may be denoted by $A \rightarrow B$. Note that if both A and B are true, then $A \rightarrow B$ is true. If A is false, then the statement "When A is true then B is also true" is not violated, regardless of whether B is true or false. But if A is true and B is false, then clearly the statement $A \rightarrow B$ is false. These facts are represented by the truth Table 8.1d.

Example 8.4

Consider two propositions A and B. We can form the truth table of the combined proposition (*NOT A*) *OR B* as given below:

A	~A	B	~A ∨ B
T	F	T	T
T	F	F	F
F	T	T	T
F	T	F	T

Here we have used the truth tables (Table 8.1a and b). Note that the result is identical to Table 8.1d. This equivalence is commonly exploited in the logic associated with knowledge-based decision making.

In logic, knowledge is represented by propositions. A simple proposition does not usually make a knowledge base. Many propositions connected by logical connectives may be needed. Knowledge may be processed through *reasoning*, by the application of various laws of logic including an appropriate *rule of inference*, subjected to a given set of data (measurements, observations, external commands, previous decisions, etc.) to arrive at new inferences or decisions.

8.3.2 Boolean Algebra

Boolean algebra is the algebra of two-valued logic. It is useful in the analysis and design of digital logic circuits. The two values used are 1 (corresponding to *true*) and 0 (corresponding to *false*). Accordingly, there is also a correspondence between the operations of Boolean algebra and logic. The two values in Boolean algebra may represent any type of two states (e.g., *on* or *off* state, *presence* or *absence* state, *high* or *low* state, two voltage levels in *transistor-to-transistor logic*) in practical applications.

Laws of Boolean algebra follow from the characteristics of two-valued logic. Some basic laws are given in Table 8.2. Some of these are obvious and the others may be verified using truth tables. The notation used for the basic operations is the following: Logic "NOT" is analogous to bit switching (between 0 and 1) in Boolean algebra, and is denoted by an over-bar. Logic "OR" is denoted by Boolean "+" and logic "AND" is denoted by Boolean "·". For "Exclusive OR" (XOR), the Boolean notation is "$\oplus$".

TABLE 8.2

Some Laws of Boolean Algebra and Two-Valued Logic

Name of Property/Law	Statement of Property/Law
Commutativity	$a+b=b+a$
	$a \cdot b = b \cdot a$
Associativity	$(a+b)+c=a+(b+c)$
	$(a \cdot b) \cdot c = a \cdot (b \cdot c)$
Distributivity	$a \cdot (b+c) = (a \cdot b) + (a \cdot c)$
	$a+(b \cdot c) = (a+b) \cdot (a+c)$
Absorption	$a+(a \cdot b) = a$
	$a \cdot (a+b) = a$
Idempotency	$a+a=a$
	$a \cdot a = a$
Exclusion	$a+\bar{a}=1$
	$a \cdot \bar{a} = 0$
De Morgan's laws	$\overline{a \cdot b} = \bar{a} + \bar{b}$
	$\overline{a+b} = \bar{a} \cdot \bar{b}$
Boundary conditions	$a+1=1$
	$a \cdot 1 = a$
	$a+0=a$
	$a \cdot 0 = 0$

Note: When the order of performing the operations is not explicitly indicated (e.g., through the use of parentheses), the "NOT" operations are performed before the "AND" operations, and the "AND" operations are performed before the "OR" operations.

8.3.2.1 Sum and Product Forms

Conversion of a Boolean expression from the "sum" form to the "product" form, and *vice versa*, may be accomplished by using De Morgan's laws and the fact that performance of two negations will result in the original expression.

First, start with $\overline{a+b} = \overline{a} \cdot \overline{b}$ and negate both sides. We have $a+b = \overline{\overline{a} \cdot \overline{b}}$.

Similarly, start with $\overline{a \cdot b} = \overline{a} + \overline{b}$ and negate both sides. We have $a \cdot b = \overline{\overline{a} + \overline{b}}$.

These results are useful in the analysis and design of logic circuits.

8.4 Combinational Logic Circuits

A digital circuit converts digital inputs into digital outputs. There are two types of logic devices, classified as either *combinational logic* or *sequential logic*. Combinational logic devices are static where the present inputs completely (and uniquely) determine the present outputs without using any past information (history) or memory. In contrast, the output of a sequential logic device depends on the past values of the inputs as well as the present values. In other words, they depend on the time sequence of the input data, and hence some form of *memory* would be needed. The present section deals with the realization of combinational logic circuits. Sequential logic realizations will be discussed in a later section.

8.4.1 Logic Gates

Logic gates are the basic circuit elements found in IC circuits that are used in digital systems. A logic gate has one or more logical inputs and only one logical output. Each input line or output line of the gate can have one of two states: *true* (represented by the binary digit 1) and *false* (represented by the binary digit 0). It is easy to see how a combination of switches can be used to form a logic gate. For example, two switches connected in series can serve as an AND gate because current passes through the circuit only when both switches are closed. Output of the gate is the state of the circuit. Hence, by denoting the closed-circuit state as true (bit 1) and the open-circuit state as false (bit 0), we can obtain a truth table for the AND gate, as shown in Figure 8.2a. Note that the gate has two inputs, which are the states of the two switches. Similarly, an OR gate can be formed by connecting two switches in parallel, as shown in Figure 8.2b. In this OR gate, the output state is considered true (bit 1) when both inputs are true simultaneously. Alternatively, in an XOR, the output state is taken as false (bit 0) when both inputs are true, and the output is true when only one of the inputs is true. Mechanical switches and relays are not suitable for logic gates in digital circuits. Solid-state switches are the preferred variety. Semiconductor elements such as diodes and transistors can function as solid-state switches. In digital systems, these elements are present in the IC form and not in their discrete component form.

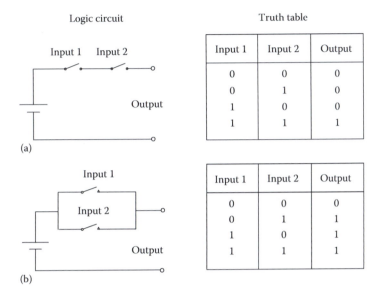

Logic circuit | Truth table

Input 1	Input 2	Output
0	0	0
0	1	0
1	0	0
1	1	1

(a)

Input 1	Input 2	Output
0	0	0
0	1	1
1	0	1
1	1	1

(b)

FIGURE 8.2
Examples of basic logic gates formed using switches: (a) AND gate and its truth table and (b) OR gate and its truth table.

The three basic logic gates are AND, OR, and NOT. The three gates NAND, XOR, and NOR can be constructed from the first three gates, but all six of these gates may be considered basic. American National Standards Institute (ANSI) symbols for these six logic gates are shown in Figure 8.3. A circle at the end of a logic signal line represents an inverter (negation). This is a simplification of the complete symbol of a NOT gate (triangle and a circle, as shown in Figure 8.3d). The truth tables give the state of the output of each gate for various states of the inputs.

Example 8.5

An OR gate obeys the Boolean operation $a+b$. For an XOR, denoted by $a \oplus b$, the same relation holds, except when both $a=1$ and $b=1$, in which case the output becomes 0. It follows that XOR obeys the Boolean relation $c = a \oplus b = (a+b) \cdot \overline{(a \cdot b)}$. This follows from the fact that the term $\overline{(a \cdot b)} = 0$ when both $a=1$ and $b=1$, and it is equal to 1 otherwise. Obtain a Boolean relation for XOR using the truth table in Figure 8.3c and show that this is equivalent to the above result. Give a digital circuit using AND, OR, and NOT gates only to realize XOR.

Solution
From the truth table of XOR, as given in Figure 8.3c, we have

$$c = \overline{a} \cdot b + a \cdot \overline{b}$$

The expression given in the example may be expanded as follows:

$$c = (a+b) \cdot \overline{(a \cdot b)} = (a+b) \cdot (\overline{a} + \overline{b}) \quad \text{(De Morgan)}$$

$$= a \cdot \overline{a} + a \cdot \overline{b} + b \cdot \overline{a} + b \cdot \overline{b}$$

$$= a \cdot \overline{b} + b \cdot \overline{a} \quad (\text{since } x \cdot \overline{x} = 0)$$

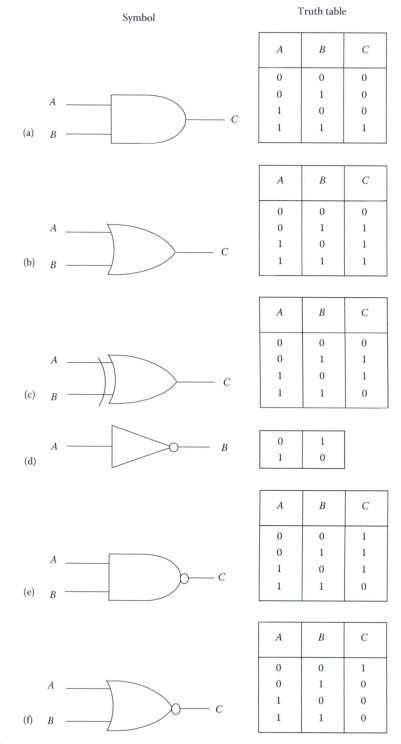

FIGURE 8.3
ANSI symbols and truth tables for the six basic logic gates: (a) AND, (b) OR, (c) XOR, (d) NOT, (e) NAND, (f) NOR.

This result is identical to what we obtained from the truth table.

A realization of the XOR gate is done by following the governing Boolean relation, as shown in Figure 8.4.

It can be shown that the NAND gate is functionally complete. That is, the three basic gates AND, OR, and NOT can be implemented with NAND gates alone. For example, Figure 8.5 shows a realization of OR using NAND gates alone. Similarly, it can be shown that the NOR gate is functionally complete.

A gate symbol can have many (more than two) input lines. Then, the logic combination that produces the output should be interpreted accordingly. For example, if an AND gate has three inputs, then the output is true only when all three inputs are true.

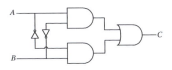

FIGURE 8.4
A realization of XOR using AND, OR, and NOT gates.

8.4.2 IC Logic Families

A logic family is identified by the nature of the solid-state elements (resistors, diodes, bipolar transistors, MOSFETS, etc.) and power levels used in the integrated circuit (IC). Examples include *resistor-transistor logic* (RTL), *diode-transistor logic* (DTL), *emitter-coupled logic* (ECL), *transistor-transistor logic* (TTL), and *complementary-symmetry metal-oxide semiconductor logic* (CMOS). Currently,

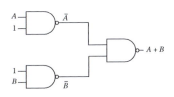

FIGURE 8.5
A realization of OR using NAND gates alone.

TTL and CMOS are the most popular logic families. CMOS devices are particularly attractive for their lower power dissipation (particularly useful in battery powered applications), smaller size (higher circuit density on chip), and better noise immunity properties. But they are less robust and more easily damaged by electrical fields than TTL devices. Each logic family uses logic gates as basic elements. Two examples are shown in Figure 8.6. A TTL NAND gate is shown in Figure 8.6a. In this case, *A* and *B* are the inputs and *C* is the output. It can be shown that this circuit element satisfies the truth table given in Figure 8.3. Specifically, the output will be at a low voltage level (logic-0 or logical FALSE) only when both inputs are simultaneously at a high voltage level (logic-1 or logical TRUE).

For an output signal of a TTL device, the low value corresponds to a voltage in the range 0–0.4 V, and the high value corresponds to a voltage in the range 2.4–5 V. For an input signal of a TTL device, the low value corresponds to a voltage in the range 0–0.8 V, and the high value corresponds to a voltage greater than 2 V. These voltage ranges provide a high degree of immunity to noise because logic-0 and logic-1 are separated by about 2 V. Furthermore, even if a noise level of 0.4 V is added to an output of a logic device, its logic state will remain the same at the input of another logic device. Figure 8.6a shows only two input lines, but more than two inputs could be connected to the emitter of the input transistor, if so desired. The power supply voltage is usually limited to the range 4.75–5.25 V in a TTL circuit. A 5 V power supply is shown in Figure 8.6a.

A NAND gate in the C-MOS logic family is shown in Figure 8.6b. Several advantages of CMOS over TTL were mentioned before. A further advantage of CMOS logic circuits is that they can use any power supply in the voltage range of 3–15 V. For example, a CMOS circuit can operate with a 12 V automobile battery, without having to use special circuitry to reduce the supply voltage, which is the case for a TTL circuit. The logic levels of CMOS circuits change with the supply voltage.

Sometimes, two or more logic families are used in the same digital device (e.g., computer). When connecting two IC units that belong to different logic families (e.g., one is CMOS and the other TTL), it is necessary to make sure, as when interconnecting any two devices, that the two units are compatible. In particular, operating voltages of the two

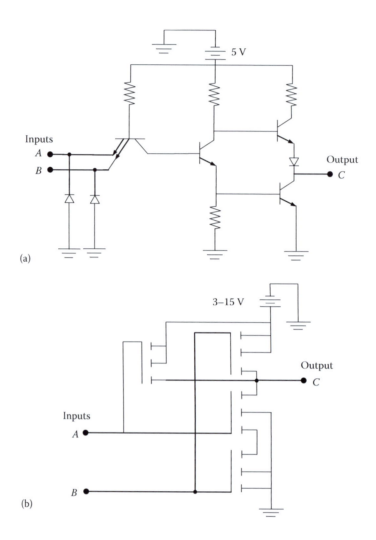

FIGURE 8.6
Typical logic gates from TTL and CMOS logic families: (a) a TTL NAND gate and (b) a CMOS NAND gate.

units have to be compatible and operating currents of the two units have to be compatible as well. In short, the two units should have matching impedances. Otherwise, interface circuitry (e.g., *level shifters* for voltage compatibility and *buffers* for current compatibility) has to be used between the two IC units that are interconnected.

8.4.3 Design of Logic Circuits

When developing a digital logic solution for a mechatronic application, preliminary work is needed to first define the specific function of the hardware, performance specifications, whether a solution is commercially available (off-the-shelf) at an acceptable cost, and so on. If a digital device has to be custom developed, then, the logic requirements of the required implementation have to be fully understood. The subsequent main steps in realizing the logic circuit for the particular application are given below:

1. Identify the inputs and outputs of the circuit for the particular application.
2. State the logic that connects the inputs and outputs and express it as a Boolean relation.
3. Minimize/optimize the relation.
4. Using the basic logic gates, sketch the realization that will satisfy the minimal Boolean relation.

In a practical realization, it is important to minimize the cost and complexity. Accordingly, a minimal realization that uses the least number of basic gates would be preferred. This topic will be addressed under Karnaugh maps (K-maps). Also, it is desirable to use least number of types of logic gates (e.g., all NOR, all NAND). We now illustrate the design/realization of combinational logic circuits for several practical applications.

8.4.3.1 Multiplexer Circuit

A digital multiplexer selects one digital input channel from a group and connects it to the output channel (i.e., reads the input channel). Consider the case of a two-input multiplexer. The inputs are denoted by x and y. The output is a. The control signal for input selection is c. The logic function of the circuit may be expressed as follows:

$$a = x \quad \text{if } c = 0$$

$$a = y \quad \text{if } c = 1$$

This logic may be translated into the following Boolean relation:

$$a = x \cdot \bar{c} + y \cdot c$$

Its realization is shown in Figure 8.7.

8.4.3.2 Adder Circuits

Logic gates are used in the *arithmetic and logic unit* (ALU) of a microprocessor to perform various processing operations. A basic arithmetic operation that is performed by the ALU is *addition*. A simple logic circuit, which performs the addition of two binary digits (bits), is shown in Figure 8.8a. This is called a *half adder*. This circuit contains two OR gates, one AND gate, and one NOT gate. The added two digits are denoted by A and B (the inputs to the logic circuit). The sum of the two digits is denoted by S, and the carry to the next higher place is denoted by C. Clearly, the truth table given in Figure 8.8b agrees with the rules of binary addition of two bits. Note that $\bar{A}$ is the complement of A. A *full adder* circuit performs binary addition on three bits.

Example 8.6

An electronic switch that uses digital logic is to be developed for switching the lights on and off in an art gallery. The switch has to be turned on at 7:00 p.m. and turned off at 6:00 a.m. Also, if there

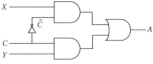

FIGURE 8.7
A two-input digital multiplexer.

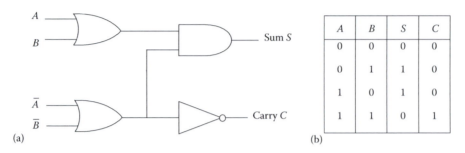

FIGURE 8.8
A half-adder: (a) logic circuit and (b) truth table.

A	B	S	C
0	0	0	0
0	1	1	0
1	0	1	0
1	1	0	1

are people in the gallery and the light that enters the gallery from outside is inadequate, the lights have to be turned on regardless of the time of the day. Assume that a sensor detects the presence of people in the gallery and produces a logic state of high (1). Also, there is sensor that detects the level of light entering the gallery from outside and generates a logic state of high (1) when the light level is inadequate. Furthermore, a digital clock produces logic high state when the time is between 6:00 a.m. to 7:00 p.m. Design a logic circuit that uses NAND gates only to operate the switch.

Solution
The logic state of the people sensor is denoted by p, that of the light sensor is denoted by l, and that of the time sensor is denoted by t. The output of the logic circuit is denoted by s, which is high (1) when the switch has to be on. The logic of the operation is, the switch has to be on when there are people in the gallery "and" the light entering from outside is not adequate, "or" if the time is not between 6:00 a.m. to 7:00 p.m. This statement translates into the Boolean relation

$$s = p \cdot l + \bar{t}$$

Now, in order to determine a realization with NAND gates only, we proceed as follows:

$$s = p \cdot l + \bar{t} = \overline{\overline{p \cdot l + \bar{t}}} = \overline{\overline{p \cdot l} \cdot t} \quad \text{(by double negation and application of the De Morgan law)}$$

It is seen that this relation can be realized using two NAND gates, as shown in Figure 8.9.

8.4.3.3 Active-Low Signals

Logic devices often use *npn* transistors (or, *n*-channel transistors) at their output terminals. This is because these transistors can handle larger currents and can operate at higher switching rates than *pnp* transistors (or *p*-channel transistors). Fairly high current levels are needed to operate devices like motors and solenoids, and hence switching devices based on *npn* transistors are desirable in such applications. A logic device based on this principle is shown in Figure 8.10. Here, the base B is the input to the transistor and the collector C represents the output. The emitter E is grounded. The collector has to be internally connected to a suitable resistor and a power supply v_{cc} (typically, 5 V). If the collector is not internally connected in this manner, we have an *open-collector* output, which is not desirable. In that case, the collector has to be

FIGURE 8.9
A logic circuit operating lights in an art gallery.

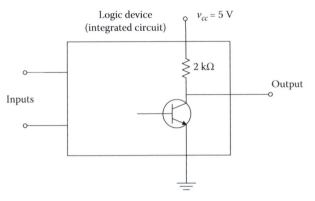

FIGURE 8.10
A logic device with active-low output.

connected externally through a *pull-up resistor* to the power source v_{cc}. This resistance has to be sufficiently high (e.g., 2 kΩ). Otherwise, the resulting high currents will increase the power dissipation and, furthermore, the output voltage (which has to be small for the logic-0 state) will increase.

The *npn* transistor in Figure 8.10 operates like a semiconductor switch. This can be explained as follows. When the input voltage to the transistor (i.e., base voltage with respect to the emitter, v_{be}) is small (<0.7 V), there will not be any current through the collector (and hence through R_c). Then the transistor is in its *off* state, and the output v_o will be close to v_{cc}. This is a high-voltage state for the output. The output impedance of the transistor is high (several kΩ) in this state. When the input to the transistor is greater than 0.7 V, the transistor is forward biased and is turned *on*. A finite current will conduct through the collector, thereby saturating the transistor, resulting in a significant voltage drop across R_c. As a result, the output voltage v_o will be close to zero (about 0.2 V). This is a low-voltage state for the output. The output impedance of the transistor is low in this state. In summary: when the transistor is turned off (logic-0), the output signal is high; when the transistor is turned on (logic-1), the transistor is saturated and the output signal is low (with a reasonable amount of current to drive a device). In other words, the logic device has an *active-low* output.

A TTL logic device (as shown in Figure 8.10, for example) has some disadvantages over a CMOS logic device in view of the faster switching speed, higher fan-out capability (i.e., the number of devices it can drive at the output), and increased immunity to noise, of the latter. To realize these advantages, the logic gates should have small output impedances regardless of whether the output is at logic-0 or logic-1. This is essential so that at fast switching rates the output transistor will be capable of both supplying and sinking large currents. This is not the case for the circuit in Figure 8.10. Here, when the input voltage (at the base) is small (logic-0), the transistor will not conduct (off), as indicated before, and the output will be at a high voltage (logic-1). In this state, the output impedance will be high (e.g., 1.4 kΩ). When the input (base) voltage is high enough (logic-1), the transistor will be turned on and the voltage at the collector will drop to a small value (logic-0). Then the output impedance (resistance between the collector of the saturated transistor and the ground) will be small.

In the developments of previous sections, we have assumed *active-high* signals, where the high voltage level represents logic-1 (logical TRUE) and the low voltage level represents logic-0 (logical FALSE). This is the default case. In other words, if active-low (denoted by signal L) is not specified, we assume active-high signals (denoted by *signal H*). Hence, a signal denoted by A is identical to $A.H$. But, as indicated before, active-low signal usage is appropriate in some practical applications. Note that $signal.L$ can be converted into $signal.H$, and vice versa, by simply adding an inverter (NOT gate). For example, if A denotes an active-high signal, then $A = A.H$; $\bar{A} = A.L$; $\overline{A.L} = A$.

Example 8.7

Using NAND gates and NOT gates (inverters) only, develop a logic circuit to realize $C = A \oplus B$ assuming that both input signals (A, B) are available in the active-high form and the output C is also needed in the active-high form signals $A.L$, $B.L$ are available and the desired output from the circuit is $C.L$.

Solution
Since $A \oplus B = A \cdot \bar{B} + \bar{A} \cdot B$ we have

$$C = \overline{\overline{A \cdot \bar{B}} + \overline{\bar{A} \cdot B}} = \overline{\overline{A \cdot \bar{B}} \cdot \overline{\bar{A} \cdot B}} \quad \text{(double negation and De Morgan)}$$

This is implemented in the active-high form in Figure 8.11a, and in the active-low form in Figure 8.11b.

Compare these implementations with what is given in Figure 8.4.

8.4.4 Minimal Realization

Boolean expressions of logic functions may contain some redundancy. Their implementation in such a form will require an increased number of logic gates, and will not be desirable (with respect to cost, complexity, reliability, power consumption, etc.). Procedures are available to reduce (and minimize) logic functions before they are implemented. In particular, various reduction/simplification operations of Boolean algebra, as given in Table 8.2 and illustrated in various examples in the previous sections, may be employed. The use of *K-maps* is a popular and convenient procedure for determining the minimal realization of a logic function, which is explained next.

8.4.4.1 Karnaugh Map Method

Logic function minimization using K-maps is based on the property $A + \bar{A} = 1$ and its extensions such as $AB + A\bar{B} = A$ and $\bar{A}\bar{B} + \bar{A}B + AB + A\bar{B} = 1$. We have seen that any Boolean (logic) function may be expressed as a truth table with respect to its variables, in a one-dimensional array. A K-map is a convenient version of a truth table where a matrix-type map is used, which is more convenient when the number of variables is large. Each row or column of the matrix represents a variable group (typically, a pair). The variable groups are arranged such that only one of the variables is

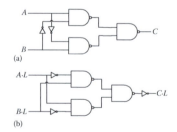

FIGURE 8.11
Implementation of XOR using NAND gates using: (a) active-high signals and (b) active-low signals.

complemented from one group to the adjacent group (i.e., *adjacency* is preserved, only one bit-switching between two adjacent groups). The reduction procedure is summarized below:

1. Separate the variables of the logic function into two groups, one group representing the rows of a matrix and the other representing the columns of the matrix, as follows: If there are only two variables, each group will contain one variable; if there are three variables, one group will have two variables and the other will have the remaining one; if there are four variables, each group will have two variables, and so on.

2. Mark the rows and columns of the matrix with various combinations of logic states for the two groups such that adjacent rows or adjacent columns correspond to only one bit switching (i.e., complementation of just one variable).

3. Mark the cells of the matrix with 1s corresponding to the terms of the logic function (the unmarked cells have 0s).

4. Identify all rectangular blocks that are filled (i.e., no 0s), starting from the blocks with the largest number of cells and ending at the remaining single cells, as follows: Blocks with an even number of rows and columns, blocks with one row and an even number of columns, blocks with one column and an even number of rows, single cells. In this procedure, allow for *wrap-around* (i.e., the first and the last rows are linked, and the first and the last columns are linked, in a cylindrical manner). *Note*: A subset of a union of already marked blocks should not be marked as a new block.

5. Reduce each block into a minimal form by using the reduction formulas of Boolean algebra.

Example 8.8

Consider the logic function $A \cdot (\bar{B} + \bar{C}) + \bar{A} \cdot \bar{B} \cdot C + A \cdot B \cdot C \cdot \bar{D}$. Using a K-map, reduce this function and express it in the

(i) Sum of product form
(ii) Product of sum form

Solution
The given expression may be expanded as: $Y = A \cdot \bar{B} + A \cdot \bar{C} + \bar{A} \cdot \bar{B} \cdot C + A \cdot B \cdot C \cdot \bar{D}$.
Its K-map is shown below.

Y:

	$\bar{C}\bar{D}$	$\bar{C}D$	CD	$C\bar{D}$
$\bar{A}\bar{B}$			1	1
$\bar{A}B$				
AB	1	1		1
$A\bar{B}$	1	1	1	1

B brackets rows $\bar{A}B$ and AB; A brackets rows AB and $A\bar{B}$.

(i) First, we partition the K-map, as shown below, to form four 4-element blocks:

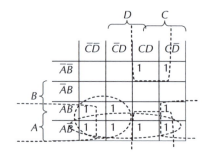

Accordingly, we write the reduced form: $Y = A \cdot \bar{B} + A \cdot \bar{C} + \bar{B} \cdot C + A \cdot \bar{D}$.

This is in the minimal, sum of product form. *Note:* In arriving at this expression, the K-map was partitioned into four blocks each having four elements. The first term corresponds to the bottom 1×4 row. The second term corresponds to the 2×2 block at the bottom left corner. The third term corresponds to the wrapped-around 2×2 block formed by the last two elements in the bottom row and the last two elements of the top row. The last term corresponds to the wrapped-around 2×2 block formed by the last two elements of the first column and the last two elements of the last column.

(ii) The 0-cells in the K-map above gives the truth-value terms for $\bar{Y}$. Hence, we can form the K-map for $\bar{Y}$ as given below.

$\bar{Y}$:

	$\bar{C}\bar{D}$	$\bar{C}D$	CD	$C\bar{D}$
$\bar{A}\bar{B}$	1	1		
$\bar{A}B$	1	1	1	1
AB			1	
$A\bar{B}$				

This K-map can be partitioned into two 4-cell blocks and one 2-cell block, giving the minimal expression: $\bar{Y} = \bar{A} \cdot B + \bar{A} \cdot \bar{C} + B \cdot C \cdot D$.

Now negate (complement) the expression to get

$$Y = \overline{\bar{A} \cdot B + \bar{A} \cdot \bar{C} + B \cdot C \cdot D} = \overline{\bar{A} \cdot B} + \overline{\bar{A} \cdot \bar{C}} \cdot \overline{B \cdot C \cdot D} = \overline{\bar{A} \cdot B} \cdot \overline{\bar{A} \cdot \bar{C}} \cdot \overline{B \cdot C \cdot D}$$

$$= (A + \bar{B}) \cdot (A + C) \cdot (B + C + D)$$

This is the minimal function in the product of sum form.

8.5 Sequential Logic Devices

The logic circuits considered in the previous section do not have feedback paths. Hence, when the inputs are removed, the outputs also disappear. Such logic circuits (called combinational logic circuits) are static (i.e., algebraic) in nature. If an output of a logic circuit is fed back, it is possible to maintain an output logic state even after the inputs are removed. Furthermore, then the output of the circuit will depend on inputs that were applied earlier

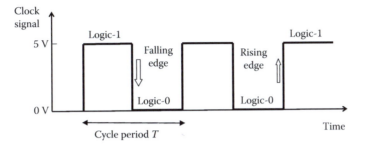

FIGURE 8.12
A clock signal for edge-triggered synchronous operation of a logic device.

and, consequently, on the time sequence of the previous inputs/outputs. Such feedback logic circuits are called *sequential logic circuits*. They are dynamic devices. Their present output depends on the past history and timing of the inputs.

A clock signal is available as the time reference for operation of a sequential logic circuit. If the operation is synchronized with the clock signal and the actions are triggered by it, we have a *synchronous* operation. If the actions are triggered by the inputs (non-periodic) and not by the clock signals, we have an *asynchronous* operation. Edge-triggered devices use the transitions in clock pulses for triggering. Positive *edge-triggered* devices use the rising edge (i.e., transition from 0 to 1) of a clock pulse for triggering. Negative edge-triggered devices use the falling edge (i.e., transition from 1 to 0) of a clock pulse for triggering. This nomenclature is illustrated in Figure 8.12. Edge trigger is needed in synchronous operation. For asynchronous operation, the *level-trigger* is used where triggering does not occur in a periodic manner but rather depends on the level of the triggering signal.

Examples of sequential logic devices include flip-flops, latches, shift registers, counters, and trigger devices with memory. In fact, the property of "memory" is an important consideration in sequential logic devices. Furthermore, complex digital systems like microcontrollers/microprocessors and state machines are formed using basic sequential circuits such as flip-flops as the hardware building blocks. Flip-flops are called bistable devices because they can assume two and only two stable output states (0 or 1). There are many types of flip-flops. A latch is a flip-flop that is able to latch onto a binary state. Up-counters generate a binary number sequence where each number is generated by incrementing the previous number by 1. Similarly, down-counters generate a binary number sequence in the consecutive descending order. ICs are used to perform important functions such as data storage (memory) and processing (ALU) in a microcontroller (miniature digital computer with a microprocessor). Solid-state memory devices (e.g., buffers, registers) and the logic elements that perform data processing (e.g., division) in microcontrollers employ sequential logic ICs. Several basic devices that fall into the category of sequential logic circuits are discussed next.

8.5.1 RS Flip-Flop

An RS flip-flop (or, *reset-set flip-flop*) is shown in Figure 8.13a. This circuit uses two cross-coupled NAND gates. Alternatively, two cross-coupled NOR gates may be used. The truth table (function table) for an RS flip-flop, as given in Figure 8.13b, is obtained from the truth table of a NAND gate. Under normal operating conditions, Q and $\bar{Q}$ are double-rail outputs, and the latter output is the complement of the former. It is seen from Figure 8.13b, however, that when both inputs are 0, both outputs become 1. This condition violates the

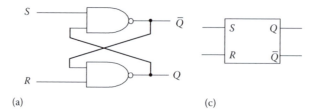

(a)

(c)

Applied input (data)		Resulting output		
S	R	Q	$\bar{Q}$	Explanation
0	0	1	1	Not acceptable
0	1	0	1	Independent of previous output
1	0	1	0	
1	1	Q_o	$\bar{Q}_o$	Retains state (same as previous output)

(b)

FIGURE 8.13
RS flip-flop formed by using NAND gates: (a) logic circuit, (b) truth table, (c) ANSI symbol.

requirement and is unacceptable. Furthermore, it is seen from the truth table that when one input is the complement of the other input, the output state simply becomes equal to the input state, irrespective of the previous state. In other words, the input data are directly transferred to the output ("data transfer"). In this mode, when the output becomes 0 (i.e., when the input is 0), it is called a "reset" operation, and when the output becomes 1 (i.e., when the input is 1), it is called a "set" operation. When both inputs to the RS flip-flop are 1, the resulting output state becomes identical to the previous output state (denoted by Q_o). Hence, the device has memory. The resulting output, as given in Figure 8.13b, is the final "settled" output in cases where it is different from the previous output. The ANSI symbol for an RS flip-flop is given in Figure 8.13c. The property that the present output of a flip-flop depends on the previous output implies that the unit has memory (can remember the state history). Hence, a flip-flop is used as a basic element in semiconductor memory units. Since the RS flip-flop can assume two stable states (01 and 10), it is a *bistable* device. A device such as an oscillator that does not have a stable state is known as an *astable* device.

8.5.2 Latch

A latch circuit is able to retain the previous output state when triggered. Triggering may be done using either an enable signal (asynchronous operation) or a clock signal (synchronous operation). Figure 8.14a shows a latch circuit formed using an RS flip-flop. Its truth table is given in Figure 8.14b, which can be easily verified using the truth tables of the RS flip-flop (Figure 8.13b), the NAND gate, and the NOT gate. The symbol of the latch is shown in Figure 8.14c.

8.5.3 JK Flip-Flop

In a JK flip-flop, J and K (similar to R and S in an RS flip-flop) denote the input data. But, unlike an RS flip-flop, a JS flip-flop does not have an unacceptable state of input

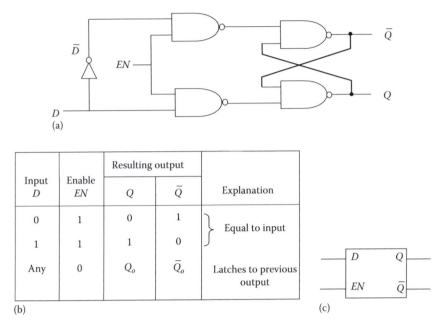

FIGURE 8.14

A latch circuit formed using an RS flip-flop and NAND gates: (a) logic circuit, (b) truth table, (c) symbol.

data. Furthermore, a JS flip-flop can provide a "toggle" action where the output is the complement of the previous output. A toggle action is important in "counting" operations. (A toggle will change the count by 1. Both up count and down count are possible.) *Note*: An RS flip-flop does not provide a toggle action.

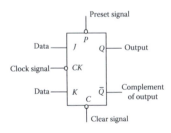

FIGURE 8.15

The symbol of a JK flip-flop.

TABLE 8.3

Truth Table for Asynchronous Operation of a JK Flip-Flop

$\bar{P}$	$\bar{C}$	Q	$\bar{Q}$
0	0	Not used	
0	1	1	0
1	0	0	1
1	1	Synchronous operation	

The ANSI symbol of a JK flip-flop is shown in Figure 8.15. As before, Q is the output that results due to the input data J and K. Again, $\bar{Q}$ denotes the complement of Q and is also available as an output. The action of a JK flip-flop can be triggered either by external signals (specifically, a "Preset" signal P and a "Clear" signal C) or by the clock signal CK. The former is the *asynchronous* operation and the latter the *synchronous* operation. Typically, active-low signals are needed for P and C (i.e., logic-0 activates the operation), and the falling edge of a clock pulse is used for triggering the device in the synchronous operation. The truth table for the asynchronous operation of a JS flip-flop is given in Table 8.3. *Note*: The values of the complement signals $\bar{P}$ and $\bar{C}$ are listed to emphasize the active-low operation. The circles that end these signal lines in Figure 8.15, which denote "NOT" operations, further indicate the use of active-low signals. The state of $\bar{P}=1$, $\bar{C}=1$ is not used in a JK flip-flop. When $\bar{P}=0$, $\bar{C}=1$, the output Q is "set" to 1 (which is the value of P). Similarly, when $\bar{P}=1$, $\bar{C}=0$, the output Q is "cleared" (i.e., becomes 0, which is the value of $\bar{C}$). When $\bar{P}=1$, $\bar{C}=1$, the JK flip-flop goes into synchronous operation where triggering is done by the clock signal CK. The circle ending the

TABLE 8.4

Truth Table of a JK Flip-Flop

Clock Signal	Applied Input		Resulting Output		
CK	J	K	Q	$\bar{Q}$	Explanation
↓	0	0	Q_o	$\bar{Q}_o$	Retains state (same as previous output; memory)
↓	0	1	0	1	} Data transfer independent of previous output
↓	1	0	1	0	(clear, set)
↓	1	1	$\bar{Q}_o$	Q_o	Toggles state (complement of previous output; i.e., counting)

line of the clock signal in Figure 8.15 indicates that the device is triggered by the falling edge of a clock pulse.

The function table (truth table) for a JK flip-flop in clocked (synchronous) operation is given in Table 8.4. *Note*: The falling-edge triggering is used (denoted by ↓). As usual, Q_o denotes the previous output (i.e., the output prior to trigger).

Since a JK flip-flop has the capabilities of memory and toggle, it has many applications (e.g., counters and other types of flip-flops such as D flip-flop and T flip-flop). In designing a circuit for a particular application, one may start with the desired output sequence from the circuit and proceed backwards to develop a suitable circuit that will achieve this output. In this approach to design, what is particularly useful is the "Excitation Table" of the flip-flop. This table gives the required input (data) for a given combination of previous output (before trigger) and new output (after trigger). The excitation table for a JK flip-flop is given in Table 8.5.

TABLE 8.5

Excitation Table of a JK Flip-Flop

Q_n	Q_{n+1}	J	K
0	0	0	Any
0	1	1	Any
1	0	Any	1
1	1	Any	0

8.5.4 D Flip-Flop

A D flip-flop (or, *data flip-flop*) has a single input (or, data) D. Its output Q is equal to the input data value. A D flip-flop may be interpreted as a special case of a JK flip-flop with $J = D$ and $K = \bar{D}$. (Hence, there is no need to have the input K.) The symbol of a D flip-flop is shown in Figure 8.16. The clock signal is used in synchronous operation. Typically, the rising-edge trigger is used, as shown in Figure 8.16. (*Note*: No circle at the end of the CK line entering the flip-flop.) Asynchronous operation is possible with the "Clear" signal C. In Figure 8.16, an active-low signal is used for C, as typical. Specifically, when $C = 0$, the output is cleared (i.e., $Q = 0$) regardless of the input data. The function table (truth table) of a clocked D flip-flop is given in Table 8.6. The excitation table is given in Table 8.7.

8.5.4.1 Shift Register

A shift register shifts the stored data in a word, one bit at a time to the right, as triggered by incoming clock pulses. A set of D flip-flops connected in series, with the first flip-flop receiving a 0 as the input data and the subsequent flip-flops receiving the output of the previous flip-flop as their input data, can serve as a shift register. Edge-trigger using a clock signal will cause the originally stored (output) data of the

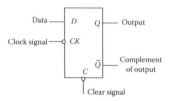

FIGURE 8.16
The symbol of a D flip-flop.

TABLE 8.6

Function Table (Truth Table) of a D Flip-Flop

Clock Signal	Data Input	Resulting Output		Explanation
CK	D	Q	$\bar{Q}$	
↑	0	0	1	} Data transfer independent of previous output (clear, set)
↑	1	1	0	

flip-flops to be shifted to the right, one at a time. At the end of the shifting sequence, all the bits will be 0 (i.e., data will be cleared).

TABLE 8.7

Excitation Table of a D Flip-Flop

Q_n	Q_{n+1}	D
0	0	0
0	1	1
1	0	0
1	1	0

8.5.5 T Flip-Flop and Counters

A T flip-flop (or, *toggle flip-flop*) is a JK flip-flop with $J=1$ and $K=1$. Also, it is assumed that $\bar{P}=1$. Hence, the signal lines for J, K, and P are not marked on the symbol of the T flip-flop, as shown in Figure 8.17a. The clock signal is in fact the data signal of a T flip-flop, and is denoted by T. When $\bar{C}=1$ it should be clear from the function table for a JK flip-flop (Tables 8.3 and 8.4) that synchronous (i.e., clocked) toggle operation takes place in the T flip-flop. The output can be cleared ($Q=0$) simply by setting $\bar{C}=0$ (asynchronous mode).

In the device shown in Figure 8.17a, the falling edge (i.e., transition from 1 to 0) of T results in a *toggle* (i.e., changing from 0 to 1 or 1 to 0) of the output Q. This property enables a group of T flip-flops to function as either a binary counter or a frequency divider (by multiples of 2). A 3 bit binary counter using T flip-flops is shown in Figure 8.17a. The corresponding timing diagram is shown in Figure 8.17b. The outputs Q_0, Q_1, and Q_2 form the 3 bit word $[Q_2Q_1Q_0]_2$. Bit changing occurs at every clock period, triggered by the falling edge of a clock pulse. From the timing diagram, it is clear that the 3 bit output changes in the sequence 000, 001, 010, 011, 100, 101, 110, and 111 at the clock frequency. Hence, we have a 3 bit up-counter. Also, from the timing diagram it is clear that the frequency of Q_0 is 1/2 the clock frequency, the frequency of Q_1 is 1/4 the clock frequency, and the frequency of Q_2 is 1/8 the clock frequency. Hence, these outputs can also serve as frequency-divider signals.

In a counter that uses the straight binary code (as in Figure 8.17b), there can be more than 1 bit switching at a time. For example, when advancing from 001 to 010, there will be 2 bit switches, and when advancing from 011 to 100, there will be 3-bit switches. Due to signal delays in the hardware, these switches will not happen simultaneously. As a result, an intermediate value, which will be different from the correct value, will result. For example, when advancing from 011 to 100, if the middle bit is switched first, the intermediate result will be 1, which is different from the correct result (decimal 4). Such a condition is called a *hazard*. It can be avoided by suing a code other than the straight binary code, specifically, a gray code, as mentioned earlier in the chapter.

Example 8.9

A 3 bit binary counter is to be designed using three D flip-flops and basic logic gates. In order to avoid hazards, a gray scale is used, where the required sequence is 000, 001, 011, 010, 110, 111, 101, and 100. *Note*: In this sequence, only one bit will switch for each increment in the count. Develop a suitable circuit for this counter.

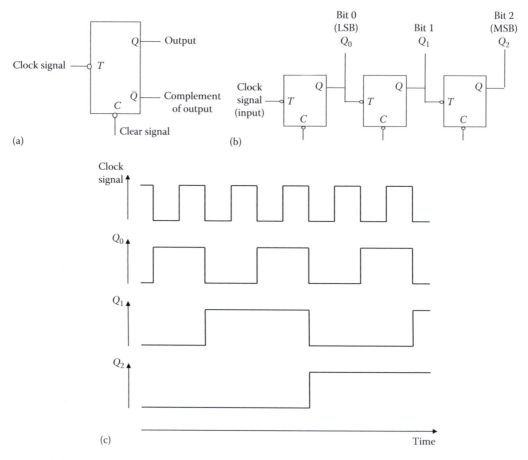

FIGURE 8.17
(a) The symbol of a T flip-flop, (b) a 3 bit binary up-counter using T flip-flops, (c) the timing diagram for the up-counter.

Solution

The output from the three flip-flops is denoted by $[Q_2Q_1Q_0]$. The required output sequence may be represented either by a timing diagram or as a table. The data inputs to the three flip-flops are denoted by $[D_2D_1D_0]$. The values of the input data, which are necessary to achieve the required changes in the three bits at each clock trigger, may be conveniently determined using the excitation for a D flip-flop (Table 8.7). This information is given in Table 8.8.

From Table 8.8 (truth table) we can express the necessary input data at clock trigger in terms of the output that has to be present at (or, just before) clock trigger in order to achieve the required results. Specifically we have:

$$D_0 = \bar{Q}_0\bar{Q}_1\bar{Q}_2 + Q_0\bar{Q}_1\bar{Q}_2 + \bar{Q}_0Q_1Q_2 + Q_0Q_1Q_2$$

$$D_1 = Q_0\bar{Q}_1\bar{Q}_2 + Q_0Q_1\bar{Q}_2 + \bar{Q}_0Q_1\bar{Q}_2 + \bar{Q}_0Q_1Q_2$$

$$D_2 = \bar{Q}_0Q_1\bar{Q}_2 + \bar{Q}_0Q_1Q_2 + Q_0Q_1Q_2 + Q_0\bar{Q}_1Q_2$$

What is related here is the output Q that is present "prior to" a clock trigger, and the necessary input data D that will generate the required output after the clock trigger. In other words, the

TABLE 8.8

Input–Output Values for the Gray-Code Counter

Output Just before Clock Trigger			Output Just after Clock Trigger			Input Data at Clock Trigger		
Q_2	Q_1	Q_0	Q_2	Q_1	Q_0	D_2	D_1	D_0
0	0	0	0	0	1	0	0	1
0	0	1	0	1	1	0	1	1
0	1	1	0	1	0	0	1	0
0	1	0	1	1	0	1	1	0
1	1	0	1	1	1	1	1	1
1	1	1	1	0	1	1	0	1
1	0	1	1	0	0	1	0	0
1	0	0	0	0	0	0	0	0

above three relations give the necessary *feedback paths* in the circuit. We now simplify (minimize) these expressions using K-maps, as follows:

D_0:

D_1:

D_2:

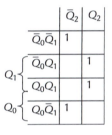

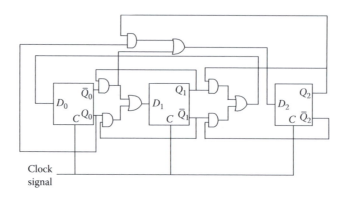

FIGURE 8.18
A gray-code counter using D flip-flops.

The minimized expressions are:

$$D_0 = Q_1Q_2 + \bar{Q}_1\bar{Q}_2$$

$$D_1 = \bar{Q}_0Q_1 + Q_0\bar{Q}_2$$

$$D_2 = \bar{Q}_0Q_1 + Q_0Q_2$$

The circuit shown in Figure 8.18 should be now developed using these minimal expressions.

8.5.6 Schmitt Trigger

Switching or triggering elements are useful in both analog and digital circuitry. If switching is ideal, then the slightest noise in a signal can create undesirable chatter near the switching region, producing erroneous results. Hence, it is desirable to have some hysteresis in the switching element so that the "switch-on" signal level is slightly higher than the "switch-off" signal level, thereby making the switch insensitive to noise. The Schmitt trigger is a solid-state switch that has this type of desirable hysteresis. A bipolar transistor circuit for the Schmitt trigger is shown in Figure 8.19a. The input/output characteristic of

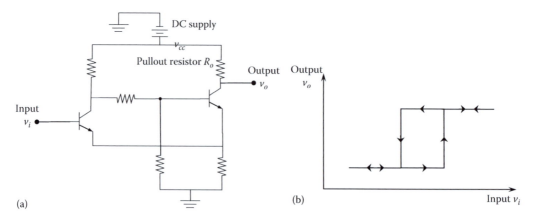

FIGURE 8.19
Schmitt trigger: (a) a circuit using bipolar transistors and (b) quasi-static characteristic curve.

the circuit under *quasi-static* conditions (i.e., assuming that the input varies very slowly) is shown in Figure 8.19b. Typically, six Schmitt triggers are included in a single IC package. The symbol for a logic gate with hysteresis is its usual ANSI symbol, with a hysteresis curve marked inside the symbol.

8.6 Practical Considerations of IC Chips

Logic circuitry in digital hardware devices and in various parts of a microcontroller (microprocessor, memory, interface hardware, etc.) is usually present in the IC form. Elements such as logic gates and bistable circuits such as flip-flops are building blocks for an IC, but these elements are present in the monolithic form and not in the discrete form discussed in Chapter 2. In this section, let us examine the manufacture and several practical considerations of digital IC chips. This insight is useful, even though not always essential, to a mechatronic engineer.

In a digital system, data storage and processing are performed using logic circuitry. These circuits consist of logic gate elements. The logic gates are arranged in an IC circuit in a suitable manner during the manufacture of the IC so that the IC will accomplish the required digital function. A typical IC chip contains a large number of microelectronic circuit elements manufactured on a small slice of silicon (smaller than a fingernail). Each circuit element consists of basic semiconductor elements such as diodes and transistors. These semiconductor elements are not present as conventional *discrete components* with leads and wiring for interconnection, but rather in the *monolithic form*. The entire IC is produced through a delicate manufacturing process and it is not possible to remove or replace, for example, a transistor in the IC without destroying the entire IC.

8.6.1 IC Chip Production

ICs (microelectronic circuits) are not manufactured by starting with discrete semiconductor elements and connecting them together to form a circuit. Such a production method would not only be expensive and time consuming but would also result in much larger chip size and poor reliability due to the large number of circuit connections that would be needed. The circuit density of an IC chip used in digital systems can be expressed in terms of the number of elementary logic circuits (logic gates) present on the chip. Four classifications are common, as given in Table 8.9. Note that the numbers given are approximate and represent the order of magnitude only. The important thing to note, however, is that

TABLE 8.9

IC Chip Classification according to Circuit Density

IC Chip Type	Number of Basic Logic Gates Present
Small-scale integration (SSI)	Less than 10
Medium-scale integration (MSI)	10–1,000
Large-scale integration (LSI)	1,000–10,000
VLSI	$10,000–1 \times 10^6$

it is practically impossible to produce a VLSI chip (e.g., a RAM memory chip containing 100,000 logic gates) using discrete transistors.

ICs are produced by various methods. In all such methods, the required circuit is first designed on paper or computer screen and then the entire circuit is produced on a semi-conductor chip (silicon, GaAs, etc.) by several steps of a delicate and carefully controlled manufacturing process. To give an idea of the various steps involved, consider the production of a monolithic IC chip using *planar diffusion technology*. A flat piece of doped silicon crystal that is polished and cleaned is used as the substrate on the surface of which the IC is formed. Suppose that the substrate has a layer of *n*-type silicon. The method of forming the IC is somewhat similar to that for manufacturing a discrete bipolar transistor (see Chapter 2).

The surface of the silicon wafer is oxidized to form a coating of silicon dioxide (insulation). Then, a chemical coating that is sensitive to light is applied on the oxidized surface. Next, a light beam is focused onto the coating through a mask. The mask has a window structure (openings) which conforms to the required (designed) circuit pattern. Areas exposed to light become insoluble to a chemical solution (developer), which is used to wash off the unexposed chemical. Next, hydrofluoric acid is used to remove the silicon dioxide layer underneath the window pattern produced by removing the unexposed coating. Next, the element is placed in a heated diffusion furnace, and a stream of gas containing an acceptor-type dopant is applied. This forms a pattern of *p*-type regions underneath the windows. Then, as in the case of a discrete transistor, it is possible to form a pattern of *n*-type regions over this by a systematic process of oxidation, chemical coating, masked exposure to light, developing, and diffusion using a donor-type dopant. Finally, electrical connections have to be made between various semiconductor-elements formed this way on the silicon crystal. To accomplish this, the same process of photomasking is used first to open up the areas that need connection. Then aluminum vapor is used to deposit aluminum on these exposed areas, thereby making electrical connection. There is no guarantee that an IC element manufactured in this manner will function correctly. Hence, the element has to be tested for proper operation, preferably before packaging.

8.6.2 Chip Packaging

Once an IC is formed on a silicon wafer, the wafer has to be cut into the correct shape and then packaged. The packaging material could be plastic or ceramic. Typically, a packaged chip is rectangular in shape with pins protruding out. Some of these pins serve as leads for information signals (data in and out) and others as power leads.

A popular packaging geometry is shown in Figure 8.20. This is known as the *dual-in-line package* (DIP). The pins of a DIP protrude down from the two longitudinal edges of the housing. Typically, up to 48 pins can be provided in a single package. Alternatively, an out-line package has lead pins protruding outward from the edges in the same plane as the chip underside. About 64 pins can be provided with such a package. With increased complexity of IC chips, efforts are continuing to increase the number of pins per package.

The assembly of an IC chip on a printed-circuit board (PC board) is usually a simple matter of pushing the pins of the chip through a corresponding set of holes (socket). Typically, soldering would be necessary to hold the chip in place and make electrical connections. In addition to IC chips, a PC board may contain a few discrete elements such as resistors,

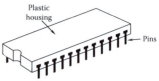

FIGURE 8.20
A DIP IC.

capacitors, and additional discrete diodes and transistors, and auxiliary devices such as switches and perhaps a socket for a ribbon cable, all permanently interconnected in a required manner using conductive lines deposited on the PC board.

8.6.3 Applications

Solid-state logic devices are commonly used in mechatronic applications. Often, these devices are embedded and distributed/integrated throughout a mechatronic system. Examples include the control of dc motors using chopper circuits, the control of ac motors using variable-frequency drives, and intelligent instruments with embedded microcontrollers. GaAs appears superior to silicon for high frequency and high-speed devices. Electron mobility in GaAs field-effect transistors (FET) is and order of magnetic higher than in silicon FETs. Also, the power dissipation (active power) is lower for a GaAs device. Typical data rates are silicon: <1000 Mbps; GaAs: 500–2500 Mbps; and CMOS: over 1500 Mbps. It is advantageous to put logic gates and power-control devices on the same chip, in view of lower cost, improved reliability, less wiring, and lower number of components.

8.7 Microcontrollers

A *microcontroller* (historically called a "microcomputer") is a dedicated, special-purpose, and miniature digital computer that is typically "embedded" in the application system (e.g., mechatronic system). Even though lower in speed and more costly in comparison to the hardware implementations (hardwired digital devices having fixed functionality), microcontrollers have the flexibility and more complex functionality that is offered by software programmability.

As for any digital computer, the minimum hardware requirement for a microcontroller is a *central processing unit* (CPU) for processing digital data, a *memory* for storing data, and *input/output circuitry* (I/O) for communication with other devices. It is a miniature computer on a single "chip" (or, IC) that typically handles special-purpose real-time functions of a specific application and does not require many of the functionalities of a general-purpose computer (e.g., desktop PC—personal computer or a laptop computer). The CPU of a microcontroller is the *microprocessor*.

An IC chip is simply a packaged IC. It can represent a microprocessor, a memory unit, or some other circuit used in a digital system, or even an analog circuit. The use of very large-scale integration (VLSI) chips not only for data processing (microprocessor) but also for memory (RAM, ROM, PROM, etc.) and interfacing (I/O) hardware was responsible for the advent of the microcontroller. Today, microcontrollers and microprocessors have become an integral part of many mechatronic devices and systems with such components as smart sensors, digital signal processors, intelligent controllers, microelectromechanical systems (MEMS), and on-board diagnostic units.

8.7.1 Microcontroller Architecture

A microcontroller consists of a microprocessor, memory unit, and input/output hardware. These devices interact through multiline electrical signal paths (conductors) known as buses. Three types of buses can be identified. A data bus carries the data needed at various

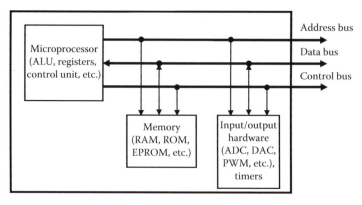

FIGURE 8.21
Typical architecture of a microcontroller.

locations in the microcontroller. An address bus carries the address (digital name) of a memory location or hardware component that needs to be accessed by the microprocessor. A control bus carries the commands issued by the microprocessor that controls the operation of the microcontroller system. In addition, a "power bus" will be available, which will provide power to various components in the microcontroller. A schematic representation of a microcontroller is given in Figure 8.21. More than one memory unit may exist in a microcontroller. In particular, a *data register* in a microprocessor is a memory in a general sense.

8.7.1.1 Microcontroller Operation

The processor unit of a microcontroller system is the microprocessor. A microprocessor performs its intended functions using *instructions* and *data*, both provided through *programs* and external commands (using various I/O means). Instructions (programs) provided by the vendor (manufacturer) or developed by the user and stored in the memory are known as software. In particular, vendor-provided software that is stored in a *read-only-memory* (ROM) unit and that cannot be generally altered by the user are called *firmware*. Also, for specialized functions, some programs with fixed procedures may be permanently converted into digital hardware, with the associated advantages of fast speed, lower cost (in mass production), and convenience of operation.

All information (*data, addresses, control instructions, operation codes*, etc.) within a microcontroller is present in the binary form as, for example, voltage levels, voltage pulses, or on/off states of digital logic elements. Consider the operation of the microcontroller system shown in Figure 8.21. For example, suppose that the microprocessor wants to read a piece of data from memory for the purpose of numerical processing (addition, multiplication, etc.). The microprocessor knows (it should) the address of the particular location in the memory of the microcontroller at which the piece of data is stored. It will place this address code (as voltage pulses, for example) on the address bus and will also place the control signal corresponding to "Read from memory" on the control bus. In response, the memory will place the piece of data on the data bus, from which the microprocessor will read the data into one of its registers. Next, consider storing in memory a piece of data processed by the microprocessor. First, the microprocessor will place the piece of data (perhaps currently located in its *accumulator* or *data register*) on the data bus, and will place on the address bus the address code of the location in the memory at which the piece of data is to be stored. Finally, the microprocessor will place the "Write in memory" control signal on the control bus. This will store the piece of data at the proper address location in the memory of the microcontroller. This process will not automatically erase the piece of data generated inside the microprocessor (which is present in the data register). Data input from other (peripheral) devices and data output to other devices can also be handled in the same manner.

Now consider the task of data processing in a microcontroller (by a microprocessor). A computer program (software) to perform the processing task is stored in the memory of the microcontroller. The particular step of the program (i.e., *instruction*) that is being executed is selected according to the value in the *program counter*. This value is the *instruction address*. Suppose that the task, according to the computer program, is to add the data stored in the memory address location *B* to the data at the address location *A* and store the result at the address location *C*. The instruction address (picked up according to the program counter) will identify an instruction in the program that will "instruct" the microprocessor to read the data at the memory location *A* of the microcontroller into the microprocessor accumulator. Consequently, this instruction address location (in the memory where the program is stored) will contain the *operation code* for "Read from memory" and the address of the memory location *A*. Similarly, the second instruction address location will contain the operation code for "Addition" and the address of the memory location *B*. This will result in the addition of the data at location *B* into what is already in the accumulator of the microprocessor. The result will remain in the accumulator. The third instruction address location will contain the operation code for "Store in memory" and the address of the memory location *C*. This will cause the data in the accumulator of the microprocessor to be placed at location *C* of the memory of the microcontroller.

8.7.2 Microprocessor

As seen before, the brain of a microcontroller is the microprocessor unit. It processes data (numerical and logical) according to instructions (computer program) supplied to the microcontroller. Let us briefly discuss the architecture of the microprocessor portion alone. Memory and input/output hardware of a microcontroller will be discussed subsequently.

A simplified structure of a microprocessor unit is shown in Figure 8.22. We have already mentioned several components included in this schematic representation. Note the interaction among various components shown in the figure. A brief description of the function of each component is given next.

8.7.2.1 Arithmetic Logic Unit

The arithmetic logic unit (ALU) of a microprocessor can perform arithmetic operations such as *addition, subtraction, multiplication,* and *division,* and logic operations such as *AND, OR, NOT, NAND,* and *NOR*. Depending on the number of *primitive operations* available with the particular microprocessor, some or all of these operations can be performed by the ALU in response to the corresponding operation codes. For instance, if multiplication is not a primitive operation for a particular microprocessor, it can be performed by several ways such as repeated addition, or addition and shift. Similarly, subtraction can be performed by using addition (for example, by the two's complement method).

Typically, data (i.e., *operands*) to be processed by the ALU are read (for instance, from memory) into the *accumulator* and the *data register* (see Figure 8.22). The operation code corresponding to the operation to be performed by the ALU is placed in the *instruction register*.

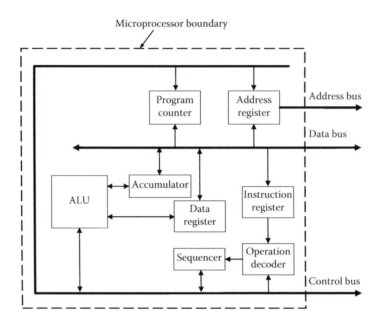

FIGURE 8.22
Schematic representation of a microprocessor.

The operation code is available from the computer program that is stored in the memory of the microcontroller. This *"opcode"* is in the program step (instruction) that is addressed according to the contents of the program counter. The operation decoder decodes the operation code. Next, a control signal is supplied to the ALU by the *sequencer* in order to perform the particular operation. Once the operation is performed, the result is stored in the accumulator and may be subsequently stored in the memory of the microcontroller or sent to an output device through the data bus.

8.7.2.2 Program Counter

During execution, the computer program is stored in the memory of the microcontroller. Each instruction (or program step) is stored at a specific memory location with an associated address. The program counter is a storage area within the microprocessor that contains the address of the instruction (program step) to be executed next. When this address is placed on the address bus by the microprocessor, the memory of the microcontroller will respond by placing the data (program step) contained in that memory location on the data bus. The address in the program counter will advance (increment) sequentially so as to execute the program in a sequential manner unless instructions such as jump or branch are encountered. In the case of a branch instruction, the contents of the program counter will change to reflect the program step to which the execution is branched off.

8.7.2.3 Address Register

The address of a memory location of the microcontroller or peripheral device that needs to be accessed by the microprocessor is contained in a temporary register known as the address register. When this address is placed on the address bus, the data in the corresponding memory location will be placed on the data bus, or the addressed hardware unit will be "enabled," perhaps by placing some data on the data bus.

8.7.2.4 Accumulator and Data Register

These are registers in the microprocessor that can temporarily hold data to be processed by its ALU. Typically, when a program step is addressed, the data to be processed in that instruction are placed on these registers. Furthermore, results produced by the ALU are temporarily stored in these registers to be placed on the data bus, for example, for subsequent storage in the memory of the microcontroller.

8.7.2.5 Instruction Register

Instructions such as an operation code contained in a program step (stored in the memory and accessed by the microprocessor) are temporarily stored in an instruction register, awaiting decoding.

8.7.2.6 Operation Decoder

The operation decoder interprets an operation code stored in the instruction register of the microprocessor. The decoded information is then passed on to a sequencer to be sent to the ALU. In this manner, the ALU finds out which operation is to be performed on the data (i.e., operands) that are stored in the data registers.

8.7.2.7 Sequencer

It is usually not possible for the ALU to process data immediately after the operation decoder interprets the opcode. The sequencer controls the sequence in which the ALU performs its operations. The sequencer may receive control signals from peripheral devices or a "Ready for action" signal from the ALU.

8.7.3 Memory

A microprocessor must have access to the data that it processes and the instructions (computer program) that inform it on how the data should be processed. Computer programs (software) and data are stored in the microcontroller memory for access during processing. Memory is different from mass storage media such as disks, which are off-line devices of a digital computer. During operation, the microcontroller may transfer data (and programs) stored in an external storage medium into its memory, and may transfer processed data currently stored in the memory back to the storage medium, through the I/O interface. It is the contents of the memory that can be directly accessed by the microprocessor.

8.7.3.1 RAM, ROM, PROM, EPROM, and EEPROM

The memory of a microcontroller may be either integral with it or available as separate memory chips. There are many categories of memory of a microcontroller (or any digital computer). Read and write memory (popularly termed random access memory) is denoted by RAM. The user can read the contents of a RAM and also write (store) information (data, programs) into a RAM. Hence, user generated software is typically stored in a RAM chip of a computer.

Read only memory is denoted by ROM. The user can read contents of a ROM but he cannot modify the contents; the user cannot write into a ROM. It does not lose its contents when the power is removed. Typically, the permanent software provided by the manufacturer (i.e., firmware) is stored in a ROM.

Sometimes, the user (not the manufacturer) may wish to store information permanently in memory. A programmable read only memory or PROM is used for this purpose. With a special writing device, the user is able to write into a blank PROM once, but the stored contents cannot be modified thereafter.

One may wish to have permanent memories that can be modified as well (occasionally), under special circumstances. Erasable programmable read only memory (EPROM) is useful in that case. The contents of an EPROM can be erased using specialized equipment (e.g., by applying ultraviolet light for 30 min). Then, new information can be stored again on the EPROM (permanently, until erased using the special equipment).

Electrically erasable PROM is denoted by EEPROM. This is also denoted by EAPROM (electrically alterable PROM). This memory can be used permanently as a ROM, but it can be modified occasionally by using electrical currents (rather than ultraviolet light) of fairly high voltage produced by special circuits. Advantages of EEPROM include the fact that carefully controlled modification of memory contents (say, erasure of a selected portion of the contents) is possible and that the memory chip does not have to be removed from the PC board for modification. Disadvantages include the slowness of the modification process and the complexity of the electrical circuitry that would be required in the process.

8.7.3.2 Bits, Bytes, and Words

Information within a microcontroller is stored, processed, and transferred as *binary digits* or bits. Data storage in computer memory may be achieved using a grid of bistable elements. A memory chip that can store 1024 bits (2^{10} bits) is a 1 kb memory chip. Note that RAM chips of hundreds of megabits (Mb) in capacity (*Note*: 1 Mb = 1024 kb) are commonly available, and the cost is decreasing as well. This is a rapidly growing area.

A basic microcontroller handles information grouped into 8 bits simultaneously. Such a group of 8 bits is termed a *byte*. In computer terminology, a collection of 1024 bytes is a one-*kilobyte* or simply 1 K. Also, 1 *megabyte* (or 1 Mb) is equal to 1024 K.

The number of bits that can be manipulated by a computer in one operation is called the *word size* of the computer. One word of data is stored in each register of a microprocessor and it can be processed by the ALU in a single operation. This usually is also the size of data that can be placed on the data bus of the microcontroller. Today 16 bit microcontrollers are common while 32 bit microcontrollers (i.e., data word = 32 bits) are available as well.

8.7.3.3 Volatile Memory

If the contents of a memory are automatically erased when the power source of the computer is turned off, the memory is said to be volatile. Many types of RAM chips have volatile memories. ROM chips should maintain their contents under power-off conditions as well. Hence, they should have *nonvolatile* memories.

8.7.3.4 Physical Form of Memory

Any bistable element can store a bit of data. It follows that a variety of physical devices can serve as computer memory. The density of data storage (which governs the physical

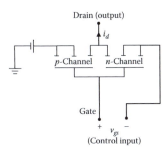

Drain (output)

i_d

p-Channel   *n*-Channel

Gate

+ v_{gs} −
(Control input)

FIGURE 8.23
CMOS (complementary symmetry MOSFET) memory element.

size), speed of data transfer, cost, and compatibility with the overall microcontroller system are the deciding factors in the development of microcontroller memory.

Semiconductor memory is widely used in microcontrollers, but magnetic bubble memory technology is used as well. A *semiconductor memory* is simply an IC chip consisting of many bistable circuit elements (e.g., flip-flops), each element representing a bit of data. The technique used to store data in a memory chip depends on the particular type of semiconductor memory.

One type of memory chip uses metal-oxide-semiconductor field-effect transistor (MOSFET) elements (see Figure 8.23) to store bits of data. The level of electric charge between the gate lead and the channel lead is used to represent a bit. A level of charge above a threshold value will represent 1 and a charge level below the threshold value will represent 0. Hence, a 1 bit can be stored by energizing the gate of the MOSFET at the correct memory location. Since this charge gradually leaks off, it has to be periodically refreshed (say, at a frequency of 1 kHz) in order to maintain its value. Hence, this type of RAM is called *dynamic RAM* (or, *DRAM*), and is implemented using dynamic circuitry. Note that this type of memory is volatile and cannot be used as ROM.

A *static* memory, unlike dynamic memory, does not have to be periodically refreshed. A common static RAM uses flip-flop elements. As discussed before, a flip-flop element consists of logic gates made of semiconductor elements, and it will maintain its output state even after the inputs are removed. A static RAM of this type tends to be more expensive than a dynamic RAM, because more complex circuitry is involved. An advantage is, however, that a static RAM does not require circuitry for continuously refreshing the inputs.

A semiconductor memory that is commonly used as EPROM employs MOSFET elements to permanently store electrical charges. Each MOSFET has an additional (floating) gate that is insulated (using a silicon dioxide layer) from the rest of the transistor. When a high voltage (e.g., 25 V) is applied between the regular gate lead and the drain lead, the floating gate will be charged, but this charge will be trapped by the insulation when the voltage is removed. Hence, a 1 bit will be retained permanently at that memory location. The chip can be erased by exposing it to ultraviolet light for about a half an hour. This process makes the silicon dioxide insulation of the floating gate temporarily conducting, thereby discharging the gate. New data can be stored again in the chip by charging the floating gates as before.

8.7.3.5 Memory Access

Each memory location has a unique address so that it can be accessed by the microprocessor. The address of a memory location is certainly not the same as the information (data) stored at that location. Typically, each memory location can store one data word. Each memory location can be connected to the data bus through a *data buffer* using a control signal. A specific memory location to be connected to the data bus is chosen depending on the address on the address bus. This address is decoded and the corresponding memory location is activated. Depending on the control signal, the contents that are temporarily stored in the buffer are then stored at the memory location, or *vice versa*. A schematic representation of this scheme of memory access (read/write) is shown in Figure 8.24. Note that the buffer has three states (*read, write,* and *no action*). The access scheme can be extended to

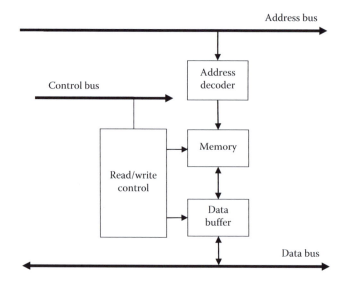

FIGURE 8.24
Memory access scheme.

the case where there are more than one memory chip to be accessed. In this case, first, the "chip select" control is activated and information on the address bus is used to choose the proper memory chip. Next, write control or read control is activated as in the single chip case to perform the appropriate operation.

Consider the static RAM chip shown in Figure 8.25. This is a $32\,K \times 8$ semiconductor (CMOS) RAM. Specifically, it has $32\,K$ memory locations, each holding 8 bits of memory. Since $32\,K = 32 \times 2^{10} = 2^{15}$, we need a 15 bit address to identify all $32\,K$ locations of memory. These address bits are denoted by A0–A14 in Figure 8.25. When a particular (8 bit) memory location is addressed using the corresponding combination of 15 bit addresses, the contents of the memory (8 bits) can be accessed by the eight I/O pins denoted by I/O1 to I/O8. In particular, when the pin WE is activated (using logic-0; because "active-low"), the eight I/O pins can be used to send 8 bits of data to be stored at the addressed memory location. Note that the "overbar" in $\overline{WE}$ denotes "active-low." When the lead OE is activated (using logic-0, because "active-low"), the contents (8 bits) of the addressed memory location can be read from the eight I/O leads. The particular memory chip is selected by the processor by sending a logic low signal to the lead CS, which is the "chip select" lead (which is active-low).

Example 8.10

Consider a RAM chip having 256 memory locations. Discuss different ways this memory can be structured and accessed using the address. What is the minimum word size of the memory address that is required?

Solution
One possible way of arranging the 256 locations of memory is as a single row having 256 locations. In this case, an 8 bit address would be needed (*Note*: $2^8 = 256$). All 8 bits of the address word are used simultaneously to define the 256 locations along the row. Alternatively, the memory could be arranged as a matrix of two rows, each row having 128 locations. In this case, 1 bit of the address is used to denote the row number. For example, value 0 of this bit could represent first row and, then, value 1 would represent the second row. Now, 7 bits are necessary to denote the

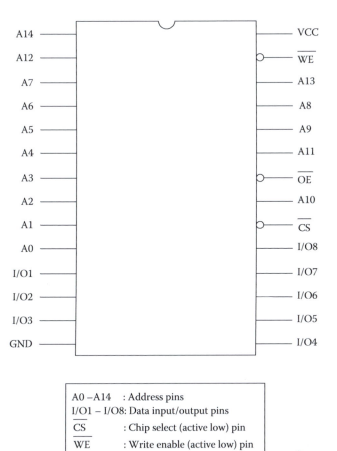

A14		VCC
A12		$\overline{WE}$
A7		A13
A6		A8
A5		A9
A4		A11
A3		$\overline{OE}$
A2		A10
A1		$\overline{CS}$
A0		I/O8
I/O1		I/O7
I/O2		I/O6
I/O3		I/O5
GND		I/O4

A0–A14 : Address pins
I/O1 – I/O8: Data input/output pins
$\overline{CS}$: Chip select (active low) pin
$\overline{WE}$: Write enable (active low) pin
$\overline{OE}$: Output enable (active low) pin
VCC : Power supply pin
GND : Ground pin

FIGURE 8.25
A $32K \times 8$ RAM chip.

memory location along the chosen row (*Note*: $2^7 = 128$). Here again, an 8 bit word is required as the memory address. Yet, another possibility is a memory matrix having 4 rows and 64 locations on each row. In this case, 2 bits are required to represent the row number, and 6 bits ($2^6 = 64$) are necessary to define the location along each row. Again, we need an 8 bit word. Next, consider a matrix of 8 rows, each row having 32 memory locations. In this case, 3 bits of the address word have to be allocated for defining the row number and 5 bits ($2^5 = 32$) are needed for defining the location along each row. Once again, an 8 bit address would be needed. These four memory arrangements are the optimal structures, and in each case, an 8 bit address is required. Other structures are not optimal. For example, consider a structure having 3 rows. In this case, 86 memory locations are needed on each row. Since 2 bits are needed to define the row number and 7 bits for defining the memory location on each row, we need a 9 bit address for this structure.

8.7.3.6 Memory Card Design

Several memory chips can be placed in a single memory card to form a memory of larger capacity, as needed. Consider a $mK \times p$ memory chip. This has mK memory locations, each location being able to hold p bits of data. Suppose that $m = 2^r$. Since, $1\,\text{K} = 2^{10}$, it is noted that there are 2^{10+r} memory locations, and as a result, $10 + r$ bits are needed to address these locations.

When several memory chips are placed in a memory card, an address is needed to select a specific chip, and a further address is needed for a memory location in that chip. A decoder may be used to convert the chip selecting the address to the required $\overline{\text{CS}}$ logic of the selected chip. The I/O will then match the addressed memory location of the selected chip.

Example 8.11

Suppose that a required number of $2\,\text{K} \times 8$ EPROM chips and $2\,\text{K} \times 8$ RAM chips are available. Construct a memory card containing $4\,\text{K} \times 8$ EPROM and $4\,\text{K} \times 8$ RAM.

Solution
We have to use two $2\,\text{K} \times 8$ EPROM chips and two $2\,\text{K} \times 8$ RAM chips in order to meet the requirement. Note that a 2 bit address is needed to select one of these four chips. A further 11 bits are needed to select a memory location in the selected chip. Let us denote these 13 bits by $A_0, A_1, \ldots, A_{12}$. The bits A_{11} and A_{12} are used to select a chip and the bits $A_0, A_1, \ldots, A_{10}$ are used to address a memory location of the selected chip. This arrangement is shown in Figure 8.26.

The values of A_{11} and A_{12} have to be chosen such that when a particular chip is addressed, its $\overline{\text{CS}}$ is set to 0 while the $\overline{\text{CS}}$ of the other chips are all set to 1. This is the case for "active-low" logic, as discussed earlier in the chapter. A 2 to 4 decoder may be used for this purpose, as indicated in Table 8.10. A logic gate realization of this decoder is shown in Figure 8.27.

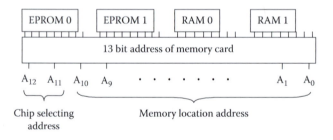

FIGURE 8.26
A memory card containing two EPROM chips and two RAM chips.

TABLE 8.10

Chip Selecting Logic

		$\overline{\text{CS}}$ of				
A_{12}	A_{11}	EPROM 0	EPROM 1	RAM 0	RAM 1	Select
0	0	0	1	1	1	EPROM 0
0	1	1	0	1	1	EPROM 1
1	0	1	1	0	1	RAM 0
1	1	1	1	1	0	RAM 1

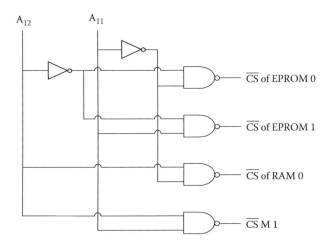

FIGURE 8.27
A 2 to 4 decoder for chip selection.

8.7.4 Input/Output Hardware

Returning to the microcontroller arrangement shown in Figure 8.21, note that so far we have discussed the microprocessor and memory in some detail. In this section, we will examine the input/output considerations of a microcontroller.

Input/output pins form the link between a microcontroller and a *peripheral device* such as a display, keyboard, or a system being monitored or controlled (e.g., a mechatronic device such as a motor). However, it is often not possible to directly connect a peripheral device to the microcontroller for reasons such as the following:

1. If devices are directly connected to a microcontroller, it will not know from which devices or memory location the data are coming from (during input), and it will not be able to "select" a device to supply data or operate the device (during output).

2. Data may be received from several devices simultaneously. If a device does not wait until the microcontroller is ready to accept data from that device, the system cannot function properly. In monitoring and control of a mechatronic system, often, many response variables from the system may have to be read and more than one control signal may have to be generated. The two items mentioned in item 1 above are applicable here as well.

3. Data from the external device may be *analog signals*, which have to be converted into the appropriate digital form before entering the microcontroller.

4. Speed of data processing of the microcontroller will be different (often faster) than the speed at which data are handled (generated, displayed, used to operate a device, etc.) by an external device. Hence, *synchronization* (or *proportioning*) of data rates would be necessary, particularly in real-time monitoring and control.

5. Some computations (e.g., fast Fourier transform) require blocks of data, not just one data sample. Hence, a means of *buffering* the data would be needed. Furthermore, an external device might generate data in a *bit-serial* manner, as a sequence of pulses, whereas a microprocessor processes words (e.g., 16 bits or 32 bits) of data and the microcontroller memory stores words of data as well. A means for this serial-parallel conversion is needed.

6. Voltage levels and currents of signals handled by external devices are often different from what are compatible with a microcontroller (TTL, CMOS, etc.). Furthermore, the *impedance* of a peripheral device has to be matched with that of the microcontroller (see Chapter 4). Otherwise, signal distortion due to loading will result.

The input/output requirements will depend on many factors including the characteristics of the external device that is connected to the microcontroller (e.g., analog or digital data, data rate, signal level/power, serial or parallel data), the nature of application (e.g., automatic control, real-time process monitoring, data acquisition and logging for off-line processing), and the number of input and output signal channels.

8.7.4.1 Microcontroller Pin-Out

A microcontroller chip has many pins (e.g., 40 pin DIP). Many of these pins are grouped into "ports," which serve input/output (interface) functions. The rest serve other miscellaneous purposes (e.g., voltage, clock). The pins are designated for various functions (e.g., input or output; analog-to-digital conversion; capture/compare/pulse-width modulation—CCP). The port operation of a microcontroller is controlled by the microprocessor through its registers.

8.7.4.1.1 Input Pins

Depending on the nature of the device and compatibility (voltage, power, impedance, etc.) a device may be directly connected to the microcontroller pins. This may be the case with TTL or CMOS devices. Sometimes external discrete elements may have to be connected with the device (e.g., a pull-up resistor for open-collector TTL or an open-drain CMOS device).

8.7.4.1.2 Output Pins

Compatible single devices (e.g., TTL or CMOS devices) may be directly connected to the output pins of a microcontroller. In some cases, external discrete elements may have to be connected to the device (e.g., a pull-up resistor for open-collector TTL or open-drain CMOS device). Additional hardware (e.g., buffer, external power source, external digital-to-analog converter—DAC) may be required to drive multiple devices through the same output pin (i.e., fan-out).

The input/output process of a microcontroller can take place in many ways. Three main ways are as follows:

1. Programmed input/output
2. Interrupt input/output
3. Direct memory access (DMA)

The three methods are illustrated in Figure 8.28 and are described next.

8.7.4.2 Programmed I/O

In this case, data transfer takes place under the control of a program running in the micro-controller. This is illustrated in Figure 8.28a. First, the program selects the proper device

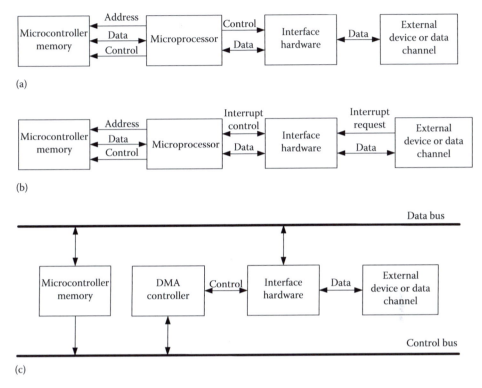

FIGURE 8.28
Three methods of input/output data transfer: (a) programmed I/O; (b) interrupt I/O; (c) DMA.

to be accessed and the microprocessor sends a control signal to activate the corresponding interface hardware (pin) and, perhaps, to inform whether it is a data input (i.e., read) operation or a data output (i.e., write) operation. In an input operation, data are transferred from the external device into the data register (or accumulator) of the microprocessor, which may eventually be stored in the microcontroller memory. In an output operation, data are transferred from the data register of the microprocessor to the external device. Since there might be many devices or data channels interfaced with a microcontroller, there should be a way to pick the proper device or channel for data transfer. A common method used for this, in programmed I/O, is known as *Memory Mapped I/O*. This method is illustrated in Figure 8.29. In this case, each data channel (or external device) is treated by the microprocessor as a memory location, and is assigned an address. Clearly, this address must be unique and not an address already used for a memory location. To select a data channel, the microprocessor places the corresponding address on the address bus. The address decoder interprets the address and activates the proper I/O line (pin). Also, the microprocessor sends a control signal to instruct whether it is an input operation or an output operation. In the case of an output operation, the microprocessor places the data on the data bus. The external device then picks up this data. In an input operation, the interface hardware places the data on the data bus and the microprocessor picks up that data. The microprocessor should have a way to find out whether an external device is ready to accept data from the microcontroller or whether a device has data ready to be read into the microcontroller. One method of doing this is by *polling*. In this method, the microprocessor (when not engaged in an important processing activity) will periodically scan the I/O channels

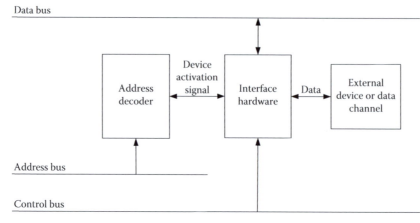

FIGURE 8.29
Memory-mapped I/O.

(pins) to see if any particular channel (or device) is ready for an input/output activity. This readiness may be indicated by a single-bit *status register* (e.g., 1 bit denoting "ready" and 0 bit denoting "not ready"). Some scheme of *priority assignment* would be needed to select a channel when more than one channel is ready. Polling is a slow and somewhat wasteful technique of device selection. It has to scan all channels in each cycle of polling and it uses up the microprocessor even when the input/output channels are not ready. A method of selecting an I/O device, which does not have these disadvantages, is the use of *interrupts*, as described next.

8.7.4.3 Interrupt I/O

This method of input/output data transfer is illustrated in Figure 8.28b. In this method, an external device, when ready for an input/output activity, sends an *interrupt request* signal to the microcontroller. The microprocessor suspends its current activity (after completing the execution of the instruction that it is currently executing) and performs the I/O activity and then returns to the original (interrupted) activity. Like in the case of polling, some priority assignment method is needed to handle situations where there are several interrupt requests simultaneously. For example, the microcontroller can have more than one interrupt request line that is occupied according to some priority (for example, on the first-come-first-served basis). The microprocessor services the interrupt requests one at a time, according to that priority. The basic steps of servicing an interrupt request are as follows: (1) the microprocessor completes execution of the current instruction; (2) it saves the contents of its registers in the microcontroller memory; (3) it sends an *interrupt acknowledge* signal through a control line to the external device—this initiates the data transfer process; (4) the microprocessor sends a *service complete* signal, which will disable the particular I/O line, and will resume the interrupted activity.

8.7.4.4 Direct Memory Access Method

This method of input/output data transfer is illustrated in Figure 8.28c. Unlike the two previous methods, this method needs very little microprocessor activity, and it is suitable for fast transfer of bulk data. In this method, I/O data transfer takes place directly between the microcontroller memory and the external device through a data bus, under the control of a *DMA controller*. The DMA controller temporarily suspends the current operation of the microprocessor, generates the control signals that are needed to perform the data transfer operation, and reactivates the microprocessor at the completion of the data transfer. This is known as *cycle stealing*, because the DMA controller steals an operating cycle of the microprocessor.

8.7.4.5 Handshaking Operation

During data input, it is desirable for the microcontroller to be notified when the external device has placed data in the buffer of its I/O port. Similarly, the microcontroller should notify the external device when the microcontroller has read the data in the I/O port buffer. This type of two-way signaling is known as *handshaking*. For example, a voltage level could be set high when the buffer is full and subsequently set low when the microcontroller reads the buffer. This type of handshaking signal for data input is shown in Figure 8.30a. Similar handshaking will take place at data output as well. For example, the output handshaking signal is set high when the microprocessor has stored the processed data in the output buffer. This signal will be set low when the external device receives the contents of the buffer. The associated handshaking signal is shown in Figure 8.30b.

The handshaking function may be accomplished by using a *status register* as well. A status register consists of a data byte or word. One bit location of the register is assigned to each I/O port. Then the bit value can be used in the handshaking operation. For example, during data input at I/O port n, the bit at the nth location of the status register is set to 1 when the data buffer is full. The bit value at that location is set to 0 again when the microprocessor has completed reading the contents of that port.

8.7.4.6 Clock, Counter, and Timer

Clocks or timing signals are crucial to the operation of a microcontroller. Clock signal is typically a *pulse sequence* at a known constant frequency, generated, for example, by a quartz crystal oscillator. Many clock signals are used in a microcontroller. In particular, the microprocessor clock is a high frequency (e.g., hundreds of MHz) pulse signal that is used to time, coordinate, and synchronize various activities of the microprocessor. In addition, clock signals are used to time or count events, synchronize input/output data transfer, and for generating control signals and interrupts in the operation of sample-and-hold circuits, ADC, DAC, multiplexers, and other interface hardware. For example, analog to digital conversion can be controlled (or synchronized with other activities) by triggering the ADC using a clock signal. The frequency of an ADC trigger signal may be in the MHz range. Variable clock frequencies can be obtained by using programmable timers. Timing of the input/output operations is particularly crucial in real-time microcontroller applications such as digital control and online monitoring of mechatronic systems.

| (a) | Input buffer full | Buffer read by microprocessor | (b) | Output buffer full | Buffer read by external device |

FIGURE 8.30
A handshaking signal: (a) for data input and (b) for data output.

8.7.5 Microcontroller Programming and Program Execution

Microcontroller application is a software implementation. A microcontroller cannot perform a task unless it is completely and correctly instructed on how to perform that task. In other words, the task has to be "programmed." The computer program (software) contains the necessary instructions to perform the task step-by-step. A microcontroller is programmed using its *instruction set*. It is a set of commands or *operation codes*, written as *mnemonics*. Each operation step of a microcontroller requires an "operation code" along with, when necessary, *data* (or *operands*) on which the operation is performed. Different brands and models of microcontrollers will have different instruction sets.

8.7.5.1 *Instruction Set, Operation Codes, and Mnemonics*

Within a microcontroller, the *instructions* for manipulating data and performing various *operations* have to be given in the binary form. The code used to represent a computer operation or instruction is known as the *operation code* or simply *opcode*. The list of these opcodes is the *instruction set* of the microcontroller. Internally, the opcode is used in the binary form. This corresponds to the machine code or *machine language* of the microcontroller. Externally, however, it is inconvenient to use the binary representation for each code. Hence, in programming in a low-level language such as the *assembly language*, a meaningful abbreviation is used to represent each opcode. These abbreviations are known as mnemonics. For example, the operation code for division of one number by another might be 0011011 in the machine code, and the mnemonic of this operation code might be DIV. The associated program step (or instruction) should contain two addresses, the address of the operation code and the address of the data location in the memory of the microcontroller. Operation codes are usually permanently stored in ROM and the data are usually stored in RAM.

8.7.5.2 *Programming and Languages*

A microprocessor executes the programs stored in the memory of the microcontroller. These programs are present in the form of machine language (in binary code) of the microcontroller. It is rather difficult to write a program in a machine code, however. Hence, many computer languages that are more convenient for programming by humans have been developed over the years. Some of these languages are *low-level*, in the sense that they are closely related to a particular machine language and are typically machine specific, and hence not interchangeable with different types of microcontrollers (i.e., not portable). Some other types of languages are high-level and generic. Typically, a program is first written in a high-level language on PC, and properly debugged and "compiled" before downloading into the microcontrollers. A high-level language may be used to program any microcontroller provided that an appropriate *compiler*, which is a program that can convert the high-level program into the machine language of the microcontroller, is available.

Note: It is much easier to write a program in a high-level language. Debugging (error removal) is also easier. But, since a high-level program is neither machine specific nor customized, it can be inefficient and the execution speed is slower than with a program written in a low-level machine-dependent language.

8.7.5.2.1 *Assembly Language Programming*

A program in assembly language has a one-to-one correspondence to its machine language program. Specifically, each program step (or line) of a machine language program will consist of three items as follows:

Instruction address   Operation code   Data address

The main difference is that the machine language uses the binary representation for each item, and the assembly language uses a *mnemonic* for the operation code and a variable name to denote the data address. Instruction addresses are not required in an assembly program because they are automatically generated during the assembly process. A program known as *assembler* is used to convert an assembly language program into the machine language. It will properly interpret the mnemonics for the opcodes and also will assign memory addresses for the data variable names. Errors in the assembly language program are also checked and indicated during the assembly process. Only an error-free assembly program will generate an object module to be stored in memory. Assembly language is machine dependent and hence it will vary from one microcontroller to the next. In order to program a microcontroller in the assembly language, the *instruction set* of the microprocessor should be known. The mnemonics for the opcodes are given in the instruction set.

8.7.5.2.2 High-Level Languages

Due to the one-to-one correspondence between machine language and assembly language, it is complex and tiresome to program in assembly language. Furthermore, an assembly language program cannot be easily transferred (ported) from one microcontroller to another. Hence, computer programs are frequently written in a high-level language.

High-level languages also employ mnemonics for operations and variable names for data addresses. But many operations are generally combined into one program statement, and hence there is no one-to-one correspondence between a high-level language and its machine-code program. A program known as the compiler is used to convert a high-level language program (the source program) into a machine-code program (the object code).

Popular high-level languages include C, FORTRAN, PASCAL, and BASIC. A mechatronics engineer should able to program using at least one of these languages. Unlike the first three languages, BASIC (Beginner's All-purpose Symbolic Instruction Coding) uses an interpreter-compiler. In this case, the entire source program is not compiled in a single step. Instead, each BASIC program line (statement) is compiled and executed before proceeding to the next statement. This process is obviously slower than the regular compiling, but is more convenient and advantageous in online computations.

8.7.5.3 Program Execution

Suppose that a computer program is written using a high-level language on a PC, debugged, and converted into the machine language by the assembler or compiler software. This machine code program (in the binary form) is first "loaded" into the memory (RAM or ROM) of the microcontroller. *Note*: A machine language program may also be stored in a peripheral medium such as hard disk and loaded into memory only when the program is executed. This economizes the valuable memory space. Each program step has an address associated with it. This address gives the memory location where that program step is stored. Next, the microprocessor of the microcontroller should follow the instructions given in each of these program steps.

Program execution commences by some command (say, by an external signal from the keyboard or trigger from a response variable of a mechatronic system). First, the address of the first program step is read into the *program counter* of the microprocessor. This address is then placed on the *address bus*. The contents of that address location in memory are returned to the microprocessor through the *data bus*. Note that the contents are both the address of an opcode and the address of the data on which the opcode would operate. The microprocessor first places the address of the opcode on the address bus. Typically, this would be an address of a ROM location. In response, the contents of the ROM location (opcode) are returned to the microprocessor through the data bus. This is automatically stored in the *instruction register* of the microprocessor. Next, the microprocessor places on the address bus the address of the data (operand). Usually, this would be an address of a RAM location. In response, the contents of the RAM location (data) are returned to the microprocessor through the data bus. This data will be stored in the *accumulator* (a *data register*) within the microprocessor. Data are "fetched" in this manner by the microprocessor. Subsequently, the operation code stored in the instruction register is decoded by the *operation decoder* and presented to the *sequencer* (see Figure 8.22), which in turn sequences the proper operation. The actual operation takes place by means of the processing hardware of the ALU of the microprocessor. Once this program step is executed, the program counter is advanced by one. This new number in the program counter corresponds to the address of the next program step. Consequently, the microprocessor will execute the next program step as before. Throughout the operation, control signals will be sent through the *control bus* to signal various activities and states of the microcontroller.

8.7.5.3.1 Branching

A microprocessor executes program steps one by one in the order in which they are written (i.e., sequentially), unless instructed by the present program step to branch (or *jump*) to a step other than the one next to it. Normally, the program counter is incremented by 1 after executing each program step, as mentioned before. If a branch statement is encountered, the contents of the program counter have to be changed to give the address of the program step that would be executed next. Hence, a branch statement should contain that address. When a branch statement in the program is encountered, the microprocessor first stores the present contents of the program counter in a memory location (known as the stack), and then loads into the program counter the address of the program step to which the execution would be branched. Once the program steps following the branch statement are executed, the program will encounter a "return" statement. On encountering this statement, the microprocessor will load the address stored in the stack back into the program counter and continue the operation as usual.

8.7.5.4 Real-Time Processing

Computation requirements, particularly those pertaining to speed and efficiency, are usually more stringent for real-time processing. In mechatronic applications, control signals have to be generated in real time. In other words, data processing cycles have to be synchronized with a real-time clock. Processing efficiency is a major consideration in real-time computing. For example, in digital control, processing speed has to increase proportionately with the sampling rate of the measured data. *Shannon's sampling theorem* indicates that the sampling rate should be at least twice (and preferably 5 or 10 times) the maximum frequency component of interest in the measured signal.

Microcontroller priorities have to be assigned for various tasks, demands, and needs. For example, *interrupt-handling* methods should be clearly defined for the I/O channels. Resource allocation includes memory allocation, and this too is quite crucial in real-time systems.

Another consideration that is of utmost importance in real-time computing is the issue of whether *software-driven devices* or *hardware logic devices* should be used. For example, in direct digital control the control signals have to be computed within a control cycle. This time period is limited by factors such as ADC sampling period for analog sensors, pulse period for encoders, and step period (time for one step at maximum speed) for stepping motors. The software-based approach has advantages such as flexibility and the ability to implement quite complex control strategies. The hardware-based approach (a hardware controller) has advantages such as fast operating speed, simplicity, and low cost (in mass production). Depending on the task at hand, a compromise may be reached: a hybrid system consisting of software-based devices (microcontrollers) and hardware logic devices (hardwired implementations). For example, simple tasks such as timing, counting, and sequencing may be implemented by hardware logic circuitry, and control algorithms may be software programmed into a microcontroller.

8.7.6 Development of Microcontroller Applications

Microcontrollers and their development/application tools are available from a variety of suppliers. The popular microcontrollers include the ARM and Cortex families from analog devices; the 8x family from Intel; the PIC family from Microchip technology; the HC family from Motorola (Freescale); and the MSP family from Texas Instruments. A typical development kit for a microcontroller includes the following: Microcontroller chip; development board (breadboard); PC (for programming, debugging, etc.); power supply (dc); connection cables; software (compiler, debugger, simulator/emulator, etc.); reference manuals (soft- or hard-copy); miscellaneous hardware (IC chips such as Schmitt trigger, drivers, additional memory, etc.; discrete elements such as resistors, etc.). Schematic representation of a project breadboard with its components is shown in Figure 8.31. The system or device to be operated by the microcontroller (e.g., a mechatronic device such as a motor) is also indicated in the figure.

Before purchasing microcontrollers and development tools, one must carefully examine the application (e.g., operating/monitoring/controlling a mechatronic device) and

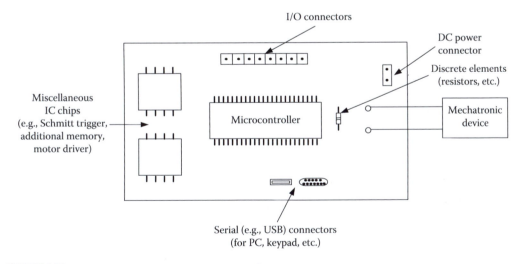

FIGURE 8.31
Microcontroller project breadboard.

understand its requirements. It is desirable to follow a systematic procedure in developing a microcontroller application. The following main steps are suggested.

1. Study the application system/device with respect to its functions, outputs (normally monitored using sensors), inputs (normally activated by a controller or external commands), and performance specifications.

2. Describe in words the functions of the microcontroller in the particular application. This will form the pseudo-program and may include graphic representations such as flowcharts and block diagrams.

3. Select a commercial microcontroller by matching its capabilities (number of I/O lines, memory, word size, speed, types of I/O capability such as PWM, UART, etc.) with the project specifications, cost, size, etc.

4. Acquire the necessary development tools for the microcontroller (software for programming, debugging, emulating, etc.; breadboard; cables; PC; power supply; reference manuals, etc.).

5. Design and draw the hardware circuit for the breadboard (including I/O requirements, additional IC chips such as Schmitt trigger and additional memory, and discrete elements such as pull-up resistors, etc.). Acquire the additional hardware.

6. Complete the project board including the system/device (mechatronic) to be operated, and test the hardware.

7. Program the application on the PC using the appropriate high-level language (e.g., C), debug, compile, debug again (emulate the application if necessary), and download the code into the microcontroller.

8. Test the operation of the overall application. If necessary, fine-tune some of the previous steps.

Problems

8.1 Discuss the reasons for using binary codes to represent information within a digital computer.

(i) Convert the following binary numbers to the decimal form: (a) 10110, (b) 0.101, (c) 1101.1101

(ii) Convert the following decimal numbers to the binary form: (a) 29, (b) 0.5625, (c) 10.3125

Why is it convenient to handle binary numbers in the hexadecimal form? Write the hex values of the six numbers given above.

8.2 Convert $(1001010.11)_2$ into a hex number. Also, convert $(6D.0F)_{16}$ into a binary number. What is the octal representation of each of these two numbers?

8.3 Convert $(18347.319)_{10}$ into the octal representation, then to binary, and from that to the hex representation.

8.4 Perform the following two's complement operations and express the result in the two's complement form:

(i) $(10010)_{2's} + (00110)_{2's}$

(ii) $(10010)_{2's} - (00110)_{2's}$

(iii) $(10011)_{2's} + (10100)_{2's}$

(iv) $(01101)_{2's} + (01011)_{2's}$

Check your result using the equivalent decimal numbers.

8.5 Carry out the following binary computations

(i) For example, $(1011)_2 \times (101)_2$

(ii) $(1101101)_2 \div (101)_2$

8.6 Determine $(010100)_{sm} + (111011)_{sm}$ and express the result in the sm form.

8.7 Explain the property of "sign extension" as applied to the coding of sign numbers for use in digital computers.

Carry out the following one's complement operations and give the answers in the one's complement form:

(i) $(01010)_{1's} + (10010)_{1's}$

(ii) $(01010)_{1's} - (10010)_{1's}$

(iii) $(10101)_{1's} + (10010)_{1's}$

Check your answers using decimal numbers.

8.8 Carry out the following binary arithmetic computations and express the result in binary. Check your answers using decimal computations.

(i) $(1101)_2 \times (110)_2$

(ii) $(11001101)_2 \div (1011)_2$

8.9 Consider the negative number $-(1101)$, which is in the binary form. It has a decimal value of -13. Determine its 2's complement representation.

8.10 Carry out the following operations and express the result in the 2's complement form. Check the results using the decimal values of the numbers.

(i) $(10001)_{2's} + (01011)_{2's}$

(ii) $(11001)_{2's} - (00011)_{2's}$

(iii) $(10001)_{2's} + (10010)_{2's}$

(iv) $(01110)_{2's} + (01101)_{2's}$

8.11 Codes are employed in digital systems to represent information such as numerical values and text (or characters). Consider the following four codes:

(a) Natural binary code

(b) A binary gray code

(c) BCD

(d) American standard code for information interchange (ASCII)

State which of these codes are used to represent numerical values, which ones are used to represent characters, and which ones for both numerical values and characters. Give one or more applications of these four codes in digital systems.

Using a 4 bit word, write the straight binary and gray codes for numbers from 0 to 15. For the starting value (zero), make the straight binary and gray codes identical. Write the BCD representation for the decimal values from 0 to 15.

8.12 (i) Convert $(456.128)_{10}$ into the BCD form.

(ii) Convert $(10000110.10010010)_{BCD}$ into the decimal form.

8.13 Demonstrate how you would apply De Morgan's law to expressions containing either products (logical AND) or sums (logical OR) of more than two terms. Illustrate your approach for

(i) $\overline{a+b+c}$

(ii) $\overline{a \cdot b \cdot c}$

Simplify the Boolean expression $\overline{(\bar{a}+b) \cdot \overline{a+(b+c)} \cdot \bar{c}}$

Note: AND operations are carried out before OR operations.

8.14 Using Boolean algebra, verify the following "Absorption Properties" of logic:

A OR A AND $B = A$

A AND (A OR B) $= A$.

Using one of these results and other Boolean properties/laws, simplify the Boolean expression

$$(a+\bar{b}) \cdot \overline{(c+\bar{d})} \cdot c.$$

Note: AND operations are carried out before OR operations.

8.15 Using Boolean algebra, verify the following "Simplification Properties" of logic:

A OR NOT A AND $B = A$ OR B

A AND (NOT A OR B) $= A$ AND B.

Note: NOT operations are performed before AND; AND operations are performed before OR.

8.16 Prepare a truth table to represent the Boolean relation $a = \overline{x \cdot y + z}$. Using this table express a in:

(i) A sum-of-product form

(ii) A product-of-sum form

Verify that your answers are equivalent to the original expression for a.

8.17 Prepare a truth table to represent the Boolean relation $a = x \cdot \bar{y} + z$. Using this table express a in a

(i) Sum-of-product form

(ii) Product-of-sum form

Verify that your answers are equivalent to the original expression for a.

8.18 The NOR gate is said to be functionally complete. Explain the meaning of this statement. Give a realization of AND using NOR gates alone.

8.19 A 2 to 4 decoder is a digital device whose input is a 2 bit binary word. The output has 4 lines, one of which is activated high (1) depending on the input value. Note that the input is one of the four (decimal) values 0, 1, 2, and 3 (or, binary 00, 01, 10, and 11). Realize a logic circuit for this decoder using AND gates and NOT gates only.

8.20 Illustrate how the Boolean function $\bar{a} \cdot b + c \cdot d$ may be implemented by using

(i) NAND gates only

(ii) NOR gates only

8.21 What are advantages of the CMOS logic family over the TTL logic family? How could the two logic families be combined in the same digital system?

A TTL logic gate is shown in Figure P8.21. The inputs to the gate are A and B, and the output is C. What logic operation does this gate provide? Explain the principle of operation of this logic gate.

8.22 A basic security system for a home operates as follows:

When the system is activated in the "home" mode, only an opening of a door or window will sound the alarm. When the system is activated in the "away" mode, any motion inside the house or opening of a door/window will sound the alarm.

The activation of the system is done manually by using a binary switch with logic-1 representing "home' and logic-0 representing "away." This logic state is denoted by H. The detection of a motion by the motion sensor is denoted by M. The opening of a door or a window is denoted by O. Activation of the alarm is denoted by A.

(a) Assuming that the logic signals H, M, O, and A are all active-high, develop a logic circuit using OR gates and NOT gates only for this alarm system.

(b) Assuming that the signals M and O are available in the active-low form (i.e., $M.L$ and $O.L$) and the output signal A is also needed in the active-form (i.e., $A.L$), modify the circuit in part (a) to realize the system (using OR and NOT gates only).

8.23 For each of the following logic functions, form the K-map and by using it obtain a minimal logic function that is equivalent:

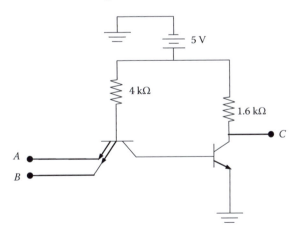

FIGURE P8.21
A basic TTL logic gate.

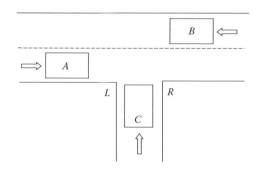

FIGURE P8.25
Turn signals for a road intersection.

(a) $\bar{A}\cdot B+A\cdot B-+\bar{A}\cdot C+A\cdot B\cdot\bar{C}+\bar{A}\cdot\bar{B}\cdot\bar{C}$
(b) $A\cdot B+\bar{C}\cdot D+C\cdot\bar{D}+\bar{A}\cdot C\cdot D+A\cdot\bar{B}\cdot C\cdot D+A\cdot\bar{B}\cdot\bar{C}\cdot\bar{D}$

8.24 Reduce the logic function $\bar{A}+A\cdot B+\bar{B}\cdot C+A\cdot B\cdot C$ by preparing a K-map for the function.

8.25 A one-way road meets a two-way road, as shown in Figure P8.25. The vehicles on the one-way road are allowed to make right turns and left turns only (i.e., it is a dead end). There are sensors to detect vehicles at positions A, B, and C, which generate high signals (logic-1) when vehicles are present and low signals (logic-0) when vehicles are not present. Logic hardware devices are to be developed to operate the right-turn signal (with active-high output R) and the left-turn signal (with active-high output L) in the one-way road. Inputs to these devices are the logic signals denoted by A, B, and C, as generated by the vehicle-detection sensors.

The logic governing the operation of the two signals is as follows: The right-turn signal is on when there are no vehicles at A and there are vehicles at C. The right-turn signal is on as well, when there are no vehicles at A, B, and C. The left-turn signal is on when there are no vehicles at A and B and there are vehicles at C.

(a) Express the logic governing the two devices

(b) Using a K-map, minimize the logic, if possible

(c) Give *circuits* for implementing the two devices using NOR gates and NOT gates only

8.26 (a) How do sequential logic devices differ from combinational logic devices?

(b) How does the asynchronous operation of a logic device differ from the synchronous operation?

(c) Give the circuit for an RS flip-flop using NOR gates only. Extend this to an asynchronous latch circuit, using NOR gates and NOT gates.

8.27 (a) Compare and contrast the following three types of flip-flops: JK flip-flop, D flip-flop, and T flip-flop. Indicate how the latter two flip-flops can be derived as special cases of a JK flip-flop. What is the practical use of the data "toggle" capability of a T flip-flop?

(b) Develop a 3 bit binary counter using D flip-flops. A counting sequence of straight binary 3 bit words (000, 001, 010, 011, 100, 101, 110, and 111) is needed.

8.28 Outline the production process of a typical IC chip. Suppose a digital control circuit is assembled using discrete elements such as bipolar junction transistors, diodes, capacitors, and resistors, instead of their monolithic versions. What are the shortcomings of such a controller in comparison to a single-board controller that uses (monolithic) IC chips?

8.29 Explain the acronyms IC, PC board, SSI, MSI, LSI, and VLSI. Give a classification for IC devices based on the logic gate density. Into what category would you put a modern 32 bit microprocessor chip?

8.30 A basic characteristic of a digital system is that many hardware components of the system are able to store and/or transfer binary data. Since a two-state element is needed to represent a binary digit (bit), this type of digital hardware should physically possess the two-state characteristic. Briefly state the two physical states associated with elements in the following types of hardware: (a) TTL circuit, (b) nonvolatile MOSFET memory, (c) magnetic bubble memory, (d) optical fiber, (e) CCD (charge-coupled device), (f) computer hard disk, (g) EAROM (electrically alterable read-only memory).

8.31 (a) What is a combinational logic circuit and what is a sequential logic circuit? Describe the use of a flip-flop or a latch as a basic element in semiconductor memory. Prepare a truth table similar to what is shown in Figure 8.13b, for an RS flip-flop that uses two cross-coupled NOR gates.

(b) An optical encoder (a motion sensor that senses displacement in steps and produces pulses correspondingly) generates two pulse sequences that are 90° out of phase. In one direction of motion (denoted by *L*), one pulse sequence (denoted by *A*) leads the other (denoted by *B*) by 90°, and in the opposite direction of motion (denoted by *R*) this pulse sequence (*A*) lags the other (*B*) by 90°. This situation is shown in Figure P8.31. Suppose that the two sequences are read into a 2 bit register, the high voltage level of each signal being represented by a 1 bit and the low (zero) voltage level being represented by a 0-bit. Verify that in the *L* direction of motion, the register value will change according to the sequence:

A	0	1	1	0	0	....
B	0	0	1	1	0	....

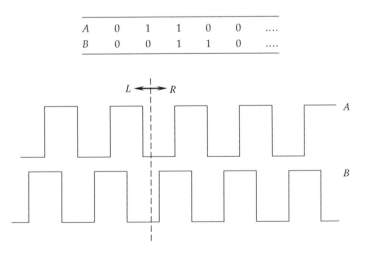

FIGURE P8.31
Two pulse sequences that are 90° out of phase as generated by a motion encoder.

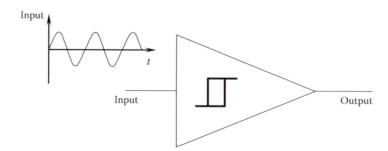

FIGURE P8.32
A Schmitt trigger excited by a sinusoidal signal.

and in the R direction of motion it will be

A	0	0	1	1	0		...	
B		0	1	1	0	0	...	

Explain a simple way to detect the direction of motion by checking the binary value in the register. Discuss a way to physically implement this direction-detection method.

8.32 What is the main advantage of including hysteresis in a switching element? What is a Schmitt trigger? Describe the operation of the Schmitt trigger circuit that employs two bipolar junction transistors as in Figure 8.19a. Even though a Schmitt trigger can be constructed using discrete elements in this manner, this device is commercially available in the monolithic form as a single IC chip. Suppose that a sinusoidal signal is applied to a Schmitt trigger, as schematically shown in Figure P8.32. Discuss the shape of the output signal. What is the output frequency?

8.33 Typically in a digital controller of a mechatronic system, the arithmetic operations, ADD and SUBTRACT, are performed using hardware. More complex operations including MULTIPLY and DIVIDE may be implemented either by hardware or by software. The software operations can be more than an order of magnitude slower. Compare the use of hardware-intensive controllers (or hard-wired controllers) with software-intensive controllers (or digital computer-based controllers) in terms of factors such as speed of processing, flexibility, possibility of using complex control algorithms, and controller cost. Would you classify a ROM-based controller that does not have programmable memory, as a hard-wired controller or as a software-intensive controller?

8.34 What are the essential hardware components in a basic microcontroller? Using a schematic diagram show the organization of such a basic computer.
    Explain the acronyms RAM, ROM, PROM, EPROM, and EEPROM (or, EAPROM). What is the difference between

(a) Volatile memory and nonvolatile memory?
(b) Static memory and dynamic memory?
(c) Semiconductor memory and magnetic bubble memory?

Which of these types of memory cannot be used as a ROM?

8.35 Explain the following methods of data transfer between a microcontroller and a peripheral device: (a) programmed I/O, (b) interrupt I/O, (c) DMA.

Which method would you use in data acquisition for real-time control? Into which of the three categories would you put memory-mapped I/O?

8.36 Explain what roles the following factors play in real-time control using a microcontroller:

(a) Word size of the microprocessor

(b) Machine cycle time of the microprocessor

(c) Instruction cycle time of the microcontroller system

(d) Number of instructions in the machine language program of the control algorithm

What are advantages and disadvantages of expanding the instruction set of a microprocessor?

The frequency of the main clock of a microprocessor is known to be 5 MHz. One machine cycle takes two clock periods, and two machine cycles are needed for one instruction. The microprocessor is used in a real-time control application. The input hardware of the controller includes a multiplexer and a bank of 10 bit ADC units. The ADC cycle time is 5 ms and the multiplexer takes 0.5 ms for one channel switch.

(a) How many input channels can the multiplexer handle optimally?

(b) What is the instruction rate in MIPS?

(c) Estimate the maximum program size for a control algorithm in K (1024 bytes) if three bytes (one byte for the opcode and two bytes for the address field) are used for one instruction.

(d) Estimate the control bandwidth.

8.37 Using a suitable schematic diagram, identify the basic interface hardware components for connecting a mechatronic system to a microcontroller for real-time control, and explain their functions. List several factors that determine the

(a) ADC rate (rate at which data are supplied into the microcontroller)

(b) DAC rate (rate at which data are sent out from the microcontroller)

(c) Control frequency (rate at which control signal is updated)

for real-time control of a mechatronic device. Explain why an input register and a buffer in series (*double buffering*) would be needed to improve the ADC rate.

A microprocessor-based real-time control loop has a single ADC with a double buffer, a control microcontroller, and a single DAC. The ADC rate at which the measured feedback signal is sampled into the input buffer is 10 k words/s. The microprocessor can read data from the input buffer at a rate of 1 M words/s. It can write data into the output register at the same rate. A typical control computation cycle involves reading the contents of the input buffer, performing a "system identification" (i.e., computation of a dynamic model) using this data, computation of the new control signal, and loading the control signal into the output register to be picked up by the DAC of the control channel. The computations alone require a processing time of 1 ms during which time the microprocessor is not available for reading data from the input buffer or for sending data to the DAC.

(a) Estimate the required minimum size of the input buffer.

(b) What is the best control frequency that can be provided by this controller?

8.38 What are the two main functions of the hardware used in controlling a stepper motor? Using a schematic diagram, explain a microcontroller-based feedback control system for a stepper motor.

A stepper motor controller uses a 12 bit counter chip (a hardware counter). If one-step of the motor corresponds to a load movement of 0.2 mm, what is the maximum load movement (stroke) that could be controlled?

8.39 What is the main function of an input/output adapter chip for a microcontroller? Consider an I/O chip having 40 pins. Pin allocation includes a power supply pin, a microcontroller clock signal pin, a reset pin, interrupt request pins, a read/write control pin, chip select pins, and register select pins. Do you think that this I/O chip

(a) Can handle bidirectional data (i.e., data transfer in both directions) in bytes?

(b) Can handle interrupts?

(c) Has handshaking capability?

8.40 Machine language program that is stored in a microcontroller memory is simply a set of "instructions." Consider an instruction that is 3 bytes long. State what information it might carry. Instruction cycle time will vary from instruction to instruction depending on the information carried by an instruction. What are the factors that contribute to determining the instruction cycle time?

Explain why the accuracy and the speed of a microcontroller can be increased by increasing the word size of the microprocessor and by addressing one word, and not one byte, at a time. How many memory locations (words in a word-addressable machine) can be addressed using a 16-bit address?

8.41 Shannon's sampling theorem tells us that the largest useful frequency component that is retained when a signal is sampled is half the sampling rate. In practice, however, we wish to make the sampling rate at least 5 times larger than the maximum frequency of interest. In a servocontroller of a robotic manipulator, suppose that the maximum frequency of interest in a measured motion signal (angular position and angular speed) is 50 Hz. What is an acceptable ADC conversion rate to be used in an associated digital controller? Suppose that 12 channels of data (six joints with a position channel and speed channel for each joint) are sampled and read into the digital controller. The control microcontroller needs 10 μs to read a data sample from one channel. Assuming 3 byte instructions on the average, the average instruction cycle time (fetch and execute) is known to be 6 μs. Estimate the program size (number of instructions) of the control algorithm.

9

Control Systems

9.1 Introduction

The purpose of control is to make a plant (i.e., the "mechatronic" system to be controlled) behave in a desired manner, according to some *performance specifications*. The overall system that includes at least the plant and the controller is called the control system. The control problem can become challenging due to such reasons as

- Complex system (many inputs and many outputs, dynamic coupling, nonlinearities, etc.)
- Rigorous performance specifications
- Unknown excitations (unknown inputs/disturbances/noise)
- Unknown dynamics (incompletely known plant)

A good control system should satisfy performance requirements concerning such attributes as *accuracy, stability, speed of response* or *bandwidth, sensitivity,* and *robustness*. This chapter will introduce the subject of control systems, which is necessary for the field of mechatronics. It is advised to study the concepts of transfer functions and block diagrams as presented in Chapter 3 before proceeding further. MATLAB® and LabVIEW® software tools that are useful in the study and implementation of control systems are outlined in Appendix D.

9.2 Control System Structure

A control system contains a controller as an integral part. The purpose of the controller is to generate control signals, which will drive the process to be controlled (the plant) in the desired manner. Actuators are needed to perform the control actions as well as to drive the plant directly. Sensors and transducers are necessary to measure output signals (process responses) for feedback control; to measure input signals for feedforward control; to measure process variables for system monitoring, diagnosis, and supervisory control; and for a variety of other purposes. Since many different types and levels of signals are present in a control system, signal modification (including signal conditioning and signal conversion) is indeed a crucial function associated with any control system. In particular, signal modification is an important consideration in component interfacing.

9.2.1 Feedback and Feedforward Control

A simplified schematic example of a feedback control system is shown in Figure 9.1. In a *feedback control system*, the control loop has to be closed, making measurements of the system response and employing that information to generate control signals so as to correct any output errors. Hence, feedback control is also known as *closed-loop control*. In digital control, a digital computer serves as the controller. Virtually any control law may be programmed into the control computer.

If the plant is stable and is completely and accurately known, and if the inputs to the plant can be precisely generated (by the controller) and applied, then accurate control might be possible even without feedback control. Under these circumstances, a measurement system is not needed (or at least not needed for feedback) and thus we have an *open-loop control* system. In *open-loop control*, we do not use current information on the *system response* to determine the control signals. In other words, there is no feedback.

Generally, the performance of a control system can be improved by measuring these (unknown) inputs and somehow using the information to generate control signals. In

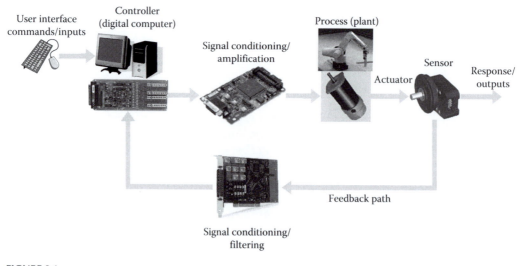

FIGURE 9.1
A feedback control system.

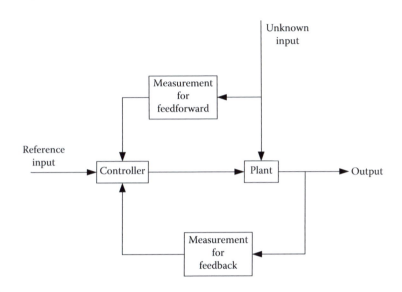

FIGURE 9.2
A system with feedback and feedforward control.

feedforward control, unknown "inputs" are measured and that information, along with the desired inputs, is used to generate control signals that can reduce errors due to these unknown inputs or variations in them. A block diagram of a typical control system that uses feedforward control is shown in Figure 9.2. In this system, in addition to feedback control, a feedforward control scheme is used to reduce the effects of a disturbance input that enters the plant. The disturbance input is measured and fed into the controller. The controller uses this information to modify the control action so as to compensate for the disturbance input by "anticipating" its effect.

Terminology
Some useful terminology introduced in this chapter is summarized below:

Plant or process: System to be controlled.

Inputs: Excitations (known, unknown) to the system.

Outputs: Responses of the system.

Sensors: The devices that measure the system variables (excitations, responses, etc.).

Actuators: The devices that drive various parts of the system.

Controller: Device that implements the control law (generates the control signal).

Control law: Relation or scheme according to which the control signal is generated.

Control system: At least the plant and the controller (may include sensors, signal conditioning, etc., as well).

Feedback control: Plant response is measured and fed back into the controller. Control signal is determined according to the error (between the desired and actual responses).

Closed-loop control: Same as feedback control. There is a feedback loop (closed loop).

Open-loop control: Plant response is not used to determine the control action.

Feed-forward control: Control signal is determined according to plant "excitation."

9.2.2 Programmable Logic Controllers

A programmable logic controller (PLC) is essentially a digital-computer-like system that can properly sequence a complex task, consisting of many discrete operations and involving several devices, which needs to be carried out in a sequential manner. PLCS are rugged computers typically used in factories and process plants to connect input devices such as switches to output devices such as valves, at high speed, at appropriate times in a task, as governed by a program. Internally, a PLC performs basic computer functions such as logic, sequencing, timing, and counting. It can carry out simpler computations and control tasks such as proportional-integral-derivative (PID) control. Such control operations are called *continuous-state control*, where the process variables are continuously monitored and made to stay very close to the desired values. There is another important class of controls, known as the *discrete-state control*, where the control objective is for the process to follow a required sequence of states (or steps). In each state, however, some form of continuous-state control might be operated, but it is not quite relevant to the discrete-state control task. Programmable logic controllers are particularly intended for accomplishing discrete-state control tasks.

There are many control systems and industrial tasks that involve the execution of a sequence of steps, depending on the state of some elements in the system and on some external input states. For example, consider an operation of turbine blade manufacture. The discrete steps in this operation might be

1. Move the cylindrical steel billets into the furnace
2. Heat the billets
3. When a billet is properly heated, move it to the forging machine and fixture it
4. Forge the billet into shape
5. Perform surface finishing operations to get the required aerofoil shape
6. When the surface finish is satisfactory, machine the blade root

Note that the entire task involves a sequence of events where each event depends on the completion of the previous event. In addition, it may be necessary for each event to start and end at specified time instants. Such *time sequencing* would be important for coordinating the operation with other activities, and perhaps for proper execution of each operation step. For example, activities of the parts handling robot have to be coordinated with the schedules of the forging machine and milling machine. Furthermore, the billets will have to be heated for a specified time, and the machining operation cannot be rushed without compromising product quality, tool failure rate, safety, etc. Note that the task of each step in the discrete sequence might be carried out under continuous-state control. For example, the milling machine would operate using several direct digital control (DDC) loops (say, PID control loops), but discrete-state control is not concerned with this except for the starting point and the end point of each task.

A process operation might consist of a set of two-state (on–off) actions. A PLC can handle the sequencing of these actions in a proper order and at correct times. Examples of such tasks include sequencing the production line operations, starting a complex process plant, and activating the local controllers in a distributed control environment. In the early days of industrial control, solenoid-operated electromechanical relays, mechanical timers, and drum controllers were used to sequence such operations. An advantage of using a PLC is that the devices in a plant can be permanently wired, and the plant operation can be modified or restructured by software means (by properly programming the PLC) without requiring hardware modifications and reconnection.

A programmable logic controller operates according to some "logic" sequence programmed into it. Connected to a PLC are a set of input devices (e.g., pushbuttons, limit switches, and analog sensors such as RTD temperature sensors, diaphragm-type pressure sensors, piezoelectric accelerometers, and strain-gage load sensors) and a set of output devices (e.g., actuators such as dc motors, solenoids, and hydraulic rams, warning signal indicators such as lights, alphanumeric LED displays and bells, valves, and continuous control elements such as PID controllers). Each such device is assumed to be a two-state device (taking the logical value 0 or 1). Now, depending on the condition of each input device and according to the programmed-in logic, the PLC will activate the proper state (e.g., on or off) of each output device. Hence, the PLC performs a switching function. Unlike the older generation of sequencing controllers, in the case of a PLC, the logic that determines the state of each output device is processed using software and not by hardware elements such as hardware relays. Hardware switching takes place at the output port, however, for turning on or off the output devices controlled by the PLC.

9.2.2.1 PLC Hardware

As noted before, a PLC is a digital computer that is dedicated to perform discrete-state control tasks. A typical PLC consists of a microprocessor, RAM and ROM memory units, and interface hardware, all interconnected through a suitable bus structure. In addition, there will be a keyboard, a display screen, and other common peripherals. A basic PLC system can be expanded by adding expansion modules (memory, I/O modules, etc.) into the system rack.

A PLC can be programmed using a keyboard or touch-screen. An already developed program could be transferred into the PLC memory from another computer or a peripheral mass-storage medium such as hard disk. The primary function of a PLC is to switch (energize or de-energize) the output devices connected to it, in a proper sequence, depending on the states of the input devices and according to the logic dictated by the program. A schematic representation of a PLC is shown in Figure 9.3. Note the sensors and actuators in the PLC.

In addition to turning on and off the discrete output components in a correct sequence at proper times, a PLC can perform other useful operations. In particular, it can perform simple arithmetic operations such as addition, subtraction, multiplication, and division on input data. It is also capable of performing counting and timing operations, usually as part of its normal functional requirements. Conversion between binary and binary-coded decimal (BCD) might be required for displaying digits on an LED panel and for interfacing the PLC with other digital hardware (e.g., digital input devices and digital output devices). For example, a PLC can be programmed to make a temperature measurement and a load measurement, display them on an LED panel, make some computations on these (input) values, and provide a warning signal (output) depending on the result.

The capabilities of a PLC can be determined by such parameters as the number of input devices (e.g., 16) and the number of output devices (e.g., 12) that it can handle, the number of program steps (e.g., 2000), and the speed at which a program can be executed (e.g., 1 M steps/s). Other factors such as the size and the nature of memory and the nature of timers and counters in the PLC, signal voltage levels, and choices of outputs, are all important factors.

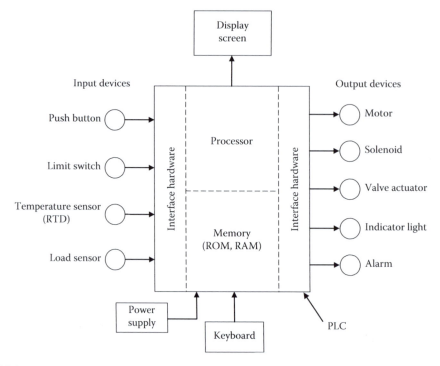

FIGURE 9.3
Schematic representation of a PLC.

9.2.3 Distributed Control

For complex processes with a large number of input/output variables (e.g., a chemical plant and a nuclear power plant) and with systems that have various and stringent operating requirements (e.g., the space shuttle), centralized direct digital control is quite difficult to implement. Some form of distributed control is appropriate in large systems such as manufacturing workcells, factories, and multi-component process plants. A distributed control system will have many users who would need to use the resources simultaneously and, perhaps, would wish to communicate with each other as well. Also, the plant will need access to shared and public resources and means of remote monitoring and supervision. Furthermore, different types of devices from a variety of suppliers with different specifications, data types, and levels may have to be interconnected. A communication network with switching nodes and multiple routes is needed for this purpose.

In order to achieve connectivity between different types of devices having different origins it is desirable to use a standardized bus that is supported by all major suppliers of the needed devices. The Foundation Fieldbus or Industrial Ethernet may be adopted for this purpose. Fieldbus is a standardized bus for a plant, which may consist of an interconnected system of devices. It provides connectivity between different types of devices having different origins. Also, it provides access to shared and public resources. Furthermore, it can provide means of remote monitoring and supervision.

A suitable architecture for networking an industrial plant is shown in Figure 9.4. The industrial plant in this case consists of many "process devices" (PD), one or more programmable logic controllers (PLC), and a distributed control system (DSC), or a supervisory controller. The pds will have direct I/O with their own components while possessing

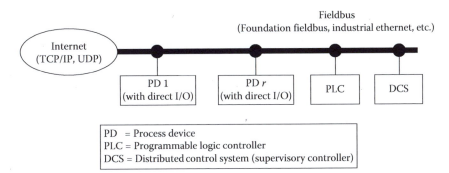

FIGURE 9.4
A networked industrial plant.

connectivity through the plant network. Similarly, a PLC may have direct connectivity with a group of devices as well as networked connectivity with other devices. The DSC will supervise, manage, coordinate and control the overall plant.

9.2.3.1 Hierarchical Control

A favorite distributed control architecture is provided by hierarchical control. Here, distribution of control is available both geographically and functionally. A hierarchical structure can facilitate efficient control and communication in a complex control system. Consider a three-level hierarchy. Management decisions, supervisory control, and coordination between plants in the overall facility may be provided by the supervisory control computer, which is at the highest level (level 3) of the hierarchy. The next lower level (intermediate level) generates control settings (or reference inputs) for each control region (subsystem) in the corresponding plant. Set points and reference signals are inputs to the direct digital controllers (DDC), which control each control region. The computers in the hierarchical system communicate using a suitable communication network. Information transfer in both directions (up and down) should be possible for best performance and flexibility. In master-slave distributed control, only downloading of information is available.

As an illustration, a three-level hierarchy of an intelligent mechatronic system (IMS) is shown in Figure 9.5. The bottom level consists of electromechanical components with component-level sensing. Furthermore, actuation and direct feedback control are carried out at this level. The intermediate level uses intelligent preprocessors for abstraction of the information generated by the component-level sensors. The sensors and their intelligent preprocessors together perform tasks of intelligent sensing. State of performance of the system components may be evaluated by this means, and component tuning and component-group control may be carried out as a result. The top level of the hierarchy performs task-level activities including planning, scheduling, monitoring of the system performance, and overall supervisory control. Resources such as materials and expertise may be provided at this level and a human–machine interface would be available. Knowledge-based decision making is carried out at both intermediate and top levels. The resolution of the information that is involved will generally decrease as the hierarchical level increases, while the level of "intelligence" that would be needed in decision-making will increase.

Within the overall system, the communication protocol provides a standard interface between various components such as sensors, actuators, signal conditioners, and controllers, and also with the system environment. The protocol will not only allow highly flexible implementations, but will also enable the system to use distributed intelligence to perform

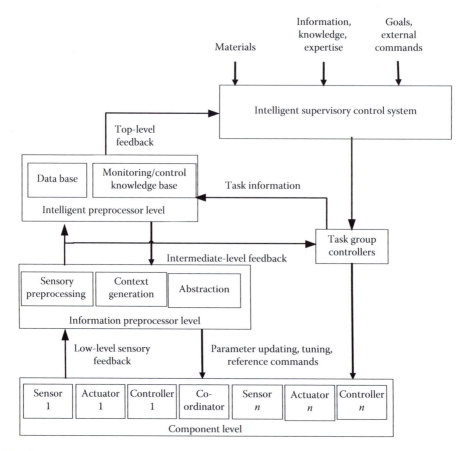

FIGURE 9.5
A hierarchical control/communications structure for an intelligent mechatronic system.

preprocessing and information understanding. The communication protocol should be based on an application-level standard. In essence, it should outline what components can communicate with each other and with the environment without defining the physical data link and network levels. The communication protocol should allow for different component types and different data abstractions to be interchanged within the same framework. It should also allow for information from geographically removed locations to be communicated to the control and communication system of the IMS.

9.3 Control System Performance

A good control system should possess the following performance characteristics:

1. Sufficiently stable response (*stability*). Specifically, the response of the system to an initial-condition excitation should decay back to the initial steady state (asymptotic stability). The response to a bounded input should be bounded (bounded-input–bounded-output—BIBO stability).

TABLE 9.1

Performance Specifications for a Control System

Attribute	Desired Value	Objective	Specifications
Stability level	High	The response does not grow without limit and decays to the desired value	Percentage overshoot, settling time, pole (eigenvalue) locations, time constants, phase and gain margins, damping ratios
Speed of response	Fast	The plant responds quickly to inputs/ excitations	Rise time, peak time, delay time, natural frequencies, resonant frequencies, bandwidth
Steady-state error	Low	The offset from the desired response is negligible	Error tolerance for a step input
Robustness	High	Accurate response under uncertain conditions (input disturbances, noise, model error, etc.) and under parameter variation	Input disturbance/noise tolerance, measurement error tolerance, model error tolerance
Dynamic interaction	Low	One input affects only one output	Cross-sensitivity, cross-transfer functions

2. Sufficiently fast response (*speed of response* or *bandwidth*). The system should react quickly to a control input or excitation.

3. Low sensitivity to noise, external disturbances, modeling errors, and parameter variations (*sensitivity* and *robustness*).

4. High sensitivity to control inputs (*input sensitivity*).

5. Low error; for example, tracking error and steady-state error (*accuracy*).

6. Reduced coupling among system variables (*cross sensitivity* or *dynamic coupling*).

As listed here, some of these specifications are rather general. Table 9.1 summarizes typical performance requirements for a control system. Some requirements might be conflicting. For example, fast response is often achieved by increasing the system gain, and increased gain increases the actuation signal, which has a tendency to destabilize a control system. Note further that what is given here are primarily qualitative descriptions for "good" performance. In designing a control system, however, these descriptions have to be specified in a quantitative manner. The nature of the used quantitative design specifications depends considerably on the particular design technique that is employed. Some of the design specifications are time-domain parameters and the others are frequency-domain parameters.

9.3.1 Performance Specification in Time Domain

Speed of response and degree of stability are two commonly used specifications in the conventional time-domain design of a control system. These two types of specifications are conflicting requirements in general. In addition, steady-state error is also commonly specified. Speed of response can be increased by increasing the gain of the control system. This, in turn, can result in reduced steady-state error. Furthermore, steady-state error requirement can often be satisfied by employing integral control. For these reasons, first we will treat speed of response and degree of stability as the requirements for performance specification in time domain, tacitly assuming that there is no steady state error. The steady-state error requirement will be treated separately.

Performance specifications in the time domain are usually given in terms of the response of an oscillatory (under-damped) system to a unit step input, as shown in Figure 9.6. First, assuming that the steady-state error is zero, note that, the response will eventually settle at the steady-state value of unity. Then the following performance specifications can be stipulated:

Peak time (T_p): Time at which the response reaches its first peak value.

Rise time (T_r): Time at which the response passes through the steady-state value (normalized to 1.0) for the first time.

Modified rise time (T_{rd}): Time taken for the response to rise from 0.1 to 0.9.

Delay time (T_d): Time taken for the response to reach 0.5 for the first time.

Two-percent settling time (T_s): Time taken for the response to settle within ±2% of the steady-state value (i.e., between 0.98 and 1.02).

Peak magnitude (M_p): Response value at the peak time.

Percentage overshoot (PO): This is defined as

$$PO = \frac{\text{peak magnitude} - \text{steady state value}}{\text{steady state value}} \times 100\% \tag{9.1a}$$

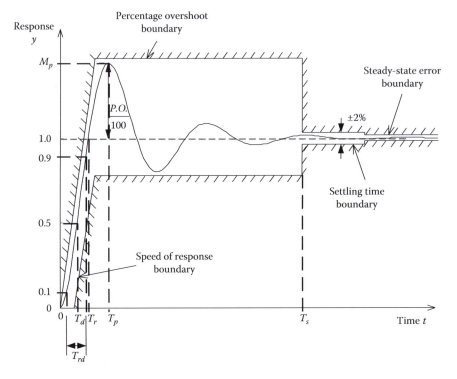

FIGURE 9.6
Performance specifications for time-domain design of a control system.

In the present case of unity of steady-state value, this may be expressed as

$$PO = 100(M_p - 1)\% \tag{9.1b}$$

Note that T_r, T_p, T_{rd}, and T_d are "primarily" measures of the speed of response whereas T_s, M_p, and PO are "primarily" measures of the level of stability. Note further that T_r, T_p, M_p, and PO are not defined for non-oscillatory responses. Simple expressions for these time-domain design specifications may be obtained, assuming that the system is approximated by a simple oscillator.

Specifications on the slope of the step-response curve (speed of response), percentage overshoot (stability), settling time (stability), and steady-state error can also be represented as boundaries to the step response curve. This representation of conventional time-domain specifications is shown in Figure 9.6.

9.3.1.1 Simple Oscillator Model

A damped simple oscillator (mechanical or electrical) may be expressed by the input–output differential equation:

$$\frac{d^2y}{dt^2} + 2\zeta\omega_n \frac{dy}{dt} + \omega_n^2 y = \omega_n^2 u \tag{9.2}$$

where
  U is the input (normalized)
  Y is the output (normalized)
  ω_n is the undamped natural frequency
  ζ is the damping ratio

Suppose that a unit step input is applied to the system. As indicated in Chapter 3, the resulting response of the oscillator, with zero initial conditions, is given by

$$y(t) = 1 - \frac{1}{\sqrt{1-\zeta^2}} e^{-\zeta\omega_n t} \sin(\omega_d t + \phi) \quad \text{for } \zeta < 1 \tag{9.3}$$

where

$$\omega_d = \sqrt{1-\zeta^2}\,\omega_n = \text{damped natural frequency} \tag{9.4}$$

$$\cos\phi = \zeta; \quad \sin\phi = \sqrt{1-\zeta^2} \tag{9.5}$$

Note: It is clear from Equation 9.3 that the steady-state value of the response (i.e., as $t \to \infty$) is 1. Hence, the steady-state error is zero. It follows that the present model does not allow us to address the issue of steady-state error.

The response given by Equation 9.3 is of the form shown in Figure 9.6. Clearly, the first peak of the response occurs at the end of the first (damped) half cycle. It follows that the *peak time* is given by

$$T_p = \frac{\pi}{\omega_d} \tag{9.6}$$

The *peak magnitude* M_p and the *percentage overshoot P.O.* are obtained by substituting Equation 9.6 into 9.3; thus,

$$M_p = 1 + \exp\left(-\zeta\omega_n T_p\right) \tag{9.7}$$

$$PO = 100\exp\left(-\zeta\omega_n T_p\right) \tag{9.8a}$$

Note: In obtaining this result we have used the fact that $\sin\phi = \sqrt{1-\zeta^2}$. Alternatively, by substituting Equations 9.4 and 9.6 into 9.8a we get

$$PO = 100\exp\left(-\pi\zeta\big/\sqrt{1-\zeta^2}\right) \tag{9.8b}$$

The *settling time* is determined by the exponential decay envelope of Equation 9.3. The 2% settling time is given by

$$\exp\left(-\zeta\omega_n T_s\right) = 0.02\sqrt{1-\zeta^2} \tag{9.9}$$

For small damping ratios, T_s is approximately equal to $4/(\zeta\omega_n)$. This should be clear from the fact that $\exp(-4) \approx 0.02$. Note further that the *poles* (*eigenvalues*) of the system, as given by the roots of the *characteristic equation* $s^2 + 2\zeta\omega_n s + \omega_n^2 = 0$ are $p_1, p_2 = -\zeta\omega_n \pm j\omega_d$. It follows that the *time constant* of the system (inverse of the real part of the dominant pole) is

$$\tau = \frac{1}{\zeta\omega_n} \tag{9.10}$$

Hence, an approximate expression for the 2% settling time is

$$T_s = 4\tau \tag{9.11}$$

The rise time is obtained by substituting $y = 1$ in Equation 9.11 and solving for t. This gives $\sin(\omega_d T_r + \phi) = 0$, or

$$T_r = \frac{\pi - \phi}{\omega_d} \tag{9.12}$$

In which the *phase angle* ϕ is directly related to the damping ratio, through Equation 9.5.

The expressions for the performance specifications, as obtained using the simple oscillator model, are summarized in Table 9.2. For a higher order system, when applicable, the

TABLE 9.2

Analytical Expressions for Time-Domain Performance
Specifications (Simple Oscillator Model)

Performance Specification Parameter	Analytical Expression (Exact for a Simple Oscillator)
Peak time T_p	π/ω_d
Rise time T_r	$(\pi - \phi)/\omega_d$
Time constant τ	$1/(\zeta\omega_n)$
2% Settling time T_s	$-\left(\ln 0.02\sqrt{1-\zeta^2}\right)\tau \approx 4\tau$
Peak magnitude M_p	$1+\exp\left(-\dfrac{\pi\zeta}{\sqrt{1-\zeta^2}}\right)$
Percentage overshoot PO	$100\exp\left(-\dfrac{\pi\zeta}{\sqrt{1-\zeta^2}}\right)$

damping ratio and the natural frequency that are needed to evaluate these expressions may be obtained from the dominant complex pole pair of the system.

In the conventional time-domain design, relative stability specification is usually provided by a limit on the percentage overshoot (*PO*). This can be related to damping ratio (ζ) using a design curve.

Example 9.1

A control system is required to have a percentage overshoot of less than 10% and a settling time of less than 0.8 s. Indicate this design specification as a region on the *s*-plane.

Solution

A 10% overshoot means (see Table 9.2): $0.1 = \exp\left(-\dfrac{\pi\zeta}{\sqrt{1-\zeta^2}}\right)$

Hence, $\zeta = 0.60 = \cos\phi$ or: $\phi = 53°$

Next, T_s of 0.8 s means (see Table 9.2): $T_s = 4\tau = 4/\zeta\omega_n = 0.8$ or: $\zeta\omega_n = 5.0$

For the given specifications, we require $\phi \leq 53°$ and $\zeta\omega_n \geq 5.0$. The corresponding region on the *s*-plane is given by the shaded area in Figure 9.7.

Note: In this example if we had specified a T_p spec as well, then it would correspond to a ω_d spec. This will result in a horizontal line boundary for the design region in Figure 9.7.

9.4 Control Schemes

In a *regulator*-type control system, the objective is to maintain the output at a desired (constant) value. In a *servomechanism*-type control system, the objective is for the output to follow a desired trajectory (i.e., a specified time response or a path with respect to time).

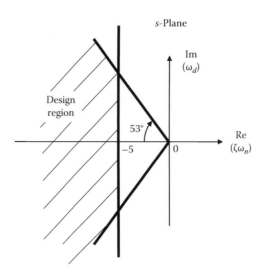

FIGURE 9.7
Design specification on the s-plane.

In a control system, in order to meet a specified performance, a suitable control method has to be employed.

In a *feedback control system,* as shown in Figure 9.8, the control loop has to be closed by making measurements of the system response and employing that information to generate control signals so as to correct any output errors. In Figure 9.8, since the feedback signal is not modified (i.e., gain = 1) before subtracting from the reference input, it represents the *"unity feedback."* A *control law* is a relationship between the controller output and the plant input. Common control modes are:

1. On–off (bang–bang) control
2. Proportional (*P*) control
3. Proportional control combined with reset (integral—*I*) action and/or rate (derivative—*D*) action (i.e., multimode or multi-term control).

Control laws for commonly used control actions are given in Table 9.3. Some advantages and disadvantages of each control action are also indicated as well in this table. Compare this information with what is given in Table 9.1.

The proportional action provides the necessary speed of response and adequate signal level to drive a plant. Besides, increased proportional action has the tendency to reduce the steady-state error. A shortcoming of increased proportional action is the degradation of

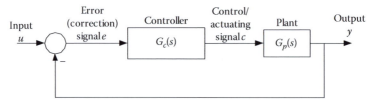

FIGURE 9.8
A feedback control system with unity feedback.

TABLE 9.3

Comparison of Some Common Control Actions

Control Action	Control Law	Advantages	Disadvantages
On–off	$\frac{c_{max}}{2}\left[\text{sgn}(e)+1\right]$	Simple Inexpensive	Continuous chatter Mechanical problems Poor accuracy
Proportional	$k_p e$	Simple Fast response	Offset error (steady-state error) Poor stability
Reset (integral)	$\frac{1}{\tau_i}\int e\,dt$	Eliminates offset Filters out noise	Low bandwidth (Slow response) Reset windup Instability problems
Rate (derivative)	$\tau_d\dfrac{de}{dt}$	High bandwidth (Fast response) Improves stability	Insensitive to dc error Allows high-frequency noise Amplifies noise Difficult analog implementation

stability. Derivative action (or rate action) provides stability that is necessary for satisfactory performance of a control system. In the time domain, this is explained by the fact that the derivative action tends to oppose sudden changes (large rates) in the system response. Derivative control has its shortcomings, however. For example, if the error signal that drives the controller is constant, the derivative action will be zero and it has no effect on the system response. In particular, derivative control cannot reduce steady-state error in a system. Also, derivative control increases the system bandwidth, which has the desirable effect of increasing the speed of response (and tracking capability) of the control system. Derivative action has the drawback of allowing and amplifying high-frequency disturbance inputs and noise components. Hence, derivative action is not practically implemented in its pure analytic form, but rather as a lead circuit, as will be discussed in a later chapter.

The presence of an *offset* (i.e., *steady-state error*) in the output may be inevitable when proportional control alone is used for a system having finite dc gain. When there is an offset, one way to make the actual steady state value equal to the desired value would be to change the set point (i.e., input value) in proportion to the desired change. This is known as *manual reset*. Another way to zero out the steady-state error would be to make the dc gain infinity. This can be achieved by introducing an integral term (with transfer function $1/s$) in the forward path of the control system (because $1/s \rightarrow \infty$ when $s=0$; i.e., at zero frequency because $s=j\omega$ in the frequency domain). This is known as *integral control* or *reset control* or *automatic reset*.

9.4.1 Feedback Control with PID Action

Many control systems employ *three-mode controllers* or *three-term controllers*, which are PID controllers providing the combined action of proportional, integral and derivative modes. The control law for proportional plus integral plus derivative (PID) control is given by

$$c = k_p\left(e + \tau_d\dot{e} + \frac{1}{\tau_i}\int e\,dt\right) \tag{9.13a}$$

or in the transfer function form

$$\frac{c}{e} = k_p\left(1 + \tau_d s + \frac{1}{\tau_i s}\right) \qquad (9.13b)$$

in which

e is the error signal (controller input)

c is the control/actuating signal (controller output or plant input)

k_p is the proportional gain

τ_d is the derivative time constant

τ_i is the integral time constant

The parameters k_p, τ_d, and τ_i are the design parameters of the PID controller and are used in controller tuning as well.

Example 9.2

Consider an actuator with transfer function

$$G_p = \frac{1}{s(0.5s + 1)} \qquad (i)$$

Design:

(a) A position feedback controller

(b) A tacho-feedback controller (i.e., position plus velocity feedback controller) that will meet the design specifications $T_p = 0.1$ and $PO = 25\%$.

Solution

(a) Position feedback

The block diagram for the position feedback control system is given in Figure 9.9a
From the standard result for a closed loop (see Chapter 3)

$$\text{Closed-loop TF} = \frac{kG_p(s)}{1 + kG_p(s)} = \frac{k}{s(0.5s+1)+k} = \frac{2k}{s^2 + 2s + 2k} \qquad (ii)$$

where k is the gain of the proportional controller in the forward path.

Note: In this case, only one parameter (k) is available for specifying two performance requirements. Hence, it is unlikely that both specifications can be met.

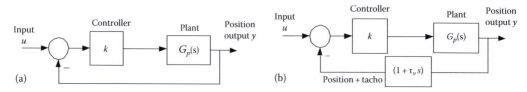

FIGURE 9.9
(a) Position feedback control system; (b) tacho-feedback control system.

To check this further note from the denominator (characteristic polynomial) of (ii) that $\zeta\omega_n = 1$ and $\omega_n^2 = 2k$. Hence,

$$\zeta = \frac{1}{\sqrt{2k}} \tag{iii}$$

And

$$\omega_d = \sqrt{\omega_n^2 - (\zeta\omega_n)^2} = \sqrt{2k - 1} \tag{iv}$$

For a given T_p we can compute ω_d using the expression in Table 9.2; k using (iv); ζ using (iii); and finally PO using Table 9.2.

Alternatively, for a given PO, we can determine ζ using Table 9.2; k using (iii); ω_d using (iv); and finally T_p using Table 9.2. These two sets of results are given in Table 9.4.

Note: For $T_p = 0.1$ we have $PO = 90.5\%$; For $PO = 25\%$ we have $T_p = 1.39$.

Hence, both requirements cannot be met with the single design parameter k, as expected.

(b) Tacho-feedback

The block diagram for the tacho-feedback system (a) is shown in Figure 9.9b.

Tachometer is a velocity sensor. Customarily, tacho-feedback uses feedback of both position and velocity. Hence, the feedback transfer function is $H = 1 + \tau_v s$, and from the standard result for a closed-loop system (see Chapter 3):

$$\text{Closed-loop TF} = \frac{kG_p(s)}{1 + kG_p(s) \times (\tau_v s + 1)} = \frac{k}{s(0.5s + 1) + k(\tau_v s + 1)}$$

$$= \frac{2k}{s^2 + 2(1 + k\tau_v)s + 2k}$$

where k is the proportional gain and τ_v is the velocity feedback parameter (*tacho gain*). By comparing with the simple oscillator TF, we note

$$\omega_n^2 = 2k \tag{v}$$

$$\zeta\omega_n = 1 + k\tau_v \tag{vi}$$

Since two parameters (k and τ_v) are available to meet the two specifications, it is likely that the design goal can be achieved. The computation steps are given below:

As before, for $T_p = 0.1$ we have $\omega_d = 10\pi$
Also, for $P.O. = 25\%$ we have $\zeta = 0.404$
Hence, $\omega_n = \omega_d/\sqrt{1 - \zeta^2} = 10.93\pi$
And $\zeta\omega_n = 0.404 \times 10.93\pi = 13.873$
Then we use (v) to compute k.
Substitute in (vi) to compute τ_v.
We get $k = 590$ and $\tau_v = 0.022$.

TABLE 9.4

Results for Position Control

T_p	ω_d	K	ζ	PO (%)
0.1	10π	494.0	0.032	90.5
1.39	2.264	3.063	0.404	25

9.4.1.1 System Type and Error Constants

There is an *offset* (i.e., *steady-state error*) when proportional control is used for a system having finite dc gain. The final value theorem (FVT) is useful in determining the steady-state error.

Final value theorem: The steady-state value of a signal $x(t)$ is given by

$$x_{ss} = \lim_{s \to 0} sx(s) \tag{9.14}$$

In which $x(s)$ is the Laplace transform of $x(t)$.

Characteristics of a system can be determined by applying a known input (*test input*) and studying the resulting response or the error of the response. For a given input, system error will depend on the nature of the system (including its controller). It follows that, error, particularly the *steady-state error*, to a standard test input, may be used as a parameter for characterizing a control system. This is the basis of the definition of error constants. Before studying that topic, we should explain the term "system type."

Consider a feedback control system with unity feedback (*Note*: A general feedback system can be normalized to a unity feedback system). The *forward transfer function* is $G(s)$.

System type: Assuming that the *feedback transfer function* has unity dc gain, the system type is defined as the number of free integrators present in the forward transfer function $G(s)$. For example, if there are no free integrators, it is a *type-zero system*. If there is only one free integrator, it is a *type-1 system*, and so on.

Position error constant K_p: This is given by

$$K_p = \lim_{s \to 0} G(s) = G(0) \tag{9.15}$$

The steady-state error is

$$e_{ss} = \frac{1}{1 + G(0)} = \frac{1}{1 + K_p} \tag{9.16}$$

Note: $G(0)$ will be finite and K_p will exist only if the system is type zero. This is termed position error constant because, for a position control system, a step input can be interpreted as a constant position input.

Velocity error constant K_v: This is given by

$$K_v = \lim_{s \to 0} sG(s) \tag{9.17}$$

The steady-state error for a unit ramp is a nonzero constant only for a type-1 system, and is given by

$$e_{ss} = \frac{1}{K_v} \tag{9.18}$$

The constant K_v is termed velocity error constant because for a position control system, a ramp position input (with Laplace transform $1/s^2$) is a constant velocity input.

TABLE 9.5

Dependence on the Steady-State Error on the System Type

Input	$u(t)\ t \geq 0$	$u(s)$	Steady-State Error		
			Type 0 System	Type 1 System	Type 2 System
Unit step	1	$\dfrac{1}{s}$	$\dfrac{1}{1+K_p}$	0	0
Unit ramp	t	$\dfrac{1}{s^2}$	∞	$\dfrac{1}{K_v}$	0
Unit parabola	$\dfrac{t^2}{2}$	$\dfrac{1}{s^3}$	∞	∞	$\dfrac{1}{K_a}$

Acceleration error constant K_a: This is given by

$$K_a = \lim_{s \to 0} s^2 G(s) \tag{9.19}$$

For a type-0 system or type-1 system, this steady-state error goes to infinity. For a type-2 system the steady-state error to a unit parabolic input (with Laplace transform $1/s^3$) is finite, however, and is given by

$$e_{ss} = \frac{1}{K_a} \tag{9.20}$$

The constant K_a is termed acceleration error constant because, for a position control system, a parabolic position input is a constant acceleration input.

How the steady-state error depends on the system type and the input is summarized in Table 9.5.

Note: For control loops with one or more free integrators (i.e., system type-1 or higher) the steady-state error to a step input would be zero. This explains why integral control is used to eliminate the offset error in systems under steady inputs, as noted previously.

9.4.2 Performance Specification Using s-Plane

The s-plane is given by a horizontal axis corresponding to the real part of s and a vertical axis corresponding to the imaginary part of s. The poles of a damped oscillator are given by the real part $-\zeta \omega_n$ and the imaginary part ω_d of the two roots. Now recall the expressions for the performance specifications as given in Table 9.2. The following facts are clear:

- A "constant settling time line" is the same as a "constant time constant line" (i.e., a vertical line on the s-plane).
- A "constant peak time line" is the same as a "constant ω_d line" (i.e., a horizontal line on the s-plane).
- A "constant percentage overshoot line" is the same as a "constant damping ratio line" (i.e., a radial line, cosine of whose angle with reference to the negative real axis is equal to the damping ratio ζ—see Equation 9.5).

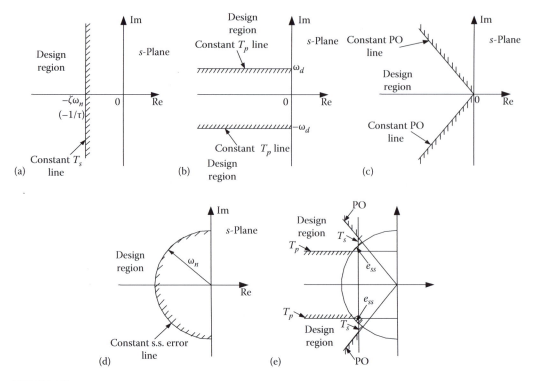

FIGURE 9.10
Performance specification on the s-plane: (a) Settling time; (b) peak time; (c) percentage overshoot; (d) steady-state error; (e) combined specification.

These lines are shown in Figure 9.10a through c. Since a satisfactory design is expressed by an inequality constraint on each of the design parameters, we have indicated the acceptable design region in each case.

Next, consider an appropriate measure on the s-plane for steady-state error. We recall that for a type-zero system, the steady-state error to a step input decreases as the loop gain increases. Furthermore, it is also known that the undamped natural frequency ω_n increases with the loop gain. It follows that for a system with variable gain: A "constant steady-state error line" is a "constant ω_n line" (i.e., a circle on the s-plane, with radius ω_n and centered at the origin of the coordinate system).

This line is shown in Figure 9.10d.

A composite design boundary and a design region (corresponding to a combined design specification) can be obtained by simply overlaying the four regions given in Figure 9.10a through d. This is shown in Figure 9.10e. *Note*: In Figure 9.10, we have disregarded the right half of the s-plane, at the outset, because it corresponds to an *unstable* system.

9.5 Stability

Stable response is a requirement for any control system. It ensures that the natural response to an initial condition excitation does not grow without bounds (or, more preferably, decays back to the initial condition) and the response to an input excitation (which itself

TABLE 9.6

Dependence of Natural Response on System Poles

Pole		Nature of Response	
Real	Negative	Transient (non-oscillatory)	Decaying (stable)
	Positive		Growing (unstable)
Imaginary (pair)		Oscillatory with constant amplitude	Steady (marginally stable)
Complex (pair)	Negative real part	Oscillatory with varying amplitude	Decaying (stable)
	Positive real part		Growing (unstable)
Zero value		Constant	Marginally stable
Repeated		Includes a linearly/polynomially increasing term (unstable if not accompanied by exponential decay)	

is bounded) does not lead to an unlimited response. Asymptotic stability and bounded-input–bounded-output (BIBO) stability are pertinent in this context. In designing a control system, the required level of stability can be specified in several ways, both in the time domain and the frequency domain. Some ways of performance specification with regard to stability in the time domain were introduced previously. The present section revisits the subject of stability, in time and frequency domains. Routh–Hurwitz method, root locus method, and Nyquist and Bode diagram methods incorporating gain and phase margins are introduced for stability analysis of linear time-invariant (LTI) systems.

The characteristic equation of an nth-order system can be expressed as

$$a_n s^n + a_{n-1} s^{n-1} + \cdots + a_0 = 0 \tag{9.21}$$

This is also the denominator of the system transfer function when equated to zero. This has n roots, which are the poles (or eigenvalues) of the system. If a root of the characteristic equation (pole) has an imaginary part, then that pole produces an oscillatory (sinusoidal) natural response in the system. Also, if the real part is negative, it generates an exponential decay and if the real part is positive, it generates an exponential growth (or, unstable response). These observations are summarized in Table 9.6 and further illustrated in Figure 9.11.

Note: The case of repeated poles should be given special care. The constants of integration will have terms such as t and t^2 (polynomial) in this case. These are growing (unstable) terms unless they are accompanied by decaying exponential terms that will counteract the polynomial growth (because an exponential decay is stronger than a polynomial growth of any order)

9.5.1 Routh–Hurwitz Criterion

The Routh test or Routh–Hurwitz stability criterion is a simple way to determine whether a system is stable (i.e., whether none of the poles have positive real parts) by examining the characteristic polynomial, without actually solving the characteristic equation for its roots. If the system is unstable, the Routh test also tells us how many poles are on the right half plane (RHP); i.e., the number of unstable poles. First, a Routh array has to be formed in order to perform the test.

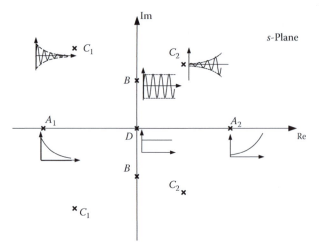

FIGURE 9.11
Pole location on the *s*-plane and the corresponding response.

9.5.1.1 Routh Array

It is possible to determine the stability of the system without actually finding these n roots, by forming a Routh array from the characteristic equation (9.21), as follows:

	First Column	Second Column	Third Column	...	
s^n	a_n	a_{n-2}	a_{n-4}	...	←First row
s^{n-1}	a_{n-1}	a_{n-3}	a_{n-5}	...	←Second row
s^{n-2}	b_1	b_2	b_3	...	←Third row
s^{n-3}	c_1	c_2	c_3	...	←Fourth row
.	.	.	.	...	
.	.	.	.	...	
s^0	h_1				←Last row

The first two rows are completed first, using the coefficients $a_n, a_{n-1}, \ldots, a_1, a_0$ of the characteristic polynomial, as shown. Note the use of alternate coefficients in these two rows. Each subsequent row is computed from the elements of the two rows immediately above it by *cross-multiplying* the elements of those two rows. For example:

$$b_1 = \frac{a_{n-1}a_{n-2} - a_n a_{n-3}}{a_{n-1}}$$

$$b_2 = \frac{a_{n-1}a_{n-4} - a_n a_{n-5}}{a_{n-1}}, \text{ etc.}$$

$$c_1 = \frac{b_1 a_{n-3} - a_{n-1}b_2}{b_1}$$

$$c_2 = \frac{b_1 a_{n-5} - a_{n-1}b_3}{b_1}, \text{ etc.}$$

And so on, until the coefficients of the last row are computed.

The Routh–Hurwitz stability criterion states that for the system to be stable, the following three conditions must be satisfied:

1. All the coefficients $(a_0, a_1, \ldots, a_n)$ of the characteristic polynomial must be positive (i.e., same sign; because, all signs can be reversed by multiplying by –1).
2. All the elements in the first column of the Routh array must be positive (i.e., same sign).
3. If the system is unstable, the number of unstable poles is given by the number of successive sign changes in the elements of the first column of the Routh array.

Example 9.3

Consider a system having the (closed loop) transfer function

$$G(s) = \frac{2(s+5)}{3s^3 + s^2 + 4s + 2}$$

Its Routh array is formed by examining the characteristic equation:

$$3s^3 + s^2 + 4s + 2 = 0$$

The Routh array is

$$
\begin{array}{ccc}
s^3 & 3 & 4 \\
s^2 & 1 & 2 \\
s^1 & b_1 & 0 \\
s^0 & c_1 & 0
\end{array}
$$

where

$$b_1 = \frac{1 \times 4 - 3 \times 2}{1} = -2$$

$$c_1 = \frac{b_1 \times 2 - 1 \times 0}{b_1} = 2$$

The first column of the array has a negative value, indicating that the system is unstable. Furthermore, since there are "two" sign changes (positive to negative and then back to positive) in the first column, there are two unstable poles in this system.

9.5.1.2 Auxiliary Equation (Zero-Row Problem)

A Routh array may have a row consisting of zero elements only. This usually indicates a *marginally stable* system (i.e., a pair of purely imaginary poles). The roots of the polynomial equation formed by the row that immediately precedes the row with zero elements will give the values of these marginally stable poles.

Example 9.4

Consider a plant $G(s) = 1/s(s+1)$ and a feedback controller $H(s) = K(s+5)/(s+3)$. Its closed-loop characteristic polynomial $(1 + GH = 0)$ is

$$1 + \frac{K(s+5)}{s(s+1)(s+3)} = 0$$

$$\rightarrow s(s+1)(s+3) + K(s+5) = 0 \rightarrow s^3 + 4s^2 + (3+K)s + 5K = 0$$

The Routh array is

s^3	1	$3+K$
s^2	4	$5K$
s^1	$\dfrac{12-K}{4}$	0
s^0	$5K$	

Note: When $K = 12$, the 3rd row (corresponding to s^1) of the Routh array will have all zero elements.

The polynomial equation corresponding to the previous row (s^2) is $4s^2 + 5K = 0$
   With $K = 12$, we have the auxiliary equation, $4s^2 + 5 \times 12 = 0$ or $s^2 + 15 = 0$ whose roots are $s = \pm j\sqrt{15}$.
   Hence, when $K = 12$ we have a marginally stable closed-loop system, two of whose poles are $\pm j\sqrt{15}$. The third pole can be determined by comparing coefficients as shown below. This remaining root has to be real (because if it is a complex root, it must occur as a conjugate pair of roots). Call it p. Then, with $K = 12$, on combining with the factor corresponding to the auxiliary equation (the complex root par) the characteristic polynomial will be $(s-p)(s^2 + 15)$. This must correspond to the same characteristic equation as given in the problem. Hence:

$$s^3 + 4s^2 + 15s + 60 = (s-p)(s^2 + 15) = 0$$

By comparing coefficients: $60 = -15p$ or $p = -4$
   Hence, the real pole is at -4, which is stable. Since the other two poles are marginally stable, the overall system is also marginally stable.

9.5.1.3 Zero Coefficient Problem

If the first element in a particular row of a Routh array is zero, a division by zero will be needed when computing the next row. This will create an ambiguity as to the sign of the next element. This problem can be avoided by replacing the zero element by a small positive element ε, and then completing the array in the usual manner.

Example 9.5

Consider a system whose characteristic equation is $s^4 + 5s^3 + 5s^2 + 25 + 10 = 0$
Let us study the stability of the system.
Routh array is

$$
\begin{array}{c|ccc}
s^4 & 1 & 5 & 10 \\
s^3 & 5 & 25 & 0 \\
s^2 & \varepsilon & 10 & 0 \\
s^1 & \dfrac{25\varepsilon - 50}{\varepsilon} & 0 & \\
s^0 & 10 & &
\end{array}
$$

Note that the 1st element in the 3rd row (s^2) should be $(5 \times 5 - 25 \times 1)/5 = 0$. But we have represented it by ε, which is positive and will tend to zero Then, the first element of the 4th row (s^1) becomes $(25\varepsilon - 50)/\varepsilon$. Since ε is very small, 25ε is also very small. Hence, the numerator of this quantity is negative, but the denominator (ε) is positive. It follows that this element is negative (and large). This indicates two sign changes in the 1st column, and hence the system has two unstable poles.

9.5.1.4 Relative Stability

Consider a stable system. The pole that is closest to the imaginary axis is the *dominant pole*, because the natural response from remaining poles will decay to zero faster leaving behind the natural response of this (dominant) pole. It should be clear that the distance of the dominant pole from the imaginary axis is a measure of the "level of stability" or "degree of stability" or "stability margin" or "relative stability" of the system. In other words, if we shift the dominant pole closer to the imaginary axis, the system becomes less stable (i.e., the "relative stability" of the resulting system becomes lower).

The stability margin of a system can be determined by the Routh test. Specifically, consider a stable system. All its poles will be on the LHP. Now, if we shift all the poles to the right by a known amount, the resulting system will be less stable. The stability of the shifted system can be established using the Routh test. If we continue this process of pole shifting in small steps, and repeatedly apply the Routh test, until the resulting system just goes unstable, then the total distance by which the poles have been shifted to the right provides a measure of the stability margin (or, relative stability) of the original system.

Example 9.6

A system has the characteristic equation: $s^3 + 6s^2 + 11s + 36 = 0$

(a) Using Routh–Hurwitz criterion, determine the number of unstable poles in the system.
(b) Now move all the poles of the given system to the right of the s-plane by the real value 1 (i.e., add 1 to every pole). Now how many poles are on the right-half plane?

Note: You should answer this question "without" actually finding the poles (i.e., without solving the characteristic equation).

Solution

(a) Characteristic equation: $s^3 + 6s^2 + 11s + 36 = 0$
Routh array:

$$
\begin{array}{ccc}
s^3 & 1 & 11 \\
s^2 & 6 & 36 \\
s^1 & \dfrac{6 \times 11 - 1 \times 36}{6} & 0 \\
s^0 & 36 &
\end{array}
$$

Since the entries of the first column are all positive, there are no unstable poles in the original system.

(b) Denote the shifted poles by $\tilde{s}$
We have $\tilde{s} = s + 1$ or $s = \tilde{s} - 1$
   Substitute in the original characteristic equation. The characteristic equation of the system with shifted poles is

$$(\tilde{s} - 1)^3 + 6(\tilde{s} - 1)^2 + 11(\tilde{s} - 1) + 36 = 0$$

or

$$\tilde{s}^3 - 3\tilde{s}^2 + 3\tilde{s} - 1 + 6\tilde{s}^2 - 12\tilde{s} + 6 + 11\tilde{s} - 11 + 36 = 0$$

or

$$\tilde{s}^3 + 3\tilde{s}^2 + 2\tilde{s} + 30 = 0$$

Routh array:

$$
\begin{array}{ccc}
\tilde{s}^3 & 1 & 2 \\
\tilde{s}^2 & 3 & 30 \\
\tilde{s}^1 & \dfrac{3 \times 2 - 1 \times 30}{3} = -3 & 0 \\
\tilde{s}^0 & 30 &
\end{array}
$$

There are two sign changes in the first column. Hence, there are two unstable poles.

9.5.2 Root Locus Method

Root locus is the locus of (i.e., continuous path traced by) the closed-loop poles (i.e., roots of the characteristic equation) of a system, as one parameter of the system (typically the loop gain) is varied. Specifically, the root locus shows how the locations of the poles of a closed-loop system change due to a change in some parameter of the loop transfer function. Hence, it indicates the stability of the closed-loop system as a function of the varied parameter.

The root locus starts from the open-loop poles (strictly speaking, loop poles). Hence, as the first step, these loop poles must be marked on the complex s-plane. Consider the feedback control structure with a forward transfer function of $G(s)$ and a feedback transfer function of $H(s)$. The overall transfer function of this system (i.e., the *closed-loop transfer function*) is

$$\frac{Y(s)}{U(s)} = \frac{G(s)}{1+G(s)H(s)} \tag{9.22}$$

The stability of the closed loop system is completely determined by the poles (not zeros) of the closed-loop transfer function (9.22).

Note: Zeros are the roots of the numerator polynomial equation of a transfer function.

The closed-loop poles are obtained by solving the characteristic equation (the equation of the denominator polynomial):

$$1+G(s)H(s) = 0 \tag{9.23a}$$

It follows that the closed-loop poles are (and hence, the stability of the closed loop system is) completely determined by the *loop transfer function* $G(s)H(s)$. It is clear that the roots of (9.23a) depend on both poles and zeros of $G(s)H(s)$. Hence, stability of a closed loop system depends on the poles and zeros of the loop transfer function.

Equation 9.23a can be rewritten in several useful and equivalent forms. First, we have

$$GH = -1 \tag{9.23b}$$

Next, since GH can be expressed as a ratio of two *monic polynomials* (i.e., polynomials whose highest order term coefficient is equal to unity) $N(s)$ and $D(s)$, we can write:

$$K\frac{N(s)}{D(s)} = -1 \quad \text{or} \quad KN(s)+D(s) = 0 \tag{9.23c}$$

in which

$N(s)$ is the numerator polynomial of the loop transfer function
$D(s)$ is the denominator polynomial of the loop transfer function
K is the loop gain

Now since the polynomials can be factorized, we can write

$$K\frac{(s-z_1)(s-z_2)\ldots(s-z_m)}{(s-p_1)(s-p_2)\ldots(s-p_n)} = -1 \tag{9.23d}$$

or

$$(s-p_1)(s-p_2)\ldots(s-p_n)+K(s-z_1)(s-z_2)\ldots(s-z_m) = 0 \tag{9.23e}$$

in which

z_i is the zero of the loop transfer function
p_i is the pole of the loop transfer function
m is the order of the numerator polynomial = number of zeros of GH
n is the order of the denominator polynomial = number of poles of GH

For physically realizable systems, we have $m \le n$. Equation 9.23c is in the "ratio-of-polynomials form" and Equation 9.23d is in the "pole-zero form." Now we will list the main rules for sketching a root locus.

Note: It is just one equation, the characteristic equation (9.23) of the closed-loop system, which generates all these rules.

9.5.2.1 Root Locus Rules

Rule 0 (Symmetry): The root locus is symmetric about the real axis on the s-plane.

Rule 1 (Number of branches): Root locus has n branches. They start at the n poles of the loop transfer function GH. Out of these, m branches terminate at the zeros of GH and the remaining $(n-m)$ branches go to infinity, tangential to $n-m$ lines called *asymptotes*.

Rule 2 (Magnitude and phase conditions):
The *magnitude condition* is

$$K \frac{\prod_{i=1}^{m}|s - z_i|}{\prod_{i=1}^{n}|s - p_i|} = 1 \tag{9.24a}$$

The *phase angle condition* is

$$\sum_{i=1}^{n} \angle (s - p_i) - \sum_{i=1}^{m} \angle (s - z_i) = \pi + 2r\pi \tag{9.24b}$$

$$r = 0, \pm 1, \pm 2, \ldots$$

Rule 3 (Root locus on real axis): Pick any point on the real axis. If (# poles − # zeros) of GH to the right of the point is odd, the point lies on the root locus.

Rule 4 (Asymptote angles): The $n - m$ asymptotes form angles

$$\theta_r = \frac{\pi + 2\pi r}{n - m}$$

$$r = 0, \pm 1, \pm 2, \ldots \tag{9.25}$$

with respect to the positive real axis of the s-plane.

Rule 5 (Break points): *Break-in points* and *breakaway points* of the root locus are where two or more branches intersect. They correspond to the points of repeated (multiple) poles of the closed-loop system. These points are determined by differentiating the characteristic equation (9.23c) with respect to s, substituting for K using (9.23c) again, as $K = -(D(s)/N(s))$. This gives

$$N(s)\frac{dD}{ds} - D(s)\frac{dN}{ds} = 0 \tag{9.26}$$

Note: The root-locus branches at a break point are equally spaced (in angle) around the break point.

Rule 6 (Intersection with imaginary axis): If the root locus intersects the imaginary axis, the points of intersection are given by setting $s = j\omega$ and solving the characteristic equation:

$$D(j\omega) + KN(j\omega) = 0 \tag{9.27}$$

This gives two equations (one for the real terms and the other for the imaginary terms).

Alternatively, the Routh–Hurwitz criterion and the auxiliary equation for marginal stability may be used to determine these points and the corresponding gain value (K).

Rule 7 (Angles of approach and departure): The *departure angle* α of root locus, from a *GH* pole, is obtained using

$$\alpha + \angle \text{ at other poles} - \angle \text{ at zeros} = \pi + 2r\pi \tag{9.28a}$$

The *approach angle* α to a *GH* zero is obtained using

$$\angle \text{ at poles} - \alpha - \angle \text{ at other zeros} = \pi + 2r\pi \tag{9.28b}$$

Note: Angles mentioned in Rule 7 are measured by drawing a line from the approach/ departure point to the other pole or zero of *GH* that is considered and determining the angle of that line measured from the positive real axis (i.e., horizontal line drawn to the right at the other pole or zero).

Rule 8 (Intersection of asymptotes with real axis): Asymptotes meet the real axis at the *centroid* about the imaginary axis, of the poles and zeros of *GH*. Each pole is considered to have a weight of +1 and each zero a weight of –1.

9.5.2.2 Steps of Sketching Root Locus

Now we list the basic steps of the normal procedure that is followed in sketching a root locus.

Step 1: Identify the loop transfer function and the parameter (gain K) to be varied in the root locus.

Step 2: Mark the poles of *GH* with the symbol (**×**) and the zeros of *GH* with the symbol (**o**) on the *s*-plane.

Step 3: Using Rule 3 sketch the root locus segments on the real axis.

Step 4: Compute the asymptote angles using Rule 4 and the asymptote origin using Rule 8, and draw the asymptotes.

Step 5: Using Rule 5, determine the break points, if any.

Step 6: Using Rule 7 compute the departure angles and approach angles, if necessary.

Step 7: Using Rule 6, determine the points of intersection with the imaginary axis, if any.

Step 8: Complete the root locus by appropriately joining the points and segments that have been determined in the previous steps.

Example 9.7

Consider the feedback control system shown in Figure 9.12. The following three types of control may be used:

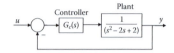

FIGURE 9.12
A feedback control system.

(a) Proportional (P) Control: $G_c = K$
(b) Proportional + Derivative (PD) Control: $G_c = K(1 + s)$
(c) Proportional + Integral (PI) Control: $G_c = K(1 + (1/S))$

Sketch the root loci for these three cases and compare the behavior of the corresponding controlled systems.

Solution

(a) The loop transfer function $GH = K/(s^2 - 2s + 2)$
     The loop poles are at $s = 1 \pm j$. The two RL branches will start from them.
     These are marked on the s-plane in Figure 9.13a.
     There are no loop zeros.
     Hence, according to Rule 3, there are no segments of the RL on the real axis.
     Since there are no GH zeros, there are two asymptotes where the RL branches will end (at infinity).
     The asymptote angles are $\pm 90°$ (Rule 4)
     The pole centroid $\bar{s} = 1 \times 2/2 = 1$
     The asymptotes intersect at this centroid (Rule 8), and can be sketched as in Figure 9.13a.

     Note: Even though obvious, the departure angle α at the pole $1 + j$ can be determined by:
     $\alpha + 90° = 180° \Rightarrow \alpha = 90°$ (Rule 7)
     The complete root locus for this case (P control) is sketched in Figure 9.13a.
     It is seen that the system is always unstable.

(b) The loop transfer function $GH = K(1 + s)/(s^2 - 2s + 2)$
     There are two loop poles, at $s = 1 \pm j$ from which the two branches of RL originate.
     There is a loop zero at -1 where one of the RL branches terminates.
     These are marked on the s-plane in Figure 9.13b.
     According to Rule 3, the RL lies on the real axis between $-\infty$ and -1, as sketched in Figure 9.13b.
     Since there are two loop poles and one loop zero, the RL has one asymptote with asymptote angle $180°$ (Rule 1 and Rule 4).
     The departure angle α from pole $1 + j$ is determined by (Rule 7)
     $\alpha + 90° - \theta = 180°$ (where $\tan \theta = \frac{1}{2}$ or $\theta = 26.6°$) $\Rightarrow \alpha = 116.6°$

     Note: The departure angle from pole $1 - j$ may be determined simply by the symmetry of the RL (Rule 0) or by (Rule 7) as
     $\alpha + (-90°) - (-\theta) = 180°$ (where $\theta = 26.6°$ as before) $\Rightarrow \alpha = 243.4°$ or $-116.6°$
     Break points (Rule 5):
     Here, $N(s) = (1 + s)$ and $D(s) = s^2 - 2s + 2$
     Hence, the break point is given by (Equation 9.26)

$$(1 + s)(2s - 2) - (s^2 - 2s + 2) = 0 \Rightarrow s^2 + 2s - 4 = 0 \Rightarrow s = -1 \pm \sqrt{5}$$

The correct break point must be on the root locus (i.e., <-1). Hence, we pick

$$s = -1 - \sqrt{5}$$

The complete root locus for the present case of PD control is sketched in Figure 9.13b.

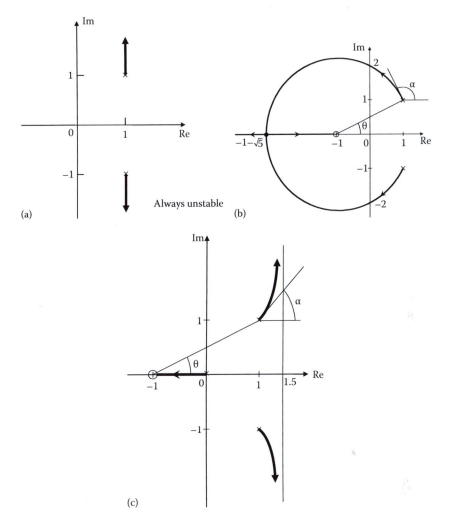

FIGURE 9.13
(a) Root locus for the system with P control; (b) root locus for the system with PD control; (c) root locus for the system with PI control.

It is seen that the system is unstable for low values of gain K, starting from 0, but becomes stable beyond a certain gain value. This is due to the inclusion of derivative control (or, a loop zero on the left half plane—LHP) that has a stabilizing effect. The gain value and the frequency of marginal stability can be determined as usual. In the present case, this is a relatively simple exercise. The closed-loop characteristic equation (9.23c) is

$$\frac{K(1+s)}{(s^2 - 2s + 2)} = -1 \Rightarrow (s^2 - 2s + 2) + K(1+s) = 0 \Rightarrow s^2 + (K-2)s + K + 2 = 0$$

Hence, for stability, we must have $K > 2$. The gain for marginal stability is $K = 2$. The corresponding characteristic equation is $s^2 + K + 2 = 0$, and the resulting marginally stable closed-loop poles are $s = \pm j2$.

(c) The loop transfer function $GH = K(1+s)/s(s^2 - 2s + 2)$
Now there are three loop poles, at $s=0$ and $1 \pm j$ from which the three branches of RL originate.

There is a loop zero at -1 where one of the RL branches terminates.

These are marked on the s-plane in Figure 9.13c.

According to Rule 3, the RL lies on the real axis from 0 to -1, as sketched in Figure 9.13b. Since there are three loop poles and one loop zero, the RL has two asymptotes with asymptote angles $= \pm 90°$ (Rule 1 and Rule 4).

The pole centroid $\bar{s} = \dfrac{1 \times 2 - 1 \times (-1)}{3 - 1} = 1.5$

The asymptotes intersect at this point (Rule 8), as sketched in Figure 9.13c.
The departure angle α from the pole $1 + j$ is determined by
$\alpha + 90° + 45° - \theta = 180°$ where $\tan \theta = \frac{1}{2}$ or $\theta = 26.6° \Rightarrow \alpha = 71.6°$

Note: As usual, the departure angle from the other (conjugate) pole $1 - j$ is determined by symmetry as $\alpha = -71.6°$.

The complete root locus for the case with PI control is sketched in Figure 9.13c.

It is seen that the system is always unstable. In fact, the system is more unstable than with P control alone, and becomes worse as the gain K is increased. This is due to the presence of the integral (I) action in the controller, which has a destabilizing effect.

9.5.3 Stability in the Frequency Domain

The concepts of transfer function and frequency domain models have been discussed in Chapter 3. Now we will specifically use the concept of *frequency transfer function* (or, *frequency response function*), where the independent variable is frequency ω radians/s or f cycles/s (or hertz), to outline some useful techniques in the stability analysis of control systems. In particular, the following may be used as measures of relative stability, in the frequency domain:

1. Peak magnitude (and associated Q-factor and half-power bandwidth)
2. Phase margin
3. Gain margin

9.5.3.1 Marginal Stability

If a dynamic system oscillates steadily in the absence of a steady external excitation, this condition represents a state of *marginal stability*. The "distance" to a state of marginal stability is a measure of the level of stability, and is called a *stability margin*.

9.5.3.2 The 1, 0 Condition

Consider a feedback control system represented by the block diagram in Figure 9.14. We have assumed unity feedback, but this can be generalized. In fact, without loss of generality we can interpret G in Figure 9.14 as the loop transfer function GH, because a system with a general feedback transfer function H can be reduced to a unity feedback system, through block diagram reduction, by placing GH as the forward transfer function.

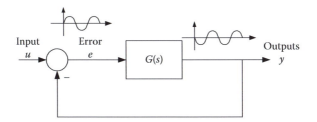

FIGURE 9.14
A feedback control system with unity feedback.

Suppose that the open-loop transfer function $G(s)$ is such that at a specific frequency of operation ω, we have

1. Magnitude $|G(j\omega)| = 1$
2. Phase angle $\angle G(j\omega) = -\pi$

Then, if an error signal of frequency ω is injected into the loop (due to noise, disturbance, initial excitation, etc.), its amplitude will not change while passing through $G(s)$, but the phase angle will reduce by π. Hence the output signal y will have the same amplitude as the error signal e, but y will "lag" e by π. Since y is fed back into the loop with a negative feedback (*Note*: −1 corresponds to a further phase lag of π) the overall phase lag in the feedback signal, when reaching the forward path of the loop, will be 2π. Since a phase change of 2π is the same as no phase change, the feedback signal will have the same amplitude as the forward signal (i.e., gain = 1) and the same phase angle as the forward signal (i.e., phase = 0). This is called the (1, 0) *condition*. Under this condition, it is clear that even in the absence of an external input u, a harmonic signal of a specific frequency ω can sustain in the loop without growing or decaying. This is a state of self-sustained steady oscillation. If such a condition of steady oscillation is possible, in the absence of a steady external input, the system is said to be marginally stable.

Note: The specific frequency ω at which this condition of steady oscillation would be feasible is itself a property of the system, and depends on system parameters.

Next, consider a system with non-unity feedback with a feedback transfer function of $H(s)$. It should be clear that in applying the (1, 0) condition for marginal stability, what matters is the overall gain and phase shift in the entire loop. Hence, in Figure 9.14 we need to consider the overall loop transfer function $G(s)H(s)$ and not the individual components. The (1, 0) condition for *marginal stability* is, at a specific frequency of operation ω:

1. Magnitude $|G(j\omega)H(j\omega)| = 1$
2. Phase angle $\angle G(j\omega)H(j\omega) = -\pi$. $\hspace{2cm}$ (9.29a)

Note: The characteristic equation of the closed-loop system $\tilde{G}$ is $GH + 1 = 0$. Hence, the stability of a closed-loop system is completely determined by the loop transfer function GH, as already concluded under the root locus method. Specifically, now we need to study the magnitude and the phase of GH $(j\omega)$, in the frequency domain. In the special case of unity feedback ($H = 1$) we need to study $G(j\omega)$. For convenience, in these studies we denote GH simply by G, keeping in mind that then G represents the loop transfer function GH.

Bode diagram (Bode plot) and Nyquist diagram (Nyquist plot) are convenient ways of graphically representing transfer functions (see Chapter 3). These plots are valuable in the stability of dynamic systems, in the frequency domain.

9.5.3.3 Phase Margin and Gain Margin

The characteristic equation of the closed-loop system is $G(s)H(s) = -1$. The system is marginally stable if one pair of roots of the characteristic equation is purely imaginary (i.e., $\pm j\omega$) while the remaining roots are not unstable. Hence, the condition for marginal stability is that there exists a frequency ω such that

$$G(j\omega)H(j\omega) = -1 \tag{9.29b}$$

In fact the two conditions given by Equation 9.29a are exactly equivalent to this single complex Equation 9.29b, because -1 has a magnitude of 1 and a phase angle of $-\pi$. It follows that if the plot of the loop transfer function GH in the complex plane (i.e., the polar plot of Imaginary $GH(j\omega)$ vs. Real $GH(j\omega)$—the Nyquist plot—see Chapter 3), as ω changes, passes through the point $(-1, 0)$, then the control system is *marginally stable*.

Gain margin: Suppose that at a particular operating frequency ω, the phase $\phi = -180°$ but the gain (magnitude) M is less than unity (i.e., $M < 1$). Then, if the external input u in Figure 9.14 is disconnected, the amplitude of the feedback signal will steadily decay. This, of course, corresponds to a stable system. The smaller the value of M, the more stable the system. Hence, a stability margin known as the *gain margin* g_m can be defined as

$$g_m = \frac{1}{|G(j\omega)H(j\omega)|} \tag{9.30a}$$

at the frequency ω where $\angle G(j\omega)H(j\omega) = -180°$

Note: If the magnitude (i.e., gain) of the transfer function $GH(j\omega)$ is increased by a factor of g_m at this frequency, then the marginal stability conditions (Equation 9.29) will be satisfied. Hence, g_m is the margin by which the gain of a stable system may be increased so that the system becomes just unstable. It follows that the larger the g_m, the better the degree of stability. It is convenient to express g_m in decibels (dB) because the transfer function magnitude (in the frequency domain) is usually expressed in dB (particularly in Bode diagrams—see Chapter 3). Then,

$$g_m = -20 \log_{10} |G(j\omega)H(j\omega)| \tag{9.30b}$$

at the frequency ω where $\angle G(j\omega)H(j\omega) = -180°$

Phase margin: Suppose that there exists some frequency ω at which the magnitude (i.e., gain) of the loop transfer function $GH(j\omega)$ is $M = 1$ (i.e., 0 db) but the phase ϕ lies between 0 and $-180°$. This frequency ω_c is called the crossing frequency or crossover frequency, because it corresponds to the point where the gain (magnitude) curve crosses the unity (0 dB) line. Since the phase angle decreases with frequency, there will be a higher frequency at which the phase is $-180°$ but the gain (magnitude) will be less than unity (because the loop transfer function magnitude usually decreases with increasing frequency, at high frequencies). This, as noted under the topic of gain margin, corresponds to a stable system. The amount by which the phase of the loop transfer function at gain $= 0$ dB, may be decreased (i.e., the phase lag increased) until it reaches the $-180°$ value, is termed phase margin (ϕ_m). Specifically, the phase margin is defined as

$$\phi_m = 180° + \angle G(j\omega)H(j\omega) \tag{9.31}$$

at the frequency ω where $|G(j\omega)H(j\omega)| = 1$.

The larger the phase margin, the more stable the system.

In summary, gain margin tells us the amount (margin) by which the gain may be increased at a phase of $-180°$, before the system becomes marginally stable; and phase margin tells us the amount (margin) by which the phase may be "decreased" (i.e., "phase lag" increased) at unity gain, before the system becomes marginally unstable. A more rigorous development of the concepts of gain margin and phase margin requires a knowledge of the *Nyquist stability criterion*, as presented in a separate section.

9.5.3.4 Bode and Nyquist Plots

As discussed in Chapter 3, Bode diagram (Bode plot) and Nyquist diagram (Nyquist plot) are convenient graphical representations of transfer functions, in the frequency domain. Specifically, the Bode plot of a transfer function $G(s)$ constitutes the following pair of curves:

Magnitude $|G(j\omega)|$ versus frequency ω

Phase angle $\angle G(j\omega)$ versus frequency ω

Note that the Bode plot requires two curves—one for gain and one for phase. These two curves can be represented as a single curve by using a so-called *polar plot* with a real axis and an imaginary axis. A polar plot is a way to represent both magnitude and phase of a rotating vector, with one curve. When phase $= 0°$, the vector (which represents the complex number) points to the right; when phase $= 90°$, the vector points up; and when phase $= -180°$, the vector points to the left, etc.

The solid curve in Figure 9.15a represents the path of the tip of the directed line (two-dimensional vector) representing the frequency transfer function (i.e., complex transfer function in the frequency domain) as the frequency varies. Any one point represents both the *amplitude* (distance to origin) and *phase* (angle measured from the positive real axis), at one given frequency. Thus, all the information in the two curves of the Bode plots (Figure 9.15b) is represented by this single curve called the *Nyquist plot* (or *polar plot* or *argand plot*).

The marginal stability condition is (a) gain $= 1$ or 0 dB, and (b) phase lag $= 180°$ at some specific operating frequency. A gain of 1 is represented by a vector of unit length. Its tip traces a unit circle with its center at the origin of the coordinate frame (see the dotted circle in Figure 9.15a). A phase of $180°$ corresponds to a horizontal vector pointing to the left from the origin (i.e., the negative real axis). The intersection of the Nyquist plot with the unity-gain circle gives the *phase* at the 0 db point; the length of the vector (distance from origin) of the point where the Nyquist plot intersects the negative side of the real axis gives the gain at the critical $180°$ phase (lag) point.

Now formal definitions for gain margin g_m and phase margin ϕ_m may be given using either the Nyquist diagram or the Bode diagram as in Figure 9.15. Consider a stable closed-loop system with the transfer function $\tilde{G} = G/(1+GH)$. First we plot the Nyquist diagram for the loop transfer function GH, for example, as shown in Figure 9.15a. The factor by which the Nyquist curve should be expanded (say, by increasing its gain) in order to make

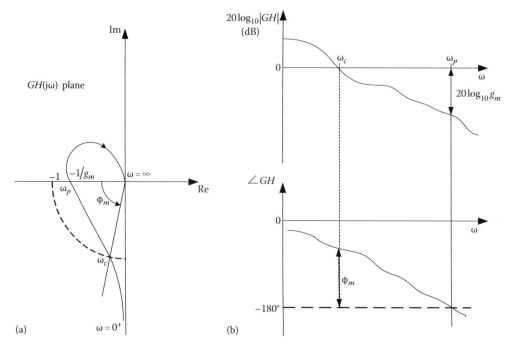

FIGURE 9.15
Definition of gain margin and phase margin: (a) using Nyquist diagram; (b) using Bode diagram.

the system marginally stable (i.e., to pass through the critical point (−1, 0)) measures the relative stability (or, *stability margin*) of the closed-loop system. The stability margins may be similarly defined using the Bode plot in Figure 9.15b.

The relative stability (stability margin) of a control system can be improved by adding a compensator so as to increase the GM. Since, in general, the gain margin of a system automatically improves when the phase margin of the system is improved, in design specifications it is adequate to consider only the phase margin.

Example 9.8

Figure 9.16a and b show Bode and Nyquist plots (of the loop transfer functions *GH*) for two systems, respectively. The one on the left is stable because the phase lag is less than 180° at the critical 0 dB (i.e., where gain = 1) point, and the gain is less than 1 at the critical phase lag point (i.e., where phase lag = 180°). Note that the amount by which the gain is less than 0 dB at the phase-crossover (−180°) point is the *gain margin*. Similarly, the amount by which the phase lag is less than 180° at the gain-crossover point (0 dB) is the *phase margin*.

9.6 Advanced Control

The emphasis of the present chapter has been on conventional control (also known as classical control) that is commonly used in mechatronic/industrial applications. It primarily deals with single-input–single-output (SISO) systems both in time domain and

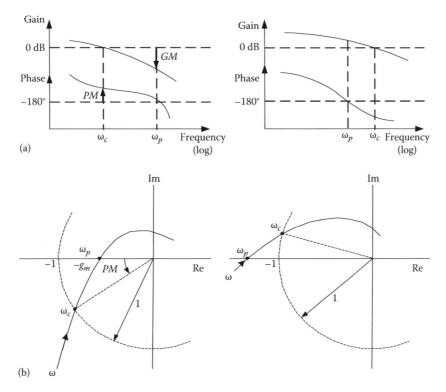

FIGURE 9.16
(a) Bode plots; (b) Nyquist plots, of a stable system (left) and an unstable system (right) (GM, gain margin; PM, phase margin).

frequency domain. What are commonly identified as modern control techniques are time-domain multivariable (multi-input–multi-output—MIMO) techniques that use the state-space representation for the system. The present section will outline some of these advanced control techniques, particularly in the categories of optimal control and modal control. Another control method that has been quite popular in mechatronic/industrial applications is fuzzy logic control, which will be presented in the subsequent section.

9.6.1 Linear Quadratic Regulator Control

In linear quadratic regulator (LQR), the objective is to minimize a cost function (maximize a performance index) and hence, this technique falls under the general category of *optimal control*. Consider a mechatronic system that is represented by the linear state-space model:

$$\dot{x} = Ax + Bu \tag{9.32}$$

Assume that all the states x are measurable and all the system modes are controllable. Then, we use the constant-gain feedback control law:

$$u = Kx \tag{9.33}$$

The choice of parameter values for the feedback gain matrix K is infinite. Therefore, we can use this freedom to minimize the cost function:

$$J = \frac{1}{2}\int_{t}^{\infty}\left[x^T Q x + u^T R u\right]d\tau \tag{9.34}$$

This is the time integral of a quadratic function in both state and input variables, and the optimization goal may be interpreted as bringing x down to zero (regulating x to 0), but without spending a rather high control effort; hence, the name linear quadratic regulation (LQR). Also, Q and R are weighting matrices, with the former being at least positive semi-definite and the latter positive definite. Typically Q and R chosen as diagonal matrices with positive diagonal elements whose magnitudes are decided based on the degree of relative emphasis that should be made on various elements of x and u. It is well known that K that minimizes the cost function (9.34) is given by

$$K = -R^{-1}B^T K_r \tag{9.35}$$

where K_r is the positive-definite solution of the matrix Riccati algebraic equation:

$$K_r A + A^T K_r - K_r B R^{-1} B^T K_r + Q = 0 \tag{9.36}$$

It is also known that the resulting closed-loop control system is stable. Furthermore, the minimum (optimal) value of the cost function (9.34) is given by

$$J_m = \frac{1}{2}x^T K_r x \tag{9.37}$$

where x is the present value of the state vector. Major computational burden of the LQR method is in the solution Equation 9.36. Other limitations of the technique arise due to the need for measuring all the state variables (which may be relaxed to some extent). Furthermore, even though stability of the controlled system is guaranteed, the level of stability that is achieved (i.e., stability margin or the level of modal damping) cannot be directly specified. Also, robustness of the control system, in the presence of model errors, unknown disturbances, and so on, may be questionable. Besides, the cost function incorporates an integral over an infinite time duration, which does not typically reflect the practical requirement of rapid vibration control.

9.6.2 Modal Control

The LQR control technique has the serious limitation of not being able to directly achieve specified levels of modal damping, which may be an important goal in control of a mechatronic system. The method of modal control accomplishes this objective through *pole placement*, where poles (eigenvalues) of the controlled system are placed at specified values so that the modes (i.e., fundamental free natural responses) of the system behave in a desired manner (with respect to stability, speed of response, etc.). Specifically, consider the plant (9.32) and the feedback control law (9.33). Then, the closed-loop system is given by

$$\dot{x} = (A + BK)x \tag{9.38}$$

It is well known that if the plant (A, B) is controllable, then a control gain matrix K can be chosen that will arbitrarily place the eigenvalues of the closed-loop system matrix $A + BK$. That means, under the given assumptions, the modal control technique can not only assign the modal damping but also the damped natural frequencies at specified values. The assumptions given above are quite stringent, but they can be relaxed to some degree. A shortcoming, however, of this method is the fact that it does not place a restriction on the control effort, as for example the LQR technique does, in achieving a specified level of modal control.

9.6.3 Nonlinear Feedback Control

Simple, linear servo control is known to be inadequate for transient and high-speed operation of complex plants. Past experience of servo control in process applications is extensive, however, and servo control is extensively used in many commercial applications (e.g., robots). For this type of control to be effective, however, nonlinearities and dynamic coupling must be compensated faster than the control bandwidth at the servo level. One way of accomplishing this is by implementing a linearizing and decoupling controller inside the servo loops. This technique is termed *feedback linearization technique* (FLT). One such technique that is useful in controlling nonlinear and coupled dynamic systems such as robots is outlined here.

Consider a mechanical dynamic system (plant) given by

$$M(q)\frac{d^2q}{dt^2} = n\left(q, \frac{dq}{dt}\right) + f(t) \tag{9.39}$$

in which

$f = [f_1 f_2 \ldots f_r]^T$ is the vector of input forces at various locations of the system

$q = [q_1 q_2 \ldots q_r]^T$ is the vector of response variables (e.g., positions) at the forcing locations of the system

$M(q)$ is the inertia matrix (nonlinear)

$n\left(q, \dfrac{dq}{dt}\right)$ is a vector of remaining nonlinear effects in the system (e.g., damping, backlash, gravitational effects)

Now suppose that we can model M by $\hat{M}$ and n by $\hat{n}$. Then, let us use the nonlinear (linearizing) feedback controller given by

$$f = \hat{M}K\left[e + T_i^{-1}\int e\,dt - T_d\frac{dq}{dt}\right] - \hat{n} \tag{9.40}$$

in which $e = q_d - q =$ error (correction) vector; $q_d =$ desired response; and K, T_i, and T_d are constant control parameter matrices. This control scheme is shown in Figure 9.17. By substituting the controller equation (9.40) into the plant equation (9.39) we get

$$M\frac{d^2q}{dt^2} = n - \hat{n} + \hat{M}K\left[e + T_i^{-1}\int e\,dt - T_d\frac{dq}{dt}\right] \tag{9.41}$$

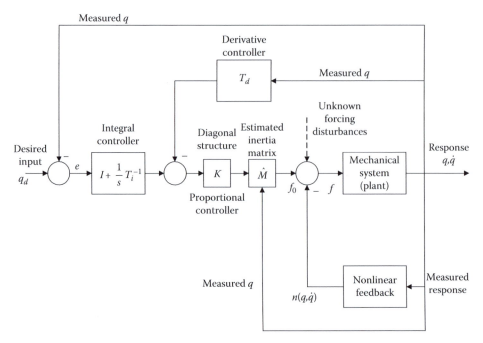

FIGURE 9.17
The structure of the model-based nonlinear feedback control system.

If our models are exact, we have $M = \hat{M}$ and $n = \hat{n}$. Then, because the inverse of matrix $\hat{M}$ exists in general (because the inertia matrix *is positive definite*), we get

$$\frac{d^2q}{dt^2} = K\left[e + T_i^{-1}\int e\,dt - T_d\frac{dq}{dt}\right]$$ (9.42)

Equation 9.42 represents a linear, constant parameter system with PID control. The proportional control parameters are given by the gain matrix K, the integral control parameters by T_i, and the derivative control parameters by T_d. It should be clear that we are free to select these parameters so as to achieve the desired response. In particular, if these three parameter matrices are chosen to be diagonal, then the control system, as given by Equation 9.42 and shown in Figure 9.17 will be uncoupled (i.e., one input affects only one output) and will not contain dynamic interactions. In summary, this controller has the advantages of linearizing and decoupling the system; its disadvantages are that accurate models will be needed and that the control algorithm is crisp and unable to handle qualitative or partially known information, learning, etc.

Instead of using analytical modeling, the parameters in $\hat{M}$ and $\hat{n}$ may be obtained through the measurement of various input–output pairs. This is called *model identification*, and can cause further complications in terms of instrumentation and data processing speed, particularly because some of the model parameters must be estimated in real time.

9.6.4 Adaptive Control

An adaptive control system is a feedback control system in which the values of some or all of the controller parameters are modified (adapted) during the system operation (in real time) on the basis of some performance measure, when the response (output) requirements are not satisfied. The techniques of adaptive control are numerous because many criteria can be employed for modifying the parameter values of a controller. According to the above definition, *self-tuning control* falls into the same category. In fact, the terms "adaptive control" and "self-tuning control" have been used interchangeably in the technical literature. Performance criteria used in self-tuning control may range from time-response or frequency-response specifications, parameters of "ideal" models, desired locations of poles and zeros, and cost functions. Generally, however, in self-tuning control of a system some form of parameter estimation or identification is performed online using input–output measurements from the system, and the controller parameters are modified using these estimated parameter values. A majority of the self-tuning controllers developed in the literature is based on the assumption that the plant (process) is linear and time invariant. This assumption does not generally hold true for complex industrial processes. For this reason, we shall restrict our discussion to an adaptive controller that has been developed for nonlinear and coupled plants.

Online estimation or system identification, which may be required for adaptive control, may be considered to be a preliminary step of "learning." In this context, *learning control* and adaptive control are related, but learning is much more complex and sophisticated than a quantitative estimation of parameter values. In a learning system, control decisions are made using the cumulative experience and knowledge gained over a period of time. Furthermore, the definition of learning implies that a learning controller will "remember" and improve its performance with time. This is an evolutionary process that is true for intelligent controllers, but not generally for adaptive controllers.

Here, we briefly describe a *model-referenced adaptive control* (MRAC) technique. The general approach of MRAC is illustrated by the block diagram in Figure 9.18. In nonadaptive feedback control the response measurements are fed back into the drive controller through a feedback controller, but the values of the controller parameters (feedback gains) themselves are unchanged during operation. In adaptive control, these parameter values are changed according to some criterion. In model-referenced adaptive control, in particular, the same reference input that is applied to the physical system is applied to a reference

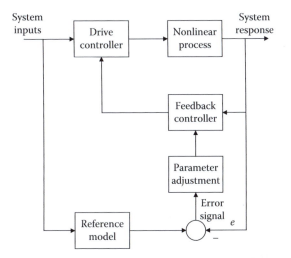

FIGURE 9.18
Model-referenced adaptive controller.

model as well. The difference between the response of the physical system and the output from the reference model is the error. The ideal objective is to make this error zero at all times. Then the system will perform just like the reference model. The error signal is used by the adaptation mechanism to determine the necessary modifications to the values of the controller parameters in order to reach this objective. Note that the reference model is an idealized model which generates a desired response when subjected to the reference input, at least in an asymptotic manner (i.e., the error converges to zero). In this sense, it is just a means of performance specification and may not possess any resemblance or analogy to an analytical model of the process itself. For example, the reference model may be chosen as a linear, uncoupled system with desired damping and bandwidth properties (i.e., damping ratios and natural frequencies).

A popular approach to derive the adaptive control algorithm (i.e., the equations expressing how the controller parameters should be changed in real time) is through the use of the MIT rule. In this method, the controller parameters are changed in the direction opposite to the maximum slope of the quadratic error function. Specifically, the quadratic function

$$V(p) = e^T We \tag{9.43}$$

is formed, where

E is the error signal vector shown in Figure 9.18
W is a diagonal and positive-definite weighting matrix
P is a vector of control parameters that will be changed (adapted) during control

The function V is minimized numerically with respect to p during the controller operation, subject to some simplifying assumptions. The details of the algorithm are found in the literature.

The adaptive control algorithm described here has the advantage that it does not necessarily require a model of the plant itself. The reference model can be chosen to specify the required performance, the objective of MRAC being to drive the response of the system toward that of the reference model. Several drawbacks exist in this scheme, however. Because the reference model is quite independent of the plant model, the required control effort could be excessive and the computation itself could be slow. Furthermore, a new control law must be derived for each reference model. Also, the control action must be generated much faster than the speed at which the nonlinear terms of the plant change because the adaptation mechanism has been derived by assuming that some of the nonlinear terms remain more or less constant.

Many other adaptive control schemes depend on a reasonably accurate model of the plant, not just a reference model. The models may be obtained either analytically or through identification (experimental). Adaptive control has been successfully applied in complex, nonlinear, and coupled systems, even though it has several weaknesses, as mentioned previously.

9.6.5 Sliding Mode Control

Sliding mode control, variable structure control, and suction control fall within the same class of control techniques, and are somewhat synonymous. The control law in this class is generally a switching controller. A variety of switching criteria may be employed. Sliding mode control may be treated as an adaptive control technique. Because the switching surface is not fixed, its variability is somewhat analogous to an adaptation criterion. Specifically, the error of the plant response is zero when the control falls on the sliding surface.

Consider a plant that is modeled by the nth order nonlinear ordinary differential equation:

$$\frac{d^n y}{dt^n} = f(y,t) + u(t) + d(t) \tag{9.44}$$

where

y is the response of interest
$u(t)$ is the input variable
$d(t)$ is the unknown disturbance input
$f(\bullet)$ is an unknown nonlinear model of the process which depends on the response vector:
$y = \left[y, \dot{y}, \cdots, \frac{d^{n-1}y}{dt^{n-1}} \right]^T$. A time-varying sliding surface is defined by the differential equation:

$$s(y,t) = \left(\frac{d}{dt} + \lambda \right)^{n-1} \tilde{y} = 0 \tag{9.45}$$

with $\lambda > 0$. Note the response error $\tilde{y} = y - y_d$, where y_d is the desired response. Similarly, $\dot{\tilde{y}} = \dot{y} - \dot{y}_d$ may be defined. It should be clear from Equation 9.45 that if we start from rest with zero initial error ($\tilde{y}(0) = 0$ with all the derivative of $\tilde{y}$ up to the n–1th derivative being zero at $t = 0$) then $s = 0$ corresponds to $\tilde{y}(t) = 0$ for all t. This will guarantee that the desired trajectory $y_d(t)$ is tracked accurately at all times. Hence, the control objective would be to force the error state vector $\tilde{y}$ onto the sliding surface $s = 0$. This control objective will be achieved if the control law satisfies

$$s\,\text{sgn}(s) \le -\eta \quad \text{with } \eta > 0 \tag{9.46}$$

where sgn(s) is the *signum function*.

The nonlinear process $f(y,t)$ is generally unknown. Suppose that $f(y,t) = \hat{f}(y,t) + \Delta f(y,t)$ where $\hat{f}(y,t)$ is a completely known function, and Δf represents modeling uncertainty. Specifically, consider the control equation:

$$u = -\hat{f}(y,t) - \sum_{p=1}^{n-1} \binom{n-1}{p} \lambda^p \frac{d^{n-p}\tilde{y}}{dt^{n-p}} - K(y,t)\,\text{sgn}(s) \tag{9.47}$$

Where $K(y,t)$ is an upper bound for the total uncertainty in the system (i.e., disturbance, model error, speed of error reduction, etc.) And:

$$\binom{n-1}{p} = \frac{(n-1)!}{p!(n-1-p)!}$$

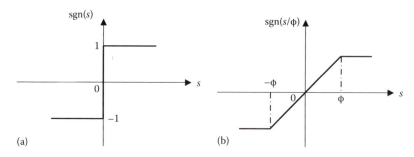

FIGURE 9.19
Switching functions used in sliding mode control: (a) signum function; (b) saturation function.

This sliding-mode controller satisfies Equation 9.46, but has drawbacks arising from the sgn(s) function. Specifically, very high switching frequencies can result when the control effort is significant. This is usually the case in the presence of large modeling errors and disturbances. High-frequency switching control can lead to the excitation of high-frequency modes in the plant. It can also lead to chattering problems. This problem can be reduced if the signum function in Equation 9.47 is replaced by a *saturation function*, with a boundary layer ±φ, as shown in Figure 9.19. In this manner, any switching that would have occurred within the boundary layer would be filtered out. Furthermore, the switching transitions would be much less severe. Clearly, the advantages of sliding mode control include robustness against factors such as nonlinearity, model uncertainties, disturbances, and parameter variations.

9.6.6 Linear Quadratic Gaussian Control

This is an *optimal control* technique that is intended for quite linear systems with random input disturbances and output (measurement) noise. Consider the *linear* system given by the set of first order differential equations (*state equations*):

$$\frac{d\boldsymbol{x}}{dt} = \boldsymbol{Ax} + \boldsymbol{Bu} + \boldsymbol{Fv} \tag{9.48}$$

and the output equations:

$$\boldsymbol{y} = \boldsymbol{CX} + \boldsymbol{w} \tag{9.49}$$

in which
$\boldsymbol{x} = [x_1 \, x_2 \, \dots \, x_n]^T$ is the state vector
$\boldsymbol{u} = [u_1 \, u_2 \, \dots \, u_r]^T$ is the vector of system inputs
$\boldsymbol{y} = [y_1 \, y_2 \, \dots \, y_m]^T$ is the vector of system outputs

The vectors v and w represent input disturbances and output noise, respectively, which are assumed to be white noise (i.e., zero-mean random signals whose *power spectral density junction is* flat) with covariance matrices $\boldsymbol{V}$ and $\boldsymbol{W}$. Also, $\boldsymbol{A}$ is called the system matrix, $\boldsymbol{B}$ the *input distribution matrix,* and $\boldsymbol{C}$ the *output formation matrix.* In linear quadratic Gaussian (LQG) control, the objective is to minimize the *performance index* (cost function):

$$J = E\left\{\int_0^\infty \left(x^T Q x + u^T R u\right) dt\right\} \tag{9.50}$$

in which Q and R are diagonal matrices of weighting and E denotes the "expected value" (or mean value) of a random process. In the LQG method, the controller is implemented as the two-step process:

1. Obtain the estimate $\hat{x}$ for the state vector x using a *Kalman filter* (with gain K_f)
2. Obtain the control signal as a product of $\hat{x}$ and a gain matrix K_0 by solving a noise-free linear quadratic optimal control problem

This implementation is shown by the block diagram in Figure 9.20. As mentioned before, it can be analytically shown that the noise-free quadratic optimal controller is given by the gain matrix:

$$K_0 = R^{-1} B^T P_0 \tag{9.51}$$

where P_0 is the positive semi-definite solution of the *algebraic Riccati equation*:

$$A^T P_0 + P_0 A - P_0 B R^{-1} B^T P_0 + Q = 0 \tag{9.52}$$

The Kalman filter is given by the gain matrix:

$$K_f = P_f C^T W^{-1} \tag{9.53}$$

where P_f is obtained as before by solving:

$$A P_f + P_f A^T - P_f C^T W^{-1} C P_f + F V F^T = 0 \tag{9.54}$$

An advantage of this controller is that the stability of the closed-loop control system is guaranteed as long as both the plant model and the Kalman filter are *stabilizable* and *detectable*. Note that if uncontrollable modes of a system are stable, the system is stabilizable.

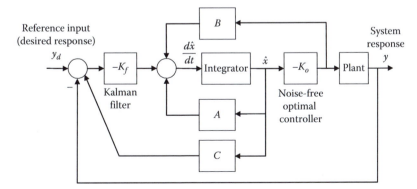

FIGURE 9.20
Linear quadratic Gaussian (LQG) control.

Similarly, if the unobservable modes of a system are stable, the system is detectable. Another advantage is the precision of the controller as long as the underlying assumptions are satisfied, but LQG control is also a model-based "crisp" scheme. Model errors and noise characteristics can significantly affect the performance. Also, even though stability is guaranteed, good stability margins and adequate robustness are not guaranteed in this method. Computational complexity (solution of two Riccati equations) is another drawback.

9.6.7 H_∞ (*H*-Infinity) Control

This is a relatively new optimal control approach that is quite different from the LQG method. However, this is a frequency-domain method. This technique assumes a linear plant with constant parameters, which may be modeled by a *transfer function* in the SISO case or by a *transfer matrix* in the MIMO case. Without going into the analytical details, let us outline the principle behind H_∞ control.

Consider the MIMO, linear, feedback control system shown by the block diagram in Figure 9.21, where I is an *identity matrix*. It satisfies the relation: $GG_c[y_d - y] = y$, or

$$[I + GG_c]y = GG_cy_d \tag{9.55}$$

Because the plant G is fixed, the underlying design problem here is to select a suitable controller G_c that will result in a required performance of the system. In other words, the closed-loop transfer matrix:

$$H = [I + GG_c]^{-1}GG_c \tag{9.56}$$

must be properly "shaped" through an appropriate choice of G_c. The required shape of $H(s)$ may be consistent with the classical specifications such as

1. Unity $|H(j\omega)|$ or large $|GG_c(j\omega)|$ at low frequencies in order to obtain small steady-state error for step inputs
2. Small $|H(j\omega)|$ or small $|GG_c(j\omega)|$ at large frequencies so that high-frequency noise would not be amplified, and further, the controller would be physically realizable
3. Adequately high gain and phase margins in order to achieve required stability levels

Of course, in theory there is an "infinite" number of possible choices for $G_c(s)$ that will satisfy such specifications. The H_∞ method uses an optimal criterion to select one of these

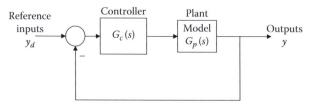

FIGURE 9.21
A linear multivariable feedback control system.

"infinite" choices. Specifically, the choice that minimizes the so-called "H_∞ norm" of the closed-loop transfer matrix $H(s)$, is chosen. The rationale is that this optimal solution is known to provide many desired characteristics (with respect to stability, robustness in the presence of model uncertainty and noise, sensitivity, etc.) in the control system.

The H_∞ norm of a transfer matrix H is the maximum value of the largest *singular value* of $H(j\omega)$, maximum being determined over the entire frequency range. A singular value of $H(j\omega)$ is the square root of an eigenvalue of the matrix $H(j\omega)H^T(j\omega)$.

The H_∞ control method has the advantages of stability and robustness. The disadvantages are that it is a "crisp" control method that is limited to linear, time-invariant systems, and that it is a model-based technique.

9.7 Fuzzy Logic Control

An intelligent controller may be interpreted as a computer-based controller that can somewhat "emulate" the reasoning procedures of a human expert in the specific area of control, to generate the necessary control actions. Here, techniques from the field of *artificial intelligence (AI)* are used for the purpose of acquiring and representing knowledge and for generating control decisions through an appropriate reasoning mechanism. With steady advances in the field of AI, especially pertaining to the development of practical *expert systems* or *knowledge systems*, there has been a considerable interest in using AI techniques for controlling complex processes. Complex engineering systems use intelligent control to cope with situations where conventional control techniques are not effective.

Intelligent control depends on efficient ways of representing and processing the control knowledge. Specifically, a knowledge base has to be developed and a technique of reasoning and making "inferences" has to be available. Knowledge-based intelligent control relies on knowledge that is gained by intelligently observing, studying, or understanding the behavior of a plant, rather than explicitly modeling the plant, to arrive at the control action. In this context, it also heavily relies on the knowledge of experts in the domain, and also on various forms of general knowledge. Modeling of the plant is implicit here. Soft computing is an important branch of study in the area of intelligent and knowledge-based systems. It has effectively complemented conventional AI in the area of machine intelligence (computational intelligence). Fuzzy logic, probability theory, neural networks, and genetic algorithms are cooperatively used in soft computing for knowledge representation and for mimicking the reasoning and decision-making processes of a human. Decision making with soft computing involves *approximate reasoning*, and is commonly used in intelligent control. This section presents an introduction to intelligent control, emphasizing fuzzy logic control.

9.7.1 Fuzzy Logic

Fuzzy logic is useful in representing human knowledge in a specific domain of application and in reasoning with that knowledge to make useful inferences or actions. The conventional binary (bivalent) logic is crisp and allows for only two states. This logic cannot handle fuzzy descriptors, examples of which are "fast" which is a *fuzzy quantifier* and "weak" which is a *fuzzy predicate*. They are generally qualitative, descriptive, and subjective and may contain some overlapping degree of a neighboring quantity, for example, some degree of "slowness" in the case of the fuzzy quantity "fast." Fuzzy logic allows for a realistic extension of binary, crisp logic to qualitative, subjective, and approximate situations, which often exist in problems of intelligent machines where techniques of artificial intelligence are appropriate.

In fuzzy logic, the knowledge base is represented by if–then rules of fuzzy descriptors. Consider the general problem of approximate reasoning. In this case, the knowledge base K is represented in an "approximate" form, for example, by a set of if–then rules with *antecedent* and *consequent* variables that are fuzzy descriptors. First, the data D are preprocessed according to

$$F_D = FP(D) \qquad (9.57)$$

which, in a typical situation, corresponds to a data abstraction procedure called "fuzzification" and establishes the membership functions or membership grades that correspond to D. Then for a fuzzy knowledge base F_K, the fuzzy inference F_I is obtained through fuzzy-predicate approximate reasoning, as denoted by

$$F_I = F_K \circ F_D \qquad (9.58)$$

This uses a *composition* operator "$\circ$" for fuzzy matching of data (D) with the knowledge base (K) is carried out, and making inferences (I) on that basis.

Fuzzy logic is commonly used in "intelligent" control of processes and machinery. In this case, the inferences of a fuzzy decision making system are the control inputs to the process. These inferences are arrived at by using the process responses as the inputs (context data) to the fuzzy decision-making system.

9.7.2 Fuzzy Sets and Membership Functions

A fuzzy set has a fuzzy boundary. The membership of an element lying on the boundary is fuzzy: there is some possibility that the element is inside the set, and a complementary possibility that it is outside the set. A fuzzy set may be represented by a membership function. This function gives the grade (degree) of membership within the set, of any element of the universe of discourse. The membership function maps the elements of the universe on to numerical values in the interval [0, 1]. Specifically,

$$\mu_A(x) : X \rightarrow [0,1] \qquad (9.59)$$

where $\mu_A(x)$ is the membership function of the fuzzy set A in the universe in X. Stated in another way, fuzzy set if A is a set of ordered pairs:

$$A = \{(x, \mu_A(x)); \; x \in X, \; \mu_A(x) \in [0,1]\} \qquad (9.60)$$

The membership function $\mu_A(x)$ represents the grade of possibility that an element x belongs to the set A. It follows that a membership function is a *possibility function* and not a probability function. A membership function value of zero implies that the corresponding element is definitely not an element of the fuzzy set. A membership function value of unity means that the corresponding element is definitely an element of the fuzzy set.

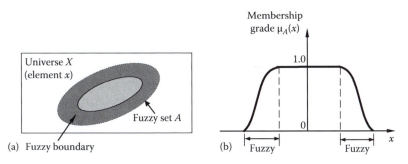

FIGURE 9.22
(a) A fuzzy set; (b) the membership function of a fuzzy set.

A grade of membership greater than 0 and less than 1 corresponds to a non-crisp (or fuzzy) membership, and the corresponding elements fall on the fuzzy boundary of the set. The closer the $\mu_A(x)$ is to 1 the more the x is considered to belong to A, and similarly the closer it is to 0 the less it is considered to belong to A. A typical fuzzy set is shown in Figure 9.22a and its membership function is shown in Figure 9.22b.

Note: A crisp set is a special case of fuzzy set, where the membership function can take the two values 1 (membership) and 0 (non-membership) only. The membership function of a crisp set is given the special name *characteristic function*.

9.7.3 Fuzzy Logic Operations

It is well known that the "complement," "union," and "intersection" of crisp sets correspond to the logical operations NOT, OR, and AND, respectively, in the corresponding crisp, bivalent logic. Furthermore, it is known that, in the crisp bivalent logic, the union of a set with the complement of a second set represents an "implication" of the first set by the second set. Set inclusion (i.e., extracting a subset) is a special case of implication in which the two sets belong to the same universe. These operations (connectives) may be extended to fuzzy sets for corresponding use in fuzzy logic and fuzzy reasoning. For fuzzy sets, the applicable connectives must be expressed in terms of the membership functions of the sets that are operated on. In view of the isomorphism between fuzzy sets and fuzzy logic, both the set operations and the logical connectives can be addressed together. Some basic operations that can be defined on fuzzy sets and the corresponding connectives of fuzzy logic are described now.

9.7.3.1 Complement (Negation, NOT)

Consider a fuzzy set A in a universe X. Its complement A' is a fuzzy set whose membership function is given by

$$\mu_{A'}(x) = 1 - \mu_A(x) \quad \text{for all } x \in X \tag{9.61}$$

The complement in fuzzy sets corresponds to the negation (NOT) operation in fuzzy logic, just as in crisp logic, and is denoted by $\bar{A}$ where A now is a fuzzy logic proposition (or a fuzzy state).

9.7.3.2 Union (Disjunction, OR)

Consider two fuzzy sets A and B in the same universe X. Their union is a fuzzy set containing all the elements from both sets, in a "fuzzy" sense. This set operation is denoted by $\cup$. The membership function of the resulting set $A \cup B$ is given by

$$\mu_{A \cup B}(x) = \max[(\mu_A(x), \mu_B(x)] \quad \forall x \in X \tag{9.62}$$

The union corresponds to a logical OR operation (called *Disjunction*), and is denoted by $A \lor B$, where A and B are fuzzy states or fuzzy propositions. The rationale for the use of *max* to represent fuzzy-set union is that, because element x may belong to one set or the other, the larger of the two membership grades should govern the outcome (union). Furthermore, this is consistent with the union of crisp sets. Similarly, the appropriateness of using *max* to represent fuzzy-logic operation "OR" should be clear. Specifically, since either of the two fuzzy states (or propositions) would be applicable, the larger of the corresponding two membership grades should be used to represent the outcome.

Even though set intersection is applicable to sets in a common universe, a logical "OR" may be applied for concepts in different universes. In particular, when the operands belong to different universes, orthogonal axes have to be used to represent them in a common membership function.

9.7.3.3 Intersection (Conjunction, AND)

Again, consider two fuzzy sets A and B in the same universe X. Their intersection is a fuzzy set containing all the elements that are common to both sets, in a "fuzzy" sense. This set operation is denoted by $\cap$. The membership function of the resulting set $A \cap B$ is given by

$$\mu_{A \cap B}(x) = \min[(\mu_A(x), \mu_B(x)] \quad \forall x \in X \tag{9.63}$$

The union corresponds to a logical AND operation (called *Conjunction*), and is denoted by $A \land B$, where A and B are fuzzy states or fuzzy propositions. The rationale for the use of *min* to represent fuzzy-set intersection is that, because the element x must simultaneously belong to both sets, the smaller of the two membership grades should govern the outcome (intersection).

Furthermore, this is consistent with the intersection of crisp sets. Similarly, the appropriateness of using *min* to represent fuzzy-logic operation "AND" should be clear. Specifically, since both fuzzy states (or propositions) should be simultaneously present, the smaller of the corresponding two membership grades should be used to represent the outcome.

9.7.3.4 Implication (If–Then)

An if–then statement (a rule) is called an "implication." A knowledge base in fuzzy logic may be expressed by a set of linguistic rules of the if–then type, containing fuzzy terms. In fact, a fuzzy rule is a *fuzzy relation*. A knowledge base containing several fuzzy rules is also a relation that is formed by combining (aggregating) the individual rules according to how they are interconnected.

Consider a fuzzy set A defined in a universe X and a second fuzzy set B defined in another universe Y. The fuzzy implication "If A then B," is denoted by $A \rightarrow B$. Note that in this fuzzy rule, A represents some "fuzzy" situation, and is the *condition* or the *antecedent* of the rule.

Similarly, *B* represents another fuzzy situation, and is the *action* or the *consequent* of the fuzzy rule. The fuzzy rule $A \to B$ is a fuzzy relation. Since the elements of *A* are defined in *X* and the elements of B are defined in Y, the elements of $A \to B$ are defined in the *Cartesian product space* $X \times Y$. This is a two-dimensional space represented by two orthogonal axes (*x*-axis and *y*-axis), and gives the domain in which fuzzy rule (or fuzzy relation) is defined. Since *A* and *B* can be represented by membership functions, an additional orthogonal axis is needed to represent the membership grade.

Fuzzy implication may be defined (interpreted) in several ways. Two definitions of fuzzy implication are

Method 1:

$$\mu_{A \to B}(x, y) = \min[(\mu_A(x), \mu_B(y)] \quad \forall x \in X, \ \forall y \in Y \tag{9.64}$$

Method 2:

$$\mu_{A \to B}(x, y) = \min[1, \ \{1 - \mu_A(x) + \mu_B(y)\}] \quad \forall x \in X, \ \forall y \in Y \tag{9.65}$$

These two methods are approaches for obtaining the membership function of the particular fuzzy relation given by an if–then rule (implication). Note that the first method gives an expression that is symmetric with respect to *A* and *B*. This is not intuitively satisfying because "implication" is not a commutative operation (specifically, $A \to B$ does not necessarily satisfy $B \to A$). In practice, however, this method provides a good, robust result. The second method has an intuitive appeal because in crisp bivalent logic, $A \to B$ has the same truth table as [(NOT *A*) OR *B*] and hence are equivalent. Note that in Equation 9.65, the membership function is upper-bounded to 1 using the *bounded sum* operation, as required (a membership grade cannot be greater than 1). The first method is more commonly used because it is simpler to use and often provides quite accurate results.

9.7.4 Compositional Rule of Inference

In knowledge-based systems, the knowledge is often expressed as rules of the form:
   "**IF** condition Y_1 is y_1 **AND IF** condition Y_2 is y_2 **THEN** action *C* is *c*."

In fuzzy knowledge-based systems (e.g., fuzzy control systems), rules of this type are linguistic statements of expert knowledge in which y_1, y_2, and *c* are fuzzy quantities (e.g., small negative, fast, large positive). These rules are fuzzy relations that employ the fuzzy implication (IF–THEN). The collective set of fuzzy relations forms the knowledge base of the fuzzy system. Let us denote the fuzzy relation formed by this collection of rules as the fuzzy set *K*. This relation is an aggregation of the individual rules, and may be represented by a multivariable membership function. In a fuzzy decision making process (e.g., in fuzzy logic control), the rule base (knowledge base) *K* is first collectively matched with the available data (context). Next, an inference is made on another fuzzy variable that is represented in the knowledge base, on this basis. The matching and inference making are done using the composition operation, as discussed previously. The application of composition to make inferences in this manner is known as the *compositional rule of inference* (CRI).

For example, consider a control system. Usually the context would be the measured outputs *Y* of the process. The control action that drives the process is *C*. Typically, both these variables are crisp, but let us ignore this fact for the time being and assume them to be fuzzy, for general consideration. Suppose that R, a fuzzy relation, denotes the control knowledge base. The method of obtaining the rule base R is analogous to model identification in conventional crisp control. Then, by applying the compositional rule of inference we get the fuzzy control action as

$$\mu_C = \max_Y \min(\mu_Y, \mu_R) \tag{9.66}$$

9.7.4.1 Extensions to Fuzzy Decision Making

Thus far, we have considered fuzzy rules of the form:

$$\text{IF } x \text{ is } A_i \text{ AND IF } y \text{ is } B_i \text{ THEN } z \text{ is } C_i \tag{9.67}$$

where A_i, B_i, and C_i are fuzzy states governing the ith rule of the rule base. This is the Mamdani approach (Mamdani system or Mamdani model) named after the person who pioneered the application of this approach. Here, the knowledge base is represented as fuzzy protocols and represented by membership functions for A_i, B_i, and C_i, and the inference is obtained by applying the compositional rule of inference. The result is a fuzzy membership function, which typically has to be *defuzzified* for use in practical tasks.

Several variations to this conventional method are available. One such version is the *Sugeno model* (or, *Takagi–Sugeno–Kang model* or *TSK model*). Here, the knowledge base has fuzzy rules with crisp functions as the consequent, of the form:

$$\text{IF } x \text{ is } A_i \text{ AND IF } y \text{ is } B_i \text{ THEN } c_i = f_i(x, y) \tag{9.68}$$

For Rule i, where f_i is a crisp function of the condition variables (antecedent) x and y. Note that the condition part of this rule is the same as for the Mamdani model (9.67), where A_i and B_i are fuzzy sets whose membership functions are functions of x and y, respectively. The action part is a crisp function of the condition variables, however. The inference $\hat{c}(x,y)$ of the fuzzy knowledge-based system is obtained directly as a crisp function of the condition variables x and y, as follows:

For Rule i, a weighting parameter $w_i(x, y)$ is obtained corresponding to the condition membership functions, as for the Mamdani approach, by using either the "min" operation or the "product" operation. For example, using the "min" operation we form:

$$w_i(x, y) = \min[\mu_{A_i}(x), \mu_{B_i}(y)] \tag{9.69}$$

The crisp inference $\hat{c}(x,y)$ is determined as a weighted average of the individual rule inferences (crisp) $c_i = f_i(x, y)$ according to

$$\hat{c}(x, y) = \frac{\sum_{i=1}^r w_i c_i}{\sum_{i=1}^r w_i} = \frac{\sum_{i=1}^r w_i(x, y) f_i(x, y)}{\sum_{i=1}^r w_i(x, y)} \tag{9.70}$$

where r is the total number of rules. For any data x and y, the knowledge-based action $\hat{c}(x,y)$ can be computed from (9.70), without requiring any defuzzification. The Sugeno model is particularly useful when the actions are described analytically through crisp functions, as in conventional crisp control, rather than linguistically. The TSK approach is commonly used in the applications of direct control and in simplified fuzzy models.

The Mamdani approach, even though popular in low-level direct control, is particularly appropriate for knowledge representation and processing in expert systems and in high-level (hierarchical) control systems.

9.7.5 Basics of Fuzzy Control

In fuzzy control, some information (e.g., output measurements) from the system to be controlled is matched with a knowledge base of control for the particular system, using CRI. The knowledge base is generally a set of (n) rules or "linguistic relation" of the form (9.67): Rule i: A_i and $B_i => C_i$.

Because these fuzzy sets are related through IF–THEN implications and because an implication operation for two fuzzy sets can be interpreted as a "minimum operation" on the corresponding membership functions, the membership function of this fuzzy relation may be expressed as

$$\mu_{Ri}(a,b,c) = \min\left[\mu_{Ai}(a), \mu_{Bi}(b), \mu_{Ci}(c)\right] \tag{9.71}$$

The individual rules in the rule-base are joined through ELSE connectives, which are OR connectives ("unions" of membership functions). Hence, the overall membership function for the complete rule-base (relation R) is obtained using the "maximum" operation on the membership functions of the individual rules:

$$\mu_R(a,b,c) = \max_i \mu_{Ri}(a,b,c) = \max_i \min\left[\mu_{Ai}(a), \mu_{Bi}(b), \mu_{Ci}(c)\right] \tag{9.72}$$

In this manner the membership function of the entire rule-base can be determined (or, "identified" in the terminology of conventional control) using the membership functions of the response variables and control inputs.

A fuzzy knowledge base is a multivariable function—a multidimensional array (a three-variable function or a dimensional array in the case of Equation 9.72) of membership function values. This array corresponds to a fuzzy control algorithm in the sense of conventional control. The control rule-base may represent linguistic expressions of experience, expertise, or knowledge of the domain experts (control engineers, skilled operators, etc.). Alternatively, a control engineer may instruct an operator (or a control system) to carry out various process tasks in the usual manner; monitor and analyze the resulting data; and learn appropriate rules of control, say by using neural networks.

Once a fuzzy control knowledge base of the form given by Equation 9.72 is obtained, we need a procedure to infer control actions using process measurements, during control. Specifically, suppose that fuzzy process measurements A' and B' are available. The corresponding control inference C' is obtained using the *compositional rule of inference*. The applicable relation is (see 9.66)

$$\mu_{C'}(c) = \sup_{a,b} \min\left[\mu_{A'}(a), \mu_{B'}(b), \mu_R(a,b,c)\right] \tag{9.73}$$

Actual process measurements are crisp. Hence, they have to be *fuzzified* in order to apply the compositional rule of inference. This is conveniently done by reading the grade values of the membership functions of the measurement at the specific measurement values. Typically, the control action must be a crisp value as well. Hence, each control inference C' must be *defuzzified* so that it can be used to control the process. Several methods are available to accomplish defuzzification. In the *mean of maxima* method, the control element corresponding to the maximum grade of membership is used as the control action. If there is more than one element with a maximum (peak) membership value, the mean of these values is used. In the *center of gravity* (or *centroid*) *method*, the centroid of the membership function of control decision is used as the value of crisp control action. This weighted control action is known to provide a somewhat sluggish, yet more robust control.

Because process measurements are crisp, one method of reducing the real-time computational overhead is to pre-compute a decision table relating quantized measurements to crisp control actions. The main disadvantage of this approach is that it does not allow for convenient modifications (e.g., rule changes and quantization resolution adjustments) during operation. Another practical consideration is the selection of a proper sampling period in view of the fact that process responses are generally analog signals. Factors such as process characteristics, required control bandwidth, and the processing time needed for one control cycle, must be taken into account in choosing a sampling period. Scaling or gain selection for various signals in a fuzzy logic control system is another important consideration. For reasons of processing efficiency, it is customary to scale the process variables and control signals in a fuzzy control algorithm. Furthermore, adjustable gains can be cascaded with these system variables so that they may serve as tuning parameters for the controller. A proper tuning algorithm would be needed, however. A related consideration is real-time or online modification of a fuzzy rule-base. Specifically, rules may be added, deleted, or modified on the basis of some scheme of *learning and self-organization*. For example, using a model for the process and making assumptions such as input–output monotonicity, it is possible during control to trace and tag the rules in the rule base that need attention. The control-decision table can be modified accordingly.

Hardware fuzzy processors (*fuzzy chips*) may be used to carry out the fuzzy inference at high speed. The rules, membership functions, and measured context data are generated as usual through the use of a control "host" computer. The fuzzy processor is located in the same computer, which has appropriate interface (input/output) hardware and driver software. Regardless of all these, it is more convenient to apply the inference mechanism separately to each rule and then combine the result instead of applying it to the entire rule base using the compositional rule of inference.

Fuzzy logic is commonly used in direct control of processes and machinery. In this case, the inferences of a fuzzy decision making system form the control inputs to the process. These inferences are arrived at by using the process responses as the inputs (context data) to the fuzzy system. The structure of a direct fuzzy controller is shown in Figure 9.23. Here, y represents the process output, u represents the control input, and R is the relation, which represents the fuzzy control knowledge base.

Example 9.9

Consider the room-comfort control system schematically shown in Figure 9.24. The temperature (T) and humidity (H) are the process variables that are measured. These sensor signals are provided to the fuzzy logic controller, which determines the cooling rate (C) that should be generated by the air conditioning unit. The objective is to maintain a particular comfort level inside the room.

A simplified fuzzy rule-base of the comfort controller is graphically presented in Figure 9.25. The temperature level can assume one of two fuzzy states (*HG, LW*), which denote high and low, respectively, with the corresponding membership functions. Similarly, the humidity level can assume two other fuzzy states (*HG, LW*) with associated membership functions. Note that the membership functions of T are quite different from those of H, even though the same nomenclature is used. There are four rules, as given in Figure 9.25. The rule base is:

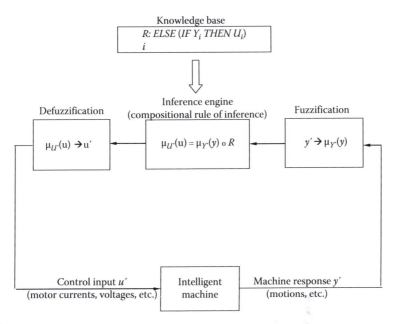

FIGURE 9.23
Structure of a direct fuzzy controller.

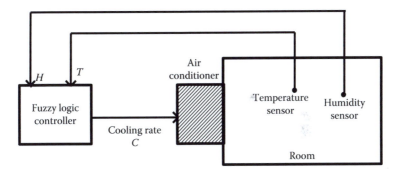

FIGURE 9.24
Comfort control system of a room.

Rule 1:		If	T	Is	HG	And	H	Is	HG	Then	C	Is	PH
Rule 2:	Else	If	T	Is	HG	And	H	Is	LW	Then	C	Is	PL
Rule 3:	Else	If	T	Is	LW	And	H	Is	HG	Then	C	Is	NL
Rule 4:	Else	If	T	Is	LW	And	H	Is	LW	Then	C	Is	NH
	And	If											

The nomenclature used for the fuzzy states is as follows:

Temperature (T)	Humidity (H)	Change in cooling rate (C)
HG = High	HG = High	PH = Positive high
LW = Low	LW = Low	PL = Positive low
		NH = Negative high
		NL = Negative low

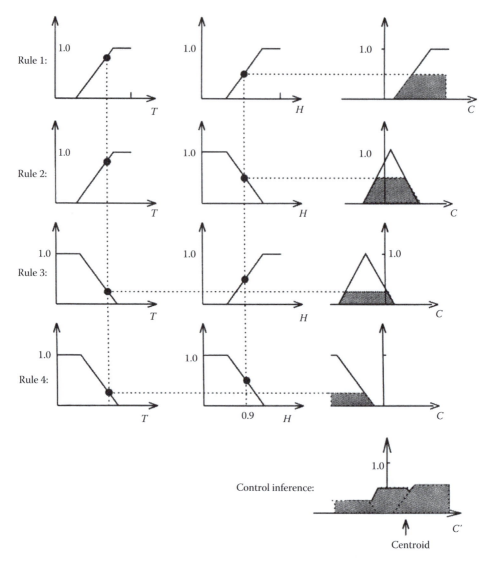

FIGURE 9.25
The fuzzy knowledge base of the comfort controller.

Application of the compositional rule of inference is done here by using individual rule-based composition. Specifically, the measured information is composed with individual rules in the knowledge base and the results are aggregated to give the overall decision. For example, suppose that the room temperature is 30°C and the relative humidity is 0.9. Lines are drawn at these points, as shown in Figure 9.25, to determine the corresponding membership grades for the fuzzy states in the four rules. In each rule the lower value of the two grades of process response variables is then used to clip (or modulate) the corresponding membership function of C (a *min* operation). The resulting "clipped" membership functions of C for all four rules are superimposed (a *max* operation) to obtain the control inference C' as shown. This result is a fuzzy set, and it must be defuzzified to obtain a crisp control action $\hat{c}$ for changing the cooling rate. The centroid method may be used for defuzzification.

9.7.6 Fuzzy Control Surface

A fuzzy controller is a nonlinear controller. A well-defined problem of fuzzy control, with analytical membership functions and fuzzification and defuzzification methods, and well-defined fuzzy logic operators, may be expressed as a nonlinear control surface through the application of the compositional rule of inference. The advantage then is that the generation of the control action becomes a simple and very fast step of reading the surface value (control action) for given values of crisp measurement (process variables). The main disadvantage is, the controller is fixed and cannot accommodate possible improvements to control rules and membership functions through successive learning and experience. Nevertheless, this approach to fuzzy control is quite popular. A useful software tool for developing fuzzy controllers is the MATLAB Fuzzy Logic Toolbox.

Example 9.10

A schematic diagram of a simplified system for controlling the liquid level in a tank is shown in Figure 9.26a. In the control system, the error (actually, correction) is given by $e =$ Desired level − Actual level.

The change in error is denoted by Δe. The control action is denoted by u, where $u > 0$ corresponds to opening the inflow valve and $u < 0$ corresponds to opening the outflow valve. A low-level direct fuzzy controller is used in this control system, with the control rule-base as given in Figure 9.26b.

The membership functions for E, ΔE, and U are given in Figure 9.26c. Note that the error measurements are limited to the interval $[-3a, 3a]$ and the Δerror measurements to $[-3b, 3b]$. The control actions are in the range $[-4c, 4c]$.

Following the usual steps of applying the compositional rule of inference for this fuzzy logic controller, we can develop a crisp *control surface* $u(e, \Delta e)$ for the system, expressed in the three-dimensional coordinate system $(e, \Delta e, u)$, which then can be used as a simple and fast controller. This method is described next.

The crisp control surface is developed by carrying out the rule-based inference for each point: $(e, \Delta e)$ in the measurement space $E \times \Delta E$, using individual rule-based inference. To demonstrate this procedure, consider a set of context data $(e_o, \Delta e_o)$, where e_o is in $[-3a, -2a]$ and Δe_o is in $[-b/2, 0]$. Then, from the membership functions and the rule base, it should be clear that only two rules are valid in this region, as given below:

R_1: if e is NL and Δe is NS then u is NL
R_2: if e is NL and Δe is ZO then u is NM

Since, in the range $[-3a, -2a]$, the membership grade of singleton fuzzification of e_o is always 1, the lower grade of the two context values is the one corresponding to the singleton fuzzification of Δe_o for both rules. Then, in applying the individual rule-based inference, the lower grade value of the two context variables is used to clip off the corresponding membership function of the control action variable U in each rule (this is a *min* operation). The resulting membership functions of U for the two applicable rules are superimposed (this is a *max* operation) to obtain the control inference U', as shown in Figure 9.27.

For defuzzification, we apply the moment method to find the centroid of the resulting membership function of control inference. From the moment method, we obtain the crisp control action as a function of e and Δe. The above procedure is repeatedly applied to all possible ranges of e $[-3a, 3a]$ and Δe $[-3b, 3b]$ to obtain the complete control surface. Also, the procedure can be implemented in a computer program to generate a control surface. A control surface with $a = 1$, $b = 2$, and $c = 0.5$ is shown in Figure 9.28.

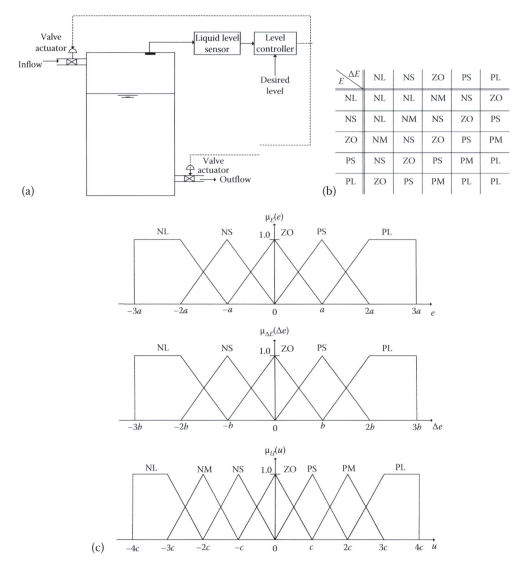

FIGURE 9.26
(a) Liquid level control system; (b) the control rule base; (c) the membership functions of error, change in error, and control action.

In the present example, what we have applied is in fact the Mamdani approach. Sugeno model (or, Takagi–Sugeno–Kang model or TSK model) could have been used as well, thereby avoiding the defuzzification step.

9.8 Digital Control

In a digital control system, a digital device is used as the controller. The digital controller may be a *hardware* device consisting of permanent logic circuitry or a *software device*—a digital computer. Hardware controllers are inexpensive and fast, but lack flexibility or

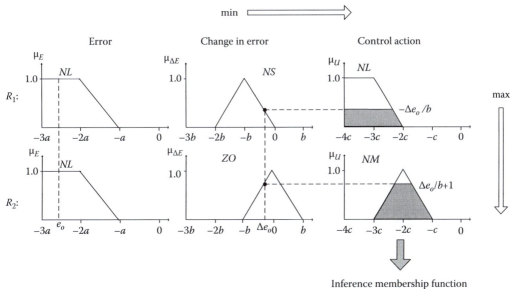

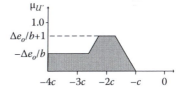

FIGURE 9.27
Individual rule-based inference for $e_o[-3a, -2a]$ and $\Delta e_o[-b/2, 0]$.

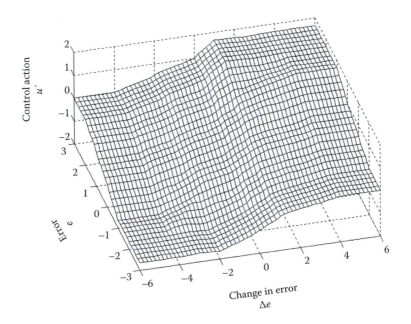

FIGURE 9.28
Control surface with $a = 1$, $b = 2$, and $c = 0.5$.

programmability. A software-based digital controller has programmable memory in addition to a central processor (see Chapter 8). The control algorithm is "programmed" into the computer memory and is used by the processor in real time to generate the control signals. The control algorithm in such a controller can be modified simply by reprogramming, without the need for hardware changes. Typically, data are sampled into a digital controller at a fixed sampling period (see Chapter 5).

9.8.1 Computer Control Systems

In a computer-based control system, a digital computer serves as the controller. A digital feedback control system is shown in Figure 9.29. The information enters into the control computer in the digital form. Signals generated by the computer are in the digital form. Typically, they have to be converted into the analog form for use in the external purposes such as driving a plant or its actuators. Virtually any control law may be programmed into the control computer. Control computers have to be fast and dedicated machines for real-time operation where processing has to be synchronized with plant operation and actuation requirements. This requires a real-time operating system. Apart from these requirements, control computers are basically no different from general-purpose digital computers. They consist of a processor to perform computations and to oversee data transfer, memory for program and data storage during processing, mass storage devices to store information that is not immediately needed, and input/output devices to read in and send out information.

9.8.2 Components of a Digital Control System

Digital control systems might utilize digital instruments and additional processors as well for actuating, signal-conditioning, or measuring functions. For example, a stepper motor that responds with incremental motion steps when driven by pulse signals can be considered a digital actuator. Furthermore, it usually contains digital logic circuitry in its drive system. Similarly, a two-position solenoid is a digital (binary) actuator. Digital flow control may be accomplished using a digital control valve. A typical digital valve consists of a bank of orifices, each sized in proportion to a place value of a binary word (2^i, $i = 0$, 1, 2, ..., n). Each orifice is actuated by a separate rapid-acting on/off solenoid. In this manner, many digital combinations of flow values can be obtained. Direct digital

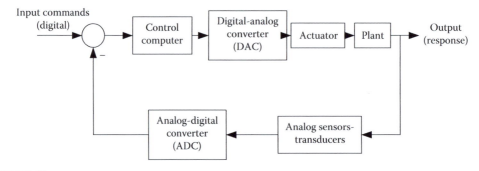

FIGURE 9.29
A digital feedback control system.

measurement of displacements and velocities can be made using shaft encoders. These are digital transducers that generate coded outputs (e.g., in binary or gray-scale representation) or pulse signals that can be coded using counting circuitry. Such outputs can be read in by the control computer with relative ease. Frequency counters also generate digital signals that can be fed directly into a digital controller. When measured signals are in the analog form, an analog front end is necessary to interface the transducer and the digital controller. Input/output interface cards that can take both analog and digital signals are available with digital controllers.

Analog measurements and reference signals have to be sampled and encoded prior to digital processing within the controller. Digital processing can be effectively used for signal conditioning as well. Alternatively, digital signal processing (DSP) chips can function as digital controllers. However, analog signals have to be *preconditioned* using analog circuitry prior to digitizing in order to eliminate or minimize problems due to *aliasing distortion* (high-frequency components above half the sampling frequency appearing as low-frequency components) and *leakage* (error due to signal truncation) as well as to improve the signal level and filter out extraneous noise (see Chapter 5). The drive system of a plant typically takes in analog signals. Often, the digital output from controller has to be converted into analog form for this reason. Both *analog-to-digital conversion* (ADC) and *digital-to-analog conversion* (DAC) can be interpreted as signal-conditioning (modification) procedures. If more than one output signal is measured, each signal will have to be conditioned and processed separately. Ideally, this will require separate conditioning and processing hardware for each signal channel. A less expensive (but slower) alternative would be to time-share this expensive equipment by using a *multiplexer*. This device will pick one channel of data from a bank of data channels in a sequential manner and connect it to a common input device.

The current practice of using dedicated, microcontroller-based (Chapter 8), and often, decentralized (distributed) digital control systems in industrial applications can be rationalized in terms of the major advantages of digital control.

9.8.3 Advantages of Digital Control

The following are some of the important advantages of digital control.

1. Digital control is less susceptible to noise or parameter variation in instrumentation because data can be represented, generated, transmitted, and processed as binary words, with bits possessing two identifiable states.

2. Very high accuracy and speed are possible through digital processing. Hardware implementation is usually faster than software implementation.

3. Digital control can handle repetitive tasks extremely well, through programming.

4. Complex control laws and signal conditioning methods that might be impractical to implement using analog devices can be programmed.

5. High reliability in operation can be achieved by minimizing analog hardware components and through decentralization using dedicated microprocessors for various control tasks.

6. Large amounts of data can be stored using compact, high-density data storage methods.

7. Data can be stored or maintained for very long periods of time without drift and without being affected by adverse environmental conditions.

8. Fast data transmission is possible over long distances without introducing excessive dynamic delays, as in analog systems.

9. Digital control has easy and fast data retrieval capabilities.

10. Digital processing uses low operational voltages (e.g., 0–12 V dc).

11. Digital control is cost effective.

9.8.4 Computer Implementation

In computer-based control systems, a suitable control algorithm has to be programmed into the memory of the control computer. A digital controller is functionally similar to its analog counterpart except that the input data to the controller and the output data from the controller are in the digital form (see Figure 9.29). The control law can be represented by a set of *difference equations*. These difference equations relate the discrete output signals from the controller and the discrete input signals into the controller. The problem of developing a digital controller can be interpreted as the formulation of appropriate difference equations that are able to generate the required control signals. Similarly, just the same way as an analog controller may be represented by a set of analog transfer functions, a digital controller may be represented by a set of *discrete transfer functions*. These discrete transfer functions, in turn, can be transformed into a set of difference equations.

Once a control law is available in the analog form, as a transfer function, the corresponding digital control law may be determined by obtaining the discrete transfer function that is equivalent to the analog transfer function. This approach is particularly useful when, for example, it is required to update (modernize) a well-established analog control system by replacing its analog compensator circuitry with a digital controller/compensator. Then the (Laplace) transfer function of the analog compensator can be obtained by testing or analysis (or both) of the compensator. The eventual objective would be to develop a difference equation to represent the analog compensator. This is a basic task in the development of a digital controller, and is conveniently handled by the *z*-transform method.

A discrete transfer function necessarily depends on the sampling period T used to convert analog signals into discrete data (sampled data). Digital control action approaches the corresponding analog control action when T approaches zero. Faster sampling rates provide better accuracy and less aliasing error, but demand smaller processing cycle times, which in turn call for efficient processors and improved control algorithms for a given level of control complexity. Faster sampling rates are more demanding on the interface hardware as well. A large word size is needed to accurately represent data. By increasing the word size (number of bits per word), the *dynamic range* and the *resolution* of the represented data can be improved and the *quantization error* decreased. Even though the processing cycle time will generally increase by increasing the word size, on average there is also a speed advantage to increasing the word size of a computer. The larger the program size (number of instructions per program) the greater the memory requirements and, furthermore, the slower the associated control cycle for a given control computer. It follows that sampling rate, processing cycle time, data word size, and memory requirements are crucial parameters that are interrelated in digital control.

Digital control is particularly preferred when the control algorithms are complex. The algorithm for a *three-point controller* (PID controller), for example, is quite simple and straightforward. Even though a PID controller can be easily implemented by analog means, or even by a *hardware digital controller*, one may decide to employ a simple microprocessor as the controller in each proportional-integral-derivative (PID) loop of a control system. The microcontroller approach has the advantages of low cost, small size, and flexibility. In particular, integration with a higher-level supervisory controller in a distributed-control environment will be rather convenient when microprocessor-based loop controllers are employed. Also, integration of PID loops with more complex control schemes such as linearizing control (nonlinear feedback control) and adaptive control will be simplified when the software-based digital control approach is used. Digital implementation of lead and lag compensators can be slightly more difficult than the implementation of three-point controllers.

Problems

9.1 You are asked to design a control system to turn on lights in an art gallery at night, provided that there are people inside the gallery. Explain a suitable control system, identifying the open loop and feedback functions, if any, and describing the control system components.

9.2 (a) Discuss possible sources of error that can make open-loop control or feedforward control meaningless in some applications.

  (b) How would you correct the situation?

9.3 In each of the following examples, indicate at least one (unknown) input that should be measured and used for feedforward control to improve the accuracy of the control system.

  (a) A servo system for positioning a mechanical load. The servomotor is a field-controlled dc motor, with position feedback using a potentiometer and velocity feedback using a tachometer.

  (b) An electric heating system for a pipeline carrying a liquid. The exit temperature of the liquid is measured using a thermocouple and is used to adjust the power of the heater.

  (c) A room heating system. Room temperature is measured and compared with the set point. If it is low, a valve of a steam radiator is opened; if it is high, the valve is shut.

  (d) An assembly robot that grips a delicate part to pick it up without damaging the part.

  (e) A welding robot that tracks the seam of a part to be welded.

9.4 Hierarchical control has been applied in many industries, including steel mills, oil refineries, chemical plants, glass works, and automated manufacturing. Most applications have been limited to two or three levels of hierarchy, however. The lower levels usually consist of tight servo loops, with bandwidths on the order of 1 kHz. The upper levels typically control production planning and scheduling events measured in units of days or weeks.

A five-level hierarchy for a flexible manufacturing facility is as follows: The lowest level (level 1) handles servo control of robotic manipulator joints and machine tool degrees of freedom. The second level performs activities such as coordinate transformation in machine tools, which are required in generating control commands for various servo loops. The third level converts task commands into motion trajectories (of manipulator end effector, machine tool bit, etc.) expressed in world coordinates. The fourth level converts complex and general task commands into simple task commands. The top level (level 5) performs supervisory control tasks for various machine tools and material-handling devices, including coordination, scheduling, and definition of basic moves. Suppose that this facility is used as a flexible manufacturing workcell for turbine blade production. Estimate the event duration at the lowest level and the control bandwidth (in hertz) at the highest level for this type of application.

9.5 The programmable logic controller (PLC) is a sequential control device that can sequentially and repeatedly activate a series of output devices (e.g., motors, valves, alarms, and signal lights) on the basis of the states of a series of input devices (e.g., switches, two-state sensors). Show how a programmable controller and a vision system consisting of a solid-state camera and a simple image processor (say, with an edge-detection algorithm) could be used for sorting fruits on the basis of quality and size for packaging and pricing.

9.6 It is well known that the block diagram in Figure P9.6a represents a dc motor, for armature control, with the usual notation. Suppose that the load driven by the motor is a pure inertia element (e.g., a wheel or a robot arm) of moment of inertia J_L that is directly and rigidly, attached to the motor rotor.

(a) Obtain an expression for the transfer function $\dfrac{\omega_m}{v_a} = G_m(s)$ for the motor with the inertial load, in terms of the parameters given in Figure P9.6a, and J_L.

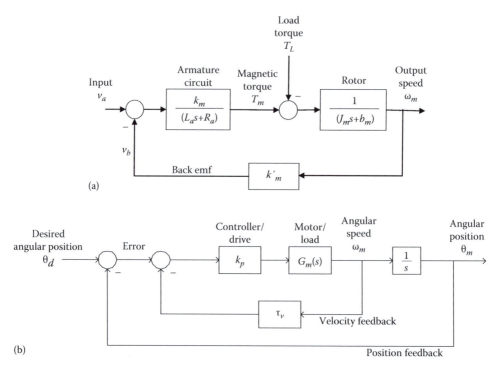

FIGURE P9.6
(a) Block diagram of a dc motor for armature control; (b) motor control with feedback of position and velocity.

(b) Now neglect the leakage inductance L_a. Then, show that the transfer function in Part (a) can be expressed as $G_m(s) = k/(\tau s + 1)$. Give expressions for τ and k in terms of the given system parameters.

(c) Suppose that the motor (with the inertial load) is to be controlled using position plus velocity feedback. The block diagram of the corresponding control system is given in Figure P9.6b, where $G_m(s) = k/(\tau s + 1)$. Determine the transfer function of the (closed-loop) control system $G_{CL}(s) = \theta_m/\theta_d$ in terms of the given system parameters (k, k_p, τ, τ_v). Note that θ_m is the angle of rotation of the motor with inertial load, and θ_d is the desired angle of rotation.

9.7 Consider a field-controlled dc motor with a permanent-magnet rotor and electronic commutation. In this case, the rotor magnetic field may be approximated to a constant. As a result, the motor magnetic torque may be approximately expressed as $T_m = k_m i_f$, where i_f = field current; k_m = motor torque constant.

Suppose that the load driven by the motor is purely inertial, with a moment of inertia J_L, which is connected to the motor rotor (of inertia J_m) by a rigid shaft. A schematic representation of this system is given in Figure P9.7a, where the field (stator) circuit is clearly shown. Note that the field resistance is R_f, the field inductance is L_f, and the input (control) voltage to the field circuit is v_f.

The mechanical dynamics of the motor system are represented by Figure P9.7b. Here, ω_m is the motor speed and b_m is the mechanical damping constant of the motor. The damping is assumed to be linear and viscous.

(a) In addition to the torque equation given above, give the field circuit equation and the mechanical equation (with inertial load) of the motor, in terms of the system parameters given in Figure P9.7a and b. Clearly explain the principles behind these equations.

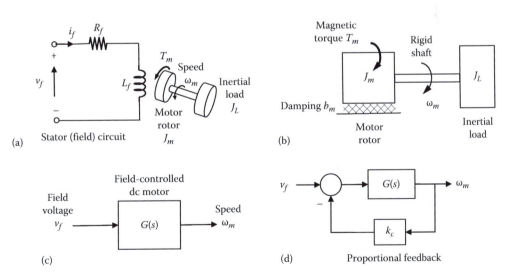

FIGURE P9.7
(a) A field-controlled dc motor with a permanent-magnet rotor; (b) mechanical system of the motor with an inertial load; (c) open-loop motor for speed control; (d) proportional feedback system for speed control.

(b) From the equations given in Part (a) above, obtain an expression for the transfer function $\dfrac{\omega_m}{v_f} = G(s)$ of the motor with the inertial load. The corresponding open-loop system is shown in Figure P9.7c.

(c) Express the mechanical time constant τ_m and the electrical time constant τ_e of the (open-loop) motor system in terms of the system parameters given in Figure P9.7a and b.

(d) What are the poles of the open-loop system (with speed as the output)? Is the system stable? Why? Sketch (*Note*: no need to derive) the shape of the output speed of the open-loop system to a step input in field voltage, with zero initial conditions. Justify the shape of this response.

(e) Now suppose that a proportional feedback controller is implemented on the motor system, as shown in Figure P9.7d, by measuring the output speed ω_m and feeding it back with a feedback gain k_c. Express the resulting closed-loop transfer function in terms of τ_m, τ_e, k, and k_c, where $k = k_m/R_f b_m$. What is the characteristic equation of the closed loop system?

(f) For the closed-loop system, obtain expressions for the undamped natural frequency and the damping ratio, in terms of τ_m, τ_e, k, and k_c. Give an expression for the control gain k_c, in terms of the system parameters τ_m, τ_e, and k, such that the closed-loop system has *critical damping*.

(g) Assuming that the closed-loop system is *under-damped* (i.e., the response is oscillatory), determine the time constants of this system. Compare them with the time constants of the open-loop system.

9.8 A dc motor with velocity feedback is given by the block diagram in Figure P9.8 (without the feedforward control path indicated by the broken lines). The input is u, the output is the motor speed ω_m, and the load torque is T_L. The electrical dynamics of the motor are represented by the transfer function $k_e/(\tau_e s + 1)$ and the mechanical dynamics of the motor are represented by the transfer function $k_m/(\tau_m s + 1)$ where s is the Laplace variable, as usual.

(a) Obtain a transfer function equation relating the output ω_m to the two inputs u and T_L in terms of the given parameters and the Laplace variable.

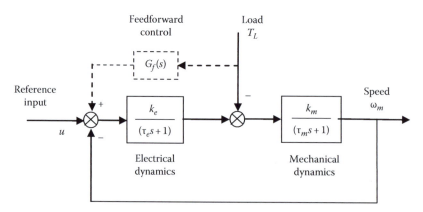

FIGURE P9.8
Control block diagram of a dc motor.

(b) Now include the feedfoward controller as shown by the broken line. Obtain an expression for the feedforward control transfer function $G_f(s)$, in terms of the given parameters and the Laplace variable, such that the effects of the load torque would be fully compensated (i.e., not felt in the system response ω_m).

9.9 Consider six control systems whose loop transfer functions (or, forward tfs with unity feedback) are given by

$$\text{(a) } \frac{1}{(s^2+2s+17)(s+5)} \qquad \text{(d) } \frac{10(s+2)}{(s^2+2s+101)}$$

$$\text{(b) } \frac{10(s+2)}{(s^2+2s+17)(s+5)} \qquad \text{(e) } \frac{1}{s(s+2)}$$

$$\text{(c) } \frac{10}{(s^2+2s+101)} \qquad \text{(f) } \frac{s}{(s^2+2s+101)}$$

Compute the additional gain (multiplication) k needed in each case to meet a steady-state error specification of 5% for a step input.

9.10 A tachometer is a device that is commonly used to measure speed, both rotatory (angular) and translatory (rectilinear). It consists of a coil that moves in a magnetic field. When the tachometer is connected to the object whose speed is to be sensed, the coil moves with the object and a voltage is induced in the coil. In the ideal case, the generated voltage is proportional to the speed. Accordingly, the output voltage of

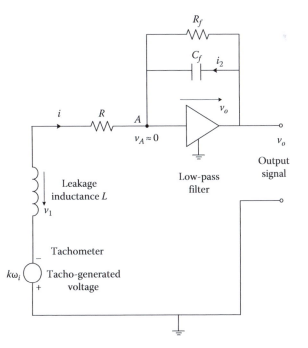

FIGURE P9.10
An approximate model for a tachometer-filter combination.

the tachometer serves as a measure of the speed of the object. High frequency noise that may be present in the tachometer signal can be removed using a low-pass filter.

Figure P9.10 shows a circuit, which may be used to model the tachometer-filter combination. The angular speed of the object is ω_i and the tachometer gain is k. The leakage inductance in the tachometer is demoted by L and the coil resistance (possibly combined with the input resistance of the filter) is denoted by R. The low-pass filter has an operational amplifier with a feedback capacitance C_f and a feedback resistor R_f. Since the operational amplifier has a very high gain (typically 10^5–10^9) and the output signal v_o is not large, the voltage at the input node A of the op-amp is approximately zero. It follows that v_o is also the voltage across the capacitor.

(a) Comment on why the speed of response and the settling time are important in this application. Give two ways of specifying each of these two performance parameters.

(b) Using voltage v_o across the capacitor C_f and the current i through the inductor L as the state variables and v_o itself as the output variable, develop a state space model for the circuit. Obtain the matrices A, B, C, and D for the model.

(c) Obtain the input–output differential equation of the model and express the undamped natural frequency ω_n and the damping ratio ζ in terms of L, R, R_f, and C_f. What is the output of the circuit at steady state? Show that the filter gain k_f is given by R_f/R and the discuss ways of improving the overall amplification of the system.

(d) Suppose that the percentage overshoot of the system is maintained at or below 5% and the peak time at or below 1 ms. Also, it is known that $L = 5.0\,\text{mH}$ and $C_f = 10.0\,\mu\text{F}$. Determine numerical values for R and R_f that will satisfy the given performance specifications.

9.11 Compare position feedback servo, tacho-feedback servo, and PPD servo with particular reference to design flexibility, ease of design, and cost.
Consider an actuator with transfer function

$$G_p = \frac{1}{s(0.5s+1)}$$

Design a position feedback controller and a tacho-feedback controller that will meet the design specifications $T_p = 0.09$ and $P.O. = 10\%$.

9.12 Consider the problem of tracking an aircraft using a radar device that has a velocity error constant of $10\,\text{s}^{-1}$. If the airplane flies at a speed of $2000\,\text{km/h}$ at an altitude of $10\,\text{km}$, estimate the angular position error of the radar antenna that tracks the aircraft.

9.13 Describe the operation of the cruise control loop of an automobile, indicating the *input*, the *output*, and a *disturbance input* for the control loop. Discuss how the effect of a disturbance input can be reduced using feedforward control.
Synthesis of feedforward compensators is an important problem in control system design. Consider the control system shown in Figure P9.13. Derive the transfer function relating the disturbance input u_d and the plant output y. If you have the complete freedom to select any transfer function for the feedforward compensator G_f, what would be your choice? If the process bandwidth is known to be very low and if G_f is a pure gain, suggest a suitable value for this gain.

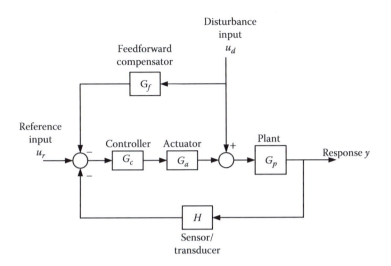

FIGURE P9.13
A feedback control system with feedforward compensation.

Suppose that a unit step input is applied to the system in Figure P9.13. For what value of step disturbance u_d will the output y be zero at steady state?

9.14 A control system with tacho-feedback is represented by the block diagram in Figure P9.14. The following facts are known about the control system:

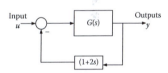

FIGURE P9.14
A control system with tacho-feedback.

1. It is a third order system but behaves almost like a second order system.

2. Its 2% settling time is 1 s, for a step input.

3. Its peak time is $\dfrac{\pi}{3}$ s, for a step input.

4. Its steady-state error, to a step input, is zero.

    (a) Completely determine the third order forward transfer function $G(s)$.

    (b) Estimate the damping ratio of the closed-loop system.

9.15 A satellite-tracking system (typically a position control system) having the plant transfer function $G_p(s) = 1/s(2s+1)$ is controlled by a control amplifier with *compensator*, having the combined transfer function $G_c(s) = K(s+1)/(\tau s+1)$ and unity feedback ($H = 1$). A block diagram of the control system is shown in Figure P9.15.

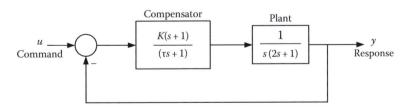

FIGURE P9.15
Block diagram of a satellite tracking system.

(a) Write the closed-loop characteristic equation (in polynomial form).

(b) Using the Routh–Hurwitz criterion for stability, determine the conditions that should be satisfied by the compensation parameter τ and the controller gain K in order to maintain stability in the closed-loop system.

(c) Sketch this stability region using K as the horizontal axis and τ as the vertical axis.

(d) When $K=5$ and $\tau=3$ find the poles (i.e., eigenvalues or roots) of the closed-loop system. What is the natural frequency of the system for these parameters values?

9.16 A system is given by the input–output differential equation.

$$\frac{d^3y}{dt^3} + 6\frac{d^2y}{dt^2} + 11\frac{dy}{dt} + 6y = 2\frac{du}{dt} + 6u$$

where $u=$ input, $y=$ output.

(a) Using Routh–Hurwitz criterion (and without solving the characteristic equation), determine how many poles of the system are on the left half plane. Is the system stable?

(b) For a unit step input, determine the steady state value of the response, by using the differential equation and explaining your rationale. Next, verify your answer using Final Value Theorem.

Suppose that all the poles of the given system are moved to the right by 1 (and the system zeros are not changed).

(c) Using Routh–Hurwitz criterion (and without actually solving the characteristic equation) determine the stability of the new system (with moved poles).

9.17 Consider the six transfer functions:

(a) $\dfrac{1}{(s^2 + 2s + 17)(s+5)}$

(b) $\dfrac{10(s+2)}{(s^2 + 2s + 17)(s+5)}$

(c) $\dfrac{10}{(s^2 + 2s + 2)}$

(d) $\dfrac{10(s+2)}{(s^2 + 2s + 2)}$

(e) $\dfrac{1}{s(s+2)}$

(f) $\dfrac{1}{s(s+2)}$

Suppose that these transfer functions are the plant transfer functions of five control systems under proportional feedback control. If the loop gain is variable, sketch the root loci of the five systems and discuss their stability.

9.18 The loop transfer function of a feedback control system is given by

$$GH = \frac{K}{s(s+1)(s+2)}$$

    (a)   Sketch the root locus of the closed-loop system by first determining the

        (i)   Location and the angles of the asymptotes

        (ii)   Break points

        (iii)   Points at which the root locus intersects with the imaginary axis and the corresponding gain value

    (b)   Fully justifying your answer, state whether the system is stable for $K = 10$.

    (c)   Suppose that a zero at −3 is introduced to the control loop so that

$$GH = \frac{K(s+3)}{s(s+1)(s+2)}$$

Sketch the root locus of the new system.

9.19 A control system has an unstable plant given by the transfer function

$$G_p(s) = \frac{1}{\left(s^2 - s + 1\right)}$$

By sketching root locus, discuss whether the plant can be stabilized using

    (a)   Proportional (P) feedback control

    (b)   Proportional plus derivative (PPD) control

    (c)   Proportional plus integral (PI) control

Are these observations intuitively clear?

9.20 Consider the feedback (closed-loop) control system shown in Figure P9.20. You are given the following loop transfer function for the system:

$$GH = \frac{1}{s^3 + 3s^2 + (K+2)s + 3K - 1}$$

where K is a control system parameter that can be varied.

    Determine the root locus of the closed-loop system as the parameter K changes from 0 to ∞. Specifically, you must first determine the

    (i)   Segments of the root locus on the real axis

    (ii)   Angles of the asymptotes and the location where the asymptotes intersect the real axis

    (iii)   Break points (as numerical expressions, which need not be evaluated)

    (iv)   Points at which the root locus intersects with the imaginary axis, if it does

    (v)   The range of values of K for which the closed-loop system is stable

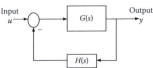

FIGURE P9.20
A feedback control system.

Note: You must give details and justify all your steps.

9.21 An interesting issue of force feedback control in robotic manipulators is discussed in the literature by Eppinger and Seering of the Massachusetts Institute of Technology. Consider the two models representing a robotic manipulator, which interacts with a workpiece, as shown in Figure P9.21a and b. In (a) robot is modeled as a rigid body (without flexibility) connected to ground through a viscous damper, and the workpiece is modeled as a mass-spring-damper system. In this case, only the rigid body mode of the robot is modeled. The robot interacts with the workpiece through a compliant device (e.g., remote center compliance—RCC device or a robot hand), which has an effective stiffness and damping. In (b), the robot model has flexibility, and the workpiece is modeled as a clamped rigid body, which cannot move. Note that in this second case a flexible (vibrating) mode as well as a rigid body mode of the robot are modeled. The interaction between the robot and the workpiece is represented the same way as in case (a). In both cases, the employed force feedback control strategy is to sense the force f_c transmitted through the compliant terminal device (the force in spring k_c), compare it with a desired force f_d, and use the error to generate the actuator force f_a. The controller (with driving actuator) is represented by a simple gain k_f. This gain is adjusted in designing or tuning the feedback control system shown in Figure P9.21c.

(a) Derive the dynamic equations for the two systems and obtain the closed-loop characteristic equations.

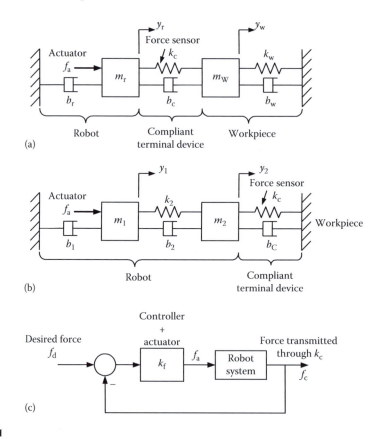

FIGURE P9.21
(a) A model for robot–workpiece interaction; (b) an alternative model; (c) a simple force feedback scheme.

(b) Give complete block diagrams for the two cases showing all the transfer functions.

(c) Using the controller gain k_f as the variable parameter, sketch the root loci for the two cases.

(d) Discuss stability of the two feedback control systems. In particular, discuss how stability may be affected by the location of the force sensor. (*Note*: The two models are analytically identical except for the force sensor location.)

9.22 The open-loop transfer function of a control system is given by

$$G(s) = \frac{s+1}{(s^2 + s + 4)}$$

(a) With this transfer function, if the loop is closed through a unity feedback, determine the phase margin. You should use direct computation rather than a graphical approach.

(b) If the input to the open-loop system is $u = 3\cos 2t$ determine the output y under steady conditions.

9.23 (a) Define phase margin (PM) and gain margin (GM) of a system. For what type of linear system, the PM and GM considerations may not be appropriate in assessing relative stability?

(b) An approximate relationship for PM in terms of damping ratio ζ is given by:

$$\phi_m = 100\zeta \text{ degrees}$$

Give the main steps of deriving this result using a damped oscillator model.

(c) A position control system, which uses a dc motor to drive an inertial load, is represented by the block diagram given in Figure P9.23.
The forward transfer function is given by $G(s) = 2/s(2s+1)$

(i) Sketch the Nyquist diagram of G. On this basis, comment on the stability of the closed loop system.

(ii) Compute the phase margin and gain margin of the closed loop system.

(iii) Determine the exact damping ratio of the closed-loop system and check whether the result agrees with the approximate relation given in Part (a).

(iv) A reference position input of $u = 3\sin t$ is applied to the system. Determine the position response y at steady state.

9.24 (i) Which control method would you recommend for each of the following applications:

(a) Servo control of a single-axis positioning table with a permanent-magnet dc motor (linear).

(b) Active control of a vehicle suspension system (linear, multivariable).

(c) Control of a rotary cement kiln (nonlinear, complex, difficult to model).

FIGURE P9.23
A position control system.

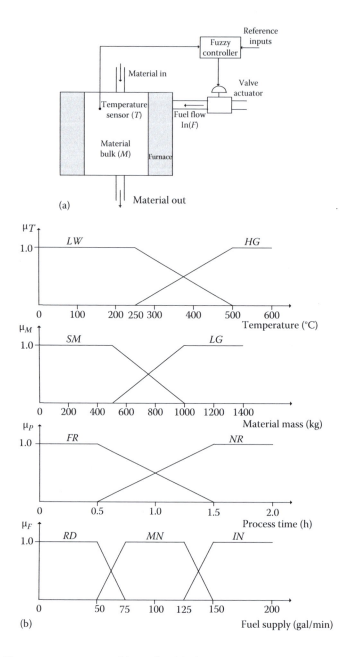

FIGURE P9.24
(a) A metallurgical heat treatment process; (b) membership functions.

   (ii) A metallurgical process consists of heat treatment of a bulk of material for
         a specified duration of time at a suitable temperature. The heater is con-
         trolled by its fuel supply rate. A schematic diagram of the system is shown in
         Figure P9.24a.

         The following fuzzy quantities are defined, with the corresponding states:

         T: Temperature of the material (*LW* = low; *HG* = high)
         M: Mass of the material (*SM* = small; *LG* = large)

P: Process termination time   (*FR*=far; *NR*=near)
F: Fuel supply rate   (*RD*=reduce; *MN*=maintain; *IN*=increase)

The membership functions of these quantities are given in Figure P9.24b. A simple rule base that is used in a fuzzy controller for the fuel supply unit is given below:

If *T* is *LW* and *P* is *FR* then *F* is *IN*
Or if *T* is *HG* then *F* is *RD*
Or if *M* is *SM* and *P* is *NR* then *F* is *MN*
Or if *M* is *LG* and *P* is *FR* then *F* is *IN*
Or if *P* is *NR* then *F* is *RD*
End if.

At a given instant, the following set of process data is available:

Temperature=300°C
Material mass=800 kg
Process operation time=1.3 h

Determine the corresponding inference membership function for the fuel supply, and a crisp value for the control action. Comment on the suitability of this inference.

9.25 Consider the experimental setup of an inverted pendulum shown in Figure P9.25.

Suppose that direct fuzzy logic control is used to keep the inverted pendulum upright. The process measurements are the angular position, about the vertical (*ANG*) and the angular velocity (*VEL*) of the pendulum. The control action (*CNT*) is the current of the motor driving the positioning trolley. The variable *ANG* takes two fuzzy states: positive large (*PL*) and negative large (*NL*). Their memberships are defined in the support set [−30°, 30°] and are trapezoidal. Specifically:

$\mu_{PL}=0$ for *ANG*=[−30°, −10°]
     = linear [0,1.0]   for *ANG*=[−10°, 20°]
     = 1.0   for *ANG*=[20°, 30°]

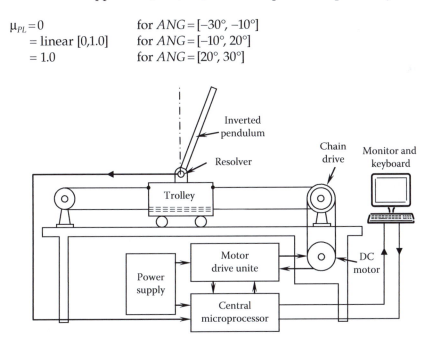

FIGURE P9.25
A computer-controlled inverted pendulum.

$\mu_{NL} = 1.0$ for $ANG = [-30°, -20°]$
      = linear [1.0, 0]   for $ANG = [-20°, 10°]$
      = 0   for $ANG = [10°, 30°]$

The variable *VEL* takes two fuzzy states *PL* and *NL*, which are quite similarly defined in the support set [−60°/s, 60°/s]. The control inference *CNT* can take three fuzzy states: Positive large (*PL*), no change (*NC*), and negative large (*NL*). Their membership functions are defined in the support set [−3A, 3A] and are either trapezoidal or triangular. Specifically:

$\mu_{PL} = 0$ for $CNT = [-3A, 0]$
     = linear [0,1.0]   for $CNT = [0.2A]$
     = 1.0   for $CNT = [2A, 3A]$
$\mu_{NC} = 0$ for $CNT = [-3A, -2A]$
     = linear [0, 1.0]   for $CNT = [-2A, 0]$
     = linear [1.0,0]   for $CNT = [0, 2A]$
     = 0   for $CNT = [2A, 3A]$
$\mu_{NL} = 1.0$ for $CNT = [-3A, -2A]$
     = linear [1.0,0]   for $CNT = [-2A, 0]$
     = 0   for $CNT = [0, 3A]$

The following four fuzzy rules are used in control:

	If	*ANG*	Is	*PL*	And	*VEL*	Is	*PL*	Then	*CNT*	Is	*NL*
Else	If	*ANG*	Is	*PL*	And	*VEL*	Is	*NL*	Then	*CNT*	Is	*NC*
Else	If	*ANG*	Is	*NL*	And	*VEL*	Is	*PL*	Then	*CNT*	Is	*NC*
Else	If	*NAG*	Is	*NFL*	And	*EL*	Is	*NFL*	Then	*CT*	Is	*LP*
End	If.											

(a) Sketch the four rules in a membership diagram for the purpose of making control inferences using individual rule-based inference.

(b) If the process measurements of $ANG = 5°$ and $VEL = 15°/s$ are made, indicate on your sketch the corresponding control inference.

9.26 Compare analog control and direct digital control for motion control in high-speed applications of industrial manipulators. Give some advantages and disadvantages of each control method for this application.

10

Case Studies in Mechatronics

Study Objectives

- Engineering design
- Mechatronic design
- Economic analysis
- Case studies
- Examples and projects

10.1 Introduction

Mechatronics is a multidisciplinary engineering field, which involves the synergistic integration of several areas such as mechanical engineering, electronic engineering, control engineering, and computer engineering. Similarly, the design and development of a mechatronic system will require an integrated approach to deal with the subsystems and subprocesses of a mixed system, specifically, an electromechanical system. As reiterated in the book, the subsystems of a mechatronic system should not be designed or developed independently without addressing the system integration, subsystem interactions and matching, and the intended operation of the overall system. Such an integrated approach will make a mechatronic design more optimal than a conventional design. In this chapter, some important issues in the design and development of a mechatronic product are highlighted. As illustrative examples, several case studies of practical mechatronic systems are provided.

10.2 Engineering Design

Engineering design is the process of developing engineering products, processes, or systems. It is different from solving mathematical and scientific problems in many ways. In particular, engineering design is not the same as solving problems of engineering science. The main differences are highlighted in Table 10.1.

The typical steps followed in the process of engineering design are listed below:

1. Identify the need for the required design.
2. Describe the objectives of the designed product or system.

TABLE 10.1

Comparison of Engineering Design and Problem Solution in Engineering Science

	Engineering Design	Engineering Science
Problem statement	Not complete or final. Can be qualitative and fuzzy	Complete and final
Involved knowledge, subjects or fields	Flexible and not conclusive	Clear from the problem
Solution	Multiple and not unique	Typically unique
Solution steps	Flexible, incompletely known and may determine the outcome	Known by expert and will not change the solution

3. Gather preliminary performance specifications (some may be qualitative).

4. Outline several conceptual designs.

5. Identify the technologies and expertise that are needed.

6. Identify available (e.g., commercial, off-the-shelf) components or subsystems for each conceptual design.

7. Carry out a feasibility study and an economic (cost–benefit) analysis for each conceptual design.

8. Establish quantitative and final performance specifications.

9. Obtain preliminary detailed designs.

10. Evaluate the designs through modeling, analysis (including economic), and the use of experts and narrow down the design alternatives (to one or two).

11. Carry out a detailed design.

12. Evaluate the detailed design through analysis, prototype development, and testing, etc.

Typically, it is necessary to revisit one or more steps several times before the final design is reached. In particular, several iterations of some or all of the indicated design steps may be necessary. In this context, it is less time consuming and more economical to make design changes early in the design process. For example, design changes in the modeling step will be less costly and faster than those after prototyping. Completing the final development, technology transfer, implementation, production, maintenance, and so on may be incorporated into the design process. Social and environmental factors are increasingly becoming important in engineering design and should be incorporated in the design process.

10.2.1 Engineering Design as an Optimization Problem

An engineering design problem may be described by the following two items:

1. Design objectives (or performance function)
2. A set of constraints

Since many design solutions may be possible that satisfy these two requirements, the best solution should be chosen. In this sense, engineering design is analogous to the "constrained optimization problem" in applied mathematics. However, a mathematical solution is typically not possible for the engineering design process primarily because the two items given above are rather complex, incomplete, fuzzy, and not analytic in general. In particular, the objective function may not be differentiable and hence its gradient may not be computable. Also, even in special cases that can use standard optimization tools, the solution may be trapped in a local optimum instead of the desired global optimum. Then, gradient-based optimization techniques will not be applicable. However, methods of soft computing such as evolutionary computing and genetic programming/algorithms may be effectively used in the optimization of engineering design.

Example 10.1

A machine to automatically and accurately remove the heads of fish, in a fish-processing plant, is to be designed. For this design problem, some of the objectives and constraints are given below.

Design objectives: Cutting accuracy better than 1.0 mm; throughput rate (fish-processing rate) of at least 2 fish/s.

Constraints: The machine should be washable at the end of a shift, the cost should not be more than $50,000, the size should not exceed 2 m (length) × 0.75 m (width) × 2 m (height), and the weight should not exceed 400 kg.

Example 10.2

Two design concepts for a machine to automatically and accurately cut off heads in fish are given in Figures 10.1 and 10.2. These original concepts have evolved into the final prototype shown in Figure 10.3.

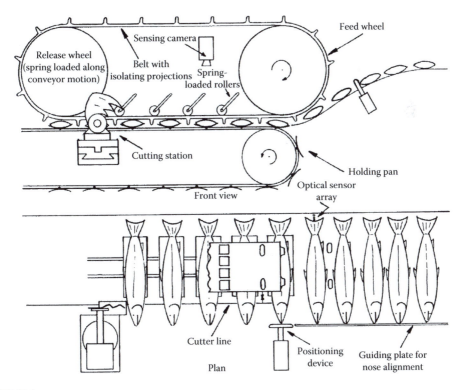

FIGURE 10.1
Design concept of a machine for fish head removal (rough sketch).

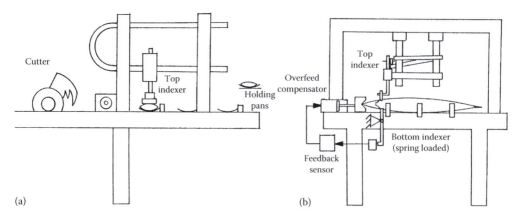

FIGURE 10.2

An alternative design concept of a machine for fish head removal (a sketch): (a) Front view; (b) side view.

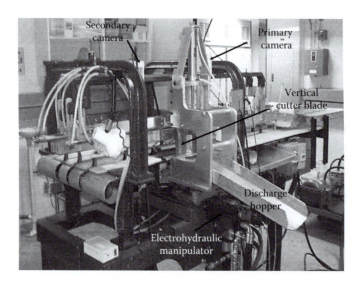

FIGURE 10.3

The final industrial prototype of the automated machine for fish head removal.

10.2.2 Useful Terminology

Some terminology that is useful in engineering design is listed below.

Design for manufacturability: The design process of a product should necessarily take into account the manufacturing aspects (e.g., parts count, ease of machining, ease of assembly, availability of material or components). Communication with the manufacturing team is important as early as possible in the design process. Some such communications may be made using computer-aided design (CAD) drawings and solid models in soft- or hard-copy form.

Concurrent design: It is desirable to simultaneously design all components or subsystems of the required system. This is known as concurrent design and is particularly relevant in multidisciplinary or mechatronic design. Essentially, concurrent design implies carrying out all aspects of design (all functions, all domains, all components, etc.) simultaneously.

According to another terminology, design through regular interaction with the manufacturing team (i.e., design for manufacturability) is also known as concurrent design. In conventional engineering design, various interactions and coupling of functions, components, effects, and so on in the designed system are taken into account by proper interaction and communication among members and teams involved in the design (and manufacturing, etc.). In modern, mechatronic design, this is incorporated concurrently into the design procedure itself.

Design for sustainability: This is also known as the design for disposability or recycling. Here, the reuse of components of the product (i.e., recycling) at the end of the product life (retirement) is considered in the design process. Also, the environmental impact of disposing the product is minimized through improved design practices.

Rapid prototyping: Using CAD models and linking them with computer-controlled production/manufacturing machinery to rapidly manufacture a product is called rapid prototyping. Design for manufacturability is directly applicable here.

Mechatronic design: Since mechatronics is a multidisciplinary field of engineering, mechatronic design should involve more than one engineering discipline; notably mechanical engineering, electrical and electronics engineering, and computer and information engineering. Here, the integrated and concurrent design of all components and all functions of the product or system should be considered. In this sense, it involves concurrent design.

10.2.3 Design Projects

Engineering design projects are carried out by teams of engineers. This is particularly so since typically multiple knowledge domain disciplines and expertise are involved. All aspects, components, or developments, however, may not need all members of the team. Regular record keeping and communication are crucial in view of this. It is important to get not only the production personnel, but also researchers and technology developers, clients/customers, marketing people, and even packaging and shipping (transportation) personnel involved as well for discussion and input so that the viewpoints, needs, objectives, and capabilities of all of them are accommodated into the design. Communication among members of the design team and with other teams (e.g., the manufacturing department) may be done verbally or by using text and graphics. A detailed report is used to document the design and to communicate the information to others. Both its technical quality and the literary quality should be high. Oral presentation (e.g., a PowerPoint presentation) is also an effective and often indispensable part of communication in engineering design.

Once the design team is established (by you and your supervisor, in consultation with potential members) depending on the needed skills, complexity of the project, logistics, and other considerations, the project may be carried out using the following steps:

1. Based on the purpose of the design and through brainstorming, think of several conceptual design ideas for the device/system.
2. Write a proposal for the project describing the designed system/device, its purpose, functional requirements, possible performance considerations, the planned approach for the design process, etc.
3. Establish (by consulting experts, information sources, industry, potential users, manufacturers, etc.) and identify performance parameters and design specifications.

4. Carry out a preliminary evaluation of the conceptual designs through feasibility considerations, modeling, analysis, computer simulation, economic (cost–benefit) analysis, component availability, etc. and narrow down the conceptual design.

5. Carry out a detailed design complete with material selection, component selection/design, etc., together with manufacturing and assembly considerations, etc.

6. Complete the engineering drawings for the design.

7. Develop the prototype.

8. Test the prototype to evaluate its performance, and make modifications if necessary.

9. Prepare a project report.

10. Make an oral presentation to your supervisor and others.

10.2.4 Quality Function Deployment

The quality function deployment (QFD) method was developed in the 1970s in Japan, specifically to integrate the customer requirements into the engineering design process. Through the use of this method, development is reduced by about 60% and the development time is reduced by one third. A large majority of companies in the industrial world (including in the United States) extensively use the QFD method.

The method involves the development of a QFD chart (or matrix), which matches the customer needs with the engineering requirements of the designed product or system, and also compares the design with competing products or systems (i.e., benchmarking).

Example 10.3

Consider the design of a handheld body massager. A preliminary QFD chart for its design is given in Figure 10.4.

10.2.5 Design Report

The preparation of a professional design report is important in any design project. The typical structure of a design report is outlined below.

Title page: The title page should include the title of the design, names of the design personnel (report authors), company/division name, coordinates (address, e-mail address, etc.), and date.

Executive summary/abstract: The executive summary/abstract gives a summary of the report in fairly nontechnical language. It outlines the design objectives, design process, and the design outcome. It may indicate major accomplishments and problems.

Table of contents: The table of contents lists at least the chapters and their main section headings (and often the subsection headings as well) with their page numbers.

List of figures: The list of figures lists the figure numbers and their titles.

List of tables: The list of tables lists the table numbers and their titles.

Nomenclature: The nomenclature gives a list of notations used in the report and their short definitions. It includes English and special (Greek) characters and symbols used to

		Motor speed	Motor power	Motor weight	Casing material density	Yield strength of casing	GE	Philips	Panasonic
	Lightweight	×	×	×					•
	Comfortable	×	×	×			•	•	
Consumer requirements	Durable				×	×	•		
	Low cost	×	×	×	×			•	•
	Low noise	×					•	•	
	Attractive			×					•

Engineering requirements (spans Motor speed, Motor power, Motor weight, Casing material density, Yield strength of casing)

Benchmarking (spans GE, Philips, Panasonic)

Engineering parameters:
- 480 rpm (Motor speed)
- 100 W (Motor power)
- 0.1 kg (Motor weight)
- $1 \times 10^3\ \mathrm{kg/m^3}$ (Casing material density)
- $1 \times 10^8\ \mathrm{N/m^2}$ (Yield strength of casing)

FIGURE 10.4
A QFD chart for a handheld body massager.

denote variables, parameters, operations, etc. in the body of the report. They should be properly ordered (in alphabetical order).

Acknowledgments: This section acknowledges the key people (other than the report authors) who have assisted in the design and report preparation.

Body of the report: The body of the report will consist of several chapters. The first chapter is typically an introductory chapter giving the purpose of the report, objectives of the design, history background work, outline of the design process, key characteristics of the design outcome, and organization of the rest of the report. The last chapter should discuss the main accomplishments of the design process, brief discussions of key issues of the design, conclusions (major inferences drawn from the design and the discussion), recommendations (the recommended course of action based on the conclusions), and possible further improvements and shortcomings.

References/bibliography: This ends the main body of the report. The References section lists all the sources of information that were used in the design, which are cited (referred to) in the report. Alternatively, sources that are cited and also other sources of information that are not cited in the report may be listed under Bibliography. Typically include the names of all the authors of the information source (books, reports, journal papers, conference papers, Web sites, etc.); the title of the source; location (city and state or country); Volume (Vol.) and issue number (No.) for papers (and some reports) or ISBN; and for books, provide a range of page numbers (pp.) and the year (and month for papers or reports) of publication.

Appendices: One or more appendices may be included after the references/bibliography section. They will include useful information, data, and procedures that are not immediately necessary in the body of the report or are too complex and will interfere with the flow of the material in the main report. Also, the material in the appendices may directly come from other sources (e.g., catalogs, handbooks), and may include computations, analytical derivations, computer source codes, drawings, and graphics.

10.2.6 Design of a Mechatronic System

A mechatronic system may be treated as a control system, consisting of a plant (which is the process, machine, device, or system to be controlled), actuators, sensors, interfacing and communication structures, signal modification devices, and controllers and compensators. The function of the mechatronic system is primarily centered at the plant. Actuators (Chapter 7), sensors (Chapter 6), and signal modification devices (Chapter 4) might be integral with the plant itself, or might be needed as components that are external to the plant for proper operation of the overall mechatronic system. Controllers (Chapter 9) are an essential part of a mechatronic system. They generate control signals to the actuators in order to operate (drive) the plant in a desired manner (Chapter 5). These various components may not be present as physically autonomous units in a mechatronic system in general, even though they may be separately identified, from a functional point of view. For example, an actuator and a sensor might be an integral part of the plant itself. The design of a mechatronic system can be interpreted as the process of integrating (physical/functional) components such as actuators, sensors, signal modification devices, interfacing and communicating structures, and controllers with a plant so that the plant in the overall mechatronic system will respond to inputs (or commands) in a desired manner.

10.3 Robotics Case Study

Robots are mechatronic devices. They may be employed either individually or within a workcell to carry out industrial tasks, service functions, and household chores. In either case, the design, development, and selection of a robot for a specific task have to be carried out by giving careful attention to practical, economic, and social considerations. Issues such as process requirements, commercial availability, cost and economic realities, time constraints, and human factors have to be taken into consideration in the process of robotization.

10.3.1 General Considerations

Applications of robots may be classified into the following broad categories of activity:

1. Point-to-point motion
2. Trajectory following
3. Local/fine manipulation

A specific task may need a combination of these activities. Following are a few examples: the mixing and dispensing of drugs (activities *a* and *c*), assisting a disabled person to walk (activities *b* and *c*), vacuum cleaning a floor (activity *b*), loading and unloading of parts to and from machine tools (activity *a*), tool replacement (activities *a* and *c*), spot welding (activity *a*), seam welding (activity *b*), die casting (activity *c*), sealant application (activity *b*), packaging (activities *a* and *c*), and product inspection (activities *a* and *c*). There are many benefits of using robots in practical applications. They include improved task flexibility, elimination of low-quality human labor, increased productivity, improved product quality, improved utilization of capital equipment, hazard reduction, robustness to external economic factors (e.g., inflation), round-the-clock and on-demand operation, better inventory management, better production planning, increased production competitiveness and flexibility, improved work environment, and overall improvement of the quality of life. Prior to robotization of a task, however, it is necessary to evaluate many factors such as appropriateness, feasibility, time constraints of production, costs, and benefits. In an industrial application, for example, it is advisable to carry out the following studies.

1. **Evaluating the plant for tasks of potential application of robotics**
   Here, the level and nature of automation (hard versus flexible automation), needed level of production flexibility (e.g., parts on demand), degree of structure in the operation and plant environment (for highly structured and fixed processes, hard automation without robots might be more appropriate), work environment (e.g., potential hazards, user friendliness), desired production rates and volumes, and the existence of similar proven applications (on site or elsewhere) should be considered.

2. **Determining the features of robots needed for the specific application**
   Important considerations in this context include payload, operating speed/ bandwidth, accuracy (repeatability, precision, resolution, etc.), work envelope (reachability), method of actuation (dc or ac servomotors, stepper motors, hydraulic, pneumatic), desired robotic structure (degrees of freedom, revolute or prismatic joints, and in what combination), end-effector requirements (gripper, hand, tools, sensors, and the characteristics of objects that are to be handled), method of control (servo, adaptive, hierarchical, etc.), instrumentation requirements (sensory, monitoring, data acquisition, fixturing, control, and coordination needs), operating environment (moisture, chemicals, fire hazards, dust, temperature, etc.), and commercial availability of the desired types of robots.

3. **Studying the robot installation requirements and their consequences in plant operation**
   Relevant considerations will include necessary utilities (dc or ac, single-phase or three-phase power, compressed air, etc.), installation timing (down-time of operations, product demands, etc.), other machinery within workcells (interfacing, communication, networking, control, etc.), plant layout (integration with raw material, product, tool flow plans, services, operator interaction and interfaces, safety, etc.), and personnel (training, programming, maintenance, operation, etc.).

4. **Performing an economic analysis**
   This is a cost–benefit evaluation, taking into account such considerations as availability of off-the-shelf units, capital investment, down-time, efficiency, wastage reduction, labor reduction, operating costs, and the rate of return on investment. Specifically, the cost of purchase and installation of the robot, useful life, maintenance, service, and utility costs, depreciation, and money cost should be considered in the analysis. It may be estimated, for example, that the approximate hourly cost of labor has increased rather exponentially from $6.00 in 1975 to $35.00

in 2004, whereas the approximate hourly cost of the operation of a robot has only increased from \$4.00 to \$8.00 during the same period. The former cost continues to rise and the latter is leveling off.

5. **Addressing human relations considerations**
   Here, the relationship between workers and management, loss of employment due to automation, massive loss of employment due to plant closures (without automation), inefficiency, worker retraining, and union representation should be considered.

10.3.2 Robot Selection

The selection of the "best" robot for a given task is a crucial step in robotic application. Clearly, the term "best" is used within the set of constraints that govern the problem; for example, cost, timing of installation and operation, and the availability of hardware and personnel. For precision tasks (e.g., manufacturing products of fine tolerance), it is important to make sure that the accuracy specifications (including repeatability and motion resolution) can be met. In such applications, structural integrity and strength (e.g., robot stiffness) also would be prime considerations. Other aspects, such as the controller and its architecture (e.g., open and user-programmable at low level), compatibility and ease of communication with other machinery and their controllers in coordinated operation (e.g., in workcells), necessary end effectors, tools, fixtures, instrumentation, ease of programming, and operator friendliness would be important. Special requirements (e.g., clean room tasks) may require custom modifications and component/system qualifications (i.e., analysis and/or testing to evaluate and determine the suitability for the specific application). A typical set of steps that would be followed in selecting an industrial robot is given below:

1. **Define the tasks, which are to be carried out by the robot**
   First, a verbal description of the task sequence would be appropriate (e.g., pick a bulb from the conveyor, inspect for faults, decide a category, place it in the appropriate bin). Next, the motion sequence should be defined in a quantitative/analytical form, for example, giving time sequences and/or trajectories for the set of actions. Also, define the time sequences for nonmotion tasks (e.g., grasp the object, release the object, wait for the part). Error tolerances should be specified as well.

2. **Develop robot specifications for the tasks**
   For example, work envelope, speed limits and cycle time, force and payload capability, repeatability, and motion resolution have to be specified for the robot.

3. **Identify the necessary mechanical structure for the robot**
   The required number of degrees of freedom, type of joint combination (e.g., SCARA robot with three revolute joints and one prismatic joint), required end-effector motions (e.g., complete or partial rotations, strokes of linear motions), and lengths of robot links have to be decided upon.

4. **Identify the sensor and actuator preferences**
   The nature of the desired drive system for each joint (e.g., backlash-free transmission or harmonic drive with ac servomotor, direct-drive joint) and the associated sensory preferences (e.g., incremental optical encoders, resolvers, joint torque sensors) have to be identified as completely as possible.

5. Identify the end-effector requirements

Depending on the expected tasks, a variety of end effectors might be needed. Simple grasping operations may need basic two-finger grippers. More sophisticated tasks would require robotic hands, tools, and custom devices with adequate dexterity, motion resolution, force resolution, and sensors (e.g., tactile, wrist torque and force, optical, and ultrasonic sensors).

6. Identify the control and programming requirements

Decide whether high-level task programming alone is adequate or whether low-level direct programming of the joint controllers would be needed. Also, depending on the expected operators and programmers, decide upon the desired difficulty level of programming. Also, communication and interfacing needs and compatibility with other interacting devices and tools have to be established; for example, when the robot is expected to be an integral part of a larger system, such as a workcell.

7. Identify the user interface needs

This is somewhat related to Item 6 above. A graphic user interface (GUI) that suits the specific application has to be considered. Generally, it has to be user friendly with simple input/output means such as touch screens, voice activation, and handwritten commanding, particularly when technologically nonsophisticated users are involved.

8. Decide on a budget and contact suppliers

If a suitable robot is commercially available within budget and within the required time frame, the selection would be straightforward. Otherwise, several iterations of robot specifications and identification would be needed and each time some of the specifications would need relaxing/modifying.

10.3.2.1 Commercial Robots

The task of selecting a robot is greatly simplified if the required specifications can be matched to those of a commercially available robot. A typical set of commercial robots and some useful specifications attributed to them are listed in Table 10.2. Since an end effector has to be chosen separately and does not usually come with the robot, the payload that is indicated in the table includes the weight of the end effector. The *repeatability* of a robot specifies how accurately a robot can reach a commanded point in space. A repeatability error may result from such factors as Coulomb friction, backlash, and poor control and is one of many factors that determine the overall *accuracy* of a robot. The accuracy itself will depend on such considerations as the speed of operation, payload, and the specific trajectory of motion. The *resolution* of a robot, like repeatability, is a lower bound for accuracy and represents the smallest motion increment that can be executed. Again, resolution may depend on factors such as friction, backlash, unknown disturbances, the resolution of digital motion transducers such as encoders, the bit size of a control command, and for a robot that uses stepper motors, the step size of incremental motion.

Often, what is given in product specifications of commercial robots is the no-load speed. A more meaningful specification is the *cycle time* for a specified pick-and-place cycle when carrying the specified payload. Specifically, by assuming a triangular speed profile, where the robot steadily accelerates from rest after the "pick" to reach the peak speed and then steadily decelerates to the "place" point, the peak velocity v_{peak} is given by

TABLE 10.2

Data for Several Commercial Robots

Robot	Mechanical Structure	Drive System	Payload (kg)	End-Effector Speed, with Payload (m/s)	Repeatability (mm)
PUMA 560	6-axis revolute	Geared dc servomotors	2.3	0.5	0.10
SEIKO RT5000	4-axis cylindrical	Gear/rack/belt/ harmonic drive, dc servomotors	5.0	2.0	0.04
SCORA-ER 14	4-axis SCARA (3 revolute + 1 prismatic)	Harmonic-drive, dc servomotors	2.0	1.9	0.05
Adept-3	4-axis SCARA (3R + 1P)	Direct-drive, dc servomotors	25.0	0.7	0.05
Pana Rob HR-50	4-axis SCARA (3R + 1P)	AC servomotors with speed reducers	5.0	1.2	0.05
Staubli RX 130	6-axis revolute	AC servomotors with speed reducers/gears	12.0	1.5	0.025
CRS Robotics A 465	6-axis revolute	DC servomotors with harmonic drives and belts	3.0	1.0	0.05
GMF Robotics M-300	4-axis cylindrical	DC servomotors	100.0	Moderate	1.0
Hitachi A4030	4-axis SCARA	DC servomotors with speed reducers	10.0	1.5	0.05

$$v_{peak} = \frac{2\Delta\ell}{\Delta t} \qquad (10.1)$$

where
Δt is the cycle time
$\Delta\ell$ is the pick-and-place distance
The corresponding acceleration (and deceleration) is given by

$$a = \frac{2v_{peak}}{\Delta t} \qquad (10.2)$$

or

$$a = \frac{4\Delta\ell}{\Delta t^2} \qquad (10.3)$$

If the payload is M, the force f_e exerted at the end effector for steady acceleration to peak speed is given by

$$f_e = Ma \qquad (10.4)$$

or

$$f_e = \frac{4M\Delta\ell}{\Delta t^2} \tag{10.5}$$

The capabilities of the robot, as specified by the cycle time, payload, peak operating speed, and the force at peak speed and steady acceleration have to match with the requirements of the robotic task.

As an example, consider a robot that is able to execute a pick-and-place operation over a distance of 0.5 m in less than 1.0 s, carrying a payload of 30 kg. Suppose that it has a resolution of ±0.01 mm and a repeatability of ±0.05 mm. We can determine the maximum operating speed of the robot when carrying a payload of 30 kg. Also, we can determine the maximum force that can be exerted by the end effector of the robot at its peak operating speed.

Here, $\Delta\ell = 0.5$ m and $\Delta t = 1.05$. Hence, from Equation 10.1 we have

$$v_{peak} = \frac{2 \times 0.5}{1.0} = 1.0 \ \text{m/s}$$

which is the maximum operating speed with the payload. From Equation 10.2,

$$a = \frac{2 \times 1.0}{1.0} = 2.0 \ \text{m/s}^2$$

From Equation 10.4, $f_e = 30 \times 2.0 = 60.0$ N.

This is the maximum force that can be exerted by the end effector.

10.3.3 Robotic Workcells

A robotic workcell is a group of machinery (generically termed *machine tools*) such as material removal devices (e.g., computer-numerical-control or CNC mills, lathes, drills, borers, cutters), material handling equipment (e.g., conveyors, gantry mechanisms, positioning tables, automated guided vehicles or AGVs) along with one or more robots working together to achieve a common task objective under the supervision and control of a *cell host* computer. Each machine tool will be controlled by its own *machine-tool controller*, but will communicate with and be coordinated by the cell host. The *cell supervisor* is a high-level program that runs the cell host. A production/manufacturing system, process plant, or factory may consist of two or more workcells, which communicate with each other through a local area network (LAN) under the supervision of a *system control computer*. Note that the machine tools and robots operate under their own controllers running their own programs (e.g., CNC parts programs and robot motion/operation sequence programs) on command, coordination, and supervision by the cell host. Parts and material movement within a workcell are guided by the cell host. During a parts run, very little communication is needed with the system control computer.

Autonomous operation is a desired feature for a robotic workcell. In this context, a workcell should be able to carry out its task without relying on external assistance. In particular, unmanned and automated operation, flexibility, and the capabilities of self-reconfiguration, self-repair, learning, and adaptation would be desirable. Smart sensors, intelligent controllers, programmable machine tools, and effective communication and control structures would be needed as well. Networking and communication protocol considerations are paramount here.

The design philosophy for robotic workcells is based primarily on flexibility and autonomy. To achieve flexibility, the use of programmable and modular components along with communication and control architectures that allow fast restructuring would be necessary. Accordingly, robotic workcells fall into the category of *flexible automation*. For autonomous operation, sensors to obtain the necessary information on each component of the workcell and its environment and intelligent control systems that can handle unfamiliar and unexpected conditions, along with the capabilities of self-reconfiguration and repair would be needed. Structured communication and control, particularly a hierarchical control architecture where the required information and control signals can be related to a specific and well-defined layer and where lower layers may be modified with little effect on the upper layers would be desirable. Allowance should be made for the ease of upgrading and expansion of the workcell. Standard and programmable components that are compatible with the plant network and the associated communication protocols and whose operation can be defined in terms of input–output characteristics would help simplify workcell integration, component replacement, and workcell modification. The following guidelines may be followed in the design and development of a robotic workcell:

1. Identify the workcell tasks and process requirements (production rates, tolerances, etc.). Reach a compromise between product flexibility and production rate.

2. Identify the machine tool and robot requirements, limiting if possible, to either existing or commercially available units.

3. The initial workcell need not be fully autonomous. Integrate humans if feasible.

4. Develop a workcell architecture. Use simple networking topology and communication protocols and existing computer technology and hardware.

5. Identify the necessary accessories such as grippers, fixtures, sensors, and instrumentation. Use simple and off-the-shelf components with simple interfacing requirements where possible.

6. Modify and enhance existing components and controllers (e.g., by adding sensors, processors, memory, software) as is required and feasible.

7. Identify the critical components of the workcell and consider the possibility of incorporating either software or hardware redundancy.

When designing a flexible production system having multiple workcells, an analysis has to be made to determine the workload demand on each component under normal operating conditions. Then, in selecting the workcell components, care has to be exercised to ensure that the component capacity is greater than or equal to the workload demand and to reach a somewhat optimal balance between these two levels. A dynamic restructuring system with proper monitoring, control, and rescheduling capabilities will be able to accomplish this. If there is excess capacity in each component, it may be possible to share common components within or between workcells, thereby releasing appropriate components that operate well below their capacity. Also, if component overloading is present, a similar procedure may be used to shed the overload into a partner component that operates below capacity.

10.3.4 Robot Design and Development

The main steps of the design and development of a customized robot may be given as follows:

1. Arrive at kinematic (motion) and dynamic (force/torque-motion) specifications for the range of tasks to be carried out by the robot. Modeling and analysis will be required.

2. Determine the geometric requirements (e.g., degrees of freedom, revolute and/or prismatic joints, length of links, motion resolution, and accuracy) for the robot, based on the task kinematic specifications. Modeling and analysis will be required.

3. Determine the geometric requirements for the end effector. Modeling and analysis will be required.

4. Determine the dynamic (forces and torque) requirements for the end effector and the robotic joints based on the dynamic specifications. Modeling and analysis will be required.

5. Select actuators (type, load, motion, and power capacities) from commercially available units to match the kinematic and dynamic requirements. Analysis and design will be required.

6. If available, direct-drive actuators cannot match the requirements. Select motion transmission units (perhaps available as integral with the actuators) to meet the requirements. Analysis and design will be required.

7. Select matching drive systems and power supplies for the actuators.

8. Select the digital control platform for the robot. This may include the control computer, input/output hardware and software, user interface, and other communication needs.

9. Apart from the sensors provided with the actuators (e.g., encoders, tachometers, resolvers), determine what other sensors are necessary for the tasks (e.g., proximity sensors, ultrasonic, optical, and vision sensors) and select them from commercially available units to meet the requirements (of task, accuracy, resolution, bandwidth, etc.).

10. Decide upon a preliminary design for the robot and carry out a model analysis/simulation exercise to validate the design. Fine tune the design (some of the previous steps may have to be repeated here) to meet the specifications.

11. Acquire/build the components for the robotic system. Assemble/integrate the robotic system.

12. Test the robot and compare with computer simulations. Carry out further improvements based on the test results.

10.3.4.1 Prototype Robot

We were given the task of developing a laboratory robot at the University of British Columbia for use in research and development, particularly in space robotics. The developed manipulator can be used to assess, through real-time experiments, the effectiveness of a variety of control procedures for their possible application to space-based systems. Robotic manipulators play an important role in space exploration because of the harsh environment in which they have to operate and the challenges associated with it. Their tasks include the capture and release of spacecraft, maneuver of the payload, and support of extra-vehicular activities (EVA). One example is the mobile servicing system (MSS), Canada's contribution to the International Space Station project.

After a preliminary study, we decided to develop a nonconventional robot consisting of multiple modules connected in series, each module consisting of a revolute (slewing) joint and a prismatic (deploying) joint. The particular robotic design is termed multi-module deployable manipulator system (MDMS). A robot of this type offers several useful characteristics with respect to the dynamics and control, over the conventional manipulator designs that involve only revolute joints:

- Reduced inertial coupling, for the same number of joints
- Better capability to overcome obstacles
- A reduced number of singular positions for a given number of joints
- Simpler decision making during task execution

We have developed a detailed, nonlinear, dynamic model for the robot based on the task requirements; carried out extensive analyses and computer simulation; designed and developed the robot; implemented a variety of control schemes; and extensively tested the prototype robot.

10.3.4.2 Robot Design

In the process of developing the present four-module manipulator, we first developed a two-module variable geometry manipulator (VGM). Through the working experience with this initial prototype, the new four-module system was developed so as to achieve the following: a higher level of rigidity in the joint connections; reduced size, weight, and inertia; reduced machine shop time; and ease of construction, assembly, and maintenance. In addition to these criteria, the mechanical and electrical components were chosen using a mechatronic approach based on their performance characteristics such as power, speed, accuracy, reliability, and robustness. A view of the developed manipulator is shown in Figure 10.5. Starting from the top of the picture, mounted underneath the wooden board is a dc motor for shoulder joint motion. Each of the subsequent elbow joints is also equipped with a dc motor for slewing motion. In between the elbow joints are the deploying links. Each link, which can vary its length, provides not only an extra degree of freedom, but also increased versatility and maneuverability to this robotic arm. In addition, each elbow joint is provided with a rolling support comprising three ball transfers, which run smoothly on a bench with a steel surface. The prototype has the following main features and specifications:

- It is a planar, 8-axis robotic manipulator with four modules, each consisting of one slewing link and one deploying link
- It uses rolling supports on a flat surface to compensate for gravity
- Maximum extension of each deploying link is 15 cm ($\approx$6 in.)
- Maximum acceleration of 0.08 m/s² at a design payload of 5 kg
- Maximum slew speed = 60°/s
- Maximum deployment speed = 4 cm/s
- PC-based control system on which a variety of control strategies could be implemented

The 15 cm extension provides a sufficient change in length for demonstrating the characteristically improved performance due to prismatic joints. The rolling supports are an

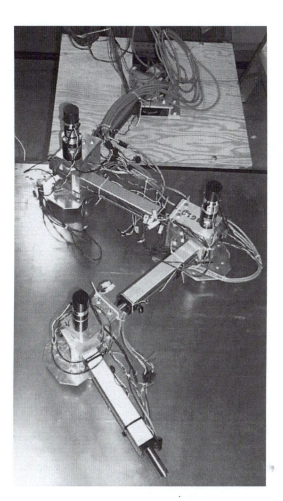

FIGURE 10.5
The MDMS prototype.

important feature because in the fully extended configuration, the manipulator induces a high bending load at each unit, particularly at the shoulder joint. The workspace of the prototype manipulator is a circle of approximately 4.5 m in diameter.

Two categories of components are employed in the construction of the manipulator: those commercially available (supplied by outside manufacturers/vendors) and those designed and machined in-house. The selection of the components in the former category is focused here. The latter type of components include machined parts like the base, motor mounts, and joint supports, and they are made of aluminum 6061 for its low mass density ($\rho_{Al} = 2710 \, \text{kg/m}^3$), strength (yield strength = 255 MPa), and high machinability.

10.3.4.3 Actuator Selection/Sizing

The manipulator was built employing four harmonic-drive dc servomotors from HD Systems Inc. for the revolute links, and four pulse power I (PPI) linear actuators from Dynact Inc. for the deploying links. Electromagnetic actuators were chosen as opposed to hydraulic or pneumatic actuators because of their simplicity, availability, lower cost, and ease of integration and control. These actuators are particularly suitable for carrying out light duty tasks in laboratory experimentations using the robot. Our older robot prototype consisting of two modules (VGM) uses dc servomotors with conventional gear transmissions. The resulting problems include increased weight, noise, and gear backlash, which contribute to the nonlinearity of the system making the control problem more challenging. Low backlash gear units are 25%–50% more expensive. Direct-drive servomotors eliminate the backlash and noise problems. However, the size and weight of the units available to us did not meet the present design requirements. Harmonic-drive actuators, which are virtually backlash free and are available in compact packages, provide high positional accuracy and stiffness. Precise motion control can be performed using them. Furthermore, these actuators have a large torque capacity and are 10%–30% less expensive than the conventional geared servomotor units.

As in the two-module VGM, the new prototype manipulator also uses a ball-screw mechanism for its prismatic (deploying) links. In view of several attractive features, direct-drive linear servomotors were considered for these links, but their main drawback is the heavy magnetic assembly, which did not satisfy our design criteria. Also they generally have a lower thrust-force to weight (size) ratio when compared with ball-screw actuators. Ball-screw actuators with drive nuts that have very low backlash are available as integral packages from Dynact Incorporated in Orchard Park, New York. Each actuator includes a dc servomotor, a ball-screw mechanism, a deploying shaft, and an optical encoder for sensing and feedback of the angular position. The machine shop time was greatly reduced as well with these integral packages. The actuator housing and the extensible shaft are made of formed and machined aircraft-quality aluminum and are rugged and lightweight. In addition, the actuators incorporate magnetic reed switches to protect them from overtravel in both extend and retract positions. The drive screws of the linear actuators are available in either ACME or ball thread form. Ball thread type screws were selected in the present prototype because of their high efficiency, high duty cycle, low friction, long design life, and high speed and load capabilities. Optional nuts with lower backlash can be chosen as well, for the ball screw.

The particular commercial models for the revolute link actuators were selected based on the torque and speed requirements. The computed design torque consists of three parts: actuating torque for the driven load, resistance torque in the bearings, and the resistance torque contributed by the ball-transfer rolling supports. It is required that the revolute joint motion is capable of achieving a speed of 60°/s. The computed design torque requirements for the four slewing joint actuators, starting from the shoulder joint, are listed in Table 10.3. With these torque values and the speed requirement, harmonic-drive actuators were selected by referring to their torque-speed curves as provided by the manufacturer (see Chapter 7) and they are listed in Table 10.4.

For the deploying links, the design thrust consists of the driving force for the payload and the force to overcome the friction between the actuated load and the workspace surface. The maximum required thrust was computed to be 92.0 N. PPI linear actuators, which correspond to the smallest direct (in-line) drive model available from Dynact, were selected. With a standard motor, the actuator can produce a maximum thrust of 550.0 N. In order to reduce the motor mass, the smaller, nonstandard motors were used instead. DC servomotors model 14201 from Pittman that are smaller and lighter but still meet the torque-speed requirements

TABLE 10.3

Computed Design Torque
for Revolute Joints

Slewing Joint Motor No.	Design Torque (N·m)
1	30
2	13
3	4
4	0.5

TABLE 10.4

Selected Revolute Joint Actuators and Their Characteristics

	Joint #1	Joint #2	Joint #3	Joint #4
Model	RFS-20-3007	RH-14C-3002	RH-11C-3001	RH-8C-3006
Rated voltage (V)	75	24	24	24
Rated current (A)	1.9	1.8	1.3	0.8
Peak current (A)	4.8	4.1	2.1	1.1
Rated output torque (N·m)	24	5.9	3.9	2.0
Rated output speed (rpm)	30	30	30	30
Max. continuous stall torque (N·m)	28	7.8	4.4	2.3
Peak torque (N·m)	84	20	7.8	3.5
Motor positioning accuracy (arc-min)	1.0	2.0	2.0	2.5
Diameter (mm)	85	50	40	33
Length (mm)	216	148	125	107
Mass (kg)	3.6	0.78	0.51	0.32
Inertia (kg·m²)	1.2	0.082	0.043	0.015
Transmission ratio	100	100	100	100

TABLE 10.5

Characteristics of a Deploying Joint Motor

Pittman Model 14201 DC Servomotor	
Rated voltage (V)	24.0
Rated current (A)	1.1
Peak current (A)	8.6
Rated output torque (N·m)	0.0244
Rated output speed (rpm)	3700
Max. continuous stall torque (N·m)	0.07
Peak torque (N·m)	0.5
Diameter (mm)	54
Length (mm)	75
Mass (kg)	1.3
Inertia (kg·m²)	1.15E−5
Transmission ratio	1

were chosen to drive the linear actuators. They are 50% lighter than the PPI standard motors. Some characteristics of this servomotor unit are given in Table 10.5.

10.3.4.4 Final Design

Having sized and selected the actuators for the MDMS, the remaining components, mounting brackets, and connectors were designed with the aid of a CAD software package. The assembled CAD model is shown in Figure 10.6. The base plate is the wooden board in Figure 10.5. Modules one to four are the PPI actuators. The whole manipulator is mounted on a base plate attached to a steel frame. Note that modules two to four have the same type of connection to the adjacent modules. Hence, the system can easily be reduced to a

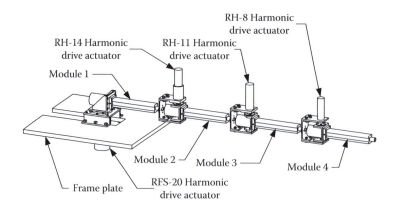

FIGURE 10.6
The CAD model of the MDMS prototype.

two- or three-module manipulator for specific experimental studies. The mount plate for the shoulder joint actuator also features four jacking screws for leveling the manipulator system. Overall, the new prototype manipulator design has improved on the rigidity and compactness over its forerunner, the VGM.

A detailed drawing of one of the revolute joints is given in Figure 10.7. The joint is constructed with two C-shape channels connected through ball bearings and machined shaft-like connectors. The module connector integrates the joint bracket to the end of the deploying shaft of the previous module through a taper pin. The connectors and the channels are machined out of aluminum. The size and the weight of the joint are minimized through careful design practices. For example, the size of the outer channel was minimized by allowing just enough clearance for the linear actuator motor and the electrical wiring (not shown in the figure) for required movement. Furthermore, the length of the channel flange was minimized; hence, flange thickness need not be too large in

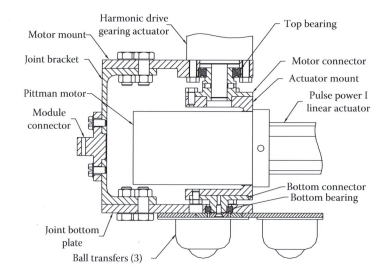

FIGURE 10.7
Details of a revolute joint.

order to realize a rigid structure. With the PPI linear actuator mounted on the inside smaller channel, the slewing link rotates with respect to the outer channel. Two deep groove ball bearings are installed, one each at the top and bottom plates, to support the slewing link. The motor connector is attached to the harmonic drive gear actuator and is secured in place with two setscrews. Two setscrews are used to prevent any possible slack during operation. Rolling support with ball transfers is attached to the bottom plate of the joint.

10.3.4.5 Amplifiers and Power Supplies

All the actuators are driven by brush type pulse-width modulated (PWM) servo amplifiers and power supplies from Advanced Motion Controls (AMC) in Camarillo, California. The amplifiers and power supplies were selected to match the actuators. For example, the harmonic drive actuator at the base has the following ratings:

- Rated voltage = 75 V
- Rated current = 1.9 A
- Peak current = 4.8 A
- Rated output torque = 24 N · m
- Rated output speed = 30 rpm
- Maximum continuous stall torque = 28 N · m
- Peak torque = 84 N · m
- Motor positioning accuracy = 1.0 arc-min
- Transmission ratio (harmonic drive) = 1:100
- Optical encoder resolution = 500 P/rev

The AMC PWM amplifier model 12A8 selected to drive this actuator has the following ratings:

- DC supply voltage = 20–80 V
- Peak current (2 s max. internally limited) = ±12 A
- Max. continuous current (internally limited) = ±6 A
- Switching frequency = 36 kHz
- Bandwidth = 2.5 kHz

Both the dc supply voltage range and the peak current meet the motor needs. The continuous current of amplifier (±6 A) is greater than the rated current of the motor. Moreover, the operating bandwidth of the amplifier, at 2.5 kHz, is far greater than what is needed for a typical robotic task (e.g., 125 Hz). The power supply, AMC model PS16L30, has the following ratings:

- Supply voltage = 120 VA
- Rated output voltage = 30 VDC
- Nominal output current = 53 A

The output voltage of the power supply matches the range of operation of the amplifier. Also, the continuous output current of the power supply (53 A) is sufficient even for eight motors (a total of 14 A) or eight amplifiers (a total of 48 A). Besides, when driving the actuator based on a command signal from the control computer, each amplifier is able to receive end-of-stroke signals from the magnetic reed switches, which are mounted on the linear actuators, in order to disconnect the power to the linear actuators.

10.3.4.6 Control System

The control system is schematically shown in Figure 10.8. The controlled motion of the prototype manipulator system is carried out using optical encoders as feedback sensors, which come integral with the actuators, a data acquisition board, and an IBM-PC compatible computer. To implement different control algorithms on this robotic manipulator system, an open architecture real-time control system has been established using an 8-axis ISA bus servo I/O card from Servo To Go, Inc. in Indianapolis, Indiana. This data acquisition board features the following functionalities:

- Eight channels of encoder input of up to 10 MHz input rate
- Eight channels of 13 bit analog output in the +10 to −10 V range
- Interval timers capable of interrupting the PC
- Timer interval programmable to 10 min in 25 µs increments
- Board base address determined automatically without a configuration file
- IRQ number is software selectable, i.e., no board jumper is required

For real-time control, the eight-axis card was originally operated under the QNX real-time operating system. The necessary drivers and function library are available in C language from Quality Real-Time Systems (QRTS) in Falls Church, Virginia. Controller programs have been written in C language. The control program also serves as the manager for

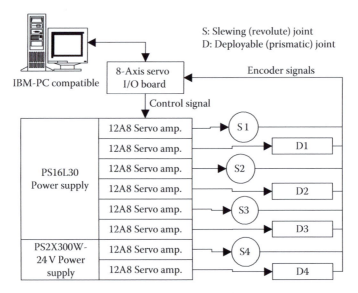

FIGURE 10.8
The robot control system.

data exchange and the coordinator of the following functions: setting the sampling rate, acquiring encoder signals, and sending out command signals to the amplifiers through the digital-to-analog conversion (DAC) channels.

The MDMS prototype uses a single host computer for the development and implementation of its control system. This enables easier debugging and faster modification of program code, while maintaining the performance of a dedicated real-time controller. This provided a cost effective option, particularly since no coprocessors or DSPs were employed. Originally, the real-time control of MDMS was achieved using the QNX real-time operating system. Even though it has many attractive and multitasking features, the range of supported hardware configurations was not as extensive as for a popular operating system such as Windows, which was found to me more user-friendly. Even though Windows NT has been used in real-time applications, it was not designed for "hard" real-time applications. Any events that are given the highest priority setting are still subject to unpredictable delays due to lower level processes running in Windows NT. Windows NT is better suited for operations that only require precision in the 100 ms range. VenturCom's Real-Time Extension (RTX) for Windows NT adds real-time capabilities down to the submillisecond range. RTX provides real-time capabilities by adding a new subsystem known as RTSS to the Windows NT architecture. It allows users to schedule events ahead of all Windows NT schedules and to create threads that are not subject to time-sliced sharing of the processor. RTX enables users to take advantage of the sophisticated GUI and connectivity options of Windows and the high-performance, reliability, and determinism of a real-time operating system at the same time on the same computer. RTX provides a set of real-time functions, allowing the developer to program a real-time controller in C/C++. These functions are similar to those available in the Win32 API, but they allow the user to set thread priority at any one of the 128 levels provided by RTX, and particularly higher priority than Windows NT schedules.

A PID controller was implanted on the MDMS, particularly using the features of the clock and timer provided by RTX. The control program was developed with two threads: the timer function and the main function. The timer function was scheduled to execute every sampling period and was given priority over the main function. The main function waited until the ESC key was hit or the robot task was completed and then zeroed the DACs and killed the process. The timer function repeated the following steps at each sampling period:

- Determined the desired output from a function or file
- Read the actual output from the encoders (motion sensors at robot joints)
- Used a PID control law to determine the input to the joint motors
- Wrote the control input value to the DACs

Rather than using the Windows drivers for the servo card, simple read and write functions were developed in RTX. This improved performance by ensuring that each function call has priority over other Windows NT schedules. This was not strictly necessary for the PID controller, but would be useful for implementing more computationally intensive controllers in the future. Figure 10.9 shows a schematic diagram of the control systems with the PID control program. The control action at time instant n is computed by

$$u_n = K_p \left[e_n + \frac{T_d(e_n - e_{n-1})}{T_s} + \frac{1}{T_i} \left\{ \frac{(e_n + e_{n-1})T_s}{2} + PInt \right\} \right] \tag{10.6}$$

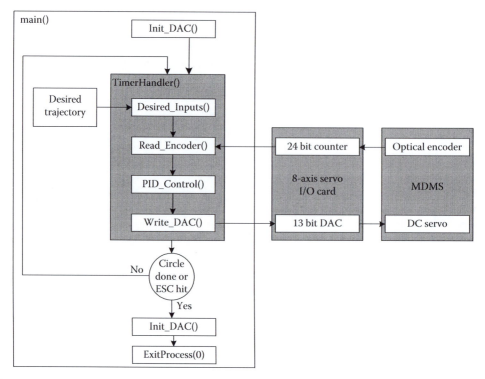

FIGURE 10.9
PID control of the MDMS.

where
T_s is the sampling time
e_n is the response error
u_n is the control action
K_p is the proportional gain
K_i is the integral gain
K_d is the derivative gain
$T_i = K_p/K_i$ is the integral action time
$T_d = K_d/K_p$ is the derivative action time
The value of *PInt* is initialized at step 0

In addition to the standard PID control, other sophisticated control schemes such as predictive control have been implemented and successfully tested using the MDMS.

10.3.4.7 Economic Analysis

Economic analysis involves a cost–benefit evaluation and will require a computation of the payback period. Figures on the initial investment, the number of people replaced by the robot, and the corresponding wage savings, productivity increase due to robotization, inflation rate, corporate tax rate, and operating expenses, such as utilities, maintenance, and insurance would be required for an analysis of this type. The procedure that is given in the next case study may be applied in the present case study as well.

10.4 Iron Butcher Case Study

The Iron Butcher, which is commonly used in the fish-processing industry for head cutting of fish, is known to be inefficient and wasteful and the resulting product quality may be unacceptable for high-end markets. We have developed two improved designs of the Iron Butcher machines for fish cutting. The two machines are similar except for the cutter design and the types of actuators used. The development of one machine is presented here. The machine, which uses a variety of sensors, actuators, and hardware for component interface and control, operates with the help of a dedicated supervisory control system. A layered architecture has been used for the system. It has several knowledge-based modules for carrying out tasks such as machine monitoring, controller tuning, machine conditioning, and product quality assessment.

10.4.1 Technology Needs

Fish processing, a multibillion dollar industry in North America alone, by and large employs outdated technology. Wastage of the useful fish meat during processing, which reaches an average of about 5%, is increasingly becoming a matter of concern for reasons such as dwindling fish stocks and increasing production costs. Due to the seasonal nature of the industry, it is not cost effective to maintain a highly skilled labor force to carry out fish-processing tasks and to operate and service the processing machinery. Due to the rising cost of fish products and also diverse tastes and needs of the consumer, the issues of product quality and products-on-demand are gaining prominence. To address these needs and concerns, the technology of fish processing should be upgraded so that the required tasks could be carried out in a more efficient and flexible manner.

A machine termed the Iron Butcher, which was originally designed at the turn of the century and has not undergone any major changes, is widely used in the fish-processing industry for carrying out the head cutting operation of various species of salmon. The fish are manually fed onto a moving conveyor at one end of the machine. A pair of pins is provided at the feeding end of the machine, with respect to which the fish should be manually positioned across the conveyor. But, since the typical feeding rate is 2 fish/s, accurate manual positioning is infeasible. Instead, a mechanical indexing mechanism is employed for automatic positioning. The main component of this mechanism is an indexer foot, which drops onto the fish body. The indexer permits finer adjustment of the lateral position of a fish with respect to the cutter blade. The indexer mechanism moves diagonally along the conveyor, on a carriage guideway, and is driven by the same actuator as for the conveyor. Accordingly, the indexer moves at the same speed in the direction of motion of the conveyor. In addition, the indexer also has a lateral motion toward the cutter edge, in view of the overall diagonal movement. During this lateral motion, the indexer foot slides over the fish body until it engages with the collarbone at the gill plate of the fish. Subsequently, due to this engagement, the indexer foot pushes the entire fish laterally toward the cutter blade, thereby accomplishing the required positioning action. Just before the fish reaches the cutter, the indexer foot lifts off and starts the return motion along the carriage guideway. The rotary knife of the cutter, then, lops off the head of the fish.

The straightforward indexing mechanism of the Iron Butcher, even though mechanically robust, can result in two common types of positioning error. An "over-feed error" results if the indexer foot becomes engaged into some location of the fish body prior to reaching the collarbone. This could result due to an external damage or some other structural nonuniformity on the fish body. Consequently, the fish would be pushed too much in the lateral direction with respect to the cutter. The subsequent cut would be wasteful, removing a chunk of valuable meat with the head. An "under-feed error" results if the indexer foot slides over the collarbone and the gill plate or pushes too much into a soft or damaged gill plate. In this case, a portion of the head would be retained with the fish body after the cut. Such fish needs to be manually trimmed in a subsequent operation, and this would represent a reduction in the throughput rate and a wastage of labor. Another shortcoming of the Iron Butcher is the poor quality of cut even with accurate positioning. The primary contributing factors are clear. The fish move continuously, not intermittently, with the conveyor and are not stationary during the cutting process. Also, the high-inertia rotary cutter has a pointed outer end, which first hooks the fish and then completes a guillotine-type cut. The combined effect of the cutter inertia, hooking engagement, and the motion of the fish during cutting is an irregular cut with excessive stressing and deformation. This will result in a lower product quality.

Motivated by the potential for improvement of this conventional machine, specifically with regard to waste reduction, productivity, and production flexibility, a project was undertaken by us to develop an innovative machine for fish cutting. The machine has been designed, integrated, tested, and refined. The test results have shown satisfactory performance. The new Iron Butcher possesses the following important features:

1. High cutting accuracy: obtained using mechanical fixtures, positioners, tools and associated sensors, actuators, and controllers, which have been properly designed and integrated into the machine.

2. Improved product quality: achieved through high-accuracy cutting and also through mechanical designs that do not result in product damage during handling and processing, along with a quality assessment and supervisory control system, which monitors the machine performance, determines the product quality, and automatically makes corrective adjustments.

3. Increased productivity and efficiency: attained through accurate operation and low wastage with a reduction in both downtime and the need for reprocessing of poorly processed fish.

4. Flexible automation: requiring fewer workers for operation and maintenance than the number needed for a conventional machine, which is possible due to the capabilities of self-monitoring, tuning, reorganization, and somewhat autonomous operation, as a result of advanced components, instrumentation, and the intelligent and hierarchical supervisory control system of the machine.

10.4.2 Final Design

With the specific objectives in mind, we have designed the industrial prototype shown in Figure 10.3. This machine has three main subsystems: (1) the conveyer system, (2) the pneumatic system, (3) and the hydraulic system. Their functions are to move the fish, position the cutter in the horizontal positioning plane, and cut the fish, respectively. A schematic representation of the machine is shown in Figure 10.10.

In the normal operation of the machine, a worker places a salmon on the feeding table, which is located at one end of the machine, with its head pointed towards the cutter and its belly facing away from the direction of the conveyer motion. The feeding table allows the salmon to slide down to the conveyer bed, which transports the fish from the feeding table end to the cutter. The conveyor, driven by an ac motor, indexes the fish in an intermittent

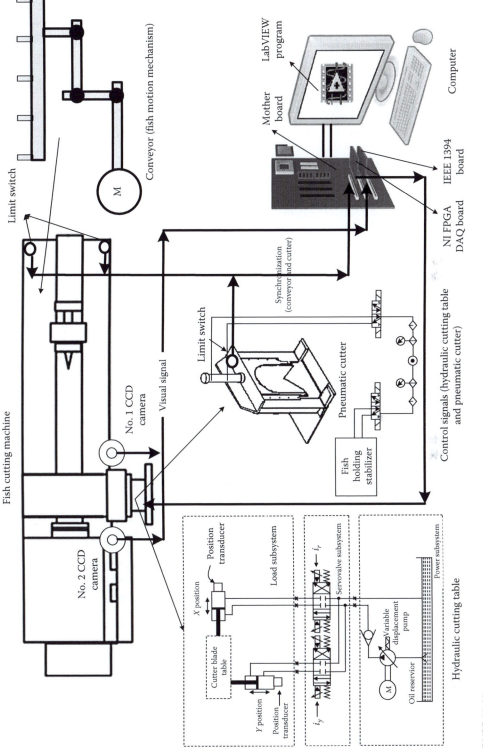

FIGURE 10.10
Schematic representation of the fish cutting machine.

manner. An image of each fish, obtained using a digital charge-coupled device (CCD) camera, is processed to determine the geometric features, which in turn establish the proper cutting location. A two-axis hydraulic drive unit positions the cutter accordingly, and the cutting blade is operated using a pneumatic actuator. The position sensing of the hydraulic manipulator is done using linear magnetostrictive displacement transducers, which have a resolution of 0.025 mm when used with a 12 bit analog-to-digital converter (ADC). A set of six gage-pressure transducers are installed to measure the fluid pressure in the head and rod sides of each hydraulic cylinder, and also in the supply lines. A high-level imaging system determines the cutting quality, according to which adjustments may be made on-line, to the parameters of the control system so as to improve the process performance. The control system has a hierarchical structure with conventional direct control at the component level (low level) and an intelligent monitoring and supervisory control system at the upper level.

The primary vision module of the machine is responsible for the fast and accurate detection of the gill position of a fish on the basis of an image of the fish as captured by the primary CCD camera. This module is located in the machine host and is comprised of a CCD camera, an IEEE 1394 board for image grabbing, a trigger switch for detecting a fish on the conveyor, and an NI FPGA data acquisition (DAQ) board for analog and digital data communication between the control computer and the electrohydraulic manipulator. The secondary vision module is responsible for the acquisition and processing of visual information pertaining to the quality of the processed fish that leaves the cutter assembly. This module functions as an intelligent sensor in providing high-level information feedback into the control module of the software. The hardware and the software associated with this module are a CCD camera at the exit end for grabbing images of processed fish and a developed image processing module based on NI LabVIEW for visual data analysis. The CCD camera acquires images of processed fish under the direct control of the host computer, which determines the proper instance to trigger the camera by timing the duration it takes for the cutting operation to complete. The image is then grabbed by the image-processing module software for further processing. In this case, however, image processing is accomplished to extract high level information such as the quality of processed fish.

10.4.3 Control System Architecture

Figure 10.11 shows the overall architecture of the control system of the fish cutting machine. The NI FPGA data acquisition board located within the host computer acquires all sensor data (position sensors, limit switches, pressure sensors, etc.) from the machine. Using this data, the computer executes the control algorithm, which determines the drive signals for the actuators (of the hydraulic table, pneumatic cutter, and the conveyor). The IEEE 1394 board receives vision signals from the CCD camera for image processing.

10.4.3.1 Motor Control Console

The motor control console (MCC) is a power box that controls the power to the hydraulic pump. This is needed since the current is very high (about 40 A) when the motor starts. Figure 10.12 shows the components of the MCC box, and Table 10.6 gives the component names.

Figure 10.13 shows the wiring diagram of the MCC. It uses a three-phase power system (208 V, 60 A, 60 Hz) and contains a main switch, an E-stop button, and a 2-way switch. The 2-way switch is used to select manual control or computer control of the hydraulic pump.

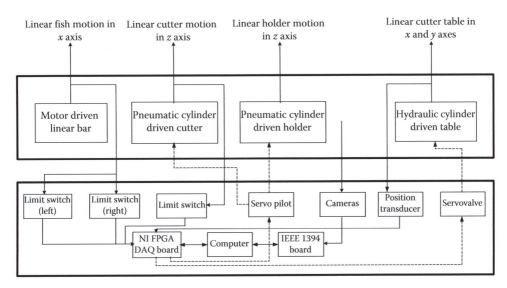

Linear fish motion in   Linear cutter motion   Linear holder motion   Linear cutter table in
      x axis   in *z* axis   in *z* axis   *x* and *y* axes

FIGURE 10.11
Control system architecture.

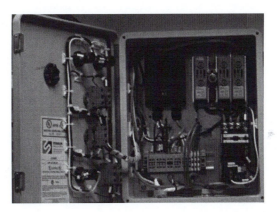

FIGURE 10.12
The MCC box.

10.4.3.2 Junction Box

The junction box shown in Figure 10.14 is the interface between the computer and the machine. It provides ±5 V, ±15 V, and ±42 mA to drive the servovalves and to power the position and pressure transducers.

An important component of the junction box is the current signal generator that is required by the hydraulic servovalve since currents in the range of −42 to 42 mA are required for the servovalves, while the output signal provided by PC (NI data acquisition board) is a voltage in the range of −10 to 10 V.

10.4.4 Hydraulic System

Figure 10.15 shows a schematic diagram of the hydraulic system. The system contains a hydraulic pump, a two degree-of-freedom (*x* and *y*) electrohydraulic manipulator (cutting table) and two position transducers (for *x* and *y* directions).

TABLE 10.6

Control Panel Components Breakdown for 36 kW, 575 VAC

Component Name	Supplier	Model Number	Quantity
Main circuit breaker	ABB	T1N030TL	1
Breaker operation mechanism	ABB	KT3VD-M	1
Operation shaft	ABB	OXP6X	1
Panel door operation handle	ABB	OHB65J6	1
Enclosure, $16 \times 14 \times 8$	Vanco	EJ16148	1
2 pole dead front fuse holder	Vanco	LPSM002	1
1 pole dead front fuse holder	Vanco	LPSM001	1
Contactor for heater hi-limit	Vanco	A30–30–11–84	1
Overload relay	Vanco	TA42DU42	1
NO contact block for pushbutton	Vanco	CBKCB10	2
NC contact block for pushbutton	Vanco	CBKCB01	2
Start-stop combo green/red/extended	Vanco	CBK2P1	1
Three-position selector switch	Vanco	TBA	1
Red, LED	Vanco	CL513R	1
Green, LED	Vanco	CL513G	1
Alum lug 1 cond 2–14 1/4 bolt	Vanco	TA2–1	2
Labels	ABOND	TBA	4
SSR	TBA	TBA	1

10.4.4.1 Physical Parameters of the Cutter Table

- Moving mass in the x direction $M_x = 32.7$ kg
- Moving mass in the y direction $M_y = 55.7$ kg
- Maximum stroke of the pistons $L = 2$ in. $= 5.08 \times 10^{-2}$ m
- Volume of the hoses connecting the valve to the x cylinder $V_{ox} = 8.9 \times 10^{-5}$ m^3
- Volume of the hoses connecting the valve to the x cylinder $V_{oy} = 1.14 \times 10^{-5}$ m^3
- Piston bore diameter $D_{bore} = 1.5$ in. $= 3.81 \times 10^{-2}$ m
- Piston rod diameter $D_{rod} = 1$ in. $= 2.54 \times 10^{-2}$ m

10.4.4.2 Hydraulic Piston Parameters

- Piston head-side area $A_1 = 1.140 \times 10^{-3}$ m^2
- Piston rod-side area $A_2 = 6.333 \times 10^{-4}$ m^2
- Area ratio $\gamma = A_1/A_2 = 1.8$

10.4.4.3 Flow Control Servovalves

Model: MCV113A6109

- Supply pressure = 3000 psi
- Return pressure = atmospheric to 300 psi
- Flow rating = 10 gpm (38 L/min) based on 1000 psi pressure drop and rated input signal
- Null pressure = 50% of the supply pressure
- Neutral leakage = less than 0.85 gpm (consisting of pilot stage quiescent flow plus second stage neural leakage with 1000 ps pressure drop and output ports blocked)

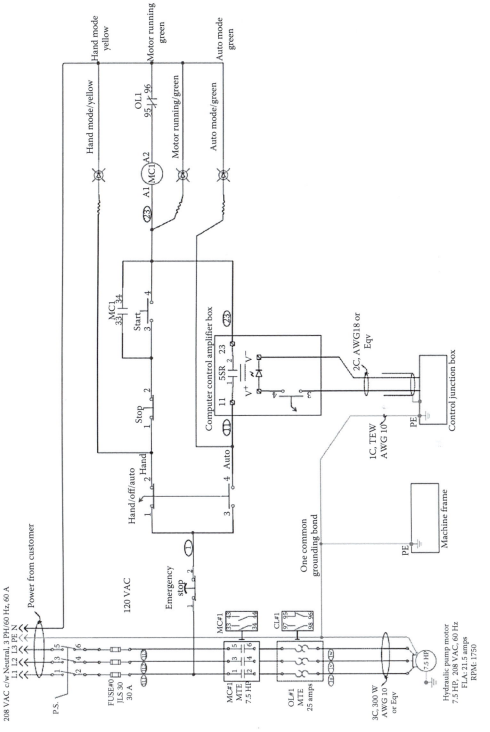

FIGURE 10.13
Wiring diagram of the MMC.

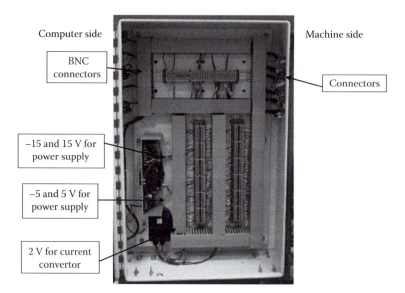

Computer side

Machine side

BNC
connectors

Connectors

−15 and 15 V for
power supply

−5 and 5 V for
power supply

2 V for current
convertor

FIGURE 10.14
The junction box.

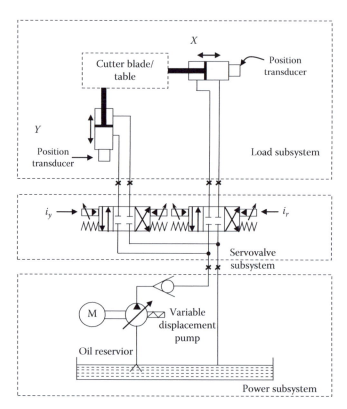

FIGURE 10.15
Schematic diagram of the hydraulic system.

- Hysteresis = 5% of the rate input signal, maximum
- Blocked output pressure rise = greater than 80% of the supply pressure with 4% of the rated input signal increase
- Frequency response = 18 Hz (−3 dB bandwidth)
- PWM signal (if used) with a frequency higher than 500 Hz is recommended
- Filtration = 10 microns in-line pressure filter

10.4.4.4 Pilot Valves

Model: MCV101A1412

Four pins, two of which are used (A and B)

Nominal full flow current = 42 mA

Coil resistant 100 Ω

10.4.4.5 Position Transducers

Temposonic™ II linear magnetostrictive displacement transducers are used (see Chapter 6). Model No. TTRCU0020; Stroke = 2.0 in. The sensor can operate continuously at pressure up to 3000 psi.

10.4.5 Pneumatic System

The pneumatic filters have the following specifications:

- Inlet pressure: 150 psig max
- Outlet pressure: 150 psig max
- Temperature: 125°F max

The lubricator has the following specifications:

- Inlet pressure: 250 psig max
- Temperature: 175°F max

10.4.6 Economic Analysis

Economic analysis involves a cost–benefit evaluation and will require the computation of the payback period. Figures on the initial investment, number of people replaced by a new machine and corresponding wage savings, productivity increase due to new technology, inflation rate, corporate tax rate, and operating expenses such as utilities, maintenance, and insurance are required for an analysis of this type. Through our knowledge of the fish-processing industry and details of the new machine, the following representative data are used:

Initial investment on a machine	= $100,000
Company expenses for machine start-up	= $10,000
First year revenues:	
Wage savings due to replaced workers	= $20,000
Increased revenues due to higher productivity	= $50,000

It has been assumed that only four workers would be needed per new machine whereas six workers are needed for the older Iron Butcher. A processing season of 3 months is assumed as well. The productivity increase given here is primarily from the increased meat recovery.

First year operating expenses (maintenance, utilities, insurance, etc.) = $20,000	
Economic parameters:	
Inflation rate	= 5%
Corporate tax rate	= 25%
Tax credit on capital expenditure	= 10%
With a planning horizon of 5 years, assume a depreciation rate	= 20%

Next, a cash flow table is completed, as shown in Table 10.7, for the data given here.

Internal rate of return:
Taking into consideration that the risk of payback is greater for the later years from the time of initial investment, the applicable equation is

$$C = \sum_{i=1}^{n} \frac{S_i}{(1+r)^n} \tag{10.7}$$

where
    C is the initial cost of new technology
    S_i is the net cash savings in the ith year
    n is the design life (planning horizon or period of amortization)
    r is the internal rate of return

Then, from the cash flow table (Table 10.7) for one machine installation, and using Equation 10.7 we have

TABLE 10.7

Cash Flow Table ($) for a Machine Installation

Item	Initial	Year 1	Year 2	Year 3	Year 4	Year 5
Capital investment tax credit (10%)	−100,000	10,000				
Machine start-up cost	−10,000					
Tax gain (25%)		2,500				
Machine depreciation (20%)		20,000	20,000	20,000	20,000	20,000
Tax loss (25%)		−5,000	−5,000	−5,000	−5,000	−5,000
Wage savings (5% inflation)		20,000	21,000	22,050	23,153	24,310
Tax loss (25%)		−5,000	−5,250	−5,513	−5,788	−6,078
Productivity increase (5% inflation)		50,000	52,500	55,125	57,881	60,775
Tax loss (25%)		−12,500	−13,125	−13,781	−14,470	−15,194
Operating expenses (5% inflation)		−10,000	−10,500	−11,025	−11,576	−12,155
Tax gain (25%)		2,500	2,625	2,756	2,894	3,039
Net savings	−110,000	72,500	62,250	64,612	67,094	69,697

$$110,000 = \frac{72,500}{(1+r)} + \frac{62,250}{(1+r)^2} + \frac{64,612}{(1+r)^3} + \frac{67,094}{(1+r)^4} + \frac{69,697}{(1+r)^5}$$

The solution for r may be obtained iteratively. An approximate value for r is obtained by using

$$r_{apr} = \frac{1}{Cn} \sum_{i=1}^{n} S_i \qquad (10.8)$$

Hence,

$$r_{apr} = \frac{72,500 + 62,250 + 64,612 + 67,094 + 69,697}{110,000 \times 5}$$

or $r_{apr} = 0.6$. A more accurate value for r would be 0.5465. Now, using $r = 0.55$, the payback period, defined as $1/r$, is approximately 1.8 years. A payback period of less than 2 years, as predicted in this analysis, is quite acceptable for the fish-processing industry.

A simplified version of Equation 10.7 as given by

$$C = S \sum_{i=1}^{n} \frac{1}{(1+r)^n} \qquad (10.9)$$

is commonly used in predicting the payback period. This approach cannot be fully justified, as the net savings figure S_i is not constant during the accounting life of the machine, as is clear from Table 10.7.

10.4.7 Networked Application

With the objective of monitoring and controlling the industrial process from remote locations, we have developed a universal network architecture, both hardware and software. The developed infrastructure is designed to perform optimally with a Fast Ethernet (100Base-T) backbone where each network device only needs a low cost network interface card (NIC). Figure 10.16 shows a simplified hardware architecture that networks two machines (the fish-processing machine and an industrial robot). Each machine is directly connected to its individual control server, which handles the networked communication between the process and the Web-server, data acquisition, sending of control signals to the process, and the execution of low level control laws. The control server of the fish-processing machine contains one or more data acquisition boards, which have ADC, DAC, digital I/O, and frame grabbers for image processing.

Video cameras and microphones are placed at strategic locations to capture live audio and video signals allowing the remote user to view and listen to a process facility and to communicate with local research personnel. The camera selected in the present application is the Panasonic Model KXDP702 color camera with built-in pan, tilt, and 21× zoom (PTZ), which can be controlled through a standard RS-232C communication protocol. Multiple cameras can be connected in a daisy-chain manner to the video-streaming

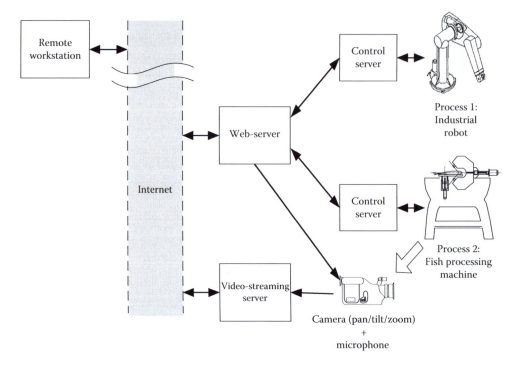

FIGURE 10.16
Network hardware architecture for the fish-processing system.

server. For capturing and encoding the audio-video (AV) feed from the camera, the Winnov Videum 1000 PCI board is installed in the video-streaming server. It can capture video signals at a maximum resolution of 640 × 480 at 30 fps, with a hardware compression that significantly reduces computational overheads of the video-streaming server. Each of the AV capture boards can only support one AV input. Hence, multiple boards have to be installed.

10.5 Exercises

10.1 Carry out a literature search and obtain two or more definitions of engineering design. Critically compare these definitions.

10.2 Establish a set of steps for carrying out an engineering design according to each of the definitions obtained in Exercise 10.1. While keeping the design objectives the same, compare these design procedures.

10.3 Compare this engineering design to another type of "design."

10.4 Consider the engineering design process of a hybrid (electric-gasoline) car. First establish the need for this vehicle based on the consumer needs, environmental consideration, economic factors, potential market, etc., particularly by comparing with a popular type of car. Develop a preliminary QFD chart for the design of this vehicle.

10.6 Projects

You are a mechatronic engineer who has been assigned the task of designing, developing, and instrumenting a mechatronic system. Giving the necessary details, describe the steps of system design, integration, testing, and the application of the projects outlined below. The design should include the structural system, electronic system, hardware, and software including sensors, actuators, motion transmission devices, power sources, controllers, signal conditioning, and interfacing needs. Off-the shelf components may be used where available and appropriate. The given specifications may be modified and additional specifications may be established, when necessary. Provide data for the main components of the system. Using suitable diagrams and sketches, describe how the overall system (plant, sensors, actuators, controllers, signal modification devices, etc.) is interconnected (integrated). Explain how the system operates (e.g., what is the purpose of the system, what commands are provided to the system, what responses and signals are generated by the system and sensed, how the control signals are generated and according to what criteria, how the plant is actuated, and what type of plant response would be achieved under proper operation).

10.6.1 Project 1: Automated Glue Dispensing System

Application of an adhesive is useful in many industrial processes; for example, automotive, wood product, and building. Consider a system for the two-dimensional application of glue. The system has to sense the application area, position the dispensing gun, and operate it, which will include simultaneous dispensing of glue and moving the dispensing head accurately with respect to the glued object (e.g., window). The speed of the system will be governed by the constraints of the glue gun (including the properties of the glue) and the moving system. The following preliminary specifications are given:

Maximum area of application: $1\,m \times 1\,m$
Speed of the dispensing unit $= 20\,cm/s$
Positioning accuracy $= \pm 0.5\,mm$
Glue application accuracy (uniformity) $= \pm 0.2\,mm$
Glue dispensing pressure $= 100\,psi\ (690\,kPa)$

The design should include the glue dispensing system as well. Also, describe in detail an application of your design.

10.6.2 Project 2: Material Testing Machine

The testing of material (e.g., tensile/compressive, bending, torsional, fatigue, impact) is important in product development, monitoring, and qualification. Testing of biological material may be needed in medical, agricultural, and food-product applications. Design a machine for *in vitro* (outside the body, in an artificial environment) testing of spine segments obtained from a human cadaver. The spine segment is mounted on a form base and various load profiles (forces and moments) are applied. The resulting motions (displacements and rotations) are measured and recoded for further analysis. Some of the test requirements are given below:

Moment increment (for a ramp test) $= 1\,N \cdot m$
Moment range $= -15$ to $+15\,N \cdot m$
Speed $= 4$ increments/s
Moment step (for a step test) $= \pm 10\,N \cdot m$
Accuracy $= 2\%$

10.6.3 Project 3: Active Orthosis

Powered prosthetic devices are increasingly used to assist deformed, disabled, or injured upper and lower limbs of humans. An orthosis is a fully integrated prosthetic device attached to a human body and assumes that the limb is not missing and the sensation of the limb is not completely lost. An active device (as opposed to a passive device) will require a power source. Sensations of temperature, pressure, and texture (e.g., tactile sensing) are particularly important in human functions. Design an active upper-limb orthosis. The head and the shoulder may be used to control the device, and two-state, multistate, or continuous commands can be provided. Functionality, reliability, convenience and comfort, speed, accuracy, cost, and appearance are important considerations in the design. The assisted functions and movements may include those of the upper arm, elbow, forearm, wrist, and fingers. You may establish the necessary design specifications for the orthotic device through self-testing, experience, and literature search.

10.6.4 Project 4: Railway Car Braking System

In braking a train, the braking forces have to be applied to the wheels rapidly, systematically, and under control. Derailments should be avoided and the braking operation should not be damaging to the train and its occupants. Under normal conditions, braking should be done while minimizing any discomfort to the passengers. Consider a multicar lightrail system (e.g., an elevated guideway transit system or a subway system). Assume that a hydraulic system is used to apply the braking force to the wheels through brake shoes, and these forces can be quite high (e.g., $3 \times 10^4\,N$). Establish suitable design specifications. Design a braking system that includes antilock braking features. Train speed, wheel-rail conditions, weather conditions, and the nature of the stop (normal or emergency) should be factored into the control system.

10.6.5 Project 5: Machine Tool Control System

Productivity, product quality, machine life, tool life, and safety will improve through the proper control of machine tools. Consider a standard vertical milling machine consisting of a positioning (x–y) table and a vertical spindle assembly, which carries the toolbit. The following parameter values and specifications are available:

Mass of the positioning table: 250 kg
Mass of the spindle assembly: 50 kg

Maximum mass of workpiece: 50 kg

Positioning accuracy: 0.01 mm

Maximum speed of the positioning table: 0.2 m/s

Maximum acceleration of the positioning table: $1.0 \, \text{m/s}^2$

Maximum cutting force: 2000 N

Operating bandwidth of the milling machine: 100 Hz

Assume that dc motors and ball screws are used to drive the positioning table. The servo rise time may be taken as 50 ms. Design a suitable control system for tool positioning and machining.

Examine whether/how the design should be modified depending on the cutting (workpiece) material (e.g., steel, aluminum, other metals and alloys, plastic, wood, rubber).

10.6.6 Project 6: Welding Robot

Considerations of productivity, flexibility, hazards, and cost have provided the motivation for using robots for industrial welding applications. The automotive industry is a good example. Both seam welding and spot welding may be carried out by robots. Design an arc-welding industrial robot for a production line of an industrial plant. Select a specific industrial application and on that basis, establish a set of specifications for the robot. The design should involve kinematics, dynamics, mechanics, electronics, control, and system integrating and plant networking aspects. The design should involve the detection of the welded part prior to positioning the welding torch. In particular, consider a six degree-of-freedom robot with three prismatic joints and three revolute joints. The prismatic joints are primarily used for gross positioning of the end effector (the welding torch) and the revolute joints are primarily used for fine manipulation (e.g., orientation and proximity adjustments) of the welding torch with respect to the welded part. Some preliminary design specifications are given below:

Maximum speed of a prismatic joint: 1.0 m/s

Maximum speed of a revolute joint: 2.0 rad/s

Maximum linear acceleration: 1 g

Linear positioning accuracy: ±0.1 mm

Angular positioning accuracy: ±1°

Payload (including the welding torch): 15 kg

Work envelope: hemisphere with 1.5 m radius

You may use ac servomotors.

10.6.7 Project 7: Wood Strander

A strander is a machine for producing wood chips (flakes) in the manufacture of strand boards (chipboards). A schematic diagram of a strander is shown in Figure P10.7.

The lumber is intermittently delivered into the cutting ring by a conveyor. The feeder pushes the lumber against the rotating ring whose blades chip the lumber.

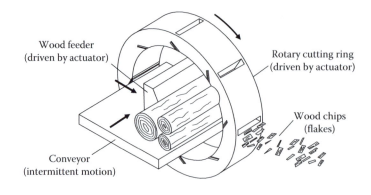

FIGURE P10.7
A strander for strand board manufacture.

Design a strander that can chip logs with a maximum diameter of 0.4 m in 0.5 m segments (which corresponds to the axial length of the cutting ring). The following preliminary specifications are provided:

Cutting ring diameter: 3.0 m
Number of cutting knives: 88
Cutting ring speed: 900 rpm
Feed velocity: 10 cm/s
Average cutting force: 5000 N
Average chip thickness: 0.3 cm
Chip thickness accuracy: 1%

10.6.8 Project 8: Automated Mining Shovel

A mining shovel is similar to an excavator used in earth removal operations. It has a boom to hold the stick, which carries the shoveling bucket. The stick is moved and manipulated by means of a cable winch mechanism. It is proposed to automate the shoveling process by incorporating appropriate sensors, actuators, controllers, and other hardware. The following partial specifications are available.

Shovel (bucket or dipper) capacity: 40 m³
Boom length: 15 m
Stick length: 7 m
Hoist speed: 2 m/s
Stick speed: 1 m/s
Stick rotation speed: 5 rpm
Dipper rotation speed: 10 rpm
Cycle time: 10 s

10.6.9 Project 9: Can-Filling Machine

Packaging measured portions of food products is an important operation in the food-processing industry. In particular, "portion control" is necessary. Consider an automated can-filling operation. It optimally portions and fills fish into cans. The overall system includes the stages of pre-filling, filling, and post-filling. An integrated approach to optimal grouping and cutting, according to a weight-based portioning criterion, is central to the advantages of the approach. The system that is considered here incorporates optimal grouping and cutting of fish, using robotic devices in order to minimize the weight of a fish portion from the target weight of a can. Devices have to be designed and developed for mechanical handling and cutting of fish, with associated sensing and control systems. Automated can filling and sensor-based post-filling inspection and integrated correction/repair, under the supervision and control of a high-level control system, are needed. The associated technology includes the fast and accurate estimation of the weight distribution of each fish, a portion optimization method, handling, conveying, and cutting devices, advanced sensor technology, and multilayered intelligent control. An important feature of the system is the weight-based optimization of the fill weight so as to minimize over-fills and under-fills of canning. The typical throughput rate is 5 cans/s.

10.6.10 Project 10: Fish-Marking Machine

The marking of juvenile fish in hatcheries, before releasing them into lakes and rivers, is an activity that is very valuable in fishery management. Samples of grown fish could be subsequently harvested and examined to collect data, which would be useful for many purposes such as predicting fish stocks, determining migration patterns of fish, and ascertaining the survival ratio of hatchery fish. A simple presence/absence type mark can provide a straightforward identification means for hatchery fish, at high speed. This project concerns the design and development of an automated machine for the spray marking of fish. The machine consists of four main modules: the feeding unit, the conveying unit, the spray-marking unit, and the pigment recirculation unit. A fluorescent pigment may be used for mass marking, which is fast and inexpensive, the fish survival is excellent, and a mark retention of 130 days or more is possible. A commercial spray gun is adopted that uses high-pressure air to embed microscopic fluorescent granules into the epidermis of fish. A pigment emulsion in water is used instead of the dry powdered pigment. A conveyor system is used to transport live juvenile fish into the spray marker, which dispenses the pigment mixed with water through a nozzle with the aid of compressed air. An agitator is used to continuously mix the container of pigment suspended in water, in order to reduce clogging of the nozzle and ensure the uniformity of marking. The marks are detected in the field by examining samples of fish under a low-power ultraviolet (UV) lamp. During examination under a UV light source, the spray-marked areas on the fish body, which contain the fluorescent particles, will shine brightly in a specific color such as red, green, or orange in the visible spectrum and these marks can be very easily detected through the naked eye or by optical means. By this method, fish can be marked at a rate of approximately 15,000/h using about one pound (0.45 kg) of fluorescent pigment per 7,000 fish. The length of the fish ranges from 38 to 52 mm. Although the particles of the fluorescent pigment spray impinge on fish body with a reasonable momentum, only a fraction of the pigment particles actually embed into the scales of fish. Much of the sprayed emulsion is collected in a container attached to the underside of the machine at the exit and is recycled, providing a degree of environmental friendliness. Water in the emulsion increases the momentum of the pigment spray during marking, thereby increasing the mark retention. Also, an emulsion facilitates the use of the system under damp conditions, thereby reducing nozzle clogging. An air pressure of 120 psi is used by the spray gun. Spraying a target (fish) positioned 30 cm from the nozzle resulted in a 8.5 cm spray region diameter and a 5 cm marking region diameter. The conveyor belt is roughly 15 cm wide. According to typical experimental conditions, one spray gun is able to cover only a 5 cm width. A triple-gun system is needed in the design to produce the required marking area on the conveyor.

10.6.11 Project 11: Machine for Grading Herring Roe

Quality is crucial for manufactured products such as automotive parts and high-end fish products such as herring roe. Vision-based systems are used in industry for inspection, quality assessment, and grading of products. Processed herring roe, which is considered a delicacy, has a lucrative market in countries like Japan. Size, shape, color, texture, and firmness are important in determining the overall quality of a skein of roe. A Grade 1 product may command double the price of a Grade 2 product. Accurate grading is quite important in this respect. Grading of herring roe is done mainly by manual labor at present. In view of associated difficulties, such as speed and maintaining a uniform product quality, machine grading has received much attention. Design a grading machine that employs intelligent sensor integration and fusion, for herring roe. In the machine, the roe skeins are arranged in a single file at the feeder and are sent through the sensory system. A camera-based sensory system, as schematically shown in Figure P10.11, may be used for high-speed sensing of size, shape, and color. Ultrasound or other methods may be used for sensing texture and firmness sensing. Roe firmness, geometric features, associated weight estimates, and color are incorporated in an intelligent sensor fusion system to arrive at a grading decision. The typical throughput for the machine is 5 skeins/s. The dimension of a skein of roe will not exceed 20 cm × 5 cm. Consider the experimental arrangement shown in Figure P10.11. Images are captured continuously through a PULNIX 6701 progressive scan camera equipped with a 12.5 mm, *f.* 1.4 Cosmicar TV lens, by a PCI frame grabber board manufactured by Matrox Genesis. Single and multiple linear laser stripes generated from the LASIRIS laser diode structured-light projector (30 mW, 670 nm) are projected onto the object. Microsoft Visual Basic 6.0 programming language was used as an object-oriented environment to develop the human machine interface. Visual C++ 6.0 programming language was used to develop the algorithm for quality assessment, as a dynamic link library (DLL) in a powerful and high-speed PC computer.

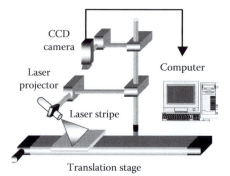

FIGURE P10.11
Laser-based sensing system.

10.6.12 Project 12: Hydraulic Control System

Component sizing is an important consideration in the design of a hydraulic control system. You are asked to design a hydraulic system for a radar positioning drive. Specifically, you must

1. Select a suitable hydraulic motor and suitable gearing to drive the load (radar).
2. Select a suitable pump for continuous hydraulic power supply.
3. Design a speed transmission unit (e.g., gear) for the motor to load (radar) coupling.
4. Determine the inlet pressure at the hydraulic motor.
5. Determine the pump outlet pressure.

The following data are given:

Load inertia $= 2000 \, \text{kg} \cdot \text{m}^2$

Maximum load speed $= 1 \, \text{rad/s}$

Maximum load acceleration $= 10 \, \text{rad/s}^2$

Wind torque $= 2000 \, \text{N} \cdot \text{m}$

Distance from pump to hydraulic motor $= 4 \, \text{m}$

Size of pipeline (steel) $= 1.25 \, \text{cm}$ OD and $1.0 \, \text{mm}$ thickness

Maximum supply pressure of fluid $= 20,000 \, \text{kPa}$

Hydraulic power loss in pipeline $= 5\%$

Pump leakage $=$ motor leakage $= 5\%$

Motor efficiency $=$ gear box efficiency $= 95\%$

Assume that the hydraulic fluid is MIL-H-5606.

Design the following:

1. A pump-controlled system in which the pump directly supplies a controlled flow to the hydraulic motor.
2. A valve-controlled system in which a valve is used between the pump and the hydraulic motor to supply a controlled flow to the motor (assume a 5% valve leakage).

You may use a mechanical engineering handbook to obtain the specifications for the pump and the hydraulic motor (usually the same specs are given for both pumps and motors) and to estimate the pressure loss in the steel piping carrying MIL-H-5606 oil. Commercially available sensors, transmission units, motors, pumps, valves, and power supplies may be used.

Comment: Consider the problem of servovalve selection for a hydraulic drive. The first step is to choose a suitable hydraulic actuator (ram or motor) that meets the load requirements. This establishes the load flow Q_L and the load pressure P_L at the operating speed of the load. The supply pressure P_s is also known. Note that under no-load conditions, the pressure drop across the servovalve is P_s, and under normal operating conditions, it is

$P_s - P_L$. In the manufacturer's specifications, the rated flow of a servovalve is given at some specified pressure drop (e.g., 7000 kPa). Since the flow is proportional to the square root of the pressure drop, we can determine the required flow rating for the valve (typically, by increasing the computed value of the rated flow by 10% to allow for leakage, fluctuations in load, etc.). This rated flow is one factor that governs the choice of a servovalve. The second factor is the valve bandwidth, which should be several times larger than the primary resonant frequency of the load for proper control. Typical information available from servovalve manufacturers includes the frequency corresponding to the 90° phase lag point of the valve response. This frequency may be used as a measure of the valve bandwidth. To select a servovalve for a valve-controlled radar drive, the first step would be to obtain a catalog with a data sheet from a well-known servovalve manufacturer. Assume that the resonant frequency of the radar system is 10 Hz.

Appendix A: Solid Mechanics

This appendix presents some results in the mechanics of solids that are useful in the analysis and design of mechanical components.

A.1 General Problem of Elasticity

A.1.1 Strain Components

By definition, for deformations u, v, and w in the Cartesian directions x, y, and z, we have the corresponding direct strain (ε) and shear strain (γ) components:

$$\varepsilon_x = \frac{\partial u}{\partial x} \qquad\qquad \gamma_{xy} = \frac{\partial v}{\partial x} + \frac{\partial u}{\partial y} = \gamma_{yx}$$

$$\varepsilon_y = \frac{\partial v}{\partial y} \qquad\qquad \gamma_{yz} = \frac{\partial w}{\partial y} + \frac{\partial v}{\partial z} = \gamma_{zy} \qquad (A.1)$$

$$\varepsilon_z = \frac{\partial w}{\partial z} \qquad\qquad \gamma_{zx} = \frac{\partial u}{\partial z} + \frac{\partial w}{\partial x} = \gamma_{xz}$$

A.1.2 Constitutive Equations

These are stress (direct stress σ or shear stress τ) versus strain (direct strain ε or shear strain γ) equations.

Assumptions:

1. Linear elastic $\Rightarrow$ stress–strain relations are 1st degree
2. Homogeneous $\Rightarrow$ uniform material
3. Isotropic $\Rightarrow$ material properties are independent of direction

We get

$$\varepsilon_x = \frac{1}{E}\left[\sigma_x - v(\sigma_y + \sigma_z)\right] \qquad\qquad \gamma_{xy} = \frac{2(1+v)}{E}\tau_{xy}$$

$$\varepsilon_y = \frac{1}{E}\left[\sigma_y - v(\sigma_z + \sigma_x)\right] \qquad\qquad \gamma_{yz} = \frac{2(1+v)}{E}\tau_{yz} \qquad (A.2)$$

$$\varepsilon_z = \frac{1}{E}\left[\sigma_z - v(\sigma_x + \sigma_y)\right] \qquad\qquad \gamma_{zx} = \frac{2(1+v)}{E}\tau_{zx}$$

Here
E is Young's modulus (of elasticity)
ν is Poisson's ratio

In addition to these six constitutive equations, we have the following equations.

A.1.3 Equilibrium Equations

From the equilibrium of a parallelepiped $\delta x \times \delta y \times \delta z$, we get

$$\sum_i \frac{\partial \sigma_{ij}}{\partial x_i} + X_j = 0 \tag{A.3}$$

We have three equilibrium equations.

Note: X_j is the body force for the j direction.

Now taking the moment about the central axis, canceling $\delta x \times \delta y \times \delta z$, and then neglecting terms of $O(\delta)$ [*Note*: body forces and derivatives contain $O(\delta)$ multiple; neglect them], we get

$$\sigma_{ij} = \sigma_{ji} \left(\text{In the absence of body moments}\right) \tag{A.4}$$

A.1.4 Compatibility Equations

Compatibility means, in addition to continuity, there are no kinks (geometric singularities). These are equations satisfied by the second derivatives of strains. They are obtained by double-differentiating (A.1) and eliminating the terms containing u, v, and w. There are six compatibility equations.

Principal stresses: maximum or minimum direct stresses. On the corresponding planes, the shear stress will be zero.

Principal strains: maximum or minimum direct strains. On the corresponding planes, the shear strains will be zero.

Note: For isotropic solids, the principal planes of stress and strain coincide.

A.2 Plane Strain Problem

1. Strains in the z direction are zero.
2. Derivatives w.r.t. z are zero; i.e., all stresses are functions of x, y; all existing strains are functions of x, y; body forces are functions of x, y.
3. Displacements in the z direction are zero (this is governed by (1))

Note: (2) and (3) → All properties are independent of z (not functions of z). Still, stresses in the z direction may not be zero.

A.2.1 Constitutive Equations

$$\varepsilon_x = \frac{1}{E}\left[\sigma_x - \nu\left(\sigma_y + \sigma_z\right)\right] \tag{A.5}$$

$$\varepsilon_y = \frac{1}{E}\left[\sigma_y - \nu\left(\sigma_z + \sigma_x\right)\right] \tag{A.6}$$

$$0 = \frac{1}{E}\left[\sigma_z - \nu\left(\sigma_x + \sigma_y\right)\right] \tag{A.7}$$

$$\gamma_{xy} = \frac{2(1+\nu)}{E}\tau_{xy} \tag{A.8}$$

$$\gamma_{yz} = \frac{2(1+\nu)}{E}\tau_{yz} \tag{A.9}$$

$$\gamma_{zx} = \frac{2(1+\nu)}{E}\tau_{zx} \tag{A.10}$$

By substituting (A.7) in (A.5) and (A.6), we get

$$\varepsilon_x = \frac{1}{E}\left[\sigma_x\left(1-\nu^2\right) - \nu\sigma_y\left(1+\nu\right)\right]$$

$$\varepsilon_y = \frac{1}{E}\left[\sigma_y\left(1-\nu^2\right) - \nu\sigma_x\left(1+\nu\right)\right]$$

A.2.2 Equilibrium Equations

$$\frac{\partial \sigma_x}{\partial x} + \frac{\partial \tau_{xy}}{\partial y} + X_x = 0 \tag{A.11}$$

$$\frac{\partial \tau_{xy}}{\partial x} + \frac{\partial \sigma_y}{\partial y} + X_y = 0 \tag{A.12}$$

$$X_z = 0 \tag{A.13}$$

Note: Compatibility equations are identically satisfied.

A.3 Plane Stress Problem

1. Stresses in the z direction are zero
2. Properties are independent of z (i.e., $\partial/\partial z = 0$)

A.3.1 Constitutive Equations

$$\varepsilon_x = \frac{1}{E}\left[\sigma_x - \nu\sigma_y\right] \tag{A.14}$$

$$\varepsilon_y = \frac{1}{E}\left[\sigma_y - \nu\sigma_x\right] \tag{A.15}$$

$$\varepsilon_z = -\frac{1}{E}\nu\left(\sigma_x + \sigma_y\right) \tag{A.16}$$

$$\gamma_{xy} = \frac{2(1+\nu)}{E}\tau_{xy} \tag{A.17}$$

$$\gamma_{yz} = \frac{2(1+\nu)}{E}\tau_{yz} = 0 \tag{A.18}$$

$$\gamma_{zx} = \frac{2(1+\nu)}{E}\tau_{zx} = 0 \tag{A.19}$$

A.3.2 Equilibrium Equations

$$\frac{\partial\sigma_x}{\partial x} + \frac{\partial\tau_{xy}}{\partial y} + X_x = 0 \tag{A.20}$$

$$\frac{\partial\tau_{xy}}{\partial x} + \frac{\partial\sigma_y}{\partial y} + X_y = 0 \tag{A.21}$$

$$X_z = 0 \tag{A.22}$$

Note: Unlike in the plane strain problems, compatibility equations are not identically satisfied in general in the plane stress problem.

A.3.3 Plane Stress Problem in Polar Coordinates

A.3.3.1 Strain Components

$$\varepsilon_r = \frac{\partial u}{\partial r}$$

$$\varepsilon_\theta = \frac{1}{r}\frac{\partial v}{\partial \theta} + \frac{u}{r} \tag{A.23}$$

$$\gamma_{r\theta} = \frac{1}{r}\frac{\partial u}{\partial \theta} + \frac{\partial v}{\partial r} - \frac{v}{r}$$

A.3.3.2 Constitutive Equations

$$\varepsilon_r = \frac{1}{E}\left(\sigma_r - \nu\sigma_\theta\right)$$

$$\varepsilon_\theta = \frac{1}{E}\left(\sigma_\theta - \nu\sigma_r\right) \tag{A.24}$$

$$\gamma_{r\theta} = \frac{2\left(1+\nu\right)}{E}\tau_{r\theta}$$

A.3.3.3 Equilibrium Equations

$$\frac{\partial\sigma_r}{\partial r} + \frac{1}{r}\frac{\partial\tau_{r\theta}}{\partial\theta} + \frac{\sigma_r - \sigma_\theta}{r} + F_r = 0$$

$$\frac{\partial\tau_{r\theta}}{\partial r} + \frac{1}{r}\frac{\partial\sigma_\theta}{\partial\theta} + \frac{2\tau_{r\theta}}{r} + F_\theta = 0 \tag{A.25}$$

A.4 Rotating Members

Rotates at angular speed ω.

A.4.1 Rotating Disks

The equilibrium equations reduce to (*Note*: axisymmetric)

$$\frac{d\sigma_r}{dr} + \frac{\sigma_r - \sigma_\theta}{r} + \rho\omega^2 r = 0 \quad \textit{Note:} \ \ \rho = \text{density in absolute units} \tag{A.26}$$

Also,

$$\varepsilon_\theta = \frac{u}{r} = \frac{1}{E}\left(\sigma_\theta - \nu\sigma_r\right)$$

Hence,

$$\frac{du}{dr} = \frac{d}{dr}\frac{r}{E}\left(\sigma_\theta - \nu\sigma_r\right) = \frac{1}{E}\frac{d}{dr}\left[r\left(\sigma_\theta - \nu\sigma_r\right)\right]$$

Also,

$$\varepsilon_r = \frac{du}{dr} = \frac{1}{E}\left(\sigma_r - \nu\sigma_\theta\right)$$

Hence,

$$\sigma_r - v\sigma_\theta = \frac{d}{dr}\left[r\left(\sigma_\theta - v\sigma_r\right)\right] \tag{A.27}$$

Introduce a function ψ so that

$$\sigma_r = \frac{\psi}{r}$$

$$\sigma_\theta = \frac{d\psi}{dr} + \rho\omega^2 r^2 \tag{A.28}$$

Substitute (A.28). Then, (A.26) will be identically satisfied. But (A.27) becomes

$$\frac{d^2\psi}{dr^2} + \frac{1}{r}\frac{d\psi}{dr} - \frac{\psi}{r^2} = -(3+v)\rho\omega^2 r \tag{A.29}$$

$$\text{or,}\quad \frac{d}{dr}\left[\frac{1}{r}\frac{d}{dr}(r\psi)\right] = -(3+v)\rho\omega^2 r$$

Solution

$$\psi = Ar + \frac{B}{r} - \frac{(3+v)}{8}\rho\omega^2 r^3 \tag{A.30}$$

From (A.28), we get

$$\sigma_r = A + \frac{B}{r^2} - \frac{(3+v)}{8}\rho\omega^2 r^2$$

$$\sigma_\theta = A - \frac{B}{r^2} - \frac{(1+3v)}{8}\rho\omega^2 r^2 \tag{A.31}$$

A.4.2 Rotating Thick Cylinders

With similar end conditions, it is assumed that plane sections remain plane, if taken sufficiently distant from ends. Hence, the strain in the direction of the axis is constant.

1. $e_z = $ constant c
2. By the symmetry and similarity of sections remote from ends, all the shear stresses are absent in the element considered.
3. By symmetry $F_\theta = 0$
4. $F_r = \rho\omega^2 r$ in the case of rotating cylinder

A.4.2.1 Strain Equations

$$e_r = \frac{du}{dr}$$

$$e_\theta = \frac{u}{r}$$

(A.32)

Hence,

$$\frac{d}{dr}(re_\theta) = e_r = r\frac{de_\theta}{dr} + e_\theta$$

or,

$$\frac{de_\theta}{dr} = \frac{e_r - e_\theta}{r}$$

A.4.2.2 Stress–Strain (Constitutive) Relations

$$e_r = \frac{1}{E}\left[\sigma_r - \nu(\sigma_\theta + \sigma_z)\right]$$

$$e_\theta = \frac{1}{E}\left[\sigma_\theta - \nu(\sigma_r + \sigma_z)\right]$$

(A.33a)

$$e_z = \frac{1}{E}\left[\sigma_z - \nu(\sigma_r + \sigma_\theta)\right]$$

A.4.2.3 Equilibrium Equations

Consider the element shown in Figure A.1.

$$\left(\sigma_r + \frac{d\sigma_r}{dr}\cdot dr\right)(r+dr)d\theta - \sigma_r \cdot r \cdot d\theta - 2\sigma_\theta \cdot dr\frac{d\theta}{2}$$

$$+ \rho\omega^2 rrd\theta \cdot dr = 0$$

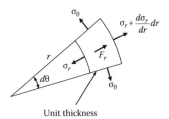

We get

$$\frac{d\sigma_r}{dr} + \frac{\sigma_r - \sigma_\theta}{r} + \rho\omega^2 r = 0 \qquad \text{(A.34)}$$

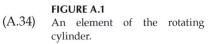

FIGURE A.1
An element of the rotating cylinder.

A.4.2.4 Final Result

$$\sigma_r = A + \frac{B}{r^2} - \frac{(3-2v)}{8(1-v)}\rho\omega^2 r^2$$

$$\sigma_\theta = A - \frac{B}{r^2} - \frac{(1+2v)}{8(1-v)}\rho\omega^2 r^2$$

(A.35a)

A.4.2.5 Temperature Stresses

Here, Equations A.34 and A.32 are the same. But (A.33a) takes the following form:

$$e_r = \frac{1}{E}\left[\sigma_r - v(\sigma_\theta + \sigma_z)\right] + \alpha T$$

$$e_\theta = \frac{1}{E}\left[\sigma_\theta - v(\sigma_r + \sigma_z)\right] + \alpha T$$

$$e_z = \frac{1}{E}\left[\sigma_z - v(\sigma_r + \sigma_\theta)\right] + \alpha T$$

(A.33b)

Note: α = the temperature coefficient of linear expansion.

A.4.3 Particular Cases of Cylinders

A.4.3.1 Case 1: Axially Restrained Ends

$$\sigma_r = A + \frac{B}{r^2}$$

$$\sigma_\theta = A - \frac{B}{r^2}$$

(A.35b)

$$\sigma_z = EC + v(\sigma_r + \sigma_\theta) = \text{constant}$$

Axial force $P = \int_a^b \sigma_z \cdot 2\pi r \cdot dr = \sigma_z \int_a^b 2\pi r \cdot dr = \pi(b^2 - a^2)\sigma_z$

Hence,

$$\sigma_z = \frac{P}{\pi(b^2 - a^2)}$$

A.4.3.2 Case 2: Thick Pressure Vessel

Internal radius $=a$
External radius $=b$
Internal pressure $=p$
Hence, the end force exerted on the cylinder is $P=\pi a^2 p$.
The boundary conditions are $\sigma_r = -p$ at $r=a$ and $\sigma_r=0$ at $r=b$.
Hence,

$$\sigma_r = -\frac{pa^2}{\left(b^2-a^2\right)}\left(\frac{b^2}{r^2}-1\right)$$

$$\sigma_\theta = \frac{pa^2}{\left(b^2-a^2\right)}\left(\frac{b^2}{r^2}+1\right) \tag{A.35c}$$

$$\sigma_z = \frac{pa^2}{\left(b^2-a^2\right)}$$

A.4.3.3 Case 3: Thin Pressure Vessel

Here $t/a \ll 1$ where thickness $t=(b-a)$

$$b^2-a^2 \to 2at \qquad b-r=t' \quad \text{where } 0<t'<t$$

Hence,

$$\sigma_r = -\frac{pt'}{t}$$

$$\sigma_\theta = \frac{pa}{t} \tag{A.35d}$$

$$\sigma_z = \frac{pa}{2t}$$

Note: σ_θ and σ_z are much larger than σ_r.

The same relations may be obtained considering the equilibrium of the sectioned cylinder.

A.4.3.4 Case 4: Rotating Cylinder with Free Ends

The boundary conditions are $\sigma_r=0$ at $r=a$ and $r=b$.

Hence, from (A.35), we get

$$A = \frac{(3-2v)}{8(1-v)}\left(a^2+b^2\right)\rho\omega^2$$

$$B = \frac{(3-2v)}{8(1-v)}a^2b^2\rho\omega^2$$

$$\sigma_z = EC + 2vA - \frac{v\rho\omega^2 r^2}{2(1-v)} = K - \frac{v\rho\omega^2 r^2}{2(1-v)}$$

End force $P = \int_a^b \sigma_z \cdot 2\pi r \cdot dr = \int_a^b \left[K - \frac{v\rho\omega^2 r^2}{2(1-v)} \right] 2\pi r \cdot dr = 0$

Hence, $K = \dfrac{v\rho\omega^2 \left(a^2 + b^2\right)}{4(1-v)}$

We get

$$\sigma_z = \frac{v}{4(1-v)} \left[1 + \frac{a^2}{b^2} - \frac{2r^2}{b^2} \right] \rho\omega^2 b^2$$

Note: $\sigma_{\theta max}$ and $\sigma_{z max}$ occur at the lowest value of r; i.e., at $r = a$.

A.5 Mohr's Circle of Plane Stress

Mohr's circle is used to obtain the stress state in a different direction (plane) at a location, given the stress state in another direction. (*Note*: Mohr's circle may be used to transform strains and moments of inertia as well, in the same manner). The principal stresses (σ_{max} and σ_{min}) occur in the stress state where the shear stress is zero, and a state of pure shear (τ_{max}) occurs in a direction that is at 45° to the principal direction. Since an angle along the Mohr's circle is double the actual angle in the physical domain, pure shear occurs at an angle 90° to the state of principal stresses.

Consider the state of plane stress in a general direction, as shown in Figure A.2a. Suppose that the direction of principal stress is at an angle θ, as shown in Figure A.2b. The Mohr's circle is shown in Figure A.2b.

The following results can be obtained:

$$\sigma_{max}, \sigma_{min} = \frac{\sigma_x + \sigma_y}{2} \pm \sqrt{\left[\frac{\sigma_x - \sigma_y}{2} \right]^2 + \tau_{xy}^2} \tag{A.36}$$

$$\tau_{max} = \sqrt{\left[\frac{\sigma_x - \sigma_y}{2} \right]^2 + \tau_{xy}^2} = \frac{\sigma_{max} - \sigma_{min}}{2} \tag{A.37}$$

$$\tan 2\theta = \frac{2\tau_{xy}}{\sigma_x - \sigma_y} \tag{A.38}$$

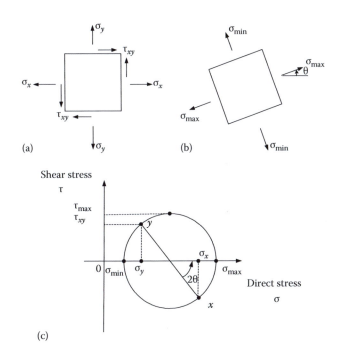

FIGURE A.2
(a) General state of plane stress; (b) state of principal stress; (c) Mohr's circle of plane stress.

A.6 Torsion

A.6.1 Circular Members

Assumptions:

1. Cross sections remain plane
2. All radii remain straight
3. No longitudinal displacement

Consider a portion of radius r of the cylinder of radius a ($a > r$), as shown in Figure A.3. On this, a line on the surface, parallel to the axis (generator) before applying T, will be a spiral after T is applied. Consider the element of length dz from it, as in Figure A.3b.

γ = shear strain = angular deformation of element

We have, $dz \cdot \gamma = r \cdot d\theta$

Hence,

$$\gamma = r \frac{d\theta}{dz} \tag{A.39}$$

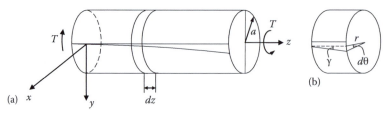

(a) x   y   dz

(b)

FIGURE A.3
(a) Circular torsion member; (b) small element.

Since identical conditions exist at every section,

$$\text{Angle of twist/unit length } \frac{d\theta}{dz} = \alpha = \text{constant} = \frac{\theta}{l} \qquad (A.40)$$

Hence, the strain distribution (whether elastic or not) is as shown in Figure A.4.

$$\text{Shear stress for elastic case: } \tau = G\gamma = G\alpha r \qquad (A.41)$$

FIGURE A.4
Shear strain distribution under torsion.

$$\text{Torque } T = \int_A r\tau\, dA$$

where
  dA is an elemental area at radius r on the shaft cross-section
  τ is shear stress

Hence,

$$T = \int_0^a r\tau 2\pi r \cdot dr \qquad (A.42)$$

By substituting (A.41) in (A.42) and using the definition of J, we get

$$T = G\alpha J \qquad (A.43)$$

where J is the polar moment of area, which is defined as

$$J = \int_A r^2\, dA \qquad (A.44)$$

By substituting (A.41) in (A.43), we get

$$\tau = \frac{Tr}{J} \qquad (A.45)$$

Hence,

$$\frac{T}{J} = \frac{\tau}{r} = G\alpha = G\frac{d\theta}{dz} \tag{A.46}$$

For conical members where $d\theta/dz$ is not a constant (because J varies from section to section),

Twist angle

$$\theta = \int \frac{T}{GJ} dz \tag{A.47a}$$

For cylindrical shafts, from (A.40)

$$\theta = \frac{Tl}{GJ} \tag{A.47b}$$

A.6.2 Torque Sensor

Treat the cylindrical torsion member shown in Figure A.3a as a torque sensor. Note that in the foregoing equations, r is any radius within the shaft cross-section and τ is the shear stress at that location. Now assume that r denotes the outer (maximum) radius and τ is the corresponding shear stress τ_{max} (i.e., use $r \equiv r_{max}$ and $\tau \equiv \tau_{max}$ to avoid the use of subscripts).

In a small square (two-dimensional) element on the cylindrical outer surface of the shaft, with one side parallel to the shaft axis and another side along the shaft circumference, the state of stress is a "pure shear," as shown in Figure A.5. Mohr's circle of the state of stress on the outer surface of the shaft is indicated in Figure A.6.

The principal stresses occur at points A and C in Figure A.6, and a state of pure shear occurs in a direction that is at 45° to the principal direction. Since an angle along the Mohr's circle is double the actual angle in the physical domain, pure shear occurs at an angle 90° to the state of principal stresses, as given by points B and D in Figure A.6.

It follows from the Mohr's circle that on the outer surface of the shaft, for an element as shown in Figure A.5, at 45° to the shaft axis, the state of stress is pure tension or compression without shear. Accordingly, x and y in Figure A.5 are directions of principal stress on the shaft surface. Note that the principal stress (tension or compression) σ is given by (see Figure A.6)

$$\sigma = \tau \tag{A.48}$$

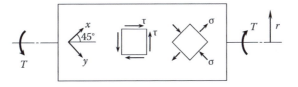

FIGURE A.5
Pure shear state of stress and the principal directions x and y in pure torsion.

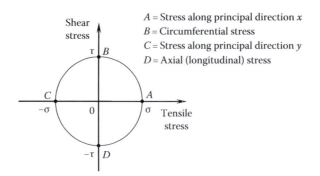

FIGURE A.6
Mohr's circle for pure torsion of a shaft.

Now, to determine the strain along the two principal directions x and y, we use stress–strain constitutive relations for a plane stress problem:

$$\varepsilon_x = \frac{1}{E}\left(\sigma_x - v\sigma_y\right) \tag{A.49}$$

$$\varepsilon_y = \frac{1}{E}\left(\sigma_y - v\sigma_x\right) \tag{A.50}$$

where
   E is Young's modulus of elasticity
   v is Poisson's ratio

Using the fact that

$$\sigma_x = -\sigma_y = \sigma \tag{A.51}$$

from Equations A.49 and A.50, we get

$$\varepsilon_x = -\varepsilon_y = \varepsilon = \frac{(1+v)}{E}\sigma \tag{A.52}$$

Now, in view of Equations A46, A.48, and A.52, we have

$$\varepsilon = \frac{(1+v)r}{EJ}T \tag{A.53}$$

or using the fact that shear modulus G is given by

$$E = 2(1+v)G \tag{A.54}$$

we can write

$$\varepsilon = \frac{r}{2GJ} T \qquad (A.55)$$

Equation A.55 is used to determine torque T from the principal strain ε, as measured using a strain gage device. This is the principle of a torque sensor using the strain gage method.

Example A.1

Determine the torques at the ends of the frustum, which are rigidly fixed, shown in Figure A.7.

Solution
Selecting the coordinates as shown in Figure A.7b, the twist of an element dz at z is

$$d\theta = \frac{T}{GJ} dz \quad \text{where } J = \frac{\pi r^4}{2} = \frac{\pi z^4 \tan^4 \alpha}{2} = Kz^4$$

Hence, $\theta = \int \frac{T}{GKz^4} dz = \int_{l}^{2l} \frac{T_2}{GKz^4} dz + \int_{2l}^{3l} \frac{T_1}{GKz^4} dz = 0$

Hence, $\left[-\frac{T_2}{3z^3} \right]_{l}^{2l} + \left[-\frac{T_1}{3z^3} \right]_{2l}^{3l} = 0$

We get

$$7T_2 + \frac{19}{27} T_1 = 0 \qquad (i)$$

From equilibrium,

$$T = T_2 - T_1 \qquad (ii)$$

Solve (i) and (ii)

$$T_1 = -\frac{189}{208} T$$

$$T_2 = \frac{19}{208_1} T$$

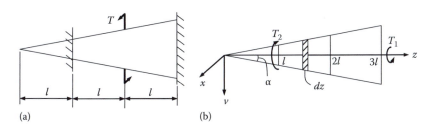

FIGURE A.7
(a) Torsion in a conical member (frustum); (b) an element of analysis in the conical member.

A.7 Beams in Bending and Shear

See Figure A.8 for a beam with a bending moment M, shear force V, and transverse deflection v at the longitudinal location x.

Bending moment $M = EI\dfrac{d^2v}{dx^2}$

Shear force $Q = \dfrac{dM}{dx}$

Shear flow $q = \dfrac{QP}{I}$

E is Young's modulus of elasticity
I is the second moment area of the beam cross section about its neutral axis of bending
P is the first moment of area of the cross section of the beam segment at which the shear flow is determined (about the neutral axis of bending)

A.7.1 Mohr's Theorems

1. For a beam transversely loaded in any manner, slope at B – slope at $A = \dfrac{1}{EI}$ [area of bending moment diagram between AB]

2. $\dfrac{1}{EI}$ [moment about a line of bending moment diagram from A to B] = Intercept made on the line by the tangents to beam at A and B

A.7.2 Maxwell's Theorem of Reciprocity

Suppose that a force system P_i has a corresponding deflection δ_i, and another P_i' has δ_i'. Then, $\displaystyle\sum P_i\delta_i' = \sum P_i'\delta_i$.

Note: This is true for moments as well, provided that slopes are used instead of deflections.

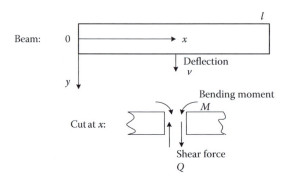

FIGURE A.8
Beam in bending and shear.

TABLE A.1

Useful Results of Beam Deflection

Beam Configuration	End Angle (Slope)	End Deflection
M	$\dfrac{Ml}{EI}$	$\dfrac{Ml^2}{2EI}$
P	$\dfrac{Pl^2}{2EI}$	$\dfrac{Pl^3}{3EI}$
w $\quad l$	$\dfrac{wl^3}{6EI}$	$\dfrac{wl^4}{8EI}$

A.7.3 Castigliano's First Theorem

$$\frac{\partial U}{\partial P_r} = \delta_r; \qquad \frac{\partial U}{\partial \delta_r} = P; \qquad \frac{\partial U}{\partial M_r} = \theta_r; \qquad \frac{\partial U}{\partial \theta_r} = M_r$$

A.7.4 Elastic Energy of Bending

For a small beam element of length ds and flexural rigidity EI, bent by moment M, the elastic energy of bending is

$$dU = \frac{1}{2} M \frac{ds}{r} = \frac{1}{2} \frac{M^2}{EI} ds \qquad (A.56)$$

where r is the radius of curvature.

Some important results of beam deflection are given in Table A.1.

A.8 Open-Coiled Helical Springs

A.8.1 Case 1: Axial Load *W*

Consider element AB of a spring. The moments at A are shown in Figure A.9.

The bending moment component tends to unwind.

The twisting couple component twists the wire inward, causing deflection.

Bending $d\phi = \dfrac{ds}{R} = \left(\dfrac{1}{R_2} - \dfrac{1}{R_1} \right) ds$

The beam bending result is $\dfrac{M}{EI} = \dfrac{1}{R} = \left(\dfrac{1}{R_2} - \dfrac{1}{R_1} \right) = \dfrac{d\phi}{ds}$

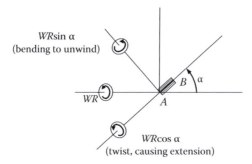

FIGURE A.9
Spring element under axial load.

Hence, $d\phi = \dfrac{M}{EI}ds$

Twisting $d\theta = \dfrac{T}{GJ}ds$

The corresponding vertical deflection $d\Delta$ is given by

$$\frac{1}{2}wd\Delta = \frac{1}{2}Md\phi + \frac{1}{2}Td\theta = \frac{1}{2}\frac{M^2}{EI}ds + \frac{1}{2}\frac{T^2}{GJ}ds = \frac{1}{2}\frac{w^2R^2\sin^2\alpha}{EI}ds + \frac{1}{2}\frac{w^2R^2\cos^2\alpha}{GJ}ds$$

or $d\Delta = w^2R^2ds\left[\dfrac{\cos^2\alpha}{GJ} + \dfrac{\sin^2\alpha}{EI}\right]$

For the full spring,

$$\Delta = w^2R^2L\left[\frac{\cos^2\alpha}{GJ} + \frac{\sin^2\alpha}{EI}\right] \tag{A.57}$$

where $L = \text{length} = \dfrac{2\pi Rn}{\cos\alpha}$

The axial winding due to $T = \dfrac{T}{GJ}ds\sin\alpha = \dfrac{wR}{GJ}\cos\alpha\sin\alpha\cdot ds$

The axial unwinding due to $M = \dfrac{M}{EI}ds\cos\alpha = \dfrac{wR}{EI}\sin\alpha\cos\alpha\cdot ds$

The overall winding $= \dfrac{wR}{GJ}\cos\alpha\sin\alpha\cdot ds - \dfrac{wR}{EI}\sin\alpha\cos\alpha\cdot ds$ in an element ds.

The overall winding is

$$\theta = wRL\sin\alpha\cos\alpha\left[\frac{1}{GJ} - \frac{1}{EI}\right] \tag{A.58}$$

A.8.2 Case 2: Axial Couple *M*

Consider the spring element shown in Figure A.10.

Axial deflection is $d\theta$.

We have

$$\frac{1}{2}M\,d\theta = \frac{1}{2}\frac{T'^2}{GJ}ds + \frac{1}{2}\frac{M'^2}{EI}ds = \frac{1}{2}\frac{M^2\sin^2\alpha}{GJ}ds + \frac{1}{2}\frac{M^2\cos^2\alpha}{EI}ds$$

Hence, $d\theta = Mds\left[\dfrac{\sin^2\alpha}{GJ} + \dfrac{\cos^2\alpha}{EI}\right]$

or,

$$\theta = ML\left[\frac{\sin^2\alpha}{GJ} + \frac{\cos^2\alpha}{EI}\right] \tag{A.59}$$

Vertical deflection due to $T' = \dfrac{T'}{GJ}\cos\alpha \cdot R \cdot ds$ downward

Vertical deflection due to $M' = \dfrac{M'}{EI}\sin\alpha \cdot R \cdot ds$ upward

Net vertical deflection downward $d\Delta = \dfrac{T'}{GJ}\cos\alpha \cdot R \cdot ds - \dfrac{M'}{EI}\sin\alpha \cdot R \cdot ds$

$$d\Delta = \frac{MR\sin\alpha\cos\alpha}{GJ}\cdot ds - \frac{MR\sin\alpha\cos\alpha}{EI}\cdot ds$$

$$d\Delta = MR\sin\alpha\cos\alpha\cdot ds\left[\frac{1}{GJ} - \frac{1}{EI}\right]$$

or

$$\Delta = MRL\sin\alpha\cos\alpha\left[\frac{1}{GJ} - \frac{1}{EI}\right] \tag{A.60}$$

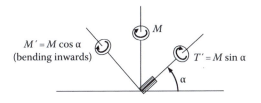

FIGURE A.10
Spring element under axial couple.

A.9 Circular Plates with Axisymmetric Loading

Assumptions:

1. Material: homogeneous; isotropic; obeys Hook's Law
2. Geometry: flat plates; deflections and slopes small; mid-plane is neutral plane; planes normal to mid-plane before deflection remain normal
3. Stresses: work within elastic limit; normal forces and central shears are negligible

By symmetry $\partial/\partial\theta = 0$, and $\gamma_{r\theta}$, $\tau_{r\theta}$, $M_{r\theta}$ are absent. From Figure A.11, deflection in the planar (*x*) direction at depth *z* from the neutral plane is $u = -z(dw/dr)$.

A.9.1 Strains

$$\varepsilon_r = \frac{du}{dr} = -z\frac{d^2w}{dr^2}$$

$$\varepsilon_\theta = \frac{u}{r} = -\frac{z}{r}\frac{dw}{dr}$$

(A.61)

A.9.2 Stresses

$$\sigma_r = \frac{E}{\left(1-v^2\right)}\left[\varepsilon_r + v\varepsilon_\theta\right] = -\frac{Ez}{\left(1-v^2\right)}\left[\frac{d^2w}{dr^2} + v\frac{1}{r}\frac{dw}{dr}\right]$$

$$\sigma_\theta = \frac{E}{\left(1-v^2\right)}\left[\varepsilon_\theta + v\varepsilon_r\right] = -\frac{Ez}{\left(1-v^2\right)}\left[\frac{1}{r}\frac{dw}{dr} + v\frac{d^2w}{dr^2}\right]$$

(A.62)

A.9.3 Moments

Considering unit width in its logical sense and not in its physical sense, we get moments per unit length at a particular point:

$$M_r = \int\left(\sigma_r \cdot 1 \cdot dz\right)z = -\frac{E}{\left(1-v^2\right)}\left[\frac{d^2w}{dr^2} + v\frac{1}{r}\frac{dw}{dr}\right]\int z^2 \cdot dz$$

$$M_\theta = \int\left(\sigma_\theta \cdot 1 \cdot dz\right)z = -\frac{Ez}{\left(1-v^2\right)}\left[\frac{1}{r}\frac{dw}{dr} + v\frac{d^2w}{dr^2}\right]\int z^2 \cdot dz$$

or,

$$M_r = -D\left[\frac{d^2w}{dr^2} + v\frac{1}{r}\frac{dw}{dr}\right]$$

$$M_\theta = -D\left[\frac{1}{r}\frac{dw}{dr} + v\frac{d^2w}{dr^2}\right]$$

(A.63)

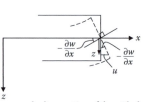

$w = w(x,y)$ = equation of the mid-plane after deflection

FIGURE A.11
Element of plate in bending.

A.9.4 Equilibrium Equations

See Figure A.12.

The shear force balance is $-\left[Q_r + \dfrac{dQ_r}{dr}\,dr\right][r+dr]\,d\theta + Q_r \cdot r \cdot d\theta = q \cdot dr \cdot r \cdot d\theta$

Hence, $Q_r \cdot dr \cdot d\theta + r\dfrac{dQ_r}{dr}\,dr \cdot d\theta = q \cdot dr \cdot r \cdot d\theta$ or,

$$\frac{d(rQ_r)}{dr} = -qr \tag{A.64}$$

The moment balance is $\left[M_r + \dfrac{dM_r}{dr}\,dr\right][r+dr]\,d\theta - M_r \cdot r \cdot d\theta - Q_r \cdot r \cdot d\theta \cdot dr - M_\theta \cdot dr \cdot d\theta = 0$

or $M_r + r\dfrac{dM_r}{dr} - M_\theta = Q_r \cdot r$

Substitute (A.63) and (A.64)

$$-D\left[\frac{d^2w}{dr^2} + \frac{\nu}{r}\frac{dw}{dr} + r\left(\frac{d^3w}{dr^3} + \frac{\nu}{r}\frac{d^2w}{dr^2} - \frac{\nu}{r^2}\frac{dw}{dr}\right) - \frac{1}{r}\frac{dw}{dr} - \nu\frac{d^2w}{dr^2}\right] = rQ_r$$

We get

$$\frac{d}{dr}\left[\frac{1}{r}\frac{d}{dr}\left(r\frac{dw}{dr}\right)\right] = -\frac{Q}{D} \tag{A.65}$$

where $Q_r = Q$
Substitute (A.64)

$$\frac{1}{r}\frac{d}{dr}\left\{r\frac{d}{dr}\left[\frac{1}{r}\frac{d}{dr}\left(r\frac{dw}{dr}\right)\right]\right\} = \frac{q}{D} \tag{A.66}$$

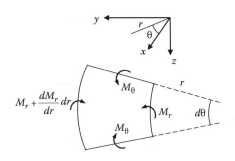

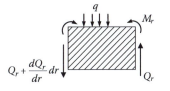

FIGURE A.12
Equilibrium condition of plate element.

A.9.5 Boundary Conditions

A.9.5.1 Fixed Edge

$$W = 0 \text{ at } r = a$$

$$\frac{dw}{dr} = 0 \text{ at } r = a \text{ and } r = 0$$

A.9.5.2 Simply Supported Edge

$$W = 0 \text{ at } r = a$$

$$M_r = -D\left[\frac{d^2w}{dr^2} + \frac{v}{r}\frac{dw}{dr}\right] = 0 \text{ at } r = a$$

$$\frac{dw}{dr} = 0 \text{ at } r = 0$$

A.9.5.3 Partially Restrained Edge

$$\frac{dw}{dr} = 0 \text{ at } r = a$$

Edge deflection of plate = deflection of edge beam

Edge rotation of plate = rotation of edge beam

Appendix B: Transform Techniques

In this appendix, we will formally introduce the Laplace transformation and the Fourier transformation and we will illustrate how these techniques are useful in the analysis of dynamic systems. The preference of one domain over another will depend on such factors as the nature of the excitation input, the type of the analytical model available, the time duration of interest, and the quantities that need to be determined.

B.1 Laplace Transform

The Laplace transformation relates the time domain to the *Laplace domain* (also called *s-domain* or complex frequency domain). The Laplace transform $Y(s)$ of a piecewise-continuous function or signal $y(t)$ is given, by definition, as

$$Y(s) = \int_0^\infty y(t)\exp(-st)\,dt \tag{B.1}$$

and is denoted using the Laplace operator $\mathcal{L}$ as

$$Y(s) = \mathcal{L}y(t) \tag{B.1*}$$

Here, s is a complex independent variable known as the Laplace variable, defined by

$$s = \sigma + j\omega \tag{B.2}$$

where σ is a real-valued constant that will make the transform (Equation B.1) finite, ω is simply frequency, and $j = \sqrt{-1}$. The real value (a) can be chosen to be sufficiently large so that the integral in Equation B.1 is finite even when the integral of the signal itself (i.e., $\int y(t)\,dt$) is not finite. This is the reason why, for example, Laplace transform is better behaved than Fourier transform, which will be defined later, from the analytical point of view. The symbol s can be considered to be a constant, when integrating with respect to t, in Equation B.1.

The inverse relation (i.e., obtaining y from its Laplace transform) is

$$y(t) = \frac{1}{2\pi j} \int_{\sigma - j\omega}^{\sigma + j\omega} Y(s)\exp(st)\,ds \tag{B.3}$$

and is denoted using the inverse Laplace operator $\mathcal{L}^{-1}$ as

$$y(t) = \mathcal{L}^{-1}Y(s) \tag{B.3*}$$

The integration in (B.3) is performed along a vertical line parallel to the imaginary (vertical) axis, located at σ from the origin in the complex Laplace plane (s-plane). For a given piecewise-continuous function $y(t)$, the Laplace transform exists if the integral in Equation B.1 converges. A sufficient condition for this is

$$\int_0^\infty |y(t)| \exp(-\sigma t) dt < \infty \tag{B.4}$$

Convergence is guaranteed by choosing a sufficiently large and positive σ. This property is an advantage of the Laplace transformation over the Fourier transformation.

B.1.1 Laplace Transforms of Some Common Functions

Now we determine the Laplace transform of some useful functions using the definition (B.1). Usually, however, we use Laplace transform tables to obtain these results.

B.1.1.1 Laplace Transform of a Constant

Suppose our function $y(t)$ is a constant, B. Then the Laplace transform is

$$\mathcal{L}(B) = Y(s) = \int_0^\infty Be^{-st} dt$$

$$= B \left. \frac{e^{-st}}{-s} \right|_0^\infty = \frac{B}{s}$$

B.1.1.2 Laplace Transform of the Exponential

If $y(t)$ is e^{at}, its Laplace transform is

$$\mathcal{L}(e^{at}) = \int_0^\infty e^{-st} e^{at} dt$$

$$= \int_0^\infty e^{(a-s)t} dt$$

$$= \frac{1}{(a-s)} e^{(a-s)t} \Big|_0^\infty = \frac{1}{s-a}$$

Note: If $y(t)$ is e^{-at}, it is obvious that the Laplace transform is

$$\mathcal{L}\left(e^{-at}\right) = \int_0^\infty e^{-st} e^{-at} dt$$

$$= \int_0^\infty e^{-(a+s)t} dt$$

$$= \frac{-1}{(a-s)} e^{-(a+s)t} \Big|_0^\infty = \frac{1}{s+a}$$

This result can be obtained from the previous result simply by replacing a with $-a$.

B.1.1.3 Laplace Transform of Sine and Cosine

In the following, the letter $j = \sqrt{-1}$. If $y(t)$ is $\sin \omega t$, the Laplace transform is

$$\mathcal{L}\left(\sin \omega t\right) = \int_0^\infty e^{-st} \left(\sin \omega t\right) dt$$

Consider the following identities:

$$e^{j\omega t} = \cos \omega t + j \sin \omega t$$

$$e^{-j\omega t} = \cos \omega t - j \sin \omega t$$

If we add and subtract these two equations, respectively, we obtain the expressions for the sine and the cosine in terms of $e^{j\omega t}$ and $e^{-j\omega t}$:

$$\cos \omega t = \frac{1}{2}\left(e^{j\omega t} + e^{-j\omega t}\right)$$

$$\sin \omega t = \frac{1}{2i}\left(e^{j\omega t} - e^{-j\omega t}\right)$$

$$\mathcal{L}\left(\cos \omega t\right) = \frac{1}{2} L\left(e^{j\omega t}\right) + \frac{1}{2} L\left(e^{-j\omega t}\right)$$

$$\mathcal{L}\left(\sin \omega t\right) = \frac{1}{2} L\left(e^{j\omega t}\right) - \frac{1}{2} L\left(e^{-j\omega t}\right)$$

We have just seen that

$$\mathcal{L}\left(e^{at}\right) = \frac{1}{s-a}; \quad \mathcal{L}\left(e^{-at}\right) = \frac{1}{s+a}$$

Hence,

$$\mathcal{L}\left(e^{j\omega t}\right) = \frac{1}{s - j\omega t}; \quad \mathcal{L}\left(e^{-j\omega t}\right) = \frac{1}{s + j\omega t}$$

Substituting these expressions, we get

$$\mathcal{L}(\cos \omega t) = \frac{1}{2}\left[\frac{1}{s - j\omega}\right] + \frac{1}{2}\left[\frac{1}{s + j\omega}\right]$$

$$= \frac{1}{2}\left[\frac{s + j\omega}{s^2 - \left(j\omega\right)^2} + \frac{s - j\omega}{s^2 - \left(j\omega\right)^2}\right]$$

$$= \frac{s}{s^2 + \omega^2}$$

$$\mathcal{L}(\sin \omega t) = \frac{1}{2j}L\left(e^{j\omega t} - e^{-j\omega t}\right)$$

$$= \frac{1}{2j}\left[\frac{1}{s - j\omega}\right] - \frac{1}{2j}\left[\frac{1}{s + j\omega}\right]$$

$$= \frac{1}{2j}\left[\frac{s + j\omega}{s^2 - \left(j\omega\right)^2} + \frac{s - j\omega}{s^2 - \left(j\omega\right)^2}\right]$$

$$= \frac{1}{2j}\left[\frac{2j\omega}{s^2 + \omega^2}\right]$$

$$= \frac{\omega}{s^2 + \omega^2}$$

B.1.1.4 Laplace Transform of a Derivative

Let us transform a derivative of a function. Specifically, the derivative of a function y of t is denoted by $\dot{y} = \left(dy/dt\right)$. Its Laplace transform is given by

$$\mathcal{L}\left(\dot{y}\right) = \int_0^\infty e^{-st}\dot{y}dt = \int_0^\infty e^{-st}\frac{dy}{dt}dt \tag{B.5}$$

Now we integrate by parts to eliminate the derivative within the integrand.

B.1.1.4.1 Integration by Parts

From calculus, we know that $d(uv) = udv + vdu$.

By integrating, we get $uv = \int udv + \int vdu$.

Hence,

$$\int u dv = uv - v du \qquad (B.6)$$

This is known as integration by parts.
In (B.5), let
$\quad u = e^{-st}$ and $v = y$
Then, $dv = dy = \dfrac{dy}{dt} dt = \dot{y} dt$

$$du = \frac{du}{dt} dt = -se^{-st} dt.$$

Substitute in (B.5) to integrate by parts:

$$\mathcal{L}(\dot{y}) = \int_{0}^{\infty} e^{-st} dy$$

$$= \int u dv = uv - \int v du$$

$$= e^{-st} y(t) \Big|_{0}^{\infty} - \int_{0}^{\infty} -se^{-st} y(t) dt$$

$$= -y(0) + s\mathcal{L}[y(t)]$$

$$= s\mathcal{L}(y) - y(0)$$

where $y(0)$ = the initial value of y. This says that the Laplace transform of a first derivative $\dot{y}$ equals s times the Laplace transform of the function y minus the initial value of the function (the initial condition).

Note: We can determine the Laplace transforms of the second and higher derivatives by repeated application with this result, for the first derivative. For example, the transform of the second derivative is given by

$$\mathcal{L}[\ddot{y}(t)] = \mathcal{L}\left[\frac{d\dot{y}(t)}{dt}\right] = s\mathcal{L}[\dot{y}(t)] - \dot{y}(0) = s\{s\mathcal{L}[y(t)] - y(0)\} - \dot{y}(0)$$

or

$$\mathcal{L}[\ddot{y}(t)] = s^2 \mathcal{L}[y(t)] - sy(0) - \dot{y}(0)$$

B.1.2 Table of Laplace Transforms

Table B.1 shows the Laplace transforms of some common functions. Specifically, the table lists functions as $y(t)$, and their Laplace transforms (on the right) as $Y(s)$ or $\mathcal{L} y(t)$. If one is

TABLE B.1

Laplace Transform Pairs

$y(t) = \mathcal{L}^{-1}\,[Y(s)]$	$\mathcal{L}\,[y(t)] = Y(s)$
B	B/s
e^{-at}	$\dfrac{1}{s+a}$
e^{at}	$\dfrac{1}{s-a}$
$\sinh at$	$\dfrac{a}{s^2-a^2}$
$\cosh at$	$\dfrac{s}{s^2-a^2}$
$\sin \omega t$	$\dfrac{\omega}{s^2+\omega^2}$
$\cos \omega t$	$\dfrac{s}{s^2+\omega^2}$
$e^{-at}\sin \omega t$	$\dfrac{\omega}{(s+a)^2+\omega^2}$
$e^{-at}\cos \omega t$	$\dfrac{s+a}{(s+a)^2+\omega^2}$
Ramp t	$\dfrac{1}{s^2}$
$e^{-at}(1-at)$	$\dfrac{s}{(s+a)^2}$
$y(t)$	$Y(s)$
$\dfrac{dy}{dt} = \dot{y}$	$sY(s) - y(0)$
$\dfrac{d^2y}{dt^2} = \ddot{y}$	$s^2Y(s) - sy(0) - \dot{y}(0)$
$\dfrac{d^3y}{dt^3} = \dddot{y}$	$s^3Y(s) - s^2y(0) - s\dot{y}(0) - \ddot{y}(0)$
$\displaystyle\int_a^t y(t)dt$	$\dfrac{1}{s}Y(s) - \dfrac{1}{s}\displaystyle\int_0^a y(t)dt$
$af(t) + bg(t)$	$aF(s) + bG(s)$
Unit step $U(t) = 1$ for $t \geq 0 = 0$ otherwise	$\dfrac{1}{s}$
Delayed step $cU(t-b)$	$\dfrac{c}{s}e^{-bs}$
Pulse $c[U(t) - U(t-b)]$	$c\left(\dfrac{1-e^{-bs}}{s}\right)$

TABLE B.1 (continued)

Laplace Transform Pairs

Impulse function $\delta(t)$	1
Delayed impulse $\delta(t-b) = U(t-b)$ 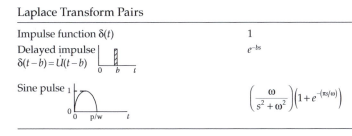	e^{-bs}
Sine pulse	$\left(\dfrac{\omega}{s^2+\omega^2}\right)\left(1+e^{-(\pi s/\omega)}\right)$

TABLE B.2

Important Laplace Transform Relations

$\mathcal{L}^{-1}F(s) = f(t)$	$\mathcal{L}\, f(t) = F(s)$
$\dfrac{1}{2\pi j}\displaystyle\int_{\sigma-j\infty}^{\sigma+j\infty} F(s)\exp(st)ds$	$\displaystyle\int_0^\infty f(t)\exp(-st)dt$
$k_1 f_1(t) + k_2 f_2(t)$	$k_1 F_1(s) + k_2 F_2(s)$
$\exp(-at)f(t)$	$F(s+a)$
$f(t-\tau)$	$\exp(-\tau s)F(s)$
$f^{(n)}(t) = \dfrac{d^n f(t)}{dt^n}$	$s^n F(s) - s^{n-1}f(0^+) - s^{n-2}f^1(0^+)$ $-\cdots- f^{n-1}(0^+)$
$\displaystyle\int_{-\infty}^t f(t)dt$	$\dfrac{F(s)}{s} + \dfrac{\displaystyle\int_{-\infty}^0 f(t)dt}{s}$
t^n	$\dfrac{n!}{s^{n+1}}$
$t^n e^{-at}$	$\dfrac{n!}{(s+a)^{n+1}}$

given a function, one can get its Laplace transform from the table. Conversely, if one is given the transform, one can get the function from the table.

Some general properties and results of the Laplace transform are given in Table B.2.

In particular, note that with zero initial conditions, differentiation can be interpreted as multiplication by s. Also, integration can be interpreted as division by s.

B.2 Response Analysis

The Laplace transform method can be used in the response analysis of dynamic systems, mechatronic and control systems in particular. We will give examples for the approach.

Example B.1

The capacitor-charge equation of the *RC* circuit shown in Figure B.1 is

$$e = iR + v \tag{i}$$

For the capacitor,

$$i = C\frac{dv}{dt} \qquad\text{(ii)}$$

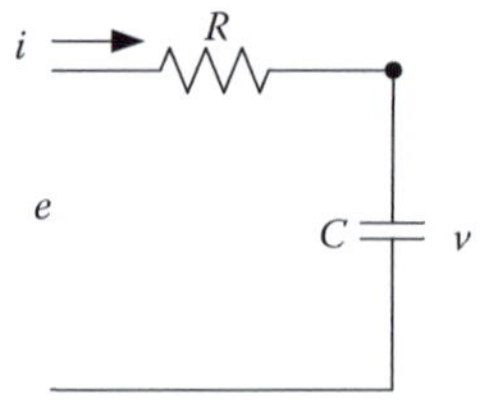

Substitute (ii) in (i) to get the circuit equation:

$$e = RC\frac{dv}{dt} + v \qquad\text{(iii)}$$

FIGURE B.1
An RC circuit with applied voltage e and voltage v across capacitor.

Take the Laplace transform of each term in (iii) with all initial conditions = 0:

$$E(s) = RCsV(s) + V(s)$$

The transfer function expressed as the output/input ratio (in the transform form) is

$$\frac{V(s)}{E(s)} = \frac{V(s)}{sRCV(s)+V(s)} = \frac{1}{sRC+1} = \frac{1}{\tau s+1} \qquad\text{(iv)}$$

where $\tau = RC$.

The actual response can now be found from Table B.1 for a given input E. The first step is to get the transform into proper form (like line 2)

$$\frac{1}{\tau s+1} = \frac{1/\tau}{s+(1/\tau)} = \frac{a}{s+a} = a\left(\frac{1}{s+a}\right)$$

where $a = 1/\tau$. Suppose that input (excitation) e is a unit impulse. Its Laplace transform (see Table B.1) is $E = 1$. Then from (iv),

$$V(s) = \frac{1}{\tau s+1}$$

From line 2 of Table B.1, the response is

$$v = ae^{-at} = \frac{1}{\tau}e^{-t/\tau} = \frac{1}{RC}e^{-t/RC}$$

A common transfer function for an overdamped second-order system (e.g., one with two RC circuit components of Figure B.1) would be

$$\frac{V(s)}{E(s)} = \frac{1}{(1+\tau_1 s)(1+\tau_2 s)}$$

This can be expressed as "partial fractions" in the from

$$\frac{A}{1+\tau_1 s} + \frac{B}{1+\tau_2 s}$$

and can be solved in the usual manner.

Example B.2

The transfer function of a thermal system is given by

$$G(s) = \frac{2}{(s+1)(s+3)}$$

If a unit step input is applied to the system, with zero initial conditions, what is the resulting response?

Solution

Input $U(s) = (1/s)$ (for a unit step)

Since $\dfrac{Y(s)}{U(s)} = \dfrac{2}{(s+1)(s+3)}$

the output (response)

$$Y(s) = \frac{2}{s(s+1)(s+3)}$$

Its inverse Laplace transform gives the time response. For this, first convert the expression into partial fractions as

$$\frac{2}{s(s+1)(s+3)} = \frac{A}{s} + \frac{B}{(s+1)} + \frac{C}{(s+3)} \tag{i}$$

The unknown A is determined by multiplying equation (i) throughout by s and then setting $s=0$. We get

$$A = \frac{2}{(0+1)(0+3)} = \frac{2}{3}$$

Similarly, B is obtained by multiplying (i) throughout by $(s+1)$ and then setting $s=-1$. We get

$$B = \frac{2}{(-1)(-1+3)} = -1$$

Next, C is obtained by multiplying (i) throughout by $(s+3)$ and then setting $s=-3$. We get

$$C = \frac{2}{(-3)(-3+1)} = \frac{1}{3}$$

Hence,

$$Y(s) = \frac{2}{3s} - \frac{1}{(s+1)} + \frac{1}{3(s+3)}$$

Take the inverse transform using line 2 of Table B.1.

$$y(t) = \frac{2}{3} - e^{-t} + \frac{1}{3}e^{-3t}$$

Example B.3

The transfer function of a damped simple oscillator is known to be of the form

$$\frac{Y(s)}{U(s)} = \frac{\omega_n^2}{\left(s^2 + 2\zeta\omega_n s + \omega_n^2\right)}$$

where
  ω_n is the undamped natural frequency
  ζ is the damping ratio

Suppose that a unit step input (i.e., $U(s) = (1/s)$) is applied to the system. Using Laplace transform tables, determine the resulting response with zero initial conditions.

Solution

$$Y(s) = \frac{1}{s} \cdot \frac{\omega_n^2}{\left(s^2 + 2\zeta\omega_n s + \omega_n^2\right)}$$

The corresponding partial fractions are of the form

$$Y(s) = \frac{A}{s} + \frac{Bs+C}{\left(s^2 + 2\zeta\omega_n s + \omega_n^2\right)} = \frac{\omega_n^2}{s\left(s^2 + 2\zeta\omega_n s + \omega_n^2\right)} \qquad \text{(i)}$$

We need to determine A, B, and C.
Multiply (i) throughout by s and set $s = 0$. We get

$$A = 1$$

Next note that the roots of the characteristic equation

$$s^2 + 2\zeta\omega_n s + \omega_n^2 = 0$$

are

$$s = -\zeta\omega_n \pm \sqrt{\zeta^2 - 1}\,\omega_n = -\zeta\omega_n \pm j\omega_d$$

These are the poles of the system and are complex conjugates. Two equations for B and C are obtained by multiplying (i) by $s + \zeta\omega_n - \sqrt{\zeta^2 - 1}\,\omega_n$, setting $s = -\zeta\omega_n + \sqrt{\zeta^2 - 1}\,\omega_n$, multiplying (i) by $s + \zeta\omega_n + \sqrt{\zeta^2 - 1}\,\omega_n$, and setting $s = -\zeta\omega_n - \sqrt{\zeta^2 - 1}\,\omega_n$. We obtain $B = -1$ and $C = -2\zeta\omega_n$. Consequently,

$$Y(s) = \frac{1}{s} - \frac{s + 2\zeta\omega_n}{\left(s^2 + 2\zeta\omega_n s + \omega_n^2\right)}$$

$$= \frac{1}{s} - \frac{s + \zeta\omega_n}{\left[\left(s + \zeta\omega_n\right)^2 + \omega_d^2\right]} - \frac{\zeta}{\sqrt{1-\zeta^2}} \cdot \frac{\omega_d}{\left[\left(s + \zeta\omega_n\right)^2 + \omega_d^2\right]}$$

where $\omega_d = \sqrt{1-\zeta^2}\,\omega_n =$ damped natural frequency.

Now use Table B.1 to obtain the inverse Laplace transform:

$$y_{step}(t) = 1 - e^{-\zeta\omega_n t}\cos\omega_d t - \frac{\zeta}{\sqrt{1-\zeta^2}}e^{-\zeta\omega_n t}\sin\omega_d t$$

$$= 1 - \frac{e^{-\zeta\omega_n t}}{\sqrt{1-\zeta^2}}[\sin\phi\cos\omega_d t + \cos\phi\sin\omega_d t]$$

$$= 1 - \frac{e^{-\zeta\omega_n t}}{\sqrt{1-\zeta^2}}\sin(\omega_d t + \phi)$$

where
$\cos\phi = \zeta = $ damping ratio
$\sin\phi = \sqrt{1-\zeta^2}$

Example B.4

The open-loop response of a plant to a unit impulse input, with zero ICs, was found to be $2e^{-t}\sin t$. What is the transfer function of the plant?

Solution
By linearity, since a unit impulse is the derivative of a unit step, the response to a unit impulse is given by the derivative of the result given in the previous example; thus

$$y_{impulse}(t) = \frac{\zeta\omega_n}{\sqrt{1-\zeta^2}}e^{-\zeta\omega_n t}\sin(\omega_d t + \phi) - \frac{\omega_d}{\sqrt{1-\zeta^2}}e^{-\zeta\omega_n t}\cos(\omega_d t + \phi)$$

$$= \frac{\omega_n}{\sqrt{1-\zeta^2}}e^{-\zeta\omega_n t}[\cos\phi\sin(\omega_d t + \phi) - \sin\phi\cos(\omega_d t + \phi)]$$

or

$$y_{impulse}(t) = \frac{\omega_n}{\sqrt{1-\zeta^2}}e^{-\zeta\omega_n t}\sin\omega_d t$$

Compare this with the given expression. We have

$$\frac{\omega_n}{\sqrt{1-\zeta^2}} = 2; \quad \zeta\omega_n = 1; \quad \omega_d = 1$$

But,

$$\omega_n^2 = (\zeta\omega_n)^2 + \omega_d^2 = 1 + 1 = 2$$

Hence,

$$\omega_n = \sqrt{2}$$

Hence,

$$\zeta = \frac{1}{\sqrt{2}}$$

The system transfer function is

$$\frac{\omega_n^2}{\left(s^2 + 2\zeta\omega_n s + \omega_n^2\right)} = \frac{2}{s^2 + 2s + 2}$$

Example B.5

Express the Laplace transformed expression

$$X(s) = \frac{s^3 + 5s^2 + 9s + 7}{(s+1)(s+2)}$$

as partial fractions. From the result, determine the inverse Laplace function $x(t)$.

Solution

$$X(s) = s + 2 + \frac{2}{s+1} - \frac{1}{s+2}$$

From Table B.1, we get the inverse Laplace transform

$$x(t) = \frac{d}{dt}\delta(t) + 2\delta(t) + 2e^{-t} - e^{-2t}$$

where $\delta(t) = $ unit impulse function.

B.3 Transfer Function

By the use of Laplace transformation, a *convolution integral* equation can be converted into an algebraic relationship. To illustrate this, consider the convolution integral that gives the response $y(t)$ of a dynamic system to an excitation input $u(t)$ with zero ICs. By definition, Equation B.1, its Laplace transform, is written as

$$Y(s) = \int_0^\infty \int_0^\infty h(\tau)u(t-\tau)d\tau \exp(-st)dt \tag{B.7}$$

Note that $h(t)$ is the *impulse response function* of the system. Since the integration with respect to t is performed while keeping τ constant, we have $dt = d(t-\tau)$. Consequently,

$$Y(s) = \int_{-\tau}^\infty u(t-\tau)\exp[-s(t-\tau)] d(t-\tau) \int_0^\infty h(\tau)\exp(-s\tau)d\tau$$

The lower limit of the first integration can be made equal to zero, in view of the fact that $u(t) = 0$ for $t < 0$. Again, by using the definition of Laplace transformation, the foregoing relation can be expressed as

$$Y(s) = H(s)U(s) \tag{B.8}$$

in which

$$H(s) = \mathcal{L}h(t) = \int_0^\infty h(t)\exp(-st)dt \tag{B.9}$$

Note that, by definition, the transfer function of a system, denoted by $H(s)$, is given by Equation B.8. More specifically, the system transfer function is given by the ratio of the Laplace-transformed output and the Laplace-transformed input with zero initial conditions. In view of Equation B.9, it is clear that the system transfer function can be expressed as the Laplace transform of the impulse-response function of the system. The transfer function of a linear and constant-parameter system is a unique function that completely represents the system. A physically realizable, linear, constant-parameter system possesses a unique transfer function, even if the Laplace transforms of a particular input and the corresponding output do not exist. This is clear from the fact that the transfer function is a system model and does not depend on the system input itself.

Note: The transfer function is also commonly denoted by $G(s)$. But in the present context, we use $H(s)$ in view of its relation to $h(t)$.

Consider the nth-order linear, constant-parameter dynamic system given by

$$a_n \frac{d^n y}{dt^n} + a_{n-1} \frac{d^{n-1} y}{dt^{n-1}} + \cdots + a_0 y = b_0 u + b_1 \frac{du(t)}{dt} + \cdots + b_m \frac{d^m u(t)}{dt^m} \tag{B.10}$$

For a physically realizable system, $m \leq n$. By applying Laplace transformation and then integrating by parts, it may be verified that

$$\mathcal{L} \frac{d^k f(t)}{dt^k} = s^k \hat{F}(s) - s^{k-1} f(0) - s^{k-2} \frac{df(0)}{dt} - \cdots + \frac{d^{k-1} f(0)}{dt^{k-1}} \tag{B.11}$$

By definition, the initial conditions are set to zero in obtaining the transfer function. This results in

$$H(s) = \frac{b_0 + b_1 s + \cdots + b_m s^m}{a_0 + a_1 s + \cdots + a_n s^n} \tag{B.12}$$

for $m \leq n$. Note that Equation B.12 contains all the information that is contained in Equation B.10. Consequently, transfer function is an analytical model of a system. The transfer function may be employed to determine the total response of a system for a given input, even though it is defined in terms of the response under zero initial conditions. This is quite logical because the analytical model of a system is independent of the initial conditions of the system.

B.4 Fourier Transform

The Fourier transform $Y(f)$ of a signal $y(t)$ relates the time domain to the frequency domain. Specifically,

$$Y(f) = \int_{-\infty}^{+\infty} y(t)\exp(-j2\pi ft)dt$$

$$= \int_{-\infty}^{+\infty} y(t)e^{-\omega t}dt$$

(B.13)

Using the Fourier operator "$\mathcal{F}$" terminology,

$$Y(f) = \mathcal{F}y(t)$$

(B.14)

Note that if $y(t)=0$ for $t<0$, as in the conventional definition of system excitations and responses, the Fourier transform is obtained from the Laplace transform by simply changing the variable according to $s=j2\pi f$ or $s=j\omega$. The Fourier transform is a special case of the Laplace transform where, in Equation B.2, $\sigma=0$:

$$Y(f) = Y(s)\big|_{s=j2\pi f}$$

(B.15)

or

$$Y(\omega) = Y(s)\big|_{s=j\omega}$$

(B.16)

The (complex) function $Y(f)$ is also termed the (continuous) *Fourier spectrum* of the (real) signal $y(t)$. The inverse transform is given by

$$y(t) = \int_{-\infty}^{+\infty} Y(f)\exp(j2\pi ft)df$$

(B.17)

$$\text{or } y(t) = \mathcal{F}^{-1} Y(f)$$

Note that according to the definition given by Equation B.13, the Fourier spectrum $Y(f)$ is defined for the entire frequency range $f(-\infty, +\infty)$, which includes negative values. This is termed the *two-sided spectrum*. Since in practical applications it is not possible to have "negative frequencies," the *one-sided spectrum* is usually defined only for the frequency range $f(0, \infty)$.

For a two-sided spectrum to have the same amount of *power* as a one-sided spectrum, it is necessary to make the one-sided spectrum double the two-sided spectrum for $f > 0$.

If the signal is not sufficiently transient (fast-decaying or damped), the infinite integral given by Equation B.13 might not exist, but the corresponding Laplace transform might still exist.

B.4.1 Frequency-Response Function (Frequency Transfer Function)

The Fourier integral transform of the impulse-response function is given by

$$H(f) = \int_{-\infty}^{\infty} h(t)\exp(-j2\pi f t)dt \qquad (B.18)$$

where f is the *cyclic frequency* (measured in cycles/s or hertz). This is known as the frequency-response function (or frequency transfer function) of a system. The Fourier transform operation is denoted as $\mathcal{F} h(t) = H(f)$. In view of the fact that $h(t) = 0$ for $t < 0$, the lower limit of integration in Equation B.18 could be made zero. Then, from Equation B.9, it is clear that $H(f)$ is obtained simply by setting $s = j2\pi f$ in $H(s)$. Hence, strictly speaking, we should use the notation $H(j2\pi f)$ and not $H(f)$. But for the notational simplicity, we denote $H(j2\pi f)$ by $H(f)$. Furthermore, since the angular frequency $\omega = 2\pi f$, we can express the frequency response function by $H(j\omega)$ or simply by $H(\omega)$ for the notational convenience. It should be noted that the frequency-response function, like the (Laplace) transfer function, is a complete representation of a linear, constant-parameter system. In view of the fact that both $u(t) = 0$ and $y(t) = 0$ for $t < 0$, we can write the Fourier transforms of the input and the output of a system directly by setting $s = j2\pi f = j\omega$ in the corresponding Laplace transforms.

Then, from Equation B.8, we have

$$Y(f) = H(f)\, U(f) \qquad (B.19)$$

Note: Sometimes for notational convenience, the same lowercase letters are used to represent the Laplace and Fourier transforms as well as the original time-domain variables.

If the Fourier integral transform of a function exists, then its Laplace transform also exists. The converse is not generally true, however, because of poor convergence of the Fourier integral in comparison with the Laplace integral. This arises from the fact that the factor $\exp(-\sigma t)$ is not present in the Fourier integral. For a physically realizable, linear, constant-parameter system, $H(f)$ exists even if $U(f)$ and $Y(f)$ do not exist for a particular input. The experimental determination of $H(f)$, however, requires system stability. For the nth-order system given by Equation B.10, the frequency-response function is determined by setting $s = j2\pi f$ in Equation B.12 as

$$H(f) = \frac{b_0 + b_1 j2\pi f + \cdots + b_m (j2\pi f)^m}{a_0 + a_1 j2\pi f + \cdots + a_n (j2\pi f)^n} \qquad (B.20)$$

This, generally, is a complex function of f, which has a magnitude denoted by $|H(f)|$ and a phase angle denoted by $\angle H(f)$.

B.5 The s-Plane

We have noted that the Laplace variable s is a complex variable with a real part and an imaginary part. Hence, to represent it we will need two axes at right angles to each other—the real axis and the imaginary axis. These two axes are from a plane, which is called the *s*-plane. Any general value of *s* (or any variation or trace of *s*) may be marked on the *s*-plane.

B.5.1 An Interpretation of Laplace and Fourier Transforms

In the Laplace transformation of a function $f(t)$, we multiply the function by e^{-st} and integrate with respect to *t*. This process may be interpreted as determining the "components" $F(s)$ of $f(t)$ in the "direction" e^{-st} where *s* is a complex variable. All such components $F(s)$ should be equivalent to the original function $f(t)$.

In the Fourier transformation of $f(t)$, we multiply it by $e^{-j\omega t}$ and integrate with respect to *t*. This is the same as setting $s = j\omega$. Hence, the Fourier transform of $f(t)$ is $F(j\omega)$. Furthermore, $F(j\omega)$ represents the components of $f(t)$ that are in the direction of $e^{-j\omega t}$. Since $e^{-j\omega t} = \cos \omega t - j\sin \omega t$, in the Fourier transformation what we do is to determine the sinusoidal components of frequency ω of a time function $f(t)$. Since *s* is complex, $F(s)$ is also complex and so is $F(j\omega)$. Hence, they all will have a real part and an imaginary part.

B.5.2 Application in Circuit Analysis

The fact that sin ωt and cos ωt are 90° out of phase is further confirmed in view of

$$e^{j\omega t} = \cos \omega t + j\sin \omega t \qquad (B.21)$$

Consider the *R-L-C* circuit shown in Figure B.2. For the capacitor, the current (*i*) and the voltage (*v*) are related through

$$i = C\frac{dv}{dt} \qquad (B.22)$$

If the voltage $v = v_o \sin \omega t$, the current $i = v_o \omega C \cos \omega t$. Note that the magnitude of v/i is $1/\omega C$ (or $1/2\pi fC$ where $\omega = 2\pi f$; *f* is the cyclic frequency and ω is the angular frequency). But *v* and *i* are out of phase by 90°. In fact, in the case of a capacitor, *i* leads *v* by 90°. The equivalent circuit resistance of a capacitance is called *reactance* and is given by

$$X_C = \frac{1}{2\pi fC} \qquad (B.23)$$

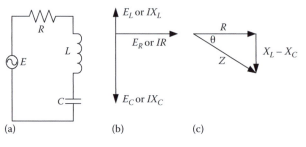

FIGURE B.2
(a) Series *R-L-C* circuit; (b) phases of voltage drops; (c) impedance triangle.

$$= \frac{1}{\omega C} \tag{B.24}$$

Note that this parameter changes with the frequency.

We cannot add the reactance of the capacitor and the resistance of the resistor algebraically; we must add them vectorialy because the voltages across a capacitor and resistor in series are not in phase, unlike in the case of a resistor. Also, the resistance in a resistor does not change with frequency. In a series circuit, as in Figure B.2, the current is identical in each element, but the voltages differ in both amplitude and phase; in a parallel circuit, the voltages are identical, but the currents differ in amplitude and phase.

Similarly, for an inductor we have

$$v = L\frac{di}{dt} \tag{B.25}$$

The corresponding reactance is

$$X_L = \omega L = 2\pi f L \tag{B.26}$$

If the voltage (E) across R in Figure B.2a is in the direction shown in Figure B.2b (i.e., pointing to the right), then the voltage across the inductor L must point upwards (90° leading) and the voltage across the capacitor C must point down (90° lagging). Since the current (I) is identical in each component of a series circuit, we see the directions of IR, IX_L, and IX_C as in Figure B.2b giving the impedance triangle shown in Figure B.2c.

To express these reactances in the s domain, we simply substitute s for $j\omega$:

$$-jX_C = \frac{1}{sC}$$

$$jX_L = sL$$

The series impedance of the RLC circuit can be expressed as

$$Z = R + jX_L - jX_C = R + sL + \frac{1}{sC}$$

In this discussion, note the use of $\sqrt{-1}$ or j to indicate a 90° phase change.

Appendix C: Probability and Statistics

In this appendix, we will review some important concepts in probability and statistics.

C.1 Probability Distribution

C.1.1 Cumulative Probability Distribution Function

Consider a random variable X. The probability that the random variable takes a value equal to or less than a specific value x is a function of x. This function, denoted by $F(x)$, is termed *cumulative probability distribution function*, or simply *distribution function*. Specifically,

$$F(x) = P[X \le x] \tag{C.1}$$

Note that $F(\infty) = 1$ and $F(-\infty) = 0$ because the value of X is always less than infinity and can never be less than negative infinity. Furthermore, $F(x)$ has to be a monotonically increasing function, as shown in Figure C.1a, because the negative probabilities are not defined.

C.1.2 Probability Density Function

Assuming that the random variable X is a continuous variable and, hence, $F(x)$ is a continuous function of x, the probability density function $f(x)$ is given by the slope of $F(x)$, as shown in Figure C.1b. Thus,

$$f(x) = \frac{dF(x)}{dx} \tag{C.2}$$

Hence,

$$F(x) = \int_{-\infty}^{x} f(x) dx \tag{C.3}$$

Note that the area under the density curve is unity. Furthermore, the probability that the random variable falls within two values is given by the area under the density curve within these two limits. This can be easily shown using the definition of $F(x)$ and $f(x)$:

$$P[a < X \le b] = F(b) - F(a) = \int_{-\infty}^{b} f(x) dx - \int_{-\infty}^{a} f(x) dx = \int_{a}^{b} f(x) dx \tag{C.4}$$

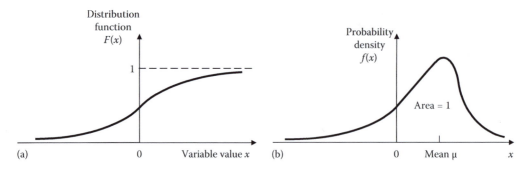

FIGURE C.1
(a) A cumulative probability distribution function; (b) a probability density function.

C.1.3 Mean Value (Expected Value)

If a random variable X is measured repeatedly a very large (infinite) number of times, the average of these measurements is the *mean value* μ or *expected value* $E(X)$. It should be easy to see that this may be expressed as the weighted sum of all the possible values of the random variable, each value being weighted by the associated probability of its occurrence. Since the probability that X takes the value x is given by $f(x)\delta x$, with δx approaching zero, we have

$$\mu = E(X) = \lim_{\delta x \to 0} \sum xf(x)\delta x$$

Since the right-hand side summation becomes an integral in the limit, we get

$$\mu = E(X) = \int_{-\infty}^{\infty} xf(x)dx \tag{C.5}$$

C.1.4 Root-Mean-Square Value

The mean square value of a random variable X is given by

$$E(X^2) = \int_{-\infty}^{\infty} x^2 f(x)dx \tag{C.6}$$

The root-mean-square (rms) value is the square root of the mean square value.

C.1.5 Variance and Standard Deviation

The variance of a random variable is the mean square value of the deviation from the mean. This is denoted by $\text{Var}(X)$ or σ^2 and is given by

$$\text{Var}(X) = \sigma^2 = \int_{-\infty}^{\infty} (x-\mu)^2 f(x)dx \tag{C.7}$$

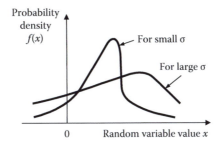

FIGURE C.2
Effect of standard deviation on the shape of a probability density curve.

By expanding Equation C.7, we can show that

$$\sigma^2 = E(X^2) - \mu^2 \tag{C.8}$$

The standard deviation σ is the square root of the variance. Note that the standard deviation is a measure of the statistical "spread" of a random variable. A random variable with a smaller σ is less random and its density curve will exhibit a sharper peak, as shown in Figure C.2.

Some thinking should convince you that if the probability density function of random variable X is $f(x)$, then the probability density function of any (well-behaved) function of X is also $f(x)$. In particular, for constants a and b, the probability density function of $(aX + b)$ is also $f(x)$. Note, further, that the mean of $(aX + b)$ is $(a\mu + b)$. Hence, from Equation C.7, it follows that the variance of aX is

$$\text{Var}(aX) = \int_{-\infty}^{\infty} (ax - a\mu)^2 f(x) dx = a^2 \int_{-\infty}^{\infty} (x - \mu)^2 f(x) dx$$

Hence,

$$\text{Var}(aX) = a^2 \text{Var}(X) \tag{C.9}$$

C.1.6 Independent Random Variables

Two random variables, X_1 and X_2, are said to be independent if the event "X_1 assumes a certain value" is completely independent of the event "X_2 assumes a certain value." In other words, the processes that generate the responses X_1 and X_2 are completely independent. Furthermore, the probability distribution of X_1 and X_2 are completely independent. Hence, it can be shown that for independent random variables X_1 and X_2, the mean value of the product is equal to the product of the mean values. Thus,

$$E(X_1 X_2) = E(X_1) E(X_2) \tag{C.10}$$

for the independent random variables X_1 and X_2.

Now, using the definition of variance and Equation C.10, it can be shown that

$$\text{Var}(X_1 + X_2) = \text{Var}(X_1) + \text{Var}(X_2) \tag{C.11}$$

for independent X_1 and X_2.

C.1.7 Sample Mean and Sample Variance

Consider the N measurements $\{X_1, X_2, ..., X_N\}$ of the random variable X. This set of data is termed a *data sample*. It generally is not possible to extract all information about the probability distribution of X from this data sample. We are able, however, to make some useful *estimates*. One would expect that the larger the data sample, the more accurate these statistical estimates would be.

An estimate for the mean value of X would be the *sample mean* $\bar{X}$, which is defined as

$$\bar{X} = \frac{1}{N} \sum_{i=1}^{N} X_i \tag{C.12}$$

An estimate for variance would be the *sample variance* S^2, given by

$$S^2 = \frac{1}{(N-1)} \sum_{i=1}^{N} (X_i - \bar{X})^2 \tag{C.13}$$

An estimate for standard deviation would be the *sample standard deviation*, S, which is the square root of the sample variance.

One might be puzzled by the denominator $N - 1$ on the right-hand side of Equation C.13. Since we are computing an "average" deviation, the denominator should have been N. But in that case, with just one reading ($N = 1$), we get a finite value for S, which is not correct because one cannot talk about a sample standard deviation when only one measurement is available. Since, according to Equation C.13, S is not defined (0/0) when $N = 1$, this definition of S^2 is more realistic. Another advantage of Equation C.13 is that this equation gives an *unbiased estimate* of variance. This concept will be discussed next. Note that if we use N instead of $N - 1$ in Equation C.13, the computed variance is called the *population variance*. Its square root is the *population standard deviation*. When $N > 30$, the difference between the sample variance and the population variance becomes negligible.

C.1.8 Unbiased Estimates

Note that each term X_i in the sample data set $\{X_1, X_2, ..., X_N\}$ is itself a random variable just like X because the measured value of X_i contains some randomness and is subjected to chance. In other words, if N measurements were taken at one time and then the same measurements were repeated, the values would be different from the first set since X was random to begin with. It follows that $\bar{X}$ and S in Equations C.12 and C.13 are also random variables. Note that the mean value of $\bar{X}$ is

$$E(\bar{X}) = E\left[\frac{1}{N} \sum_{i=1}^{N} X_i \right] = \frac{1}{N} \sum_{i=1}^{N} E(X_i) = \frac{N\mu}{N}$$

Hence,

$$E(\bar{X}) = \mu \tag{C.14}$$

We know that $\bar{X}$ is an estimate for μ. Also, from Equation C.14, we observe that the mean value of $\bar{X}$ is μ. Hence, the sample mean $\bar{X}$ is an *unbiased estimate* of the mean value μ. Similarly, from Equation C.13, we can show that the mean value of S^2 is

$$E(S^2) = \sigma^2 \tag{C.15}$$

assuming that X_i are independent measurements. Thus, the sample variance S^2 is an unbiased estimate of variance σ^2. In general, if the mean value of an estimate is equal to the exact value of the parameter that is being estimated, the estimate is said to be unbiased. Otherwise, it is a *biased estimate*.

Example C.1

An instrument has a response X that is random with a standard deviation σ. A set of N independent measurements $\{X_1, X_2, \ldots, X_N\}$ is made and the sample mean $\bar{X}$ is computed. Show that the standard deviation of $\bar{X}$ is $\sigma/\sqrt{N}$.

Also, a measuring instrument produces a random error whose standard deviation is 1%. How many measurements should be averaged in order to reduce the standard deviation of error to less than 0.05%?

Solution
To solve the first part of the problem, start with Equation C.12 and use the properties of variance given by Equations C.9 and C.11:

$$\mathrm{Var}(\bar{X}) = \mathrm{Var}\left[\frac{1}{N}(X_1 + X_2 + \cdots + X_N)\right] = \frac{1}{N^2}\mathrm{Var}(X_1 + X_2 + \cdots + X_N)$$

$$= \frac{1}{N^2}\left[\mathrm{Var}(X_1) + \mathrm{Var}(X_2) + \cdots + \mathrm{Var}(X_N)\right] = \frac{N\sigma^2}{N^2}$$

Here, we used the fact that X_i are independent.

Hence,

$$\mathrm{Var}(\bar{X}) = \frac{\sigma^2}{N} \tag{C.16}$$

Accordingly,

$$\mathrm{Std}(\bar{X}) = \frac{\sigma}{\sqrt{N}} \tag{C.17}$$

For the second part of the problem, $\sigma = 1\%$ and $\sigma/\sqrt{N} < 0.05\%$. Then,

$$\frac{1}{\sqrt{N}} < 0.05 \rightarrow N > 400$$

Thus, we should average more than 400 measurements to obtain the specified accuracy.

C.1.9 Gaussian Distribution

Gaussian distribution, or *normal distribution*, is probably the most extensively used probability distribution in engineering applications. Apart from its ease of use, another justification for its widespread use is provided by the *central limit theorem*. This theorem states that a random variable that is formed by summing a very large number of independent random variables takes Gaussian distribution in the limit. Since many engineering phenomena are consequences of numerous independent random causes, the assumption of normal distribution is justified in many cases. The validity of Gaussian assumption can be checked by plotting data on *probability graph paper* or by using various tests such as the *chi-square test*.

The Gaussian probability density function is given by

$$f(x) = \frac{1}{\sqrt{2\pi}\sigma} \exp\left[-\frac{(x-\mu)^2}{2\sigma^2}\right] \tag{C.18}$$

Note that only two parameters, mean μ and standard deviation σ, are necessary to determine a Gaussian distribution completely.

A closed algebraic expression cannot be given for the cumulative probability distribution function $F(x)$ of Gaussian distribution. It should be evaluated by numerical integration. Numerical values for the normal distribution curve are available in tabulated form with the random variable X being normalized with respect to μ and σ according to

$$Z = \frac{X - \mu}{\sigma} \tag{C.19}$$

Note that the mean value of this normalized variable Z is

$$E(Z) = E\left[\frac{(X-\mu)}{\sigma}\right] = \frac{[E(X)-\mu]}{\sigma} = \frac{(\mu-\mu)}{\sigma}$$

or

$$E(Z) = 0 \tag{C.20}$$

and the variance of Z is

$$\mathrm{Var}(Z) = \left[\frac{\mathrm{Var}(X-\mu)}{\sigma}\right] = \frac{\mathrm{Var}(X-\mu)}{\sigma^2} = \frac{\mathrm{Var}(X)}{\sigma^2} = \frac{\sigma^2}{\sigma^2}$$

or

$$\mathrm{Var}(Z) = 1 \tag{C.21}$$

Furthermore, the probability density function of Z is

$$f(z) = \frac{1}{\sqrt{2\pi}} \exp\left(\frac{-z^2}{2}\right) \tag{C.22}$$

What is usually tabulated is the area under the density curve $f(z)$ of the normalized random variable Z for different values of z. A convenient form is presented in Table C.1, where the area under the $f(z)$ curve from 0 to z is tabulated up to four decimal places for different positive values of z up to two decimal places. Since the density curve is symmetric about the mean value (zero for the normalized case), values for negative z do not have to be tabulated. Furthermore, when $z \to \infty$, area A in Table C.1 approaches 0.5. The value for $z = 3.09$ is already 0.4990. Hence, for most practical purposes, area A may be taken as 0.5 for z values greater than 3.0. Since Z is normalized with respect to σ, it follows that $z = 3$ actually corresponds to three times the standard deviation of the original random variable X. Hence, for a Gaussian random variable, most of the values will fall within $\pm 3\sigma$ about the mean value. It can be stated that approximately

- 68% of the values will fall within $\pm\sigma$ about μ
- 95% of the values will fall within $\pm 2\sigma$ about μ
- 99.7% of the values will fall within $\pm 3\sigma$ about μ

This can be easily verified using Table C.1.

C.1.10 Confidence Intervals

The probability that the value of a random variable would fall within a specified interval is called a *confidence level*. As an example, consider a Gaussian random variable X that has mean μ and standard deviation σ. This is denoted by

$$X = N(\mu, \sigma) \tag{C.23}$$

Suppose that N measurements $\{X_1, X_2, \ldots, X_N\}$ are made. The sample mean $\overline{X}$ is an unbiased estimate for μ. We also know that the standard deviation of $\overline{X}$ is $\sigma\sqrt{N}$.

Now consider the following normalized random variable:

$$Z = \frac{\overline{X} - \mu}{\sigma / \sqrt{N}} \tag{C.24}$$

This is a Gaussian random variable with zero mean and unity standard deviation. The probability p that the values of Z fall within $\pm z_o$

$$P(-z_o < Z \le z_o) = p \tag{C.25}$$

can be determined from Table C.1 for a specified value of z_o. Now substituting Equation C.24 in C.25, we get $P\left(-z_o < \dfrac{\overline{X} - \mu}{\sigma / \sqrt{N}} \le z_o\right) = p \to$

$$P\left(\overline{X} - \frac{z_o \sigma}{\sqrt{N}} \le \mu < \overline{X} + \frac{z_o \sigma}{\sqrt{N}}\right) = p \tag{C.26}$$

TABLE C.1

A Table of Gaussian Probability Distribution

$$f(z) = \frac{1}{\sqrt{2\pi}} \exp(-z^2/2)$$

$$A = \int_0^z f(z)dz$$

Area A

z	0.00	0.01	0.02	0.03	0.04	0.05	0.06	0.07	0.08	0.09
0.0	0.0000	0.0040	0.0080	0.0120	0.0160	0.0199	0.0239	0.0279	0.0319	0.0359
0.1	0.0398	0.0438	0.0478	0.0517	0.0557	0.0596	0.0636	0.0675	0.0714	0.0753
0.2	0.0793	0.0832	0.0871	0.0910	0.0948	0.0987	0.1026	0.1064	0.1103	0.1141
0.3	0.1179	0.1217	0.1255	0.1293	0.1331	0.1368	0.1406	0.1443	0.1480	0.1517
0.4	0.1554	0.1591	0.1628	0.1664	0.1700	0.1736	0.1772	0.1808	0.1844	0.1879
0.5	0.1915	0.1950	0.1985	0.2019	0.2054	0.2088	0.2123	0.2157	0.2190	0.2224
0.6	0.2257	0.2291	0.2324	0.2357	0.2389	0.2422	0.2454	0.2486	0.2517	0.2549
0.7	0.2580	0.2611	0.2642	0.2673	0.2704	0.2734	0.2764	0.2794	0.2823	0.2852
0.8	0.2881	0.2910	0.2939	0.2967	0.2995	0.3023	0.3051	0.3078	0.3106	0.3233
0.9	0.3159	0.3186	0.3212	0.3238	0.3264	0.3289	0.3315	0.3340	0.3365	0.3389
1.0	0.3413	0.3438	0.3461	0.3485	0.3508	0.3531	0.3554	0.3577	0.3599	0.3621
1.1	0.3643	0.3665	0.3686	0.3708	0.3729	0.3749	0.3770	0.3790	0.3810	0.3830

z	0.00	0.01	0.02	0.03	0.04	0.05	0.06	0.07	0.08	0.09
1.2	0.3849	0.3869	0.3888	0.3907	0.3925	0.3944	0.3962	0.3980	0.3997	0.4015
1.3	0.4032	0.4049	0.4066	0.4082	0.4099	0.4115	0.4131	0.4147	0.4162	0.4177
1.4	0.4192	0.4207	0.4222	0.4236	0.4251	0.4265	0.4279	0.4292	0.4306	0.4319
1.5	0.4332	0.4345	0.4357	0.4370	0.4382	0.4394	0.4406	0.4418	0.4429	0.4441
1.6	0.4452	0.4463	0.4474	0.4484	0.4495	0.4505	0.4515	0.4525	0.4535	0.4545
1.7	0.4554	0.4564	0.4573	0.4582	0.4591	0.4599	0.4608	0.4616	0.4625	0.4633
1.8	0.4641	0.4649	0.4656	0.4664	0.4671	0.4678	0.4686	0.4693	0.4699	0.4706
1.9	0.4713	0.4719	0.4726	0.4732	0.4738	0.4744	0.4750	0.4758	0.4761	0.4767
2.0	0.4772	0.4778	0.4783	0.4788	0.4793	0.4799	0.4803	0.4808	0.4812	0.4817
2.1	0.4821	0.4826	0.4830	0.4834	0.4838	0.4842	0.4846	0.4850	0.4854	0.4857
2.2	0.4861	0.4864	0.4868	0.4871	0.4875	0.4878	0.4881	0.4884	0.4887	0.4890
2.3	0.4893	0.4896	0.4898	0.4901	0.4904	0.4906	0.4909	0.4911	0.4913	0.4916
2.4	0.4918	0.4920	0.4922	0.4925	0.4927	0.4929	0.4931	0.4932	0.4934	0.4936
2.5	0.4938	0.4940	0.4941	0.4943	0.4945	0.4946	0.4948	0.4949	0.4951	0.4952
2.6	0.4953	0.4955	0.4956	0.4957	0.4959	0.4960	0.4961	0.4962	0.4963	0.4964
2.7	0.4965	0.4966	0.4967	0.4968	0.4969	0.4970	0.4971	0.4972	0.4973	0.4974
2.8	0.4974	0.4975	0.4976	0.4977	0.4977	0.4978	0.4979	0.4979	0.4980	0.4981
2.9	0.4981	0.4982	0.4982	0.4983	0.4984	0.4984	0.4985	0.4985	0.4986	0.4986
3.0	0.4987	0.4987	0.4987	0.4988	0.4988	0.4988	0.4989	0.4989	0.4989	0.4990

Note that the lower limit has the "≤" sign and the upper limit has the "<" sign within the parentheses. These have been used for mathematical precision, but for practical purposes, either ≤ or < may be used in each limit. Now, from Equation C.26, it follows that the confidence level is p that the actual mean value μ would fall within $\pm z_o \, \sigma/\sqrt{N}$ of the estimated (sample) mean value $\bar{X}$.

Example C.2

The angular resolution of a resolver (a rotary motion sensor) was tested sixteen times, independently, and recorded in degrees as follows:

0.11,	0.12,	0.09,	0.10,	0.10,	0.14,	0.08,	0.08
0.13,	0.10,	0.10,	0.12,	0.08,	0.09,	0.11,	0.15

If the standard deviation of the angular resolution of this brand of resolvers is known to be $0.01°$, what are the odds that the mean resolution would fall within 5% of the sample mean?

Solution
To solve this problem, we assume that resolution is normally distributed. The sample mean is computed as

$$\bar{X} = \frac{1}{16}(0.11 + 0.12 + \cdots + 0.11 + 0.15) = 0.10625$$

In view of Equation C.26, we must have

$$\frac{z_o \sigma}{\sqrt{16}} = 5\% \text{ of } \bar{X}$$

Hence,

$$\frac{z_o \times 0.01}{\sqrt{16}} = \frac{5}{100} \times 0.10625$$

or

$$z_o = 2.125$$

Now, from Table C.1,

$$P(-2.125 < Z < 2.125) = 2 \times \frac{(0.4830 + 0.4834)}{2} = 0.9664$$

C.2 Sign Test and Binomial Distribution

The sign test is useful in comparing the accuracies of two similar instruments. First, measurements should be made on the same measurand (i.e., input signal to the instrument) using the two devices. Next, the readings of one instrument are subtracted from the corresponding readings of the second instrument and the results are tabulated. Finally, the probability of getting the number of negative signs (or positive signs) equal to what is present in the tabulated results is computed using *binomial distribution*.

Before discussing binomial distribution, let us introduce some new terminology. First, *factorial r* (denoted by $r!$) of an integer r is defined as the product

$$r! = r \times (r-1) \times (r-2) \times \cdots \times 2 \times 1 \tag{C.27}$$

Now, suppose that there are n distinct articles that are distinguishable from one another. The number of ways in which r articles could be picked from the batch of n, giving proper consideration to the order in which the r articles are picked (or arranged), is called the number of *permutations* of r from n. This is denoted by ${}^nP_r$, which is given by

$$ {}^nP_r = n \times (n-1) \times (n-2) \times \cdots \times (n-r+2) \times (n-r+1) = \frac{n!}{(n-r)!} \tag{C.28}$$

This can be easily verified, since the first article can be chosen in n ways and the second article can be chosen from the remaining $(n-1)$ articles in $(n-1)$ ways and can be kept next to the first article, and so on.

If we disregard the order in which the r articles are picked (and arranged), the number of possible choices of r articles is termed the number of *combinations* of r from n. This is denoted by ${}^nC_r$. Now, since each combination can be arranged in $r!$ different ways (if the order of arrangement is considered), we have

$$ {}^nC_r \times r! = {}^nP_r \tag{C.29}$$

Hence, using Equation C.28, we get

$$ {}^nC_r = \frac{n \times (n-1) \times (n-2) \times \cdots \times (n-r+2) \times (n-r+1)}{r!} = \frac{n!}{(n-r)!\,r!} \tag{C.30}$$

With the foregoing notation, we can introduce binomial distribution in the context of the sign test. Suppose that n pairs of readings are taken from the two instruments. If the probability that a difference in reading would be positive is p, then the probability that the difference would be negative is $1 - p$. Note that if the systematic error in the two instruments is the same and if the random error is purely random, then $p = 0.5$.

The probability of getting exactly r positive signs among the n entries in the table is

$$ p(r) = {}^nC_r p^r (1-p)^{n-r} \tag{C.31}$$

To verify Equation C.31, note that this event is similar to picking exactly r items from n items and constraining each picked item to be positive (having probability p) and also constraining the remaining $(n-r)$ items to be negative (having probability $1 - p$). Note that r is a discrete variable that takes the values $r = 1, 2, \ldots, n$. Furthermore, it can be easily verified that

$$ \sum_{r=1}^{n} p(r) = \sum_{r=1}^{n} {}^nC_r p^r (1-p)^{n-r} = (p+1-p)^n = 1 \tag{C.32}$$

Hence, $p(r)$, $r=1, 2, ..., n$, is a discrete function that resembles a continuous probability density function $f(x)$. In fact, $p(r)$ given by Equation C.31 represents *binomial probability distribution*. Using Equation C.31, we can perform the sign test. The details of the test are conveniently explained by means of an example.

Example C.3

To compare the accuracies of two brands of differential transformers (DTs, which are displacement sensors), the same rotation (in degrees) of a robot arm joint was measured using both brands, DT1 and DT2. The following 10 measurement pairs were taken:

DT1	10.3	5.6	20.1	15.2	2.0	7.6	12.1	18.9	22.1	25.2
DT2	9.8	5.8	20.0	16.0	1.9	7.8	12.2	18.7	22.0	25.0

Assuming that both devices are used simultaneously (so that backlash and other types of repeatability errors in manipulators do not enter into our problem), determine whether the two brands are equally accurate at the 70% level of significance.

Solution
First, we form the sign table by taking the difference of the corresponding measurements:

DT1 – DT2	0.5	−0.2	0.1	−0.8	0.1	−0.2	−0.1	0.2	0.1	0.2

Note that there are six positive signs and four negative signs. If we had tabulated DT2 − DT1, however, we would get four positive signs and six negative signs. Both these cases should be taken into account in the sign test. Furthermore, more than six positive signs or fewer than four positive signs would make the two devices less similar (in accuracy) than what is indicated by the data. Hence, the probability of getting six or more positive signs or four or fewer positive signs should be computed in this example in order to estimate the possible match (in accuracy) of the two devices.

If the error in both transducers is the same, we should have

P (positive difference) $= p = 0.5$

This is the hypothesis that we are going to test. Using Equation C.31, the probability of getting six or more positive signs or four or fewer negative signs is calculated as

1 − probability of getting exactly 5 positive signs

$$= 1 - {}^{10}C_5 (0.5)^5 \times (0.5)^5 = 1 - \frac{10!}{5!5!} \times (0.5)^{10} = 1 - 0.246 = 0.754$$

Note that the hypothesis of two brands being equally accurate is supported by the test data at a level of significance over 75%, which is better than the specified value of 70%.

C.3 Least Squares Fit

Instrument *linearity* may be measured by the largest deviation of the input/output data (or calibration curve) from the least squares straight-line fit of data. Since many algebraic expressions become linear when plotted to a logarithmic scale, a linear (straight-line) fit is generally more accurate if log–log axes are used. The linear least squares fit can be thought of as an estimation method because it "estimates" the two parameters of an input/ output model, the straight line, that fits a given set of data such that the squared error is a minimum. The estimated straight line is also known as the *linear regression line* or *mean calibration curve*.

Consider N pairs of data $\{(X_1, Y_1), (X_2, Y_2), \ldots, (X_N, Y_N)\}$ in which X denotes the *independent variable* (input variable) and Y denotes the *dependent variable* (output variable).

Suppose that the estimated linear regression is given by

$$Y = mX + a \qquad \text{(C.33)}$$

For the independent variable value X_i, the dependent variable value on the regression line is $(mX_i + a)$, but the actual (measured) value of the dependent variable is Y_i. Hence, the sum of squared error for all data points is

$$e = \sum_{i=1}^{N} \left(Y_i - mX_i - a\right)^2 \qquad \text{(C.34)}$$

We have to minimize e with respect to the two parameters m and a. The required conditions are $\partial e/\partial m = 0$ and $\partial e/\partial a = 0$. By carrying out these differentiations in Equation C.34, we get $\sum_{i=1}^{N} X_i\left(Y_i - mX_i - a\right) = 0$ and $\sum_{i=1}^{N} \left(Y_i - mX_i - a\right) = 0$.

By dividing the two equations by N and using the definition of sample mean, we get

$$\frac{1}{N} \sum X_i Y_i - \frac{m}{N} \sum X_i^2 - a\bar{X} = 0 \qquad \text{(i)}$$

$$\bar{Y} - m\bar{X} - a = 0 \qquad \text{(ii)}$$

By solving these two simultaneous equations for m, we obtain

$$m = \left(\frac{1}{N} \sum_{i=1}^{N} X_i Y_i - \bar{X}\bar{Y}\right) \Bigg/ \left(\frac{1}{N} \sum_{i=1}^{N} X_i^2 - \bar{X}^2\right) \qquad \text{(C.35)}$$

The parameter a does not have to be explicitly expressed because from Equations C.33 and ii, we can eliminate a and express the linear regression line as

$$Y - \bar{Y} = m\left(X - \bar{X}\right) \qquad \text{(C.36)}$$

Note from Equation C.33 that a is the Y-axis intercept (i.e., the value of Y when $X = 0$) and is given by

$$a = \bar{Y} - m\bar{X} \qquad \text{(C.37)}$$

Example C.4

Consider the capacitor circuit shown in Figure C.3.

First, the capacitor is charged to voltage v_o using a constant dc voltage source (switch in position 1); then it is discharged through a known resistance R (switch in position 2). Voltage decay during discharge is measured at known time increments. Three separate tests are carried out. The measured data are as follows:

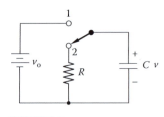

FIGURE C.3
A circuit for the least squares estimation of capacitance.

Time t (s)	0.1	0.2	0.3	0.4	0.5
Voltage v (V)					
Test 1	7.3	2.8	1.0	0.4	0.1
Test 2	7.4	2.7	1.1	0.3	0.2
Test 3	7.3	2.6	1.0	0.4	0.1

If the resistance is accurately known to be 1000 Ω, estimate the capacitance C in microfarads (μF) and the source voltage v_o in volts.

Solution

To solve this problem, we assume the well-known expression for the free decay of voltage across a capacitor to be

$$v(t) = v_o \exp\left[-\frac{t}{RC}\right] \tag{i}$$

Take the natural logarithm of Equation (i):

$$\ln v = -\frac{t}{RC} + \ln v_o \tag{ii}$$

With $Y = \ln v$ and $X = t$, Equation (ii) represents a straight line with a slope

$$m = -\frac{1}{RC} \tag{iii}$$

and the Y-axis intercept

$$a = \ln v_o \tag{iv}$$

Using all the data, the overall sample means can be computed. Thus,

$$\bar{X} = 0.3 \quad \text{and} \quad \bar{Y} = -0.01335$$

$$\frac{1}{N}\sum X_i Y_i = -0.2067 \quad \text{and} \quad \frac{1}{N}\sum X_i^2 = 0.11$$

Now substitute these values in Equations C.35 and C.37. We get

$$m = -10.13 \quad \text{and} \quad a = 3.02565$$

Next, from Equation (iii), with $R = 1000$, we have

$$C = \frac{1}{10.13 \times 1000} \, F = 98.72 \, \mu F$$

From Equation (iv),

$$v_o = 20.61 \, V$$

Note that in this problem, the estimation error would be tremendous if we did not use log scaling for the linear fit.

The least squares curve fitting is not limited to a linear (i.e., straight-line) fit. The method can be extended to a polynomial fit of any order. For example, in a *quadratic fit*, the data are fitted to a second-order (i.e., quadratic) polynomial. In that case, there are three unknown parameters, which would be determined by minimizing the quadratic error.

Appendix D: Software Tools

Modeling, analysis, design, data acquisition, and control are important activities within the field of mechatronic engineering. Computer software tools and environments are available for effectively carrying out these activities, both at the learning level and at the professional application level. Several such environments and tools are commercially available. A selected few, which are particular useful for the tasks related to this book, are outlined here.

MATLAB®* is an interactive computer environment with a high-level language and tools for scientific and technical computation, modeling and simulation, design, and control of dynamic systems. Simulink®* is a graphical environment for the modeling, simulation, and analysis of dynamic systems, and is available as an extension to MATLAB. LabVIEW* is a graphical programming language and a program development environment for data acquisition, processing, display, and instrument control.

D.1 Simulink®

A computer simulation of a dynamic model using Simulink is outlined in Chapter 3. Simulink is a graphic environment that uses block diagrams. It is an extension of MATLAB.

D.2 MATLAB®

The MATLAB interactive computer environment is very useful in computational activities in mechatronics. Computations involving scalars, vectors, and matrices can be carried out and the results can be graphically displayed and printed. MATLAB toolboxes are available for performing specific tasks in a particular area of study such as control systems, fuzzy logic, neural network, data acquisition, image processing, signal processing, system identification, optimization, model predictive control, robust control, and statistics. User guides, and web-based online help are provided by the parent company, MathWorks, Inc., and various other sources. What is given here is a brief introduction to get started in MATLAB for tasks that are particularly related to controls and mechatronics.

D.2.1 Computations

Mathematical computations can be done by using the MATLAB command window. Simply type in the computations against the MATLAB prompt "≫" as illustrated next.

---

* MATLAB and Simulink are registered trademarks and products of The MathWorks, Inc. LabVIEW is a product of National Instruments, Inc.

D.2.1.1 Arithmetic

An example of a simple computation using MATLAB is given below.

```
>> x=2; y=-3;
>> z=x^2-x*y+4
z=14
```

In the first line, we have assigned values 2 and 3 to two variables x and y. In the next line, the value of an algebraic function of these two variables is indicated. Then, MATLAB provides the answer as 14. Note that if you place a semicolon (;) at the end of the line, the answer will not be printed/displayed.

Table D.1 gives the symbols for common arithmetic operations used in MATLAB.

The following example shows the solution of the quadratic equation $ax^2 + bx + c = 0$:

```
>> a=2;b=3;c=4;
>> x=(-b+sqrt(b^2-4*a*c))/(2*a)
x =
-0.7500+1.1990i
```

The answer is complex, where i denotes $\sqrt{-1}$. Note that the function sqrt( ) is used, which provides the positive root only. Some useful mathematical functions are given in Table D.2.

D.2.1.2 Arrays

An array may be specified by giving the start value, increment, and the end value limit. An example is given below.

```
>> x=(.9:-.1:0.42)
x =
0.9000 0.8000 0.7000 0.6000 0.5000
```

TABLE D.1

MATLAB Arithmetic Operations

Symbol	Operation
+	Addition
−	Subtraction
*	Multiplication
/	Division
^	Power

TABLE D.2

Useful Mathematical Functions in MATLAB

Function	Description
abs( )	Absolute value/magnitude
acos( )	Arc-cosine (inverse cosine)
acosh( )	Arc-hyperbolic-cosine
asin( )	Arc-sine
atan( )	Arc-tan
cos( )	Cosine
cosh( )	Hyperbolic cosine
exp( )	Exponential function
imag( )	Imaginary part of a complex number
log( )	Natural logarithm
log10( )	Log to base 10 (common log)
real( )	Real part of a complex number
sign( )	Signum function
sin( )	Sine
sqrt( )	Positive square root
tan( )	Tan function

Note: MATLAB is case sensitive.

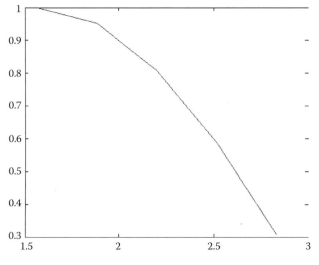

FIGURE D.1
A plot using MATLAB.

The entire array may be manipulated. For example, all the elements are multiplied by π as shown below:

```
>> x=x*pi
x =
2.8274 2.5133 2.1991 1.8850 1.5708
```

The second and the fifth elements are obtained by

```
>> x([2 5])
ans =
2.5133 1.5708
```

Next, we form a new array y using x and then plot the two arrays, as shown in Figure D.1.

```
>> y=sin(x);
>> plot(x,y)
```

A polynomial may be represented as an array of its coefficients. For example, the quadratic equation $ax^2 + bx + c = 0$ as given before, with $a = 2$, $b = 3$, and $c = 4$, may be solved using the function "roots" as below.

```
>> p=[2 3 4];
>> roots(p)
ans =
-0.7500+1.1990i
-0.7500-1.1990i
```

The answer is the same as what we obtained before.

D.2.2 Relational and Logical Operations

Useful relational operations in MATLAB are given in Table D.3. Basic logical operations are given in Table D.4.

TABLE D.3

Some Relational Operations

Operator	Description
<	Less than
<=	Less than or equal to
>	Greater than
>=	Greater than or equal to
= =	Equal to
~=	Not equal to

Consider the following example:

```
>> x=(0:0.25:1)*pi
x =
0 0.7854 1.5708 2.3562 3.1416
>> cos(x)>0
ans =
1 1 1 0 0
>> (cos(x)>0)&(sin(x)>0)
ans =
0 1 1 0 0
```

TABLE D.4

Basic Logical Operations

Operator	Description
&	AND
|	OR
~	NOT

In this example, first an array is computed. Then the cosine of each element is computed. Next, it is checked whether the elements are positive (a truth value of 1 is sent out if true and a truth value of 0 is sent out if false). Finally the "AND" operation is used to check whether both corresponding elements of two arrays are positive.

D.2.3 Linear Algebra

MATLAB can perform various computations with vectors and matrices (see Chapter 9). Some basic illustrations are given here.

A vector or a matrix may be specified by assigning values to its elements. Consider the following example:

```
>> b=[1.5 -2];
>> A=[2 1;-1 1];
>> b=b'
b =
1.5000
-2.0000
>> x=inv(A)*b
x =
1.1667
-0.8333
```

In this example, first a second order row vector and a 2×2 matrix are defined. The row vector is transposed to get a column vector. Finally, the matrix-vector equation $Ax = b$ is solved according to $x = A^{-1}b$. The determinant and the eigenvalues of A are determined by

```
>> det(A)
ans =
3
>> eig(A)
```

```
ans =
1.5000+0.8660i
1.5000-0.8660i
```

Both eigenvectors and eigenvalues of *A* are computed as

```
>> [V,P]=eig(A)
V =
0.7071 0.7071
-0.3536+0.6124i -0.3536-0.6124i
P =
1.5000+0.8660i 0
0 1.5000-0.8660i
```

Here, the symbol V is used to denote the matrix of eigenvectors. The symbol P is used to denote the diagonal matrix whose diagonal elements are the eigenvalues.

Useful matrix operations in MATLAB are given in Table D.5 and several matrix functions are given in Table D.6.

TABLE D.5

Some Matrix Operations in MATLAB

Operation	Description
+	Addition
−	Subtraction
*	Multiplication
/	Division
∧	Power
'	Transpose

TABLE D.6

Useful Matrix Functions in MATLAB

Function	Description
det( )	Determinant
inv( )	Inverse
eig( )	Eigenvalues
[,]=eig( )	Eigenvectors and eigenvalues

D.2.4 M-Files

The MATLAB commands have to be keyed in on the command window, one by one. When several commands are needed to carry out a task, the required effort can be tedious. Instead, the necessary commands can be placed in a text file and edited as appropriate (using text editor), which MATLAB can use to execute the complete task. Such a file is called an M-file. The file name must have the extension "m" in the form *filename.m*. A toolbox is a collection of such files for use in a particular application area (e.g., control systems, fuzzy logic; see Chapter 9). Then, by keying in the M-file name at the MATLAB command prompt, the file will be executed. The necessary data values for executing the file have to be assigned beforehand.

D.3 Control Systems Toolbox

There are several toolboxes with MATLAB that can be used to analyze, compute, simulate, and design control problems. Both time-domain representations and frequency-domain representations can be used. Also, both classical and modern control problems can be handled. The application is illustrated here through several conventional control problems (see Chapter 9).

D.3.1 Compensator Design Example

The speed control system shown in Figure D.2a has a compensator, a controller and amplifier gain *K*, and a dc motor with transfer function $1/(10s+1)$ that are in the forward path and a low-pass filter of transfer function $1/(0.1s+1)$ that is in the feedback path. The signal *y* (system output) from the speed sensor is conditioned by the filter and compared with the speed command *u* (system input). The resulting error signal is fed into the

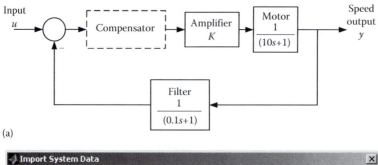

(a)

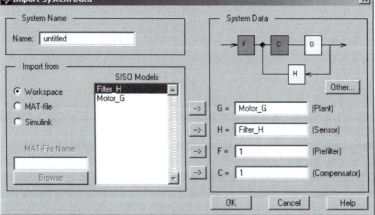

(b)

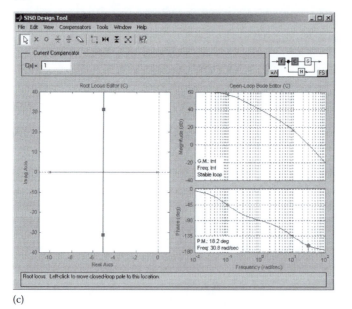

(c)

FIGURE D.2

(a) Compensator design for a velocity servo; (b) importing the model into the SISO Design Tool; (c) root locus and Bode plots for the motor model.

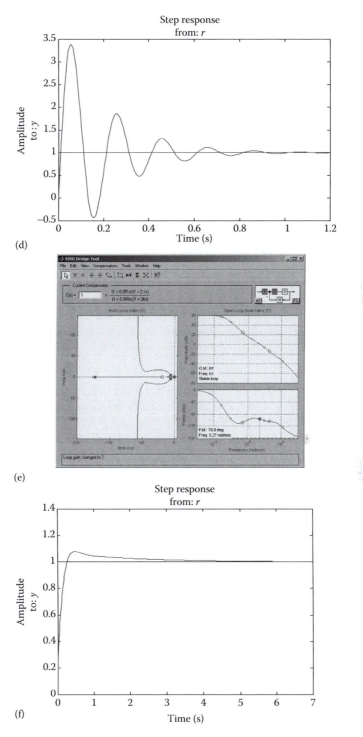

FIGURE D.2 (continued)

(d) closed-loop step response of the motor system without compensation; (e) root locus and Bode plots of the compensated system; (f) closed-loop step response of the compensated system.

control amplifier. The controller may be tuned by adjusting the gain *K*. Since the required performance was not achieved by this adjustment alone, it was decided to add a compensator network into the forward path of the control loop. The design specifications are (1) steady-state accuracy of 99.9% for a step input and (2) percentage overshoot of 10%.

Design a lead compensator and a lag compensator to meet these design specifications.

The MATLAB single-input-single-output (SISO) Design Tool is used here to solve this problem.

D.3.1.1 Building the System Model

Build the transfer function model of the motor and filter, in the MATLAB workspace, as follows:

```
Motor_G=tf([999], [10 1]);
Filter_H=tf([1], [0.1 1]);
```

To open the SISO Design Tool, type

```
sisotool
```

at the MATLAB prompt (≫).

D.3.1.2 Importing Model into SISO Design Tool

Select **Import Model** under the **File** menu. This opens the **Import System Data** dialog box, as shown in Figure D.2b.

Use the following steps to import the motor and filter models:

1. Select Motor_*G* under **SISO Models**.
2. Place it into the **G** Field under **Design Model** by pressing the right arrow button to the left of **G**.
3. Similarly, import the filter model.
4. Press **OK**.

Now, the main window of the SISO Design Tool will show the root locus and Bode plots of open loop transfer function *GH* (see Figure D.2c). As given in the figure, the phase margin is 18.2°, which occurs at 30.8 rad/s (4.9 Hz).

The closed-loop step response, without compensation, is obtained by selecting **Tools → Loop responses → closed-loop step** from the main menu. The response is shown in Figure D.2d. It is noted that the phase margin is not adequate, which explains the oscillations and the long settling time. Also the *P.O.* is about 140%, which is considerably higher than the desired one (10%) and is not acceptable.

D.3.1.3 Adding Lead and Lag Compensators

To add a lead compensator, right click the mouse in the white space of the Bode magnitude plot, choose **Add Pole/Zero** and then **lead** in the right-click menu for the open-loop Bode diagram. Move the zero and the pole of the lead compensator to get a desired phase margin about 60°.

To add a lag compensator, choose **Add Pole/Zero** and then **lag** in the right-click menu for the open-loop Bode diagram. Move the zero and the pole of the lag compensator to get a desired phase angle of about −115° at the crossing frequency, which corresponds to a phase margin of 180° − 115° = 65°.

With the added lead and leg compensators, the root locus and Bode plots of the system are shown in Figure D.2e. The closed-loop step response of the system is shown in Figure D.2f.

D.3.2 PID Control with Controller Tuning

Consider a unity-feedback control system where the forward path has the controller and a plant of transfer function $1/s(s^2 + s + 4)$. The SISO Design Tool of the MATLAB control systems toolbox is used. First, build the transfer function model of the given system (call it Mill).

```
Mill_G=tf([1], [1 1 4 0]);
Filter_H=tf([1], [1]);
```

Now import the system model into the SISO Design Tool.

D.3.2.1 *Proportional Control*

First, by trial and error we determine the proportional gain that will make the system marginally stable. As seen in Figure D.3a, when $K = 4$, the gain margin is just below 0 dB, which makes the system unstable. The response of the system is shown in Figure D.3b.

According to Ziegler–Nochols controller settings, we can choose the proper proportional gain as $K_p = 0.5 \times 4 = 2$. The corresponding system response is shown in Figure D.3c.

D.3.2.2 *PI Control*

Note that the period of oscillations (ultimate period) is

$$P_u = \frac{2\pi}{\omega_n} = \frac{2\pi}{2} = \pi \text{ s}$$

Hence, from the Ziegler–Nichols settings, we have the following settings for a PI controller:

$$K_p = 0.45 \times 4 = 1.8$$

$$\tau_i = 0.83\pi = 2.61 \text{ s}$$

Hence, the PI controller transfer function is

$$K_p\left(1 + \frac{1}{\tau_i s}\right) = \frac{K_p \tau_i s + K_p}{\tau_i s} = \frac{4.68s + 1.8}{2.61s} = 0.214 \frac{2.6s + 1}{s}$$

Insert this controller into C in the SISO Design Tool. The corresponding system Bode plot and the step response are shown in Figure D.3d and e, respectively.

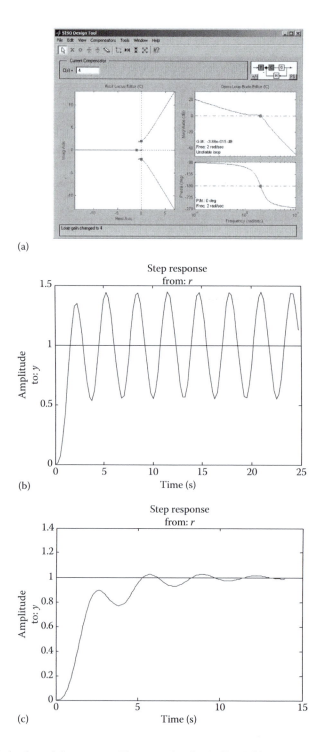

(a)

(b)

(c)

FIGURE D.3

(a) Root locus and Bode plots of the system with proportional gain $K_p = 4$; (b) step response of the closed-loop system with $K_p = 4$; (c) step response of the closed-loop system with $K_p = 2$.

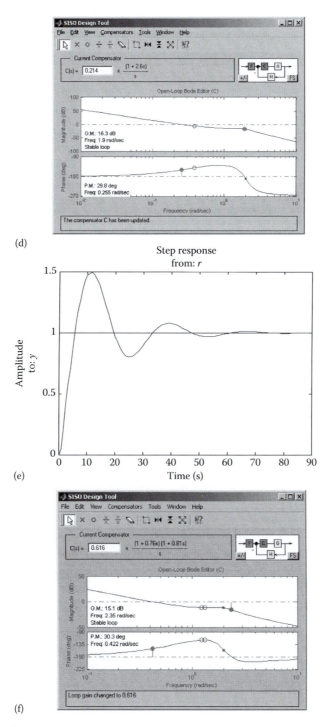

(d)

(e)

(f)

FIGURE D.3 (continued)

(d) Bode plot of the system with PI control; (e) step response of the system with PI control; (f) Bode plot of the system with PID control.

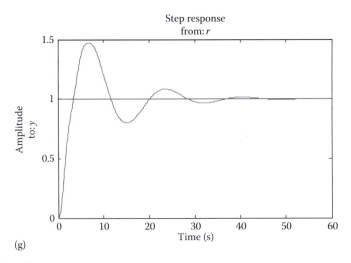

(g)

FIGURE D.3 (continued)

(g) step response of the system with PID control.

D.3.2.3 *PID Control*

From the Ziegler–Nichols settings, we pick the following parameters for a PID controller:

$$K_p = 0.6 \times 4 = 2.4$$

$$\tau_i = 0.5\pi = 1.57\,\text{s}$$

$$\tau_d = 0.125\pi = 0.393\,\text{s}.$$

The corresponding transfer function of the PID controller is

$$K_p\left(1 + \frac{1}{\tau_i s} + \tau_d s\right) = \frac{K_p \tau_i \tau_d s^2 + K_p \tau_i s + K_p}{\tau_i s} = \frac{1.48 s^2 + 3.768 s + 2.4}{1.57 s} = \frac{0.94 s^2 + 2.4 s + 1.53}{s}$$

Use the MATLAB function **roots** to calculate the roots of the numerator polynomial.

```
R=roots([0.94 2.4 1.53]);
R=-1.3217
  -1.2315
```

Hence, the transfer function of the PID controller is

$$\frac{(s+1.32)(s+1.23)}{s} = 0.616\frac{(0.76s+1)(0.81s+1)}{s}$$

Insert this controller into *C* of the SISO Design Tool. The corresponding Bode plot and the step response of the controlled system are shown in Figures D.3f and g.

D.3.3 Root Locus Design Example

Consider a control system (steel rolling mill with unity feedback). The forward path has a controller represented by gain K, a lead compensator $(s+z)/(s+p)$, and the mill whose transfer function is $1/s(s+5)$. We use the SISO Design Tool to design a suitable lead compensator. First, build the transfer function model for the rolling mill with no filter:

```
Mill_G=tf([1], [1 5 0]);
Filter_H=tf([1], [1]);
```

Then, import the system model into the SISO Design Tool. The root locus and the step response of the closed-loop system are shown in Figure D.4a and b. From Figure D.4b, it is seen that the peak time and the 2% settling time do not meet the design specifications.

To add a lead compensator, right click in the white space of the root locus plot. Choose *Add Pole/Zero* and then *lead* in the right-click menu. Left click on the root locus plot where we want to add the lead compensator.

Now we have to adjust the pole and zero of the lead compensator and the loop gain so that the root locus passes through the design region. To speed up the design process, turn on the grid setting for the root locus plot. The radial lines are constant damping ratio lines and the semicircular curves are constant undamped natural frequency lines (see Chapter 9).

On the root locus plot, drag the pole and zero of the lead compensator (cross or circle symbol on the plot) so that the root locus moves toward the design region. Left click and move the closed-loop pole (small square box) to adjust the loop gain. As you drag the closed-loop pole along the locus, the current location of that pole, the system damping ratio, and the natural frequency will be shown at the bottom of the graph.

Drag the closed-loop pole into the design region. The resulting lead compensator, the loop gain, and the corresponding root locus are shown in Figure D.4c. The step response of the compensated closed-loop system is shown in Figure D.4d.

D.4 Fuzzy Logic Toolbox

The use of fuzzy logic in intelligent control has been discussed in Chapter 9. The fuzzy logic toolbox of MATLAB is quite useful in this regard. By using it we create and edit fuzzy decision-making systems (for control and other applications) by means of interactive graphical tools or command-line functions. Simulink can be used to simulate the developed fuzzy system. The time workshop can create portable C code from a Simulink environment for use in real-time and non-real-time applications. The toolbox also provides source codes in C for implementing a stand-alone fuzzy inference engine. The stand-alone C-code fuzzy inference engine can read an FIS file (the file format for saving the fuzzy engine in MATLAB). In other words, it is able to parse the stored information, to perform fuzzy inference directly, or it can be embedded in other external applications. The design process of a fuzzy decision making system involves the following general steps, as discussed in Chapter 9: input data, fuzzification, implication (or fuzzy rules), aggregation (or composition), and inference defuzzification.

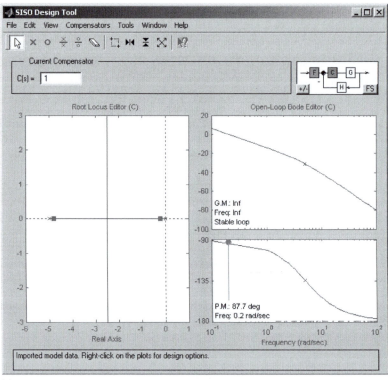

(a)

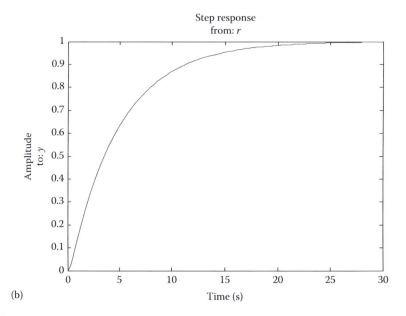

(b)

FIGURE D.4
(a) Root locus and Bode plots of the rolling mill system without compensation; (b) step response of the closed-loop system without compensation.

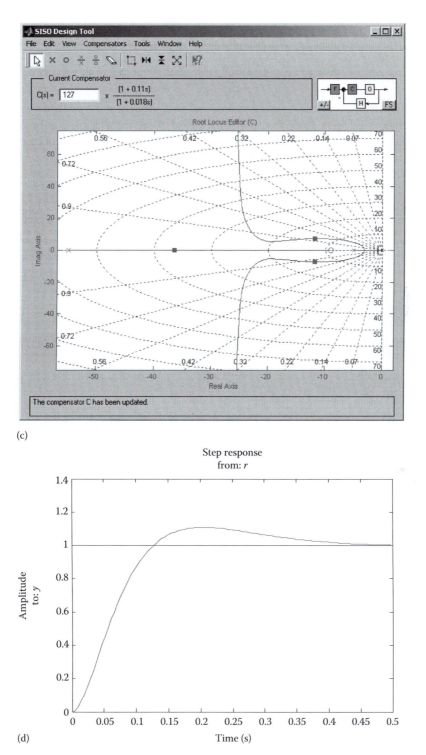

(c)

(d)

FIGURE D.4 (continued)

(c) root locus of the compensated system; (d) step response of the compensated closed-loop system.

D.4.1 Graphical Editors

There are five primary graphic user interface (GUI) tools for building, editing, and observing fuzzy inference systems in the MATLAB fuzzy logic toolbox: the FIS editor, the membership function editor, the rule editor, the rule viewer, and the surface viewer. The FIS editor handles the high-level issues for the system; e.g., number of inputs, outputs, and names. The membership function editor is used to define the shapes of the membership functions associated with each variable. The rule editor is used for editing the rules in the fuzzy knowledge base, which describes and defines the knowledge of the application (control knowledge in the case of fuzzy control). The rule viewer and the surface viewer are used for observing (not editing) the designed FIS. Try the example of tipping in a restaurant by clicking on the file menu and loading FIS from the disk. The tipper.fis is located at: **./Matlabr12/toolbox/fuzzy/fuzzydemos/tipper.fis**

1. **FIS editor**
   The FIS editor displays general information about a fuzzy inference system. Double-click on an icon to open and carry out editing related to that particular item and save the results.

2. **Membership function editor**
   The membership function editor shares some features with the FIS editor. It is a menu-driven interface that allows the user to open/display and edit the membership functions for the entire fuzzy inference system; specifically the membership functions of inputs and outputs.

3. **Rule editor**
   The rule editor contains an editable text field for displaying and editing rules. It also has some landmarks similar to those in the FIS editor and the membership function editor, including the menu bar and the status line. The pop-up menu **Format** is available from the pull-down menu **Options** in the top menu bar. This is used to set the format for the display.

4. **Rule viewer**
   The Rule Viewer displays a roadmap of the entire fuzzy inference process. It is based on the fuzzy inference diagram. The user will see a single figure window with seven small plots nested in it. The two small plots across the top of the figure represent the antecedent and the consequent of the first rule. Each rule is a row of plots and each column is a variable.

5. **Surface viewer**
   This allows the user to view the overall decision-making surface (the control surface, as discussed in Chapter 9). This is a non-fuzzy representation of the fuzzy application and is analogous to a look-up table albeit continuous.

D.4.2 Command Line–Driven FIS Design

A pre-designed FIS may be loaded into the MATLAB workspace by typing:

```
>> myfis = readfis('name_of_file.fis')
```

Typing the **showfis(myfis)** command will enable us to see the details of the FIS. Use the **getfis** command to access information of the loaded FIS. For example,

```
≫ getfis(myfis)
≫ getfis(myfis, 'Inlabels')
≫ getfis(myfis, 'input', 1)
≫ getfis(myfis, 'output', 1)
```

The command **setfis** may be used to modify any property of an FIS. For example,

```
≫ setfis(myfis, 'name', 'new_name');
```

The following three functions are used to display the high-level view of a fuzzy inference system from the command line:

```
≫ plotfis(myfis)
≫ plotmf(myfis, 'input', input_number)
or plotmf(myfis, 'output', output_number)
≫ gensuf(myfis)
```

To evaluate the output of a fuzzy system for a given input, we use the following function:

```
≫ evalfis([input matrix], myfis)
```

For example, **evalfis([1 1], myfis)** is used for single input evaluation and **evalfis([1 1; 2 3] myfis)** is used for multiple input evaluation.

Note that we may directly edit a previously saved .fis file, besides manipulating a fuzzy inference system from the toolbox GUI or from the MATLAB workspace through the command line.

D.4.3 Practical Stand-Alone Implementation in C

The MATLAB fuzzy logic toolbox allows you to run your own stand-alone C programs directly without the need for Simulink. This is made possible by a stand-alone fuzzy inference engine that reads the fuzzy systems saved from a MATLAB session. Since the C source code is provided, you can customize the stand-alone engine to build fuzzy inference into your own code. This procedure is outlined in Figure D.5.

D.5 LabVIEW*

LabVIEW or Laboratory Virtual Engineering Workbench is a product of National Instruments. It is a software development environment for data acquisition, instrument control, image acquisition, motion control, and presentation. LabVIEW is a complied graphical environment, which allows the user to create programs graphically through wired icons similar to creating a flowchart. Unlike text-based programming languages where instructions determine program execution, LabVIEW uses dataflow programming where the flow of data determines the execution order.

---

* For details see the LabVIEW User Manual Glossary and G programming Reference Manual Glossary, which are available online at http://www.ni.com/pdf/manuals/320999b.pdf and http://www.ni.com/pdf/manuals/321296b.pdf, respectively.

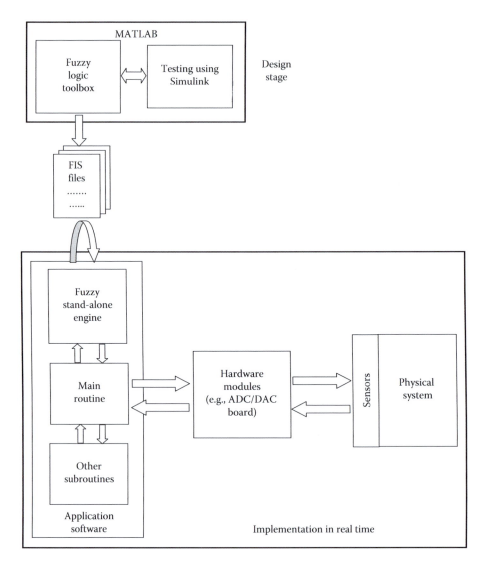

FIGURE D.5
Target implementation of a fuzzy system.

D.5.1 Introduction

LabVIEW programs, called virtual instruments (VIs), use icons to represent subroutines. It is similar to flow charting codes. The LabVIEW development environment uses the graphical programming language G.

D.5.2 Some Key Concepts

Block diagram: A block diagram is a pictorial description or representation of a program or algorithm. In a G program, the block diagram consists of executable icons called nodes and wires that carry data between the nodes.

G programming: G is a convenient graphical data flow programming language on which LabVIEW is based. G simplifies scientific computation, process monitoring and control, and applications of testing and measurement.

Control: The control is a front panel object such as a knob, push button, or dial for entering data to a VI interactively or by programming.

Control terminal: The control terminal is linked to a control on the front panel, through which input data from the front panel passes to the block diagram.

Front panel: The front panel is an interactive user interface of a VI. The front panel appearance imitates physical instruments, such as oscilloscopes and multimeters.

Indicator: The indicator is a front panel object that displays output, such as a graph or turning on an LED.

Waveform chart: A waveform chart is an indicator that plots data points at a certain rate.

While loop: The while loop is a loop structure that repeats a code section until a given condition is met. It is comparable to a Do loop or a Repeat-Until in conventional programming languages.

Wire: The wire is a data path between nodes.

D.5.3 Working with LabVIEW

As a software-centered system, LabVIEW resides in a desktop computer, laptop, or PXI as an application where it acts as a set of VIs providing the functionality of traditional hardware instruments such as oscilloscopes. Comparing physical instruments with fixed functions, LabVIEW VIs are flexible and can easily be reconfigured to different applications. It is able to interface with various hardware devices such as GPIB, data acquisition modules, distributed I/O, image acquisition, and motion control making it a modular solution. This utility is shown in Figure D.6.

A VI contains the following three parts:

- **Front panel**—a user interacts with the VI through the front panel
- **Block diagram**—the code that controls the program
- **Icon/connector**—means of connecting a VI to other VIs

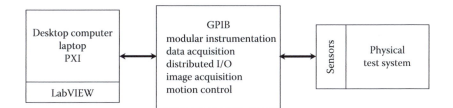

FIGURE D.6
Modular solution of LabVIEW.

D.5.3.1 Front Panel

Upon launching LabVIEW, you will be able to create or open an existing VI where the layout of the graphical user interface can be designed. Figure D.7 shows the front panel of the simple alarm slide control (alarmsld.lib) example included with LabVIEW suite of examples. This is the first phase in developing a VI. Buttons, indicators, I/O, and dialogs are placed appropriately. These control components are selected from the "controls palette," which contains a list of pre-built library or user-customized components.

A component is selected from the controls palette by left clicking the mouse on the particular control icon and can be placed on the front panel by left clicking again. Then the component can be resized, reshaped, or moved to any desired position. A component property such as visibility, format, precision, labels, data range, or action can be changed by right-clicking, with the cursor placed anywhere on the selected component, to bring up the pop-up menu.

D.5.3.2 Block Diagrams

After designing the GUI in the front panel, the VI has to be programmed graphically through the block diagram window in order to implement the intended functionality of the VI. The block diagram window can be brought forward by clicking on the "Window" pull menu and selecting "Show Diagram." For every control component created on the front panel, there is a corresponding terminal automatically created in the block diagram window. Figure D.8 shows the block diagram for the alarm slide control example provided with LabVIEW.

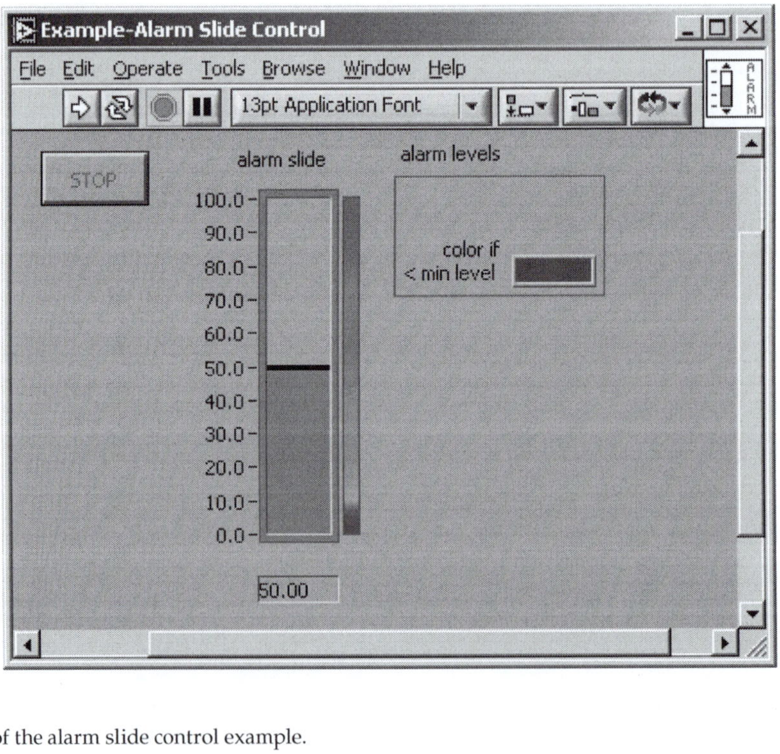

FIGURE D.7
Front panel of the alarm slide control example.

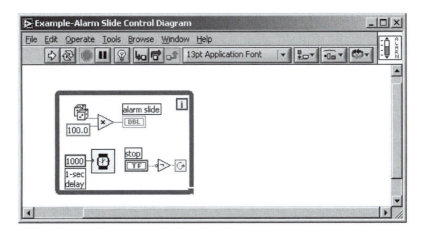

FIGURE D.8
Block diagram of the alarm slide control example.

The terminal is labeled automatically according to the data type of each control. For example, the stop button has a terminal labeled TF, which is a Boolean type. The vertical level indicator has a DBL type terminal, indicating a double-precision number. Other common controls with a DBL terminal include various numeric indicators, sliders, and graphs.

LabVIEW uses the G programming language to implement the functionality of a VI. It provides an extensive library of basic conditional and looping structures, mathematical operators, Boolean operators, comparison operators, and more advanced analysis and conditioning tools provided through the functions palette. A function may be placed on the block diagram window similar to how a control component is placed on the front panel. Depending on the required flow of execution, they are then wired together using the connect wire tool in the tools palette. In order to wire two terminals together, first click on the connect wire icon in the tools palette, then move the cursor to the input/output hotspot of one terminal, left-click to make the connection, and then move the cursor to the output/input hotspot of the other terminal and left-click again to complete the connection. The corresponding control component on the front panel can be selected by double-clicking on the terminal block.

The general flow of execution is to first acquire the data, then analyze followed by the presentation of results. The terminals and functional components are wired in such a way that data flows from the sources (e.g., data acquisition) to the sinks (e.g., presentation). LabVIEW executes its G programming code in data flow manner, executing an icon as data becomes available to it through connecting wires.

The dice terminal is a random number generator and its output is multiplied by a constant using the multiplier operator (see Figure D.8). The multiplication result is connected to the input of the alarm slide, which will show up as the level in the vertical indicator on the front panel during VI execution. The gray box surrounding the terminals is the while loop in which all the flow within the gray box will run continuously until the loop is terminated by the stop button with the corresponding Boolean terminal. When the stop terminal is true, the while loop terminates upon reading a false through the not operator. The wait terminal (watch icon) controls the speed of the while loop. The wait terminal input is given in milliseconds. In the figure, the loop runs at an interval of 1 s since a constant of 1000 is wired to the wait terminal. In order to run the VI, left click on the arrow icon on the top rows of icons or click on "Operate" and then select "Run." No compilation is required.

Note the remove broken wire command found in the edit pull-down menu. This command cleans up the block diagram of any unwanted or incomplete wiring. The debugging pop-up window that appears when an erroneous VI is executed is very helpful in trouble-shooting the VI. Double-clicking on the items in the errors list will automatically highlight the problematic areas, wires, or terminals in the diagram.

D.5.3.3 Tools Palette

LabVIEW has three main floating palettes for creating VIs. They are the tools palette, controls palette, and functions palette. The tools palette, shown in Figure D.9, is the general editing palette with tools for editing components in the front panel and block diagram panel, modifying the position, shape and size of components, labeling, wiring of terminals in the block diagram panel, debugging, and coloring. When manipulating the front panel and the block diagram panel, note which tool icon is selected. For example, the values of a control or terminal cannot be selected or edited when the positioning icon is selected.

D.5.3.4 Controls Palette

Figure D.10 shows the controls palette, which contains the pre-built and user-defined controls to create a graphical user interface. This palette will be available when the front panel is selected. If it is not showing, click on the "Window" pull-down menu and select the "Show Controls Palette" option. The figure shows the main group of top-level components available in its pre-built library. Clicking on the appropriate top-level icons will bring up the subpalettes of the available controls and indicators. To go back to the top-level icons, click on the up arrow icon on the top-left of the controls palette.

D.5.3.5 Functions Palette

When the block diagram panel is selected, the functions palette is shown as in Figure D.11, enabling you to program the VI. The functions palette contains a complete library of

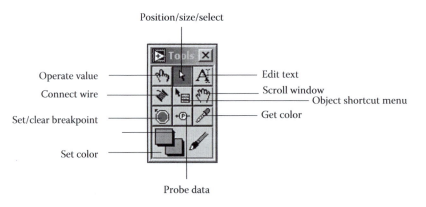

FIGURE D.9
LabVIEW tools palette.

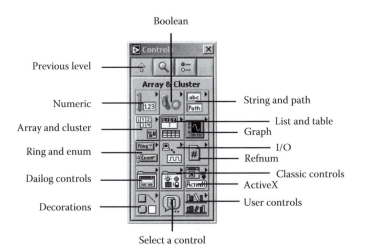

FIGURE D.10
LabVIEW controls palette.

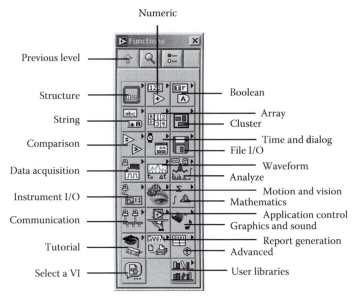

FIGURE D.11
LabVIEW functions palette.

necessary operations for developing the functionality of the VI. Similar to the controls palette, the top-level icons show the grouping of different sub-functions available for the programmer. Several commonly used groups are indicated below.

- Structures: The structures icon consists of the usual programming language sequences, conditional statements, and conditional loops. These structures are in the form of boxes where the terminals within the boxes are executed when the statements or loops are invoked. In addition, there is a formula node where custom text-based formulas can be included if you prefer the traditional text-based equations. There are also variable declaration nodes where local and global variables can be declared.

- Numeric: The elementary operators, such as summation, subtraction, multiplication, division, and power, are grouped under this icon.
- Boolean: This icon contains the Boolean operators required for logic manipulation.
- Array: The array grouping consists of tools for array manipulation.
- Comparison: Operators for numerical comparison, which provide Boolean outputs, are found under this icon.
- Analyze: This icon contains the more advanced analysis tools such as FFT spectrum, power spectrum, filters, triggering, and waveform generation.
- Mathematics: The tools for mathematical manipulation such as calculus, statistics and probability, linear algebra, optimization, and numeric functions are found under this icon.

Index